AF855109

Macroclimate and plant forms: An introduction to predictive modeling in phytogeography

Tasks for vegetation science 1

Series Editor
HELMUT LIETH
University of Osnabrück, F.R.G.

ELGENE OWEN BOX

Macroclimate and plant forms: An introduction to predictive modeling in phytogeography

WITHDRAWN OCCIDENTAL COLLEGE LIBRARY

DR W. JUNK PUBLISHERS THE HAGUE/BOSTON/LONDON 1981

581.542 B788m

OCCIDENTAL COLLEGE LIBRARY

Distributors

for the United States and Canada

Kluwer Boston, Inc.
190 Old Derby Street
Hingham, MA 02043
USA

for all other countries

Kluwer Academic Publishers Group
Distribution Center
P.O. Box 322
3300 AH Dordrecht
The Netherlands

Library of Congress Cataloging in Publication Data

Box, Elgene O.
Macroclimate and plant forms.

(Tasks for vegetation sciences; 1)
Bibliography: 16 pp.
1. Vegetation and climate–Mathematical models. 2. Phytogeography–Mathematical models. 3. Prediction theory.
I. Title. II. Series.
QK754.5.B69 581.5′42 81–12368
AACR2

ISBN 90 6193 941 0 (this volume)
ISBN 90 6193 897 X (series)

Cover design: Max Velthuijs

Copyright © 1981 Dr W. Junk Publishers, The Hague.

All rights reserved. No part of this publication may be reproduced, stored in a retrieval system, or transmitted in any form or by any means, mechanical, photocopying, recording, or otherwise, without the prior written permission of the publishers Dr W. Junk Publishers, P.O. Box 13713, 2501 ES The Hague, The Netherlands.

PRINTED IN THE NETHERLANDS

Preface by the T:VS Series editor

The first volume of any scientific book series is always something special. Long deliberations and negotiations have finally borne fruit. A new scientific series, Tasks for Vegetation Science, presents itself with the first 3 volumes almost simultaneously, setting in content and format standards for the books to come. Volumes 1 and 3 deal with modelling attempts of the structure and function of vegetation. Volume 1, this volume, contains the attempt to predict vegetation structure from various climatic parameters. This book will probably serve as a milestone in the long list of previous attempts to correlate vegetation structures with climatic parameters. Aridity indices, EP-values, bioclimatic thresholds of various sorts have been tried in the past with increasing success to predict vegetation types.

All these attempts were of limited value, however, since considerable margins of uncertainties or exceptions for greater portions of the world remained. Previous workers in this field together with the many critics always demanded better solutions after they had presented their models. They demanded better data for meteorological observations, closer networks for vegetation data, the inclusion of more environmental parameters, better systems for vegetation classification to name only a few.

It was therefore a noteworthy attempt of Dr. Box to select such a difficult task for his thesis. Dr. Box, a mathematician by early training, possessed the mathematical tools seldom available to a vegetation scientist to undertake a new attempt towards the prediction of vegetation structure from environmental, climatic parameters.

Since our time has generally better information than our colleagues in the past, it is only natural that we try to model relations between vegetation parameters and environmental parameters more precisely, if such relations, indeed, exist.

To show the vegetation types, geographical areas, and parameters where such relations exist and where they do not or only weakly exist was the purpose of Dr. Box' study. His approach has entered a new field of numeric vegetation modelling, the use of information systems. Since the approach is the first of its kind in vegetation science, one cannot expect perfection in the predictive power of his model set. However, the new method once presented is open for discussion, for refinement and for intensive testing.

For the series editor it is especially gratifying to present the volume as the first one in a row of volumes with similar tasks. The author of this book worked for a long time together with the series editor while he developed the basic structure of his information system called 'ECOSIEVE'.

Once perfect, the predictive power of this model has considerable theoretical and practical value for management purposes and is therefore very timely in its appearance. As an intensive study, combining skills in vegetation science and mathematics and computer science, it will certainly set an example for the next student generation of vegetation science. The use of information systems as well as multifactorial regression analysis has just begun to yield results in vegetation analyses of various kinds. Our plan is to present in future volumes of this series more new techniques and concepts of this kind useful for our science to the vegetation scientists.

Volume 3, by Dr. Ross, dealing with biophysical modelling on ecosystems, is already in print. Several more books, presenting new methods and special treatments of important ecosystems, are in various stages of completion.

With the first volume appearing, the new series T:VS is on its way. It is our hope that the series will serve our profession in the future and that the first volume will have a good start.

Osnabrück, March 1981 H. Lieth

Foreword

This study arose out of the old question of what actually determines vegetation structure and distributions. Is climate the overriding control, as one would suppose from reading the more geographically oriented literature? Or is climate only incidental, as suggested by more site and/or taxon-oriented writers? The question might be phrased more realistically: How much does climate control vegetation processes, structures, and distributions?

It seemed to me, as an ambitious doctoral student, that one way to attempt an answer might be to try to predict world vegetation from climate alone and then compare the predicted results with actual vegetation patterns. If climatic data were sufficient to reproduce the world's actual vegetation patterns, then one could conclude that climate is the main control. This book represents an expanded, second-generation version of that original thesis. It presents world-scale vegetation and ecoclimatic models and a methodology for applying such models to predict vegetation and for evaluating model results. This approach also provides a means of geographical simulation of vegetation patterns and changes, which represent necessary data inputs in other fields such as atmospheric chemistry and biogeochemical cycling.

It has been fairly well accepted that climatic and other environmental conditions are associated with the evolution of particular aspects of plant form (convergent evolution). The particular configurations of plant size, photosynthetic surface area and structure (e.g. sclerophylly, stomatal 'resistance'), and their seasonal variations represent what one can recognize fairly readily as distinct growth forms. These aspects of form, however, also determine the potential rates and annual totals of photosynthesis, respiration and water loss of plants and thereby set the climatic limits on their water and energy budgets and thus their distributional limits in the real world. This logic is perhaps best called 'ecophysiognomic' and can be stated as an initial hypothesis:

> Plant form and structure are evolutionary results of climate (among other factors) but also largely determine the current distributions and viability of plants through their effects on plant water and energy budgets.

If true, this implies both an ecology and a biogeography at the level of plant life forms – a fundamentally ecological life-form approach to vegetation, which not only should provide new insights and conceptual tools but perhaps also make vegetation study and appreciation somewhat more accessible to those without special training in botanical Latin and taxonomy.

The study presented herein represents an attempt to discern world patterns in the relationship between plant form and vegetation structure (physiognomy and morphology) on the one hand and environmental, mainly climatic conditions on the other. The focus is on plant water and energy budgets and how they and related physiological processes are related to and limited by climatic conditions (in different types of climates) working through the particular combinations of form characters represented by different plant life forms. The project attempts to organize much detail and many smaller explanations into a world-scale framework and initial empirical model, as the basis for subsequent, more analytical work on relations between plant form and physiological limits. The proposed world set of plant life forms is related to climate by means of climatic envelopes, which are used with a world climatic data-base to predict world life-form distributions. These are interpreted as vegetation types and compared with actual vegetation patterns and site data, with generally good results. (The validation suffers some from insufficient descriptions of local vegetation which also provide climatic data. Readers are invited to send local vegetation and climatic data for comparison with predicted vegetation profiles.)

As the title states, the book is intended as an introduction. In developing the vegetation classification the first criterion was to cover the main types of terrestrial plants in the world (but not necessarily all taxa). In describing climate the main emphasis was on covering the most important ecological aspects. For sake of manageability, both vegetation and climate had to be described with as few units as possible. Since the purpose is to discern basic, general relationships, if they exist, sweeping generalizations and simplifications are sometimes made. These are then pushed as far as possible in the modeling in order to see where they break down when exposed to the full range of variation in earth's climatic conditions. Criticism is expected. Only by this sort of generalization, however, can one put together a world-scale model which can be expected to answer general questions. The model represents perhaps the first serious attempt to predict fairly detailed vegetation profiles, including seasonal aspects, and distributions for all parts of the world's land area. It is also probably one of the first vegetation studies, at any geographic scale, to compare predicted and actual vegetation profiles rigorously as a means of assessing model accuracy and thus the validity (and adequacy) of the underlying assumptions.

The study draws on several sources and methods which are not commonplace in traditional American botany and plant ecology (as judged at least by recent journal articles). Among these are a comparative, geographic outlook over large areas (involving literature in other languages); a fairly rigorous quantitative approach (involving necessarily large amounts of climatic data); a focus on life forms as the fundamental plant unit (providing better ecological and physiognomic resolution); geographical and mathematically simple ecological modeling; and rigorous scrutiny of predicted results. The result of this synthesis is a quantitative geographical methodology which can be used both for simulation/prediction and as a geographic alternative and complement to traditional statistical hypothesis-testing. It nevertheless represents an approach which must be superficial, at least initially, and which requires further development and improvements.

This book is dedicated to Heinrich Walter for his many and enormous conceptual and geographic contributions to the ecological study of world vegetation; and to Helmut Lieth, who made the author's admission to doctoral study in ecology at an American university possible, provided early guidance, and encouraged the development of latent world-scale interests. Special thanks is also due the Kernforschungs-anlage (Nuclear Research Center) in Jülich, Germany, where much of the computerized basis for this and other research was initially developed. Work in Jülich and Hohenheim (with Walter) also permitted the author to observe many vegetation types of Europe, North Africa and Atlantic islands which could not have been visited otherwise. (More extensive ecological study during 1981–82 in the Old and New World tropics is anticipated.) Grateful acknowledgement is due the computer center of the University of North Carolina, and also those of Duke University and the University of Georgia, for computing time and other assistance; and to the UNC Photographic Laboratory for photography of the largest computer maps.

The manuscript was typed by the superb office staff at the University of Georgia geography department: Pat Wall, Mildred Bell, Anne Berryman, Kathy Smith, Sharon Mullins, Dawn Tolbert, and Sue Werner, under the direction of Yvonne Pommerville, office manager. No department ever had a more helpful office staff.

Most importantly, this project would not have been even remotely possible without the perseverance of the author's wife, who worked rotating shifts to provide daily transportation, all living expenses, and some school expenses while the author was a doctoral student.

Athens (Georgia), 31 May 1981 Elgene Owen Box

Contents

APPENDICES

List of tables

List of world maps

Climate maps

Vegetation maps

CHAPTER 1

Introduction

Man has probably always recognized vegetation as a most conspicuous and characteristic aspect of landscape. Not only does vegetation provide food and other resources, but it is also responsible for the general appearance of landscapes, providing both individual landmarks and perhaps the most basic means of distinguishing and characterizing different regions. Certain rough correlations between climate and vegetation types were appreciated certainly by the peoples of classical times and probably long before that. Modern botanists and ecologists certainly recognize the correlation between climate and vegetation types but view climate, as the main determinant of vegetation types and distributions, with varying degrees of skepticism, preferring to look instead for individualistic properties of species, particular community or association affinities, complex biotic interactions, or edaphic or other overriding site factors. These non-climatic factors certainly may be present and play interesting roles at the species level. It is the basic premise of this study, however, that their effects are super-imposed on a general climatic situation and are of secondary importance in determining the general structural, functional and geographical patterns of land vegetation.

Plant and vegetation geography have been studied most commonly at two levels:

1. The taxonomic level, involving distributions of individual species, genera and families.
2. The formation level, involving distributions of particular vegetation formations and formation types.

The vegetation units in the first case are entirely taxonomic, while those of the second are mainly physiognomic and ecological.

Between these two levels is that of a more general concept of plant types, at which the vegetation unit is particular types of plants composing the vegetation but at which the plant types are ecological and/or physiognomic rather than taxonomic units. The level of physiognomic plant types, i.e. growth forms, is at least as important as the taxonomic and formation levels for the following reasons:

1. Plant growth forms provide the basic structural components of vegetation stands, thus also the most obvious level of subdivision both for describing and explaining vegetation structure.
2. The primary physiological processes of plants are largely controlled by such aspects of plant form as the amount of respiring phytomass (plant size), amount and configuration of photosynthetic and transpiring surface (leaf form and size), and seasonal variations in form (e.g. leaf habit).
3. Since plant species tend to cluster into characteristic general forms, plant form provides a useful means of getting at general principles of plant-environment relations without becoming mired in taxonomic detail.

The idea of ecological plant types (groups of taxa with similar ecological requirements) leads directly to that of ecophysiognomic life forms, i.e. growth forms which show a more or less obvious adaptation to environmental conditions due to similar form-based water and energy budgets.[1] Since the mechanisms of environmental limitation operate primarily at this level, life forms would appear to be the most appropriate unit for studying the ecological aspects of vegetation geography.

In mapping vegetation and other larger ecological distributions (e.g. ecosystems and biomes) geobotanists and biogeographers have generally focused on more or less physiognomically defined vegetation formation types. These are usually characterized by one or more dominant or co-dominant life forms, such as the evergreen and summergreen needle-leaved trees of the boreal forest or the larger trees, shrubs and stem-succulents of warm semi-deserts and open woodlands. Formations, however, generally also include other important and lesser life forms which cannot always be recognized in more general formation approaches. For the boreal forest such additional forms regularly include the summergreen birches plus summergreen and smaller evergreen shrubs and herbs, and various ferns, mosses and lichens. These different synusiae are constituent parts of most boreal forests and as such make their own particular contributions to vegetation and ecosystem structure and functional integration. The individual roles of the constituent life forms are important to any system. The actual mix of forms within different formations of the same general type (e.g. forests, woodlands, semi-deserts) may be less consistent, however, because of reduced dominance by a single form (e.g. semi-deserts) or because of various historical (e.g. Australia) or other ecological factors. Although tropical rainforests, mediterranean oak forests, and warm-temperate 'laurel forests' are all dominated by broad-leaved evergreen trees, for example, their physiognomy, structure, and functional patterns all differ widely because of both the general size of the plants and the mix of constituent subdominant and understorey forms. Perhaps more importantly, however, the life-form approach provides a means of interpreting vegetation change along gradients at some ecological level higher than that of individual species and genera, thus hopefully providing more insight into mechanisms of environmental limitation.

It has been known for a long time that important relations exist between climate and terrestrial vegetation patterns. General levels and seasonal patterns of solar radiation, air and soil temperature, water availability (precipitation) and runoff and potential water loss (potential evaporation) determine the basic conditions of existance for land plants. Plants in turn must adapt to environmental conditions through modification of form and surface structure, physiological processes, and seasonal habits. In many respects the most important of these is the general growth form, since it is through its amount of surface area and its area-volume ratios that the other processes are constrained to operate. The rates of basic physiological processes and thus the seasonal and annual water and energy budgets of plants are directly governed by such aspects of plant form as general plant size (respiring biomass), growth form (potential leaf area and canopy height), leaf size and 'hardness' (potential water loss), and leaf duration. Climatic limits to the adequacy of particular combinations of leaf and plant size, shape, surface resistance and duration, as measured by positive water and energy balances, can thus be seen as the most important mechanism of environmental limitation. The logic of environmental (primarily climatic) limitation and its applicability to distributions of such features as soils, vegetation, animals and even human land-use patterns have been recognized since the early 19th Century, if not before. The main obstacles to quantitative application of environmental limitation principles have generally involved problems of scale: micro-environments, spatial heterogeneity, and above all the availability of sufficient quantitative data for all important environmental factors.

The usefulness of world and regional-scale vegetation models has also been well recognized in plant geography since the early 19th Century. Such models have generally been based on environmental limitation, correlating distributional limits of vegetation types with various values for important

[1]The term 'growth form' is used throughout this book as a purely physiognomic concept, the combination of a particular 'structural type' (e.g. tree, shrub), leaf type (form but not size or texture), and seasonal habit. The term 'life form' involves the assumption of certain environmental relationships as well as a bit more physiognomic detail (namely leaf size and texture). The actual life form of a plant or species is not always known, particularly as regards functional characteristics such as seasonal habit. As a result, the term 'life form', as used herein, does not refer to any particular classification system unless explicitly stated or implied by context. The structural types recognized in this study are shown in Table 1, p. 14.

macroclimatic factors. Predictive modeling, i.e the rigorous application (extrapolation) of quantitative models to environmental data (at sites other than those used to construct the model) in order to predict actually occurring patterns, can be particularly useful in plant geography and plant-environment relations, since it provides a ready means of testing the validity of the model and the understanding behind it. This was not done until recently because of the lack of sufficient world climatic data and of a sufficiently convenient means of generating and analyzing the results. The existence of large computers, computerized climatic data-bases, and computer mapping systems, however, now means that ecological hypotheses can be turned into quantitative models and then into maps within a few days. If world maps can be generated, representing the full range of climatic variation found on earth (and presumably much of that from past climates), they can often be interpreted quickly by someone familiar with world patterns in order to evaluate the hypotheses behind the maps.

It is hoped that the life-form approach will provide a more versatile and accurate basis for analyzing and synthesizing plant-environment relations and vegetation geography. In order to assess the degree to which meaningful vegetation units can be predicted from climate, an empirical model was developed which relates the world's geographically most important plant life forms to macroclimatic conditions. This model differs from earlier models in at least three important ways:

1. It treats plant life forms rather than formation types, as the units of vegetation. These life forms can be combined to provide characterizations of particular formations and formation types.
2. It includes more vegetation units (77 life forms plus sub-types) than most models. This provides much greater potential for resolution of vegetation types.
3. It includes more predictive climatic variables (eight) than most other models. This permits greater resolution of seasonal patterns which are reflected in vegetation phenology and seasonal physiognomy.

The model is used to predict and characterize the natural vegetation at 1225 terrestrial test-sites worldwide. These predictions are then compared with known vegetation distributions and detailed descriptions of the actual vegetation composition at a set of validation sites not used for developing the model.

The model described herein is a more analytical, enlarged version of an earlier 'synthetic' model (Box 1978a, 1981c), which focused only on dominant life forms and which attempted to include some aspects of vegetation structure (e.g. forest vs. open woodland) within the vegetation units. The present 'analytical' version separates plant form and stand structure entirely and includes the many understorey forms omitted from the first model. Both models relate plant life forms entirely to macroclimatic variables by means of climatic envelopes composed of limiting values determined from correlations between distributional limits and local climatic conditions. This was felt to be a necessary first step because such empirical environmental correlates sometimes represent the only readily available type of appropriate ecological data but will permit generation of readily interpretable results which can be used to develop models based on more integrative environmental factors. The model was specifically designed for world-level applicability and evaluation, i.e. for generation of world-scale vegetation predictions and rigorous comparison with actual vegetation. Vegetation predictions at particular sites are generated from the model using the general screening program ECOSIEVE (Box 1979a, 1981b). World results were computer-mapped using SYMAP (Harvard University) plus base-map, climatic data, and auxiliary software described elsewhere (Box 1975, 1978a, 1979a).

CHAPTER 2

Modeling ecological structure and function

One of the central problems in ecological modeling involves the coupling of continuous and discrete submodels. Most functional processes, such as production and transpiration, and many related environmental and state phenomena, such as air and leaf temperature, are mathematically continuous in the sense that their values vary continuously as the values of environmental determinants vary, at least on the time-scales usually employed in ecology. These processes are continuous because they represent well buffered feedback systems which are not overpowered by even rapid environmental changes. Mathematically this means that most functional processes and many state phenomena can be related to environmental determinants by continuous, often exponentially-based (i.e. feedback-based) mathematical functions. Annual net primary productivity, for example, has been successfully related to annual actual evapotranspiration and to annual precipitation by simple saturation curves and to mean annual temperature by a simple sigmoid curve, providing two bases for predictive world maps (Lieth & Box 1972).

Structural phenomena, on the other hand, generally represent balances among several basic functional processes and thus are often more complex. Some structural aspects, such as amount of leaf area or standing biomass, may require much more complex mathematical forms but still vary continuously, if less smoothly, with variations in environmental determinants. Other structural phenomena, however, such as species occurrence (distribution) and community composition, plus many behavioral phenomena, may involve processes which are discrete, combinatorial, hierarchical, probabilistic or iterative (state-dependent) in nature and which require quite different mathematics. Plant and vegetation distributions, at a given time, are best described as discrete (e.g. maps with discrete boundaries) or as probabilistic. Vegetation composition, whether described determinately or probabilistically, is a combinatorial phenomenon, involving interacting combinations of plant types or taxa. The processes which determine competitive success and vegetation succession (including productivity and leaf area) are continuous over the range of environmental conditions involved, but the structural dynamics of vegetation succession, in terms of life-form composition, is hierarchical in nature. Coupling the different types of models required to simulate these various aspects of vegetation structure, dynamics and distribution is certainly one of the main problems encountered in vegetation modeling and one which probably has no single best solution.

A. World vegetation models in general

Most quantitatively expressed vegetation-environment models can be thought of as consisting of three components:

1. a vegetation classification system characterizing plant and vegetation diversity (vegetation submodel);
2. a set of environmental variables important in describing the effective environment and in limiting vegetation distributions (environmental submodel); and

3. some sort of quantitative values (e.g. environmental limiting values) which relate the vegetation types to the environmental variables and represent an expression of plant-environment ecological relationships (ecological model).

The first quantitatively expressed world vegetation classification system and environmental model was probably that of Schimper (1898). Quantitative world and regional models have been presented also by Rübel (1930), Holdridge (1947), Troll & Paffen (1964), Mather & Yoshioka (1966), Hueck & Seibert (1972), and others. The classification of vegetation-related world climate-diagram types by Walter & Lieth (1960–67) represents a useful and elegant semi-quantitative system. Each model recognizes certain environmentally delimited vegetation types and provides appropriate, empirically determined limiting or critical values relative to important climatic variables.

The primary data-bases for modeling plant-environment relations are the various accumulated climate and other physical data and the available vegetation descriptions (including maps) and ecological interpretations for all parts of the world. Over the last century climatologists have been able to collect an enormous volume of climatic data for most parts of the world (though some important climatic factors have been poorly represented in data-collecting programs). It has been possible to construct detailed maps of macroclimatic conditions for most of the world. Plant data, on the other hand, have not been so complete, especially in tropical and other less accessible areas. Classification of plants has generally been approached on the three levels of:

1. plant species (taxonomy), based primarily on the relatively invariable reproductive parts of plants;
2. plant life forms, based on plant ecophysiognomy;
3. vegetation formations and formation types, based on the physiognomy of plant stands, which depends also on that of the dominant species.

Because of the need to be able to identify all plants relative to some classification system, primary emphasis was necessarily placed on the taxonomic effort. Coordinated data on other aspects of plant form, on process rates, on phenology, and on their environmental situations have generally been neglected until quite recently and are still inadequate. Fortunately, a few geobotanists have provided enough detailed, coordinated physiognomic and climatic data for various parts of the world that one can begin to piece together world ecophysiognomic patterns and attempt to model them predictively.

B. The current model

The model presented herein, being an initial attempt, necessarily reflects a very simplistic modeling structure. The purpose of the project was to produce a model which, without attempting to simulate major ecophysiological mechanisms in detail, can:

1. Determine, at global scale, to what extent plant form and vegetation structure are determined by climatic factors and which factors are most important.
2. Provide a world-scale modeling framework for geoecologic hypothesis-testing, further model development, and study of vegetation response to changes in climate.

This is attempted by examining the available world climatic and vegetation data and relating the distributional limits of plant life forms to what appear to be their most important climatic determinants. The model developed is essentially descriptive (empirical) and geographic in nature. Its output is a spectrum of plant life forms which can be used to characterize the vegetation at the climate site. The model is applied to a world climatic data-base in order to predict world distributions of plant forms and vegetation formation types. The predicted global results are used to check the model and can be expected to provide information on questions such as the following:

1. To what extent are the world's large-scale vegetation patterns really determined by climatic conditions? (Is climate the overriding factor, as has always been supposed?)
2. Which vegetation occurrences appear to be climatically anomalous?
3. To what extent do similar plant types and vegetation structures in different regions and possibly dissimilar climates represent the same ecophysiognomic response?
4. To what extent are certain vegetation types actually limited to known regions by climatic factors rather than by other proposed factors,

such as historical development, floristic relations, biological interactions, soil conditions, or man's influences?

Information on questions such as these, along with an initial world catalog of ecophysiognomic plant types, can be expected to provide a more unified, albeit initial theoretical framework for plant-environment studies than has previously been available. World patterns of both the predictive ecoclimatic variables and predicted vegetation are presented as computer-printed world maps. Validation of the model is attempted by comparing actual and predicted vegetation at particular sites. The sensitivity of the plant types to changes in particular climatic variables is also studied.

The logic of the life-form approach places certain constraints on the methodology, the most important of which can be summarized briefly as follows:

1. The definition of plant types must be based on plant features, i.e. there can be no vague environmental or taxonomic groupings such as 'semi-desert', which may consist of many life forms while retaining some uniformity as a formation type.
2. The model must be based on environmental limitation of major physiological processes, such as water uptake and transpiration, stomatal behavior, cold-hardiness, phenological phases, production and respiration, growing season length, etc., all operating through the particular characters of form.
3. The set of plant types must cover the full range of variation in plant form found on earth.
4. The plant types must also be global in concept, i.e. must cover ecological relations in all major occurrences of the types.
5. The model must be applied equally to all climatic situations.

The first two requirements are fairly obvious and state essentially that the model must be ecophysiognomic, which is to say that it must relate plant forms to the environmental factors most important in determining plant form. The third requirement states the necessary criterion for any general model. The last two requirements are equally important, however, since only by examining each disjunct occurrence of a particular plant form in different parts of the world can one hope to ascertain to what extent these occurrences represent the same biological answer to similar but disjunct environmental situations. A world-scale study does not guarantee that all possible ecophysiognomic manifestations will be covered, since some may have been precluded by historical or other reasons. World-scale simulation is, however, the best one can do without going to a quite different, less geographic approach.

Both available environmental data and the logic of 'first things first' strongly suggest that an initial model be based on macroclimatic factors and seek to characterize macro-level ecoclimates corresponding to the occurrence of particular plant forms. Macroclimatic data are also the only environmental data provided with almost all ecological vegetation descriptions and so represent the best basis for an initial world model. Although soil, historical, anthropogenic, biological, and other factors also influence plant distributions and vegetation structure, climate is the most important factor on most sites, and some of the other factors, such as soil conditions, may also be under macroclimatic control to a significant degree.

The global nature of the study requires that the model be able to predict life-form spectra which can be used to characterize both world vegetation distributions and the vegetation at particular sites. Since macroclimatic data are readily available for thousands of sites, these can be used with the model to generate implied world vegetation distributions which can be used to check the model.

Finally, the available ecological data strongly suggest that the model focus on environmental limitation patterns as the most feasible approach. The global scale requires a generalized but quantitative approach. Despite the several models of world vegetation which have been presented, a systematic, quantitative, global correlation of physiognomic forms (including seasonal aspects) with major ecoclimatic factors has never been attempted. The existing models are either too restricted geographically (e.g. Mather & Yoshioka) or involve vegetation units which are much too general or vague (e.g. Holdridge) to describe ecosystem structure or plant-environment relations.

C. Model structure and technical considerations

The basic structure of the model is that of descriptive vegetation and climate submodels related to each other through a deterministic ecological model. The vegetation model consists of a set of ecophysiognomic plant forms identified in Table 2, each well defined physiognomically. The climatic model consists of a minimal set of macroclimatic variables which are particularly important in determining plant form and vegetation structure and which represent a model of the main ecological aspects of climate. These variables can be thought of as defining an environmental hyperspace in which the occurrences of the plant types can be represented by environmental envelopes relative to the hyperspace variables. The ecological model consists of the estimated climatic limiting values for the plant types, which define their climatic envelopes, plus a growth-form dominance hierarchy expressing the ability of the forms to outcompete each other.

The main problems associated with this approach include the following:

1. The large number of hyperspace dimensions (ecoclimatic variables) which may be required to describe the effects of environmental limitation adequately.
2. Factor interaction between the variables, and the need for mathematical simplicity in the model.
3. The rigidity of envelope boundaries, with no mechanism for environmental variation resulting from factors not included in the model.

All three problems must be viewed, however, with reference to the so-called 'law of diminishing returns'. The first problem is basically one of model design. It was found that, for world-scale vegetation patterns, the main climatic effects could be represented adequately with eight variables. The problems of factor interaction and envelope boundaries require more discussion.

One manifestation of factor interaction and general complexity in both biological and environmental systems is that limiting values for environmental factors can be mathematically complex. Environmental limits can also vary with recent environmental conditions, as in the case of cold-hardening. Factor interaction between environmental variables can be expressed mathematically either through environmental limits which are variable-valued functions of other factors, resulting in curved or state-dependent envelope boundaries, or through the use of compound variables which combine highly interactive factors into critical variables, for which limiting values are usually much more uniform. The latter is much simpler mathematically. A central example in plant-environment relations can be seen on scattergrams of vegetation occurrence plotted against mean annual temperature and average annual precipitation (e.g. Lieth 1956; Whittaker 1975). Temperature limits are typically very nearly linear, but precipitation limits increase rapidly with increasing temperature, because of the increased evaporative demands. By contrast, the models of Holdridge (1947) and Mather & Yoshioka (1966) use a moisture index expressing the relationship between precipitation and potential water loss (evapotranspiration). The result is that the shape of the envelopes is greatly simplified, approaching an orthogonal form. Many authors have recognized both the simplified mathematical properties and greater scientific insight associated with such incisive compound variables as actual evapotranspiration and moisture indices (e.g. Major 1963; Lieth & Box 1972; Carter & Mather 1966; Holdridge 1947; Budyko 1975; Meentemeyer & Elton 1977). The choice of environmental variables can also result in artificial interaction, as in the case of temporally related monthly and annual values. This can be useful in representing different aspects of limitation by the same factor (e.g. total annual and seasonal moisture), but one must then understand that individual limiting values may not remain accurate if removed from the total configuration of variables.

The problem of rigid envelope boundaries can be overcome somewhat by generosity in estimation of limiting values, coupled with some degree of redundancy in interrelated variables. Limiting values must be chosen so as to include the effects of normal microclimatic variations. A probabilistic approach to environmental limits could also be employed but may require additional, less available data. The environmental-envelope approach has been used in many types of ecological studies, including studies of plant and animal habitats and environmental niches at the species level (e.g. Twomey 1936; Bodenheimer 1938; Fry 1947; Hart 1952; Went 1957; Gates 1969) and other quantitative models of world and regional vegetation (e.g. Lieth 1956; Holdridge 1947; and Mather & Yoshioka 1966). Each of these vegetation models involves only two environmental dimensions (the third in Holdridge's model being redundant). This number of variables is quite insufficient to represent seasonality. An apparent growing interest in environmental-envelope models is well illustrated by a recent formalized nomenclature and ecological theory of environmental envelopes by Dobson (in press) and by a recent cataloguing by Duke & Terrell (1975) of the climatic limits of some 1000 economic plants. To some extent, rigid limits can be softened by coupling with spatial environmental indices such as potential cover by the particular plant type.

A major consideration in designing an initial model was that the model be as simple as possible both structurally and mathematically. In order to make the model applicable without requiring a computer, all climatic limits are constrained to be constant-valued. Factor interactions are expressed through compound variables within an orthogonal system of environmental factors. The climatic envelopes include their limiting values. The prediction of plant types at a given site is then performed by a screening operation which involves comparing site climatic values, for each variable, with the climatic limits of each life-form candidate. Although this can be done without a computer (see Appendix F), a general computer program ECOSIEVE was developed to facilitate application of the model to larger data-bases (Box 1978a, 1979a, 1981b). ECOSIEVE sets up the user-specified screening structure, generates and prints the results for the sites provided, can provide optional cross-tabulation (i.e. distributions) by plant type and optional, user-designable output files for use with other programs, and can be coupled with user-designed successional or other models for further analysis of results. Each site is treated separately by ECOSIEVE, but some site interaction can be provided by coupling with computer mapping programs which interpolate values between sites. The spectrum of life forms predicted to occur at a site can be interpreted as a vegetation formation type by means of a life-form dominance model to be described later. This model can be coupled with ECOSIEVE and with the mapping programs.

As a result of earlier research (e.g. Lieth & Box 1972; Box 1975, 1978b), the author had helped develop a sufficient world computer-cartographic capability and climatic data-base (1225 terrestrial sites) to permit world-scale extrapolation, mapping and testing of quantitative vegetation models. As background for development of the present model, the vegetation types predicted by the earlier models of Rübel, Holdridge, Troll, and Mather & Yoshioka were generated for the 1225 climatic sites by ECOSIEVE and mapped for comparison with actual vegetation patterns (Box, unpublished). Due to omission of important environmental factors and/or insufficient discrimination of vegetation types, only Troll's model comes close to reproducing actual vegetation patterns over most of the world.

CHAPTER 3

Ecological classification of world vegetation

The area of plant ecology which deals with the relation of plant structure (including its seasonal variation) to environmental conditions has been called ecophysiognomy (Mooney 1974). One of the most conspicuous (and gratifying) features of plant structure is that it tends to converge to certain successful structural-functional forms, which can be called ecophysiognomic forms or life forms. These life forms provide adequate form-based water and energy budgets, life cycles, and other aspects of strategy for survival in their respective environments and can generally be described structurally by particular combinations of characters such as general growth form, size, leaf type, and seasonal habit. These characters are not independent, however, and compensation of one character by another implies that it is the combination, i.e. the life form, which is the basic unit and which must be related to environmental conditions. Because of the central importance of plant form, understanding the relation of form to environmental conditions suggests a means of interpreting plant-environment relations in general. A logical foundation for this approach would appear to be an initial, putative list of environmentally correlated plant forms. In order to ensure that we understand as much as possible about the relations between form and function, such a list must be based on the full range of plant-form variation found on earth. Although world classifications of growth forms, life forms and vegetation types have existed for some time, most suffer from some combination of insufficient scope, weakness in certain regions, vagueness, or lack of ecological conceptualization or applicability. Even the life-form classification of Raunkiaer (1934) is not very useful in arid areas (Zohary 1952; Orshan 1953; Walter 1973). Its elaboration by Ellenberg & Mueller-Dombois (1967b) is the best available classification, employing both growth-form and seasonality criteria. This geographic classification still requires some adaptation for use with a climate-based geographic world model.

A. Vegetation description and vegetation data

Development of a world classification of ecophysiognomic plant forms requires not only general familiarity with the vegetation of all parts of the earth but also more information on the physiognomic characters of at least the most important plants of these regions than is necessary for other types of world vegetation classification. The most important physiognomic characters of plants include the basic growth form, plant size, leaf form and size, leaf duration (or other index of seasonal activity), and some indicator of the more detailed nature of gas-exchanging surfaces. This last item, which represents the plant's control over water loss and related CO_2 intake ('stomatal resistance'), can be described crudely by leaf-texture terms such as sclerophyllous, malacophyllous, or coriaceous. Descriptions of individual species can best be obtained from local manuals (floras) where these exist but must be interpreted with reference to the species' importance in the structure of the local vegetation. Since the model is to predict vegetation type as well as plant distributions, information on cover and structure is needed also. The same plants, depending on available water, may grow so closely together that both canopies and root systems overlap or may grow in much more open stands (with little difference in the form characters of the individual plants). As a result, in order to obtain information on the effective mix of ecophysiognomic characters of the entire vegetation stand, one must seek data from a variety of more general, integrative and geographical sources, including vegetation manuals, vegetation descriptions, photographs, vegetation classifications, and maps.

Complete physiognomic description of local or regional vegetation requires information on species composition, species structural characters, formation structure (synusiae and dominance), and degree of ground cover and amount of leaf area by each synusia. Successional information often provides additional insight. Formation structure and ground cover are often suggested adequately by formation-type descriptors such as savanna, forest, open woodland, or steppe and by photographs. The formation approach has also acted, however, to limit description by identifying only the current or potential dominant synusia. The vegetation of the Carolina piedmont (USA), for example, has been characterized by various authors as Southern Pine Forest, Summergreen Broad-Leaved Forest, Oak-Pine Forest, or Southern Mixed Forest, without adequate reference to the other synusiae or to the fact that the pines are successional and remain indefinitely only on poorer sites, in open stands, and along forest edges. Descriptions of vertical structure and the lower synusiae are not always readily available for all parts of the world and must often be pieced together from individual data.

Horizontal variation in vegetation structure, due to such factors as topography, soil and microclimatic heterogeneity, is generally beyond the scope of the current project but cannot be ignored completely. Problems of gradual transitions in flat country, sharper boundaries in mountains, and various types of mosaics are well known in phytogeography. It is also well known that environmental heterogeneity becomes much more important in less favorable climates, as opposed to mesic climates where forest trees can attain complete dominance and produce more uniform microclimates. Situations of co-dominance by several life forms do not always represent ecotones, since there are often certain degrees of regularity in the mixtures of life forms and of randomness to some spatial patterns, as in the warm semi-deserts. Walter (1975), with reference to the semi-desert vegetation of Soviet Middle Asia, has provided a useful and graphic description of gradual transitions, macromosaics and more complicated interdigitations along environmental gradients.[1] Only the first of these can be predicted without reference to the environmental heterogeneity of the particular site. True physiognomic ecotones generally involve environmental limitation of at least one dominant life form.

In trying to develop any world vegetation classification system it is useful to examine existing classifications. Vegetation has been classified and mapped, with increasing refinement and detail, using three main bases: physiognomy, environmental conditions, and floristics. Physiognomic classifications (e.g. von Humboldt 1807; Schimper 1898; Rübel 1930; Küchler 1967) employ the most obvious features of vegetation and best convey an impression of vegetation structure and appearance. 'Ecological' classifications, employing primarily climatic conditions (e.g. Merriam 1898; Holdridge 1947), point out general vegetation-environment relationships but provide less information about the vegetation and can be used to cover areas where the vegetation is insufficiently known. Floristic criteria are often employed as convenient surrogates for physiognomy (e.g. *Coniferae, Graminiae*). The relative merits of the remarkably few basic approaches to vegetation classification and mapping are well summarized by Küchler (1951, 1967), Whittaker (1962, 1975), Shimwell (1971), Mueller-Dombois & Ellenberg (1974), and de Laubenfels (1975). A useful bibliography of vegetation maps was compiled by Küchler (1965–1968) and is being updated.

In reality, most vegetation classification and mapping systems employ aspects of more than one approach. Two recent classification systems proved to be especially useful for ecophysiognomic studies. The life-form and associated 'physiognomic-ecological' vegetation classifications of Ellenberg & Mueller-Dombois (1967a, 1967b, reprinted as appendices in Mueller-Dombois & Ellenberg 1974)

[1]These patterns are described more fully, with examples of actual vegetation, by Walter & Box (in press).

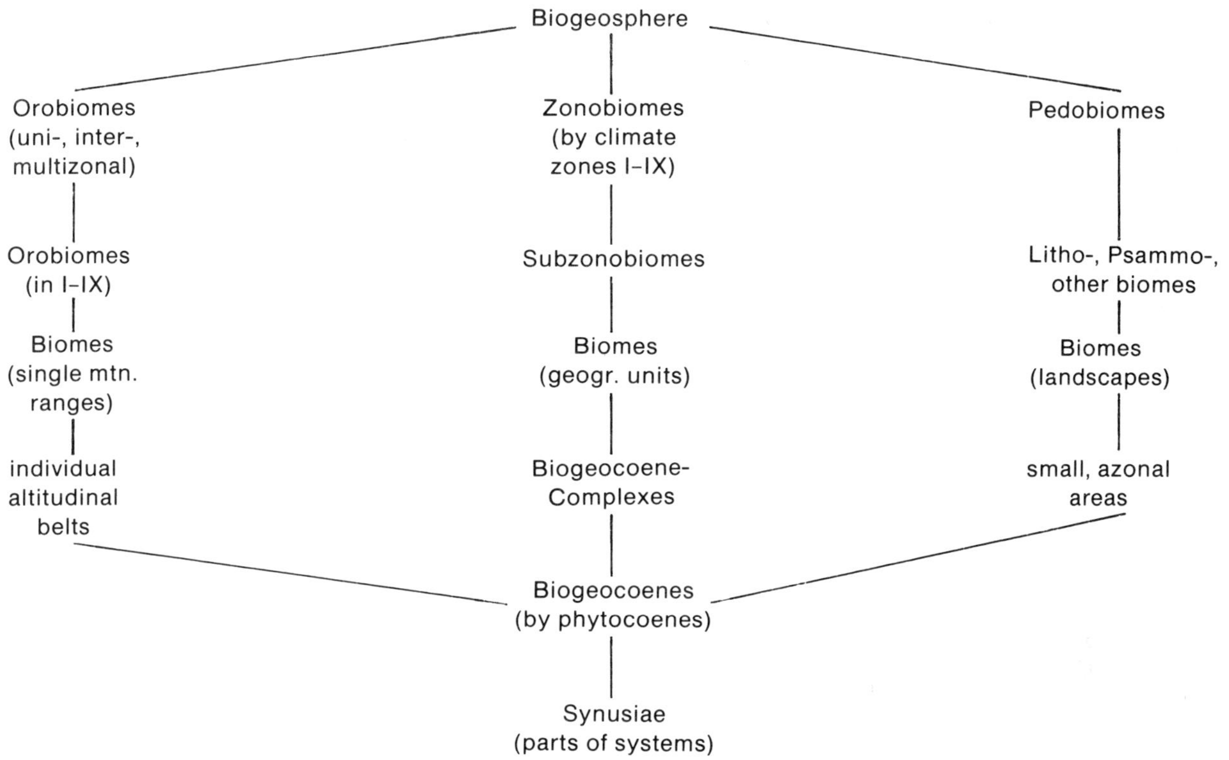

Fig. 1. Schematic representation of the global vegetation classification system of H. Walter (1975).
The biogeosphere (land areas of the world) is divided into climatically determined zonobiomes, within which there may also be edaphically determined pedobiomes and mountain ranges constituting orobiomes. Each can be subdivided on down to the level of individual phytocoenes. The Walter system provides a framework and nomenclature for classification of terrestrial vegetation units of all levels, including a systematic basis for distinguishing edaphic and other variants from the basic, climatically determined types. Classification can be based on Walter's own ten ecoclimatic zones or on other climatic classifications. Walter's ecoclimatic types, denoted by Roman numerals (see also Walter & Lieth 1960–67), are:

- I – equatorial, with diurnal seasonality
- II – tropical, with summer rain
- III – subtropical arid
- IV – mediterranean, with moist winters and dry summers
- V – warm-temperate (maritime; this zone should be divided into east-coast and west-coast sub-types)
- VI – typical temperate, with short frost period (nemoral)
- VII – arid temperate, with cold winters (continental)
- VIII – cold temperate, with cool summers (boreal)
- IX – arctic/antarctic
- X – highlands (divided into altitudinal belts but with reference also to the corresponding lowland climate zone).

The system is described in detail by Walter (1975). An English-language summary (Walter & Box 1976) is also available. A description of world zonobiomes and the more important pedobiomes and orobiomes has been presented by Walter (1977).

Table 1. Structural types of terrestrial plants and obvious ecophysiognomic bases for ecological subdivision.

The main structural types among terrestrial plants are listed in the left column along with what appear to be the most important ecophysiognomic bases for differentiation into ecological types (life forms). The central importance of leaf type (including size) in regulating energy and water budgets makes it the starting-point for subdivision of many basic forms. Seasonal habit (e.g. evergreen vs. deciduous) is also of central importance to all forms except those which show no variation (e.g. rosette-trees, which are always evergreen). The term structure can refer to branching structure, means of leaf deployment, and/or general plant (including leaf) shape. Morphs are different growth forms which can be assumed by a given taxon (usually species) under different environmental conditions, such as flagged and matt krummholz forms. The greatest variety in ecological types and forms (as well as species) may be shown by the forbs, which can be classified immediately into hemi-cryptophytic, geophytic, therophytic, and perhaps other (including larger) subtypes, each with a different ecological significance.

Structural type	Ecophysiognomic bases for subdivision
Trees	leaf type, seasonal habit, size
Dwarf-trees	leaf type, seasonal habit
Rosette-trees	size, leaf type
Rosette-treelets	general structure (palmiform, fern, xeric)
Arborescents	leaf type, seasonal habit, water source
Krummholz	leaf type, morphs, seasonal habit
Shrubs	leaf type, seasonal habit, size
Dwarf-shrubs	seasonal habit, leaf type, morphs
Rosette-shrubs	general structure
Stem-succulents	size, structure
Graminoids	structure, size, morphs, seasonal habit
Forbs	Raunkiaer life form, structure, leaf type, seasonal habit
Vines	structure, seasonal habit
Ferns	seasonal habit
Epiphytes (incl. hemi-epiphytes)	structure, leaf type, root type
Thallophytes	stand structure, substrate, plant type, seasonality

provided the life-form basis for the classification presented below. The life-form classification is a further development of that of Raunkiaer (1934), with the inclusion of such forms as tuft-trees, bottle-trees, and vines, and the useful distinction between woody, suffrutescent, herbaceous and succulent forms. The other especially useful system is the comprehensive framework for vegetation classification developed by Walter (1975, see also Walter & Box 1976). This system provides for classification of climatically determined, azonal (edaphic) and mountain vegetation units from biome scale on down to individual local synusiae (see Fig. 1). In combination with Walter's own (1968, 1973, 1976, 1977, or Walter & Lieth 1960–67) or any other adequate classification of world eco-climates, the system provides a means of partitioning the world into regions which can be expected to have different mixes of plant life forms. This was especially useful in mountainous areas, as seen in the example of the alpine belt, which takes quite different forms in the different (Walter) climate zones:

- I – páramo (rosette-treelets, grasses, cushion-shrubs)
- II – wet puna (grassland)
- III – dry puna (grass and cushion-plant semi-desert)
- IV – sclerophyllous cushion-shrubs and grasses
- VI – alpine mats
- VII – montane-alpine steppes
- VIII – montane tundra (herbs and dwarf-shrubs) with permafrost.

It is rare to find the necessary physiognomic data reported in world, regional or even local vegetation descriptions. It is even more uncommon to find those data coördinated with sufficient local climatic data, even in some recent volumes of the International Biological Program. Collecting and synthesizing these data from various sources is a major task in itself. By far the most useful vegetation descriptions in the literature are those which have followed the example of Heinrich Walter to include both site climate-diagrams or comparable monthly climatic data and vegetation description by life form, stratum, synusia, or other structural basis. Such descriptions include the following: Walter (1968, 1973, 1974, 1977), Knapp (1965, 1973), Hueck (1966), Cabrera (1968), Cuatrecasas (1968), many from the journal *Vegetatio,* and others.[2] It appears that the volumes of the new 'Ecosystems of the World' series will be quite useful, but monthly rather than quarterly data (as in Specht 1979) would be more useful.

B. Ecophysiognomic characters of terrestrial vegetation

It has been hypothesized already that the main links between plant form and plant function are the plant water and energy balances, both of which are directly linked to external environmental conditions through such features of plant form as surface area, surface (stomatal) resistance to gas exchange, and plant size. As a result, the most important ecophysiognomic characters of terrestrial plants are those which describe the various aspects of plant size, effective leaf area and structure, their effects on adjacent external microclimates, and their temporal variations. Six plant characters appear to be adequate to describe the main aspects of form which determine plant water and energy budgets:

1. structural type (without reference to the other characters, e.g. tree)
2. relative plant size (relative to others of the same structural type)
3. leaf type (form, e.g. broad, narrow)
4. relative leaf size (relative to others of the same leaf type)
5. leaf surface structure (e.g. sclerophyllous)
6. seasonal photosynthetic habit (e.g. evergreen)

The term 'leaf' is understood here to include phyllodes, cladodes, and other primary photosynthetic surfaces. The six characters are described in more detail in the following sections.

1. Structural type

Structural type is the most important and comprehensive aspect of general plant form. It is used here to refer to the most basic structural features, primarily size, branching pattern, and the distinc-

[2]Most of these volumes are in German. It is interesting to note that the most comprehensive geoecologic volume on the vegetation of all of North America, until the appearance of Daubenmire's (1978) work, was written by a German (Knapp 1965) and is not even available in a North American language! This same is true for South America (Hueck 1966).

tion between woody, herbaceous, and succulent forms (without reference to seasonal foliage patterns or the other characters). Structural types are thus such things as trees, shrubs, and graminoids. The structural types recognized in this study are shown in Table 1, along with ecophysiognomic bases for further classification.

Trees are generally the tallest and longest-lived forms. They have the largest leaf area, require the most water, may have the deepest roots, and are perhaps the ultimate K-strategists in that, despite requiring longer to reach maturity, they then form a stable structure and are relatively insensitive to most normal environmental variations. In the new plant-strategy system of Grime (1979) trees are seen as often the most successful competitors, since they can effectively shade all shorter forms except when occurring as isolated individuals. Small trees (also called treelets) can occur as obligate understorey forms or as facultatively smaller open-woodland or savanna forms.

Rosette-trees are not true dendritic trees but rather form their 'trunks' by superposition of branches which grow from a terminal bud to form a terminal rosette (tuft) and eventually die, leaving a contribution to a permanent stem. The terminal branches are actually leaf-fronds, and there is no secondary branching. Included are palms and tree ferns as well as representatives of a variety of other taxonomic groups with large rosette forms, such as yuccas (*Liliaceae*), papyas (*Caricaceae*), and the tropical alpine species of *Espeletia* and *Senecio* (*Compositae*).

Intermediate between tree and shrub forms are two somewhat more variable types: arborescents (de Laubenfels 1975) and krummholz. Arborescents may grow to tree size but are essentially overgrown shrubs, becoming arborescent if sufficient water is available. Examples include the mallee eucalypts of Australia and the mesquites (*Prosopis*) of the Americas. Both of these are phreatophytes. Krummholz can be distinctly creeping (e.g. junipers near tree-line), erect but wind-formed (flagged), or some intermediate combination. In any case the effective environmental factor is nearly constant wind which causes the low-growing and often intricately branched structures. Cushion-shrubs in particular are common to almost all windy environments. The height of krummholz in cold-winter climates is determined primarily by the winter snow cover, which protects against high winds and ice-blasting.

Shrubs may have much the same strategy and structure as trees but have much lower crowns (branching from very near ground level), less leaf area, less standing (respiring) biomass to support, but do not necessarily occur in drier environments and can occur under tree canopies where light and excess moisture permit. Rosette-shrubs are trunkless analogs of rosette-trees, to which they generally bear the same ecological relationship as just described for shrubs and trees.

Stem-succulents are generally leafless forms with fleshy, photosynthetic, expandable stems capable of storing water. These forms often have a woody internal infrastructure. Stem-succulents include the arborescent cacti and euphorbs, plus a wide variety of shorter barrel, cushion, or flat-stemmed forms.

Graminoids are efficient, narrow-leaved herbs with small flowers, well-developed root systems, and an ability to respond quickly to favorable growing conditions. They can form swards by sending out lateral shoots, grow in bunches or larger clumps, or grow in large tussocks. Dense lateral growth is favored both by killing winter temperatures and by grazing. The tussock form, best developed in maritime climates of the Southern Hemisphere, appears to be an adaptation for protection against wind and near-freezing temperatures.

The remaining types rarely attain dominance. Forbs are herbs which are neither graminoids nor ferns. Many subdivisions of form are possible, but most have relatively large, broad, malacophyllous leaves. Undifferentiated small herbs are often ephemeral and are found in extreme climates. They are generally so small that the particular plant and leaf form is less important than the plant size. Vines are woody or annual herbaceous plants which have weak stems, elongate rapidly, and support themselves by climbing or sprawling over other vegetation. Ferns (excluding tree ferns) are small, herbaceous, generally poikilohydrous plants which can often form dense ground synusiae. Epiphytes are plants which grow on other plants and include, for present purposes, hemi-epiphytes (pseudo-lianas) and semi-parasitic but photosynthetic forms. Thallophytes are non-vascular cryptogams (*sensu* Ellenberg & Mueller-Dombois) and include mosses, liverworts, lichens and algae.

The principal advantage gained by the assumption of a particular structural form is an appropriate size and shape (ratio of photosynthetic surface to respiring volume) for environmental conditions. The amount of water required by plants, without reference to particular water-conserving mechanisms, generally increases with the size of the plant. On the other hand, larger plants may be better adapted, through deeper root systems, to obtain more water and may also gain competitive advantages through greater size.

The total amount of water available can often be used to predict the structural types which will be present, although this pattern can be complicated considerably by such factors as length and dependability of rainy seasons and by various morphologic adaptations collectively referred to as xeromorphy. Also, a short but reliable wet season, providing enough water for tree growth, often produces a grassland instead, since the water may be intercepted very quickly and effectively near the surface and may not penetrate in sufficient quantities beyond the reach of grass roots. On the other hand, in the Middle Asian sand deserts, a leafless tree (*Haloxylon ammodendron*) forms the dominant vegetation because its deeper roots and adequate regulation of water loss permit it to reach sufficient water at depths which are too great for grasses and shrubs but at which the water is protected from direct evaporation.

Temperature can also control the size and type of plants present, primarily at the cold end of the spectrum, through the amount of summer warmth available at plant height. The air is generally warmest in the first few decimeters above the surface. Thus, as summers become shorter toward the poles or cooler in polar maritime climates, sufficient warmth for wood production at tree height disappears, then sufficient warmth for shrubs and krummholz. Finally only reptant dwarf-shrubs (chamaephytes) and herbs with short life-cycles find enough warmth near the surface for their annual development. Perennial evergreen species may grow significantly only in favorable years. This successive environmental limitation of structural types results to a considerable extent from the size of the plants involved, without major variation in temperature requirements.

2. Relative plant size

Relative plant size refers to the size (height) of the particular form in relation to other forms of the same structural type. The combination of structural type and relative plant size provides measures of actual plant size and (to some extent) total leaf area. Inclusion of this character also permits distinction, for example, between:

1. tall (including emergent), normal, and small (e.g. understorey) trees and rosette-trees;
2. large, normal and dwarf shrubs;
3. tall and short grasses; and
4. arborescent and non-arborescent stem-succulents.

Four sizes are recognized: tall (large), normal, short, and dwarf. Relative plant size is also generally limited by water and temperature. Increased height usually means greater leaf-surface or other photosynthetic area but also greater transpiring area, increasing potential water loss. This must be balanced against metabolic and competitive requirements, such as respiration and the need for more incident solar radiation. The factors which determine the height of vegetation in general are quite complex and are not yet well understood.

3. Leaf type

Leaf type refers to the general form of the leaf or other photosynthetic organ, such as phyllodes, cladodes or photosynthetic stems. Four types are recognized: broad, narrow, graminoid, and absent. Narrow leaves include true needles (as in pines), flattened linear leaves (as in *Podocarpus*), and cupressoid scales. Most narrow-leaved species are conifers, though other species with similar structure are also included (e.g. *Casuarina*).

In general, broad leaves offer both greater photosynthetic and greater transpiring areas and, thus, are found in more favorable environments. Narrow leaves (often harder than broad leaves) are smaller, offer better resistance against desiccation from both heat and cold, and are generally found in environments with at least some period of more extreme conditions. These generalizations can be modified, however, by leaf size and hardness.

4. Relative leaf size

Relative leaf size refers to sizes of the leaf or other photosynthetic organ in relation to those of other plants of the same leaf type. Thus, a broad megaphyll and the long needles of *Pinus palustris* can both be considered to be large leaves, suggesting certain favorable aspects of the climates where they are found. It seems useful to distinguish four leaf sizes: large (macrophyll and larger), normal (mesophyll), small (microphyll), and very small (nanophyll and smaller). Small leaves would include those of boreal birches, boreal short-needled conifers, and the microphylls of drier tropical climates (e.g. *Acacia,* including most phyllode types). Very small leaves would include the reduced, scale-like leaves of most *Cupressaceae* and the very small leaves or other organs (e.g. *Zygophyllaceae*) of many cold-climate and warm semi-desert shrubs and herbs. Leaf-size classes were suggested by those of Raunkiaer (1934).

There have been many theories concerning the factors which determine leaf size. Earlier interpretations are well summarized by Shimwell (1971). Gates (1968) assumed that leaf sizes evolved to regulate leaf temperature and developed a mathematical energy-balance equation relating the size and temperature of broad leaves to ambient conditions. Taylor (1975) used this approach in attempting to define optimal leaf forms and sizes for broad-leaved plants. A similar explanation for narrower leaves and needles does not exist. Parkhurst & Loucks (1972) maintained that Gates' approach ignores the adaptability of thermal aspects of photosynthesis and developed a leaf-size model based on water-use efficiency. Most recently, Givnish & Vermeij (1976) have improved on the Parkhurst-Loucks approach by considering the additive difference between photosynthetic gains and transpirational losses, rather than their ratio. Because of data requirements, none of these models can be used quantitatively in the current approach.

In general, larger leaves both assimilate and transpire more. Leaf size, however, is connected through energy and water relations to several other morphologic characters. Total leaf area can be affected by number of leaves as well as by leaf size. Thoday (1931) pointed out that small leaf size can facilitate both water diffusion within the leaf and more uniform illumination of a greater depth of plant foliage. Gates' results showed that smaller leaves often transpire less because they overheat less. Microphylly may be coupled with sclerophylly or less extreme cuticular hardening ('leathery' leaves) in order to reduce water loss further. Total leaf area, however, may remain high, as in the case of dense mediterranean woodlands. The large leathery leaves of tropical humid and even perhumid areas seem to represent an intermediate situation in which high leaf area is desirable but water loss must be controlled, as least for a few hours around midday.

5. Leaf structure

Leaf structure refers to the general 'hardness' of leaves or other photosynthetic surface and serves as an index of the plant's ability to control its rates of water loss and gas exchange. Leaf-structure adaptations are primarily anatomical and involve such aspects as amount of sclerenchymous tissue, stomate morphology, and degree of succulence. Six basic surface types (with considerable variation) can be readily distinguished.

1. Most broad deciduous leaves are malacophyllous (herbaceous), like the leaves of birches or maples. These are typically thin and soft but can become more reinforced with age, as in the case of beeches and summergreen oaks.
2. Coriaceous (leathery) leaves have thickened cuticles and are somewhat tougher but are still quite pliable. These are designed to retard water loss but not to withstand winter or nighttime cold. Tropical evergreen leaves, even in rainforests, are typically coriaceous, as are the narrow leaves or needles of summergreen larches and some temperate-maritime conifers.
3. Sclerophyllous leaves are mechanically reinforced with sclerenchyma and are both hard and stiff. These are designed to restrict water loss over longer periods, including colder winters, and are typical of temperate-zone evergreens, especially in mediterranean climates and in such taxa as *Eucalyptus,* evergreen *Quercus, Ericaceae, Ilex* (holly), and boreal conifers.
4. Succulent leaves or other parts have leathery or harder surfaces to retard water loss combined with internal cell structures designed to store water. These generally do not withstand prolonged freezing temperatures (some excep-

tions) and can be vulnerable to rotting in wetter areas.

5. Woody photosynthetic parts (e.g. stems) can be referred to as ligneous and can be found on evergreen, deciduous and leafless plants. Ligneous surfaces show the lowest photosynthesis rates but reduce water loss to almost zero during dry periods and can withstand cold winters.
6. Pubescent leaves or other parts are typically rather soft but somewhat fleshy and are covered with more or less dense, often white pubescence, which reduces water loss by mitigating the drying effects of wind and direct insolation. Such surfaces are typical of less extreme dry climates.

In general, deciduous leaves are malacophyllous and rarely truly harder. Evergreen leaves, on the other hand, are typically coriaceous in the tropics (lowlands), sclerophyllous where frost occurs, and rarely ever malacophyllous. 'Softer' leaves generally both transpire and photosynthesize more rapidly. They may, however, have little or no ability to regulate water loss by closing their stomates, since they would quickly overheat without the cooling effects of transpiration. 'Harder' succulent and ligneous surfaces can reduce transpiration almost completely during dry periods but permit only very low annual photosynthetic rates. There are, of course, many transitional and perhaps other basic forms. In some situations, such as tropical-rainforest canopy and understorey leaves, one can find wide variations in leaf morphology and structure on individual plants.

6. Photosynthetic habit

Photosynthetic habit refers to the seasonal photosynthetic activity of a plant (or population), as measured by obvious features such as leaf tenure or the dying back of aerial shoots. Seven basic types can be readily identified:

1. Evergreen plants (or populations) effectively retain their photosynthetic organs in active condition throughout the year. True evergreens retain their leaves or other green parts for more than one year, but plants which lose their old leaves in 'spring' when new ones are produced are also considered to be effectively evergreen.
2. Semi-evergreen ('tardily deciduous') plants are evergreen where conditions permit but may lose most or all of their leaves as the unfavorable season progresses.
3. Raingreen plants lose their leaves or annual shoots in response to a dry season and produce new ones at the beginning of a wetter season, as in tropical summer-rain climates. The old leaves are generally not killed directly by the dryness but rather close their stomates and cease production, turning yellow rather than brown.
4. Summergreen plants lose their leaves or annual shoots in response to a cold winter and produce new ones in the spring, as in temperate-zone and polar climates. Summergreen leaves and annual shoots generally are 'killed' by autumn nighttime frost, but leaves in particular may continue to produce and accumulate sugars for several days or even weeks after ceasing translocation, turning brilliant yellow or red before withering and being dropped.
5. Suffrutescent plants have perennial woody bases but produce annual herbaceous or slightly lignified shoots (with or without leaves) which are discarded at some time after the end of the growing season. The perennial base may or may not have some secondary foliage.
6. Marcescent plants (generally graminoids or other erect herbs) produce annual herbaceous or slightly lignified shoots (e.g. tropical grasses) which die at the end of the growing season but may remain standing throughout the unfavorable season and perhaps longer. The photosynthetic habit of these plants can often best be described as seasonal (no distinction between summergreen and raingreen) or opportunistic, and the term marcescent refers only to the persistence of the dead shoots.
7. Ephemeral plants can be annuals or perennials (e.g. geophytes) but produce annual herbaceous shoots which generally appear quickly in response to favorable conditions but which may endure for only a few weeks or even days.

In general, plants are evergreen if they have adaptations (sclerophylly, microphylly, etc.) which permit them to survive unfavorable periods and are deciduous or otherwise seasonal if they do not. Thus the vegetation of equatorial, mediterranean, and warm-temperate maritime climates is mostly evergreen, while that of tropical summer-rain, subtropical arid, typical temperate and polar climates is mostly seasonal. The exception to the above generalization is the boreal conifers, which are evergreen despite severe winters in order to make full use of the short, cool summer growing season. Boreal conifers survive the cold winter by a combination of sclerophylly, microphylly, and physiological changes (cold-hardening and winter dormancy) which almost eliminate respiration in winter. The situation can also be quite complicated in mediterranean and sub-mediterranean arid climates, where rainy and warm seasons overlap the least. Suffrutescent, summer-deciduous, semi-evergreen and spring-ephemeral forms can be especially important in such climates and can become dominant where total moisture is not sufficient for evergreen plants.

In some cases, especially those of summergreen and ephemeral plants, seasonal habits may be closely tied to other, mandatory life-cycle functions. Summergreen trees have been grown, for example, in nonseasonal, continuously favorable environments, but after a few years the life-cycle functions become confused (continuous blooming; simultaneous blooming foliation, and defoliation on different parts of the same plant) and the plants die. In other cases, notably that of raingreen plants, seasonal deciduousness is often facultative (at least over periods of a few years), with leaf-fall not occurring if water remains sufficient. This can be observed in many botanical gardens in tropical climates with seasonal drought.

Obligatorily evergreen, raingreen, and summergreen habits, plus facultative deciduousness, represent four different life-form physiologies. In many cases the actual character, and thus the true life form, is very difficult to determine and may remain unknown. Distinction between summergreen and raingreen life forms, for example, can be especially difficult in areas such as subtropical mountains, where the two regions often merge. Distinction between obligate and facultative evergreens can be equally difficult in rainforests.

C. Life forms of world terrestrial vegetation

The main criteria for the classification of plant life forms were discussed earlier and include provisions for covering the variation existing on earth, for definition in terms of ecologically significant physiognomic characters, and for identification of convergent forms which are typical of particular environmental situations. Fortunately for the ecologist, plant form does tend to cluster into successful, definable types. The life forms presented here and in most other classifications are intended to characterize these ecophysiognomic nodes, while at the same time realizing that there are many transitional forms which may be very hard to classify.[3]

The ecophysiognomic life forms presented here are postulated to represent structural-functional forms which have evolved as the result of adaptation to particular climatic conditions. Among other advantages, such as necessary simplification, this sort of ecophysiognomic generalization of plant detail serves to focus on the most ecologically important relationships and to overcome partly the lack of data on particular species, morphs and ecotypes. The life-form classification was patterned after those of Rübel (1930), Whittaker (1975), and Ellenberg & Mueller-Dombois (1967a, 1967b). The types were identified by studying many types of plant and vegetation data at various scales and grouping species which are physiognomically and physiologically similar and occur in similar climatic situations. Life forms generally involve groups of particular species, but some species can occur in more than one form (morph), such as various *Juniperus* species in tree, shrub and krummholz forms. Furthermore, the life form is not known for all species, especially when it depends on functional characteristics such as deciduousness.

The life-form classification is presented in Table 2, along with examples of the individual types and a textual summary of the system. More detailed descriptions of the individual forms, including general environmental relations, geographical distribution, prototypic taxa, and relevant literature, are provided in Appendix A. Identification of characteristic species or genera is especially useful but also problematic. Some very important species span two entire ecoclimates and must be considered to belong to both life forms simultaneously, for example *Pinus sylvestris* (boreal

[3]The extent of species clustering into life forms and the relative numerical importance of transitional forms will be studied in a forthcoming paper (Box, in preparation) in which the entire flora of the Carolinas, as compiled by Radford et al. (1968), is classified by life form.

Table 2. Life forms of world terrestrial vegetation.

The life forms (numbered) are listed in the left column, grouped by general growth form or structural type (e.g. broad-leaved trees) and with sub-types in some cases (e.g. warm-temperate, mediterranean, and rainforest broad-evergreen trees). Some forms are divided further into ecoclimatic sub-types with less physiognomic difference (e.g. lowland and montane tropical rainforest trees). Examples of each life form (or sub-form) are provided in the right column. The 77 life forms (90 counting all sub-types) cover the full range of variation in terrestrial plant form, including both potential formation dominants and understorey forms, which are usually more generalized. The life forms were conceived primarily as ecologically significant combinations of certain physiognomic characters, including general structural type, size, leaf form and size, and seasonal habit. Some life forms thus include very few species while others involve many hundreds (even thousands) of species and still considerable variation in less obvious aspects of both form and function. Each life form is related to various annual and seasonal aspects of temperature and water balance regimes by means of estimated tolerance limits (Table 7). Predicted world distributions and importance of the life forms, based on their tolerance limits and hypothesized form-based dominance relationships, are shown in Table 12 and in Maps 10–22. Estimates of potential changes in distribution with changes in climate are presented in Tables 17–19.

Plant form	Examples
Trees (Broad-leaved)	
Evergreen	
1. Tropical Rainforest Trees (lowland, montane)	*Lauraceae, Rubiaceae*
2. Tropical Evergreen Microphyll Trees	*Leguminosae, Meliaceae, Simaroubaceae*
3. Tropical Evergreen Sclerophyll Trees	*Eucalyptus*
4. Temperate Broad-Evergreen Trees	
a. Warm-Temperate	*Quercus virginiana*
b. Mediterranean	*Quercus ilex, Arbutus, Olea europaea*
c. Temperate Rainforest	*Magnoliaceae, Lauraceae*
Deciduous	
5. Raingreen Broad-Leaved Trees	
a. Monsoon mesomorphic (lowland, montane)	*Tectona, Dipterocarpaceae*
b. Woodland xeromorphic	*Acacia, Adansonia, Caesalpinaceae*
6. Summergreen Broad-Leaved Trees	
a. typical-temperate mesophyllous	*Quercus, Acer, Fagus*
b. cool-summer microphyllous	*Betula, Populus, Nothofagus*
Trees (Narrow and needle-leaved)	
Evergreen	
7. Tropical Linear-Leaved Trees	*Podocarpus, Agathis*
8. Tropical Xeric Needle-Trees	*Juniperus procera, Widdringtonia*
9. Temperate Rainforest Needle-Trees	*Tsuga, Thuja, Sequoia*
10. Temperate Needle-Leaved Trees	
a. Heliophilic Large-Needled	*Pinus taeda, P. caribbea*
b. Mediterranean	*Cedrus, Cupressus, Pinus pinea*
c. Typical Temperate	*Pinus strobus, P. ponderosa*
11. Boreal/Montane Needle-Trees	*Picea, Abies*
Summergreen	
12. Hydrophilic Summergreen Needle-Trees	*Taxodium, Metasequoia*
13. Boreal Summergreen Needle-Trees	*Larix, Pseudolarix*

Table 2 (continued).

Plant form	Examples
Small and dwarf trees	
14. Tropical Broad-Evergreen Small Trees	rainforest understorey, *Leguminosae*
15. Tropical Broad-Evergreen Dwarf-Trees	'campo cerrado' treelets
16. Cloud-Forest Small Trees	*Podocarpus, Ericaceae*
17. Temperate Broad-Evergreen Small Trees (typical, cool-maritime)	*Ilex, Nothofagus, Berberis*
18. Broad-Raingreen Small Trees	*Leguminosae*
19. Broad-Summergreen Small Trees	*Prunus, Nothofagus, Betula tortuosa*
20. Needle-Leaved Small Trees	*Juniperus, Actinostrobus*
Rosette-trees	
21. Palmiform Tuft-Trees	palms, *Caricaceae*
Rosette-treelets	
22. Palmiform Tuft-Treelets	understorey palms, cycads
23. Tree Ferns	*Cyatheaceae, Dicksoniaceae*
24. Tropical Alpine Tuft-Treelets	*Senecio, Espeletia*
25. Xeric Tuft-Treelets	*Yucca, Dracaena, Xanthorrhoea*
Arborescents	
26. Evergreen Arborescents	mallee eucalypts
27. Raingreen Thorn-Scrub	*Acacia, Commiphora*
28. Summergreen Arborescents	*Prosopis, Salix*
29. Leafless Arborescents	*Haloxylon, Calligonum*
Krummholz	
30. Needle-Leaved Treeline Krummholz	*Picea, Abies, Juniperus*
Shrubs	
31. Tropical Broad-Evergreen Shrubs	*Coffea, Rubiaceae, Ericaceae*
32. Temperate Broad-Evergreen Shrubs	
a. Mediterranean	*Proteaceae, Quercus dumosa, Rhamnus*
b. Typical Temperate	*Ilex, Ligustrum*
c. Broad-Ericoid (perhumid)	*Rhododendron*
33. Hot-Desert Evergreen Shrubs	*Zygophyllaceae, Acacia aneura* (mulga)
34. Leaf-Succulent Evergreen Shrubs/Treelets	*Crassula argentea*
35. Cold-Winter Xeromorphic Shrubs	*Artemisia*
36. Summergreen Broad-Leaved Shrubs	
a. mesomorphic	*Rosa, Vaccinium*
b. xeromorphic	'deciduous chaparral', šibljak
37. Needle-Leaved Evergreen Shrubs	*Juniperus communis*
Dwarf-shrubs	
38. Mediterranean Dwarf-Shrubs	*Thymus, Salvia, Eriogonum*
39. Temperate Evergreen Dwarf-Shrubs (typical, maritime heath)	heath and arctic/alpine *Ericaceae*
40. Summergreen Tundra Dwarf-Shrubs	*Betula nana, Salix reptans*
41. Xeric Dwarf-Shrubs	*Ephedra, Anabasis, Retama*
Cushion-shrubs	
42. Perhumid Evergreen Cushion-Shrubs	*Azorella selago*
43. Xeric Cushion-Shrubs	puna/Patagonian hard cushions

Table 2 (continued).

Plant form	Examples
Rosette-shrubs	
44. Mesic Rosette-Shrubs	understorey and ground palms
45. Xeric Rosette-Shrubs	*Agave, Yucca, Aloë*
Stem-succulents	
46. Arborescent Stem-Succulents	*Carnegiea gigantea, Euphorbia candelabrum*
47. Typical Stem-Succulents	unbranched barrel cacti, *Mammillaria*
48. Bush Stem-Succulents	branched *Opuntia* spp.
Graminoids	
49. Arborescent Grasses	bamboos
50. Tall Cane-Grasses	*Imperata, Arundinaria*
51. Typical Tall Grasses	*Andropogon, Festuca,* prairie grasses
52. Short Sward-Grasses	*Cynodon dactylon, Bouteloua gracilis*
53. Short Bunch-Grasses	*Festuca, Stipa, Agropyron*
54. Tall Tussock-Grasses	pampas and Patagonian grasses (e.g. *Stipa*)
55. Short Tussock-Grasses	puna grasses, *Festuca novae-selandiae*
56. Sclerophyllous Grasses	'spinifex' (*Triodia*), *Scleropoa*
57. Desert Grasses	*Aristida* (wire grass), *Stipa*
Forbs	
58. Tropical Evergreen Forbs	*Cannaceae, Begonia, Zingiberaceae*
59. Temperate Evergreen Forbs	*Gaultheria, Chimaphila, Hexastylis*
60. Raingreen Forbs	*Leguminosae, Compositae*
61. Summergreen Forbs	forest dicots, geophytes, *Compositae*
62. Succulent Forbs	*Portulacca, Sedum, Sempervivum*
Undifferentiated small herbs	
63. Xeric Cushion-Herbs	*Saxifraga, Dryas, Draba*
64. Ephemeral Dry-Desert Herbs	annuals, dwarf-geophytes, graminoids
65. Summergreen Cold-Desert Herbs	dwarf-geophytes, graminoids
66. Raingreen Cold-Desert Herbs	geophytes, graminoids
Vines and lianas	
67. Tropical Broad-Evergreen Lianas	*Ficus, Calamus,* stranglers
68. Broad-Evergreen Vines	*Philodendron, Lonicera, Smilax*
69. Broad-Raingreen Vines	*Leguminosae, Ipomoea*
70. Broad-Summergreen Vines	*Vitis, Parthenocissus, Rhus radicans*
Ferns	
71. Evergreen Ferns	rainforest ferns (e.g. *Polypodium*)
72. Summergreen Ferns	temperate ferns (e.g. *Aspidiaceae*)
Epiphytes	
73. Tropical Broad-Evergreen Epiphytes	bromeliads, orchids, aroids, cacti
74. Narrow-Leaved Epiphytes	ferns, mosses, *Tillandsia*
75. Broad-Wintergreen Epiphytes	'mistletoes' (*Loranthaceae*)
Thallophytes	
76. Mat-Forming Thallophytes	forest and tundra mosses, folious lichens
77. Xeric Thallophytes	crustose lichens

and temperate needle-trees) and *Pseudotsuga menziesii* (temperate and sub-mediterranean needle-trees). Also, should linear-leaved *Adenostoma fasciculatum* and *Rosmarinus officinalis* be called needle-leaved or mediterranean evergreen shrubs? What should one do with *Casuarina, Tamarix* and de Laubenfels' 'overgrown bushes'? Most problematic of all, as one might expect, is probably the genus *Eucalyptus,* which shows such minimal physiognomic variation in different climates and which clearly can thrive in many areas of the world into which it has been introduced only recently.

The current 'analytical' classification differs from an earlier, shorter 'synthetic' classification (Box 1978a) by having completely separated plant form and stand physiognomy (forest vs. open-woodland trees) and by including some important larger forms (e.g. palms and tropical conifers) and the understorey forms (forbs, vines, epiphytes, thallophytes) which were omitted from the earlier classification. Some earlier forms were also divided, merged and/or renamed.

Life forms are grouped in Table 2 by general growth form, but this does not always imply similar ecology, as in the case of phreatophytic and non-phreatophytic arborescents. Although the forms are given names intended to suggest regional, seasonal

Table 3. Plant ecophysiognomic diversity within structural classes.

Structural class (1)	Total number of life forms (2)	Number of growth forms (3)	Structural class (1)	Total number of life forms (2)	Number of growth forms (3)
Trees	19	5	Stem-succulents	3	3
Small trees	7	4	Graminoids	9	7
Rosette-trees and treelets	5	3	Forbs	5	4
Arborescents	4	4	Undifferentiated herbs	4	4
Krummholz (incl. Cushion-shrubs)	3	3	Vines and lianas	4	4
Shrubs	10	6	Ferns	2	2
Dwarf-shrubs	4	4	Epiphytes	3	3
Rosette-shrubs	2	2	Thallophytes	2	2

The basic structural classes of plants are shown in column 1, along with the number of corresponding life forms identified in the model, column 2. The number of growth forms (combinations of structural type and size, leaf type, and seasonal habit) represented among the life forms of each structural class is shown in column 3. Ecophysiognomic diversity within structural classes, as expected, appears to be best represented among the larger forms.

Table 4. Life-form diversity of ecoclimatic regions.

Region (1)	No. of life forms (2)	Main growth forms (3)
Tropics (excluding alpine)	18	trees, rosette-trees, etc.
Tropical alpine	2	tuft-treelet, small herb
Subtropical (including arid and mediterranean)	20	shrubs, succulents, rosettes, etc.
Tropical and temperate (mild winters, mainly S. Hem.)	10	trees, rosette-trees, ferns, cushions, etc.
Temperate zone	25	trees, shrubs, grasses, forbs, etc.
Boreal/temperate montane	3	needle-trees, krummholz
Temperate and polar/alpine	3	dwarf-shrub, herbs
Polar/temperate alpine	2	dwarf-shrub, small herb
Global	3	thallophytes, bunch-grass

The regions were determined by the ranges of the life forms involved. Each life form in Table 2 (including sub-types) was assigned to a single region of which it is most characteristic. The number of life forms assigned to each region is shown in column 2, and the most common growth forms in column 3. The greater diversity in the temperate zone probably reflects the greater diversity of temperate ecoclimates combined with the greater dominance by tree forms in the tropics.

or other obvious environmental situations, each form is well-defined physiognomically by means of the six physiognomic characters described earlier. These physiognomic characters are listed in Appendix A along with a more detailed description of each form, including geographical distribution, prototypic taxa, and relevant literature.

The distribution of life forms and growth forms among the basic structural classes is shown in Table 3. The numbers of growth forms represent different configurations of leaf type and seasonal habit among the life forms of the respective structural classes. Relatively low numbers of basic growth forms suggest a wider variety of leaf sizes and structures, as among the trees in particular. The apparent lack of such diversity among smaller forms reflects greater environmental constraints. As niches become smaller, however, smaller forms can be very diverse in specific detail, as can be seen from the number of graminoid and forb species in most florae.

The regional affinities of the various life forms are summarized in Table 4. Each life form was assigned to a single region throughout which it occurs most typically. The high number of temperate forms reflects the fact that the temperate zone is in the middle of the spectrum and reflects the greater ecoclimatic diversity and complexity of the temperate zone. Table 4 shows that the life-form classification is balanced geographically and that no region is grossly over- or under-represented.

The Box life-form classification was developed largely from the Raunkiaer-based classification of Ellenberg & Mueller-Dombois (1967b, 1974). The correspondence of these three systems is summarized in Table 5. The Box classification is not intended to be as strictly systematic (i.e. as precise) as the others but rather more generalized and readily applicable to geographic and ecologic description of vegetation and ecosystem structure. This implies omission of ecologic detail in some cases, especially the smaller forms, while seeking to cover the earth's main forms within a system of manageable size.

Table 5. Correspondence of the Raunkiaer, Ellenberg & Mueller-Dombois, and Box systems of terrestrial plant life forms.

The basic life forms of Raunkiaer (1934) plus some later additions within the Raunkiaer framework (e.g. Braun-Blanquet 1928) are listed in the first column. The main subdivisions of the system of Ellenberg & Mueller-Dombois (1967b, 1974), which represents an elaboration of the Raunkiaer system, are shown in the column 2. These subdivisions are always subdivisions of the corresponding Raunkiaer form shown in column 1. Column 3 shows the life-form groups of the Box system as presented in this monograph (see Table 2). The basic life forms or life-form groups of the Raunkiaer and Ellenberg/Mueller-Dombois systems are retained in the Box system (generally with common names, e.g. trees and shrubs) except for the hemi-cryptophytes, geophytes and therophytes, which are divided into graminoids, forbs, small herbs, and ferns. Further division of these commonly recognized general forms as hemi-cryptophyte, geophyte and therophyte forms and sub-forms is being pursued locally using initially the flora of Carolina (Box, in preparation).

Raunkiaer system (1)	Ellenberg & Mueller-Dombois (2)	Box system (3)
1. Phanerophytes	1. Trees, Shrubs, Krummholz (mega-, meso-, micro-, and nano-phanerophytes)	Trees (broad and narrow-leaved) Small and Dwarf Trees Arborescents Krummholz Shrubs
	2. Tuft-trees	Rosette (tuft)-trees and treelets
	3. Bottle-trees (normal, palmiform, leaf-succulent, aphyllous forms)	(not recognized separately)

Table 5.

Raunkiaer system (1)	Ellenberg & Mueller-Dombois (2)	Box system (3)
	4. Tall Succulents (single-stemmed and caespitose forms)	Stem-Succulents: Arborescent plus some typical and bush forms
	5. Herbaceous Phanerophytes	Graminoids: Arborescent
	graminoid	Forbs: various (height not differentiated)
	forb	
	(not explicitly recognized)	Rosette-Shrubs
2. Chamaephytes	1. Woody Dwarf-Shrubs (frutescent)	Dwarf-Shrubs Cushion-Shrubs
	2. Semi-Woody Dwarf-Shrubs (suffrutescent)	Dwarf-Shrubs (semi-shrubs) Cushion-Shrubs
	3. Herbaceous Chamaephytes (caespitose, reptant, pulvinate, scapose)	Graminoids: various Forbs: various
	4. Low Succulents (stem-, leaf-, and root-succulents)	Stem-Succulents: some typical and bush forms Rosette-Shrubs: smaller forms Forbs: larger Succulent Forbs
	5. Poikilohydrous Chamaephytes (generally xeric ferns)	Ferns: some xeric forms
3. Hemi-Cryptophytes (generally herbaceous throughout, with some partial lignification in old stems)	1. Caespitose 2. Reptant 3. Scapose 4. Aquatic	Graminoids (differentiated by size and form) Forbs, Small Herbs, and Ferns (differentiated by seasonality)
4. Geophytes (herbaceous)	1. Root-budding 2. Bulbous 3. Rhizome-Geophytes 4. Aquatic	(not recognized separately)
5. Therophytes (annuals and other herbs completing their life cycles within one favorable growing period)	1. Caespitose 2. Reptant 3. Scapose 4. Aquatic 5. Succulent	(not recognized separately except for some Ephemeral Desert Herbs)
Not recognized originally: 6. Lianas (including vines)	1. phanerophytic and chamaephytic 2. hemi-cryptophytic 3. geophytic 4. therophytic	Vines and Lianas (differentiated by form and seasonality)
7. Hemi-Epiphytes (epiphytic lianas)	roots dying, strangling, or descending	Vines and Lianas (differentiated by form and seasonality)
8. Epiphytes (vascular)	1. facultative epiphytes 2. obligative epiphytes	Epiphytes (differentiated by leaf form)
9. Thallophytes (non-vascular cryptogams)	thallo-chamaephytes thallo-hemi-cryptophytes thallo-therophytes thallo-epiphytes	Thallophytes (mat-forming vs. xeric)

CHAPTER 4

Modeling the effective environment

Modeling the effective environment, perhaps more than any other aspect of ecological modeling (*sensu lato*), represents the point at which modeling becomes as much an art as a science. Unlike the modeling of well-defined processes, amounts, forms and other phenomena, modeling the effective environment involves choosing, out of the entire set (continuum) of environmental factors, those which are most important and then choosing or constructing variables which best and most economically represent the most important spatial, temporal or other particular aspects of those factors. The primary hypothesis behind the current project is that plant form and vegetation structure are determined mainly by general climatic conditions. This means that the effective environment to be modeled is those aspects of climate which most influence plant and vegetation water and energy budgets, life-cycle processes, ecological interactions, and which are most important in limiting species and life-form distributions. The model must at the same time be simple and understandable.

A. Climate data

Data on climate phenomena are generally more readily available than are many important types of vegetation data. Of the many climatic variables monitored, the best records are for various aspects of annual regimes of temperature and precipitation. Various studies relating vegetation distributions to climatic factors (e.g. Rübel 1930; Schimper & von Faber 1935; Lieth 1956; Troll & Paffen 1964) have shown that, to a remarkable extent, it is the general levels and patterns of these macroclimatic phenomena which are best correlated with the distributions of particular plant and vegetation types. This greatly simplifies the problem of data-collection. The other important factor (e.g. Thornthwaite 1948; Major 1963; Mather & Yoshioka 1966) is evaporation, which must be estimated since few and often only questionable measurements are available. Many other climatic factors can be important in particular situations. Climatic phenomena are closely interrelated, however, and to a large extent these other factors, such as cloudiness, wind, solar radiation, and frost occurrence, plus diurnal patterns of the primary factors, can be expressed through various aspects of annual temperature, precipitation, and evaporation regimes.

Temperature is usually the easiest to treat, since the important aspect, ambient air temperature near the ground, is relatively uniform over large areas. Means of daily and monthly temperatures averaged over available periods of measurement usually converge within a few years to relatively smooth, characteristic annual patterns. Although intra-annual patterns may vary from year to year, more general trends are persistent, and it is these general trends which are most important to plants.

Precipitation can be a much more local and erratic phenomenon. It is actually soil moisture rather than precipitation, however, which is of greater importance to plant water budgets. Depending on soil depth, texture, porosity, and general rainfall levels and temporal patterns, soil moisture supplies may be much less variable than is actual precipitation.

The aspect of evaporation which is perhaps of greatest importance to plants is the climatic potential rate of water loss, generally called potential evapotranspiration (PET), which is a function of the energy supplied and the saturation deficit of the air near the vegetation. The latter can be effectively increased by wind, which removes transpired water vapor from the plants' vicinity and prevents a reduction of drying power. PET can be estimated with reasonable accuracy for periods of not less than a month by various methods based on air temperature alone, provided that the climate is not unusually windy. Although evaporative demands can change drastically within very short periods, most plants are well enough adapted to their local PET regimes that they can regulate their stomatal behavior adequately to withstand large variations.

When the author first became interested in ecological modeling (1972), records of monthly mean temperature and average precipitation were available for most parts of the world covering periods of measurement ranging from 10 to nearly 100 years. By 1977 records covering 3–20 years of measurement were appearing in print for most of the areas not represented earlier. The most comprehensive source of world climate data available at the time was the *Climate-Diagram World Atlas* (Walter & Lieth 1960–67), which includes data for about 8000 stations. Various compendia of digital monthly data were or became available, but these covered too few stations and could be used only for individual sites not available in the climate-diagram atlas. Computerized files of world climate data (e.g. Spangler & Jenne 1979) were not yet available or known. Some error is involved both in printing and in reading the non-digital monthly values on climate diagrams. Comparison of corresponding monthly climate-diagram and digital data from various sources showed, however, that the error involved with climate diagrams is no greater than the normal variation in digital values based on different 30-year periods of measurement. Monthly climate-diagram values can generally be read with an error of ±1 °C or ±2 mm at most.

As data-collection began it had not yet been determined which variables would be used in the final model. This could not be determined until results were generated using several combinations of variables. The data to be collected were thus chosen so that monthly values of annual temperature, precipitation and PET could be estimated from them if necessary. Because of the large number of values needed (potentially about 40 for each site), it was necessary to reduce the number of values actually read. This was done by recording only all local (monthly) maxima and minima of the annual temperature and precipitation curves and interpolating the other monthly values by means of cosine curve segments (see Box 1978a, Appendix F). The data collected initially included:

1. Highest monthly mean temperature (TMAX)
2. Lowest monthly mean temperature (TMIN)
3. Other local extrema in the annual mean-temperature curve
4. Highest average monthly precipitation (PMAX)
5. Lowest average monthly precipitation (PMIN)
6. Other local extrema in the annual precipitation curve.

Each datum was paired with the time of its occurrence, expressed on a continuous scale from 0 to 12 representing the twelve months. Extrema do not always occur, however, at monthly mid-points. As a result, many extrema may be better represented by the continuous climate-diagram curves than by the digital monthly data available from other sources.

Data for annual mean temperature and average precipitation, plus approximate geographic coordinates, were already available in coded form for about 1000 sites from production of the 'Miami Model' world map of estimated annual net primary productivity (Box, Lieth & Wolaver 1971). These sites were supplemented in 1976 by an additional 237 sites located in poorly represented areas and in areas of high topographic complexity. Geographic coordinates were refined from those of the Miami Model data, and annual temperature and precipitation values were scrutinized for errors. Errors in data transcription were located quite effectively by the interpolation routine used to generate the monthly values not read directly, since the iterative algorithm does not converge if data are not consistent (Box 1978a). Both the hard data and the amplified file including the interpolated monthly data were computerized for future use. The amplified file, called CLIMDAT1, contains annual and monthly data for 1225 sites covering the world's land areas and is described in more detail in Appendix E.[1] Only one interpolated monthly value (precipitation of the warmest month) was included directly in the final model. Estimated annual PET (divided by annual precipitation to provide an annual moisture index) was also included in the final model and was estimated from the interpolated monthly temperature curve using the method of Thornthwaite & Mather (1957).

B. Selection of ecoclimatic variables

In general, ambient temperature, incident solar radiation, and water availability appear to be the most important climatic factors affecting plants. Temperature levels determine potential metabolic and water loss and uptake rates. Infrequent temperature extremes may also constitute insurmountable limits to plant adaptation and distribution. Solar radiation provides the necessary energy for plant growth but can often be represented through surrogate temperature variables. Light-compensation points are generally low, and many plants assimilate more on cloudy days than under direct sunshine. (See, for example, Pisek & Tranquillini 1954; Lieth 1960; Daubenmire 1974; Larcher 1976.) Water availability determines the degree to which plants can meet the evaporative and physiologic demands posed by their particular structures, thereby influencing plant structure as well as plant productivity.

Earlier climatic models of both vegetation types (e.g. Rübel 1930; Holdridge 1947; Troll & Paffen 1964; Mather & Yoshioka 1966) and plant productivity (e.g. Rosenzweig 1968; Lieth & Box 1972) have generally been based on some measure of temperature and water availability. The most readily available and generally useful data on temperature and water availability are provided by measurements of air temperature and precipitation. Six variables are needed to represent even the most basic aspects of annual temperature and precipitation regimes. These represent, for each factor, the annual mean or total and the average maximum and minimum (monthly) values during the year. This representation of general annual levels and extreme period provided the initial basis for the model.

The first configuration of predictive variables for which world vegetation results were generated (using ECOSIEVE, see section 6.B and Appendix F) involved the above-mentioned six temperature and precipitation variables and the months in which the precipitation extremes occur. The model was used with 13 biome-level vegetation units (Lieth 1975) and predicted general biome distributions reasonably well over about half of the world's land area. Overestimation of forested area and poor distinction between mediterranean and other vegetation types, however, immediately confirmed expectations that both quantitative and temporal relationships between precipitation and water demand (evapotranspiration) would have to be expressed in order to improve prediction.

The relation between water supply and demand was approached through estimation of potential evapotranspiration (PET), since it (unlike actual evapotranspiration, AET) is a function of climate alone and would not require any prior assumptions concerning the nature of the vegetation cover. PET was introduced into the model in the form of an annual moisture index given by annual precipitation divided by annual PET. Annual PET was estimated by the method of Thornthwaite and Mather (1957), since this method requires only air temperature data and has been widely used over many parts of the world (e.g. Thornthwaite 1952; van Hylckama 1956; Pelton et al. 1960; Carter & Mather 1966; Mather 1974). It is well recognized that the Thornthwaite estimate generally underestimates PET in tropical and windy areas. It is necessary, however, to have a recognized standard for purposes of comparison, and the Thornthwaite estimate is as accurate and as widely used as any, especially in the northern temperate zone. A geographical comparison of world PET estimates by the Holdridge (1959) and Thornthwaite methods has been performed (Box 1979c), and further worldwide evaluation of estimation methods is in progress (Box, in preparation).

World vegetation patterns predicted by Rübel's (1930) model using ECOSIEVE and the world data-base (Box, unpublished) had shown that particular extreme months are not very effective pre-

[1]The computerized file of world climatic data compiled by Spangler & Jenne (1979) is being modified by the author to produce a larger ecoclimatic data-base of somewhere over 2000 sites. This file should be available by late 1981.

dictors of even seasonal vegetation types. A number of other expressions for precipitation seasonality were tried, including precipitation of the warmest month, of the warmest three-month period, and of the warmer half-year as a fraction of the annual total. Precipitation amounts of fixed periods or months were found not to be as useful as hoped, since the length and timing of dry seasons vary significantly even in similar climates. The most useful single predictor was found to be the precipitation of the warmest month, which combines both precipitation and maximum water demand in one variable.

In an attempt to increase predictive accuracy, as well as to simplify the model, a number of other approaches were also tried. These were primarily of three types:

1. A more integrative, growing season-based approach which focused on positive plant requirements (e.g. sufficient growing-season length and warmth) instead of the negative aspect represented by environmental limitation.
2. Attempts to describe ecoclimates more completely through measures of the plant microenvironment (e.g. leaf temperature, soil water, AET) and/or inclusion of more climate factors (e.g. temperature extremes, wind speed).
3. Attempts to describe more accurately environmental limitation by the basic ecoclimatic factors already recognized, primarily by inclusion of more aspects (variables) of these factors.

Attempts of the third type involved additional limiting variables such as the annual range of mean temperature, the precipitation fraction during the summer half-year, and summer and winter moisture indices. The annual range of mean temperature was found to be useful in the initial 'synthetic' model for separating tropical and extra-tropical forms and replaced mean annual temperature, which appeared to be superfluous.

Studies of the first two types (often combined) involved development of:

1. A soil-moisture simulation model SOLWAT applicable to the full range of naturally occurring soil conditions (Box 1981a). SOLWAT has since been used for a variety of other studies.
2. A model PHOSYN for simulating plant metabolic rates and production balances from environmental conditions for periods ranging from hours to years (unpublished).
3. A simple routine PERIOD for delimiting intervals during which a particular criterion is met (e.g. precipitation exceeding PET) and then intersecting the intervals for different elements to estimate more complex periods, such as (climatic) growing seasons (see SOLWAT documentation).

Other environmental variables tried in predictive configurations involving these models include length of the (longest) dry season, length of the climatic growing season, depth of aerated soil, leaf temperature and related actual transpiration, climatically estimated actual evapotranspiration, minimum soil water levels, and the length of the period with a positive climatically-estimated production balance.

These approaches provided useful quantitative estimates of plant-environment relations in different climates. One related but simpler approach involving temperature, precipitation, and PET variables and biome-level vegetation units provided the basis for world maps of estimated annual energy fixation by the vegetation cover (Box 1976) and estimated annual photosynthetic efficiency (Box 1977). It was not possible, however, to refine these integrative and microenvironmental approaches sufficiently to predict plant-form distributions accurately at the world scale. The main reason is the much greater data requirements of these methods. In addition, more simulative, integrative approaches are complicated by the fact that plant activity may not track environmental conditions in a uniform way throughout the year but responds also to genetic programming in quite complex ways, such as winter dormancy (e.g. Levitt 1972; Walter 1968; Tranquillini 1959, 1979). In general, the 'negative' environmental-limitation approach was found to be more accurate and more readily applicable than 'positive' growing-season or other more analytical or simulative approaches, which usually require estimation of the basic microenvironmental data such as soil water or leaf temperature. Average wind speed, absolute temperature extremes, and average diurnal temperature variation were identified as important additional factors but could not be investigated in detail at world scale because of the much greater data requirements. The results of the model constructed without these factors suggest that they are of secondary importance in most situations.

C. The ecoclimatic variables and their significance

The climate model finally employed consists of a set of eight macroclimatic variables which together are intended to express what appear to be the most important aspects of annual temperature, precipitation and evaporation regimes as they affect plant energy and water budgets. Special attention was payed to factors such as warm-season water availability and degree of winter cold which appear to delimit distributions of particular plant types. The ecoclimatic variables were selected by trying various combinations and comparing both predicted plant types at particular sites and predicted world distributions with known vegetation distributions. Particular reference was made to relations between climate and vegetation at a set of 113 sites chosen to represent the different occurrences of Köppen (1931) climate types throughout the world. A particular effort was made to include variables which express concurrent seasonal variations in both climate and physiognomy, factors which are generally poorly represented in earlier models (e.g. Holdridge 1947). The variables employed (with their computing abbreviations) are:

1. Mean temperature of the warmest month (°C) (TMAX)
2. Mean temperature of the coldest month (°C) (TMIN)
3. Annual range of monthly mean temperatures (°C) (DTY)
4. Average annual precipitation (mm) (PRCP)
5. Annual moisture index (annual precipitation divided by the Thornthwaite estimate of annual potential evapotranspiration) (MI)
6. Highest average monthly precipitation (mm) (PMAX)
7. Lowest average monthly precipitation (mm) (PMIN)
8. Average precipitation of the warmest month (mm) (PMTMAX)

Temperature is used as a surrogate for solar radiation since the two factors are closely related quantitatively except for a time-lag of about a month. Factor interactions involving the effects of temperature on precipitation requirements are expressed through the annual moisture index and the precipitation of the warmest month. Mather & Yoshioka (1966), Lieth & Box (1972), and others have shown that more integrative environmental variables, such as actual evapotranspiration, might provide similar results with fewer predictive variables. Such more integrative variables must be computed, however, and were rejected from this initial model, since they render non-computerized use of the model impossible and introduce additional error into the environmental data. Six of the eight variables involve directly measured data. PMTMAX is measured directly when it coincides with PMAX or PMIN and is interpolated otherwise. The moisture index MI involves PRCP and annual PET estimated from the interpolated monthly temperature values. World distributions of the eight variables, plus Thornthwaite's estimate of annual potential evapotranspiration, were obtained by computer-mapping the values at the 1225 test-sites (see section 6.B and Table 11). The resulting world distributions are shown on Maps 1–9.

Results of models involving these eight variables plus other above-mentioned variables (requiring no additional data) consistently showed little if any increase in prediction accuracy. Each of the eight variables chosen, however, appeared to be necessary and to have a particular ecological significance in most environmental situations.

The lower limit for TMAX represents the minimum temperature requirement during the growing season, which is important in probably all situations. It also serves as a rough measure of available solar energy. The upper limit for TMAX serves to estimate the point at which respiration begins to outweigh production or at which some form of metabolic collapse may occur. This is especially important for plants adapted to cooler environments.

Lower limits for TMIN specify minimum mean temperatures which can be tolerated and provide some estimate of the lowest tolerable absolute temperatures. Upper limits represent low temperatures which may be required to induce dormancy and vernalization in seasonal plant types. TMIN is the most important variable for separating evergreen and summergreen plants.

DTY is used primarily to distinguish only slightly seasonal tropical types from seasonal extra-tropical types which may occur in otherwise similar thermal situations. It is a somewhat artificial method of distinction, since it is generally not well known to what extent tropical vegetation types could exist under more seasonal temperature regimes and vice versa. It seems probable, for example, that most warm-season temperate-zone grasses came from the raingreen tropical zone, adapting their already possessed drought-dormancy to the physiological drought of the temperate-zone cold season. Moreover, for example, there is a widespread affinity between the vegetation of moist tropical mountains and the maritime temperate climates of the Southern Hemisphere (Troll & Lauer 1978). The results of the current improved model cast some doubt on the need for this variable.

The total average annual precipitation PRCP, apart from its relation to potential water loss (PET), is a measure of the total amount of water available for storage by the soil and subsequent use by vegetation. Since PRCP represents measured data, it can serve in warmer climates as a backup of MI limits when PET is underestimated. In cooler climates, limits on PRCP express the minimum annual water requirement more accurately than does PET or MI, since these variables do not include the effects of runoff.

The annual moisture index MI expresses the important relation between potential water loss and total water available (PRCP), i.e. the general wetness or dryness of a climate (without expressing seasonality). This is the most useful single variable for predicting the dominant plant structural types. The utility of annual MI as a predictive variable relies on the ability of the soil to store rain or snowmelt from wetter seasons for use during climatically drier seasons. This can be appreciated when one observes, for example, that the annual MI values of raingreen and mediterranean evergreen forests and woodlands, with their extreme dry seasons, can be only slightly lower than the MI values for summergreen and boreal (evergreen) forests and woodlands, with much less extreme, if any, dry seasons.

The lower limit for PMAX represents the minimum amount of water needed at one time for beginning or sustaining growth. This is most important in climates with low, seasonal or irregular precipitation. In dry climates the upper limit for PMAX may represent maximum amounts of stored soil water, beyond which both succulent structures and drought-conditioned life-cycle functions are either impaired or quickly out-competed by other forms.

The lower limit for PMIN represents the greatest degree of drought which can be tolerated, especially by evergreen vegetation. The upper limit represents the degree of drought which may be required by raingreen plants to induce dormancy or other life-cycle functions. This is the most important variable for separating evergreen and raingreen vegetation.

PMTMAX attempts to express the degree of summer drought in a single, easily valuated variable. The accuracy of PMTMAX as a measure of drought assumes a quasi-sinusoidal course in both temperature and precipitation patterns, a requirement which is usually met. In most climates, precipitation reaches either a maximum (polar, boreal, typical-temperate, temperate-continental, and tropical climates) or a minimum (mediterranean climates) at the time of highest temperatures. Climates transitional between temperate and mediterranean can be problematic. PMTMAX is the main variable for separating mediterranean from summergreen forms.

CHAPTER 5

The ecological model: life-form limitation, cover and dominance

In perhaps most modeling situations, including correlation models and many simpler systems models, one simply relates the target phenomena to environmental factors and does not bother with formalities about modeling the important aspects of environment or target phenomena. The distinction between models of structure and of environment, however, the two then linked by a process or 'ecological' model, can be very useful when the target phenomenon is more complex than a single quantity, since this permits one to develop structural and environmental complexity separately and then to focus on their interactions. World vegetation models (as stated in Chapter 2) have always done this, but without saying so, the ecological model being simply the set of particular environmental (climatic) values used to represent distributional limits. The ecological model presented herein has two components:

1. A traditional but much larger set of climatic correlates for the plant types, which define their macroclimatic envelopes in the environmental space employed.
2. A life-form dominance hierarchy, which permits one to interpret predicted sets of life forms as resulting in particular successional patterns and ultimately in a particular climatic-climax formation type.

The effective environment in the second component is no longer macroclimatic conditions but is instead the most important ecological aspects of the environment within the plant stand, namely the vertical shading structure and the soil water status. The ecological sub-model relating the now reduced set of potential life forms to environmental conditions within the stand is a simple hierarchy but represents a type of model which may be of more use in ecology than has generally been appreciated.

The ecological-model components together are intended to represent, in a simple way, an empirical quantification of the great body of existing literature and accepted theories concerning the environmental relations, distributions, and successional relations of the world's terrestrial plant types. Since open and closed stands are fundamentally different in their successional and dominance relationships, a simple estimator of potential vegetative cover is included, as is also a measure of potential life-form importance based on proximity to environmental limits. Synthesis of these various estimates and sub-models in order to interpret potential (climatic-climax) vegetation structure is treated in the final part of this chapter.

A. Ecological data

Ecological data, in our case, involve observations of relationships between vegetational and environmental phenomena. Such data may include:

1. Qualitative and quantitative observations on mechanisms of environmental limitation of plant growth.
2. Large-scale descriptions of vegetation distributions in relation to climatic factors.
3. Coordinated, detailed climatic and vegetation data at particular sites.
4. Analyses of vegetation changes along environmental gradients.

5. Quantitative models and correlations relating vegetation distributions to climatic factors.
6. Observations of plant successional relationships and patterns.

Possible effects of non-climatic factors must always be considered in attempting to interpret plant-climate relationships.

The first step in the collection of ecological data involved cataloguing known mechanisms of environmental limitation of species and life-form distributions (Table 6). Plant-environment relations can be approached both through the negative aspect of environmental limitation and through the complementary, positive aspect of the particular environmental requirements of plants. To some extent Table 6 reflects both aspects. The list was intended to be as complete as possible but also manageable. Instrumental in its construction were, in particular, works by Levitt (1972), Larcher (1976), Odum (1971), Whittaker (1975), Daubenmire (1974), Lieth (1974), Lieth & Whittaker (1975), Gates & Schmerl (1975), Krebs (1972), Walter (1968, 1973), Strain & Billings (1974), Precht et al. (1973), and Kozlowski (1971). Table 6 was used as a basis for identifying the most important ecoclimatic factors and for collecting and interpreting quantitative ecological data, such as correlations between distribution boundaries and climatic values.

Quantitative data had to be collected, of course, before the most important limiting mechanisms could be postulated. Quantitative estimates of physiological tolerance limits to various temperature and water-related situations are given, in particular, by Levitt (1972), Larcher (1976), Daubenmire (1974), and in various collections of symposium proceedings in the areas of physiological ecology and photosynthesis research. Most useful, however, were the ecophysiological descriptions of world and regional vegetation primarily by Walter and colleagues. These provided the main ecological and geographic basis for estimation of quantitative limits. These geoecological volumes include Rübel (1930), Walter (1968, 1973, 1974, 1977), Knapp (1965, 1973), Hueck (1966), Horvát et al. (1974), Zohary (1973), and Walter & Box (in press). The ecological works on regional vegetation were supplemented by many individual papers and by simultaneous observation of vegetation from maps and of the corresponding climatic data. Coordinated climatic and vegetation data, both at particular sites and occasionally in the form of gradient analyses, were obtained from numerous individual sources. These are listed in Appendix A, both by plant type and by geographic region.

Finally, an initial conceptual framework and initial estimates for many climatic limits were provided by various quantitative vegetation models which had appeared earlier. The first quantitative descriptions were presented by Warming (1895, see also 1909), Schimper (1898, see also 1935), and Rübel (1930). When Rübel's climatic characterizations of 18 vegetation types were projected (using ECOSIEVE and the 1225 test sites) onto a world base-map, however, less than half of the world's land area was covered (Box, unpublished). The next and perhaps most familiar model was that of Holdridge (1947), who related 19 largely environmental and only moderately physiognomic vegetation types to average annual precipitation, a moisture index (PET/precipitation), and what he called biotemperature (an annual average based only on above-freezing monthly mean temperatures). Holdridge's climatic envelopes appeared to be fairly accurate in the tropics when projected onto a world base-map (Box, unpublished) but can distinguish only general ecological types, not plant size, seasonal habit or leaf type. Lauer (1952) related vegetation zones in East Africa to length of the dry season, with generally good results, and then compared the relations with those in tropical South America. Troll has presented a quantitative climatic characterization of 37 vegetation types along with a world map (Troll & Paffen 1964). The most sophisticated models are those of Mather & Yoshioka (1966), who relate 20 physiognomic vegetation types (some representing individual dominant genera or species) to annual evapotranspiration (potential or actual) and the Thornthwaite (1952) annual moisture index composed of largely seasonal aridity and humidity components.

B. The environmental-limitation model

The model relating potential life-form occurrence to environmental conditions involves a macroclimatic envelope for each life form, within which the form occurs and outside of which it does not. The envelopes are composed of an upper and a

Table 6. Selected important climatic and non-climatic mechanisms of environmental limitation of terrestrial plant distributions.

Temperature
- winter cold
  - direct damage by degree of cold
  - required dormancy and vernalization
  - unseasonal temperatures (without cold-hardening, etc.)
  - production-respiration balance (evergreens)
- summer heat
  - lethal temperatures
  - length of warm season
  - sufficient warmth
  - production-respiration balance
- diurnal variations
  - sudden cold (without cold-hardening)
  - required thermoperiodism
- simultaneous extremes of different plant parts
- annual regime and life-cycle phases

Moisture
- drought
  - atmospheric (high saturation deficit, rapid dehydration)
  - soil (water-holding capacity relative to PET)
  - physiological (reduced uptake at low temperatures, etc.)
  - length of seasonal dry period(s)
  - required dormancy and aestivation
- surfeit
  - rotting
  - flooding and reduced aeration
- alternation of drought and surfeit
- sufficient degree and length of wet period(s)
- diurnal aspects involving saturation deficit, leaf morphology, transport, and water supply
- soil-moisture regime and life-cycle phases
- vulnerability to disease due to drought or surfeit

Other climatic factors
- wind
  - increased moisture loss and desiccation
  - mechanical damage
  - effects on surface temperatures
- insolation
  - effects on soil and plant temperatures
  - sufficient insolation
  - solar declination
  - photoperiodism

Non-climatic factors
- soil (depth, texture, nutrients, water-holding capacity, available water capacity, depth to water table, parent material, mobility, pH, etc.)
- topography (soil and surface heterogeneity, slope, aspect, effect on soil catenas, effect on water table, snow accumulation, microrelief)
- fire (frequency, type)
- geologic history (frequency of disruptions, recolonization time since last disruption, equilibration time since last disruption)
- biological interactions (root competition, light competition, allelopathy, life-history strategy, positive interactions, etc.)

The above list is intended to serve as a guide to the identification of critical environmental factors and the estimation of envelope dimensions, especially for the more important, macroclimatically based factors. Many aspects of environmental limitation have been grouped under more general headings, and many more could be added.

lower limiting value with respect to each of the eight ecoclimatic variables. (This can be thought of as defining 'rectangular' hypervolumes within the eight-dimensional ecoclimatic hyperspace.) This amounts to 16 limiting values for each of 90 life forms or sub-forms, a total of 1440 values. Many upper limits, especially for precipitation variables, may be outside the actual range of climatic conditions known on earth and are thus left unspecified and effectively infinite.

The limiting values generally represent correlations between observed plant or vegetation distributional limits and corresponding climatic values. Such correlates, however, can represent a variety of ecological as well as physiological limiting mechanisms. The ranges defined by the limiting values employed should be wide enough to represent not only truly physiological limits (assuming the ecological be narrower) but also to include a certain amount of compensation in more favorable microclimates and on more favorable (azonal) sites. When limiting values appeared to differ in different parts of the world, the wider limits were always employed except when the form concerned was known to be occuring in a special situation (e.g. near a river in a dry climate).

Initial estimates for most finite limits were made using ecological data described above or were estimated from general relationships when more specific data were not available. The more general data were supplemented, when possible, by data at particular representative sites. Gradient analyses, where available, were particularly useful but require careful interpretation. Data were sought from each regional occurrence of each plant type.

Once a complete set of limits was constructed or the set totally revised, ECOSIEVE was employed to generate the corresponding predictions for the 1225 test sites, both for each site and cross-tabulated to provide a list of potential sites for each plant type. These results were used to identify problem situations and were compared with known vegetation and plant distributions in order to improve the climatic limits as far as possible. Development of the final set of limits required about 10 iterations. Only the results predicted by the final set of limiting values were mapped.

The estimated macroclimatic envelopes of the 90 plant types, plus three types of extreme desert, can be represented by a large table of the limiting values, as shown in Table 7. The plant types are ordered in Table 7 by structural type (growth form) and, within classes, by general latitudinal order of occurrence from equator to poles, as also in Table 2.

Although the climatic limits were essentially empirically determined, certain observations concerning general patterns can be made which are not unexpected and which represent the basis for a quantitative theory.

1. Mean temperature during the warmest month (lower limit for TMAX) must be at least 10° (15° to over 20° for tropical types) for all tree, shrub and vine forms with the exception of the sub-antarctic small evergreen trees (*Nothofagus*) at 7°. Most xeromorphic forms (e.g. stem-succulents, xeric shrubs and scrub) require 15°, and the tall grasses (except tall tussocks) require 10°. The smaller growth forms, as groups, generally do not require such high minimal temperatures: dwarf-shrubs 6° (12° for most forms), cushion-shrubs only 3°–5°, ferns 10°, epiphytes 8°–16°, shorter grasses and most forbs 0°–10°, and thallophytes –2° to 0 °C.
2. Most forms do not tolerate mean temperatures (upper limit of TMAX) higher than 30°–35° and the limit may be much lower, especially for forms adapted to colder climates with lower threshold temperatures (minimum warmth requirements).
3. Deciduous forms, especially summergreens, tolerate lower temperatures (lower limit for TMIN) than do similar evergreen forms. Summergreen forms in particular generally require lower temperatures (upper limit for TMIN). Smaller, lower-growing evergreen plants (e.g. dwarf-shrubs) extend further into cold-winter climates than do larger evergreens (e.g. trees), due primarily to reduced wind, greater protection by snow or litter, and proximity to the warmer ground surface.
4. Some general structural types, such as rosette-trees (except perhaps tropical alpine dicots) and stem-succulents (except certain *Opuntia* spp.) appear to be generally frost-sensitive. Others, such as epiphytes and broad-evergreen trees, are frost-sensitive only in certain forms (e.g. tropical rainforest and eucalyptoid trees) and frost-tolerant in others (e.g. wintergreen epiphytes, subpolar *Nothofagus*).

5. Other characters being equal, larger plants (and some with only larger leaves) require more water at all times while foliated than smaller plants. This can be altered, however, by sclerophylly, by dissimilar total leaf area, and by microphylly alone. Water requirments can be reduced most by deciduousness or aphylly.
6. The annual (Thornthwaite PET-based) moisture index MI must be at least 0.9 for rainforest trees, 0.8 for broad-deciduous trees (except Xeric Raingreen Trees with long dry seasons), 0.75 for broad-evergreen trees (except eucalyptoids), and 0.6 for narrow or needle-leaved evergreens. The limits may be slightly lower for the corresponding small-tree forms. Limits may be significantly higher, however, generally around 1.0, for occurrence in closed stands. Threshold MI values are generally high for vines (0.9–1.0, except raingreen), epiphytes (1.0–1.1, except wintergreen), ferns (1.1), and evergreen forbs (0.9) but are lower for tall grasses (0.7–0.8), short sward and tussock-grasses (0.5), most shrubs (0.05–0.75), and seasonal, desert and other ephemeral herbs (0.01–0.2). Upper MI limits seem to be appropriate only for sclerophyllous, microphyllous, succulent or other generally xeromorphic forms.
7. Deciduous plant types require either a cold period (summergreens) or a period of moisture stress (raingreens). Evergreen plant types, on the other hand, are limited by such periods unless their effects can be overcome by sclerophylly, microphylly, dormancy or such. Short (1–2 month) dry periods can be tolerated by most evergreen types, even in tropical rainforests. Values (lower limits) of PMIN higher than 20 mm seem required only by tall tussock-grasses (25 mm but usually windy environments), evergreen and tree ferns (30 mm), temperate broad-leaved rainforest trees (50 mm), and subpolar broad-evergreen trees (80 mm, but probably only to ensure an even temperature).

Table 7. Estimated limiting macroclimatic values for terrestrial plant life forms.

The estimated macroclimatic envelopes for the 90 plant forms and subforms of Table 2 and for three types of extreme desert are defined by ranges for the eight ecoclimatic variables, as shown in the computer-printed table. The ecoclimatic variables are indicated by the following abbreviations:

TMAX	=	highest monthly mean temperature (°C)
TMIN	=	lowest monthly mean temperature (°C)
DTY	=	annual range of mean monthly temperatures (°C)
PRCP	=	average annual precipitation (mm)
MI	=	annual moisture index (PRCP/annual potential evapotranspiration estimated by the Thornthwaite method)
PMAX	=	highest average monthly precipitation (mm)
PMIN	=	lowest average monthly precipitation (mm)
PMTMAX	=	average precipitation of the warmest month (mm).

Asterisks denote open-ended variable ranges, i.e. unspecified and presumably unimportant (e.g. unattained) limiting values. The values are to some extent context-dependent and may not be accurate if applied individually. Some values also represent compromises between actual outer limits in unusual climates (e.g. monsoonal) and more commonly occurring limiting values (especially for PMTMAX). The estimates of limiting values were derived by iteration using the ECOSIEVE-generated world distributions implied by particular sets of limiting values. The accuracy of the limiting values can best be assessed by referring to the predicted world life-form distributions (Maps 10–22), the predicted results for 74 independent validation sites (Tables 14 and 15, plus Appendices C and D), and the predicted results for 100 potentially well known sites from the world data-base (Appendix B).

Table 7. Environmental limits (continued).

	TMAX	TMIN	DTY	PRCP	MI	PMAX	PMIN	PMTMAX
TROPICAL RAINFOREST TREES								
MAXIMUM	30.0	28.0	12.0	******	******	******	******	******
MINIMUM	20.0	18.0	0.0	1100.	0.90	150.	5.	10.
TROPICAL MONTANE RAINFOREST TREES								
MAXIMUM	24.0	23.0	12.0	******	******	******	******	******
MINIMUM	15.0	8.0	0.0	700.	0.90	100.	5.	10.
TROPICAL EVERGREEN SCLEROPHYLL TREES								
MAXIMUM	35.0	30.0	28.0	3000.	2.50	2000.	100.	2000.
MINIMUM	15.0	8.0	0.0	400.	0.50	50.	0.	10.
TROPICAL EVERGREEN MICROPHYLL-TREES								
MAXIMUM	32.0	30.0	25.0	******	******	******	******	******
MINIMUM	15.0	7.0	0.0	400.	0.75	75.	0.	10.
WARM-TEMPERATE BROAD-EVERGREEN TREES								
MAXIMUM	30.0	20.0	22.0	******	******	******	150.	******
MINIMUM	20.0	7.0	7.0	500.	0.75	80.	15.	50.
MEDITERRANEAN BROAD-EVERGREEN TREES								
MAXIMUM	32.0	15.0	30.0	******	2.50	******	75.	75.
MINIMUM	20.0	5.0	5.0	500.	0.80	60.	0.	0.
TEMPERATE BROAD-RAINFOREST TREES								
MAXIMUM	28.0	18.0	22.0	******	******	******	******	******
MINIMUM	13.0	5.0	4.0	600.	1.10	75.	50.	60.
MONSOON BROAD-RAINGREEN TREES								
MAXIMUM	35.0	28.0	15.0	******	3.00	******	30.	******
MINIMUM	23.0	14.0	0.0	800.	0.90	100.	0.	20.
MONTANE BROAD-RAINGREEN TREES								
MAXIMUM	25.0	22.0	20.0	******	******	******	30.	******
MINIMUM	15.0	8.0	0.0	400.	0.80	90.	0.	20.
XERIC RAINGREEN TREES								
MAXIMUM	40.0	30.0	25.0	******	1.00	******	25.	******
MINIMUM	24.0	10.0	0.0	400.	0.35	75.	0.	10.
SUMMERGREEN BROAD-LEAVED TREES								
MAXIMUM	30.0	17.0	50.0	******	******	******	******	******
MINIMUM	15.0	-20.0	8.0	500.	0.90	60.	3.	30.
BOREAL BROAD-SUMMERGREEN TREES								
MAXIMUM	25.0	5.0	60.0	******	******	******	******	******
MINIMUM	13.0	-30.0	8.0	100.	0.80	40.	10.	30.
TROPICAL LINEAR-LEAVED TREES								
MAXIMUM	28.0	25.0	12.0	******	******	******	******	******
MINIMUM	15.0	5.0	0.0	500.	1.00	50.	5.	10.
TROPICAL XERIC NEEDLE-TREES								
MAXIMUM	30.0	28.0	20.0	2000.	1.10	300.	75.	300.
MINIMUM	15.0	8.0	0.0	300.	0.70	75.	0.	10.
HELIOPHILIC LONG-NEEDLED TREES								
MAXIMUM	33.0	25.0	35.0	******	2.50	******	100.	******
MINIMUM	20.0	2.0	2.0	400.	1.00	40.	20.	75.
SUB-MEDITERRANEAN NEEDLE-TREES								
MAXIMUM	28.0	15.0	25.0	******	3.00	******	100.	100.
MINIMUM	14.0	-1.0	4.0	400.	0.85	75.	0.	0.
TEMPERATE RAINFOREST NEEDLE-TREES								
MAXIMUM	25.0	15.0	15.0	******	******	******	******	******
MINIMUM	12.0	-2.0	4.0	400.	1.40	80.	20.	25.
TEMPERATE NEEDLE-TREES								
MAXIMUM	24.0	15.0	30.0	2000.	3.00	300.	100.	300.
MINIMUM	15.0	-6.0	4.0	300.	0.65	30.	3.	3.
BOREAL/MONTANE SHORT-NEEDLED TREES								
MAXIMUM	22.0	3.0	60.0	******	******	******	******	******
MINIMUM	11.0	-30.0	10.0	100.	0.60	25.	5.	10.
SWAMP SUMMERGREEN NEEDLE-TREES								
MAXIMUM	28.0	20.0	30.0	******	******	******	******	******
MINIMUM	22.0	3.0	7.0	1000.	1.00	100.	20.	100.
BOREAL SUMMERGREEN NEEDLE-TREES								
MAXIMUM	22.0	3.0	90.0	******	3.00	******	50.	******
MINIMUM	12.0	-60.0	15.0	100.	0.40	30.	0.	25.
TROPICAL BROAD-EVERGREEN DWARF-TREES								
MAXIMUM	35.0	30.0	15.0	******	2.00	******	40.	******
MINIMUM	20.0	15.0	0.0	1000.	0.80	100.	1.	50.
TROPICAL BROAD-EVERGREEN SMALL TREES								
MAXIMUM	32.0	28.0	20.0	******	******	******	******	******
MINIMUM	15.0	8.0	0.0	700.	0.90	75.	5.	10.
TROPICAL CLOUD-FOREST DWARF-TREES								
MAXIMUM	20.0	18.0	6.0	******	******	******	******	******
MINIMUM	10.0	5.0	0.0	500.	1.00	50.	5.	20.

Table 7. Environmental limits (continued).

	TMAX	TMIN	DTY	PRCP	MI	PMAX	PMIN	PMTMAX
TEMP. BROAD-EVERGREEN SMALL TREES								
MAXIMUM	32.0	15.0	25.0	******	******	******	******	******
MINIMUM	12.0	2.0	4.0	400.	0.75	60.	5.	10.
SUBPOLAR BROAD-EVERGREEN SMALL TREES								
MAXIMUM	15.0	10.0	15.0	******	******	******	******	******
MINIMUM	7.0	2.0	2.0	500.	2.00	150.	80.	80.
BROAD-RAINGREEN SMALL TREES								
MAXIMUM	35.0	30.0	25.0	******	******	******	30.	******
MINIMUM	15.0	8.0	0.0	300.	0.25	50.	0.	10.
BROAD-SUMMERGREEN SMALL TREES								
MAXIMUM	30.0	17.0	60.0	******	******	******	******	******
MINIMUM	11.0	-30.0	8.0	150.	0.70	40.	1.	20.
DWARF-NEEDLE SMALL TREES								
MAXIMUM	30.0	18.0	40.0	******	3.00	300.	100.	200.
MINIMUM	18.0	-5.0	10.0	200.	0.45	25.	0.	5.
PALMIFORM TUFT-TREES								
MAXIMUM	40.0	30.0	30.0	******	******	******	******	******
MINIMUM	16.0	11.0	0.0	400.	0.70	50.	0.	10.
PALMIFORM TUFT-TREELETS								
MAXIMUM	35.0	30.0	30.0	******	******	******	******	******
MINIMUM	14.0	10.0	0.0	400.	0.70	50.	0.	5.
TREE FERNS								
MAXIMUM	22.0	22.0	12.0	******	******	******	******	******
MINIMUM	10.0	10.0	0.0	500.	1.20	75.	30.	50.
TROPICAL ALPINE TUFT-TREELETS								
MAXIMUM	15.0	10.0	5.0	******	2.00	******	100.	******
MINIMUM	1.0	0.0	0.0	200.	1.00	50.	0.	0.
XERIC EVERGREEN TUFT-TREELETS								
MAXIMUM	35.0	30.0	25.0	1500.	1.00	300.	50.	300.
MINIMUM	18.0	8.0	0.0	100.	0.25	50.	0.	0.
EVERGREEN GIANT-SCRUB								
MAXIMUM	32.0	28.0	25.0	******	2.00	******	100.	******
MINIMUM	20.0	8.0	0.0	100.	0.25	20.	0.	0.
RAINGREEN THORN-SCRUB								
MAXIMUM	40.0	30.0	25.0	1500.	1.00	1000.	20.	1000.
MINIMUM	15.0	8.0	0.0	100.	0.08	30.	0.	10.

	TMAX	TMIN	DTY	PRCP	MI	PMAX	PMIN	PMTMAX
SUMMERGREEN GIANT-SCRUB								
MAXIMUM	35.0	17.0	60.0	******	******	******	******	******
MINIMUM	12.0	-30.0	8.0	150.	0.40	40.	1.	20.
LEAFLESS XEROMORPHIC LARGE-SCRUB								
MAXIMUM	40.0	30.0	60.0	600.	0.80	100.	30.	100.
MINIMUM	15.0	-20.0	0.0	50.	0.05	5.	0.	0.
NEEDLE-LEAVED TREELINE KRUMMHOLZ								
MAXIMUM	13.0	3.0	70.0	******	******	******	100.	******
MINIMUM	10.0	-35.0	5.0	75.	0.60	25.	2.	15.
TROPICAL BROAD-EVERGREEN SHRUBS								
MAXIMUM	32.0	30.0	20.0	******	******	******	******	******
MINIMUM	10.0	6.0	0.0	400.	0.70	75.	0.	5.
MEDITERRANEAN EVERGREEN SHRUBS								
MAXIMUM	30.0	18.0	35.0	******	2.00	******	100.	125.
MINIMUM	16.0	5.0	5.0	300.	0.45	60.	0.	0.
BROAD-ERICOID EVERGREEN SHRUBS								
MAXIMUM	25.0	10.0	35.0	******	******	******	******	******
MINIMUM	13.0	-10.0	0.0	400.	1.40	35.	2.	30.
TEMPERATE BROAD-EVERGREEN SHRUBS								
MAXIMUM	30.0	20.0	50.0	******	******	******	******	******
MINIMUM	15.0	-2.0	5.0	200.	0.60	40.	1.	30.
HOT-DESERT EVERGREEN SHRUBS								
MAXIMUM	40.0	30.0	40.0	1000.	0.75	150.	20.	150.
MINIMUM	22.0	6.0	5.0	50.	0.08	10.	0.	0.
LEAF-SUCCULENT EVERGREEN SHRUBS								
MAXIMUM	32.0	25.0	20.0	1000.	0.80	300.	30.	300.
MINIMUM	15.0	12.0	0.0	100.	0.15	10.	0.	1.
COLD-WINTER XEROMORPHIC SHRUBS								
MAXIMUM	32.0	6.0	60.0	800.	0.90	200.	30.	60.
MINIMUM	15.0	-20.0	10.0	50.	0.10	15.	0.	0.
BROAD-SUMMERGREEN MESIC SHRUBS								
MAXIMUM	30.0	17.0	60.0	******	******	******	******	******
MINIMUM	14.0	-30.0	8.0	300.	0.70	30.	1.	20.
XERIC SUMMERGREEN SHRUBS								
MAXIMUM	35.0	12.0	60.0	800.	1.00	100.	30.	100.
MINIMUM	14.0	-30.0	8.0	100.	0.25	30.	1.	10.

Table 7. Environmental limits (continued).

	TMAX	TMIN	DTY	PRCP	MI	PMAX	PMIN	PMTMAX
NEEDLE-LEAVED EVERGREEN SHRUBS								
MAXIMUM	35.0	15.0	65.0	******	2.00	300.	100.	300.
MINIMUM	10.0	-30.0	0.0	100.	0.40	40.	0.	0.
MEDITERRANEAN DWARF-SHRUBS								
MAXIMUM	35.0	15.0	35.0	******	1.20	******	50.	50.
MINIMUM	16.0	3.0	5.0	150.	0.15	35.	0.	0.
TEMPERATE EVERGREEN DWARF-SHRUBS								
MAXIMUM	24.0	18.0	50.0	******	******	******	******	******
MINIMUM	6.0	-20.0	0.0	100.	0.70	30.	5.	25.
MARITIME HEATH DWARF-SHRUBS								
MAXIMUM	23.0	10.0	18.0	******	******	******	******	******
MINIMUM	12.0	-1.0	5.0	500.	1.00	60.	20.	20.
SUMMERGREEN TUNDRA DWARF-SHRUBS								
MAXIMUM	18.0	3.0	80.0	******	******	******	******	******
MINIMUM	6.0	-60.0	5.0	100.	0.70	20.	10.	10.
XERIC DWARF-SHRUBS								
MAXIMUM	40.0	30.0	60.0	1000.	0.80	100.	30.	100.
MINIMUM	12.0	-20.0	0.0	50.	0.05	5.	0.	0.
PALMIFORM MESIC ROSETTE-SHRUBS								
MAXIMUM	35.0	30.0	30.0	******	******	******	******	******
MINIMUM	15.0	3.0	0.0	500.	0.90	60.	5.	20.
XERIC ROSETTE-SHRUBS								
MAXIMUM	40.0	30.0	30.0	1500.	1.20	200.	50.	200.
MINIMUM	20.0	3.0	1.0	100.	0.10	20.	0.	2.
MESIC EVERGREEN CUSHION-SHRUBS								
MAXIMUM	18.0	10.0	15.0	******	******	******	******	******
MINIMUM	3.0	1.0	0.0	500.	1.30	60.	5.	40.
XERIC CUSHION-SHRUBS								
MAXIMUM	35.0	25.0	20.0	1000.	0.90	200.	30.	150.
MINIMUM	5.0	-5.0	0.0	50.	0.10	10.	0.	0.
ARBORESCENT STEM-SUCCULENTS								
MAXIMUM	40.0	30.0	35.0	1500.	0.80	250.	20.	250.
MINIMUM	22.0	8.0	0.0	90.	0.07	15.	0.	10.
TYPICAL STEM-SUCCULENTS								
MAXIMUM	40.0	30.0	40.0	1500.	0.60	250.	20.	250.
MINIMUM	20.0	7.0	0.0	50.	0.05	10.	0.	0.

	TMAX	TMIN	DTY	PRCP	MI	PMAX	PMIN	PMTMAX
BUSH STEM-SUCCULENTS								
MAXIMUM	40.0	30.0	40.0	2000.	1.50	300.	100.	300.
MINIMUM	10.0	0.0	0.0	50.	0.05	10.	0.	0.
ARBORESCENT GRASSES								
MAXIMUM	30.0	25.0	30.0	******	******	******	******	******
MINIMUM	10.0	4.0	0.0	100.	0.80	75.	5.	20.
TALL CANE-GRAMINOIDS								
MAXIMUM	35.0	30.0	35.0	******	3.00	******	300.	******
MINIMUM	10.0	1.0	0.0	500.	0.70	100.	0.	0.
TALL GRASSES								
MAXIMUM	35.0	23.0	60.0	******	******	******	300.	******
MINIMUM	10.0	-20.0	0.0	350.	0.70	60.	5.	30.
SHORT SWARD-GRASSES								
MAXIMUM	35.0	25.0	60.0	******	******	******	******	******
MINIMUM	5.0	-50.0	0.0	100.	0.50	40.	2.	25.
SHORT BUNCH-GRASSES								
MAXIMUM	40.0	30.0	100.0	******	3.00	******	300.	******
MINIMUM	2.0	-60.0	0.0	100.	0.20	30.	0.	0.
TALL TUSSOCK-GRASSES								
MAXIMUM	23.0	10.0	20.0	******	******	******	******	******
MINIMUM	5.0	-1.0	2.0	500.	0.80	50.	25.	30.
SHORT TUSSOCK-GRASSES								
MAXIMUM	25.0	10.0	20.0	1000.	1.20	300.	75.	300.
MINIMUM	2.0	-1.0	2.0	100.	0.50	40.	10.	25.
SCLEROPHYLLOUS GRASSES								
MAXIMUM	40.0	30.0	30.0	1000.	1.20	150.	30.	75.
MINIMUM	20.0	10.0	0.0	100.	0.10	25.	0.	10.
DESERT-GRASSES								
MAXIMUM	40.0	30.0	100.0	1000.	0.50	300.	20.	300.
MINIMUM	3.0	-60.0	0.0	40.	0.02	10.	0.	0.
TROPICAL EVERGREEN FORBS								
MAXIMUM	35.0	30.0	15.0	******	******	******	******	******
MINIMUM	3.0	2.0	0.0	500.	0.90	75.	5.	40.
TEMPERATE EVERGREEN FORBS								
MAXIMUM	30.0	18.0	50.0	******	******	******	******	******
MINIMUM	5.0	-5.0	5.0	400.	0.90	50.	10.	20.

Table 7. Environmental limits (continued).

	TMAX	TMIN	DTY	PRCP	MI	PMAX	PMIN	PMTMAX
RAINGREEN FORBS								
MAXIMUM	35.0	30.0	20.0	******	******	******	50.	******
MINIMUM	10.0	7.0	0.0	200.	0.20	30.	0.	10.
SUMMERGREEN FORBS								
MAXIMUM	35.0	15.0	100.0	******	******	******	******	******
MINIMUM	10.0	-50.0	6.0	100.	0.20	30.	0.	0.
SUCCULENT FORBS								
MAXIMUM	40.0	30.0	40.0	1500.	1.20	200.	50.	200.
MINIMUM	18.0	5.0	0.0	10.	0.05	10.	0.	0.
XERIC CUSHION-HERBS								
MAXIMUM	30.0	25.0	100.0	1000.	1.00	300.	50.	300.
MINIMUM	5.0	-60.0	0.0	40.	0.20	5.	0.	0.
EPHEMERAL DESERT HERBS								
MAXIMUM	40.0	28.0	100.0	500.	0.50	200.	15.	200.
MINIMUM	1.0	-20.0	0.0	30.	0.01	10.	0.	0.
SEASONAL COLD-DESERT HERBS								
MAXIMUM	15.0	0.0	******	******	******	******	******	******
MINIMUM	0.0	-100.0	3.0	10.	0.10	5.	0.	0.
RAINGREEN COLD-DESERT HERBS								
MAXIMUM	15.0	10.0	20.0	******	3.00	******	30.	******
MINIMUM	1.0	-3.0	0.0	100.	0.20	30.	0.	0.
TROPICAL BROAD-EVERGREEN LIANAS								
MAXIMUM	30.0	30.0	15.0	******	******	******	******	******
MINIMUM	18.0	10.0	0.0	500.	1.00	75.	2.	30.
BROAD-EVERGREEN VINES								
MAXIMUM	32.0	30.0	25.0	******	******	******	******	******
MINIMUM	10.0	4.0	0.0	400.	0.90	75.	1.	30.
BROAD-RAINGREEN VINES								
MAXIMUM	35.0	30.0	25.0	******	******	******	30.	******
MINIMUM	15.0	10.0	0.0	200.	0.25	30.	0.	10.
BROAD-SUMMERGREEN VINES								
MAXIMUM	30.0	12.0	50.0	******	******	******	******	******
MINIMUM	15.0	-5.0	8.0	400.	1.00	60.	2.	40.
TROPICAL BROAD-EVERGREEN EPIPHYTES								
MAXIMUM	30.0	30.0	12.0	******	******	******	******	******
MINIMUM	12.0	10.0	0.0	500.	1.10	75.	10.	30.
NARROW-LEAVED EPIPHYTES								
MAXIMUM	32.0	30.0	18.0	******	******	******	******	******
MINIMUM	8.0	5.0	0.0	500.	1.00	75.	10.	30.
BROAD-WINTERGREEN EPIPHYTES								
MAXIMUM	30.0	28.0	30.0	******	******	******	******	******
MINIMUM	16.0	0.0	0.0	400.	0.60	50.	0.	2.
EVERGREEN FERNS								
MAXIMUM	30.0	28.0	15.0	******	******	******	******	******
MINIMUM	10.0	5.0	0.0	500.	1.10	75.	30.	50.
SUMMERGREEN FERNS								
MAXIMUM	30.0	12.0	60.0	******	******	******	******	******
MINIMUM	10.0	-20.0	8.0	300.	1.10	50.	2.	50.
MAT-FORMING THALLOPHYTES								
MAXIMUM	30.0	30.0	130.0	******	******	******	******	******
MINIMUM	0.0	-100.0	0.0	20.	0.80	10.	1.	1.
XERIC THALLOPHYTES								
MAXIMUM	40.0	30.0	150.0	******	2.00	******	100.	******
MINIMUM	-2.0	-100.0	0.0	1.	0.01	1.	0.	0.
NON-SEASONAL COLD DESERT								
MAXIMUM	3.0	3.0	3.0	******	3.00	******	******	******
MINIMUM	0.0	-2.0	0.0	10.	0.05	2.	0.	0.
DRY DESERT								
MAXIMUM	******	******	******	200.	0.25	25.	5.	25.
MINIMUM	0.0	******	0.0	0.	0.0	0.	0.	0.
ICE DESERT								
MAXIMUM	6.0	-1.0	******	******	******	******	******	******
MINIMUM	******	******	0.0	1.	0.10	0.	0.	0.

C. The cover model

Vegetative cover (percent of ground surface covered) is essentially a function of water availability in relation to evaporative demands. This can be seen from the fact that even tundras and grasslands, growing not more than 0.5 m high, can produce seasonal cover values as high as those of forests, effectively 100%. Larcher (1976) and various IBP studies have provided scattered data which suggest the annual moisture index (MI) as an adequate and convenient initial estimate of maximum seasonal cover, up to 1.0. MI values greater than 1.0 are interpreted as 100% cover, indicating closed forests except in climates too cold for trees. MI values less than 1.0 are interpreted as indicating open stands, from open woodlands (0.75–1.0) and savannas (0.6–0.8) on down to semi-desert scrub (below 0.5). These values are also suggested by a variety of other sources, including Holdridge (1947), Budyko (1956), and Mather (1974). The correspondence of MI values with general formation classes is summarized in Table 8.

D. The dominance model

Those plant growth forms which produce the greatest total cover above that of competing forms generally make the greatest contributions to the general character of plant stands and can be thought of as dominant forms. In order to predict dominant forms in a given situation, the forms were arranged in a dominance hierarchy based on leaf area and height of the respective forms, with some consideration also of amount of lateral branching. On cleared sites free of non-climatic complications (euclimatopes) secondary succession can be

Table 8. Composition and cover requirements for vegetation formation classes.

Predicted life-form spectra can be interpreted as prediction of particular vegetation formation types by means of criteria concerning forms predicted, their expected dominance relationships (see Table 9), and expected vegetative cover. In closed stands (forests, dense shrubland, arctic/alpine grasslands) the dominant forms and thus the vegetation formation type are determined primarily by the dominance hierarchy. As stands become more open, the estimated cover, mix of forms predicted, and proximity to environmental limits become more important determinants of stand structure and formation type.

Formation class	Forms required	MI range
Rainforest (evergreen)	rainforest trees understorey trees shrubs, forbs or ferns	>1.1 (tropics) >1.3 (elsewhere)
Forest	trees or rosette-trees (not small or dwarf)	>1.0 (closed) 0.90–1.10 (open)
Woodland (open)	trees or rosette-trees (any forms)	0.75–1.00
grassy woodland or savanna forest	tall grass	0.75–0.90
forest-tundra	boreal tree and tundra forms	temperature-limited
Savanna (forest-steppe in temperate zone)	trees or rosette-trees (generally not conifers) sward and/or bunch-grass	0.60–0.80
Grassland	any grass form (except bamboo), trees unimportant	
steppe	short bunch or sward-grass	>0.30
Shrubland	any shrub form, trees unimportant	>0.30
shrub-steppe	small shrubs with grass	(open)
Scrub	various forms (usually mixed), no canopy trees grass cover not continuous	<0.50
Tundra	tundra/temperate dwarf-shrubs and/or forbs short grasses	>0.30
Semi-desert	various forms, no canopy trees	>0.30
Desert	no higher plants	⩽0.10
cold-desert	cold-desert herbs plus thallophytes only	temperature-limited

expected to proceed up the dominance hierarchy, ending eventually with dominance by the highest ranked forms climatically possible at the site. The lower forms are not necessarily eliminated but are less prominent and generally reduced in number. The general vegetation structure and formation type are determined by the dominant forms.

The hierarchy (Table 9) consists of six levels (with sub-levels). Dominance is primarily a function of growth form and relative plant size (since taller forms usually produce greater cover) and of growth rate. Evergreen forms may or may not outcompete corresponding deciduous forms in a given environment and so are usually placed in the same dominance level. (An exception is the slow-growing boreal larches, which dominate only in areas too cold for evergreens.) Also, forms which are low-growing but form dense covers may be more successful than taller but more scattered forms, as in the case of grasslands. Tree forms are considered to dominate smaller forms only if the estimated total cover (MI value) exceeds 0.9 (see also part F of this chapter). Inclusion in the same level of forms with overlapping climatic envelopes implies that these forms will coexist and be 'co-dominant' forms if no forms from a higher level are predicted. Such forms are considered to be complementary (e.g. Gimingham 1978) in the sense that they can coexist without competing directly. Coexisting forms should always be placed in the same dominance level if they commonly occur as co-dominants in mixed stands, their competition being insufficient to exclude one form or the other regularly. Life forms which never occur together, such as tropical and boreal forms, can conveniently be placed in the same dominance level, without implying anything about relations between them.

The dominance hierarchy is not intended as an ecological theory but rather as a convenient, empirically based means of predicting which forms will be most important in a given life-form mix and environmental situation. Strictly speaking, the

Table 9. Hierarchy of potential life-form dominance in vegetation stands.

Taller, leafier plant forms are expected to dominate smaller forms in a plant stand in the sense that they may reduce or exclude other forms through more successful competition for sunlight and water and thereby determine the general structure and character of the stand. Potential life-form dominance, in this sense, is summarized by the hierarchy below, in which potential dominance increases upward. This hierarchy is employed to predict dominant and sub-dominant forms in a stand in order to identify and characterize the vegetation formation type. The hierarchy is not intended to predict relative importance of forms in quite open formations or below the top two or three dominance levels present in closed formations, since these forms may be influenced more by other aspects of environment than competition for water and light.

Level	Life forms
6	Forest trees
	a. tropical rainforest trees
	b. other climax canopy trees
	c. smaller canopy trees (stressful environments), tuft-trees, taller sub-climax trees
	d. other successional canopy trees
5	Small-trees and tall grasses
4	Mesic tuft-treelets, large shrubs and arborescents, sward and tall-tussock grasses
3	Xeric tuft-treelets, xeric shrubs and bushes, mesic small-shrubs, short bunch and tussock-grasses, arborescent stem-succulents
2	Leafless and leaf-succulent scrub, dwarf-shrubs, forbs, ferns
1	Small herbs of extreme environments, leafless dwarf-shrubs, thallophytes
0	Plant-less deserts

Never dominant: vines and epiphytes
Edaphically constrained: phreatophytes (e.g. arborescents, swamp trees and scrub)
Biologically constrained: bamboos (vegetative colonies)
Extremely slow-growing (restricted to undisturbed areas): larger stem- and leaf-succulents

dominance hierarchy is applicable only to closed and only moderately open stands (including grasslands). Since it is based largely on plant size, however, it also suggests the most conspicuous forms even in the most open formation. The dominance hierarchy is empirical but does not lack a theoretical basis. The entire hierarchy can be generated by a simple algorithm employing only the physiognomic characters (see Box 1978a) of the individual life forms.

Most trees are placed in the top level (level 6) because they produce the greatest leaf area at the greatest height. Level 6 is subdivided, however, because of observable successional and other dominance relationships between tree types. Tropical rainforest trees are placed above all other because of their clearly dominant position in closed tropical forests. Most other climax canopy trees are in level 6b, since probably all of these forms can occur in more or less evenly mixed forests, such as the sub-boreal mixed forest of broad-leaved summergreens and evergreen conifers. Some canopy trees which do not commonly become dominants, have edaphic constraints, or which for some other reason would cause problems if placed with climax canopy trees are placed in level 6c. These include the tropical evergreen sclerophyll and microphyll trees, the swamp (*Taxodium*) and boreal (*Larix*) summergreen needle-trees, and the palms. These can be important forest components but usually dominate only where other trees are much reduced. Larches, for example, are slower-growing at maturity than evergreen boreal conifers (Tranquillini 1963) and form extensive forests only where evergreens are absent, in eastern Siberia. Two climax canopy tree types of generally shorter stature, the cloud-forest and sub-antarctic broad-leaved evergreen trees, are also placed in level 6c. Other successional canopy or sub-canopy trees, such as boreal summergreens (e.g. *Betula, Populus*), are placed in level 6d.

The various understorey and open-woodland smaller trees, except the slower-growing dwarf-needle (juniperoid) trees, are placed in level 5. With them in this level are the tall grasses with which the small trees often co-exist in open woodlands and savannas.

Level 4 includes tall-tussock and short sward-forming grasses, the juniperoid small trees, and a variety of other forms including understorey tuft-treelets and larger shrub forms, including arborescents. Of the shrubs and bushes, only raingreen thorn-scrub and mediterranean evergreen shrubs cover large areas as dominants, though summergreen shrubs can dominate locally (e.g. 'deciduous chaparral' of western USA).

Level 3 includes many smaller and more xeric analogs of the forms in level 4, namely the xeric tuft-treelets, xeric shrubs, smaller mesic shrubs, and short bunch and tussock-grasses. Arborescent stem-succulents are also included in this level. These forms often co-exist in warm semi-deserts and in some steppes.

Even smaller and more xeromorphic shrubs and scrub are found in level 2, including the slow-growing leafless and leaf-succulent scrub forms. These are placed with the forbs and ferns, which occur in a wide variety of environments and may occur in semi-deserts.

The smallest and most extreme forms, such as ephemerals and thallophytes, are placed in the lowest level. Outside the dominance hierarchy are certain other forms which may be dependent on other plants for support (vines and epiphytes), biologically constrained (e.g. bamboos, which occur mainly in localized colonies), or otherwise not capable of dominating large areas, except perhaps under special conditions.

E. Proximity to environmental limits

As environments become more stressful, and vegetation more open, physical conditions become equally and then more important than potential life-form dominance relationships in determing actual stand structure and life-form importance. Proximity to environmental limits may result in physiological stresses and eventual environmental limitation by any of a great variety of factors or mechanisms, the most important of which were summarized in Table 6.

In order to examine proximity to environmental limits in particular cases, the processing program ECOSIEVE was equipped with an optional algorithm to identify the closest environmental limit of each plant type possible at a site. In order to standardize for the different environmental factors and to overcome the problem of unspecified limits in some cases, the 'distance' between the site value and the closer (upper or lower) limiting value for an environmental variable is always expressed as the raw distance divided either by the greatest corresponding raw distance for any plant type at the site (with respect to the same variable) or by the plant's own tolerance range (if both limits are specified), whichever is smaller. The standardized 'distance' or 'proximity' is thus always a value between one and zero, with values approaching one occurring usually only for lower forms with wide tolerance ranges. The environmental variable with the closest limiting value and the corresponding (fractional) distance to that value are identified by ECOSIEVE and printed next to the life-form name on the printout of site results (see Table 10 and Appendices B and C). In this way one can always see which forms are occurring very near environmental limits and thus presumably not growing as vigorously as other forms at the site.

This measure of life-form 'fitness' in a given environmental situation provides additional information for use in interpreting model predictions. Fitness values of zero or very nearly zero, in particular, suggest that the plant form in question is occurring at an environmental limit and will not be abundant or important in stand structure at the site. The fitness measure is intended to eliminate forms occurring near limits from dominance considerations. (The computerized model ignores those potential dominants which have fitness measures lower than 0.1.) It is not intended to imply relative importance or potential cover of forms which are occurring well within their environmental limits (fitness values of 0.1 or greater), since this depends more on the attributes of the particular life forms.

F. Interpreting vegetation formations

Interpreting the results at a particular site as occurrence of a particular vegetation formation type is not and perhaps should not be a computerized operation, despite our desire for rigorous prediction free of subjectivity.

Interpreting model results involves at least the following operations:

1. identification of dominant forms;
2. interpretation of total stand cover and the cover (importance) of particular components;
3. interpretation of vegetation strata; and
4. selection of appropriate adjectives and nouns to describe the vegetation formation type.

The last of these requires the other three and involves both identification of the most appropriate formation class (forest, woodland, etc.) and selection of a minimal set of appropriate descriptors for the mix of dominant forms and their seasonal habits.

Formation class is a function of dominant growth forms, vegetation strata and the degree of cover provided by each. At the level of resolution required here, each is predicted sufficiently by the annual moisture index MI, which expresses the general level of water availability for robust or less dense vegetation growth. Assuming that the appropriate life forms (growth forms) have been predicted to occur (and are not occurring as scattered individuals at environmental limits), one can recognize formation classes by the criteria indicated in Table 8.

Closed stands (including forests, dense shrublands, and dense tundras or meadows) generally require that MI be at least 1.0. In such cases, the dominant forms and the formation type can be determined primarily by the dominance hierarchy (Table 9). As stands become more open, the estimated cover, mix of predicted forms, and proximity to environmental limits become more important determinants of stand structure and formation type. Grasses immediately become more important because of their ability to spread rapidly, form more or less dense ground covers, and prevent invasion by seedlings of larger forms (especially when grazed regularly). Shrubs also become more important, since they are ecologically much like trees, can form dense covers, but have much reduced total transpiring surface areas. Trees may disappear rapidly or remain as well-spaced individuals.[1] Other forms such as stem-succulents, rosette-treelets and scrub, slow-growing forms (e.g. succulents in general), and forms with edaphic (e.g.

[1]Because of the slow seasonal thawing of boreal and polar soils, permitting conifers in particular to exist at unusually low MI values, the lower MI limits for boreal evergreen and especially summergreen needle-trees had to be set artificially low. This must be kept in mind when, as a result, boreal needle-trees are predicted in dry climates south of the boreal zone. One should also keep in mind that, except for pines, tall conifers rarely form savannas or open woodlands on the warmer sides of their ranges.

phreatophytes) or other constraints characteristically occur as more scattered individuals and must be interpreted specially.

All of these characteristics, as well as the particular environmental situation, must be considered when attempting to interpret the structure and composition of open stands. In such situations the potential dominance relationships may provide useful information but can also be subordinated by more direct controls by the physical environment. In open stands (MI $<$ 0.90) the model selects potentially dominant and other important forms (pluses on the ECOSIEVE listing, see Table 10) by consulting both the dominance hierarchy and a list of forms which are identified as capable of importance in open stands. This list is designed to eliminate frequently predicted slow-growing forms with wide environmental ranges, such as needle-leaved and extremely xeromorphic (e.g. leafless) shrubs and scrub. The list of forms identified as capable of importance in most open-stand situations includes all trees except the most mesophilic (hydrophilic), raingreen thorn-scrub, evergreen large scrub, all shrubs except the mesic broad-ericoids and slow-growing forms (needle-leaved, leaf-succulent, leafless), arborescent stem-succulents (because of their conspicuousness), and all graminoids except the bambuscoid arborescents. All forbs, small herbs, ferns, vines, epiphytes and thallophytes are excluded from identification as potentially dominant in open stands. They may, however, be reconsidered if no other forms are present, as in the near-deserts. Interpretation of open and closed stands will be illustrated by examples in the next section (Table 10).

Once the results have been obtained, the formation class is determined by the dominant forms. Physiognomic (e.g. leaf form) and seasonality descriptors are attached to the name of the formation class to provide a more complete characterization of formation type, e.g. broad-leaved evergreen forest or warm-temperate mixed forest. The usual conventions are followed (e.g. Küchler 1967), with 'mixed' referring to a mixture of broad and narrow leaf forms (among the dominant elements). 'Semi-deciduous' refers to a mixture of deciduous and evergreen forms of the same stratum, whereas 'semi-evergreen' refers to a deciduous forest overstorey with an evergreen tree understorey, as is common in tropical summer-rain transitional areas (e.g. Eyre 1968). Other combinations require more detailed descriptions.

CHAPTER 6

Model results

The combined ecophysiognomic model presented in the previous three sections marshalls a variety of data and submodels and produces a variety of outputs on several levels.

The environmental-limitation model is designed so that all forms which could occur climatically at a given site are retained in the screening process. Exclusions due to non-climatic factors such as soil, competition or succession must be accomplished by additional modeling components applied subsequently. Total vegetation cover is estimated from the annual moisture index (MI). The dominance hierarchy is applied after all other operations in order to predict which of the climatically and otherwise possible forms are most likely to dominate and characterize the vegetation formation.

The model is designed for application at particular real or hypothetical sites. Such sites can be anywhere in the world's land areas, and application of the model to a large world data-base may provide interesting new perspectives. At least this probably represents the best means of evaluating the results and the usefulness of the approach. The model was applied to the world climatic data-base of 1225 sites in order to predict world distributions of the individual life forms and of related vegetation types. The predictions were computer-mapped using the mapping basis described earlier (Chapter 2), as were also the world data for the eight ecoclimatic variables plus Thornthwaite PET. Before looking at predicted world results, it is necessary to consider model use and applicability in a bit more detail.

A. Model output and applicability

Model output, at each site, consists of a spectrum of plant forms which can be expected to occur under the climatic conditions specified, provided that:

1. the soil is drained, and edaphic or other factors do not significantly alter such microenvironmental conditions as water availability;
2. particular forms are not precluded by unusual chemical conditions, such as high salinity or low nutrient content;
3. climatic conditions not included in the model, such as wind and climatic variability, do not significantly alter the climatic conditions represented by the ecoclimatic variables;
4. the local vegetation is in reasonable equilibrium with current climatic conditions; and
5. the natural vegetation has not been greatly altered or destroyed, as by natural catastrophe or anthropogenic influences.

Sites which meet all criteria except the third (and perhaps the fourth) are called euclimatopes (Walter 1975), i.e. sites where the climate can exert its full influence on the natural vegetation. The dominance hierarchy and cover considerations can be applied to the predicted plant-form spectrum in order to predict, as an end-result, a vegetation formation type characterized by particular dominant forms but including also the other predicted forms.

As examples, the life-form spectra predicted for Raleigh (North Carolina piedmont) and Dar es-Salaam (coastal Tanzania) are shown in Table 10, along with the local climatic values. Raleigh represents the temperate zone and a site (MI = 1.43)

Table 10. Predicted vegetation at Raleigh (North Carolina) and Dar es-Salaam (Tanzania).

As examples of model output and its interpretation, the life-form spectra predicted for Raleigh (temperate zone, closed formation) and Dar es-Salaam (tropical zone, open formation) are shown as they appear on an ECOSIEVE computer printout. The site name, country, and geographic coordinates are listed on the first line for each site (with site sequence-number at far left). The second line for each site shows the climatic values at the site, as indicated in the heading. (DTY is omitted since it can be obtained easily by subtracting TMIN from TMAX). The life forms predicted to occur at the site are listed after the first two lines in order of decreasing dominance level (see Table 9). Within dominance level the forms are listed in order of increasing proximity to an environmental limit (i.e. decreasing 'fitness'). The variable involved in the closest environmental limit and the corresponding 'distance' (see section 5.E) to that limit are shown to the right of each life form. Life forms of the highest dominance level represented are indicated by a preceding asterisk (except when their fitness values are below 0.1) and are interpreted as dominant forms in closed stands. In open stands (MI < 0.90) other life forms predicted to be important and perhaps dominant (primarily grass and shrub forms, see section 5.F) are indicated by preceding pluses. Interpretation of the predictions at Raleigh and Dar es-Salaam is covered in the main text.

LOCATION TMAX TMIN PRCP PMAX PMIN PMTMAX MI	LAT	LONG
198. RALEIGH, N.CAROLINA USA	35.78	-78.65
25.0 5.0 1145. 145. 72. 145. 1.43		
* 1. SUMMERGREEN BROAD-LEAVED TREES	TMAX	0.33
* 2. HELIOPHILIC LONG-NEEDLED TREES	TMIN	0.13
3. TEMPERATE BROAD-RAINFOREST TREES	TMIN	0.0
4. SWAMP SUMMERGREEN NEEDLE-TREES	TMIN	0.12
5. BOREAL BROAD-SUMMERGREEN TREES	TMAX	0.0
6. TALL GRASSES	TMAX	0.40
7. BROAD-SUMMERGREEN SMALL TREES	TMAX	0.26
8. ARBORESCENT BAMBUSOID GRASSES	TMIN	0.05
9. TEMP. BROAD-EVERGREEN SMALL TREES	TMIN	0.23
10. TALL CANE-GRAMINOIDS	TMIN	0.14
11. TEMPERATE BROAD-EVERGREEN SHRUBS	TMIN	0.32
12. BROAD-SUMMERGREEN MESIC SHRUBS	TMAX	0.31
13. DWARF-NEEDLE SMALL TREES	PMTMAX	0.38
14. BROAD-ERICOID EVERGREEN SHRUBS	TMAX	0.0
15. SHORT SWARD-GRASSES	TMAX	0.37
16. SUMMERGREEN GIANT-SCRUB	TMAX	0.43
17. NEEDLE-LEAVED EVERGREEN SHRUBS	TMAX	0.40
18. SHORT BUNCH-GRASSES	TMAX	0.56
19. SUMMERGREEN FORBS	TMAX	0.40
20. SUMMERGREEN FERNS	MI	0.23
21. TEMPERATE EVERGREEN FORBS	TMAX	0.20
22. BUSH STEM-SUCCULENTS	MI	0.05
23. BROAD-EVERGREEN VINES	TMIN	0.04
24. BROAD-WINTERGREEN EPIPHYTES	TMIN	0.18
25. BROAD-SUMMERGREEN VINES	MI	0.30
26. MAT-FORMING THALLOPHYTES	TMAX	0.19
27. XERIC THALLOPHYTES	MI	0.41
664. DAR-ES-SALAAM TANZANIA	-6.80	39.28
28.0 24.0 1110. 300. 27. 77. 0.77		
* 1. TROPICAL XERIC NEEDLE-TREES	TMAX	0.13
+ 2. TROPICAL EVERGREEN SCLEROPHYLL TREES	TMAX	0.35
3. PALMIFORM TUFT-TREES	MI	0.09
4. TROPICAL EVERGREEN MICROPHYLL-TREES	MI	0.02
5. TALL CANE-GRAMINOIDS	MI	0.09
6. TALL GRASSES	MI	0.09
+ 7. BROAD-RAINGREEN SMALL TREES	TMAX	0.35
8. SHORT SWARD-GRASSES	TMAX	0.23
9. PALMIFORM TUFT-TREELETS	MI	0.09
10. TROPICAL BROAD-EVERGREEN SHRUBS	MI	0.09
11. XERIC EVERGREEN TUFT-TREELETS	MI	0.31
+12. SHORT BUNCH-GRASSES	TMAX	0.40
13. EVERGREEN GIANT-SCRUB	TMAX	0.33
14. BUSH STEM-SUCCULENTS	TMAX	0.40
15. RAINGREEN FORBS	TMAX	0.28
16. BROAD-WINTERGREEN EPIPHYTES	TMAX	0.14
17. BROAD-RAINGREEN VINES	TMAX	0.35
18. XERIC THALLOPHYTES	TMAX	0.40

at which a closed forest would be expected. The tree component at Raleigh consists of five potential canopy forms plus three small-tree forms, all of which actually occur (examples in parentheses):

1. Summergreen Broad-Leaved Trees (*Quercus, Acer,* etc.)
2. Heliophilic Needle-Trees ('southern pines')
3. Temperate Broad-Rainforest Trees (e.g. *Magnolia*)
4. Hydrophilic Summergreen Needle-Trees (*Taxodium*)
5. Boreal Broad-Summergreen Trees (e.g. *Betula*)

plus understorey forms:

1. Temperate Broad-Evergreen Small Trees (e.g. *Ilex opaca*)
2. Broad-Summergreen Small Trees (*Cornus, Prunus,* etc.)
3. Dwarf-Needle Small Trees (*Juniperus virginiana*).

Of the five candidates for formation dominance, one can see immediately that the rainforest trees are occurring at their winter temperature limit of 5° (fitness value of zero) and that the boreal and hydrophilic trees are from lower levels in the dominance hierarchy (and are also constrained climatically or edaphically). Considering the two predicted dominants (asterisked in Table 10), one sees that the heliophilic needle-trees may be approaching a winter temperature limit and thus less robust. More importantly, however, they are evergreen sclerophylls and cannot be expected to grow as rapidly at maturity as the broad-malacophyllous summergreens. These considerations suggest that the vegetation can be characterized as a summergreen broad-leaved forest with evergreen needle-trees, with the other tree forms as scattered individuals, and with a well developed understorey composed of small broad-summergreen trees, more scattered broad-evergreen and dwarf-needle small trees, and a good variety of summergreen and some evergreen shrubs, forbs, grasses and vines. The predicted formation can be called a Summergreen Broad-Leaved Forest with evergreen needle-trees. The actual vegetation in and around Raleigh is a summergreen mixed oak forest with persistent successional pines in the canopy (up to 100 years), a wide variety of other canopy and understorey broad-summergreen trees, bald cypress (*Taxodium*) in wet depressions, well developed shrub strata where light permits (especially along edges), and a wide variety of forbs, grasses, etc. Magnolias occur only rarely as natural forest components and more often as planted ornamentals. All 27 of the forms predicted actually occur naturally, even the bush stem-succulents (*Opuntia compressa*). Not predicted but actually occurring on dry sites are also xeric tuft-treelets and rosette-shrubs (*Yucca* spp.).

The situation at Dar es-Salaam is more complicated (despite having fewer forms predicted), since it is a drier site (MI = 0.77) at which open vegetation and reduced dominance by any one form are to be expected. Of the 18 forms predicted, all can be identified (Knapp 1973; Lind and Morrison 1974) as actually occurring except the xeric needle-trees (primarily *Juniperus procera*), which are generally confined more to upland and highland areas.[1] Not yet knowing this, one must interpret the predicted vegetation as an open, rather grassy semi-deciduous woodland in which evergreen sclerophyll trees and smaller raingreen trees are the main components and xeric needle-trees are scattered, along with some palms and evergreen microphyll trees but more frequent scrub forms. The MI value, the mix of predicted forms, and the proximity of many (no fewer than six) to moisture limits strongly suggest a vegetation mosaic affected by topography and other spatial variations in water availability. Since the site is in Africa rather than Australia, one must also not picture the evergreen sclerophyll trees or arborescents as tall eucalypts but rather as forms more like the raingreen trees in size. The predicted vegetation, using the criteria in Table 9, can be called a Semi-Deciduous Mixed Short-Grass Savanna-Woodland or Savanna-Woodland Mosaic, since the MI value is clearly transitional between the two formation classes. The actual vegetation along the coastal strip including Dar es-Salaam has been described by various authors (e.g. Eyre 1968; Knapp 1973; Lind & Morrison 1974; Keay 1959; Shantz & Marbut 1923) as a dry, grassy open forest-woodland-scrub mosaic, with areas of species-rich dry forest near the coast. The trees are predominantly raingreen but xeric evergreens and facultatively deciduous trees are important. The degree of

[1]Problems with this form in other regions also suggest that its upper TMAX limit (30 °C) is probably set too high and should be lowered to around 27°–28 °C.

deciduousness of many of these trees is poorly known and even less frequently reported. Important taxa include *Cynometra, Manilkara, Diospyros, Combretum, Cussonia, Erythrina, Strychnos,* and many others, plus *Adansonia digitata, Olea africana,* and a great variety of shrubs, scrub taxa, grasses, etc. The relatively high PMIN value of 27 mm at Dar es-Salaam precludes prediction of larger raingreen trees (limit 25 mm) and raingreen thorn-scrub (limit 20 mm), but these forms do occur (e.g. *Adansonia*). Thorn-scrub is the dominant vegetation behind the wetter coastal strip.

The model is designed for application to euclimatopes and, strictly speaking, is applicable only there. It can be used to provide information at other sites, however, if appropriate modifications of microenvironmental conditions can be estimated. For sites with azonally coarser or finer soils, the precipitation requirements must be increased or decreased accordingly. (This effect is buffered in dry climates by the soil itself, however, since coarser soils may reduce actual evaporation from unvegetated areas by storing the precipitation at greater depths.) In windy climates precipitation requirements are greater, since PET may be greatly increased. Waterlogged soil may preclude trees in some environments (e.g. bogs), where the local species are not adapted to anaerobic conditions, but not in others (e.g. swamps). Dry environments with mosaic-like microrelief may produce vegetation mosaics. In such cases gentle slopes may most nearly represent the actual macroclimatic conditions, while conditions on hilltops or in depressions are shifted one way or the other.

The results predicted by the model and, more importantly, the degree to which climate determines or at least provides the initial constraints on vegetation patterns and types can best be evaluated by comparing climatically predicted and known actual vegetation at particular sites. For this reason, the vegetation predictions at 100 potentially better known sites from the world data-base (as they appear on the ECOSIEVE printout and in Table 10) are reproduced in Appendix B for readers in various regions to evaluate. Vegetation predictions for an additional 74 validation sites are shown in Appendix C.

B. Generating and mapping world results

As discussed earlier, generation of world or other large-area results requires a large climatic data-base for the area and a computer. The program ECOSIEVE was developed to apply environmental-limitation models to large, computerized data-bases (Box 1978a, 1979a, 1981b) and includes provisions for coupling with user-provided adjunct routines, such as the dominance hierarchy. ECOSIEVE was used to generate all world predictions and can provide results both for the respective sites and as potential distributions, as represented by lists of sites.

Once 'final' results have been obtained it is often desirable to map them for easier interpretation and evaluation. This can be done using the program SYMAP, developed by the Laboratory for Computer Graphics and Spatial Analysis at Harvard University. SYMAP produces raster-based maps on a normal line-printer using irregularly spaced site data as input. Contour (isoline), proximal (nearest-neighbor), and conformal (choropleth) maps can be produced by the various interpolation algorithms. The use of SYMAP has been documented by Dougenik and Sheehan (1975) and locally by Reader et al. (1972), Rase (1976, in German), and various other groups. SYMAP requires computerized base-map, a data-base with site coordinates, and various user-provided instructions for map design. The mapping process can be both facilitated and expanded (e.g. map quantification by computerized planimetry) by a variety of SYMAP simplifying and adjunct routines (Box 1979a, 1979b, 1979c).

In order to display the world ecoclimatic values and predicted plant-type distributions, the values at the 1225 test-sites were computer-mapped using SYMAP, a computerized world base-map (AOWDGL, see Reader et al. 1972), and a general coordinate-conversion routine for cylindrical and pseudo-cylindrical projections (FLEXPROJ, see Box 1979a). Since a high degree of cartographic detail is not required (and would be artificial considering the nature of the data), large computer-printed maps provide an adequate and convenient medium. The maps are based on a pseudo-cylindrical projection by Robinson (1974), which is designed to reduce polar shearing. This projection represents a compromise between areal equivalence and correct shapes, and is an excellent projection for visually oriented presentations. The computerized version (AOWDGL) was traced in 1971 from a large wall map (Rand-McNally and Co., without date) by several University of North Carolina students and has been used since for a wide variety of geoecologic maps (e.g. Lieth & Box 1972, 1977; Box 1976, 1977, 1978a, 1978b, 1979a, 1979c). The area deformation of the projection increases with latitude but is such that areas in the latitude of Iceland are enlarged by a factor of only about two. The use of a non-equal area projection does not mean that planimetry cannot be performed on the map. The area-representation of AOWDGL was investigated (Box 1975) using the planimetry program MAPCOUNT (Box 1975, 1978b, 1979a) which performs the mathematics necessary to process non-equivalent projections. It was found that AOWDGL under-represents the earth's true land area by about 7%, mostly in the polar regions.

The coordinate-conversion routine FLEXPROJ permits execution-time conversion of geographic site coordinates into the required SYMAP coordinates. It has been used for most of the above-cited maps involving the Robinson projection. The coordinate conversion is based on the mathematical definition of the Robinson projection and is added to SYMAP via its subroutine FLEXIN (FLEXible INput).[2]

The use of 1225 data-sites as the basis for world maps represents a minimum requirement. This density is not sufficient to permit accurate prediction of boundaries but does permit a reasonable representation of predicted general patterns. About 200 sites were added to the original data-base of 988 sites in order to improve the spatial representation in areas of high topographic complexity. Areas of high topographic complexity and/or low site density are misrepresented most. Poorly represented areas include the southwestern part of the Arabian Peninsula, the area just east on the Caspian Sea, the Himalayan highlands and their southern face, the deep river valley region of southern China, the mountain chains of Europe, and most of Antarctica.

[2]FLEXPROJ also enables various other modifications, including coding of data-site locations and corresponding values on the same input record, which is much more convenient that the standard SYMAP input structure.

The values of the eight ecoclimatic variables plus PET were mapped as world isoline maps. The interpolation on these maps is based on a distance-squared decay function (see Shepard 1968) using from 4 to 10 sites within a variable search radius. Isoline levels are represented by shading ranging from a single dot at each print-position to a completely blackened print-field. These maps are about 131 cm (52 inches, 515 columns) wide by 65 cm (25 inches, 204 rows) high.[3] When photo-reduced, the maps present a spectrum of shades from light to black. Photography was performed by the University of North Carolina Photographic Laboratory using high-resolution 35 mm film.

Mapping the vegetation predictions is somewhat more complex, since complete physiognomic spectra cannot be shown meaningfully on a single map. One is most interested in seeing the predicted dominant forms, plus certain other features which might give a general impression of the vegetation formations to be expected. In order to make the predicted results more comprehensible, the plant forms are grouped by growth-form, with various groups presented on separate maps, as follows:

Map	*Plant types*
10	Broad-leaved evergreen trees
11	Importance of broad-leaved evergreen trees
12	Broad-leaved deciduous trees
13	Importance of broad-leaved deciduous trees
14	Narrow-leaved trees
15	Shrubs and arborescents
16	Graminoids
17	Stem- and leaf-succulents
18	Forbs and semi-desert herbs
19	Warm-evergreen (eucalyptoid) sclerophyll trees
20	Tuft-tree forms
21	Vines and lianas
22	Epiphytes

Each map contains from two to nine forms and shows the areas in which these forms are predicted to occur as important or co-important forms, plus other occurrences on some maps. Where predicted co-important forms belong to the same map (same growth form), a special symbol for multiple predic-

[3]American computers unfortunately still think in inches: 8 rows per inch vertically and 10 columns per inch horizontally.

tions is employed. Where predicted co-dominant forms belong to different maps (different growth forms), each form is shown on its own map without any special notation. Separate maps showing the occurrence of co-dominant forms (Map 24) and the number of forms predicted at each site (Map 25) are also included. Ecological sub-types which do not greatly differ in physiognomy are grouped before dominance considerations.

Graphic limitations presented by the computerized mapping procedure represent a major consideration in the production and interpretation of the maps. Multi-colored computer printouts are not yet available at most universities. Since SYMAP permits overprinting of up to four characters at each print-position, however, there is ample opportunity for the construction of graphic symbols so that the raw predicted results can be shown without any retouching. The vegetation maps employ distinctive, hopefully intuitive symbols for the various plant forms. Symbols for forests are based on tall symbols, especially capital T (broad-leaved) and a tall vertical line (narrow-leaved), with rainforests and other evergreen forests shown by darker overprint combinations. Open woodlands are represented by more open combinations, with height of vegetation suggested by the height of the symbols. Grasses, shrubs, scrub, etc. are shown by shorter and generally simpler symbols. Some symbols are necessarily used on more than one map for different vegetation types.

The maps of plant-form distributions (and the map of co-dominants) are proximal maps produced by the SYMAP nearest-neighbor algorithm. Each vegetation zone is centered on one or more data-sites, but its shape depends on distance to sites with other predicted forms. Transitions are not shown except when they result in intermediate dominant-form mixes at intermediate sites. Because of insufficient site density, the occurrence of such transitional sites often makes the mapped patterns harder to discern. This is especially true in areas where transitional zones are long and narrow, as south of the Sahara desert. Since the vegetation maps depend on recognition of individual computer-printed symbols, which would be obscured by photoreduction, the vegetation maps are presented in their computer-printed form. This also seems most appropriate, since it best represents the predictive accuracy of the results.

C. Overview of world ecoclimates

In defining climatic envelopes for different plant types one is essentially defining a set of plant ecoclimates. More general aspects of world ecoclimates, as they affect animals, human activity or other aspects of plants, can be seen by examining the ecoclimatic variables individually. This is done by means of world maps (Maps 1–9) and distributional statistics (Table 11) for the eight ecoclimatic variables plus annual PET, based on data at the 1225 test-sites.

The map of highest mean monthly temperature, TMAX (Map 1), identifies primarily the treeless polar and subpolar regions, including the plantless ice deserts of Greenland and Antarctica (TMAX $<$ 0 °C), the polar cold deserts with few plants (0° $<$ TMAX $<$ 5 °C), and the tundra areas with more complete plant covers (5° $<$ TMAX $<$ 10 °C). Because of lower site density in polar areas, these regions represent larger fractions of world land area than suggested in Table 11. Other areas can be considered to be potential tree climates, at least thermally. Higher summer temperatures indicate more energy available for plant growth, but the potential increase in plant size in warmer areas is strongly limited by water availability. Note that the world frequency distribution of maximum mean temperature is bimodal, with 43% of the stations having tropical summer temperatures ($>$25°), only 16% having subtropical, and 25% having typical temperate summer temperatures.

The map of lowest mean monthly temperature, TMIN (Map 2), separates those areas which can support year-round plant activity, provided that water is not limiting, from those areas with winter dormancy. The division outside the tropics generally falls near the 10 °C isotherm for TMIN, though extreme absolute temperatures and overall temperature regimes may cause significant differences between maritime and continental climates. The –20 °C isotherm for TMIN delimits the coldest areas of northern North America and Eurasia, within which summergreen larches become very significant vegetation components. The world frequency distribution for TMIN is also generally bimodal, reflecting distinctly different tropical and temperate core areas.

The continental climates of northern Eurasia and North America, with extremely cold winters, are suggested by those regions on Map 3 in which the annual range of mean monthly temperature, DTY, is greater than 40 °C. Otherwise the world frequency distribution is remarkably even. This map also best identifies the tropical regions (DTY < 15 °C), in which plant activity is less seasonal than diurnal and in which evergreen plants are dominant except in areas with long dry periods. The division of the land areas into seasonal north and largely steno-thermal south is most evident on this map.

The map of annual precipitation, PRCP (Map 4), best identifies the desert climates (PRCP < 10 cm) and indicates the total amount of water available. Within broad latitudinal belts it also separates wetter and drier ecoclimates. Except in the polar and boreal/austral areas, PRCP serves as a crude index of leaf size and of the general size of the largest plants. The world distribution (Table 11) may be slightly bimodal, with moderately low-precipitation areas most common (grasslands, woodlands, boreal forest), but note that the value ranges are not uniformly wide.

The world pattern of the Thornthwaite estimate of potential evapotranspiration (Map 5) indicates the potential rate of water loss by plants and, thus, the extent to which xeromorphic features may be prominent, even under high-rainfall regimes. The daily water stress and xeromorphic structure of canopy leaves even in tropical rainforests, for example, have been pointed out by many authors (e.g. Richards 1952, Walter 1973).

The map of the annual moisture index MI (Map 6) suggests the life-form possibilities, the leaf structure, and the general structure of vegetation formations. More mesic environments generally have larger life forms, softer leaves, and often greater dominance by particular forms. Grasses are an important exception to the first generalization, however, since they can form dense covers and dominate without greater size. In general, forests do not dominate where MI < 1.0 (52.6% of the sites), but this leaves nearly half the world as potential forest climates. Trees generally disappear completely, except where additional water is available, when MI falls below about 0.5 (22.7% of all sites). Except for the slight increase below MI $= 0.1$, the world frequency distribution appears to be unimodal and centered on about MI $= 1.0$ as the most common value. The map of MI is the most useful single map for estimating plant physiognomy and vegetation structure. Most of the above generalizations do not hold, however, in the treeless polar and subpolar climates.

The map of highest average monthly precipitation, PMAX (Map 7), is the most important map for locating the extreme deserts and semi-deserts. Because of the reduced water content of colder air, this generalization also holds somewhat for the polar and subpolar regions. In climates with dry seasons and deciduous or ephemeral vegetation, PMAX may also be a useful indicator of leaf size and structure. On a linear scale, the world frequency distribution (Table 11) appears to be unimodal with mode around 75 mm but with many sites receiving larger amounts.

The map of lowest average monthly precipitation, PMIN (Map 8), very effectively identifies the world's evergreen rainforest regions (PMIN $>$ 75 mm) as well as other regions with significant year-round rainfall (PMIN > 50 mm), such as the typical four-season (nemoral) temperate climates of Western Europe and eastern North America. The most important ecoclimatic difference between these climates and the winter-dry temperate climates of East Asia is evident on this map. Most plants in areas with PMIN < 10 mm are distinctly deciduous or ephemeral (48.6% of all sites, see Table 11). PMIN may also be a good index of leaf size and structure. Note, despite the appearance given by the logarithmic value ranges, that the frequency of PMIN amounts decreases steadily from driest to more favorable amounts.

The map of average precipitation of the warmest month, PMTMAX (Map 9), is very useful for separating generally mesic and generally dry climates, since the warmest months represents critical periods. The zonal differences between PMTMAX on corresponding east and west sides of land masses are very evident. Plants in low-PMTMAX areas are usually small forms or stunted larger forms, except where summer fog moderates the effect of low rainfall (e.g. northern California to British Columbia and in southwestern Australia) and produces the world's tallest plants. The areas where PMTMAX > 75 mm (less in boreal regions) are generally dominated by forests, (38% of all sites). Note what appears, to be a bimodal distribution in Table 11.

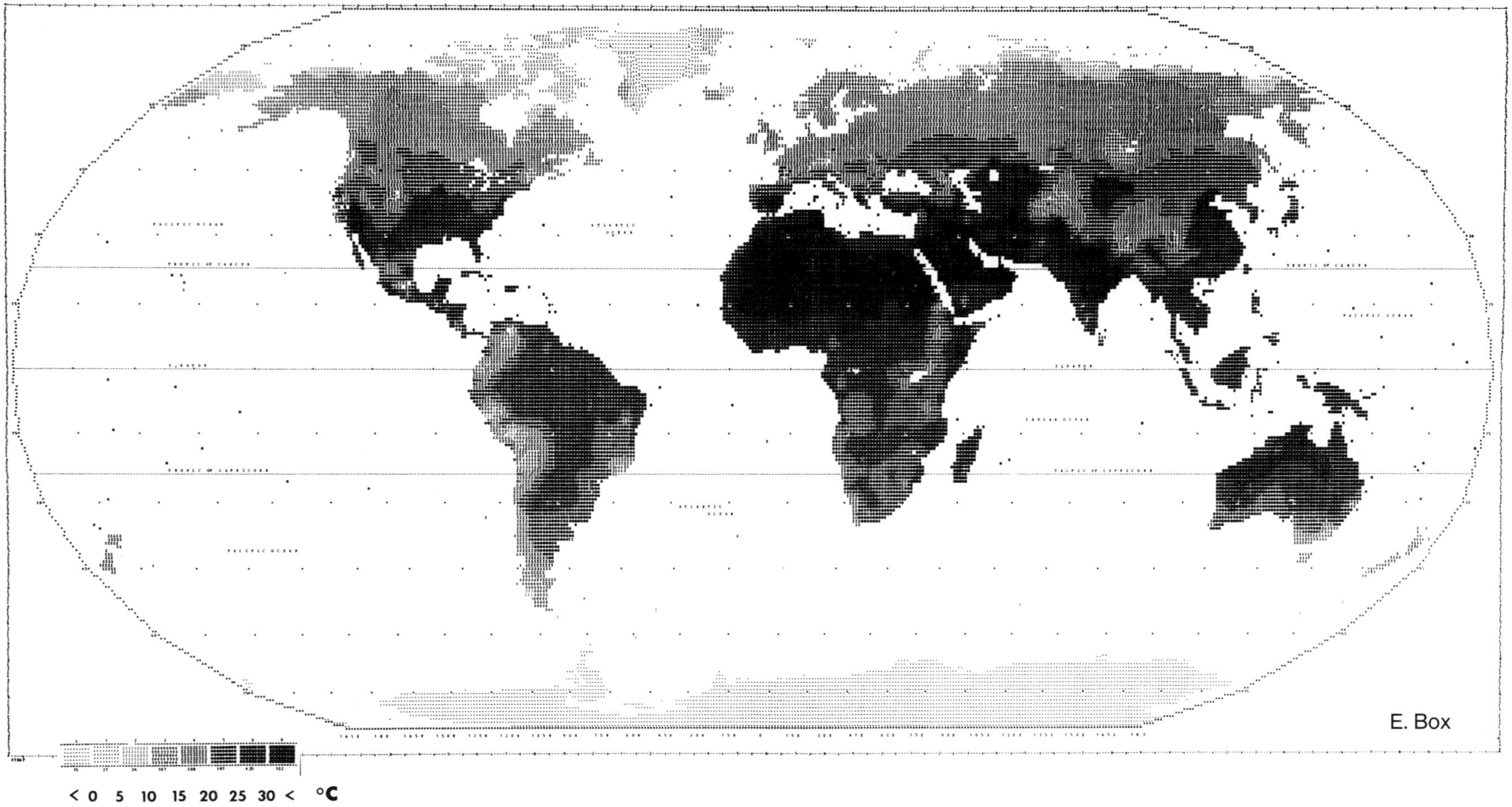

Map 1. World terrestrial pattern of highest mean monthly temperature, TMAX.
The highest mean monthly temperature serves as a measure of energy input during the year and/ or growing season. Plant production, the size of the dominant plants, and the complexity of the vegetation structure generally increase with increased energy input and ambient temperatures, provided that water or other factors are not limiting.

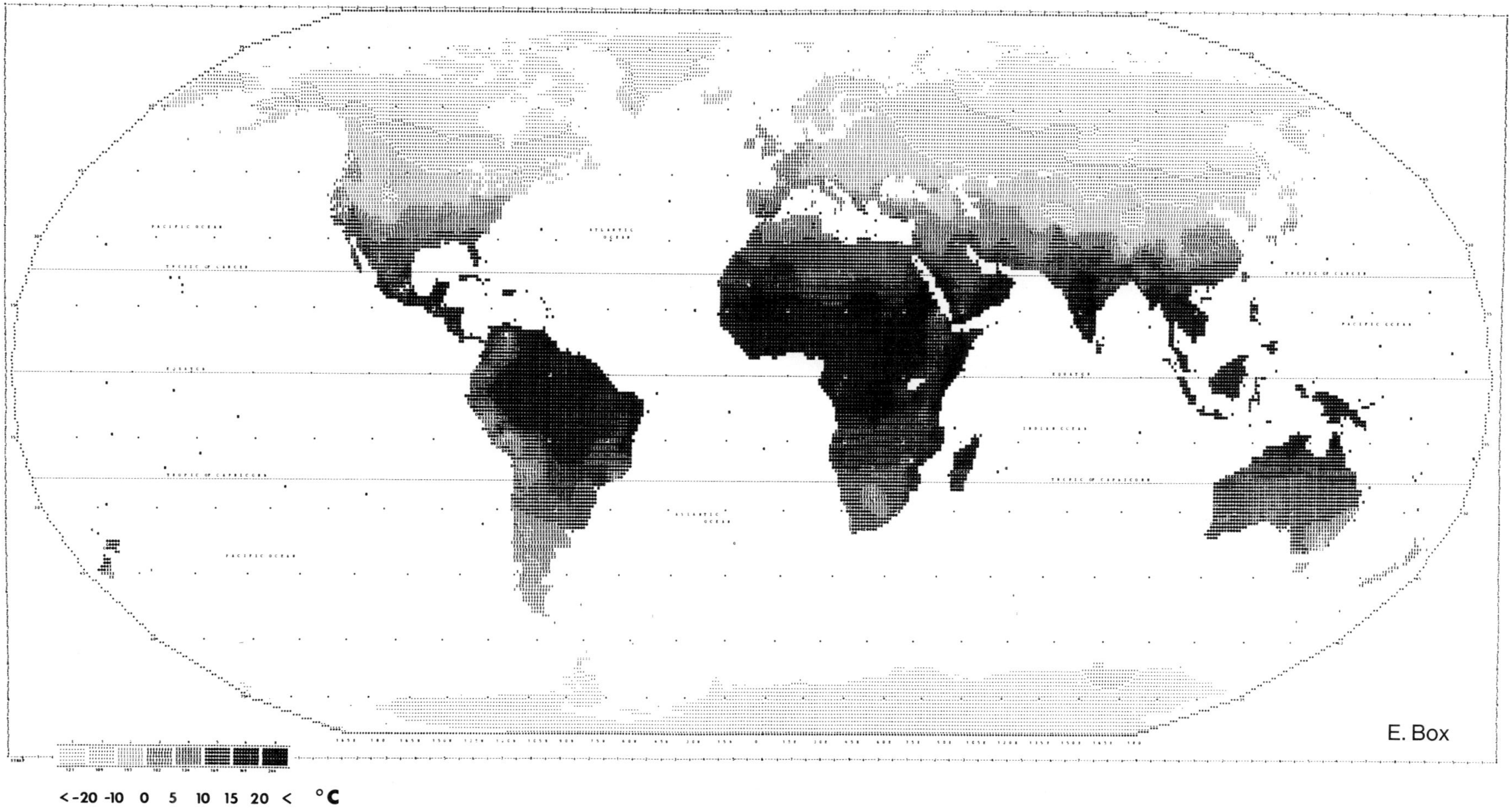

Map 2. World terrestrial pattern of lowest mean monthly temperature, TMIN.
The lowest mean monthly temperature provides an estimate of both the degree of winter cold in general and absolute minimum temperatures which must be endured. These factors represent limits to the distributions of many evergreen forms and may constitute vernalization requirements for summergreen forms.

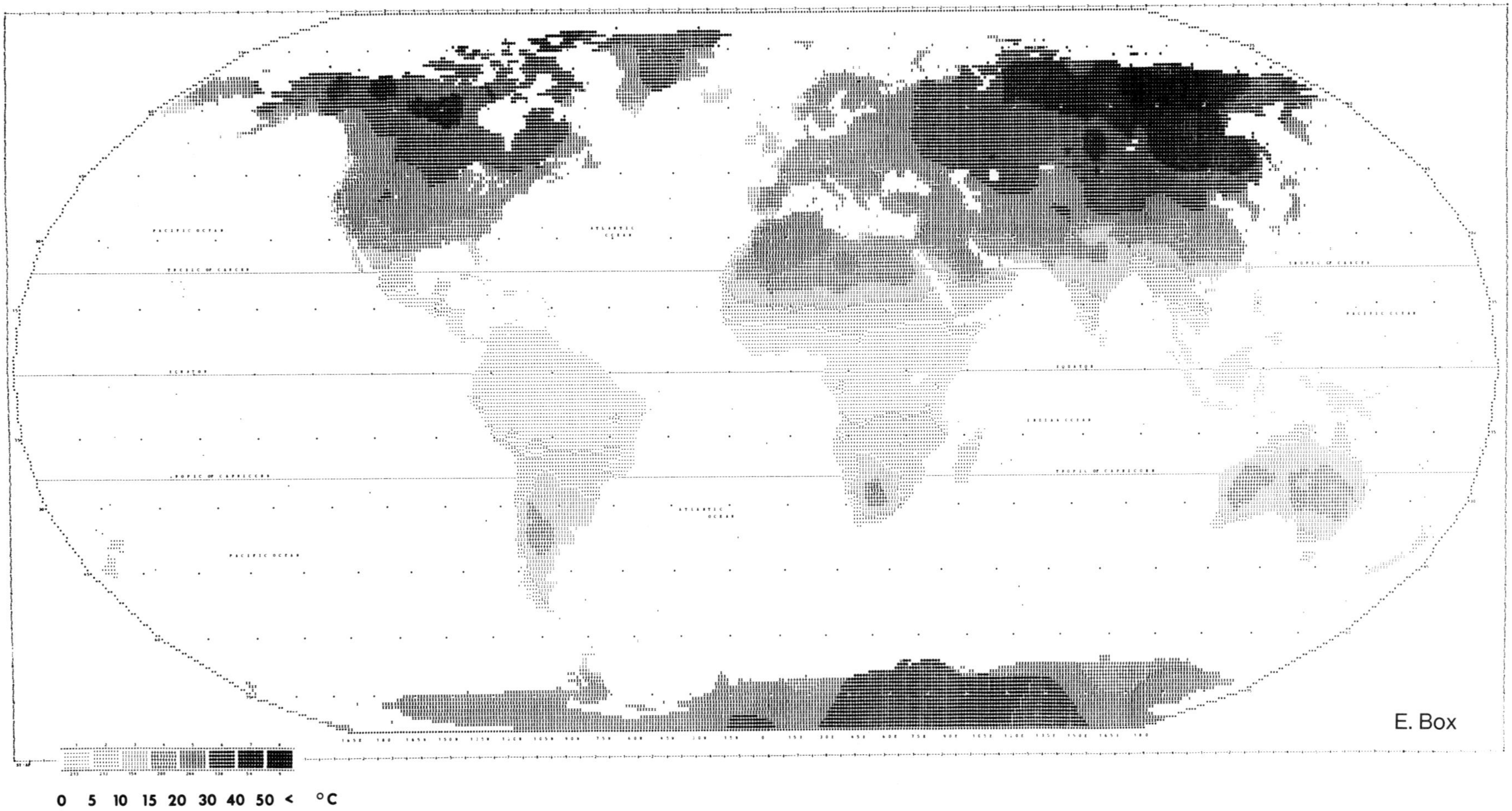

Map 3. World terrestrial pattern of annual range of mean temperature, DTY.
Annual temperature range, represented here by the range of mean temperatures, provides the main mechanism for separating tropical forms with little temperature seasonality from forms adapted to seasonal cycles. It is probably less important in the current model than in the first version.

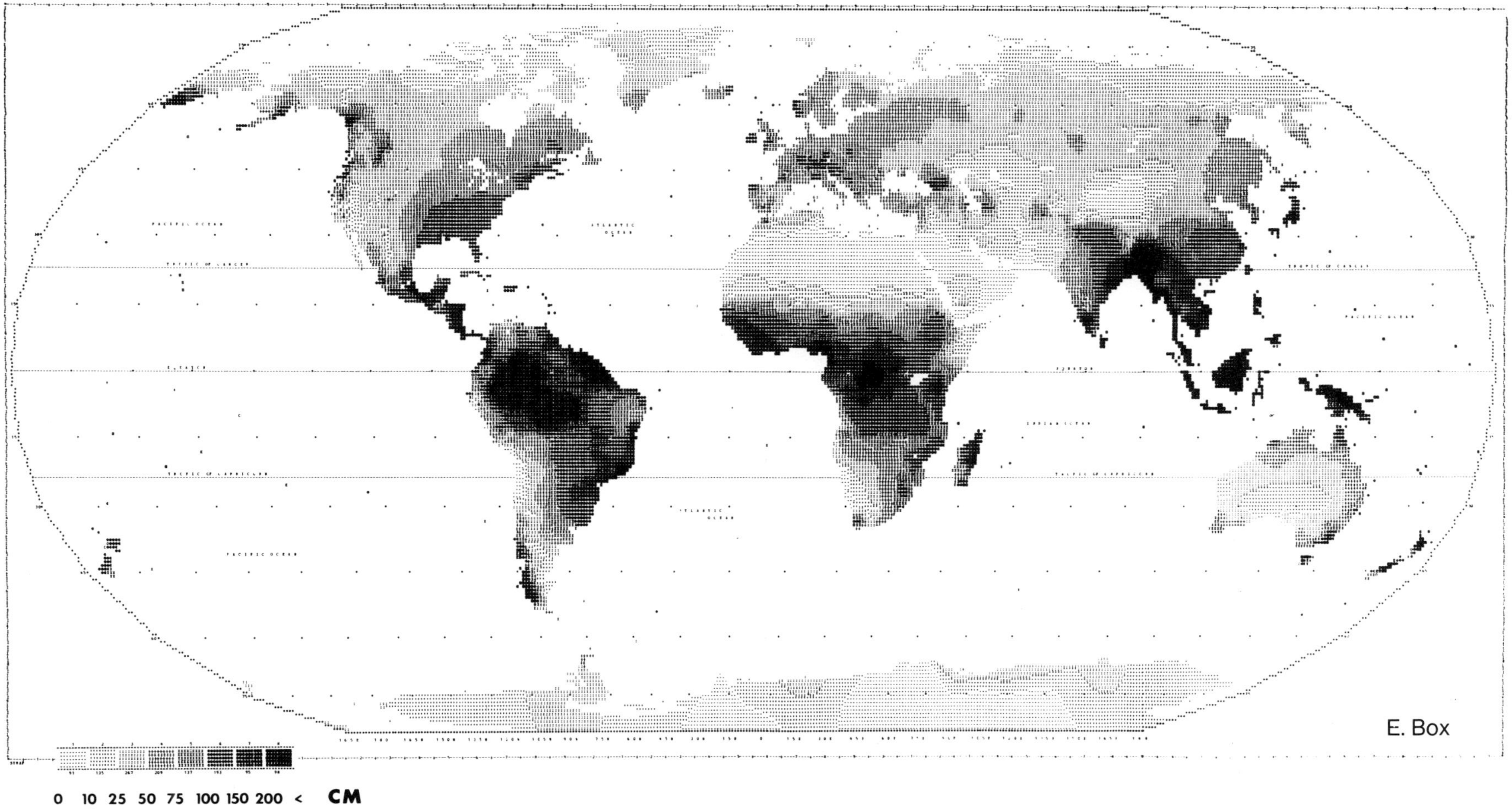

Map 4. World terrestrial pattern of average annual precipitation, PRCP.
Average annual precipitation represents the total amount of water which is (climatically) available for storage in the soil and use by plants. For a given temperature regime, plant production and vegetation structure generally increase with increased water availability. This generalization can be modified considerably by precipitation seasonality.

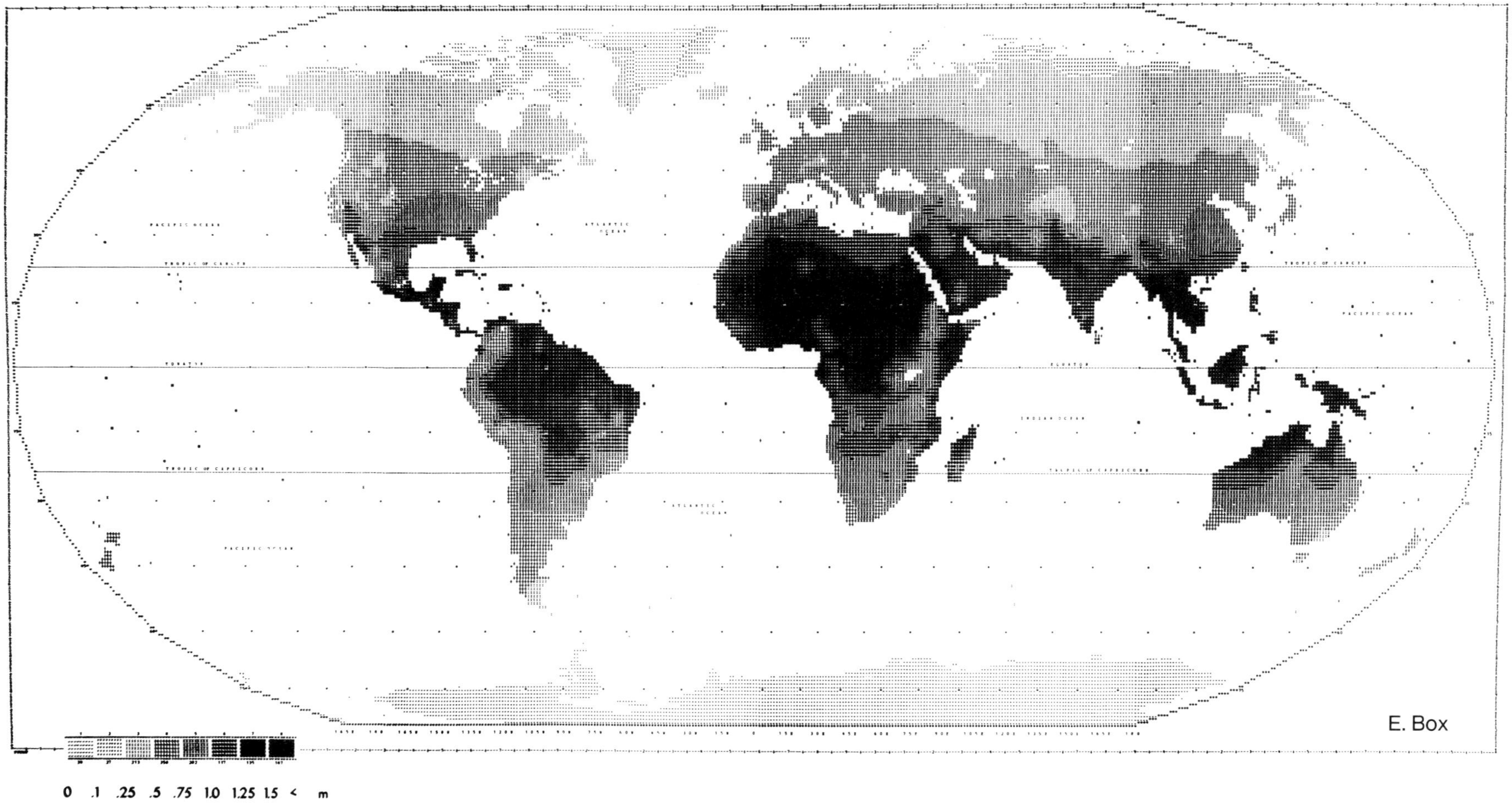

Map 5. World terrestrial pattern of average annual potential evapotranspiration (Thornthwaite estimate).
Average annual PET represents the total potential water loss by plants during a year and provides the basis for estimating annual patterns of water demand. PET over periods of at least a few weeks is generally estimated by air-temperature methods, of which perhaps the most widely used is that of Thornthwaite. The Thornthwaite estimate mapped here is based on measured monthly mean temperature values for the seasonal extrema and values interpolated for the intervening months.

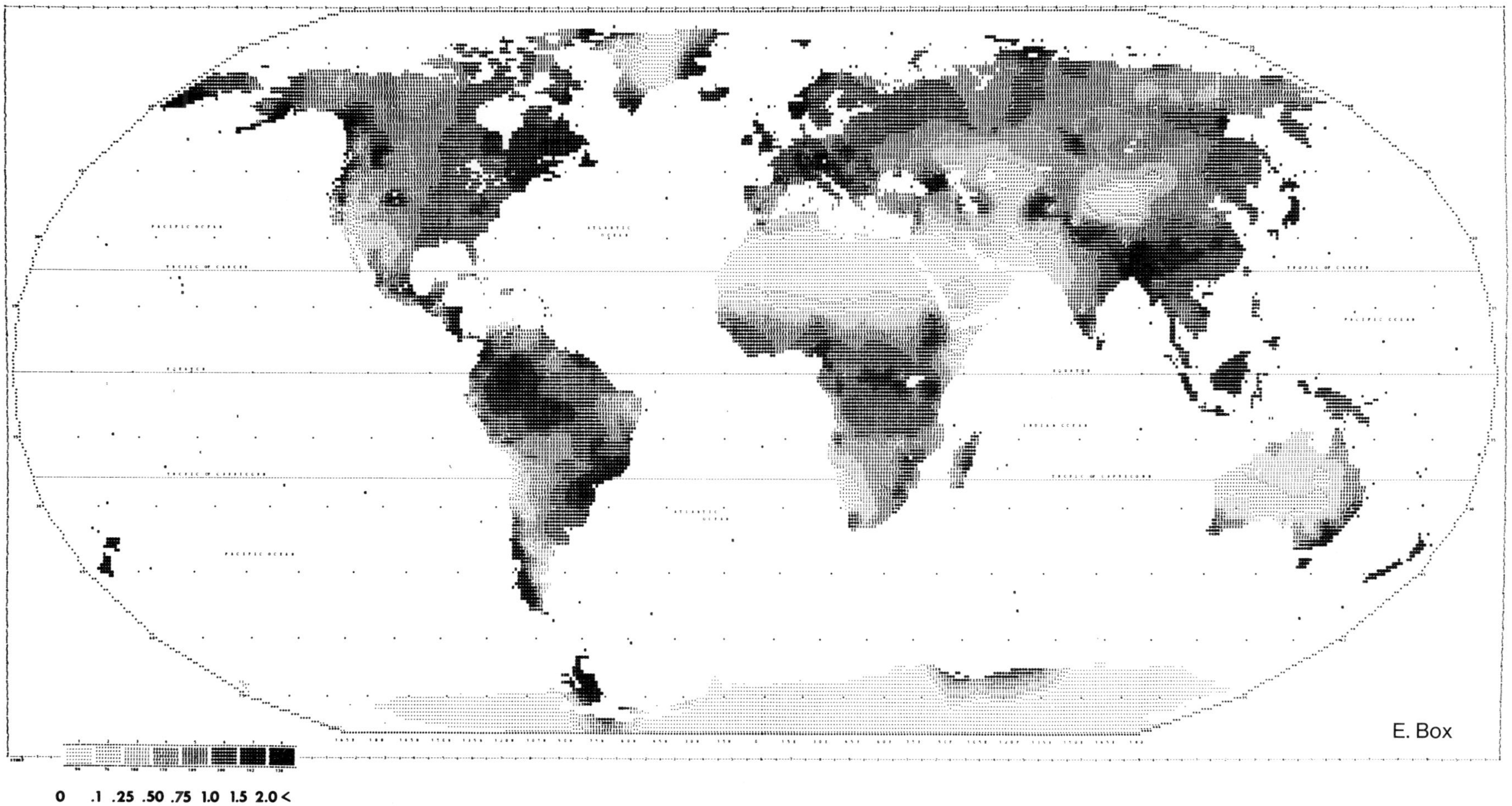

Map 6. World terrestrial pattern of the annual moisture index, MI (based on Thornthwaite's estimate of PET).
The annual moisture index MI, obtained by dividing average annual precipitation by average annual PET, serves as an index of general moisture conditions. Since it involves both energy input and available moisture, the moisture index serves as the best single indicator of the dominant growth forms, general vegetation structure, and the degree of leaf xeromorphy.

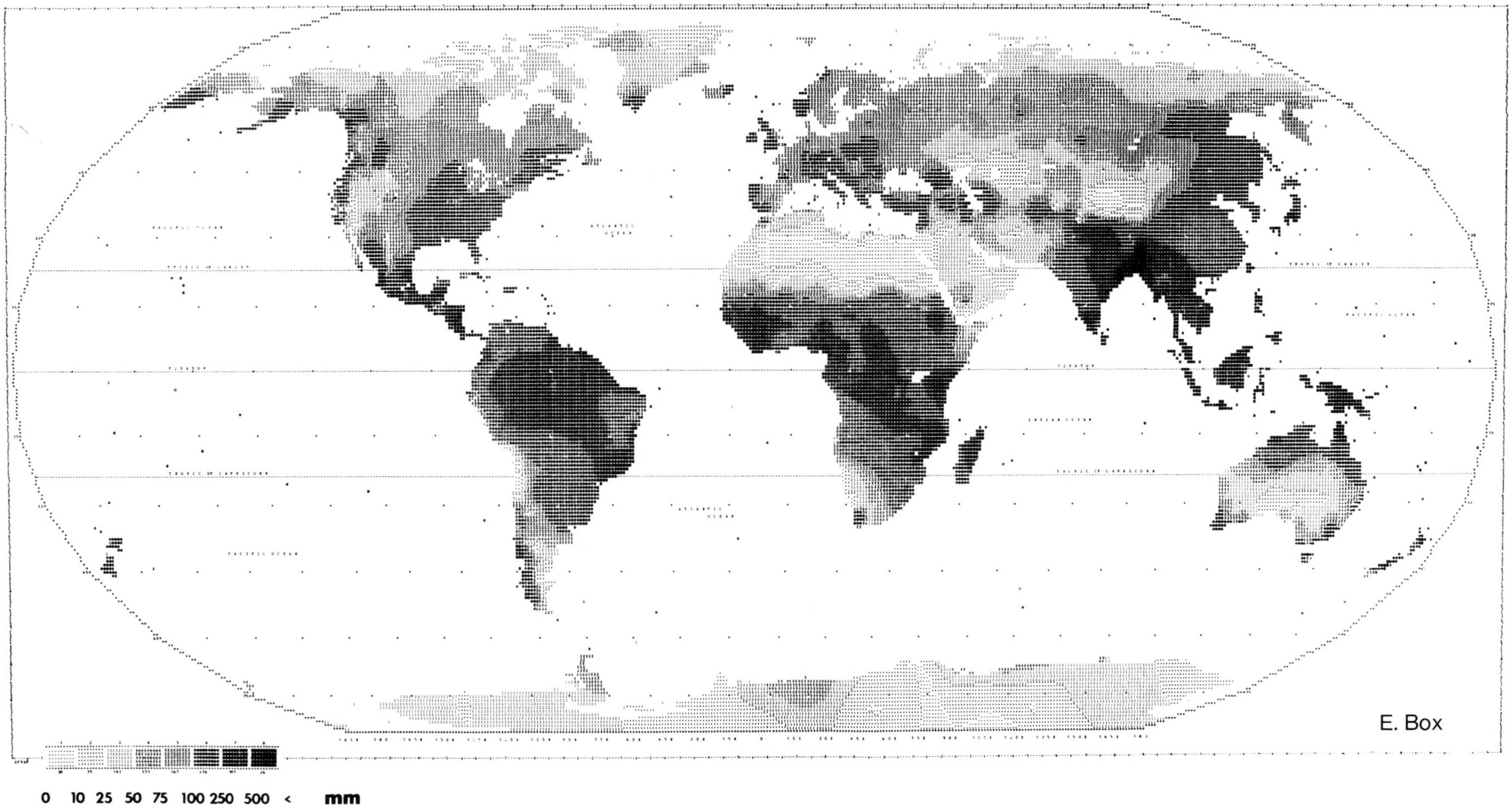

Map 7. World terrestrial pattern of highest average monthly precipitation, PMAX.
The highest average monthly precipitation provides an estimate of the amount of water which may be required, especially in drier climates, to constitute adequate growing seasons. It also indicates areas with high seasonal rainfall (seasonal rainforests).

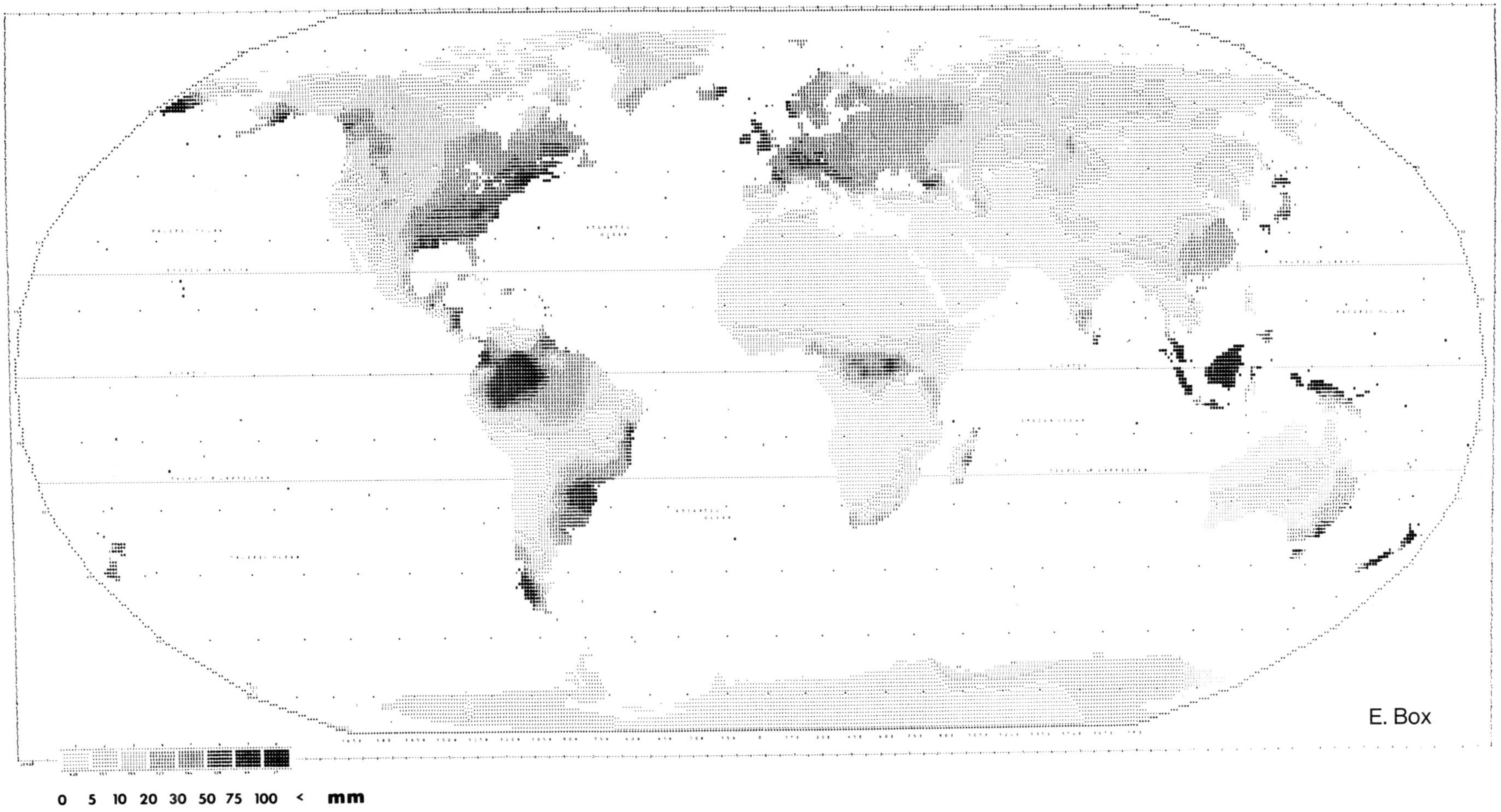

Map 8. World terrestrial pattern of lowest average monthly precipitation, PMIN.
The lowest average monthly precipitation identifies immediately the rainforest areas, which have no significant dry season. It also constitutes the mechanism for identifying areas with raingreen forms.

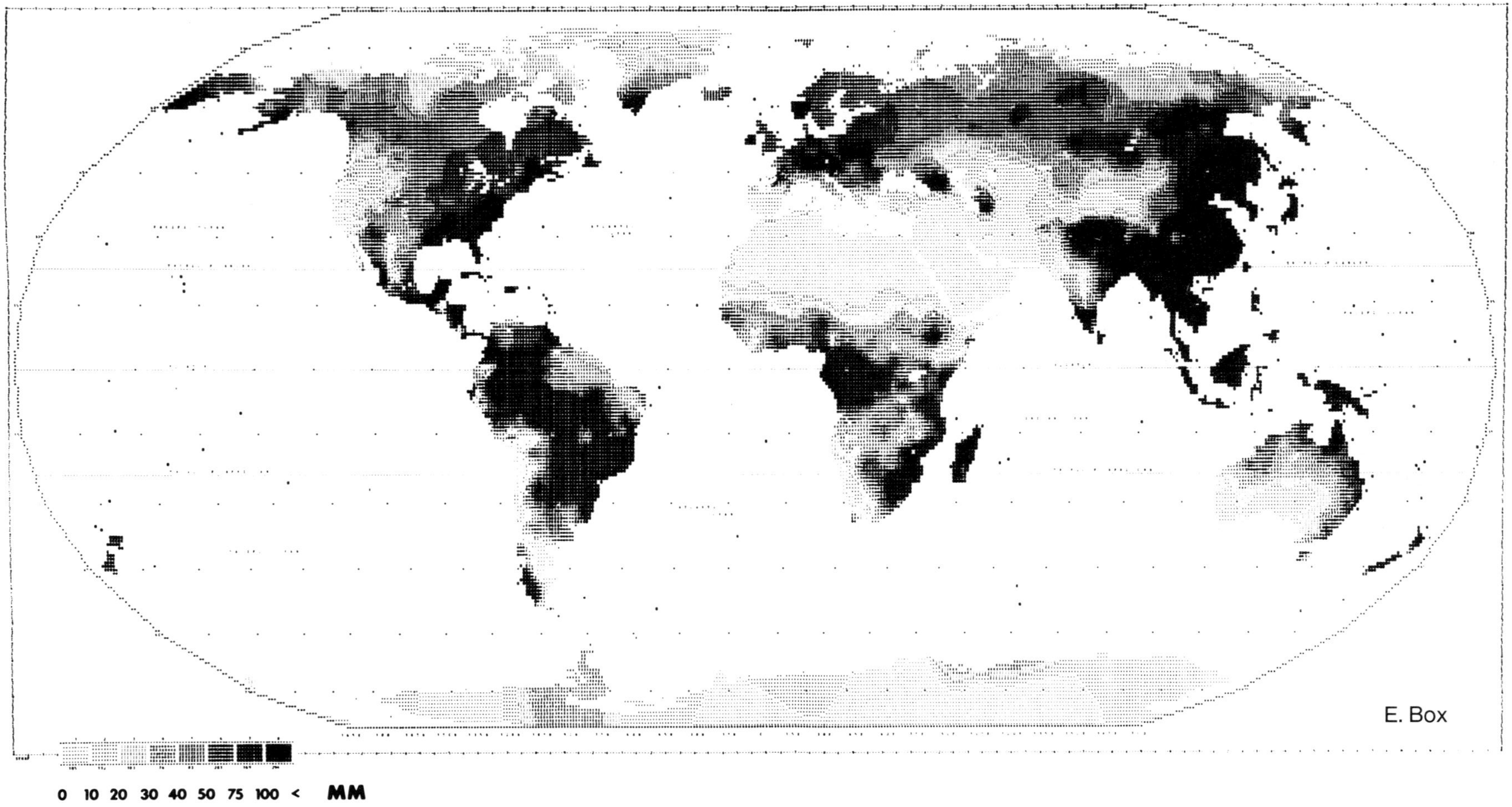

Map 9. World terrestrial pattern of average precipitation of the warmest month, PMTMAX.
The average precipitation of the warmest month serves as a useful index of precipitation seasonality in relation to growing season requirements. The map clearly identifies summer-rain areas and summerdry areas with often extreme water stress. This variable is used in particular to predict mediterranean plant forms.

Table 11. World frequency distributions of important ecoclimatic variable values.

World (terrestrial) frequency distributions, based on the values of the 1225 sites in CLIMDAT1 (the computerized world climatic data-base), are shown in separate tables for each of the eight ecoclimatic variables plus the Thornthwaite estimate of annual potential evapotranspiration. The 1225 sites are not distributed perfectly evenly throughout the world's land areas but do provide a good representation of world ecoclimatic patterns since site density is lowest in Antarctica and other extreme climates with little or no vegetation. World temperatures show a bimodal pattern with frequency peaks for all three variables in the tropical and temperate ranges and lower frequencies in the subtropical and polar ranges. Annual precipitation appears to be unimodal, with a peak in the 250–500 mm range (which is typical, however, of widely differing regions and evaporation levels). The annual balance between precipitation and potential evaporation (moisture index MI) shows a peak around the value 1.0, falling off a bit more rapidly on the wet side of this value. Seasonal precipitation values suggest a mode of about 75 mm for PMAX (highest average monthly precipitation) and suggest at least one essentially precipitation-free month per year for 34% of the world. Values for precipitation of the warmest month show a peak at 0–10 mm (mediterranean plus subtropical dry areas) and suggest a slight and much lower secondary peak around 60–70 mm. The values were obtained from the mapping program SYMAP and correspond to the isoline intervals on Maps 1–9.

TMAX

Range (°C)	No. of sites	% of total
0 °C or below	15	1.2
0°–5°	37	3.0
5°–10°	34	2.8
10°–15°	107	8.7
15°–20°	308	25.1
20°–25°	197	16.1
25°–30°	425	34.7
above 30°	102	8.3
Total	1225	100.0

TMIN

Range (°C)	No. of Sites	% of total
-20° or below	121	9.9
-20° to -10°	109	8.9
-10° to 0°	197	16.1
0–5°	102	8.3
5–10°	134	10.9
10–15°	169	13.8
15–20°	149	12.2
above 20°	244	19.9
Total	1225	100.0

DTY

Range (°C)	No. of sites	% of total
0–5°	213	17.4
5–10°	213	17.4
10–15°	154	12.6
15–20°	200	16.3
20–30°	244	19.9
30–40°	138	11.3
40–50°	54	4.4
above 50°	9	0.7
Total	1225	100.0

PRCP

Range (mm)	No. of sites	% ot total
0–100	91	7.4
100–250	135	11.0
250–500	267	21.8
500–750	209	17.1
750–1000	137	11.2
1000–1500	193	15.8
1500–2000	95	7.8
above 2000	98	8.0
Total	1225	100.0

PET

Range (mm)	No. of sites	% of total
0–100	15	1.2
100–250	23	1.9
250–500	236	19.8
500–750	350	28.6
750–1000	202	16.5
1000–1250	117	9.7
1250–1500	135	11.0
above 1500	147	12.0
Total	1225	100.0

MI

Range	No. of sites	% of total
0.0 –0.10	94	7.7
0.10–0.25	76	6.2
0.25–0.50	108	8.8
0.50–0.75	178	14.5
0.75–1.00	189	15.4
1.00–1.50	300	24.5
1.50–2.00	142	11.6
above 2.00	138	11.3
Total	1225	100.0

PMAX

Range (mm)	No. of sites	% of total
0–10	30	2.5
10–25	75	6.1
25–50	143	11.7
50–75	171	14.0
75–100	167	13.6
100–250	416	34.0
250–500	197	16.1
above 500	26	2.1
Total	1223	100.0

PMIN

Range (mm)	No. of sites	% of total
0–5	420	34.3
5–10	151	12.3
10–20	199	16.2
20–30	121	9.9
30–50	144	11.8
50–75	109	8.9
75–100	44	3.6
above 100	37	3.0
Total	1225	100.0

PMTMAX

Range (mm)	No. of sites	% of total
0–10	185	15.1
10–20	112	9.1
20–30	103	8.4
30–40	74	6.0
40–50	83	6.8
50–75	203	16.6
75–100	169	13.8
above 100	296	24.2
Total	1225	100.0

D. Life-form prediction frequency

Application of the model to the world climatic data-base provides, among other things, predictions of life-form frequency throughout the world. These frequencies were tabulated by ECOSIEVE and are listed as raw occurrences and as percentages (based on 1225 sites) in Table 12. The numbers do not represent accurate estimates of area because site density varies in different parts of the world. General estimates of area can be obtained, however, by multiplying the percentage value (last column of Table 12) by the earth's land area, about 149×10^6 km^2. Such estimates of area covered, where these can be interpreted as dominant forms and thus formation types (e.g. 16.4×10^6 km^2 for tropical rainforest trees, excluding montane), generally are close to the area estimates by Golley (1972), Lieth (1975) and Whittaker (1975) for biome-level vegetation units (e.g. 17.0×10^6 km^2 for tropical rainforest).

The highest prediction frequencies belong to the thallophytes, which can occur in most environments. The highest frequencies among the higher plant types belong to short bunch (82%) and sward (55%) grasses, summergreen forbs (52.5%), needle-leaved shrubs (43%, e.g. *Juniperus*), tall grasses (41%), and bush stem-succulents (41%, e.g. *Opuntia*). These plant types do not dominate large areas (except the grasses) but are well adapted to a wide range of largely temperate to subtropical environments, including drier situations. Next in prediction frequency is a group of forms some of which have special edaphic or other non-climatic relationships and must be interpreted differently: wintergreen epiphytes (37%), xeric cushion-herbs (34%), phreatophytic summergreen (36%) and evergreen (30%, e.g. mallee) arborescents, plus tall cane-grasses (32%) and raingreen forbs (30%). As with the first group, these forms are tolerant of various drier situations.

The most frequently predicted tree forms are the tropical sclerophyll (27%, e.g. eucalyptoid), microphyll (27%), and palmiform (24%) types, plus summergreen (30%), palmiform (26%) and raingreen (23%) small trees. Again, these forms usually are not dominants but tolerate wide ranges of conditions. Typically dominant forms which were predicted most frequently include boreal/montane evergreen needle-trees (20%), boreal summergreen needle-trees (19%), and summergreen broad-leaved trees (18%).

The least frequently predicted forms, apart from the physiographically limited tropical montane-alpine forms, include the subantarctic broad-evergreen trees (0.6%) and tree ferns (1.6%) of extremely perhumid, stenothermal climates (also physiographically limited) plus a variety of other forms: short tussock-grasses (2.6%), mediterranean broad-evergreen trees (2.9%), hydrophilic summergreen needle-trees (*Taxodium*, 3.1%), needle-leaved treeline krummholz (3.7%), evergreen cushion-shrubs (3.8%), and broad (3.9%) and narrow-leaved (3.7%) temperate rainforest trees. All of these latter forms, except krummholz, represent climates which are distinctly maritime if not perhumid.

E. Predicted plant-form distributions

Distributions of selected plant life forms and growth forms are shown on the following Maps 10–22*. Production of these maps is discussed in section 6.B. The predicted patterns shown on these maps are summarized below.

More detailed interpretations of particular patterns, including situations of co-importance and apparent discrepancy, are included in the legends for the individual maps.

Broad-leaved evergreen trees of all types (Maps 10 and 11) are especially well predicted. This is expected since these trees generally require humid, often somewhat maritime subtropical and tropical climates without extreme conditions. These situations are generally well described by the ecoclimatic variables employed.

The occurrence of broad-leaved deciduous tree types (Map 13) is generally well predicted, but the importance (Map 12) of deciduous broad-leaved trees is not always well determined in some tropical and Southern Hemisphere subtropical areas, especially relative to broad-leaved evergreen trees. Summergreen broad-leaved trees (mesophyllous variety) are predicted too far north in eastern Canada. Broad-microphyllous summergreen trees are accurately predicted in maritime subpolar areas of both hemispheres (*Betula* in northern Fenno-

*For general note on the vegetation maps, see p. 102.

scandia, Iceland, and coastal Alaska; *Nothofagus* spp. in southern Chile and New Zealand).

Narrow-leaved trees (Map 14) are less well predicted, though problems are mostly confined to particular situations. Boreal conifers appear too far north in both North America and Eurasia, though this is due to some extent to special cases and low site density. Major occurrences of the various temperate needle-leaved trees are correctly predicted on all continents (e.g. southeastern USA and adjacent Caribbean areas, southern Brazil, East Asia, sub-mediterranean Europe, the Rocky Mountains, the western Himalaya). Since envelopes for these forms were based mainly on single prototypic taxa, the results suggest, despite successional and evolutionary relationships, that these forms also possess ecophysiognomic status.

Shrubs and selected arborescents (Map 15) are well predicted in most cases. The accuracy is surprising when one considers that shrubs usually dominate in rather open stands which permit a greater variety of life forms. Shrub-dominated tundra areas are probably least well predicted and often depend as much on topography as on macroclimate.

Dominant graminoids and some other herbs (Map 16) are generally correctly located, but their areas are seldom shown to be as large as these areas are in reality. This must be due to the omission of fire, grazing, and anthropogenic influences from the model. Some problems occur in predicting dominance by sward, bunch and tussock forms. In reality, most grasslands are mixtures of these forms.

Stem and leaf-succulents (Map 17) are correctly confined to drier climates but are not well predicted, since their distributions are often determined also by topographic and historical factors. Only one stem-succulent (*Sarcostemma australis*) is reported to be native to Australia and is not important, but *Opuntia* species flourished when introduced there (e.g. Dodd 1959). Better estimation of length and extent of average and extreme dry periods might improve prediction of both these types.

Most forbs and semi-desert herbs (Map 18) are generally well predicted, though distinction between cold deserts with some scant vegetation and extreme ice or cold deserts with no vegetation is not very successful because of lack of any estimate of warm-season length or of ice and snow accumulation.

The prediction of some forms must be approached somewhat differently because of historical considerations. The most important such case is that of the warm-evergreen (eucalyptoid) sclerophyll trees. The prototypic genus was confined to Australia until transported elsewhere by man. So many eucalypts have been planted in such places as California, the Mediterranean region, and East Africa that, if introduced species can legitimately be considered in the present context, Map 19 gives a quite reasonable indication of the present worldwide distribution of eucalypts and eucalyptoids.

Palmiform and other tuft-trees (Map 20) also represent a special case because of edaphic and topographic relationships, the slow growth of the xeric forms, and other factors. Their distributions are not badly predicted by macroclimate, but dominance in seasonally flooded areas and occurrence at oases involve factors other than climate. Tropical alpine tuft-treelets are not well mapped because of the small areas and paucity of climatic data involved.

The predicted distributions of vine (Map 21) and epiphyte (Map 22) forms are special cases since these forms generally require other, usually woody and fairly large forms for support. The predicted distributions are relatively accurate and are especially interesting since one rarely sees maps for distributions of the smaller synusiae.

Table 12. World prediction frequencies of the terrestrial plant life forms.

The number of predicted occurrences worldwide (based on the world database CLIMDATI) and the corresponding percentage of the 1225 total sites are shown for each of the 90 terrestrial life forms of Table 2. The values were obtained by using ECOSIEVE to generate and tabulate the predicted forms at each site. The highest prediction frequencies are shown by thallophytes, short and tall grasses, seasonal forbs, bush stem-succulents, and several shrub and bush forms. The most frequently predicted tree forms are tropical sclerophyll, microphyll, and palmiform evergreens and the summergreen and raingreen small trees. Least frequently predicted are a variety of alpine, perhumid, Southern Hemisphere and mediterranean forms.

Plant type	No. of occurrences	Percentage
Tropical rainforest trees	133	10.9
Tropical montane rainforest trees	63	5.1
Tropical evergreen sclerophyll trees	333	27.2
Tropical evergreen microphyll trees	329	26.9
Warm-temperate broad-evergreen trees	58	4.7
Mediterranean broad-evergreen trees	36	2.9
Temperate broad-rainforest trees	48	3.9
Monsoon broad-raingreen trees	96	7.8
Montane broad-raingreen trees	67	5.5
Xeric raingreen trees	110	9.0
Summergreen broad-leaved trees	222	18.1
Boreal broad-summergreen trees	175	14.3
Tropical linear-leaved trees	173	14.1
Tropical xeric needle-trees	86	7.0
Temperate rainforest needle-trees	45	3.7
Heliophilic large-needled trees	97	7.9
Mediterranean needle-trees	157	12.8
Typical temperate needle-trees	180	14.7
Boreal/montane needle-trees	248	20.2
Hydrophilic summergreen needle-trees	38	3.1
Boreal summergreen needle-trees	236	19.3
Tropical broad-evergreen dwarf trees	89	7.3
Tropical broad-evergreen small trees	220	18.0
Cloud-forest dwarf-trees	28	2.3
Temperate broad-evergreen small trees	159	13.0
Sub-polar broad-evergreen small trees	7	0.6
Broad-raingreen small trees	284	23.2
Broad-summergreen small trees	364	29.7
Needle-leaved small trees	210	17.1
Palmiform tuft-trees	291	23.8
Palmiform tuft-treelets	313	25.6
Tree ferns	20	1.6
Tropical alpine tuft-treelets	8	0.7
Xeric tuft-treelets	184	15.0
Evergreen arborescents	368	30.0
Raingreen thorn-scrub	183	14.9
Summergreen arborescents	444	36.2
Leafless arborescents	255	20.8
Needle-leaved treeline krummholz	45	3.7
Tropical broad-evergreen shrubs	368	30.0
Mediterranean evergreen shrubs	150	12.2
Broad-ericoid evergreen shrubs	87	7.1
Temperate broad-evergreen shrubs	229	18.7

Table 12 (continued).

Plant type	No. of occurrences	Percentage
Hot-desert evergreen shrubs	131	10.7
Leaf-succulent evergreen shrubs	100	8.2
Cold-winter xeromorphic shrubs	116	9.5
Broad-summergreen mesic shrubs	353	28.8
Xeric summergreen shrubs	170	13.9
Needle-leaved evergreen shrubs	525	42.9
Mediterranean dwarf-shrubs	119	9.7
Temperate evergreen dwarf-shrubs	271	22.1
Maritime heath dwarf-shrubs	63	5.0
Summergreen tundra dwarf-shrubs	179	14.6
Xeric dwarf-shrubs	271	22.1
Mesic rosette-shrubs	221	18.0
Xeric rosette-shrubs	233	19.0
Evergreen cushion-shrubs	46	3.8
Xeric cushion-shrubs	216	17.6
Arborescent stem-succulents	140	11.4
Typical stem-succulents	180	14.7
Bush stem-succulents	503	41.1
Arborescent grasses	266	21.7
Tall cane-grasses	391	31.9
Tall grasses	504	41.1
Short sward-grasses	678	55.3
Short bunch-grasses	1002	81.8
Tall tussock-grasses	84	6.9
Short tussock-grasses	32	2.6
Sclerophyllous grasses	72	5.9
Desert-grasses	226	18.4
Tropical evergreen forbs	237	19.3
Temperate evergreen forbs	242	19.8
Raingreen forbs	371	30.3
Summergreen forbs	643	52.5
Succulent forbs	293	23.9
Xeric cushion herbs	412	33.6
Ephemeral desert herbs	190	15.5
Summergreen cold-desert herbs	167	13.6
Raingreen cold-desert herbs	24	2.0
Tropical broad-evergreen lianas	182	14.9
Broad-evergreen vines	292	23.8
Broad-raingreen vines	278	22.7
Broad-summergreen vines	134	10.9
Tropical broad-evergreen epiphytes	138	11.3
Narrow-leaved epiphytes	203	16.6
Broad-wintergreen epipytes	451	36.8
Evergreen ferns	123	10.0
Summergreen ferns	192	15.7
Mat-forming thallophytes	685	55.9
Xeric thallophytes	1040	84.9

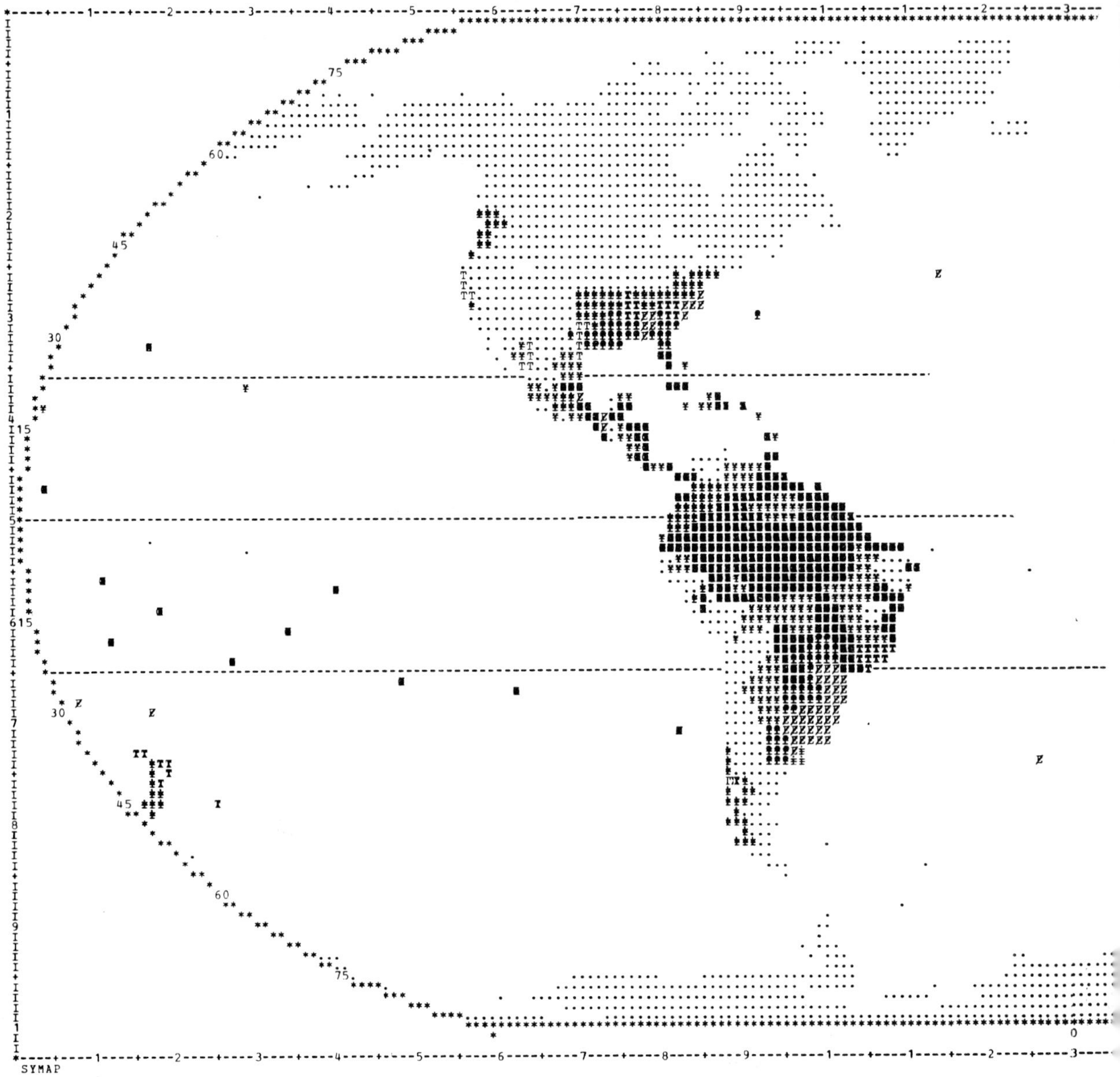

SYMAP

Map 10. Predicted distributions of most important broad-leaved evergreen tree types.

Broad-leaved evergreen trees are predicted to occur in tropical-subtropical and less extreme temperate areas with only short or no dry seasons, in mediterranean areas with sufficient total moisture, and in perhumid, stenothermal cool-temperate areas such as southern Chile (*Nothofagus*), New Zealand, and northwestern North America. The map shows the predicted dominant broad-leaved evergreen tree type in each area (except those very close to climatic limits). These types are also the formation dominants in many cases, especially in the tropical rainforest and in warm-temperate forests. Areas of predicted co-important forms (MULT) involve primarily the combinations TBRT + WTET and TBRT + TMRT (10 occurrences each). Broad-leaved evergreen trees are well predicted in general, with no major discrepancies appearing on this map. The proximity of these forms to climatic limits ('fitness') is shown on Map 11.

TRFT = Tropical RainForest Trees
EST = Evergreen Sclerophyll Trees
TEMT = Tropical broad-Evergreen Mircophyll Trees
WTET = Warm-Temperate broad-Evergreen Trees
MBET = Mediterranean Broad-Evergreen Trees
TBRT = Temperate Broad-evergreen Rainforest Trees
TEST = Tropical broad-Evergreen Small Trees (small and dwarf trees)
BEST = Temperate broad-Evergreen Small Trees (temperate, cloud-forest, and subpolar)
MULT = co-important broad-evergreen tree forms

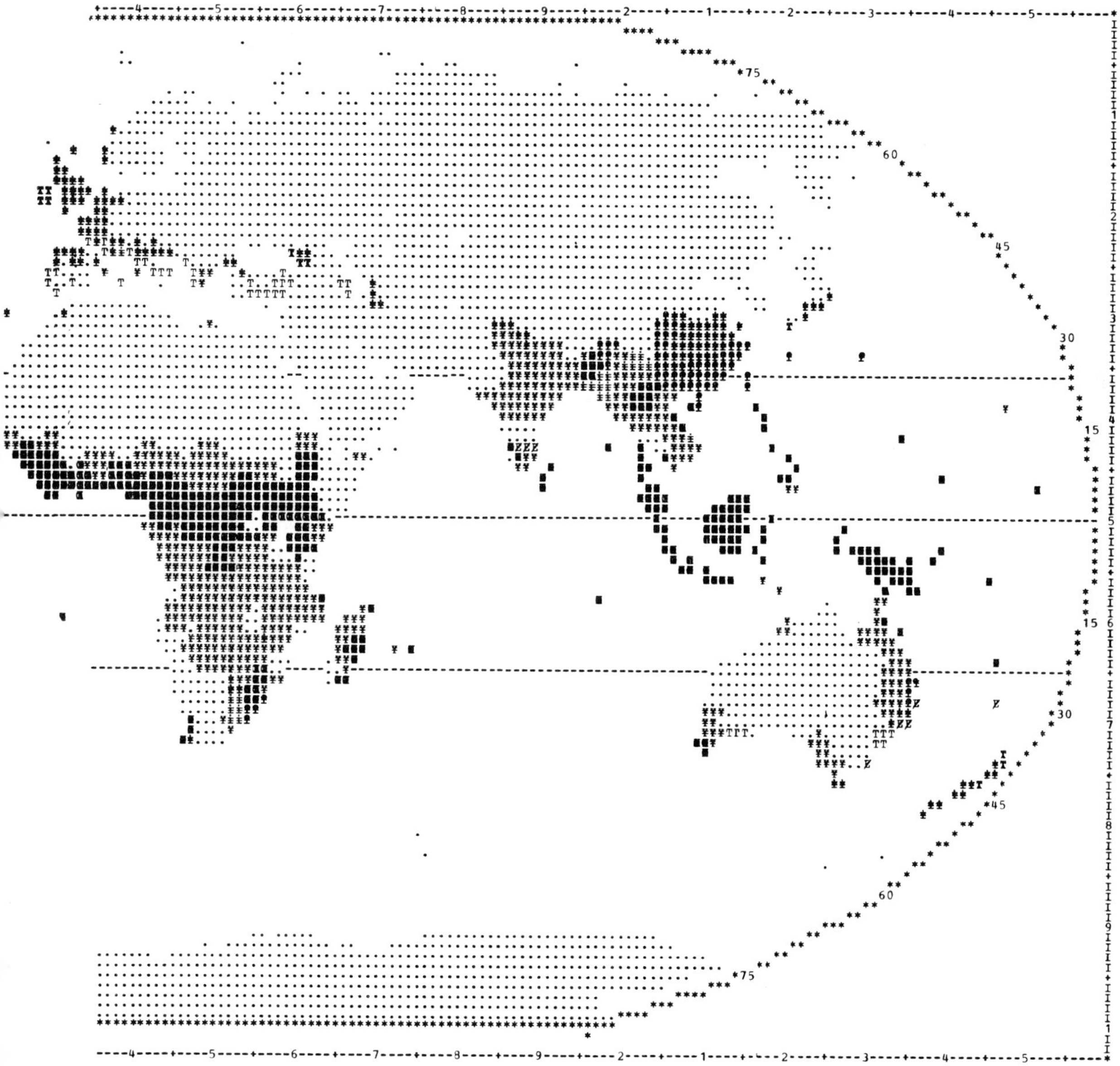

DATA VALUE EXTREMES ARE 0.0 11.00

TOTAL SUPERIMPOSED DATA POINTS IS 135. THESE OCCUR IN 60 LOCATIONS.

ABSOLUTE VALUE RANGE APPLYING TO EACH LEVEL
('MAXIMUM' INCLUDED IN HIGHEST LEVEL ONLY)

MINIMUM	BELOW	0.50	1.50	2.50	3.50	4.50	5.50	6.50	7.50	ABOVE
MAXIMUM	0.50	1.50	2.50	3.50	4.50	5.50	6.50	7.50	8.50	8.50

PERCENTAGE OF TOTAL ABSOLUTE VALUE RANGE APPLYING TO EACH LEVEL

12.50	12.50	12.50	12.50	12.50	12.50	12.50	12.50

FREQUENCY DISTRIBUTION OF DATA POINT VALUES IN EACH LEVEL

LEVEL	L	1	2	3	4	5	6	7	8	H
SYMBOLS	....	■■■■■■■■■	¥¥¥¥¥¥¥¥¥	‡‡‡‡‡‡‡‡‡	●●●●●●●●●	TTTTTTTTT	IIIIIIIII	iiiiiiiii	♣♣♣♣♣♣♣♣♣	ZZZZZZZZZ
FREQ.	795	161	138	7	29	23	14	0	92	22
FORMS	NONE	TRFT	EST	TEMT	WTET	MBET	TBRT	TEST	BEST	MULT

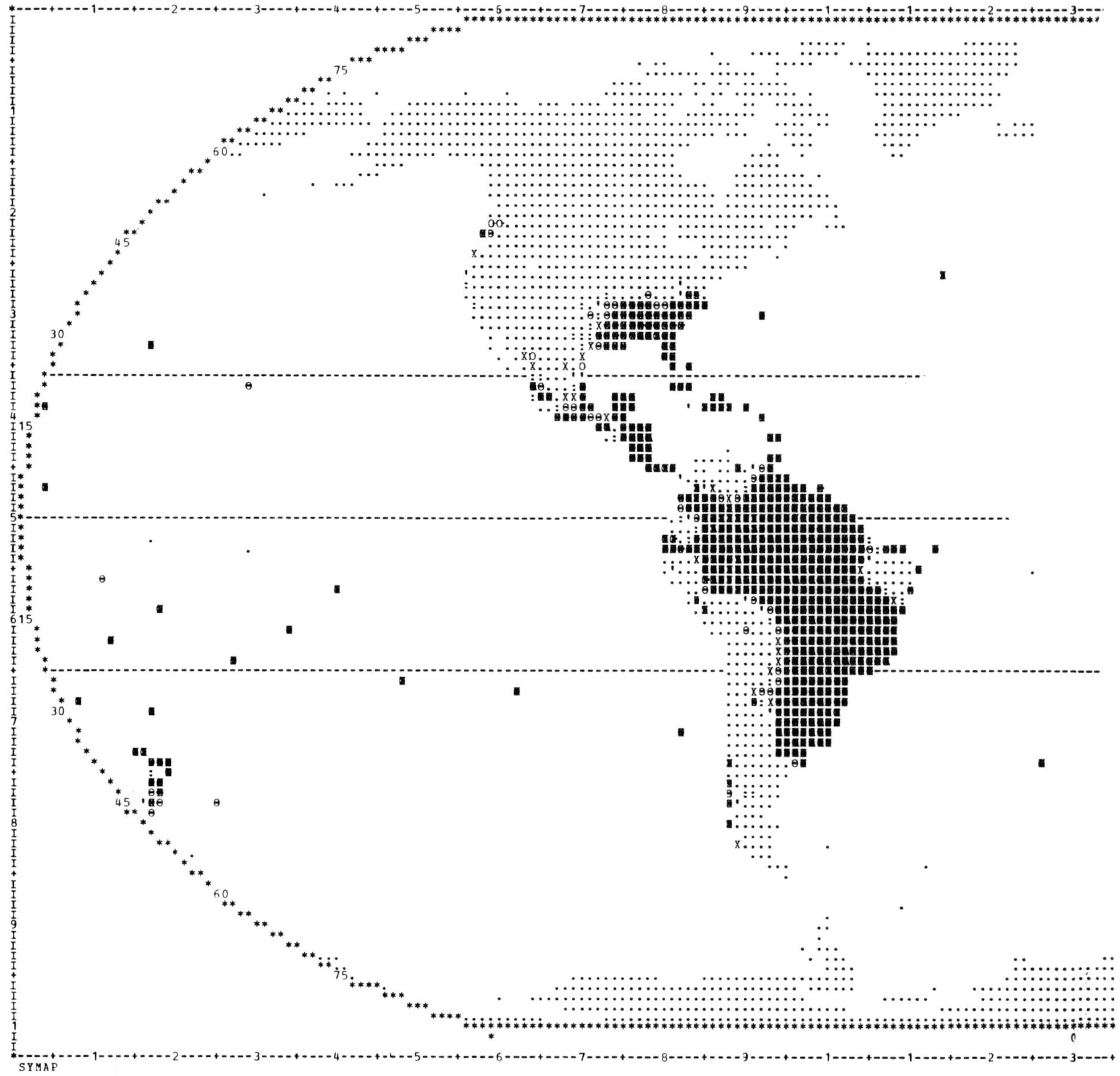

Map 11. Predicted fitness of broad-leaved evergreen tree types.

The potential 'fitness' of the locally most important broad-leaved evergreen tree types (Map 10) is suggested here by estimated proximity to respective closest climatic limit. This may also suggest potential abundance and importance in the local vegetation, although actual dominance depends also on the fitness of other tree types. The pattern shown here suggests dominance by broad-leaved evergreen trees in expected situations: tropical and warm-temperate humid climates plus some extremely maritime cool-temperate climates (e.g. southern Chile, New Zealand). The most interesting questions are suggested by the high 'fitness' values predicted in eastern African, inland southeastern USA, and southeastern South America, and the low values across northern Australia.

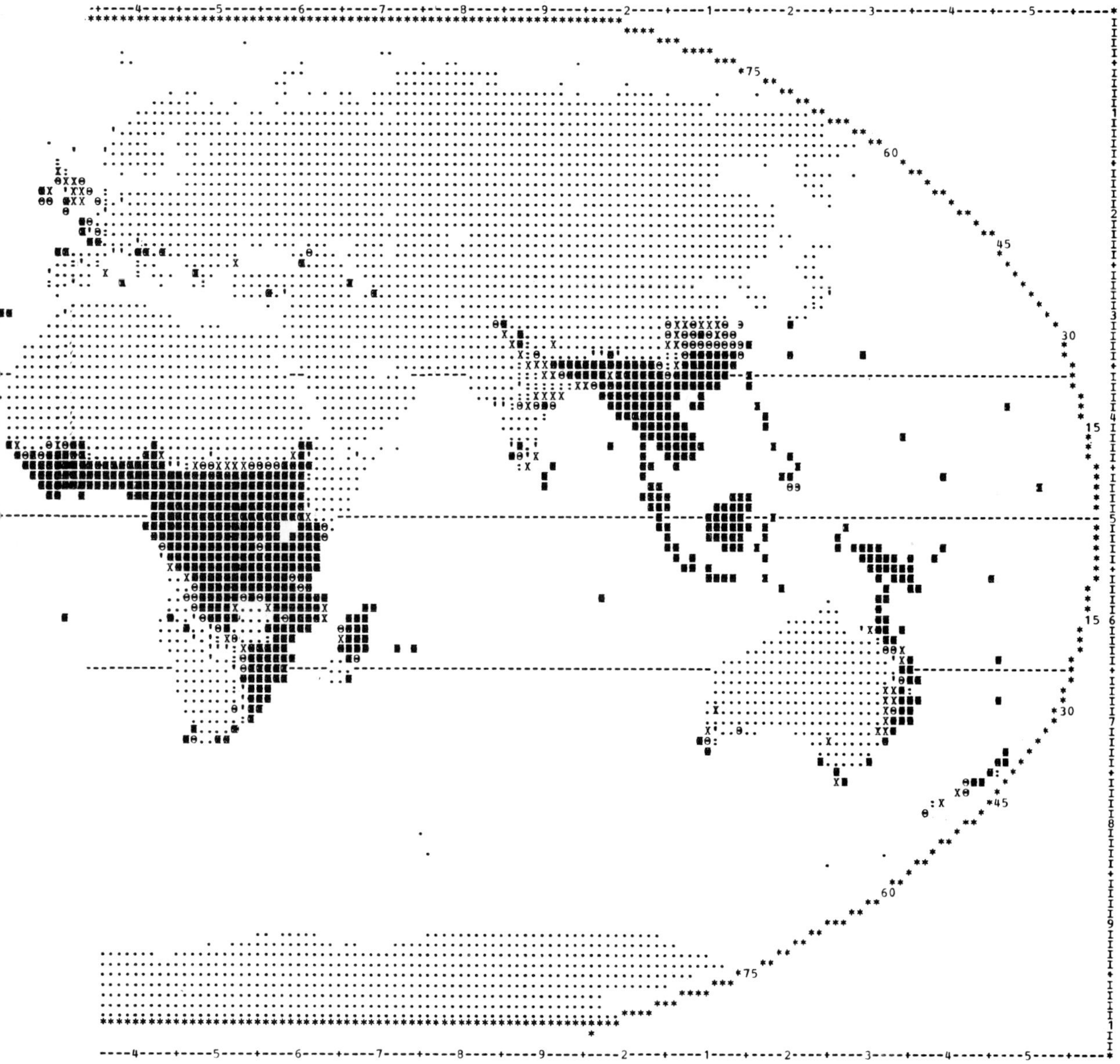

DATA VALUE EXTREMES ARE -1.00 0.50

TOTAL SUPERIMPOSED DATA POINTS IS 135. THESE OCCUR IN 60 LOCATIONS.

ABSOLUTE VALUE RANGE APPLYING TO EACH LEVEL
('MAXIMUM' INCLUDED IN HIGHEST LEVEL ONLY)

MINIMUM	BELOW	0.0	0.05	0.10	0.15	0.20	ABOVE 0.25
MAXIMUM	0.0	0.05	0.10	0.15	0.20	0.25	

PERCENTAGE OF TOTAL ABSOLUTE VALUE RANGE APPLYING TO EACH LEVEL

20.00 20.00 20.00 20.00 20.00

FREQUENCY DISTRIBUTION OF DATA POINT VALUES IN EACH LEVEL

LEVEL	L	1	2	3	4	5	H
SYMBOLS	....	''''''''	::::::::	XXXXXXXX	θθθθθθθθ	■■■■■■■■	■■■■■■■■
FREQ.	780	39	29	27	42	73	291
FITNESS	ABSENT	< 0.05	0.05-0.1	0.1-0.15	0.15-0.2	0.2-0.25	> 0.25

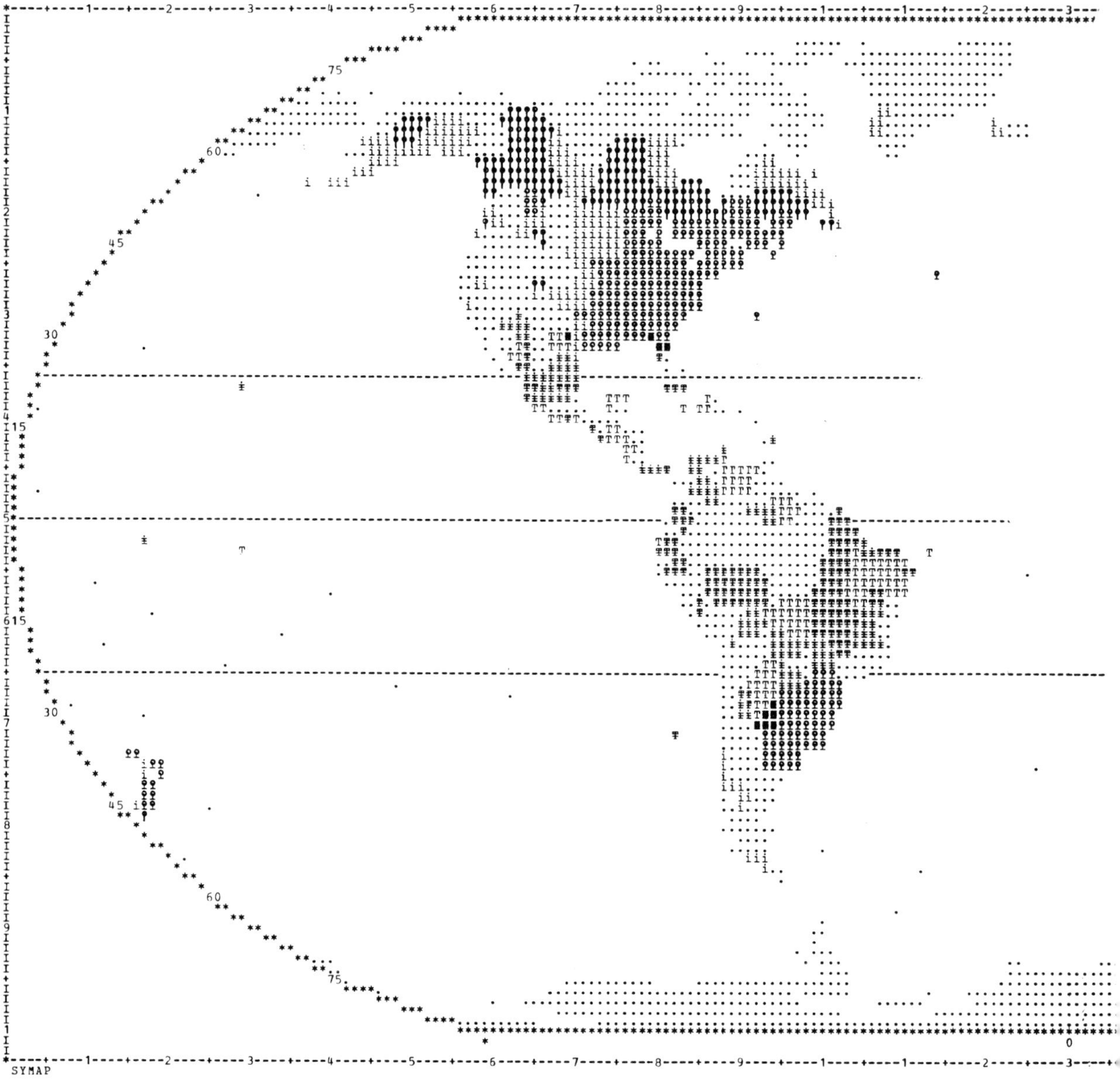

Map 12. Predicted distributions of most important broad-leaved deciduous tree types.

Broad-leaved deciduous trees, including both summergreen and raingreen types, are predicted to occur in cold/cool-winter areas of eastern North America, East Asia, and Western Europe (summergreen), in tropical summer-rain Africa, India and Southeast Asia (raingreen), and in various smaller, more scattered woodland and grassland areas. Smaller summergreen trees are also important in boreal and even some small austral climatic areas and attain formation dominance in northern Quebec, the Aleutian Peninsula and archipelago, Iceland, and Kamchatka. Smaller raingreen trees are important in dry, seasonal tropical and subtropical climates, including thorn-scrub and savanna areas mainly in Africa and tropical America. These forms are all generally well predicted. The greatest discrepancy between predicted and actual distributions involves summergreen tree types which are predicted too far north in Canada and northern Russia and probably erroneously in southern Brazil to northern Argentina and in smaller areas of southern Australia and South Africa.

RGT = mesic broad-RainGreen Trees (lowland, montane)
XRGT = Xeric broad-RainGreen Trees
SGT = SummerGreen broad-leaved Trees (temperate)
BSGT = Boreal broad-SummerGreen Trees
RGST = broad-RainGreen Small Trees
SGST = broad-SummerGreen Small Trees
MULT = co-important summergreen and raingreen tree types

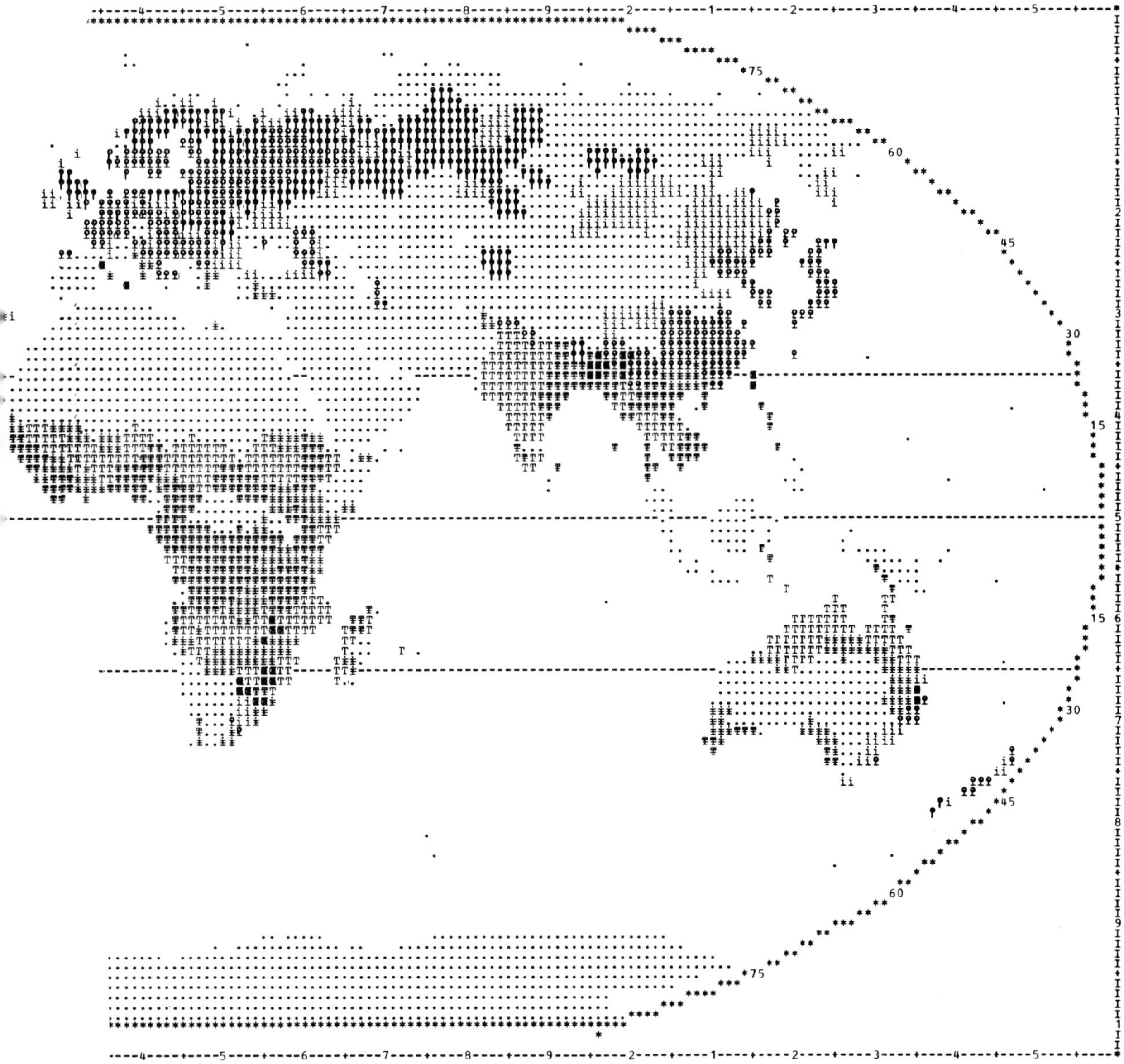

DATA VALUE EXTREMES ARE 0.0 11.00

TOTAL SUPERIMPOSED DATA POINTS IS 135. THESE OCCUR IN 60 LOCATIONS.

ABSOLUTE VALUE RANGE APPLYING TO EACH LEVEL
('MAXIMUM' INCLUDED IN HIGHEST LEVEL ONLY)

MINIMUM	BELOW	0.50	1.50	2.50	3.50	4.50	5.50	ABOVE 6.50
MAXIMUM	0.50	1.50	2.50	3.50	4.50	5.50	6.50	

PERCENTAGE OF TOTAL ABSOLUTE VALUE RANGE APPLYING TO EACH LEVEL

16.67	16.67	16.67	16.67	16.67	16.67

FREQUENCY DISTRIBUTION OF DATA POINT VALUES IN EACH LEVEL

LEVEL	L	1	2	3	4	5	6	H
SYMBOLS	....	TTTTTTTTT	TTTTTTTTT	999999999	♀♀♀♀♀♀♀♀♀	‡‡‡‡‡‡‡‡‡	iiiiiiiii	■■■■■■■■■
FREQ.	623	99	88	173	92	78	114	14
FORMS	NONE	BGT	XBGT	SGT	BSGT	BGST	SGST	MULT

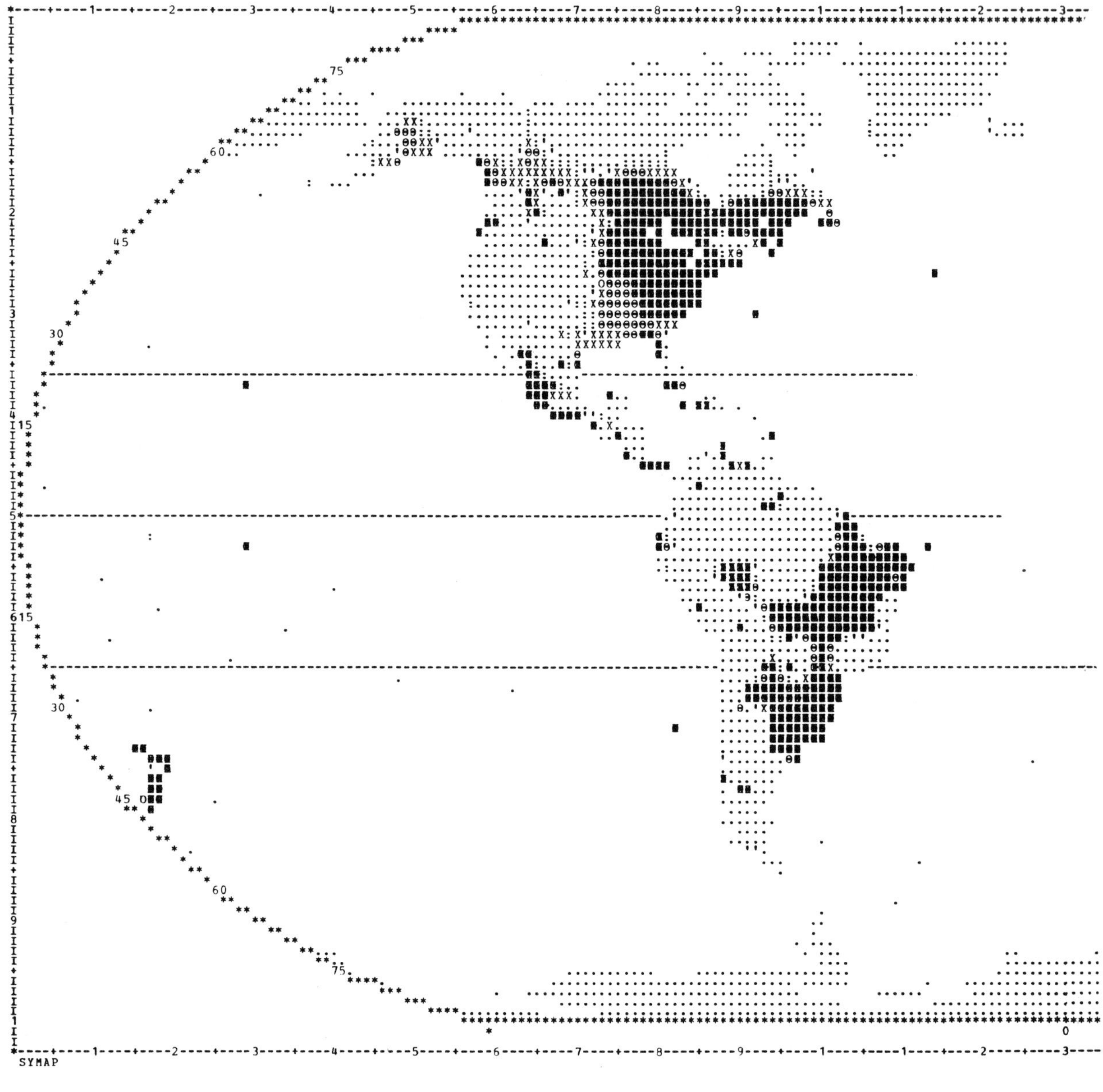

Map 13. Predicted fitness of broad-leaved deciduous tree types.

The potential 'fitness' (i.e. distance from climatic limits) of the locally most important broad-leaved deciduous tree types (Map 12) is shown on this map. Although predicted formation dominance (Map 23) depends also on other predicted tree types, the pattern shown here reflects the actual dominance of broad-leaved deciduous trees in cool-temperate and dry-season tropical climates with sufficient wet seasons. Due to the less xeromorphic nature of (malacophyllous, generally more productive) deciduous leaves, higher fitness values for broad-evergreen tree types do not imply greater importance in a given situation. (Compare Maps 11 and 13, for example, in southeastern USA, where summergreen trees are the dominant broad-leaved form). Note also the low fitness values for small summergreen trees (e.g. *Betula, Nothofagus*) in areas such as Iceland and southern Chile, where these trees can be formation dominants.

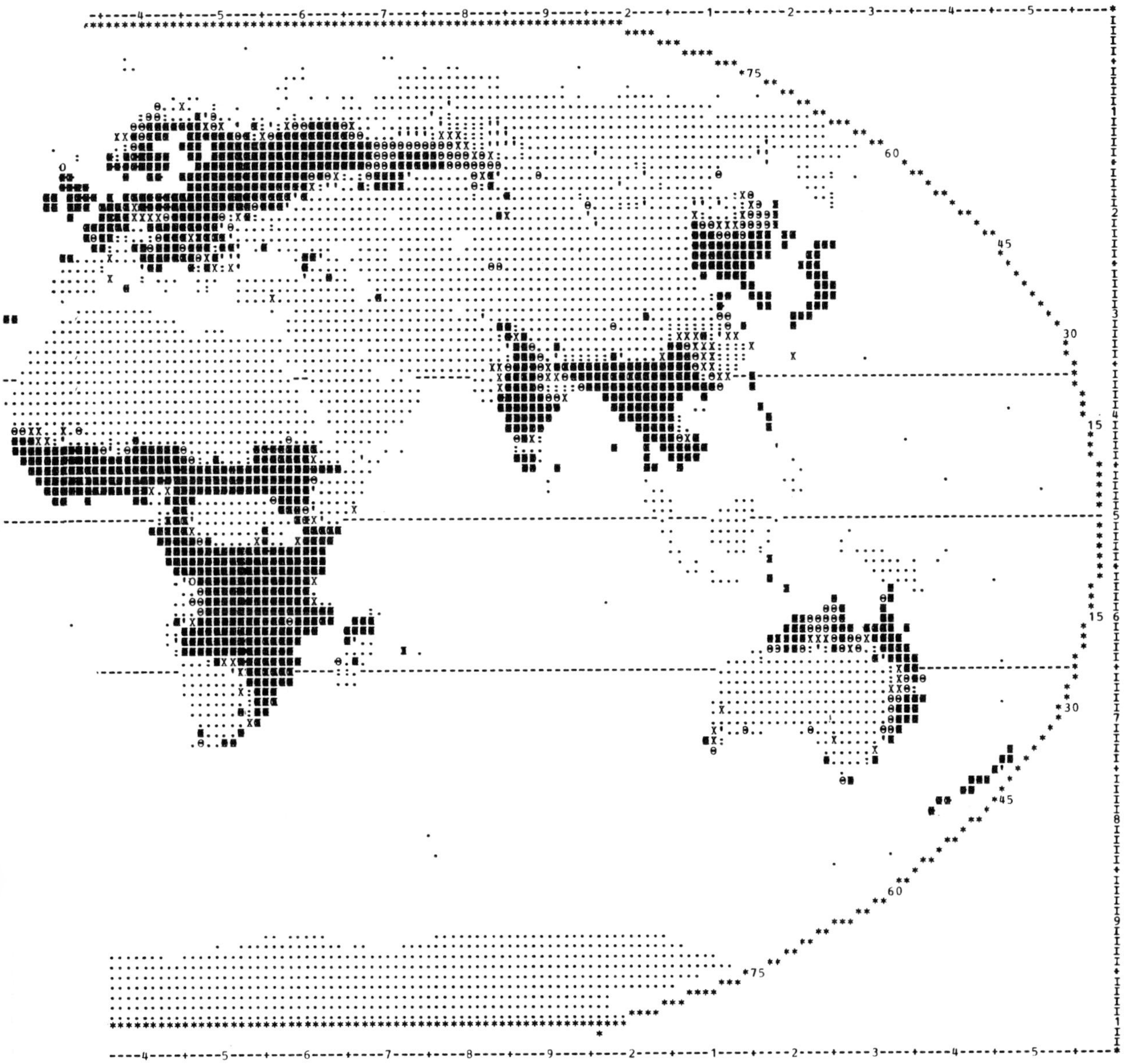

)ATA VALUE EXTREMES ARE -1.00 0.50

TOTAL SUPERIMPOSED DATA POINTS IS 135. THESE OCCUR IN 60 LOCATIONS.

ABSOLUTE VALUE RANGE APPLYING TO EACH LEVEL
('MAXIMUM' INCLUDED IN HIGHEST LEVEL ONLY)

MINIMUM	BELOW	0.0	0.05	0.10	0.15	0.20	ABOVE 0.25
MAXIMUM	0.0	0.05	0.10	0.15	0.20	0.25	

PERCENTAGE OF TOTAL ABSOLUTE VALUE RANGE APPLYING TO EACH LEVEL

20.00 20.00 20.00 20.00 20.00

FREQUENCY DISTRIBUTION OF DATA POINT VALUES IN EACH LEVEL

LEVEL	L	1	2	3	4	5	H
SYMBOLS	·········	'''''''''	:::::::::	XXXXXXXXX	θθθθθθθθθ	■■■■■■■■■	■■■■■■■■■
FREQ.	623	42	61	39	56	89	371
FITNESS	ABSENT	< 0.05	0.05-0.1	0.1-0.15	0.15-0.2	0.2-0.25	> 0.25

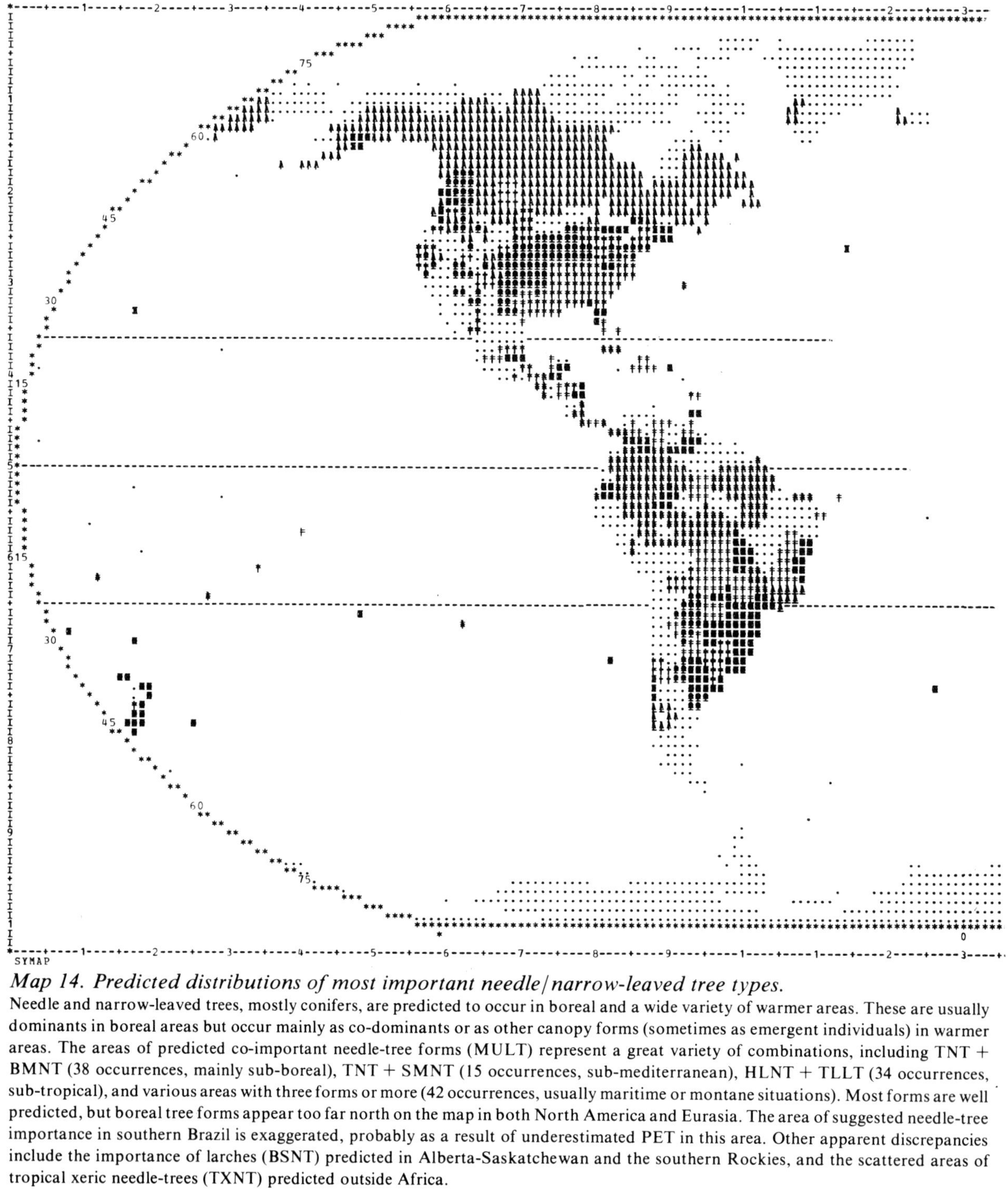

Map 14. Predicted distributions of most important needle/narrow-leaved tree types.

Needle and narrow-leaved trees, mostly conifers, are predicted to occur in boreal and a wide variety of warmer areas. These are usually dominants in boreal areas but occur mainly as co-dominants or as other canopy forms (sometimes as emergent individuals) in warmer areas. The areas of predicted co-important needle-tree forms (MULT) represent a great variety of combinations, including TNT + BMNT (38 occurrences, mainly sub-boreal), TNT + SMNT (15 occurrences, sub-mediterranean), HLNT + TLLT (34 occurrences, sub-tropical), and various areas with three forms or more (42 occurrences, usually maritime or montane situations). Most forms are well predicted, but boreal tree forms appear too far north on the map in both North America and Eurasia. The area of suggested needle-tree importance in southern Brazil is exaggerated, probably as a result of underestimated PET in this area. Other apparent discrepancies include the importance of larches (BSNT) predicted in Alberta-Saskatchewan and the southern Rockies, and the scattered areas of tropical xeric needle-trees (TXNT) predicted outside Africa.

TLLT = Tropical Linear-Leaved Trees
TXNT = Tropical Xeric Needle-Trees
HNLT = Heliophilic Needle-Leaved Trees
TNT = Temperate Needle-leaved Trees
TRNT = Temperate Rainforest Needle-Trees
SMNT = Sub-Mediterranean Needle-Trees
BMNT = Boreal/Montane Needle-Trees
BSNT = Boreal Summergreen Needle-Trees
DNST = Dwarf-Needle Small Trees
MULT = co-important narrow-leaved tree forms

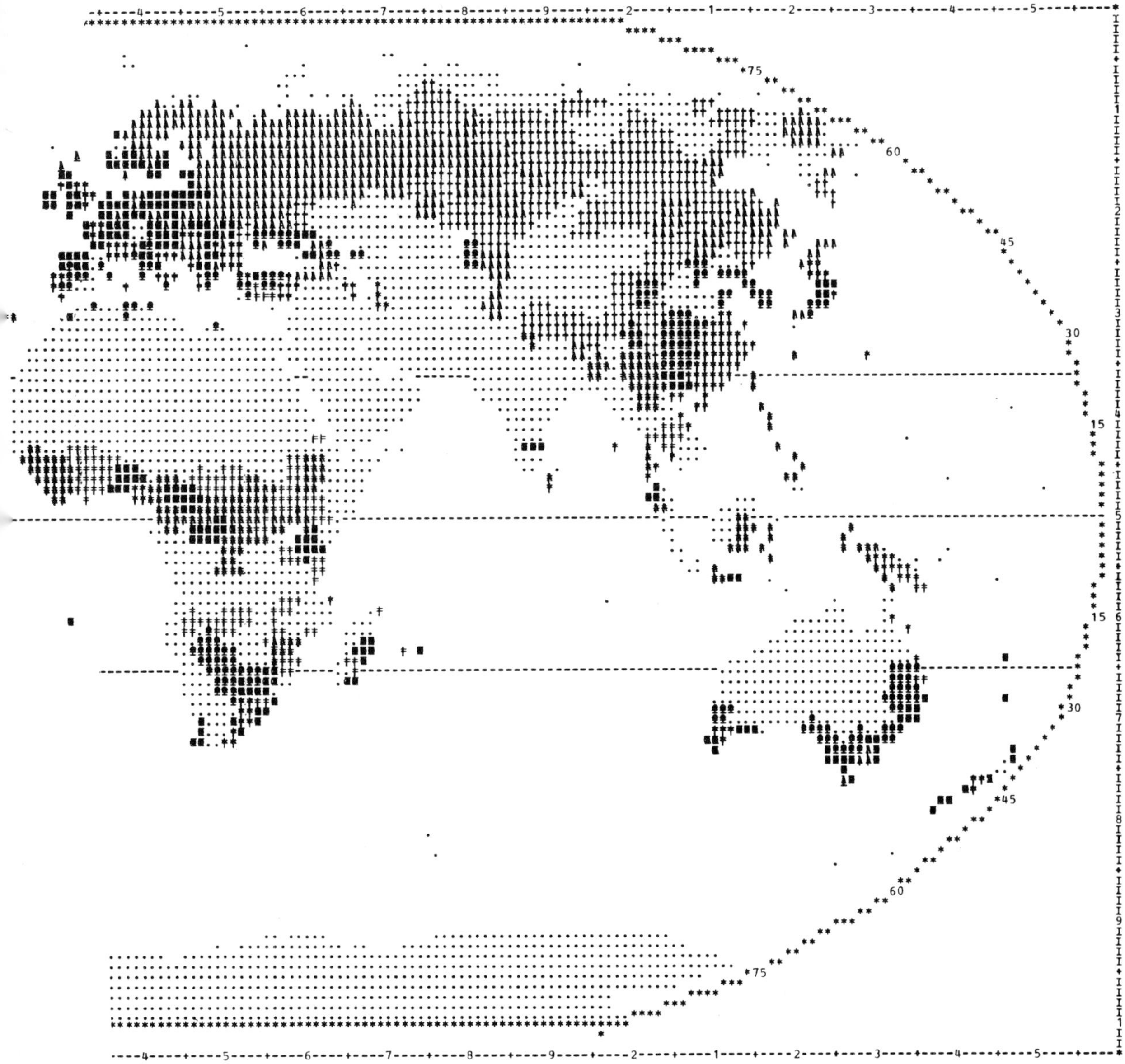

)ATA VALUE EXTREMES ARE 0.0 11.00

TOTAL SUPERIMPOSED DATA POINTS IS 135. THESE OCCUR IN 60 LOCATIONS.

ABSOLUTE VALUE RANGE APPLYING TO EACH LEVEL
('MAXIMUM' INCLUDED IN HIGHEST LEVEL ONLY)

MINIMUM	BELOW	0.50	1.50	2.50	3.50	4.50	5.50	6.50	7.50	8.50	ABOVE 9.50
MAXIMUM	0.50	1.50	2.50	3.50	4.50	5.50	6.50	7.50	8.50	9.50	

?ERCENTAGE OF TOTAL ABSOLUTE VALUE RANGE APPLYING TO EACH LEVEL

11.11	11.11	11.11	11.11	11.11	11.11	11.11	11.11	11.11

FREQUENCY DISTRIBUTION OF DATA POINT VALUES IN EACH LEVEL

LEVEL	L	1	2	3	4	5	6	7	8	9	H
SYMBOLS											
FREQ.	561	78	62	40	26	10	27	174	65	78	160
FORMS	NONE	TLLT	TXNT	HNLT	TNT	TRNT	SMNT	BMNT	BSNT	DNST	MULT

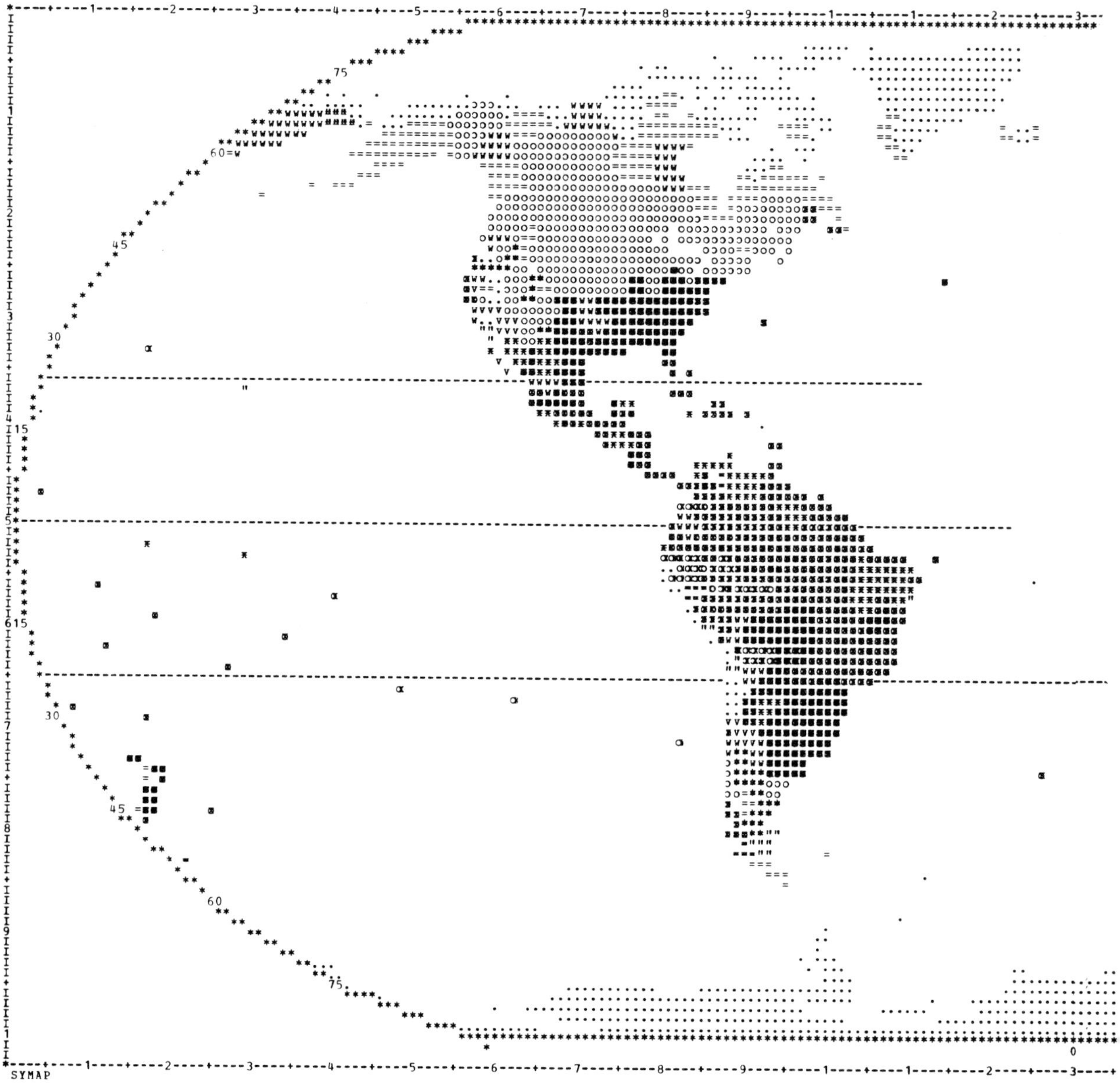

Map 15. Predicted distributions of most important shrub and similar forms.

Shrubs and similar forms, including frutescents, thorn-scrub, dwarf and cushion-shrubs, are predicted to occur over most of the non-desert world. Such forms are correctly predicted (see Map 23) to dominate large areas in western North America, Patagonia, the Middle East (including Kazakhstan), and Australia, plus smaller areas in southwestern Africa, mediterranean Europe, central Chile, Caribbean South America, Central Asia, and along the north and south borders of the Sahara desert. Shrub-tundra climates are predicted in northern Europe, southern Greenland, and in the central part of the Canadian archipelago. Of the many sites with predicted co-important shrub forms, most are combinations of evergreen and summergreen forms (127 occurrences), with raingreen plus evergreen (46 occurrences) and raingreen plus summergreen (9 occurrences) predicted less frequently. Potentially interesting occurrences of all three types were predicted in southern and southwestern Australia (3), northern Argentina (1), and the mountains of Southern Africa (2) and northern Mexico (1). Most of these areas are surprisingly well predicted when one considers that shrubs generally are most important in open stands resulting from reduction of larger forms. Predicted with least accuracy are probably the tundra forms.

BES = Broad-leaved Evergreen Shrubs
RGTS = RainGreen Thorn-Shrub
SGS = SummerGreen Shrubs
NLES = Needle-Leaved Evergreen Shrubs
HDES = Hot-Desert Evergreen Shrubs
CWXS = Cold-Winter Xeric Shrubs
TEDS = Temperate Evergreen Dwarf-Shrubs
SGDS = SummerGreen Dwarf-Shrubs
MECS = Mesic Evergreen Cushion-Shrubs
XCS = Xeric Cushion-Shrubs
MULT = co-important shrub forms

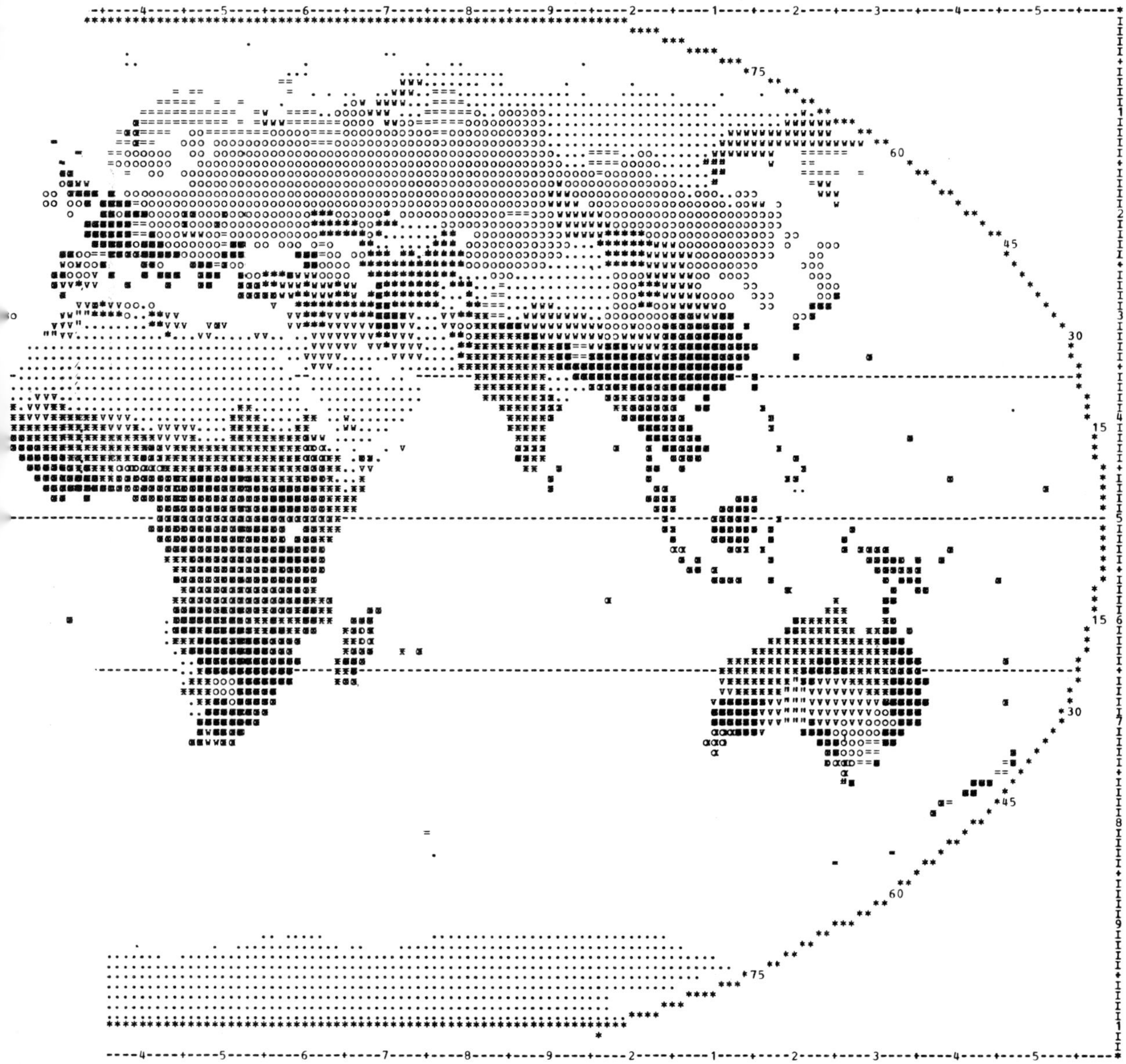

DATA VALUE EXTREMES ARE 0.0 11.00

TOTAL SUPERIMPOSED DATA POINTS IS 135. THESE OCCUR IN 60 LOCATIONS.

ABSOLUTE VALUE RANGE APPLYING TO EACH LEVEL
('MAXIMUM' INCLUDED IN HIGHEST LEVEL ONLY)

MINIMUM	BELOW	0.50	1.50	2.50	3.50	4.50	5.50	6.50	7.50	8.50	9.50	ABOVE
MAXIMUM	0.50	1.50	2.50	3.50	4.50	5.50	6.50	7.50	8.50	9.50	10.50	10.50

PERCENTAGE OF TOTAL ABSOLUTE VALUE RANGE APPLYING TO EACH LEVEL

10.00	10.00	10.00	10.00	10.00	10.00	10.00	10.00	10.00	10.00

FREQUENCY DISTRIBUTION OF DATA POINT VALUES IN EACH LEVEL

LEVEL	L	1	2	3	4	5	6	7	8	9	10	H
SYMBOLS	....	■■■■■■■■	XXXXXXXXX	OOOOOOOOO	WWWWWWWWW	VVVVVVVVV	*********	#########	=========	---------	"""""""""	■■■■■■■■
FREQ.	195	282	99	250	62	44	38	3	94	9	15	190
FORMS	NONE	BES	RGTS	SGS	NLES	HDES	CWXS	TEDS	SGDS	MECS	XCS	MULT

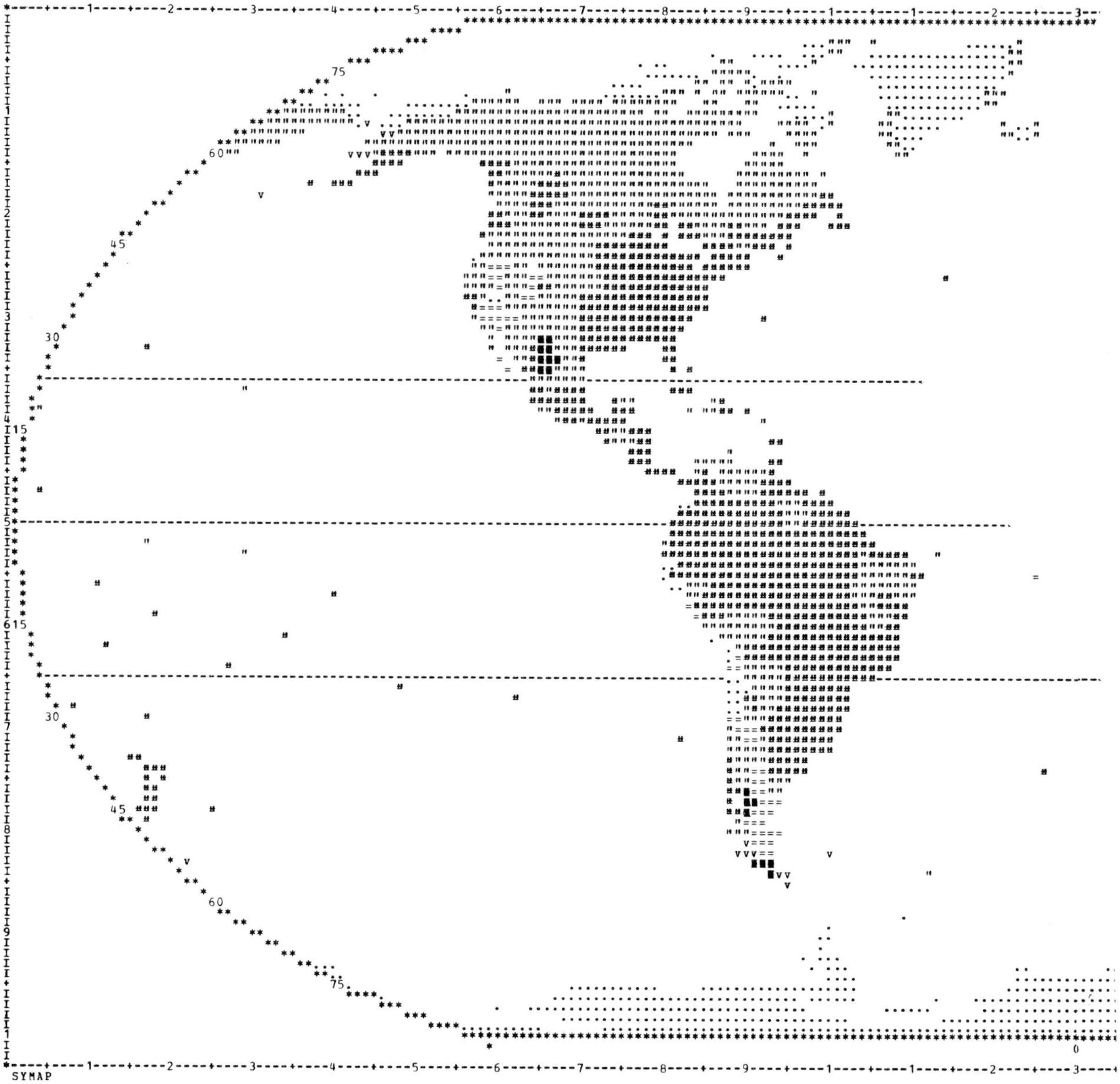

Map 16. Predicted distributions of most important graminoid types.

Graminoids (mostly grasses, with sedges also in colder climates) are predicted to occur in essentially all parts of the world except extreme deserts. Graminoids gain dominance in tropical and temperate climatic grasslands (Map 23) but may extend their areas of dominance as a result of fire, grazing, and anthropogenic influences. The areas of predicted co-important graminoid forms represent the combination DG + SCG (15 occurrences) except in southern Chile (SSG + STG). Unfortunately, no site from the Canterbury Plains short-tussock region of New Zealand is included in the data-base. The actual pattern of tall, short, and desert-grasses is shown by the predictions. Dominance by bunch, sward-forming, and tussock grass forms, on the other hand, is generally not well predicted. Because of low site density the zonal nature of the sub-Saharan grassland belts is not shown.

TG = Tall Grasses
SG = Short Grasses (bunch, sward)
TTG = Tall Tussock-Grasses
STG = Short Tussock-Grasses

SCG = Sclerophyllous Grasses
DG = Desert-Grasses
MULT = co-important graminoid forms

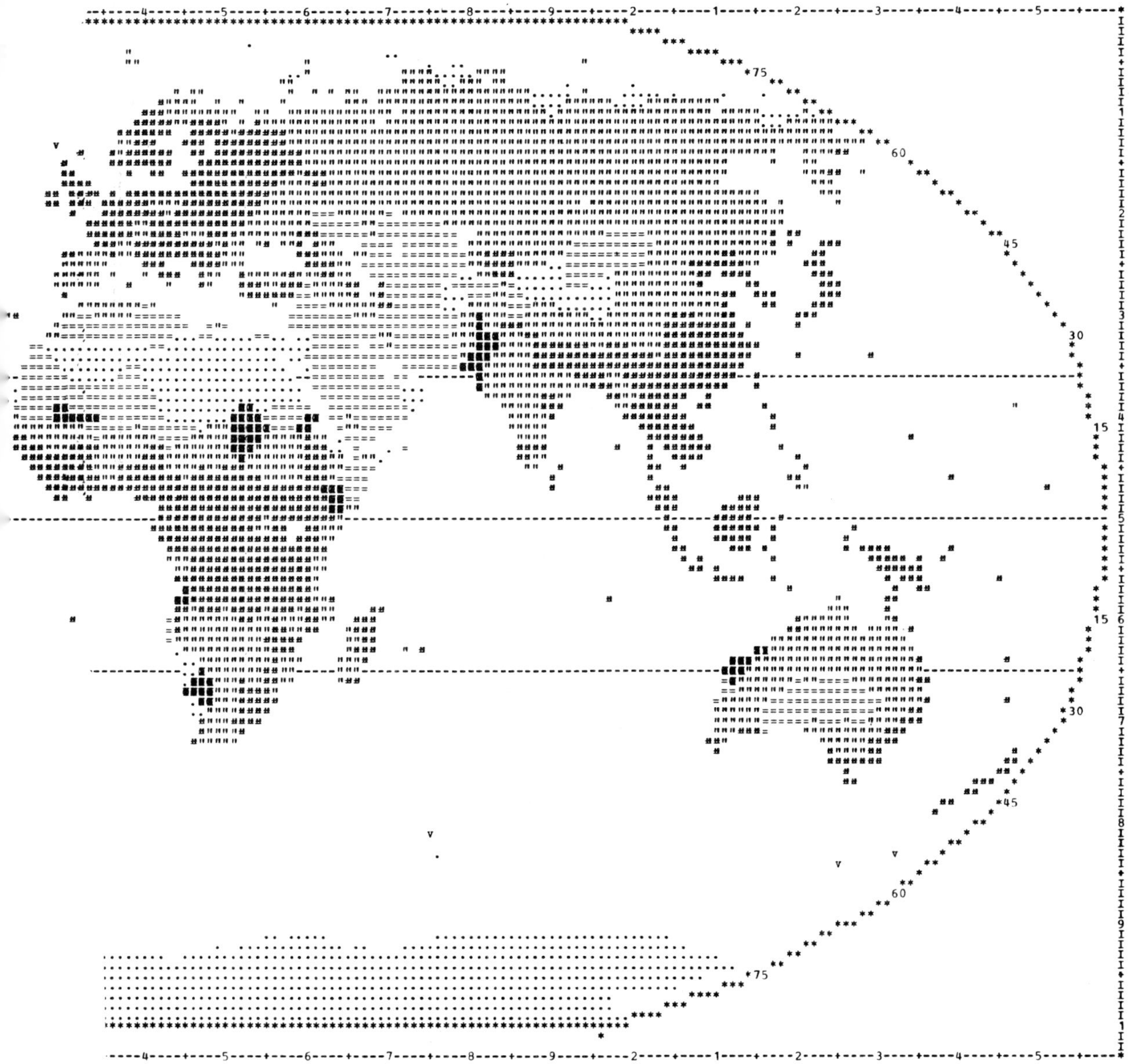

DATA VALUE EXTREMES ARE 0.0 11.00

TOTAL SUPERIMPOSED DATA POINTS IS 135. THESE OCCUR IN 60 LOCATIONS.

ABSOLUTE VALUE RANGE APPLYING TO EACH LEVEL
('MAXIMUM' INCLUDED IN HIGHEST LEVEL ONLY)

MINIMUM	BELOW	0.50	1.50	2.50	3.50	4.50	5.50	ABOVE
MAXIMUM	0.50	1.50	2.50	3.50	4.50	5.50	6.50	6.50

PERCENTAGE OF TOTAL ABSOLUTE VALUE RANGE APPLYING TO EACH LEVEL

16.67	16.67	16.67	16.67	16.67	16.67

FREQUENCY DISTRIBUTION OF DATA POINT VALUES IN EACH LEVEL

LEVEL	L	1	2	3	4	5	6	H
SYMBOLS	....	########	"""""""""	VVVVVVVVV	########	---------	=========	■■■■■■■■
FREQ.	105	544	499	12	0	0	104	17
FORMS	NONE	TG	SG	TTG	STG	SCG	DG	MULT

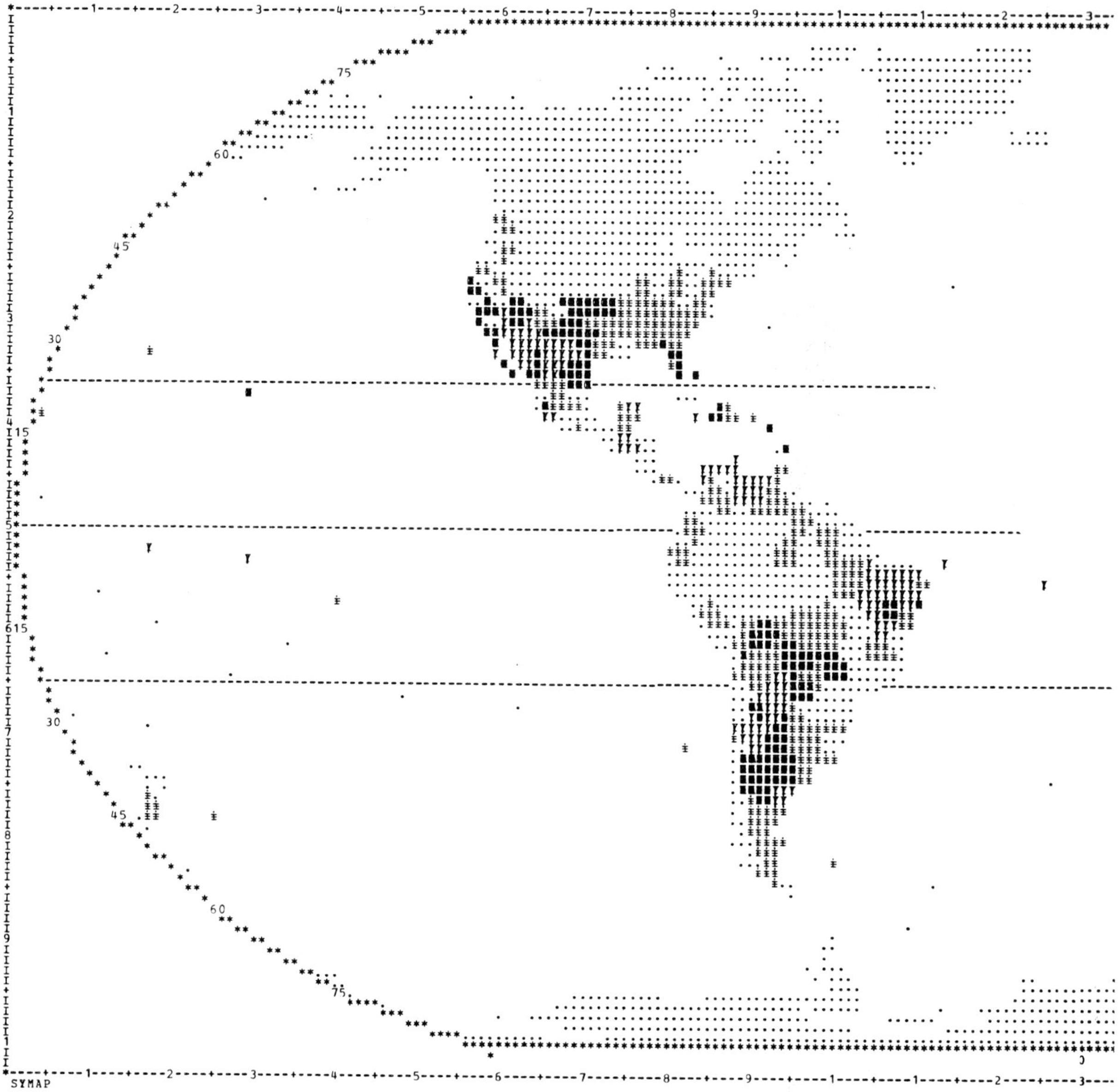

Map 17. Predicted distributions of most important stem and leaf-succulent types.
Stem-succulents and leaf-succulent xeric rosette shrubs are predicted to occur in largely frost-free climates with exception of the 'bush' form (e.g. *Opuntia*), which can withstand severe cold in some areas. All forms prefer dry climates, but some extend their ranges into surprisingly mesic climates on drier substrates. Due to the dominance structure employed, only arborescent and bush forms are predicted alone, while the other forms appear in combinations (MULT) which do not include arborescents. All of these 162 co-occurrences involve BSS: BSS + XRS (95), BSS + TSS (40), and 27 occurrences of all three (but without arborescents), mainly in southern California, Morocco-Algeria, and scattered sites in the Middle East. The prediction of a large area of arborescent stem-succulents in Australia is completely wrong, as are the smaller areas of succulents in and around the Sahara, Arabian and Iranian deserts. Off all growth forms, succulents are probably most poorly predicted by the model.

ASS = Arborescent Stem-Succulents
TSS = Typical Stem-Succulents
BSS = Bush Stem-Succulents
XRS = Xeric Rosette-Shrubs
MULT = co-important (non-arborescent) succulent forms

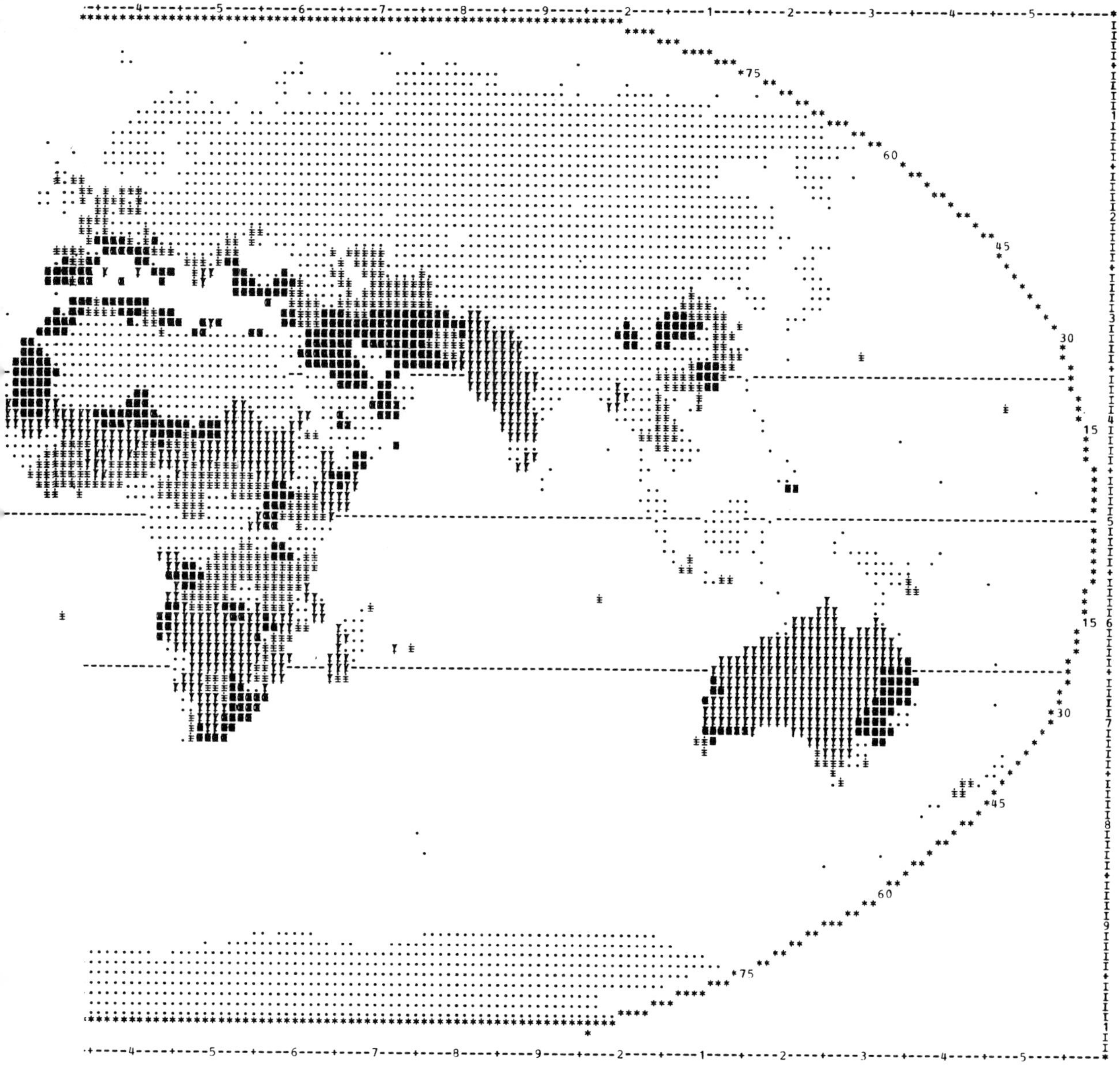

DATA VALUE EXTREMES ARE 0.0 11.00

TOTAL SUPERIMPOSED DATA POINTS IS 135. THESE OCCUR IN 60 LOCATIONS.

ABSOLUTE VALUE RANGE APPLYING TO EACH LEVEL
('MAXIMUM' INCLUDED IN HIGHEST LEVEL ONLY)

MINIMUM	BELOW	0.50	1.50	2.50	3.50	ABOVE 4.50
MAXIMUM	0.50	1.50	2.50	3.50	4.50	

PERCENTAGE OF TOTAL ABSOLUTE VALUE RANGE APPLYING TO EACH LEVEL

		25.00	25.00	25.00	25.00	

FREQUENCY DISTRIBUTION OF DATA POINT VALUES IN EACH LEVEL

LEVEL	L	1	2	3	4	H
SYMBOLS	....	YYYYYYYY	■■■■■■■■	‡‡‡‡‡‡‡‡	VVVVVVVVV	████████
FREQ.	774	140	0	205	0	162
FORMS	NONE	ASS	TSS	BSS	XRS	MULT

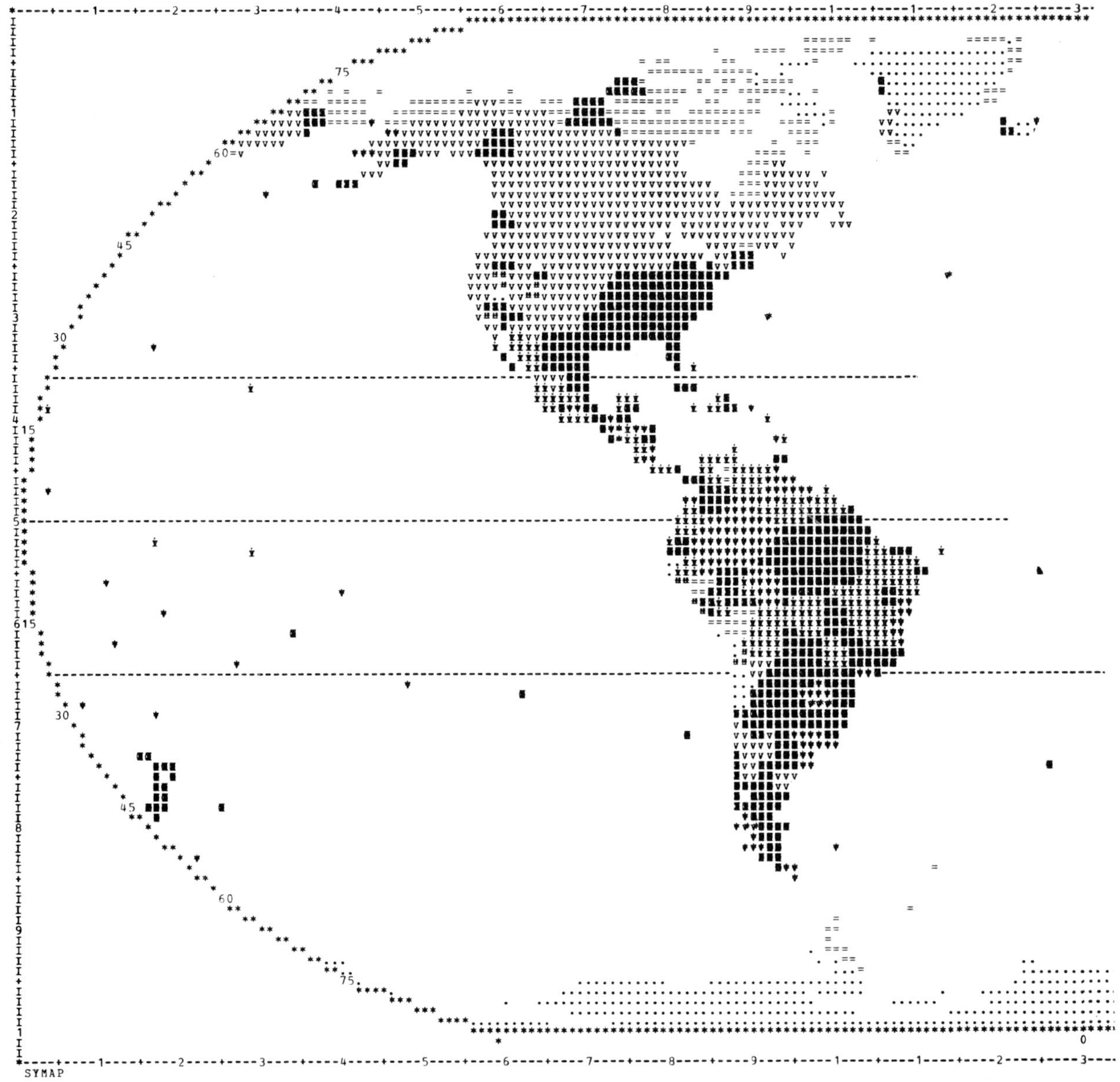

Map 18. Predicted distributions of most important forb and semi-desert herbs types.

Forbs and/or undifferentiated semi-desert herbs, like graminoids, are predicted to occur in essentially all parts of the world except extreme deserts. They gain dominance (Map 23), however, only in dry and/or cold, usually highly seasonal semi-deserts. The large areas in which the forb component is dominated by evergreen (equatorial), raingreen (tropical), or summergreen (temperate) forms are clearly visible on the map. Transition bands (MULT) of summergreen plus evergreen and/or raingreen forms are also clearly evident in the subtropics. Ephemerals are most obvious (outside the polar areas) in Middle Asia, the Near East, and around the interior Sahara desert. Evergreen forms reappear, both alone and in combinations (MULT), in some subpolar areas (e.g. Alaska, Tierra del Fuego).

TEF = Tropical Evergreen Forbs
RGF = RainGreen Forbs
SGF = SummerGreen Forbs
SCF = Succulent Cushion-Forbs
XCH = Xeric Cushion Herbs

EDH = Ephemeral Desert Herbs
SCDH = Summergreen Cold-Desert Herbs
RGDH = RainGreen cold-Desert Herbs
MULT = co-important forb/small-herb forms

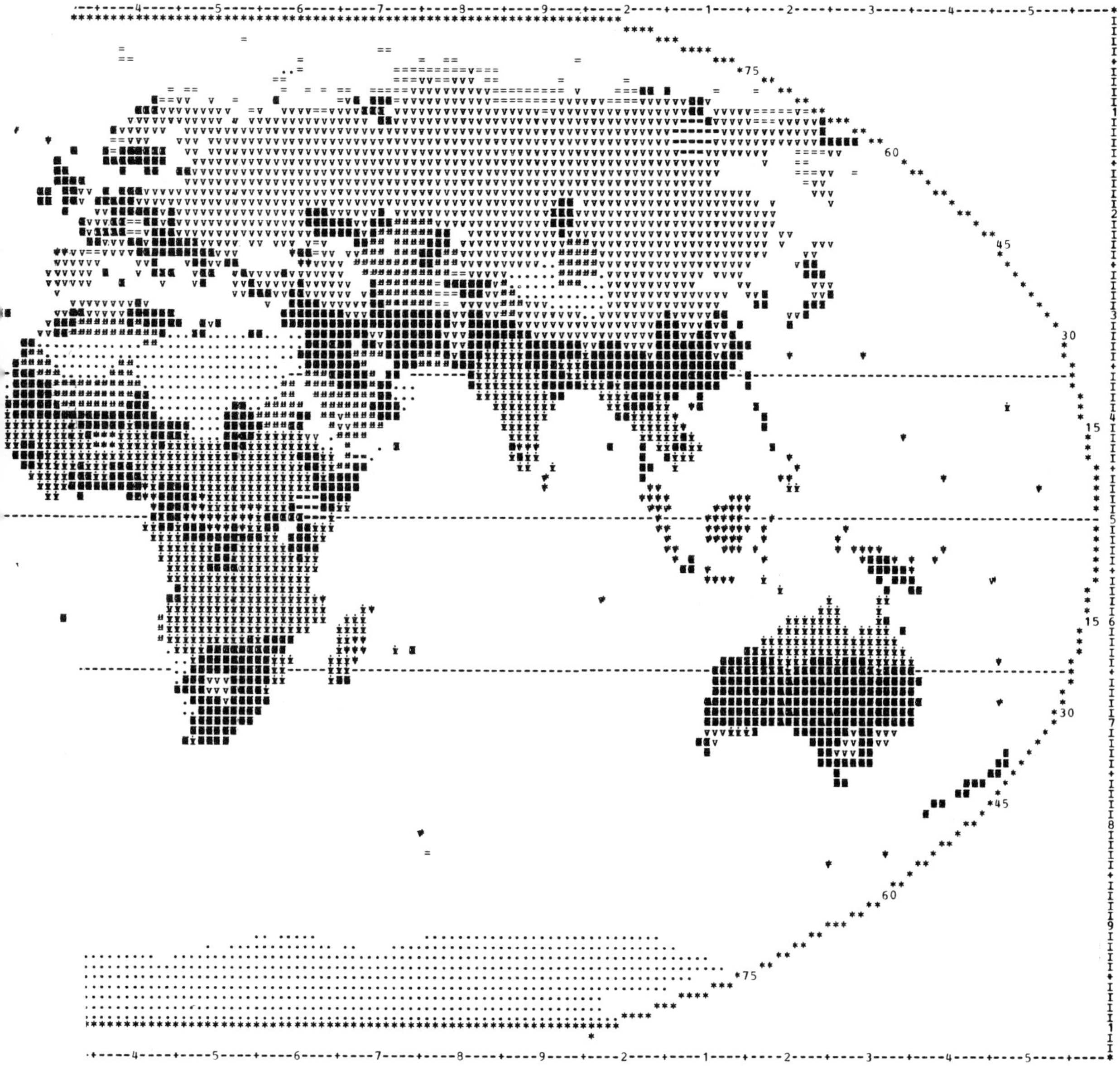

DATA VALUE EXTREMES ARE 0.0 11.00

TOTAL SUPERIMPOSED DATA POINTS IS 135. THESE OCCUR IN 60 LOCATIONS.

ABSOLUTE VALUE RANGE APPLYING TO EACH LEVEL
('MAXIMUM' INCLUDED IN HIGHEST LEVEL ONLY)

MINIMUM	BELOW	0.50	1.50	2.50	3.50	4.50	5.50	6.50	7.50	ABOVE
MAXIMUM	0.50	1.50	2.50	3.50	4.50	5.50	6.50	7.50	8.50	8.50

PERCENTAGE OF TOTAL ABSOLUTE VALUE RANGE APPLYING TO EACH LEVEL

12.50	12.50	12.50	12.50	12.50	12.50	12.50	12.50

FREQUENCY DISTRIBUTION OF DATA POINT VALUES IN EACH LEVEL

LEVEL	L	1	2	3	4	5	6	7	8	H
SYMBOLS	....	♥♥♥♥♥♥♥♥♥	ΙΙΙΙΙΙΙΙΙ	VVVVVVVVV	*********	---------	#########	=========	=========	■■■■■■■■■
FREQ.	70	102	173	405	2	3	35	80	4	407
FORMS	NONE	TEF	RGF	SGF	SCF	XCH	EDH	SCDH	RGDH	MULT

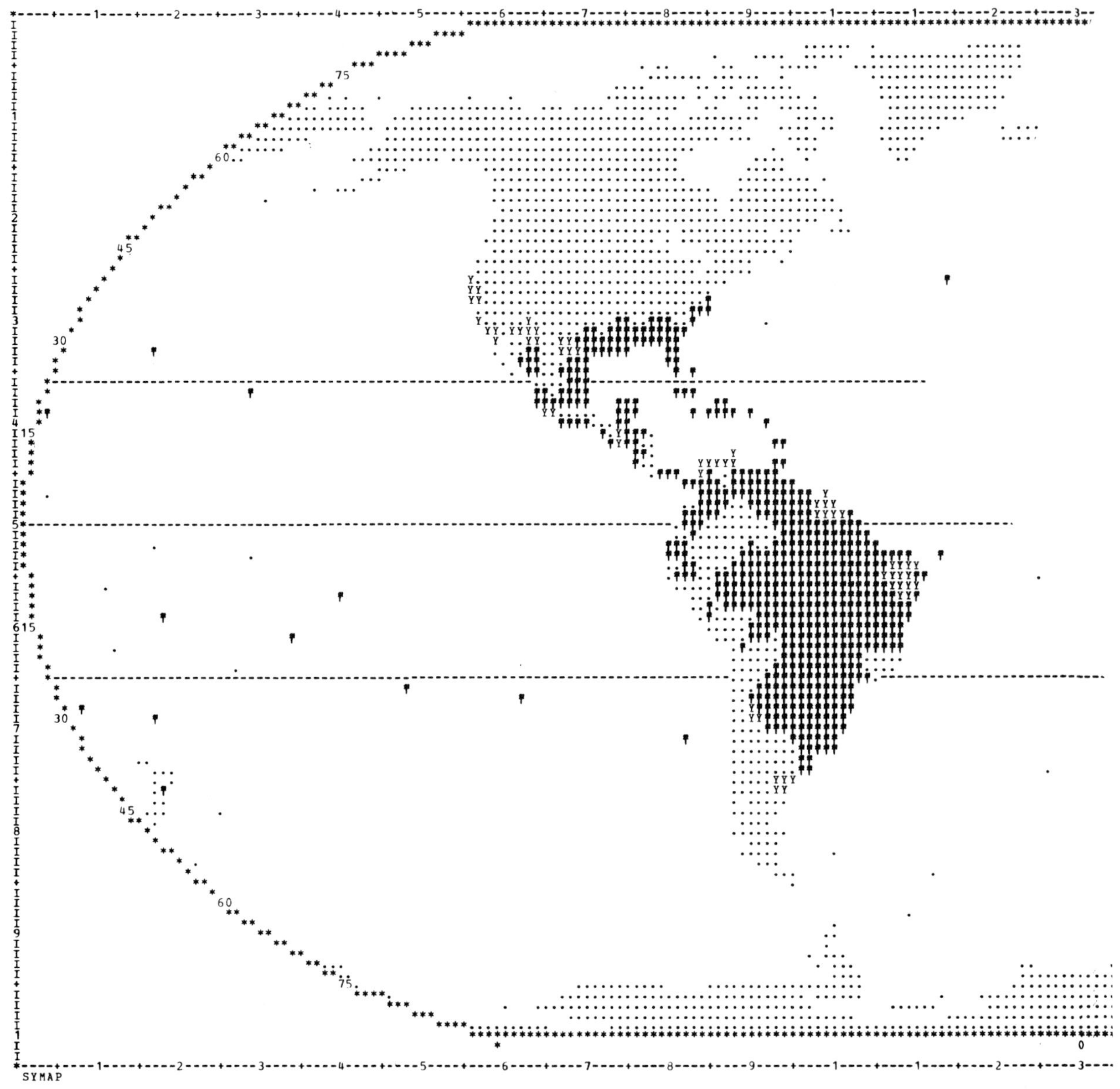

Map 19. Predicted distributions of eucalyptoid trees and scrub.

Eucalyptoid (evergreen sclerophyll) trees are separated from other broad-leaved evergreen tree forms because of the unique historical development of the prototypic genus, which was confined naturally to Australia. Both forest and woodland forms are predicted to be well adapted to large areas of tropical Africa, India, South America, Central America, southeastern USA, and Southeast Asia, as well as appropriate areas of Australia. The successful introduction of eucalypts in these areas outside Australia suggests that the predictions of eucalyptoid forms are generally accurate. Convergent non-eucalypt forms can also be identified in some of these other areas and appear especially well developed in eastern and southern Africa.

EST = Eucalyptoid Sclerophyll Trees
ESA = Eucalyptoid Sclerophyll Arborescents (e.g. mallee)

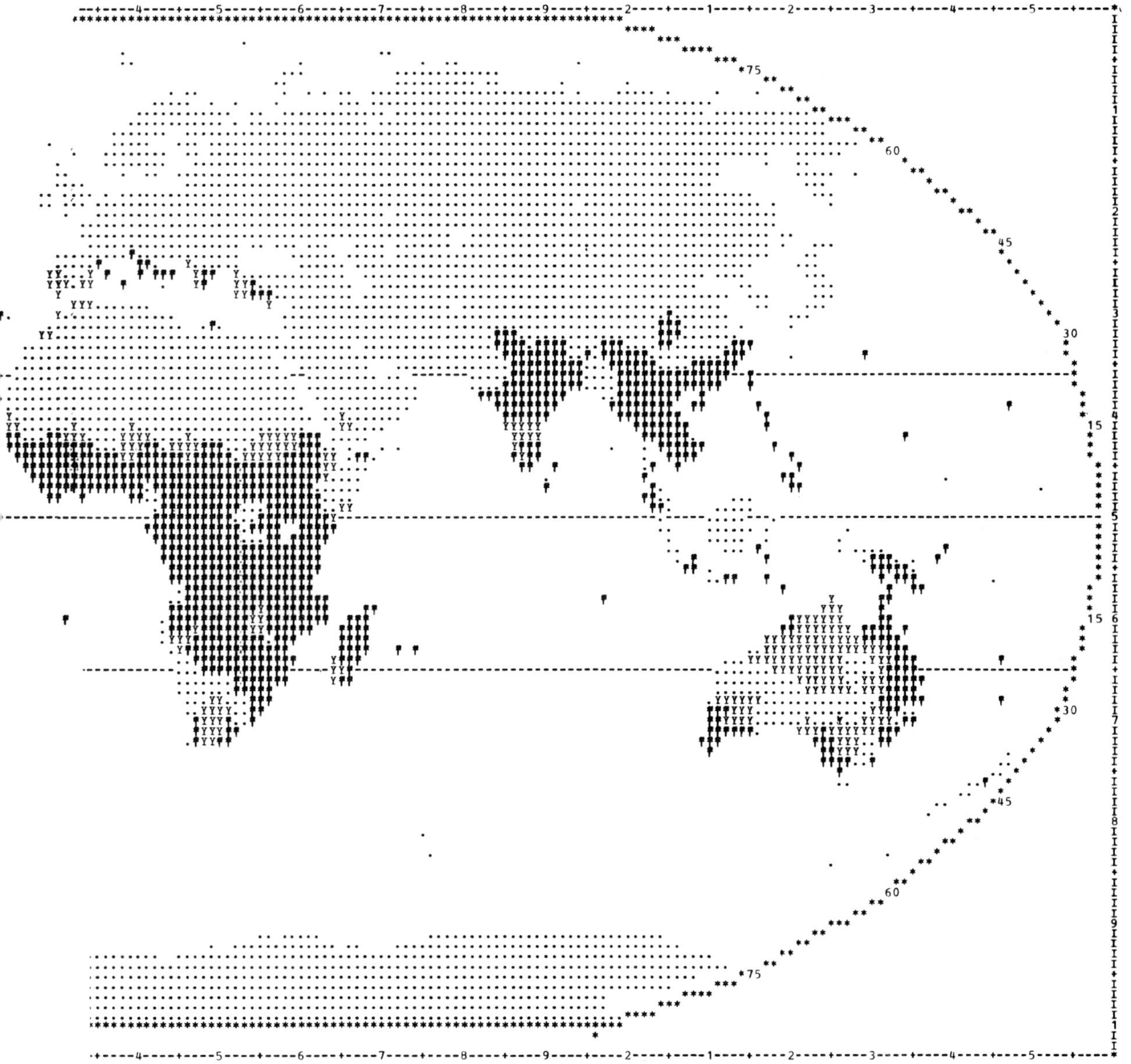

DATA VALUE EXTREMES ARE 0.0 2.00

TOTAL SUPERIMPOSED DATA POINTS IS 135. THESE OCCUR IN 60 LOCATIONS.

ABSOLUTE VALUE RANGE APPLYING TO EACH LEVEL
('MAXIMUM' INCLUDED IN HIGHEST LEVEL ONLY)

MINIMUM	BELOW	0.50	1.50
MAXIMUM	0.50	1.50	2.50

PERCENTAGE OF TOTAL ABSOLUTE VALUE RANGE APPLYING TO EACH LEVEL

50.00 50.00

FREQUENCY DISTRIBUTION OF DATA POINT VALUES IN EACH LEVEL

LEVEL	L	1	2
SYMBOLS	....	████████	YYYYYYYYY
FREQ.	863	338	80
FORMS	NONE	EST	ESA

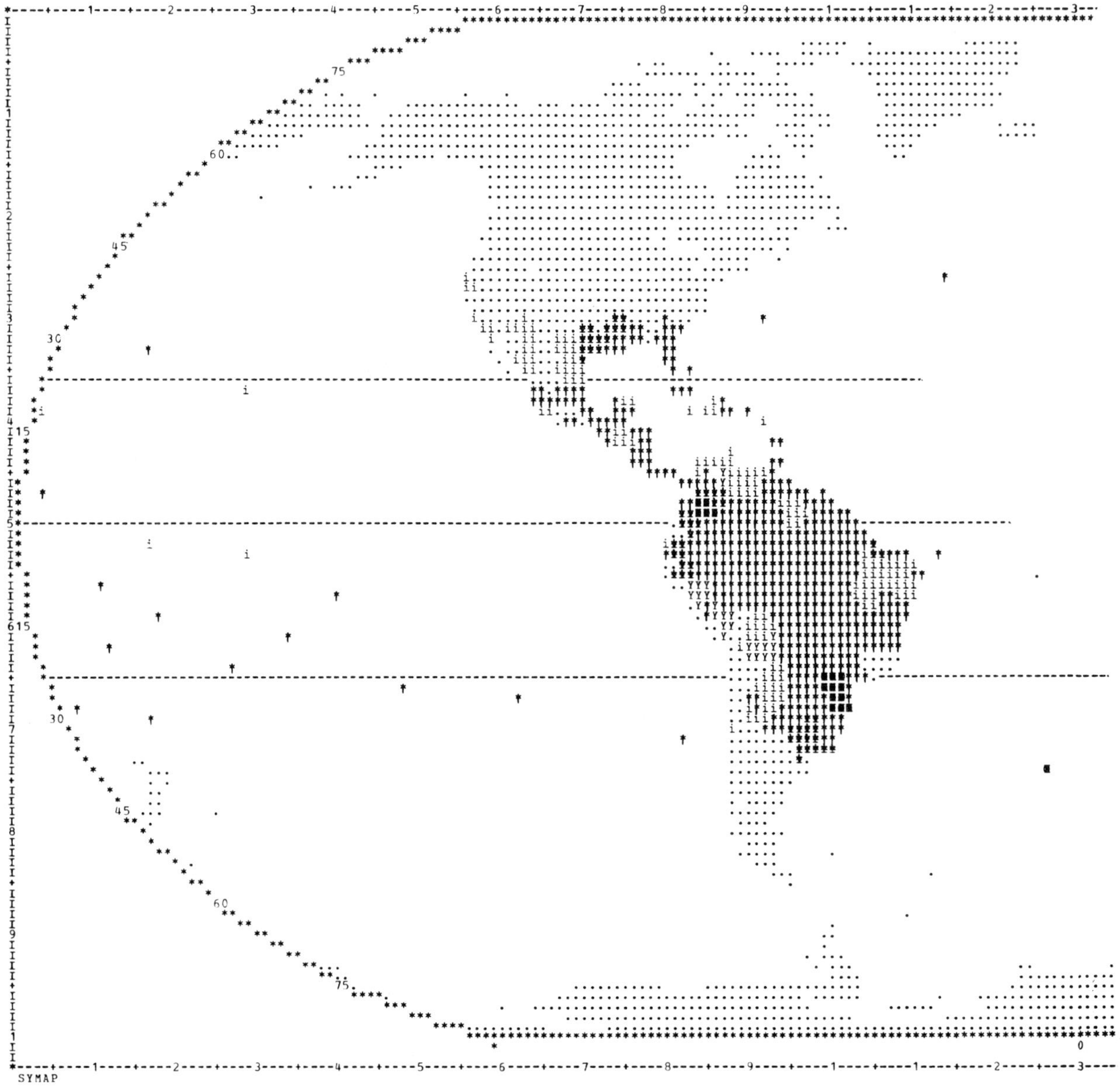

Map 20. Predicted distributions of most important tuft-tree and tuft-treelet types.

Evergreen tuft-trees and treelets, including tree ferns, are restricted to mostly frost-free climates of the tropics and subtropics, where sufficient water is available. More mesic forms are found in humid tropical and subtropical climates and more xeric forms (e.g. XETT) in the drier regions of the subtropics. The few occurrences of predicted co-important forms all involve the combination PTTL + TFN (conspicuously absent from New Zealand and Chile). Natural occurrences of the most important forms are generally well predicted, but these forms, especially palms, can also occur along water courses, at oases, in seasonally flooded savanna and swamp areas, and in other situations where climatic dryness is compensated by available groundwater. The single prediction of tree ferns as the most important tuft-tree form (Uraguay) is clearly erroneous. The occurrence of tropical alpine tuft-treelets is not well represented by the map because of the small mountain-top areas which they occupy and the paucity of such sites in the data-base, especially in Africa.

PTT = Palmiform Tuft-Trees
PTTL = Palmiform Tuft-TreeLets
TFN = Tree FerNs
TATT = Tropical Alpine Tuft-Treelets
XETT = Xeric Evergreen Tuft-Treelets
MULT = co-important tuft-tree forms

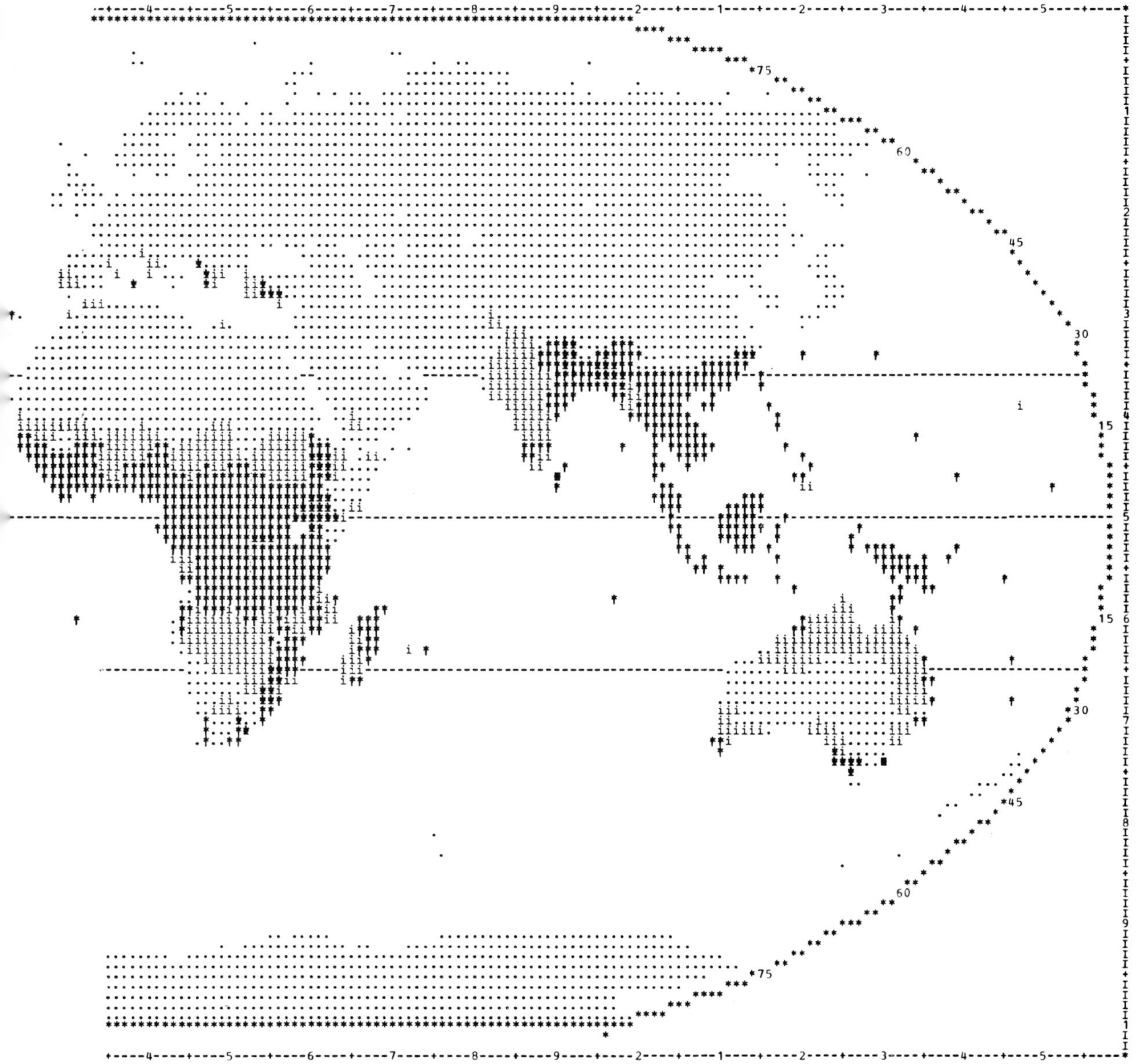

DATA VALUE EXTREMES ARE 0.0 11.00

TOTAL SUPERIMPOSED DATA POINTS IS 135. THESE OCCUR IN 60 LOCATIONS.

ABSOLUTE VALUE RANGE APPLYING TO EACH LEVEL
('MAXIMUM' INCLUDED IN HIGHEST LEVEL ONLY)

MINIMUM	BELOW	0.50	1.50	2.50	3.50	4.50	ABOVE 5.50
MAXIMUM	0.50	1.50	2.50	3.50	4.50	5.50	

PERCENTAGE OF TOTAL ABSOLUTE VALUE RANGE APPLYING TO EACH LEVEL

20.00 20.00 20.00 20.00 20.00

FREQUENCY DISTRIBUTION OF DATA POINT VALUES IN EACH LEVEL

LEVEL	L	1	2	3	4	5	H
SYMBOLS	....	[illegible]	[illegible]	[illegible]	YYYYYYYYY	iiiiiiiii	[illegible]
FREQ.	826	264	31	1	8	144	7
FORMS	NONE	PTT	PTTL	TFN	TATT	XETT	MULT

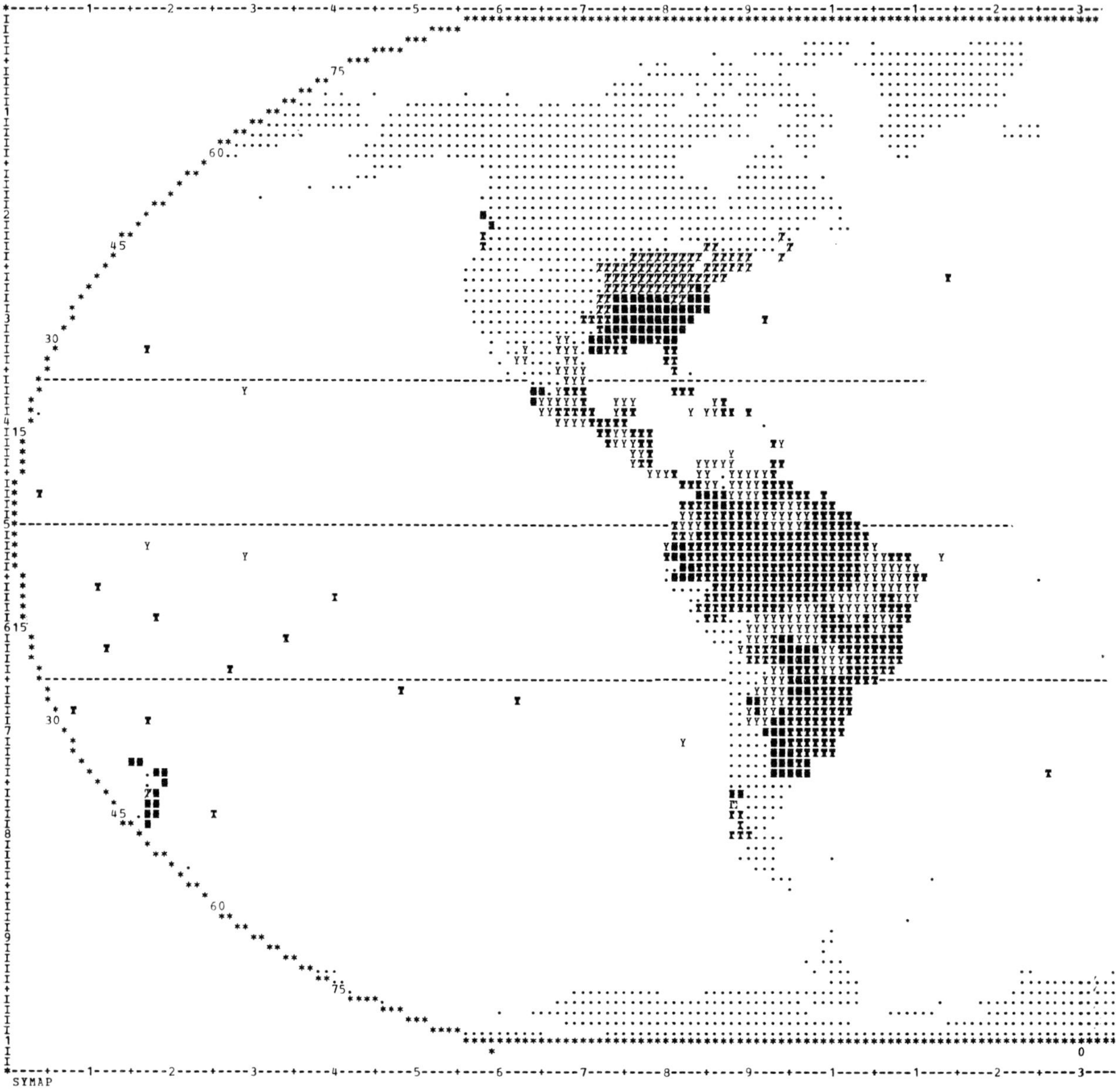

Map 21. Predicted distributions of most important vine and liana types.

Vines and lianas are generally found in wetter climates which support other woody vegetation over which the vines can climb or sprawl. Evergreen lianas (larger woody climbers) are generally restricted to the frost-free tropics where rapid growth is most easily accomplished. Smaller evergreen vines, almost always woody, occur well into the temperate zones but are largely confined to more maritime climates with moderated winter temperatures and available winter moisture. Summergreen and raingreen vines are common, respectively, in cold-winter and tropical summer-rain climates with varying degrees of total moisture. They become less important as total moisture decreases and/or the growing season becomes shorter. The areas (MULT) of predicted co-important forms represent combinations of summergreen and evergreen vines in the temperate zone (54 occurrences, e.g. southeastern USA) and of raingreen and evergreen vines in the tropics (27 occurrences, e.g. India and Africa).

TBEL = Tropical Broad-Evergreen Lianas
BEGV = Broad-EverGreen Vines
BRGV = Broad-RainGreen Vines
BSGV = Broad-SummerGreen Vines
MULT = co-important vine/liana forms

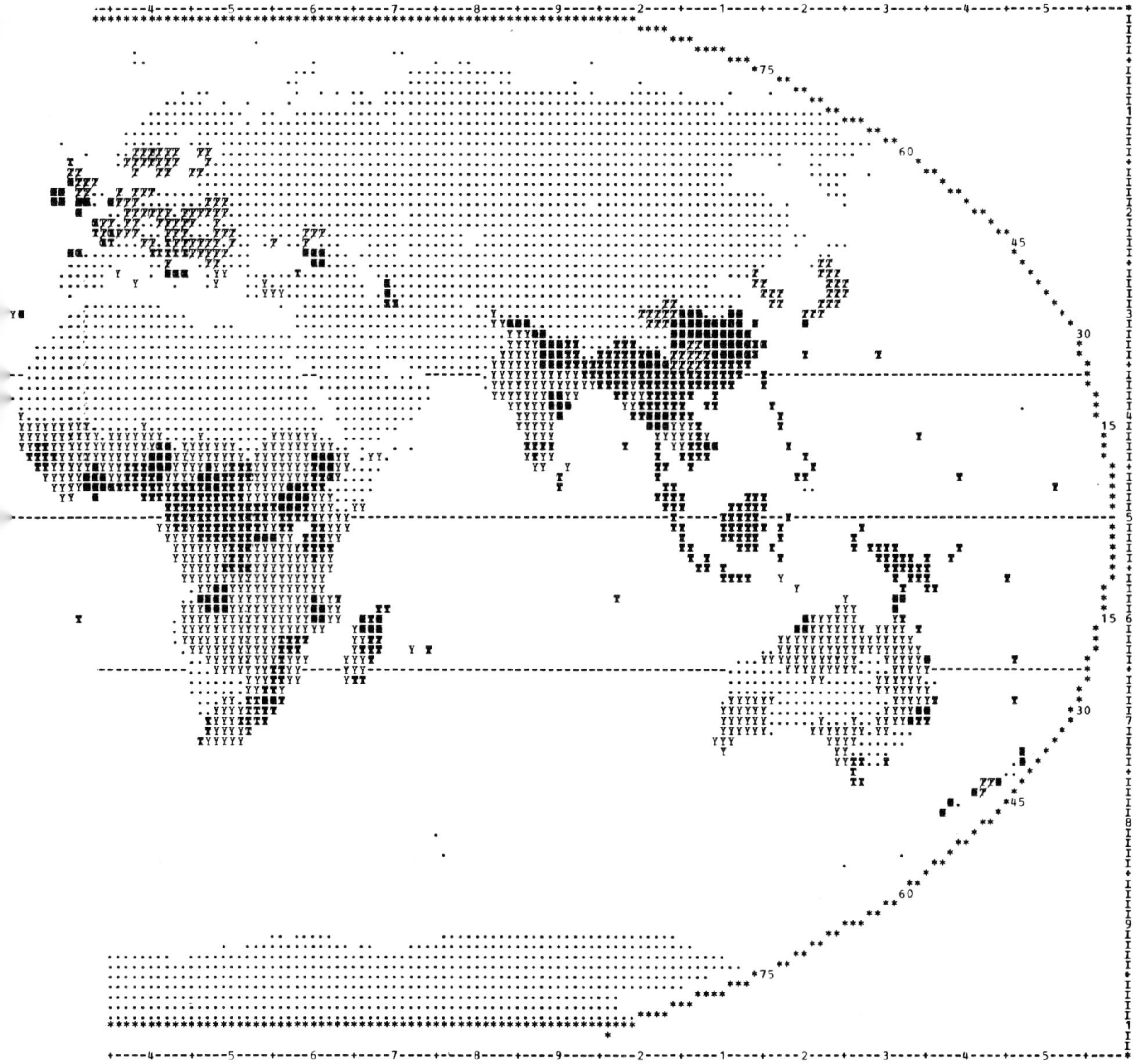

DATA VALUE EXTREMES ARE 0.0 11.00

TOTAL SUPERIMPOSED DATA POINTS IS 135. THESE OCCUR IN 60 LOCATIONS.

ABSOLUTE VALUE RANGE APPLYING TO EACH LEVEL
('MAXIMUM' INCLUDED IN HIGHEST LEVEL ONLY)

MINIMUM	BELOW	0.50	1.50	2.50	3.50	ABOVE 4.50
MAXIMUM	0.50	1.50	2.50	3.50	4.50	

PERCENTAGE OF TOTAL ABSOLUTE VALUE RANGE APPLYING TO EACH LEVEL

25.00	25.00	25.00	25.00

FREQUENCY DISTRIBUTION OF DATA POINT VALUES IN EACH LEVEL

LEVEL	L	1	2	3	4	H
SYMBOLS	....	TTTTTTTTT	IIIIIIIII	YYYYYYYYY	ZZZZZZZZZ	■■■■■■■■■
FREQ.	720	188	34	181	77	81
FORMS	NONE	TBEL	BEGV	BRGV	BSGV	MULT

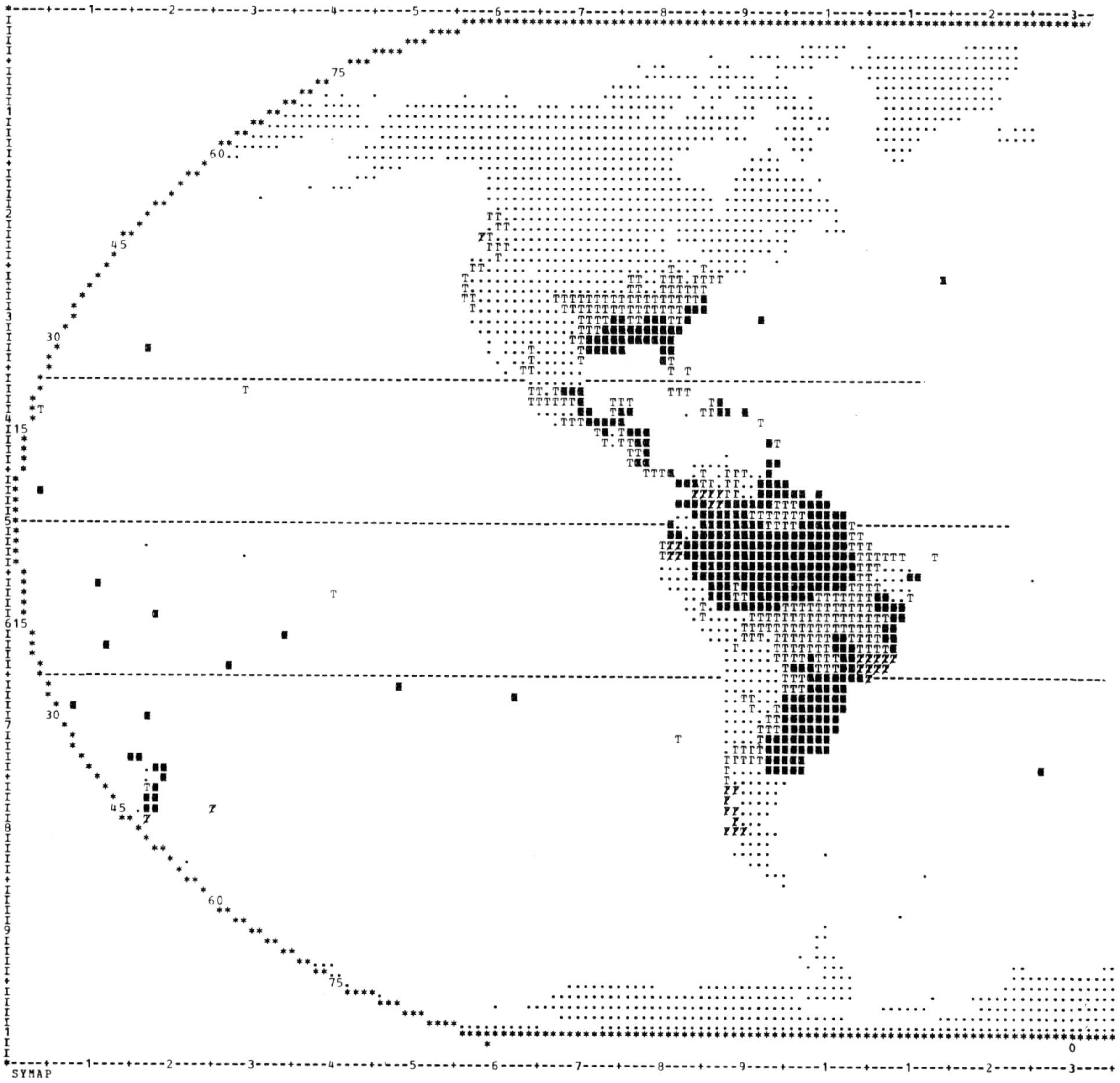

Map 22. Predicted distributions of major epiphyte types

Epiphytes (evergreen forms only) are generally restricted to tropical and relatively mild temperate climates with sufficient moisture for woody host plants. Wintergreen epiphytes (e.g. mistletoe) extend into fairly cold and/or dry areas but continue to be active during the unfavorable period. Broad and narrow-leaved (evergreen) epiphytes occur in generally frost-free climates, the former requiring more humidity to support its larger leaf area. Tropical broad-leaved epiphytes are not shown to occur separately here because they occur in climates which generally support the other types as well. Areas of predicted co-occurrence usually involve all three forms (141 occurrences) but also NLE + BWGE (53 occurrences, generally subtropical) and only 2 occurrences of NLE + TBEE without BWGE (Bogota and Quito).

TBEE = Tropical Broad-Evergreen Epiphytes
NLE = Narrow-Leaved Epiphytes

BWGE = Broad-WinterGreen Epiphytes
MULT = co-important epiphyte forms

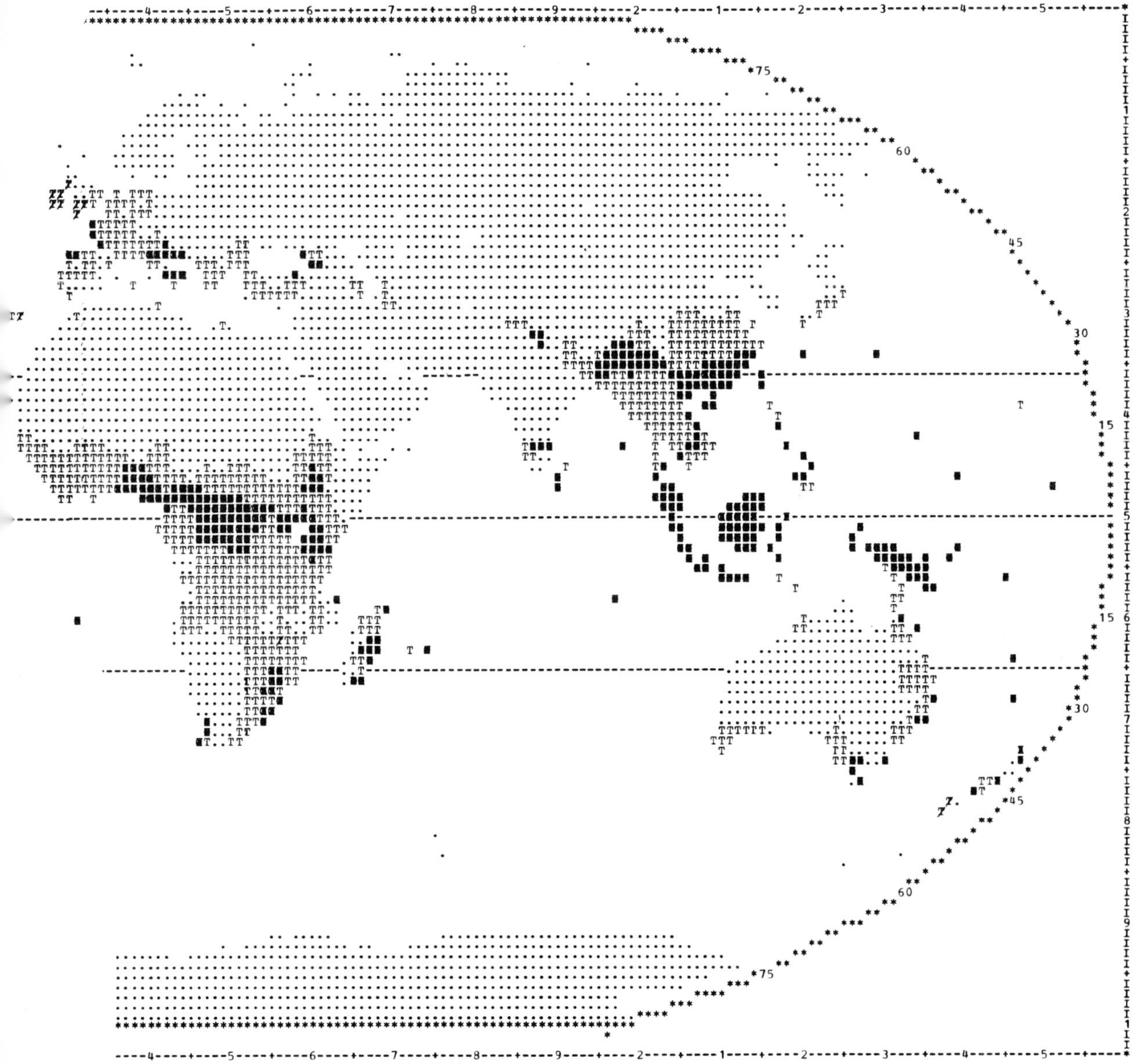

DATA VALUE EXTREMES ARE 0.0 11.00

TOTAL SUPERIMPOSED DATA POINTS IS 135. THESE OCCUR IN 60 LOCATIONS.

ABSOLUTE VALUE RANGE APPLYING TO EACH LEVEL
('MAXIMUM' INCLUDED IN HIGHEST LEVEL ONLY)

MINIMUM	BELOW	0.50	1.50	2.50	ABOVE 3.50
MAXIMUM	0.50	1.50	2.50	3.50	

PERCENTAGE OF TOTAL ABSOLUTE VALUE RANGE APPLYING TO EACH LEVEL

33.33 33.33 33.33

FREQUENCY DISTRIBUTION OF DATA POINT VALUES IN EACH LEVEL

LEVEL	L	1	2	3	H
SYMBOLS	....	IIIIIIII	ZZZZZZZZ	TTTTTTTT	■■■■■■■■
FREQ.	799	0	17	268	197
FORMS	NONE	TBEE	NLE	BWGE	MULT

F. Predicting world vegetation types

Describing, predicting and mapping vegetation types can be approached in a variety of ways using (among other criteria):

1. formation classes based on dominant growth forms (the traditional approach);
2. different combinations of constituent plant types (a more complete measure of vegetation structure); and
3. the number of constituent plant types present, with some considerations of relative importance.

These last two criteria lead to the concepts of physiognomic ecotones and form diversity of vegetation, which provide information not only on plant-type diversity but also on the general structural and functional diversity of entire ecosystems.

The simplest and perhaps the only way to present the world vegetation predictions in a single map is to map the most obvious single attribute, which is usually the dominant growth form. This is done in Map 23. The dominant form or forms are determined from the dominance hierarchy in the case of closed stands and from a combination of potential dominance, cover, and fitness in open stands. As one can see by examining Map 23, the dominant structural type is well predicted in almost all ecoclimate core areas and generally becomes problematic only in transitional areas. Among the maps, Map 23 is perhaps the best single index of model accuracy.

Identification of different combinations of predicted life forms, i.e. of different life-form spectra, was approached by means of a relatively simple enumeration program called NCOMBS (Box, unpublished). Out of 1225 sites there were 853 different combinations predicted, many differing, however, only with respect to understorey synusiae. The potential usefulness of this approach is illustrated in Table 13, in which tropical montane rainforest sub-types are differentiated according to different combinations of tree life forms. Interpretation of Table 13 and the world results is limited by the fact that they reflect only predicted presence or absence, with no consideration of importance. Addition of importance criteria would permit a type of life form-based ordination which may be more closely tied to environmental conditions than are current ordination methods.

One seldom sees maps of ecotones, so the map of predicted physiognomic ecotones and other regions of multiple but less distinct physiognomic forms (Map 24) presents an interesting pattern. The darker shaded areas may involve co-dominance by different structural types (mainly the more familiar ecotones between biomes, such as forest-steppe and forest-tundra) or prediction of two or more co-dominant forms of the same structural type (e.g. broad and narrow-leaved trees). These latter indicate areas of physiognomic diversity and coexistence within areas usually recognized as more or less uniform biomal units. Such areas can be thought of as physiognomic ecotones even though they may retain considerable ecological unity, e.g. northern mixed conifer-hardwood forests.

G. Physiognomic diversity of vegetation

Since the plant types used in the model represent well defined ecophysiognomic forms, the number of forms predicted at each site represents a measure of the physiognomic diversity of the vegetation. An initial prediction of world physiognomic diversity among dominant and sub-dominant forms (using results from the earlier model, Box 1978a) is presented on Map 25. The earlier model contained only those understorey forms which do become dominant somewhere in the world (e.g. tundra shrubs in boreal forest understorey) and so is slightly biased toward the cooler regions. Since some forms differ physiognomically more than others, it was necessary to group similar forms. The groupings used are the same as those for Maps 10–22 and generally represent commonly recognized growth forms.

The increase in number of forms along gradients of increasing aridity has been documented several times (see, for example, Shreve 1936; Whittaker & Niering 1965; Whittaker 1975; Mooney & Dunn 1970), and many authors have suggested that there can always be a greater variety of forms in open vegetation stands, where light is not limiting. One thus would expect the highest form diversity in semi-deserts and open woodlands (especially those of warmer climates) and the lowest form diversity in the climatic core areas of the various physiognomic forest types.

In looking at Map 25 one can see that this is

Table 13. Differentiation of tropical montane rainforest sub-types according to different combinations of constituent life forms.

The use of life-form combinations as a basis for perhaps greater resolution of vegetation types is illustrated using various predicted combinations which involve tropical montane rainforest trees. The different combinations of tree forms are shown in the left column with the corresponding vegetation types juxtaposed in the right column. The numbers in parentheses indicate the actual numbers of combinations (involving all life forms) which were predicted. (These numbers closely reflect the actual numbers of sites involved, since only two combinations occur at more than one site.) Among the 1225 sites in the world data-base, 60 different combinations of all life forms (63 sites) were predicted which involve tropical montane rainforest trees. All 60 also involve tropical evergreen microphyll and small trees. Of the 60, 16 also involve cloud-forest trees and can be called Mid-Montane Cloud Forest sites. Twelve of the 60 involve lowland rainforest trees and can be called Lower-Montane Rainforest sites. The remaining 32 sites involve neither cloud-forest nor lowland trees and represent the true Mid-Montane Rainforest. Further subdivisions are possible, as shown, involving admixtures of sclerophyll, raingreen, temperate summergreen, linear-leaved, and palmiform trees or small trees. Note that the combinations and frequencies reflect only presence or absence of the various forms but not their importance. Consideration of importance (e.g. proximity to environmental limits) will require further development of the identification program (NCOMBS).

Life-form combinations	Corresponding vegetation type
Tropical montane rainforest trees (60)	
(always with evergreen microphyll and small trees)	
with cloud-forest trees (16)	Mid-montane cloud forest
with neither cloud nor lowland trees (32)	Mid-montane rainforest
with sclerophyll trees (27)	drier mid-montane forest
with raingreen trees (13)	semi-deciduous mid-montane forest
with sclerophyll and raingreen trees (12)	
with neither sclerophyll nor raingreen trees (4)	
with summergreen trees (17)	subtropical semi-deciduous mid-montane forest
without linear-leaved trees (3)	
without palmiform tuft-trees/treelets (5)	
with lowland rainforest trees (12)	Lower-montane rainforest
(always with palmiform and linear-leaved trees)	
with sclerophyll trees (8)	drier lower-montane forest
with raingreen trees (5)	semi-deciduous lower-montane forest
with neither sclerophyll nor raingreen trees (3)	

indeed the pattern which is predicted. The highest physiognomic diversity is predicted for the subtropical-tropical semi-deserts of Australia, Chihuahua, Sonora-Arizona, and southern Africa. The lowest diversity is predicted for areas where a particular life form is well established, as in the tropical rainforests, and in the extreme climates of the dry and cold deserts, where total diversity is low.

Not so readily expected are the form-diversity levels predicted for boreal areas (generally imagined to be monotonously uniform), which are shown as higher than those of summergreen forest areas such as eastern North America and western Europe. In boreal areas one can readily identify three tree forms which occur throughout most of the region: the evergreen and summergreen needle-trees and the summergreen broad-microphyll (betuloid) trees. These tree forms are complemented by various shrub and herb forms, making a total of five forms, even though the evergreen conifers may exclude the other forms almost completely in older stands. In the nemoral (typical four-season temperate) regions the dominant summergreen broad-leaved trees are generally complemented by at most one other tree form, the temperate evergreen needle-trees. Shrubs

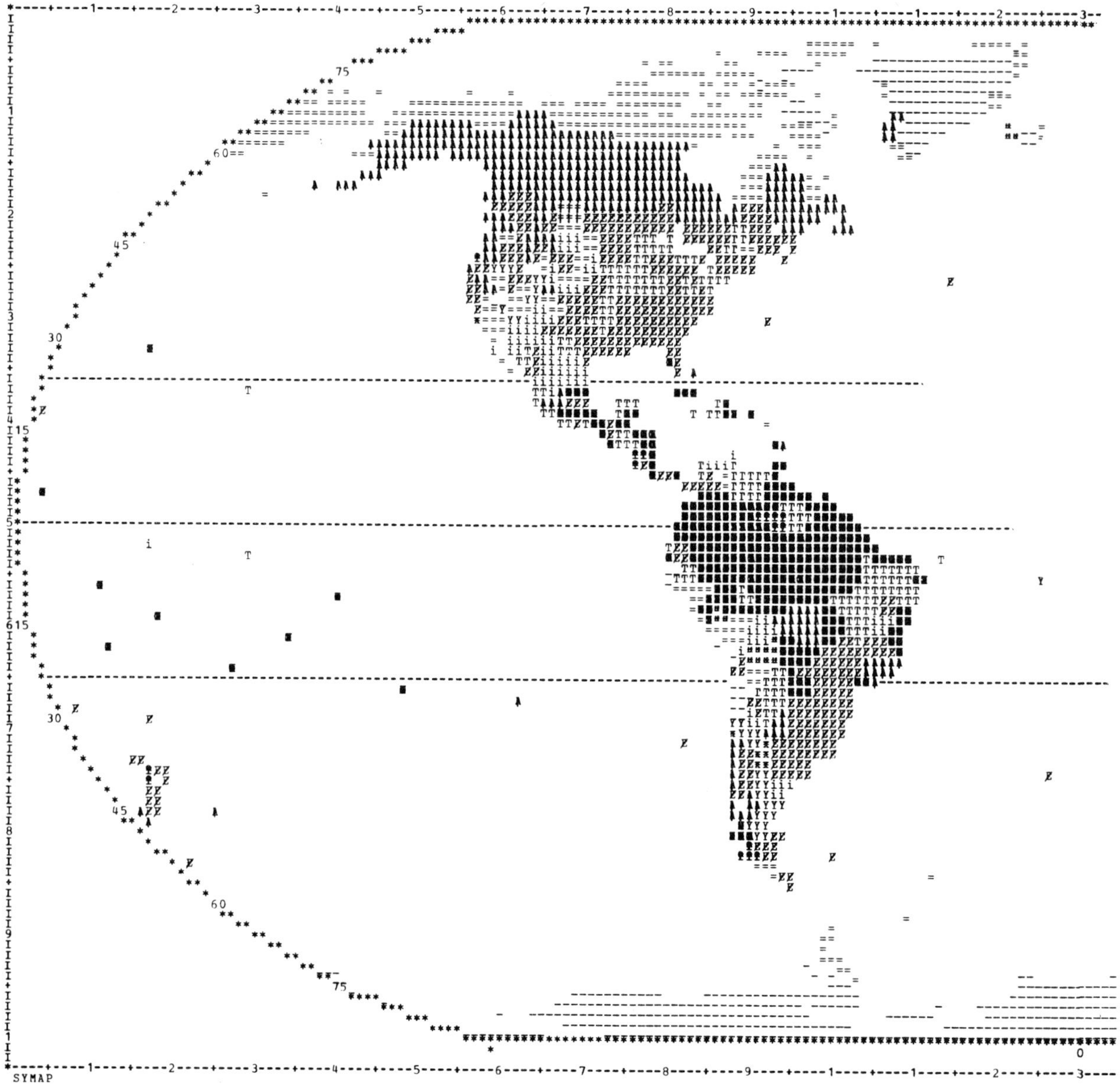

Map 23. Predicted dominant growth forms.

Tree forms are predicted as formation dominants in all humid and some moderately subhumid areas with sufficient summer warmth. Broad-leaved trees are favored in climates which are very productive at least seasonally (humid tropical and warm-summer temperate climates). Narrow and needle-leaved trees, mainly conifers and often more sclerophyllous, are favored in climates with at least seasonal stress but in which longer growing seasons must be realized to provide sufficient annual net production. In drier climates, in which transpiring leaf area and respiring biomass may need to be reduced, trees disappear and shrubs may dominate, especially where their deeper root systems provide an advantage in dry seasons over herbaceous forms. Where precipitation is low but reliable (during a sufficiently warm season), graminoids may attain dominance, with this dominance extended into adjacent areas by fire, grazing, etc. Succulent, rosette, and smaller herbaceous forms can attain dominance only in certain less favorable climates and occur more commonly as sub-dominants or understorey forms. The areas of predicted co-dominant growth forms (physiognomic ecotones) are treated in more detail on Map 24.

TRF = Tropical RainForest (broad-evergreen trees)
BEGT = other Broad-leaved EverGreen Trees
DECT = broad-leaved DECiduous Trees
NEGT = Narrow/Needle-leaved EverGreen Trees
NSGT = Needle-leaved SummerGreen Trees
T/A = small trees, Tuft-treelets and Arborescents

SCRB = xeric SCRuB (including desert shrubs and succulents)
SHRB = SHRuBs and other frutescent forms (non-desert)
TG = Tall Grasses
HERB = short grasses and other HERBaceous forms
ECOT = physiognomic ECOTones (co-dominant growth forms)

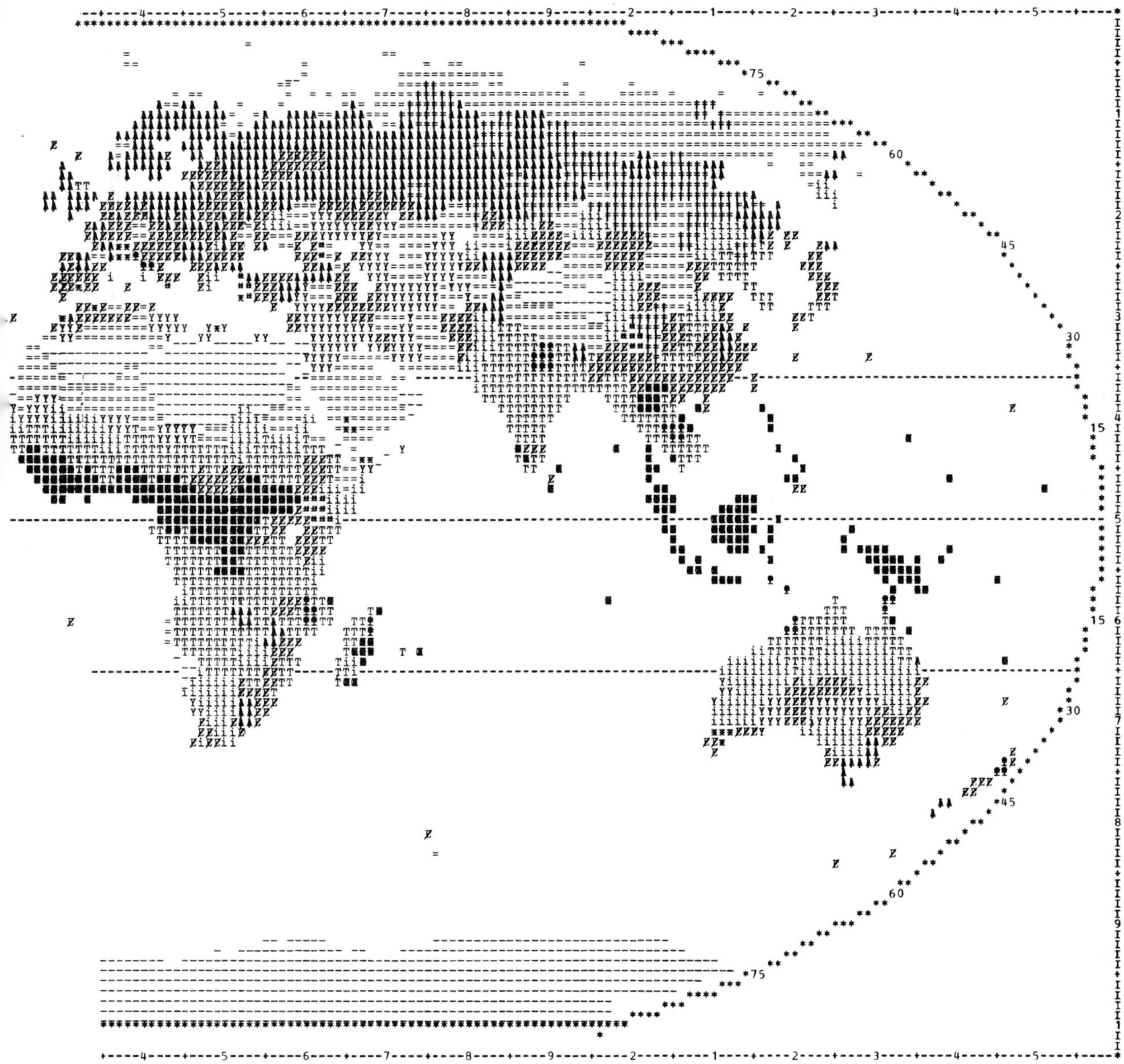

DATA VALUE EXTREMES ARE 0.0 11.00

TOTAL SUPERIMPOSED DATA POINTS IS 135. THESE OCCUR IN 60 LOCATIONS.

ABSOLUTE VALUE RANGE APPLYING TO EACH LEVEL
('MAXIMUM' INCLUDED IN HIGHEST LEVEL ONLY)

MINIMUM	BELOW	0.50	1.50	2.50	3.50	4.50	5.50	6.50	7.50	8.50	9.50	ABOVE 10.50
MAXIMUM	0.50	1.50	2.50	3.50	4.50	5.50	6.50	7.50	8.50	9.50	10.50	

PERCENTAGE OF TOTAL ABSOLUTE VALUE RANGE APPLYING TO EACH LEVEL

10.00	10.00	10.00	10.00	10.00	10.00	10.00	10.00	10.00	10.00

FREQUENCY DISTRIBUTION OF DATA POINT VALUES IN EACH LEVEL

LEVEL	L	1	2	3	4	5	6	7	8	9	10	H
SYMBOLS	---------	■■■■■■■■	●●●●●●●●	TTTTTTTT	▲▲▲▲▲▲▲▲	++++++++	iiiiiiii	YYYYYYYY	********	########	========	ZZZZZZZZ
FREQ.	68	142	16	147	209	31	89	49	25	8	204	293
FORMS	NONE	TRF	BEGF	DECF	NEGT	NSGT	T/A	SCRB	SHRB	TG	HERB	ECOT

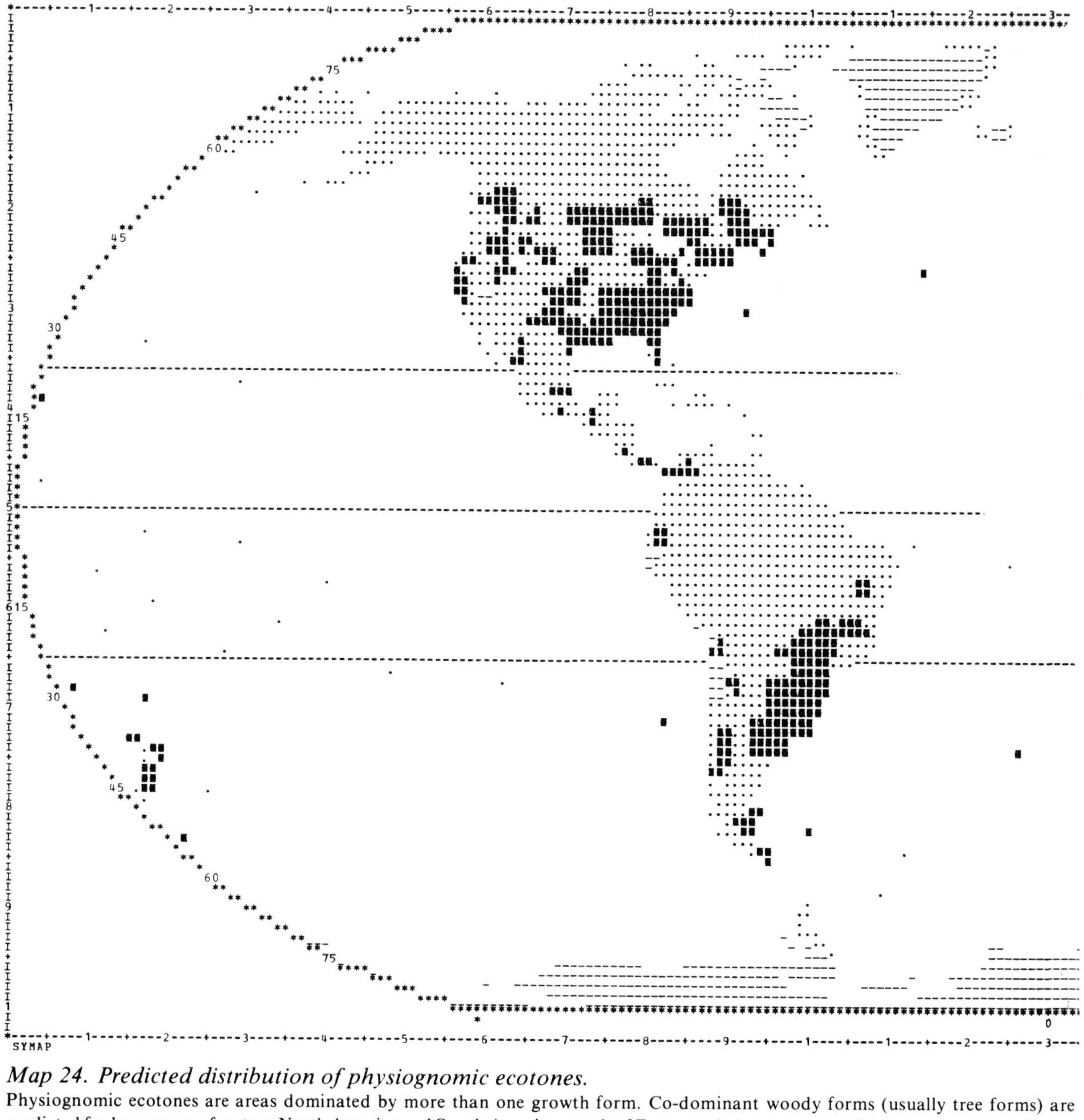

Map 24. Predicted distribution of physiognomic ecotones.

Physiognomic ecotones are areas dominated by more than one growth form. Co-dominant woody forms (usually tree forms) are predicted for large areas of eastern North America and South America, much of Europe, a belt around the African equatorial rainforest, parts of the Middle East (non-trees), eastern India, northern Japan, and large areas of East Asia and southern Australia (non-trees). The areas of co-dominant tree forms are often semi-deciduous mixed forests but may contain only subordinate admixtures of the second form. Areas of more than two co-dominant tree types (94 occurrences) are predicted especially in southeastern USA, southern Brazil, northeastern Argentina (often erroneously), western and southern Europe, southern Africa, southern China, eastern Australia, and New Zealand. Areas of co-dominant evergreen and deciduous broad-leaved trees, without narrow-leaved trees, are predicted at only a very few sites, mainly in the Far East. Areas of co-dominant broad and narrow/needle-leaved trees (91 occurrences) are predicted much more commonly, mainly in sub-boreal, sub-mediterranean, east-coast subtropical, and tropical-montane areas. Among non-tree co-dominants, areas involving more than two basic growth forms (29 occurrences, commonly xeric shrubs and bunch-grasses) are predicted especially in temperate-continental (cold-winter) semi-desert areas. Other predicted ecotones generally involve one woody and one non-woody, usually graminoid form, i.e. savannas, shrub-steppes, or very open woodlands.

SINGLE = SINGLE dominant form
ECOT = physiognomic ECOTones

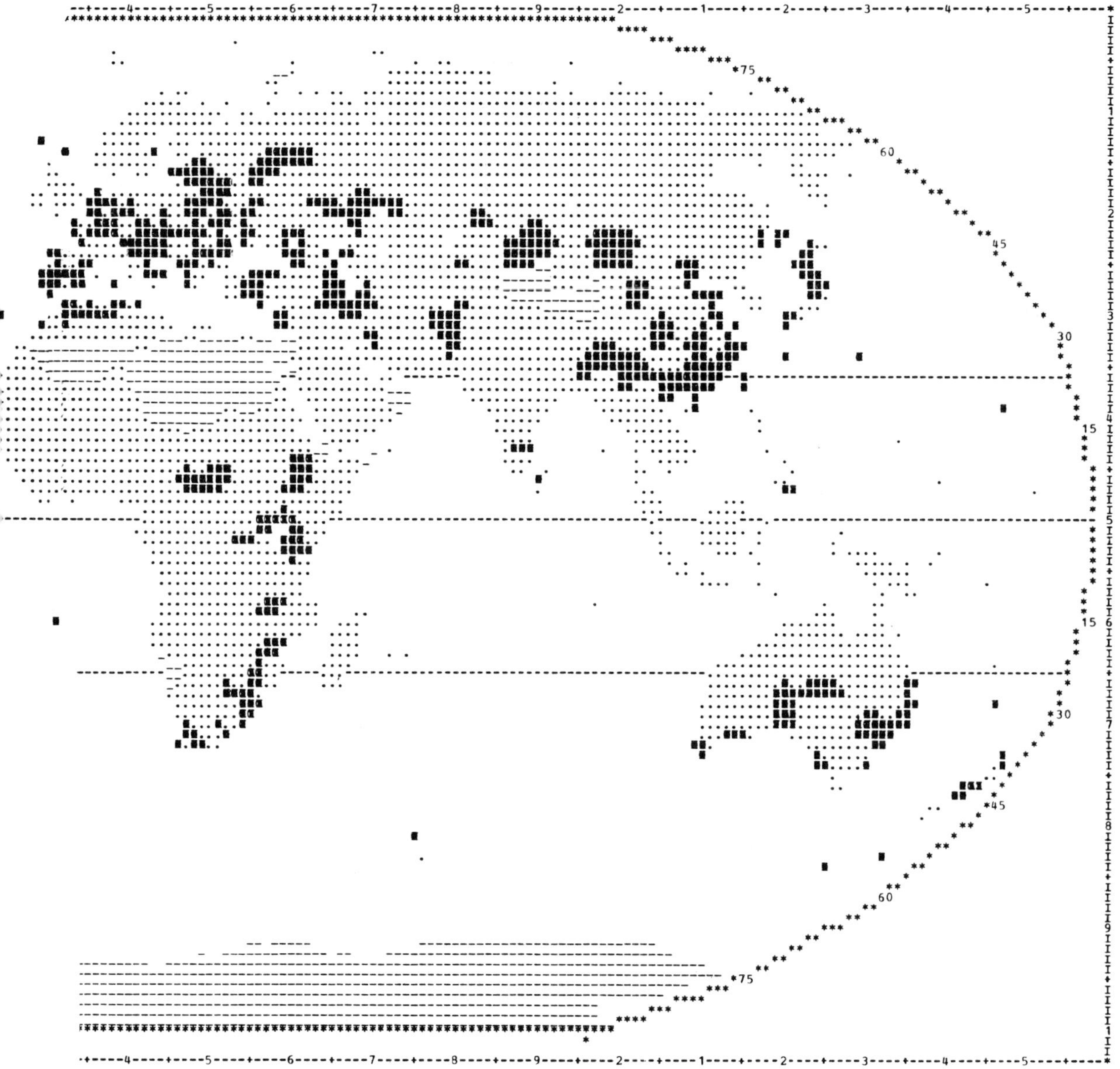

DATA VALUE EXTREMES ARE 0.0 11.00

TOTAL SUPERIMPOSED DATA POINTS IS 135. THESE OCCUR IN 60 LOCATIONS.

ABSOLUTE VALUE RANGE APPLYING TO EACH LEVEL
('MAXIMUM' INCLUDED IN HIGHEST LEVEL ONLY)

MINIMUM	BELOW	0.50	1.50	2.50	3.50	4.50	5.50	6.50	7.50	8.50	9.50	ABOVE 10.50
MAXIMUM	0.50	1.50	2.50	3.50	4.50	5.50	6.50	7.50	8.50	9.50	10.50	

PERCENTAGE OF TOTAL ABSOLUTE VALUE RANGE APPLYING TO EACH LEVEL

10.00	10.00	10.00	10.00	10.00	10.00	10.00	10.00	10.00	10.00

FREQUENCY DISTRIBUTION OF DATA POINT VALUES IN EACH LEVEL

LEVEL	L	1	2	3	4	5	6	7	8	9	10	H
SYMBOLS	----------	::::::::	::::::::	::::::::	::::::::	::::::::	::::::::	::::::::	::::::::	::::::::	::::::::	■■■■■■■■
FREQ.	68	142	16	147	209	31	89	49	25	8	204	293
FORMS	NONE	SINGLE	SINGLE	SINGLE	SINGLE	SINGLE	SINGLE	SINGLE	SINGLE	SINGLE	SINGLE	ECOT

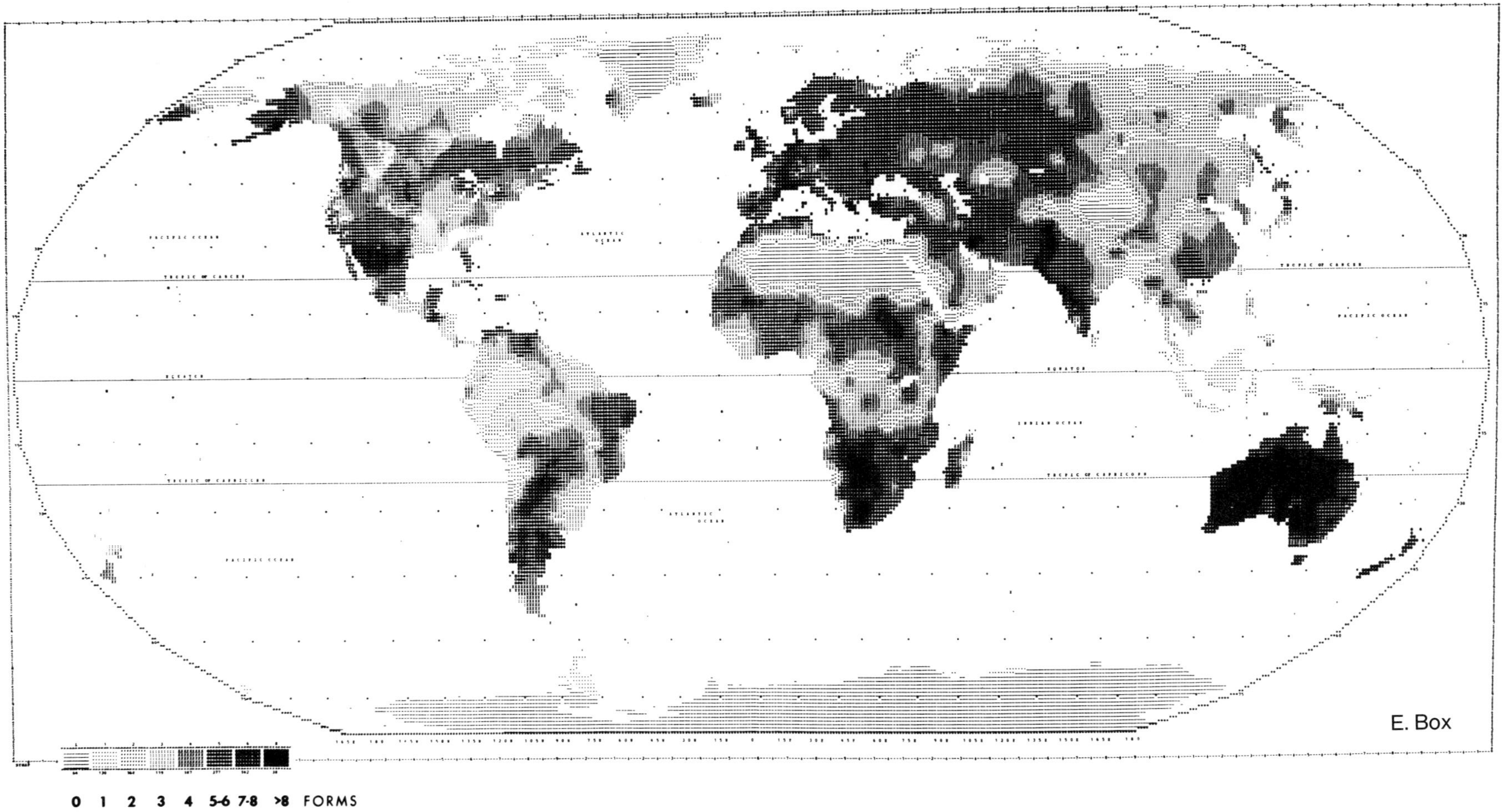

Map 25. Predicted physiognomic diversity of dominant vegetation.

and forbs occur also, but the physiognomic diversity in the tree synusia is actually higher in the boreal forest.

Maps of total (all growth forms) and higher-synusial physiognomic diversity predicted by the expanded current model are being prepared. The largest numbers of life forms (not grouped into growth forms) predicted by the current model and the corresponding sites include the following:

46 at Montevideo (Uruguay)

42 at Buenos Aires (Argentina), Jonkershoek (South Africa), and Gabo Island (Victoria)

41 at Ceres (Argentina), Rio Grande do Sul (Brazil), Pilgrim's Rest and Ladysmith (South Africa), Napoli (Italy), and Santa Cruz das Flores (Azores)

40 at Porto Alegre (Brazil), Tampa (Florida), and Southport (Queensland).

These sites are characteristically subtropical, near-coastal (or montane in South Africa), and receive at least some summer rainfall. Many of these sites are in regions which are relatively poorly predicted by the model (see validation and Appendices B–D).

Attempts to study the physiognomic diversity of vegetation, including seasonal aspects, have been very few. Unfortunately, most of these studies are not directly comparable with the results in Map 25 because they are restricted to particular plant characters, vegetation synusiae, or geographic regions. Vareschi, for example, has done considerable work on leaf shapes and sizes in the tropics and has suggested (Vareschi 1973) that the greatest diversity in tropical leaf forms is to be expected in the high-montane cloud forests. One recent diversity study which suffers less from the above-mentioned restrictions is that of Givnish (1975). By plotting life-form diversity against rainfall unpredictability for various world vegetation types, Givnish suggests (as have Whittaker, May and others) that environmental variability leads to greater structural and functional diversity. The predicted pattern of physiognomic diversity shown in Map 25 seems to be consistent with this suggestion.

H. General observations and problems

One can see immediately from the vegetation maps and frequency statistics that the predicted vegetation patterns are generally reasonable. Tree, shrub and most grass forms appear to be most accurately predicted, while problems exist with slow-growing succulent forms in particular. Natural ecotones appear to cover about 30% of the earth's land area. (Lieth, personal communication, has also suggested this figure.) The expected general patterns of dominant vegetation type (Map 23) and physiognomic diversity (Map 25) are also predicted. A particular affinity between humid tropical mountains and Southern Hemisphere maritime climates is seen in the prediction of cloud-forest trees and tropical linear-leaved trees in both regions (Appendices B and C). Projection of climatic envelopes of some forms based on single-region prototypes suggests ecophysiognomic equivalents in other parts of the world. These include the Himalayan and Rocky Mountain conifer forests as equivalents and the *Araucaria* and 'Southern pine' forests (admixtures) of southern Brazil and the southeastern U.S.A.

A number of contradictions between predicted and actual vegetation can also be identified. Most can be explained by the exclusion from the model of the appropriate limiting mechanisms.

The most obvious discrepancies seem to be the following:

1. Tropical dry evergreen woodlands (e.g. 'campos cerrados' of Brazil) are not separated from dry and wet raingreen forests.
2. Raingreen trees are predicted for large areas of Australia dominated by evergreen eucalypts.
3. Arctic tundra areas are shown as dry steppes; tropical alpine areas, on the other hand, may be shown as being too wet.
4. Arborescent and other stem-succulents do not occur in all dry environments and the model does not correctly determine in which they do occur.
5. Succulents are predicted for large parts of Australia and North Africa, but few species are native there and they are unimportant.
6. Tropical subalpine cloud forests are not distinguished from the extremely maritime evergreen rainforests of southern Chile and New Zealand.

Lack of predictive mechanisms involving length and degree of dry and favorable seasons, diurnal and interannual variation, and soil water situations can be cited as responsible for most of these discrepancies. The apparent overestimation of dryness of arctic environments and underestimation of dryness in tropical mountains, however, suggests that the Thornthwaite estimate of potential evapotranspiration is not accurate in these and other extreme regions. It has already been suggested (chapter 4) that the Thornthwaite estimate generally underestimates tropical PET and may overestimate in arctic areas, at least relative to the Holdridge (1959) estimate (see also Box, in preparation).

Prediction is complicated in some areas by non-climatic factors. The actual vegetations in the first two cases above ('campos cerrados' and northern Australia) appear to be partial reflections of low soil nutrient supplies (Rawitscher 1948; Cole 1960). A number of authors (e.g. Beard 1953; Beadle 1966; Walter 1973) have pointed out the particular edaphic problems of many savanna and open-woodland areas. Anomalous evergreenness has frequently been interpreted as an adaptation to low nutrient levels (e.g. Monk 1966; Beard 1976). Various authors (e.g. Walter 1973; Beard 1976) have also observed that, within arid regions, there do not appear to be any obvious climatic correlates for the distributions of stem-succulents. They do seem to prefer arid regions with humidity or with more than one rain period per year, which would exclude most of the Sahara-Arabian-Turanian belt.

General note on the vegetation maps

Predicted vegetation distributions and structure are represented in three ways:

1. Most important life form(s) within general structural classes (Maps 10, 12, 14–22).
2. 'Fitness' (proximity to closest climatic limit) of the locally most important life form(s) (Maps 11 and 13).
3. Dominant growth form and physiognomic ecotones (Maps 23 and 24).

Predicted formation dominants are shown only on Map 23. The term 'important' refers to 'dominance' within a given structural category.

Because of low site density the patterns appearing on all climatic and vegetation maps will be unreliable in some areas. This problem is much worse on the vegetation maps because of the nearest-neighbor mapping algorithm which must be used for typological maps. Whereas the weighting effects of the contour algorithm produce generally reasonable isolines, the nearest-neighbor algorithm distorts the shapes of nearly all life-form zones, depending on the proximities of adjacent sites with different life forms. One must look at the centers of individual life-form areas and ignore the exact computer-produced boundaries.

There are at least two other computer-generated artifacts to be kept in mind:

1. What appear to be excessively large life-form areas, including extensions too far into polar or arid regions, may be due to low site density combined with nearest-neighbor interpolation. This is the case especially in northern Canada and Alaska and in Northern Russia and northeastern Siberia.
2. Apparent conflicts in the spatial extent of corresponding life-form (Maps 10 and 12) and fitness zones (Maps 11 and 13) are due to the use of the different interpolation algorithms. The nearest-neighbor algorithm reduces the reach of adjacent absence zones and causes the life-form areas of Maps 10 and 12 to appear larger.

Symbols appearing on the vegetation maps are identified by abbreviations (see symbolism strip accompanying each map) which are explained in the legend for each map.The symbols for the fitness maps (Maps 11 and 13) represent intervals for proximity to climatic limits, as explained in the main text.

CHAPTER 7

Evaluation of model results

Formal validation has not been a traditional method for evaluating geobotanical models. World-scale models have usually been designed as correlations between vegetation patterns and environmental factors, with the emphasis much more on geographic description than on prediction and verification. Geobotanists have often relied on colleagues working or visiting in other parts of the world for evaluation of world-scale classifications, maps, and rudimentary models. In order to assess both the hypothetical basis and the implications of the model presented herein, however, a systematic means of checking the accuracy of vegetation predictions is needed. This need can be met by the methodology for formal validation of complex systems models developed recently by systems ecologists (see, for example, Caswell 1976). For predictive vegetation models this sort of rigorous validation implies direct comparison of predicted and actually occurring vegetation at particular sites not used for developing the model. These validation sites should be ecologically appropriate (free of unusual non-climatic influences but not necessarily euclimatopes), and there should be enough sites in different regions and climates to test all parts of the model. Model sensitivity to unusual conditions, including climate change, should also be studied.

A. Site comparison of predicted and actual vegetation

Where both vegetation descriptions and climate data are available, the best method for evaluating vegetation predictions is to compare predicted and actual vegetation at selected actual sites. Site comparisons permit one to compare predictions with actual vegetation observed without geographic or ecologic generalization. Such comparisons can be made within the normal framework of formal validation provided that the predicted plant-form spectra can be presented in a form which can be compared directly with actual vegetation formations. This can be done by using the cover and dominance models to interpret plant-form spectra as formation types. Comparison of actual and predicted individual synusiae (life forms) can be made by comparing the full spectra (Appendix C) with species lists (Appendix D; see also Box 1978a, Table 8).

There are certain unavoidable criteria for the choice of validation sites:

1. The actual vegetation must be natural or semi-natural in the sense that it should be free of large-scale geologic or anthropogenic physical disruptions, such as pavement or agricultural use, or maintenance in a particular state, as by mowing.
2. The sites should be free of overriding edaphic or other non-climatic complications.
3. The set of sites used for validation must represent as many climatic and vegetation regions as possible, since the ecoclimatic envelopes of the individual plant forms are independent of each other. Sites from both core areas and transitional vegetation should be included.
4. An adequate vegetation description, including species lists, must be available for each site,

either from the literature or from personal observations.

5. The necessary macroclimatic data must be available for the site.
6. The site must not be included in the data-base used for developing the model.

Personal observation is the best method for assessing actual vegetation but is not possible everywhere, since few geobotanists have seen the whole world. Finding site descriptions which include complete climatic data meant that descriptions had to be obtained from a variety of sometimes less traditional botanical sources.

In order to diminish whatever bias may exist in any particular method of vegetation description, it is perhaps better to select sites from a number of sources, such as:

1. Examples of local vegetation and associated climate from books on the vegetation of large areas. This represents the best way of covering large areas and numbers of vegetation types.
2. Detailed studies of local vegetation, in which vegetation and climate are described in detail for a specific site.
3. Sites observed personally.
4. Sites selected by the availability of illustrative material such as photographs (but with the necessary climate data).

Sites of the first type may tend to represent core areas, which are often easier to predict but which can cover large areas. The other sources need not present this bias.

About 85 sites were selected initially for the attempted validation. Half were from North America but all continents except Antarctica were represented. The sites were chosen primarily from sources 1 and 3 and were supplemented by studies of local vegetation in order to include situations not otherwise included. The sites from books were selected mainly from the volumes on the vegetation of individual large land-masses by Knapp (1965, 1973), Hueck (1966), and Walter (1968, 1973). These volumes were used because, as observed earlier, they are almost the only works covering large areas which present both sufficient physiognomic information and the required climatic data. These sources are supplemented by Küchler's (1964) map and manual of the potential vegetation of the USA and by other maps in other areas (e.g. UNESCO 1969; Emberger 1938; Küchler 1977). Most sites involve at least two sources of vegetation data. More specific site descriptions were always sought but could not always be found. The initial set of sites was culled to remove sites which involve major non-climatic problems or for which sufficient information on actual vegetation could not be found. The remaining data-base of 65 sites, of which the author had observed 23 personally, was used for validation of the initial model (Box 1978a). These sites remain quite instructive for validation since development of the current model involved primarily addition of new forms and only widening of climatic limits and since the original 65 sites were deliberately set aside and not consulted until the current model was completed. The only reason for retaining the original 65 sites was the time required for obtaining so many new site descriptions. Nine new sites were added, however, in order to improve representation of various regions, climate types and known types of actual vegetation. The nine new sites were obtained primarily from recent journal papers (mainly *Vegetatio*) and are all in the Eastern Hemisphere.[1]

The final data-base for validation consists of 74 sites, of which 30 are in North America and 23 are in mountainous areas (excluding high plateaus). Seven pairs of neighboring sites within the same general climate type are included (all in the Western Hemisphere) in order to illustrate effects of local variation. Other distributional aspects of the validation sites are presented in Table 15.

Validation data for each site include a list of the most important species occurring at and near each site and a formation name provided by the data source but modified if necessary to conform to the nomenclature used herein (Table 8). The names of actual formations contain the names of dominant genera where possible. Of course it is easier to generalize for the less well-known sites. For a few sites where the natural vegetation has been largely destroyed (e.g. São Paolo, Nanyuki) one must rely on the description of potential vegetation. The vegetation descriptions for most sites not observed

[1]Two of these sites were obtained at the last minute from the 'Ecosystems of the World' volume on heathlands by Specht (1979), which had just appeared. This and the other volumes of this series promise to be excellent world-scale sources of both descriptive and quantitative vegetation data.

Table 14. Predicted and actual vegetation at 74 validation sites.

Probably the best way to evaluate a predictive vegetation model is to compare corresponding predicted and actual vegetation at a geographically representative set of validation sites not used in developing the model. This was done using 74 sites (from all continents except Antarctica) chosen primarily by availability of the necessary paired site or regional vegetation descriptions (with species lists) and climatic data.

The sites are listed vertically in the left column, along with general location and annual moisture index (MI), which provides the basis for naming predicted formations. (Geographic coordinates and site climatic values, as well as the complete spectra of predicted forms, are given in Appendix C).

Predicted formation type and important constituent forms are shown in the center column. Some relatively similar forms are grouped into a single line in order to conserve space. Predicted dominant forms are indicated by an asterisk in closed formations (MI > 0.9) or a plus in open formations (MI < 0.9). The other forms are listed generally in order of decreasing dominance and, within dominance levels, by increasing proximity to environmental limits. The predicted life-form spectra are interpreted and the corresponding formations named according to criteria described in Tables 8 and 9 (sections 5.C–5.F).

Juxtaposed in the right column is the corresponding actual vegetation, as determined from descriptions and species lists identified and summarized more completely in Appendix D. The actual vegetation shown consists of a formation name (generally with dominant species or genera) and a short list of the most important actually occurring taxa.

Sites at which the actual dominant forms and only they are predicted to occur and to dominate (i.e. the exact combination of the dominant forms, the 'strong criterion') are indicated in the left column by two asterisks before the site name. Sites at which the actual dominant forms are all predicted to occur (but at which the other conditions for the strong criterion are not met) are indicated by a single asterisk before the site name. A summary of prediction accuracy for the validation sites is presented in Table 15.

Site	Predicted vegetation	Actual vegetation
**Fairbanks (interior Alaska, 134 m) MI = 0.74	Boreal Semi-Open Needle Forest with summergreens: *Boreal/montane short-needled trees Boreal summergreen needle-trees Broad-summergreen shrubs Needle-leaved evergreen shrubs Short bunch and sward-grasses Summergreen forbs	Boreal *Picea-Betula* forest, with *Larix, Populus, Salix, Arctostaphylos, Empetrum, Vaccinium,* and various tundra herbs.
**Vancouver (British Columbia) MI = 2.25	Diverse Needle-Leaved Rainforest with summergreens: *Boreal/montane short-needled trees *Temperate needle-trees *Sub-mediterranean needle-trees *Temperate rainforest needle-trees *Summergreen broad-leaved trees Boreal summergreen needle-trees Boreal broad-summergreen trees	*Thuja-Tsuga-Pseudotsuga* Rainforest with *Pinus, Abies, Populus, Acer, Prunus, Arbutus, Berberis, Rhododendron, Vaccinium, Rubus,* forbs, ferns, etc.
**Banff (Alberta, 1397 m) MI = 0.98	Boreal/Montane Needle Forest with summergreens: *Boreal/montane short-needled trees Boreal summergreen needle-trees Broad-summergreen small trees Needle-leaved evergreen shrubs Summergreen shrubs and tundra dwarf-shrubs Tall and short grasses Summergreen forbs	Boreal-Montane transitional *Picea-Abies* forest, with *Pinus, Larix, Betula* and understorey shrubs and herbs.

Table 14 (continued).

Site	Predicted vegetation	Actual vegetation
*Saskatoon (Saskatchewan) MI = 0.70	Short-Grass Prairie with scattered shrubs and needle-trees: +Short bunch and sward-grasses Boreal summergreen needle-trees Boreal/montane short-needled trees Broad-summergreen small trees Needle-leaved evergreen shrubs Xeric summergreen shrubs Leafless shrubs and dwarf-shrubs Summergreen forbs	Northern Grove Belt and *Festuca* Prairie: *Festuca, Koeleria, Danthonia,* et al. with *Populus tremuloides, Rosa arkansasa,* and prairie forbs.
*Swift Current (Saskatchewan) MI = 0.71	Short-Grass Prairie with scattered needle-trees and shrubs: +Short bunch and sward-grasses Boreal summergreen needle-trees Boreal/montane short-needled trees Broad-summergreen small trees Tall grasses Needle-leaved evergreen shrubs Xeric summergreen shrubs Cold-winter and leafless xeric shrubs Summergreen forbs	Northern limit of Northern Short and Mixed-Grass Prairie, with *Bouteloua, Stipa, Buchloë* et al. plus *Atriplex, Artemisia, Opuntia* and prairie forbs.
**Sioux Lookout (Ontario) MI = 1.44	Boreal Forest with summergreens: *Boreal/montane short-needled trees Summergreen broad-leaved trees Boreal summergreen needle-trees Boreal broad-summergreen trees	Central Boreal Forest: *Picea-Abies,* with *Betula, Populus, Larix, Pinus, Vaccinium,* evergreen and summergreen herbs.
**Lac Albanel (Québec) MI = 1.60	Boreal Forest with summergreens: *Boreal/montane short-needled trees Summergreen broad-leaved trees Boreal summergreen needle-trees Boreal broad-summergreen trees	Eastern Boreal Forest: *Picea-Abies,* with *Larix, Pinus, Thuja, Betula, Populus, Vaccinium,* evergreen and summergreen herbs.
*Olympia (Washington, on Puget Sound) MI = 2.05	Temperate Mesic Needle-Forest with Broad-Evergreen Understorey: *Temperate needle-trees *Sub-mediterranean needle-trees Temp. broad-evergreen small trees Tall and short graminoids Various forbs, ferns, etc.	Tall, dense *Pseudotsuga* rainforest with *Tsuga, Thuja, Pinus, Abies, Arbutus, Acer, Berberis, Prunus, Rhododendron,* evergreen and summergreen herbs.
**Davis (central valley of California) MI = 0.49	Short Bunch-Grass Steppe with various forbs and mostly smaller shrubs: +Short bunch-grasses Bush stem-succulents Leafless and mediterranean dwarf-shrubs Summergreen forbs Xeric cushion-shrubs and herbs Succulent forbs Needle-leaved, hot-desert and mediterranean evergreen shrubs	California Bunch-Grass Prairie with many forbs: *Stipa, Aristida, Festuca,* etc. with spring ephemerals, prairie forbs, *Salvia, Artemisia,* and scattered *Opuntia.*

Table 14 (continued).

Site	Predicted vegetation	Actual vegetation
*Tanbark Flat (southern California, 863 m in San Gabriel Mtns.) MI = 0.92	Mediterranean Evergreen Shrubland with trees: *Mediterranean evergreen shrubs Needle-leaved evergreen shrubs Sub-mediterranean needle-trees Mediterranean broad-evergreen trees Dwarf-needle small trees Tropical broad-evergreen shrubs Mediterranean dwarf-shrubs Xeric evergreen tuft-treelets and scrub Tall and short grasses Bush stem-succulents Summergreen forbs	*Adenostoma* (needle-leaved) Chaparral with *Arctostaphylos, Ceanothus, Rhamnus, Quercus* scrub, *Rhus, Salvia,* grasses and forbs.
**West Yellowstone (Montana - NW Wyoming, 2035 m) MI = 1.29	Montane Needle Forest with summergreens: *Boreal/montane short-needled trees Boreal summergreen needle-trees Boreal broad-summergreen trees	Montane *Picea-Abies* forest with *Pseudotsuga, Pinus, Larix, Populus, Salix, Vaccinium, Cornus,* forbs etc.
*Logan (northern Utah, 1455 m) MI = 0.67	Mixed Semi-Deciduous Montane Shrubland: +Short bunch-grasses +Needle-leaved evergreen shrubs +Cold-winter xeromorphic shrubs +Dwarf-needle small trees +Xeric summergreen shrubs Summergreen forbs Leafless scrub	Semi-deciduous montane 'deciduous chaparral': *Cercocarpus, Quercus, Arctostaphylos, Ceanothus, Purshia, Amelanchier, Acer, Rhus,* with grasses and forbs.
**Cody (northern Wyoming, 1519 m) MI = 0.44	Short Bunch-Grass Steppe with numerous xeric shrubs: +Short bunch-grasses +Xeric summergreen shrubs +Cold-winter xeromorphic shrubs Leafless xeric dwarf-shrubs Summergreen forbs Xeric cushion-herbs Desert-grasses Ephemeral desert herbs	*Artemisia*-Short Grass Steppe, with *Agropyron, Festuca, Stipa* et al., *Purshia, Lupinus, Atriplex, Opuntia.*
*Flagstaff (northern Arizona, 2070 m) MI = 1.04	Temperate Montane Needle-Forest with summergreens: *Boreal/montane short-needled trees *Temperate needle-trees Summergreen broad-leaved trees Boreal summergreen needle-trees Boreal broad-summergreen trees	Southern Rocky Mtn. *Pinus ponderosa* forest, with other *Pinus* spp., *Populus, Ceanothus* and *Quercus* scrub, bunch-grasses and steppe forbs.
*Phoenix (southern Arizona, 339 m) MI = 0.18	Hot Semi-Desert with Stem-Succulents: +Hot-desert evergreen shrubs +Arborescent stem-succulents Xeric rosette-shrubs Other stem-succulents Succulent forbs Leafless xeric shrubs Desert-grasses Ephemeral desert herbs	Colorado Valley *Larrea-Franseria* (90–95%) Semi-Desert, with scattered desert grasses, stem-succulents, dwarf-shrubs and abundant summer and winter ephemerals. (Larger stem-succulents restricted now to surrounding mountains).

Table 14 (continued).

Site	Predicted vegetation	Actual vegetation
**Tucson (southern Arizona, 739 m) MI = 0.29	Hot Semi-Desert with Mixed Shrubs and Stem-Succulents: +Raingreen thorn-scrub +Xeric summergreen shrubs +Evergreen arborescents +Hot-desert evergreen shrubs +Arborescent stem-succulents Xeric rosette-shrubs Typical and bush stem-succulents Mediterranean and leafless dwarf-shrubs Xeric cushion-shrubs Xeric evergreen tuft-treelets Short bunch and desert-grasses Summergreen, raingreen and succulent forbs Ephemeral desert herbs	Upland *Larrea-Cercidium-Opuntia* Semi-Desert (diverse) with tall cacti, plus *Prosopis, Olneya, Franseria, Fouquiera, Carnegiea, Echinocereus, Ephedra,* and abundant desert grasses, summer ephemerals and fewer winter ephemerals.
*Santa Fe (northern New Mexico, 1929 m) MI = 0.60	Mixed Xeric Shrub-Steppe with Needle-Trees: +Short bunch and sward-grasses +Needle-leaved evergreen shrubs +Summergreen scrub and xeric shrubs Cold-winter xeromorphic shrubs Leafless xeric shrubs +Boreal summergreen needle-trees Dwarf-needle small trees Summergreen forbs	Open Juniper-Piñon Woodland with small shrubs, turf and bunch grasses but fewer forbs: *Juniperus, Pinus, Cercocarpus, Ceanothus, Quercus scrub, Artemisia, Chrysothamnus, Agropyron, Bouteloua, Buchloë.*
*Colorado Springs (east base of Rocky Mtns., 1855 m) MI = 0.64	Mixed Xeric Shrub-Steppe with Needle Trees: +Short bunch-grasses +Needle-leaved evergreen shrubs +Summergreen scrub and xeric shrubs Leafless xeric shrubs +Short sward-grasses Boreal/montane short-needled trees +Boreal summergreen needle-trees Dwarf-needle small trees Summergreen forbs	Southern Short and Mixed-Grass Prairie with *Bouteloua, Muhlenbergia, Aristida, Stipa, Andropogon* et al. plus scattered *Juniperus, Yucca glauca, Opuntia* and various steppe forbs.
**Custer (Black Hills/ S. Dakota, 1622 m) MI = 0.94	Open, Grassy Needle-Tree Woodland: *Boreal/montane short-needled trees Boreal summergreen needle-trees Broad-summergreen small trees Needle-leaved evergreen shrubs Broad-summergreen shrubs +Short bunch and sward-grasses Tall grasses Summergreen forbs	Black Hills Pine Forest (locally open), with *Prunus, Arctostaphylos, Juniperus, Symphoricarpus, Artemisia, Cercocarpus, Rhus trilobata, Elymus, Poa, Agropyron, Koeleria, Bouteloua* and various forbs.
*Oelrichs (S. Dakota, south of Black Hills, 1017 m) MI = 0.74	Short Mixed-Grass Steppe with Shrubs: +Short bunch and sward-grasses Summergreen forbs +Needle-leaved evergreen shrubs +Summergreen scrub and xeric shrubs Temperate needle-trees Leafless xeric shrubs	Short Mixed-Grass Steppe: *Agropyron-Stipa* with *Koeleria, Buchloë, Bouteloua* et al. plus many steppe forbs and scattered small shrubs.

Table 14 (continued).

Site	Predicted vegetation	Actual vegetation
*Oklahoma City (central Oklahoma) MI = 0.93	Summergreen Grassy Open Woodland: +Summergreen broad-leaved trees and small trees Sub-mediterranean needle-trees +Summergreen scrub and shrubs Needle-leaved evergreen shrubs Dwarf-needle small trees Temp. broad-evergreen small trees Temperate broad-evergreen shrubs Bush stem-succulents +Tall grasses +Short bunch and sward-grasses Summergreen forbs	Southern Tall-Grass Prairie: red-clay *Stipa-Koeleria* association with scattered *Quercus* and *Carya* spp., *Andropogon, Bouteloua, Agropyron* et al., numerous steppe shrubs and semi-shrubs (e.g. *Amorpha*) and numerous prairie forbs.
**Bartlesville (northeastern Oklahoma, 218 m) MI = 1.08	Dry Summergreen Forest (semi-open): *Summergreen broad-leaved trees Broad-summergreen small trees Dwarf-needle small trees Broad-evergreen and summergreen shrubs Tall grasses (incl. cane-grasses) Short sward-grasses Summergreen forbs	*Quercus* Woodland - Tall Grass Mosaic on fine shale, with closed oak coves: *Quercus marilandica* and *Q. stellata, Carya, Cercis, Celtis, Prunus; Andropogon, Bouteloua, Elymus* et al., plus numerous prairie forbs.
**South Bend (northern Indiana, 221 m) MI = 1.42	Summergreen Broad-Leaved Forest: *Summergreen broad-leaved trees Broad-summergreen small trees (incl. boreal) Temperate needle-trees Tall grasses Summergreen forbs	Summergreen Oak-Hickory Forest (western limit): *Quercus, Carya, Acer rubrum, Fraxinus, Tilia, Ulmus, Prunus, Cercis,* with many grasses and forbs.
**Columbus (Ohio, 243 m) MI = 1.43	Summergreen Broad-Leaved Forest: *Summergreen broad-leaved trees Broad-summergreen small trees (incl. boreal) Temperate needle-trees Tall grasses Summergreen forbs	Summergreen Beech-Maple Forest: *Fagus, Acer saccharum, Fraxinus, Quercus, Carya, Betula, Prunus, Aesculus* et al., with *Lonicera canadensis* and many forbs.
**Bar Harbor (Acadian coast of Maine) MI = 2.20	Sub-Boreal Mixed Semi-Deciduous Forest: *Temperate needle-trees *Summergreen broad-leaved trees *Boreal/montane short-needled trees Boreal broad-summergreen trees Summergreen and evergreen shrubs Tall grasses	Spruce-Northern Hardwood Forest on glacial till and sand: *Picea, Abies, Pinus strobus, Thuja, Tsuga, Betula, Populus, Acer, Quercus, Fagus,* with *Vaccinium,* other ericaceous and summergreen shrubs, grasses, sedges and forbs.
**Trenton (New Jersey, on Delaware River) MI = 1.46	Summergreen Broad-Leaved Forest: *Summergreen broad-leaved trees Temperate needle-trees Broad-summergreen small trees Boreal broad-summergreen trees Tall grasses	Appalachian Summergreen *Quercus-Castanea* Forest with *Liriodendron, Acer, Fagus* et al., plus *Pinus, Betula, Tsuga, Prunus, Cornus, Vaccinium, Kalmia,* summergreen shrubs, grasses and forbs.

Table 14 (continued).

Site	Predicted vegetation	Actual vegetation
**Mt. Mitchell (western N. Carolina, 2 miles ssw of summit, 1989 m) MI = 3.56	Montane Needle-Forest with summergreens: *Boreal/montane short-needled trees Summergreen broad-leaved trees Broad-summergreen small trees Boreal broad-summergreen trees Evergreen and summergreen mesic shrubs Tall and short grasses Summergreen and evergreen forbs	Appalachian Fir-Spruce Forest: *Abies fraseri, Picea rubens*, with *Sorbus, Prunus, Betula, Acer, Vaccinium, Rhododendron, Rubus* and mostly northern forbs and graminoids.
**Charleston (S. Carolina coast) MI = 1.20	Warm-Temperate Mixed Semi-Summergreen Forest: *Warm-temperate broad-evergreen trees *Summergreen broad-leaved trees *Heliophilic long-needled trees Temperate broad-rainforest trees Tropical sclerophyll and microphyll trees Palmiform tuft-treelets Broad-evergreen small trees Broad-summergreen small trees Swamp summergreen needle-trees Arborescent grasses Tall and short grasses Evergreen and summergreen forbs and vines Narrow-leaved and wintergreen epiphytes	Lowland Warm-Temperate Mixed Forest: *Quercus virginiana, Q. laurifolia, Q. alba, Q. phellos, Pinus taeda, Liquidambar, Nyssa, Magnolia* with *Ilex, Persea, Cornus, Sabal palmetto, Bumelia, Vaccinium, Myrica, Baccharis, Arundinaria, Tillandsia*, many vines, grasses, forbs, and *Taxodium* in swamps.
**Savannah (Georgia, 25 km from coast) MI = 1.20	Warm-Temperate Mixed Semi-Summergreen Forest: *Warm-temperate broad-evergreen trees *Summergreen broad-leaved trees *Heliophilic long-needled trees Tropical sclerophyll and microphyll trees Palmiform tuft-treelets Broad-evergreen small trees Broad-summergreen small trees Swamp summergreen needle-trees Arborescent grasses Tall and short grasses Evergreen and summergreen forbs and vines Narrow-leaved and wintergreen epiphytes	Coastal-Plain Pine-Hardwood Forest: *Quercus virginiana, Q. phellos, Pinus, Liquidambar, Nyssa, Magnolia* with *Persea, Ilex, Cornus, Prunus, Sabal palmetto, Vaccinium, Myrica, Arundinaria, Tillandsia*, many vines, grasses, forbs, and *Taxodium* swamps.
*Birmingham (north-central Alabama, 182 m) MI = 1.57	Warm-Temperate Mixed Semi-Summergreen Forest: *Heliophilic long-needled trees *Warm-temperate broad-evergreen trees *Summergreen broad-leaved trees *Temperate broad-rainforest trees Tropical evergreen microphyll-trees Temp. broad-evergreen small trees Broad-summergreen small trees Swamp summergreen needle-trees Arborescent grasses Tall and short grasses Evergreen and summergreen forbs and vines Wintergreen epiphytes	Southern Pine-Summergreen Oak Forest: *Quercus* (summergreen, with evergreen), *Pinus, Carya, Nyssa* et al., with *Cornus, Cercis, Prunus, Vaccinium, Taxodium* along rivers, and many shrubs, grasses, forbs and vines.

Table 14 (continued).

Site	Predicted vegetation	Actual vegetation
**Xilitla (Mexico, 1035 m in Sierra Madre Oriental) MI = 2.58	Subtropical Montane Semi-Summergreen Forest with needle-trees: *Summergreen broad-leaved trees *Warm-temperate broad-evergreen trees *Tropical linear-leaved trees Tropical montane rainforest trees Tropical evergreen microphyll-trees Tropical broad-evergreen small trees Swamp summergreen needle-trees Palmiform tuft-trees Broad-summergreen small trees Broad-evergreen lianas and vines Arborescent grasses Tall and short grasses Evergreen and raingreen forbs Broad and narrow-leaved epipytes	Tropical Montane Mixed Semi-Summergreen Rainforest with Holoarctic Species: *Quercus, Carpinus, Carya, Liquidambar, Pinus, Cornus, Podocarpus,* with *Turpinia, Clethra, Vaccinium, Eugenia, Garrya, Ilex, Rhus, Parthenocissus,* and many holarctic herbs, including ferns.
**Mazatlan (Mexico, Pacific coast at Tropic of Cancer) MI = 0.70	Dry Raingreen Woodland and Dense Scrub: *Raingreen thorn-scrub *Tropical evergreen sclerophyll trees *Xeric raingreen trees *Broad-raingreen small trees Xeric evergreen tuft-treelets Bush stem-succulents Evergreen arborescents Leaf-succulent evergreen shrubs Short bunch-grasses Raingreen forbs and vines	Lowland Dry Raingreen Thorn Forest: *Ipomoea arborescens, Acacia, Zizyphus, Bauhinia, Ceiba, Prosopis, Caesalpinia, Jatropha, Cordia, Mimosa,* and many others, plus *Cassia, Croton, Fouquiera, Cereus, Pachycereus,* and many grasses, forbs and vines.
**Quibdo (Colombia, west base of Andes) MI = 6.10	Wet, Diverse Tropical Rainforest: *Tropical rainforest trees Palmiform tuft-trees and understorey Tropical evergreen microphyll-trees Tropical broad-evergreen small trees Tropical evergreen lianas and vines Broad and narrow-leaved epiphytes Evergreen forbs and ferns	Equatorial Coastal Rainforest: tall, diverse, multistorey forest with *Terminalia, Brosimum, Ficus, Hura, Hevea, Cedrela, Tabebuia* and countless others, plus riverine palm thickets and swamps, and numerous lianas, epiphytes, ferns, and smaller trees and arborescent shrubs.
*Barinas (Venezuela, east base of Cordillera de Mérida) MI = 1.14	Tropical Semi-Deciduous Forest: *Tropical evergreen sclerophyll trees Palmiform tuft-trees and treelets *Tropical evergreen microphyll-trees *Monsoon broad-raingreen trees *Tropical rainforest trees Raingreen and evergreen understorey trees, forbs, and vines	Tradewind Raingreen Forest with scattered evergreens: *Spondias, Pterocarpus, Bombacopsis, Sapium, Hura, Brosimum, Terminalia, Swietenia* et al., with understorey palms and arborescent shrubs, many grasses, forbs, and epiphytes (*Ficus, Phoradendron,* etc.).
**Obidos (Brazil, Amazon Basin) MI = 1.12	Tropical Semi-Deciduous Dry Rainforest: Tropical evergreen sclerophyll trees Palmiform tuft-trees Tropical evergreen microphyll-trees *Monsoon broad-raingreen trees *Tropical rainforest trees Raingreen and evergreen understorey trees, forbs, and vines	Dry Tropical Rainforest with Campos Cobertos (islands of drier open forest): *Curatella, Anacardium, Bowdichia, Vochysia, Byrsonima* et al., with *Cassia, Eugenia, Crotalaria,* tall and short grasses, and many other species.

Table 14 (continued).

Site	Predicted vegetation	Actual vegetation
**Quixeramobim (Brazil, north- western Caatingas) MI = 0.43	Raingreen Thorn-Woodland: *Raingreen thorn-scrub *Arborescent stem-succulents *Broad-raingreen small trees and occasional larger trees Xeric evergreen tuft-treelets Xeric rosette-shrubs Typical and bush stem-succulents Evergreen arborescents Short and desert grasses Raingreen forbs	True Caatinga (Raingreen Thorn-Woodland with palms and tall stem-succulents): *Zizyphus, Mimosa, Cassia, Caesalpinia, Acacia, Cavanillesia* et al., with *Cocos, Copernicia,* ground bromeliads, *Cereus, Opuntia, Capparis, Euphorbia, Jatropha,* short grasses but few forbs.
*Belo horizonte (Brazil, 200 km. north of Rio de Janeiro, 857 m.) MI = 1.72	Diverse Tropical Upland Semi-Evergreen Forest: *Tropical linear-leaved trees *Montane broad-raingreen trees *Tropical montane rainforest trees Tropical evergreen microphyll trees Tropical broad-evergreen small and dwarf trees Tropical evergreen sclerophyll trees Broad-raingreen small trees Palmiform tuft-trees Tropical broad-evergreen lianas Arborescent grasses Tall and short grasses Evergreen and raingreen forbs and vines Wintergreen epiphytes	Campos Cerrados-Mixed Forest Transition: dwarf evergreen woodland (*Kielmeyera, Byrsonima, Machaerium* et al.) with dense grass cover, small palms, arborescent shrubs, and numerous forbs, plus mixed semi-raingreen forest mainly in valleys.
**São Paolo (southern Brazil, 740 m) MI = 1.63	Diverse Subtropical Upland Semi-Deciduous Forest: *Montane broad-raingreen trees *Tropical linear-leaved trees *Temperate needle-trees *Tropical montane rainforest trees *Temperate rainforest needle-trees *Warm-temperate broad-evergreen trees Heliophilic long-needled trees Tropical evergreen microphyll trees Tropical evergreen sclerophyll trees Broad-evergreen small trees Broad-raingreen small trees Palmiform tuft-trees Tropical broad-evergreen lianas Arborescent grasses Tall and short grasses Evergreen, raingreen and summergreen forbs Broad and narrow-leaved epiphytes Evergreen and raingreen vines	Subtropical Species-Rich Semi-Deciduous Forest: *Cedrela, Balfourodendron, Hymenaea, Centrolobium, Machaerium, Piptadenia,* and many others, with numerous palms, dense underbrush, and numerous lianas, vines and epiphytes.
*Villa Nougues (Tucumán province of northwestern Argentina, 1388 m) MI = 2.22	Subtropical Montane Mixed Semi-Deciduous Forest: *Tropical montane rainforest trees *Temperate rainforest needle-trees *Montane broad-raingreen trees *Tropical linear-leaved trees *Summergreen broad-leaved trees Broad-raingreen small trees	Southern Andes Subalpine Alder Belt (near lower limit): *Alnus jorullensis, Juglans, Podocarpus, Prunus, Sambucus, Ilex, Berberis,* with mainly northern shrubs and forbs.

Table 14 (continued).

Site	Predicted vegetation	Actual vegetation
	Tropical microphyll and sclerophyll trees Summergreen and evergreen small trees Temperate needle-trees Arborescent, tall and short grasses Evergreen and deciduous forbs and vines	
Isla Victoria (Lago Nahuel Huapi, 900 m in southern Andes, 41°S) MI = 3.22	Temperate Needle-Leaved Rainforest with summergreens: *Temperate rainforest needle-trees Broad-summergreen small trees Temp. broad-evergreen trees Boreal broad-summergreen trees Evergreen and summergreen shrubs Tall and short grasses Evergreen and summergreen forbs	Patagonian Lower-Montane *Nothofagus* Rainforest, with *Eucryphia, Drimys, Libocedrus, Podocarpus, Saxegothea,* understorey *Weinmannia, Laurelia,* and many ferns, epiphytes, and often bamboos.
*Bariloche (Argentina, 825 m, near Lago Nahuel Huapi) MI = 2.16	Temperate Montane Needle-Forest: *Boreal/montane short-needled trees Temperate rainforest needle-trees Sub-mediterranean needle-trees Evergreen and summergreen small trees Summergreen scrub and shrubs Evergreen dwarf-shrubs Tall cane-graminoids Short sward and bunch-grasses Evergreen, summergreen and raingreen forbs	Patagonian Lower-Montane *Libocedrus* Forest: *L. chilensis* (to 90%), with *Lomatia, Diostea, Schinus* and scattered *Araucaria,* plus *Berberis, Azara,* and steppe dwarf-shrubs and grasses.
**Valdivia (southern Chile, 40° S near coast) MI = 4.18	Diverse Temperate Mixed Rainforest: *Temperate rainforest needle-trees *Temperate broad-rainforest trees *Tropical montane rainforest trees *Tropical linear-leaved trees Tropical evergreen microphyll-trees Broad-evergreen small trees Arborescent grasses Tall and short grasses Evergreen forbs, vines, ferns and epiphytes	Temperate Species-Rich Broad-Macrophyll Rainforest (near transition to summergreen forest): *Aextoxicon, Eucryphia, Laurelia, Drimys* et al., with *Lomatia, Nothofagus, Fitzroya, Weinmannia, Cariaria,* plus numerous lianas, epiphytes, bamboos, forbs and mosses.
Ushuaia (mainland of Tierra del Fuego) MI = 1.23	Treeline Krummholz-Tundra Transition: *Tall turf and tussock-grasses *Summergreen tundra dwarf-shrubs *Temperate evergreen dwarf-shrubs Needle-leaved shrubs and krummholz Short bunch and sward grasses Summergreen and evergreen forbs Mat-forming thallophytes	Subantarctic Summergreen *Nothofagus* Forest (near southern limit), with scattered evergreen *Nothofagus,* plus *Berberis, Ribes, Pernettya, Escallonia, Chusquea* and other graminoids, and various, mostly summergreen forbs.
*Thingvellir (Iceland, 103 m) MI = 2.97	Herb-Dwarf Shrub Tundra with Krummholz and scattered trees: *Short sward and bunch-grasses Needle-leaved treeline krummholz Boreal/montane short-needled trees Broad-summergreen small trees Evergreen and summergreen dwarf-shrubs and forbs Summergreen ferns Thallophytes	Iceland Maritime 'Mo' Tundra (herbaceous with dwarf-shrubs and mosses): *Festuca, Carex, Agrostis, Juncus, Poa* et al., with *Salix, Empetrum, Calluna, Arctostaphylos,* various forbs and many mosses and lichens.

Table 14 (continued).

Site	Predicted vegetation	Actual vegetation
**Abisko (northern Sweden, 388 m) MI = 0.74	Mixed Semi-Summergreen Forest-Tundra: Broad-summergreen small trees +Summergreen tundra dwarf-shrubs +Short bunch and sward-grasses Needle-leaved shrubs and krummholz Summergreen forbs and xeric herbs	Summergreen Fell-Birch Forest-Tundra: *Betula tortuosa, B. callosa,* with *Populus, Pinus* on moors, *B. nana, Empetrum, Vaccinium, Arctostaphylos, Loiseluria,* plus grasses, forbs, mosses and lichens.
*Lund (Dalby Söderskog) (southern Sweden) MI = 1.12	Sub-Boreal Mixed Semi-Deciduous Forest: *Boreal/montane short-needled trees *Temperate needle-trees *Summergreen broad-leaved trees Boreal summergreen needle-trees Boreal broad-summergreen trees Evergreen and summergreen shrubs and herbs	Summergreen *Ulmus-Quercus* Forest (undisturbed since 1916), with *Fagus, Fraxinus, Corylus, Crataegus, Prunus, Acer, Lonicera,* and various grasses and forest forbs.
*Mikulov/Weinviertel (northeastern Austria) MI = 0.95	Open Needle Tree-Mixed Shrub Mesic Woodland: +Temperate needle-trees +Boreal/montane short-needled trees +Boreal summergreen needle-trees +Broad-summergreen small trees Dwarf-needle small trees Broad-summergreen mesic shrubs Xeric summergreen shrubs Needle-leaved evergreen shrubs Temperate broad-evergreen shrubs Tall grasses Short bunch and sward grasses Summergreen forbs	Summergreen Woodland and Scrub with dry grassland mosaic: *Quercus, Carpinus, Pinus, Betula, Robinia* with *Festuca, Agropyron, Bromus, Poa, Convolvulus, Salvia, Artemisia* and various steppe forbs.
**Ain-Draham (coastal mountains of northern Tunisia, 739 m) MI = 2.01	Mediterranean Semi-Deciduous Montane Forest: *Summergreen broad-leaved trees *Mediterranean broad-evergreen trees Sub-mediterranean needle-trees Dwarf-needle small trees Broad-summergreen small trees and scrub Broad-evergreen and summergreen shrubs Tall graminoids Short sward and bunch-grasses Summergreen forbs and ferns Wintergreen epiphytes	Mediterranean Semi-Deciduous Montane Oak Forest: *Quercus faginea* ssp. *baetica* (SG), *Qu. suber* (EG), with *Sorbus, Prunus, Acer, Ilex, Crataegus, Erica arborea, Cytisus, Rubus,* and various grasses, mediterranean forbs, and epiphytic ferns and mosses.
**Azrou (Middle Atlas of Morocco, 1250 m) MI = 1.11	Mediterranean Evergreen Montane Forest: *Mediterranean broad-evergreen trees Sub-mediterranean needle-trees Temp. broad-evergreen small trees Tropical evergreen microphyll-trees Mediterranean shrubs and dwarf-shrubs Tall and short grasses Summergreen, raingreen and succulent forbs Wintergreen epiphytes	Mediterranean Montane Evergreen Oak Forest (near transition to cedars): *Quercus ilex* with *Sarothamnus, Cytisus, Crataegus, Cistus, Thymus, Rosa, Buxus, Pistacia,* and many vines (*Lonicera, Smilax,* et al.) grasses, dwarf and semi-shrubs, and forbs.

Table 14 (continued).

Site	Predicted vegetation	Actual vegetation
**Ksar-es-Souq (Saharan side of Middle Atlas, 1060 m) MI = 0.13	Hot-Desert Shrub Steppe: +Hot-desert evergreen shrubs Leafless scrub and dwarf-shrubs Typical stem-succulents Bush stem-succulents Xeric rosette-shrubs Succulent forbs Ephemeral desert herbs Desert-grasses	Sub-Saharan Plateau Shrub-Bunch Grass Steppe: *Artemisia, Anabasis, Ephedra, Stipa tenacissima, Aristida,* with other xeric shrubs and many, often yellow-flowered forbs.
**Damanhur (Egypt, area between Cairo and Alexandria) MI = 0.09	Sparse Hot-Desert Evergreen Scrub with ephemerals: Hot-desert evergreen shrubs Leafless scrub Small stem-succulents Succulent forbs Desert-grasses Ephemeral desert herbs Dry desert	*Artemisia-Thymelaea* Semi-Desert (10% cover), with *Aristida, Panicum, Zygophyllum, Asphodelus, Pituranthos* and various geophytes and ephemerals.
*Mahadday-Weyne (Somalia, 70 km north of Mogadisho) MI = 0.28	Raingreen Thorn-Scrub with Tall Stem-Succulents: *Raingreen thorn-scrub and small trees *Arborescent stem-succulents Typical and bush stem-succulents Xeric rosette-shrubs and tuft-treelets Leaf-succulent evergreen shrubs Xeric cushion-shrubs Short bunch and desert grasses Evergreen arborescents Raingreen and succulent forbs Raingreen vines Xeric cushion-herbs Ephemeral desert herbs	Dry Raingreen Thorn-Scrub: *Commiphora, Acacia, Capparidaceae, Anacardiaceae, Grewia* et al., plus short *Chrysopogon, Aristida, Sporobolus, Cenchrus,* etc., including scrub and forbs.
**Asmera (Eritrea, northern Ethiopia, 2372 m) MI = 0.59	Semi-Deciduous Scrub with some trees: +Tropical evergreen sclerophyll trees +Raingreen thorn-scrub +Broad-raingreen small trees +Short bunch-grasses Xeric evergreen tuft-treelets Bush stem-succulents Leaf-succulent and xeric cushion shrubs Raingreen forbs	Dry Thorn-Scrub, Dwarf-Shrub and Montane Grassland mosaic: *Acacia, Juniperus procera, Olea, Sanseveria, Celtis africana,* et al., *Acokanthera, Buxus, Euphorbia, Dracaena,* with *Festuca, Agrostis, Pentaschistis* and various forbs.
**Kericho (Western Kenya, 2042 m) MI = 2.31	Tropical Montane Rainforest: *Tropical montane rainforest trees *Tropical linear-leaved trees Tropical evergreen microphyll-trees Palmiform tuft-trees Tropical cloud-forest small-trees Tropical evergreen sclerophyll trees Treelets, tree ferns, shrubs, bamboos, other grasses, forbs, lianas and vines, epiphytes.	Equatorial Montane Rainforest: *Ocotea, Podocarpus,* with *Aningeria, Casearia, Ficus, Pygeum, Schefflera,* and many others, plus understorey trees, *Croton, Dracaena, Ficalhoa,* and numerous vines, forbs and epiphytes.

Table 14 (continued).

Site	Predicted vegetation	Actual vegetation
**Nanyuki (central Kenya, 1947 m) MI = 0.93	Tropical Dry Montane Mixed Forest (open) with scrub: *Tropical xeric needle-trees Montane broad-raingreen trees Tropical evergreen microphyll-trees Tropical evergreen sclerophyll trees Broad-raingreen small trees Palmiform tuft-trees and treelets Raingreen thorn-scrub Tall grasses, including some bamboo Short bunch and sward grasses Bush stem-succulents Tropical broad-evergreen shrubs Palmiform mesic rosette-shrubs Tropical evergreen and raingreen forbs and vines Wintergreen epiphytes	Region of potential Dry Montane Conifer Forest (open, with thorn-scrub and savanna mosaic): *Juniperus procera, Olea, Podocarpus, Croton, Cussonia, Acacia, Commiphora, Acanthus,* with *Chrysopogon, Aristida* and other grasses, plus forbs.
**Makurdi (Southeastern Nigeria) MI = 0.83	Raingreen Savanna Woodland with evergreens: +Xeric raingreen trees Broad-raingreen small trees Tropical evergreen sclerophyll trees Raingreen thorn-scrub Tall cane-grasses Short bunch-grasses Xeric tuft-treelets and rosette-scrub Evergreen scrub and shrubs Palmiform tuft-trees and treelets Bush stem-succulents Raingreen forbs and vines	Raingreen Tree Savanna (wet savanna zone): *Daniellia, Lophira, Parinari, Terminalia, Prosopis, Gardenia* et al., *Hyparrhenia, Andropogon, Pennisetum* and other tall grasses, with numerous forbs and shrubs
*Lusaka (central Zambia, 1278 m) MI = 0.89	Raingreen Woodland-Open Forest with evergreens: +Broad-raingreen small trees +Tropical evergreen sclerophyll trees Tropical xeric needle-trees Palmiform tuft-treelets and trees Tropical evergreen microphyll-trees Xeric evergreen tuft-treelets Raingreen and evergreen scrub Xeric raingreen trees Tropical and mediterranean evergreen shrubs Bush stem-succulents Xeric rosette-shrubs Tall cane-grasses Short bunch-grasses Raingreen and xeric forbs Raingreen vines Wintergreen epiphytes	Semi-Open Raingreen Miombo Forest: *Brachystegia, Julbernardia, Isoberlinia, Berlinia* and other legumes, *Parinari, Uapaca, Diospyros, Monotes, Faurea,* various shrubs, and many grasses and especially dry-season forbs.
**Maun (Ngamiland, northern Botswana, 942 m) MI = 0.43	Raingreen Woodland and Thorn-Scrub +Raingreen thorn-scrub Broad-raingreen small and larger trees Leafless xeric shrubs Mediterranean dwarf-shrubs Xeric evergreen tuft-treelets and scrub	Raingreen Mopane Open Woodland with sparse undergrowth: *Colophospermum, Acacia, Kirkia, Terminalia,* with *Grewia, Combretum, Capparidaceae,* and relatively few grasses and forbs.

Table 14 (continued).

Site	Predicted vegetation	Actual vegetation
	Bush, typical and tall stem-succulents Leaf-succulent, cushion and hot-desert shrubs Short bunch and sclerophyllous grasses Desert-grasses Raingreen and summergreen forbs	
*Kroonstad (Orange Free State, 1348 m) MI = 0.84	Diverse Savanna Woodland with Shrubs: +Short bunch and sward grasses +Tall Grasses Dwarf-needle small trees +Tropical evergreen sclerophyll trees Broad-raingreen small trees +Tropical xeric needle-trees +Temperate needle-trees +Tropical evergreen microphyll-trees Temp. broad-evergreen small trees Sub-mediterranean needle-trees Montane broad-raingreen trees Mediterranean broad-evergreen trees Xeric evergreen tuft-treelets Evergreen and deciduous arborescents Needle and broad-evergreen shrubs Summergreen shrubs Bush stem-succulents Xeric rosette and cushion shrubs Raingreen and summergreen forbs	Region of Tall, Forb-Poor High Veld: *Themeda, Heteropogon, Setaria, Eragrostis, Andropogon, Hyparrhenia* et al., plus *Elephantorrhiza, Hypoxis, Aster, Indigofera* and other forbs, and with invading shrubs if not grazed.
*Cathedral Peak (Little Berg, in Natal Drakensberg, 1860 m) MI = 2.21	Diverse Subtropical Montane Rainforest (seasonal): *Tropical montane rainforest trees *Temperate needle-trees *Montane broad-raingreen trees *Tropical linear-leaved trees Tropical evergreen sclerophyll trees Tropical evergreen microphyll-trees Broad-evergreen small trees Broad-raingreen small trees Arborescent grasses Tall cane and typical grasses Evergreen broad and needle-leaved shrubs Various forbs	Montane Heathland (Fynbos): *Passerina, Philippia, Widdringtonia, Erica, Macowania, Protea, Anthospermum, Rhus discolor,* et al., with *Asparagus, Senecio, Euphorbia, Diospyros, Stoebe,* and many other shrubs, plus *Encephalartos, Polystichum, Cymbopogon,* and *Berkheya.*
**Hell-Bourg (Réunion, 935 m) MI = 3.31	Tropical Montane Rainforest with Raingreens: *Tropical montane rainforest trees Montane broad-raingreen trees Temperate rainforest needle-trees Tropical linear-leaved trees Tropical evergreen microphyll-trees Palmiform tuft-trees Tropical cloud-forest small-trees	Montane Evergreen Rainforest: *Cassine, Diospyros, Dodonaea, Naxia, Olea, Ocotea, Sideroxylon, Terminalia, Weinmannia* and many others, with understorey trees, lianas and vines, epiphytes, etc.
*Borisovka (southern Russia, near Ukrainian border) MI = 0.94	Semi-Open Mixed Semi-Deciduous Forest: *Boreal/montane short-needled trees *Summergreen broad-leaved trees Boreal summergreen needle-trees Broad-summergreen small trees Needle-leaved evergreen shrubs	Summergreen *Quercus-Tilia* Forest (preserve) near forest-steppe transition, with *Acer, Ulmus, Malus, Crataegus,* and various shrubs, spring geophytes, summer forbs and grasses (*Poa, Festuca gigantea*).

Table 14 (continued).

Site	Predicted vegetation	Actual vegetation
	Summergreen scrub and shrubs Short and tall grasses Summergreen forbs	
**Kerki (eastern Kara-Kum, Turkmenistan, near Afghan border) MI = 0.17	Cold-Winter Xeromorphic Semi-Desert Scrub: +Leafless xeromorphic large-scrub +Cold-winter xeromorphic shrubs Xeric dwarf-shrubs Bush stem-succulents Desert-grasses Ephemeral desert herbs	Kara-Kum Leafless and Nanophyllous Semi-Desert Scrub: *Ammodendron, Haloxylon,* with *Calligonum, Salsola, Astragalus, Ephedra, Artemisia,* other shrubs and semi-shrubs, *Carex, Poa bulbosa, Allium, Aristida, Tulipa, Iris, Gagea,* and many annual herbs.
Chatkal Mtns. (Kirghiz-Uzbek border, 1433 m) MI = 1.63	Temperate Montane Steppe with Needle-Scrub: Needle-leaved evergreen shrubs Short bunch-grasses Summergreen forbs	Montane Summergreen *Juglans* Belt, with *Prunus, Acer, Malus, Crataegus* et al., *Melica, Festuca, Agropyron, Carex* and other graminoids, and various forbs.
*Najaf Depression (central Iraq, Euphrates Valley) MI = 0.06	Leafless Desert Scrub with succulents: +Leafless xeromorphic scrub +Small stem-succulents +Desert-grasses Succulent forbs Ephemeral desert herbs Dry desert	*Salsola-Zygophyllum* Shrub Steppe (5–15% cover) with saline marshes, including *Nitraria, Seidlitzia, Halocnemum, Limonium, Bienertia* and other halophytes.
Maimana (northwestern Afghanistan, 615 m) MI = 0.55	Upland Shrub-Steppe with scattered small needle-trees: +Dwarf-needle small trees +Cold-winter xeromorphic shrubs +Short bunch-grasses Bush stem-succulents Leafless shrubs and dwarf-shrubs Needle-leaved evergreen shrubs Summergreen forbs Xeric cushion-herbs	Summergreen *Pistacia* Savanna-Woodland (2–5 m, 5–40% cover), with *Amygdalus, Cerasus, Cercis, Rosa, Ephedra, Artemisia, Acantholimon, Poa bulbosa, Salvia, Compositae, Carex, Anemone, Bromus, Astragalus* and many others.
*Kotgai (eastern mountains of Afghanistan, 2450 m) MI = 0.94	Temperate Montane Dry Needle-Forest: *Temperate needle-trees *Boreal/montane short-needled trees Dwarf-needle small trees Needle-leaved evergreen shrubs Short bunch-grasses Summergreen forbs	Montane *Cedrus deodara* Forest, with *Juniperus, Pinus, Quercus, Rosa, Cotoneaster, Berberis, Lonicera, Salvia, Phlomis, Astragalus,* various grasses and forbs.
Mussoorie (northern India, 1500 m) MI = 3.32	Montane Evergreen Shrubs and Forbs: *Tropical broad-evergreen shrubs Broad-wintergreen epiphytes Summergreen forbs Mat-forming thallophytes	Montane Temperate Evergreen Oak Forest: *Quercus incana,* with *Cedrus* and *Pinus* on dry exposures, tree understorey of *Rhododendron, Lyonia, Persea,* and others, *Virburnum, Berberis, Hedera* and other climbers, *Arundinaria,* and various other grasses, forbs and ferns.

Table 14 (continued).

Site	Predicted vegetation	Actual vegetation
Yüjlin (Ordos region, central China, 1121 m) MI = 0.72	Mixed Scrub-Bunch Grass Steppe +Needle-leaved evergreen shrubs +Summergreen arborescents Broad-summergreen mesic shrubs +Short bunch-grasses +Short sward-grasses Summergreen forbs Xeric cushion-herbs	Cold-Winter Xeric Shrub Steppe (15–40% cover) on aeolian sand: *Artemisia ordosica, A. frigida, Caragana, Pycnostelma, Stipa,* with *Lasiagrostis, Carex, Allium, Peganum,* and others.
*Mitchell Plateau Camp (northwestern Australia) MI = 1.15	Monsoon Semi-Deciduous Forest: *Monsoon broad-raingreen trees Broad-raingreen small trees Tropical evergreen sclerophyll trees Tropical evergreen microphyll-trees Palmiform tuft-trees Tall cane and typical grasses Short grasses, forbs	Evergreen *Eucalyptus* Woodland with patches of raingreen monsoon forest: *E. tetrodonta, E. miniata, E. nesophila,* with *Livistona, Erythrophloeum, Terminalia, Cycas, Grevillea,* et al., plus monsoon forests of *Zizyphus, Albizia, Cochlospermum, Pouteria, Wrightia* et al.
*Wiluna (interior Western Australia, 27° S) MI = 0.25	Mixed Hot Semi-Desert Scrub: +Hot-desert evergreen shrubs +Raingreen thorn-scrub Leafless shrubs and dwarf-shrubs +Arborescent and smaller stem-succulents Xeric rosette-shrubs Sclerophyllous grasses Short bunch and desert grasses Xeric cushion-shrubs Raingreen, summergreen and succulent forbs Ephemeral desert-herbs	Evergreen Mulga Scrub: *Acacia aneura* (leafless, with polymorphic phyllodes), with other *Acacia, Eremophila, Cassia, Hakea,* smaller *Eucalyptus, Triodia, Chenopodiaceae* on saline sites, and ephemeral carpets of *Weitzia aurea* and *Helipterum.*
*Perth (coastal south-western Australia) MI = 1.14	Mediterranean-Subtropical Dry Rainforest: *Tropical evergreen sclerophyll trees *Mediterranean broad-evergreen trees *Sub-mediterranean needle-trees *Tropical linear-leaved trees Temperate needle-trees Tropical montane rainforest trees Tropical evergreen microphyll-trees Broad-raingreen small trees Broad-evergreen small trees Palmiform tuft-trees and treelets Tall cane-grasses Short bunch and sclerophyllous grasses Various evergreen shrubs Seasonal forbs Wintergreen epiphytes	Evergreen Tall Jarrah Forest: *Eucalyptus marginata,* with scattered other eucalypts, an understorey of *Casuarina* and *Banksia,* 'grass trees' (*Kingia, Xanthorrhoea*), many *Proteaceae, Myrtaceae, Leguminosae* and *Epacridaceae,* and various herbs, including orchids and insectivorous *Drosera.*

Table 14 (continued).

Site	Predicted vegetation	Actual vegetation
*Kieth (Dark Island) (South Australia) MI = 0.68	Diverse Mixed Subtropical Semi-Evergreen Scrub with scattered trees: +Needle-leaved evergreen shrubs +Xeric summergreen shrubs +Broad-raingreen small trees +Mediterranean evergreen shrubs +Dwarf-needle small trees +Raingreen thorn-scrub +Tropical evergreen sclerophyll trees +Xeric evergreen tuft-treelets Leafless scrub Evergreen arborescents Temperate needle-trees Mediterranean dwarf-shrubs Hot-desert evergreen shrubs Xeric rosette-shrubs Bush stem-succulents Xeric cushion-shrubs Short bunch-grasses Raingreen and summergreen forbs	Open Dry Evergreen Heath (1–2 m) with mallee: *Banksia ornata, Casuarina pusilla, Leptospermum myrsinoides,* and *Xanthorrhoea australis,* with scattered taller *Eucalyptus,* plus *Phyllota, Hibbertia, Leucopogon* (dwarf-shrubs), bunch-graminoids, and a variety of forbs (including *Liliaceae* and *Orchidaceae*).
*Alexandra (Canterbury Plains, South Island of New Zealand) MI = 0.63	Short Tussock and Spreading-Grass Steppe with xeric shrubs: +Short bunch-grasses +Short tussock-grasses +Short sward-grasses Needle-leaved evergreen shrubs Xeric summergreen shrubs and scrub Xeric cold-winter and cushion-shrubs Leafless dwarf-shrubs Bush stem-succulents Summergreen forbs Possible evergreen needle-trees and broad-leaved shrubs	Low-Tussock Grassland (valley site): *Festuca novae-zelandiae, Poa caespitosa,* with *Agropyron, Discaria, Aciphylla, Leucopogon, Carex, Danthonia, Agrostis,* and various meadow and steppe forbs, including European ruderals.

personally, however, are supported in their various sources by photographs taken near the sites.

After the actual vegetation had been determined for the validation sites, the site macroclimatic data were coded, other values generated, and the data put through ECOSIEVE to generate the model's predicted vegetation at each validation site. These predicted life-form spectra are listed in Appendix C. The cover and dominance models were then applied to predict a formation type for each site, which was given a name based on the predicted dominant and subdominant components and predicted cover, as described in Table 8. The predicted vegetation for each site, with predicted dominant and subdominant forms, is presented alongside the corresponding actual vegetation in Table 14. Predicted dominant forms (center column) are denoted by an asterisk, with other forms listed considered to be potentially important also. (Lesser forms are shown in the complete listing in Appendix C).

Two criteria were used for comparing predicted and actual vegetation:

1. All actually occurring dominant forms be predicted by the model.
2. All but only actually occurring dominant forms (i.e. the exact combination of dominant forms) be predicted by the model.

The latter criterion is quite strict and, if met, can be interpreted as prediction of the actual vegetation

Table 15. Summary of vegetation prediction accuracy at the validation sites.

The 74 validation sites are grouped three times in the left column according to three criteria: region, terrain type, and annual moisture index (MI). The number of sites and the prediction statistics corresponding to each grouping of sites are shown in the other four columns. For description of the strong (correct combination of dominant forms) and weak (prediction only) criteria for prediction accuracy, please refer to the text or to Table 14. Prediction accuracy appears to be highest in the Americas and Africa and in the higher and lower (but not intermediate) MI ranges. The apparently similar prediction accuracy in lowland and in mountainous areas may be misleading since all but one of the 11 mountainous sites satisfying the strong criterion also involve high MI values (>0.93). The relatively small numbers of sites in the various groupings, however, do not permit definite conclusions about relative vegetation predictability.

Site class	No. of sites	Sites meeting		
		Strong criterion (**)	Weak criterion (*)	Neither
North America	30	18	12	0
South America (incl. Central America)	13	7	4	2
Europe (excl. USSR)	4	1	3	0
Africa	14	10	4	0
USSR	3	1	1	1
Asia (excl. USSR)	5	0	2	3
Australia and Oceania	5	0	5	0
Lowland/upland sites	51	26	22	3
Mountainous sites	23	11	9	3
MI > 1.6	20	11	6	3
1.2 < MI < 1.6	9	7	1	1
0.95 < MI < 1.2	10	4	6	0
0.8 < MI < 0.95	9	3	6	0
0.6 < MI < 0.8	12	3	8	1
0.3 < MI < 0.6	6	5	0	1
MI < 0.3	8	4	4	0
Total	74	37 (50%)	31 (42%)	6 (8%)

formation type.[2] Sites at which only the first (weak) criterion is met are indicated in Table 14 by a single asterisk before the site name. Sites at which the strong second criterion is met are indicated by two asterisk before the site name. Sites meeting the strong criterion automatically satisfy the weak criterion also. A summary of sites meeting the criteria is presented in Table 15.

As one can see in Tables 14 and 15, the actual dominant plant form or forms are predicted by the model (weak criterion: one or two asterisks before the site name) in 68 of the 74 cases (92%). The actual vegetation formation type is predicted (strong criterion: two asterisks before the site name) in 37 of the 74 cases (50%), the number being lower because the criterion for correct prediction is much more rigorous. The actual formation type was considered to be correctly predicted only if all its dominant (and in some cases subdominant) components, and no others, were predicted both to occur and in fact to dominate. Considering these criteria and the number of environmental factors not included in the model, a figure of 50% seems remarkable. The actual dominant forms were not predicted at only 6 of the 74 sites (8%).

The distribution of correct and incorrect predictions, by region, terrain and general climatic wetness (annual moisture index MI), is summarized in

[2]Predicted forms of the highest dominance level present are excluded as predicted dominants if they are near environmental limits (fitness values well below 0.10). Other subjective considerations may also be made concerning growth rates, etc., according to criteria described in section 5.F.

Table 15. The small total number of sites in each grouping precludes definite conclusions, but the results do suggest greater correspondence between actual vegetation and normal climatic relationships (greater prediction accuracy) in the Americas and Africa. The smaller number of correct predictions in Asia and Australia reflect to at least some extent various site peculiarities at some Asian sites (e.g. aeolian sand) and the more isolated and unique development of the Australian vegetation (e.g. the importance of *Eucalyptus* and the lack of stem-succulents). Several sites were included because of their unusual seasonal patterns (e.g. Mussoorie, Yüjlin, Chatkal Mountains) or climatic extremes (e.g. Najaf, Quibdo, Damanhur). Three of these are in Asia and a fourth in the USSR.

At first glance it may appear that there is little difference in predictability between lowland and mountainous sites, each showing roughly the same pattern as for all sites. Of the 11 mountainous sites for which actual formation (strong criterion) was predicted, however, only one has an annual moisture index (MI) below 0.93. Those nine sites meeting only the weak criterion include six transitional or drier sites (0.6 < MI < 1.1). The other three, plus the three mountainous sites meeting neither criterion (Isla Victoria, Chatkal Mountains, Mussoorie), all have MI values above 1.6 (to over 3). Mountainous areas, both those which are climatically drier and those with accentuated seasonal orographic rainfall, appear to represent special problems for interpreting macroclimatic control on vegetation type, due primarily to greater availability of soil water.

The distribution of prediction results by moisture class (MI range) is perhaps most instructive since it considers vegetation stands by (potential) degree of openness. The highest prediction accuracy (>75% satisfying the strong criterion) is suggested for climates characterized by typical closed forests (1.2 < MI < 1.6, i.e. not extreme rainforests) and by rather open steppes and woodlands (including typical raingreen thorn-scrub). More interesting, however, is the decreased accuracy for the transitional climates (MI values from 0.6 to 1.2), reflecting the greater importance of fire, site characteristics, and particular biotic factors in determining the actual vegetation structure and dynamics in these areas. These results correspond to the general occurrence of vegetation core areas in the climatic ranges of humid forest (MI of 1.2–1.6) and dry open woodland-steppe (MI of 0.3–0.6). The vegetation of the driest (MI < 0.3) and perhaps also the wettest (MI > 1.6) climates may show greater influence by local vegetation history and/or by environmental factors (including climate) which were not expressed in the current model, such as soil water and topography.

B. Vegetation and climate at unusual sites

Perhaps the most can be gained from the evaluation of results by examining unusual sites, such as those at which the actual vegetation is not correctly predicted. Of the 74 validation sites, neither prediction criterion was satisfied at six sites: Isla Victoria, Ushuaia, Chatkal Mountains, Maimana, Mussoorie, and Yüjlin (see Table 14 for locations and predicted and actual vegetation). These include several sites which were deliberately chosen for the validation because of their unusual climates (Ushuaia, Mussoorie), unusual topographic situation (Yüjlin) or particularly interesting vegetation (Chatkal Mountains, Maimana). All of these sites except Ushuaia are located above 600 m. All but Maimana and Yüjlin have MI values above 1.2.

The dominant vegetation at Isla Victoria (in a lake in the southern Andes) is *Nothofagus dombeyii* plus other broad-leaved evergreen trees which can be considered rainforest trees (Hueck 1966). These were rejected by three model variables: TMIN, PMIN, PMTMAX. Temperate broad-evergreen small trees were predicted to occur and perhaps could represent *Nothofagus* in this case. At Ushuaia (Tierra del Fuego), the other site in southern South America, it is the summergreen *Nothofagus pumilio* and *N. antarctica,* plus scattered evergreen *Nothofagus,* which are not predicted (Hueck 1966). The mean temperature of the warmest month in the extremely maritime Ushuaia climate is only 10°, while the corresponding lower limit for broad-summergreen small trees was set in the model at 11°, based on conditions in Iceland, northern Fennoscandia, and Alaska. Other less extreme problems also occurred in predicting the vegetation of southern South America, some due to the extreme windiness of the region, with consequent underestimation of potential evapotranspiration. Compared with other comparable climatic regions, however, the minimum summer temperature requirements of most tree forms are unusually low in southern South America (e.g. some evergreen *Nothofagus* spp. surviving at TMAX = 7°).

The other four unsuccessful sites are in mountainous or upland regions in or around Central Asia. Each has a highly seasonal precipitation regime with at least one month of essentially no rainfall. Two of the sites are monsoonal, but the Chatkal Mountains site (Kirghiz-Uzbek border area in the USSR) and Maimana (Afghanistan) have continental mediterranean climates with little rainfall in summer. The actual vegetation at the Chatkal site and in similar elevational belts throughout the mountains of Middle Asia (USSR) and some of Central Asia (China-Mongolia) is an open woodland with *Juglans fallax* and a variety of wild fruit trees, e.g. *Prunus, Amygdalus, Malus,* plus *Crataegus, Pistacia* and others. The vegetation at Maimana is similar but without *Juglans* and much more open. These open woodlands (parklands) apparently survive the dry summers (with TMAX around 20–24°) on groundwater and some overland runoff from higher elevations. If this water subsidy could be expressed in the model, the predicted vegetation could be a summergreen woodland at Chatkal and a more open summergreen woodland/shrubland at Maimana.

The montane monsoonal site at Mussoorie, with MI = 3.32 but PMIN = 7 mm, well illustrates the shortcomings of the model design. The actually occurring evergreen trees (*Quercus incana,* with *Cedrus* and *Pinus*) and evergreen understorey (*Rhododendron, Persea, Lyonia,* etc.) are all rejected by the PMIN value. A model involving actual soil water, accumulated from the high-rainfall months, would certainly predict the rainforest which occurs.

The Chinese site Yüjlin, in the Ordos region of aeolian sand, illustrates a different problem. Although the MI value of 0.72 is not extremely low, there is very little precipitation except in summer, and spring can be very dry, especially on a sand substrate. The actually occurring cold-winter xeric shrubs (*Artemisia,* plus *Caragana,* et al.) generally occur in continental dry-summer climates and are not predicted at Yüjlin because of the summer rainfall peak. An aeolian sand substrate can easily reduce by half the water available from precipitation, however, which would then result in a reasonable simulation of the Yüjlin xeric shrub-steppe with 15–40% cover.

Mountainous areas, azonal soils, unusually extreme or seasonal climates, as well as wetlands and floodplains, are all areas where the macroclimatic approach may not provide useful results. This does not necessarily mean that the vegetation of these areas shows different environmental relationships but only that a macroclimatic model is insufficient to express the controlling environmental factors. Where water is the problem, precipitation and MI values can often be adjusted to express available water more accurately, resulting in a reasonable simulation of the actual vegetation composition and structure. Foggy mediterranean coastlines may remain problematic. Temperature, as illustrated by southern South America, may be a less easily circumvented problem. Temperature influences generally are not greatly changed by substrate or topographic factors, except in the cases of protected microclimates in mountains, insulating snow covers, and extremely low-growing vegetation in cool climates. For deciduous (seasonal) vegetation air temperature is usually the controlling temperature factor, but for some evergreen vegetation, such as the *Polylepis* treelets of the equatorial Andes, which can occur individually several hundred meters above treeline, soil temperature seems to be the controlling (enabling) factor. Wetlands can often be considered as equivalent to MI >> 1.0, though some forms may be eliminated by anaerobic soil conditions. Simulation success can also be improved, of course, by considering other environmental conditions, such as substrate peculiarities and the site's history of fire and anthropogenic influences.

C. Vegetation sensitivity to climatic variation

The sensitivity of plant and vegetation types to climatic variations can be studied in a number of ways. Some information was provided in the validation results by multiple sites in the same vegetation region. Additional information can be provided by changing climatic values (by amounts within the normal range of interannual variation) and comparing the new world vegetation predictions with those of the standard model. Perhaps most importantly, since there is some question about the accuracy of any single potential evapotranspiration estimation method worldwide, the model can be re-run with different estimates of PET in order to estimate the effects of this source of potential error. Such re-running of the model with altered environmental values represents a sort of sensitivity analysis, a technique which can be quite useful to the development and understanding especially of predictive ecological models (Caswell 1976). Results generated by modifying PET will always be difficult to interpret, however, since the model's MI limits were derived using PET values based on the Thornthwaite estimate. Interpretation of results is also complicated by difficulty in separating model sensitivity from the actual climatic sensitivity of existing vegetation.

The other PET estimation method which is widely used over large parts of the world is that of Holdridge (1959). It is also perhaps the best alternative to compare with the Thornthwaite values since Holdridge estimates consistently higher PET values throughout most of the tropics and Southern Hemisphere and consistently lower values north of about 40° N (Box, in prep.). The model represented by Table 7 was re-run with the world climate data, using the Holdridge estimate of PET, and the results were compared with those of the standard model by means of a simple program called NCHANGES (Box, unpublished). The resulting changes in prediction frequency of selected important plant forms (without dominance or other importance considerations) are shown in Table 16.

Table 16. Changes in selected life-form prediction frequencies using the Holdridge estimate of annual potential evapotranspiration (PET).

The occurrence frequencies (see also Table 12) of selected plant life forms, as predicted by the model (Table 7) for the 1225 world climatic data-sites, are shown in the first column of numbers (Std. freq.). The numbers of sites gained and lost using Holdridge's (1959) estimate of potential evapotranspiration (PET) are shown in the remaining columns, with the corresponding percentages of the original frequencies given in parentheses. Most forms gained and lost fewer than 20% of their original sites, with net gains or losses generally closer to 2–10%. The most significant changes occur in tropical and subtropical (including mediterranean) areas.

Plant type	Std. freq.	Holdridge PET	
		Gained (%)	Lost (%)
Tropical rainforest trees	133	1 (0.8)	5 (3.8)
Tropical evergreen sclerophyll trees	333	10 (3.0)	17 (5.1)
Warm-temperate evergreen trees	58	0	5 (8.6)
Mediterranean evergreen trees	36	0	13 (36.1)
Monsoon raingreen trees	96	3 (3.1)	13 (13.5)
Xeric raingreen trees	110	13 (11.8)	2 (1.8)
Summergreen broad-leaved trees	222	4 (1.8)	7 (3.2)
Tropical linear-leaved trees	173	0	28 (16.2)
Tropical xeric needle-trees	86	43 (50.0)	32 (37.2)
Heliophilic needle-trees	97	2 (2.1)	15 (15.5)
Sub-Mediterranean needle-trees	157	1 (0.6)	36 (22.9)
Temperate rainforest needle-trees	45	2 (4.4)	10 (22.2)
Boreal/Montane needle-trees	248	13 (5.2)	0
Broad-summergreen small trees	364	37 (10.2)	15 (4.1)
Palmiform tuft-trees	291	2 (0.7)	23 (7.9)
Raingreen thorn-scrub	183	23 (12.6)	1 (0.5)
Mediterranean evergreen shrubs	150	5 (3.3)	25 (16.7)
Broad-ericoid evergreen shrubs	87	23 (26.4)	7 (8.0)
Xeric summergreen shrubs	170	2 (1.2)	61 (35.9)
Tall cane-grasses	391	8 (2.0)	31 (7.9)
Short bunch-grasses	1002	7 (0.7)	65 (6.5)
Xeric cushion-herbs	412	24 (5.8)	113 (27.4)

The greatest percentage of changes occurred among tree forms, especially in the tropics. However, only one form, Tropical Xeric Needle-Trees (with only 86 occurrences out of 1225 sites originally), showed more than a 30% gain or 40% loss of predicted occurrences. Only four other forms (not all shown in Table 16) gained more than 20%, while only nine other forms lost as many as 20% of their original sites. Only three forms were not affected at all: the nothofagoid Cool-Temperate Broad-Evergreen Trees, the Tall Tussock-Grasses, and the Summergreen Cold-Desert Herbs, all from cool-temperate or polar climates. The greatest percentage changes in predicted sites were often by forms with restricted ranges and fewer occurrences among the original results, e.g. Tropical Alpine Tuft-Treelets. In general, the substitution of the Holdridge estimate of PET did not result in dramatic changes in predicted vegetation patterns, though it clearly can have a major effect at certain sites. The

Table 17. Predicted effect on selected extra-tropical plant-form distributions of 5 °C reduction in peak summer temperature.

The standard world vegetation model (Table 7 applied to climatic data for 1225 sites worldwide) was re-run with actual mean temperature of the warmest month (TMAX) reduced by 5 °C for each site (with recomputation of the remaining monthly temperatures to preserve an annual sinusoidal pattern and recomputation of the corresponding PET and MI values). The occurrence frequencies predicted by the standard model for selected extra-tropical forms (based on 1225 sites, see also Table 12) are shown in the first column of numbers (Std. freq.), along with the predicted changes in the remaining columns (sites gained or lost, with corresponding percentages of original frequency in parentheses). Only extra-tropical forms are shown, since results may be misleading where the annual variation of mean temperature is not clearly greater than 5 °C. Boreal needle-trees are seen to be reduced by 15–40% net while being displaced equatorward (14–19% new sites). Krummholz, tussock-grasses, and cool-mesic dwarf-shrubs make the largest net gains, with generally 10–16% of old sites lost except for krummholz, which is displaced completely (all old sites lost but many more gained). Large net decreases are evident for most typical-temperate and warm-temperate forms.

Plant type	Std. freq.	TMAX→TMAX-5°	
		Gained (%)	Lost (%)
Warm-Temperate evergreen trees	58	2 (3.4)	31 (53.4)
Summergreen broad-leaved trees	222	17 (7.7)	136 (61.3)
Boreal broad-summergreen trees	175	31 (17.7)	102 (58.3)
Temperate needle-trees	180	77 (42.8)	108 (60.0)
Boreal/Montane needle-trees	248	46 (18.5)	83 (33.5)
Boreal summergreen needle-trees	236	33 (14.0)	123 (52.1)
Dwarf-needle small trees	210	11 (5.2)	110 (52.4)
Needle-leaved treeline krummholz	45	130 (289)	45 (100)
Cold-winter xeromorphic shrubs	116	0	38 (32.8)
Broad-summergreen mesic shrubs	353	18 (5.1)	200 (56.7)
Temperate evergreen dwarf-shrubs	271	94 (34.7)	44 (16.2)
Summergreen tundra dwarf-shrubs	179	75 (41.9)	26 (14.5)
Mesic evergreen cushion-shrubs	46	13 (28.3)	10 (21.7)
Tall grasses	504	12 (2.4)	178 (35.3)
Short bunch-grasses	1002	10 (1.0)	237 (23.7)
Tall tussock-grasses	84	36 (42.9)	10 (11.9)
Summergreen forbs	643	8 (1.2)	152 (23.6)
Summergreen cold-desert herbs	167	165 (98.8)	27 (16.2)
Summergreen vines	134	8 (6.0)	72 (53.7)
Ice desert	58	39 (67.2)	0

main results appears to be a slight shifting of predicted ranges toward more mesic areas, especially in the tropics. The need for an improved, single, world-applicable estimation method for annual potential evapotranspiration remains evident.

World vegetation predictions were also generated by three other model modifications, each involving modification of a single climatic factor:

1. a 5° decrease in summer temperature (TMAX);
2. a 50% increase in annual potential evapotranspiration (reduction of MI by one-third); and
3. a 50% reduction of summer rainfall (PMTMAX).

Implied changes to values of related variables were also computed, such as recomputation of PET under the altered annual temperature regime. Each of these climatic modifications focuses on particular plant forms in certain regions of the world and is less instructive elsewhere.

The 5° reduction of mean summer temperature is not really meaningful where the annual range is 5° or less, as in most of the tropics. In boreal (austral) and polar areas, however, reduction of summer temperature resulted in the expected equatorward shift of most plant types, as shown in Table 17. Boreal evergreen and summergreen needle-trees are reduced in predicted frequency by 15% and 38% respectively, while treeline krummholz loses all its old sites and gains nearly three times as many new ones. Tussock-grasses and evergreen and summergreen dwarf-shrubs make net gains of 18–30%. In the temperate zones most forms show dramatic decreases.

The 50% increase in potential evapotranspiration is probably interpretable everywhere and has a much greater effect than the use of Holdridge PET, which generally differs from Thornthwaite by not more than 20–30%. Only the xeromorphic forms show net increases (see Table 18), while reductions of 20–60%, with no sites gained, are common for most tree, vine, epiphyte, mesic-shrub, and tall-grass forms. In terms of familiar vegetation formations, tropical rainforest, mediterranean forest, summergreen forest and temperate rainforest are all reduced by at least about 40%. The greatest net (percentage) gains are by short tussock (75%) and desert grasses (67%), ephemeral desert herbs (35%), xeric shrubs and scrub (13–21%), and stem-succulents (10–22%). Some xeric forms, however, show net losses, e.g. hot-desert evergreen shrubs (4.6%). No form is completely eliminated. Cool-temperate broad-evergreen trees (7 sites) and summergreen cold-desert herbs (167 sites) were unaffected by the increased PET.

The 50% reduction of summer rainfall is significant in most areas and also results in much greater (often dramatic) loss of predicted sites than gain (see Table 19). Again no form is completely eliminated. The greatest net losses are suffered (excluding the topographically limited swamp trees, 74%) by summergreen ferns (59%), temperate broad-leaved rainforest trees (52%), maritime cushion-shrubs (50%), heliophilic needle-trees (48%), and certain other mesic forms, while others, however, are only slightly affected. The only real gains are made by mediterranean and sub-mediterranean forms. Nineteen forms were totally unaffected, including short bunch-grasses, tropical alpine and xeric tuft-treelets, various xeric scrub forms (including hot-desert evergreen shrubs), the non-arborescent stem-succulents, and most forb forms. The apparent relative indifference of summergreen and raingreen forms, as well as mesic tropical evergreen forms, to reduced summer rainfall is perhaps the most interesting of the sensitivity results. It suggests either that rainfall is still high after 50% reduction (which is often the case in the tropics) or, mainly in summergreen areas, that the plants are well adapted to high variability in summer rainfall. Temperate evergreen forms appear to be more sensitive than the corresponding summergreen forms.

Table 18. Predicted effect on selected plant-form distributions of 50% increase in annual potential evapotranspiration (PET).

The standard world vegetation model (Table 7 applied to world climatic data) was re-run with annual potential evapotranspiration (PET) increased by 50% for each site. This is equivalent to a one-third reduction of the annual moisture index (MI). The occurrence frequencies predicted by the standard model for selected, generally more important forms (based on 1225 sites, see also Table 12) are shown in the first column of numbers (Std. freq.) along with the predicted changes in the remaining columns (sites gained or lost, with corresponding percentages of original frequency in parentheses). Most mesic forms are reduced by at least 25%, with no sites gained and the effect apparently greatest among trees and mesic evergreen forms. Some xeric forms show net gains and are shifted toward the more mesic areas. Even some xeric forms show net losses (e.g. hot-desert evergreen shrubs), suggesting that they are occurring close to dryness limits. No form is eliminated completely, but tropical xeric needle-trees are displaced by 85–90%.

Plant type	Std. freq.	PET→1.5 × PET	
		Gained (%)	Lost (%)
Tropical rainforest trees	133	0	52 (39.1)
Tropical evergreen sclerophyll trees	333	10 (3.0)	66 (19.8)
Mediterranean broad-evergreen trees	36	0	22 (61.1)
Monsoon raingreen trees	96	0	60 (62.5)
Summergreen broad-leaved trees	222	0	100 (45.0)
Tropical xeric needle-trees	86	73 (84.9)	78 (90.7)
Temperate needle-trees	180	0	49 (27.2)
Temperate rainforest needle-trees	45	0	22 (48.9)
Boreal/Montane needle-trees	248	0	65 (26.2)
Cloud-forest trees	28	0	10 (35.7)
Palmiform tuft-trees	291	0	79 (27.1)
Xeric evergreen tuft-treelets	184	48 (26.1)	20 (10.9)
Raingreen thorn-scrub	183	35 (19.1)	4 (2.2)
Broad-ericoid evergreen shrubs	87	0	50 (57.5)
Hot-desert evergreen shrubs	131	14 (10.7)	20 (15.3)
Broad-summergreen mesic shrubs	353	0	132 (37.4)
Xeric summergreen shrubs	170	45 (26.5)	12 (7.1)
Temperate evergreen dwarf-shrubs	271	0	124 (45.8)
Typical stem-succulents	180	51 (28.3)	11 (6.1)
Tall grasses	504	0	119 (23.6)
Short tussock-grasses	32	26 (81.3)	2 (6.3)
Tropical broad-evergreen lianas	182	0	88 (48.4)
Tropical broad-evergreen epiphytes	138	0	70 (50.7)
Evergreen ferns	123	0	58 (47.2)
Dry desert (no vegetation)	77	1 (1.3)	0

Table 19. Predicted effect on selected plant-form distributions of 50% reduction of summer precipitation. Precipitation.

The standard world vegetation model (Table 7 applied to world climatic data via ECOSIEVE) was re-run with average warmest-month precipitation (PMTMAX) reduced by 50% at each site (with re-estimation of the other monthly precipitation values in order to preserve the same annual total and moisture index). The occurrence frequencies predicted by the standard model for selected, mostly temperate-zone or raingreen forms (based on 1225 sites) are shown in the first column of numbers (Std. freq.) along with the predicted changes in the remaining columns (sites gained or lost, with corresponding percentages of original frequency in parentheses). Mediterranean forms show clear gains (often with no losses), but all other forms show mild to more serious reductions. The losses are greatest for temperate summer-rain rainforest forms and heliophilic needle-trees (40–60%) but are generally much less (8–22%) for summergreen and raingreen trees and shrubs, suggesting that these forms, despite malacophylly, are well adapted to withstand wide year-to-year variations in growing-season rainfall. Xeric scrub forms are also reduced (generally 15–30%), as are the graminoids (20% or less, except for sclerophyllous grasses). Wintergreen epiphytes are only slightly affected, and nineteen forms are not affected at all (not shown but identified in the main text).

Plant type	Std. freq.	PMTMAX→PMTMAX/2	
		Gained (%)	Lost (%)
Tropical rainforest trees	133	0	2 (1.5)
Warm-temperate evergreen trees	58	0	12 (20.7)
Mediterranean evergreen trees	36	22 (61.1)	0
Temperate broad-leaved rainforest trees	48	0	25 (52.1)
Monsoon raingreen trees	96	0	8 (8.3)
Summergreen broad-leaved trees	222	0	33 (14.9)
Heliophilic needle-trees	97	0	47 (48.5)
Temperate rainforest needle-trees	45	0	5 (11.1)
Boreal summergreen needle-trees	236	0	51 (21.6)
Cool-temperate broad-evergreen trees	7	0	3 (42.9)
Raingreen Thorn-scrub	183	0	33 (18.0)
Mediterranean evergreen shrubs	150	35 (23.3)	0
Temperate broad-evergreen shrubs	229	0	51 (22.3)
Broad-summergreen mesic shrubs	353	0	27 (7.6)
Xeric summergreen shrubs	170	0	29 (17.1)
Cold-winter xeromorphic shrubs	116	20 (17.2)	0
Arborescent stem-succulents	140	0	33 (23.6)
Tall grasses	504	0	88 (17.5)
Short sward-grasses	678	0	123 (18.1)
Sclerophyllous grasses	72	18 (25.0)	22 (30.6)
Raingreen vines	278	0	21 (7.6)
Summergreen vines	134	0	50 (37.3)
Wintergreen epiphytes	451	0	9 (2.0)
Summergreen ferns	192	0	114 (59.4)

CHAPTER 8

Conclusions and next steps

Plant form and function, vegetation structure, environmental factors, and ecosystemic functions are of course highly complex phenomena interrelated by perhaps countless constraining and generative feedback mechanisms. The particular species composition of a system depends not only on physical conditions and evolutionary adaptations but also on complex biotic interactions. These complex interrelationships represent one of the primary and most intriguing foci of ecology. As is generally the case with feedback systems, however, certain processes are more immediately limiting or otherwise more important than others, resulting in a hierarchical ordering of both processes and effects, with subordination of much detail.

The first observation to be made from the foregoing results is that even a rather simplistic model with considerable ecological generalization can predict the general features of vegetation structure and world vegetation distributions with surprising accuracy. Despite the complexity of ecological systems, species interactions, and plant-environment relations, the results suggest that the basic features of plant form and world vegetation (in the absence of overwhelming non-climatic complications) are determined primarily by the general levels and mean seasonal patterns of temperature and the climatic water balance, with relatively little reliance on other factor interactions. The spectrum of life forms composing a vegetation appears to vary much less with climate than does species composition, and it is the particular combination of basic life forms, not the detailed species composition, which provides the basic structural and functional framework of an ecosystem.

More specific conclusions include the following:

1. Fairly specific plant forms and vegetation structures, including seasonal aspects, can usually be determined from the most basic, readily available climatic data. Sites on which this may not hold include waterlogged, foggy or windy, extremely rocky, agricultural, mechanically disrupted, fire-maintained, heavily grazed, chemically disrupted, mobile-substrate, oasis and some coastal and riverine sites. Climate, however, usually determines many of the forms found even on these sites, with actual vegetation corresponding either to shifted climatic conditions or to a subclimax stage with climatic dominants excluded. Except for agricultural and other maintained lands, these non-climatically controlled sites cover small fractions of the earth's land area.

2. Since occurrence of the ecophysiognomic forms is well described by their macroclimatic envelopes, it appears that climate not only 'determines' physiognomy but that plant forms are limited to their particular ecoclimates to a considerable degree. This corresponds to a certain degree of what can be called ecophysiognomic status. Well-known forms which have not always been accorded such status include, in particular, various groups of pines and other conifers. The evolutionary and successional status of such forms does not preclude ecophysiognomic status.

3. Despite climatic variation and differing configurations of climatic values, the macroclimatic limits of the plant forms of Table 2 appear to be similar throughout the world. In some situations,

however, related factors not explicitly included in the model seem to set the actual limits. These include soil moisture, absolute temperature range, growing-season length and total thermal input, and wind.

4. Most vegetation types appear to be well adapted to their local climatic regimes, with the inherent variabilities, and not extremely sensitive to minor modifications. Interannual climatic variability does not appear to be a major determining factor for plant form in most climate types. (Some exceptions to both these generalizations were suggested.)

5. Natural life-form ecotones appear to cover about 30% of the earth's land area. Some mixes of life forms show considerable consistency and merit formation-type status.

6. Physiognomic diversity of vegetation appears to increase as tree forms disappear and to be negatively related to warm-season precipitation. The highest form-diversity (excluding less important understorey forms) is to be expected in the warm semi-deserts.

7. There appears to be little difference in ecophysiognomic significance between some widely separated and apparently dissimilar climates, such as:

(a) tropical high-mountain (cloud-belt) and temperate perhumid (extremely maritime) climates.
(b) correspondingly wet polar and temperate alpine climates.
(c) temperate summergreen climates and the montane summer-rain climates (with frost) of tropical margins.

These parallels have been recognized for some time through species and phenological similarities but should be studied also using integrative ecophysiognomic indices.

8. The generally good results of the site comparisons relative to full life-form spectra support further the suggestion that biotic interactions (among plants) are not as important to general vegetation structure as to species composition.

The results emphasize the importance of ecophysiognomic form in plant-environment relations and suggest that ecophysiognomy may provide a theoretical and geographical framework for studying plant-environment relations. One of the most useful results of such general geographic studies is that they often effectively point out areas where more data or research is needed. Tropical mountains, windy and maritime environments are suggested by the current study. Additional climatic data should express some aspects of windiness, diurnal temperature variations, absolute temperature extremes, and interannual variations.

As a methodology the model demonstrates the following:

1. A combined life form-environmental limitation approach to the study of vegetation structure, function and geography.
2. A general computer-based framework and methodology for simulating vegetation and ecosystem phenomena and their spatial and temporal variations.
3. A quantitative but at the same time descriptive, analytical (as opposed to purely statistical), and validation-oriented approach to ecological modeling.
4. A framework for world-scale environment based modeling which may be of some use in fields other than botany and ecology.

As one can see from the relatively large amount of model output, the life-form approach to vegetation geography provides a combinatorial basis for greater resolution in formation description without requiring long lists of species. The environmental basis permits extrapolation, prediction and hypothesis-testing under conditions of changing macro- or micro-climate. Applications range from prediction of vegetation under different climatic conditions (e.g. during the Ice Ages or after greenhouse-effect climatic change) to geographic testing of ecophysiological hypotheses. The geographical basis provides the necessary feedback for iterative refinement of estimates of environmental limits. Subsequent models, however, should seek to relate vegetation to fewer, more integrative environmental variables in order to identify more critical mechanisms and general patterns of environmental limitation. Environmental-limitation and other such discrete models can be coupled with continuous process models in order to provide more complete descriptions of system structure and function and more insight into ecological strategies. This combined structural-functional approach, with geoenvironmental basis, begins to get at what Innis (1975) and Haefner (1978) have called 'ecological grammars.'

True ecophysiognomic studies and models require a more analytical (mechanismal) approach which seeks to simulate quantitatively and to integrate the metabolic, transport, and other physiological processes of particular life forms (plant or animal). Five major environmental factors not included in the present model were identified earlier as being necessary to eliminate the main discrepancies between actual vegetation and that predicted by the macroclimatic approach. These factors are:

1. soil moisture regime
2. seasonal freezing and thawing of the soil
3. solar declination
4. annual ranges of absolute temperatures
5. delimitation of the growing season, including some measure of total solar input and warmth.

In the course of developing the model, other models and submodels of various environmental phenomena (for use within an ECOSIEVE framework or independently) were also developed. These include in particular a soil moisture simulator (SOLWAT) and a simple routine (PERIOD) for delimiting favorable and unfavorable seasons which can be used, along with more complete data on temperature extremes and climatic variabilities, for a more integrative approach to modeling life-form ecology. Refinement of the current model may well represent diminishing returns, but estimating the life-form limiting values for more integrating variables, such as soil moisture level or actual evapotranspiration, might provide interesting insights.

Models predicting ecological phenomena from environmental conditions have a variety of applications in other areas. The geographical framework provided by ECOSIEVE and an adequate climatic data-base permit application of the study of plant-environment relations over large areas and under conditions of changing climate. Two applications were immediate. The primary energetics of ecological systems involves not only primary productivity but also the caloric equivalent of the biomass. A similar ECOSIEVE-based model was used to generate a world vegetation model (using vegetation units for which values of biomass energy content were available), which was then used as the basis for maps of estimated average annual energy fixation by natural vegetation (Box 1976) and of estimated average annual photosynthetic efficiency (Box 1977). The second application involves estimating changes in global terrestrial vegetation patterns, including agricultural implications, which might result from likely changes in the earth's climate. The main problem here is estimating spatial variations in the course of climatic change. Other potential applications are to be found in biogeochemistry, environmental management and various fields of geography. Such applications require geographic descriptions of more general aspects of vegetation and often require area estimates. When computer-maps can be produced with sufficient resolution, they can also be quantified by computerized planimetry (Box 1978b, 1979a) to provide estimated world or regional budgets or inventories.

APPENDIX A

Descriptions of the plant types

The plant life forms employed in the model are listed, with examples, in the main text (Table 2). They are described in this appendix in more detail, including environmental relations, physiognomic characters, prototypic and other characteristic taxa, and relevant literature. A list of the forms, with physiognomic characters, is included. Sources of vegetation data relevant to particular life forms are cited with the respective forms in the text of the appendix. General references, especially descriptions of regional vegetation, are listed by region at the end of the appendix.

Plant form	Plant size	Leaf size	Leaf (Stem) structure
Trees (Broad-leaved)			
Evergreen			
1. Tropical Rainforest Trees (lowland, montane)	tall, med.	large-med.	cor.
2. Tropical Evergreen Microphyll Trees	medium	small	cor.
3. Tropical Evergreen Sclerophyll Trees	med.-tall	medium	scler.
4. Temperate Broad-Evergreen Trees			
a. Warm-Temperate Evergreen	med.-small	med.-small	scler.
b. Mediterranean Evergreen	med.-small	small	scler.
c. Temperate Broad-Leaved Rainforest	medium	med.-large	scler.
Deciduous			
5. Raingreen Broad-Leaved Trees			
a. Monsoon mesomorphic (lowland, montane)	medium	med.-small	mal.
b. Woodland xeromorphic	small-med.	small	mal.
6. Summergreen Broad-Leaved Trees			
a. typical-temperate mesophyllous	medium	medium	mal.
b. cool-summer microphyllous	medium	small	mal.
Trees (Narrow and needle-leaved)			
Evergreen			
7. Tropical Linear-Leaved Trees	tall-med.	large	cor.
8. Tropical Xeric Needle-Trees	medium	small-dwarf	cor.-scler.
9. Temperate Rainforest Needle-Trees	tall	large-med.	cor.
10. Temperate Needle-Leaved Trees			
a. Heliophilic Large-Needled	medium	large	cor.
b. Mediterranean	med.-tall	med.-dwarf	cor.-scler.
c. Typical Temperate	medium	medium	cor.
11. Boreal/Montane Needle-Trees	medium	small	cor.-scler.

Plant form	Plant size	Leaf size	Leaf (Stem) structure
Summergreen			
12. Hydrophilic Summergreen Needle-Trees	tall-med.	large-med.	mal.-cor.
13. Boreal Summergreen Needle-Trees	medium	medium	mal.-cor.
Small and dwarf trees			
14. Tropical Broad-Evergreen Small Trees	small	med.-small	cor.
15. Tropical Broad-Evergreen Dwarf-Trees	dwarf	large-med.	cor.
16. Cloud-Forest Small Trees	small	small	cor.-scler.
17. Temperate Broad-Evergreen Small Trees	small	small	scler.
18. Broad-Raingreen Small Trees	small	small	mal.
19. Broad-Summergreen Small Trees	small	small-med.	mal.
20. Needle-Leaved Small Trees	small	dwarf	cor.-scler.
Rosette-trees (evergreen)			
21. Palmiform Tuft-Trees	medium	large	cor.-scler.
Rosette-Treelets (evergreen)			
22. Palmiform Tuft-Treelets	small	large-med.	cor.-scler.
23. Tree Ferns	small	med.-small	cor.-mal.
24. Tropical Alpine Tuft-Treelets	small-dwarf	large-med.	cor.-scler.
25. Xeric Tuft-Treelets	small-dwarf	large-med.	scler.
Arborescents			
26. Evergreen Arborescents	med.-tall	small	scler.
27. Raingreen Thorn-Scrub	medium	small	mal.
28. Summergreen Arborescents	med.-small	small	mal.
29. Leafless Arborescents	med.-small	–	(lign.)
Krummholz			
30. Needle-Leaved Treeline Krummholz	medium	small-dwarf	scler.
Shrubs			
31. Tropical Broad-Evergreen Shrubs	med.-tall	large-med.	cor.
32. Temperate Broad-Evergreen Shrubs			
a. Mediterranean	medium	small	scler.
b. Typical Temperate	medium	small-med.	scler.
c. Broad-Ericoid (perhumid)	med.-tall	large	scler.-cor.
33. Hot-Desert Evergreen Shrubs	small	small	scler.
34. Leaf-Succulent Evergreen Shrubs/Treelets	small-dwarf	med.-small	succ.
35. Cold-Winter Xeromorphic Shrubs	small-dwarf	small-dwarf	pub.-mal.
36. Summergreen Broad-Leaved Shrubs			
a. mesomorphic	medium	med.-large	mal.
b. xeromorphic	med.-small	small-med.	mal.-cor.
37. Needle-Leaved Evergreen Shrubs	med.-small	dwarf	scler.
Dwarf-shrubs			
38. Mediterranean Dwarf-Shrubs	dwarf	small-dwarf	pub.-mal.
39. Temperate Evergreen Dwarf-Shrubs	dwarf	small-dwarf	scler.-cor.
40. Summergreen Tundra Dwarf-Shrubs	dwarf	small	mal.
41. Xeric Dwarf-Shrubs	dwarf	dwarf, –	mal., (lign.)
Cushion-shrubs			
42. Perhumid Evergreen Cushion-Shrubs	dwarf	small-dwarf	scler.
43. Xeric Cushion-Shrubs	dwarf	dwarf, –	mal., scler., (lign.)

Plant form	Plant size	Leaf size	Leaf (Stem) structure
Rosette-shrubs (evergreen)			
44. Mesic Rosette-Shrubs	medium	large	cor.-scler.
45. Xeric Rosette-Shrubs	small	med.-large	scler.-succ.
Stem-succulents (evergreen)			
46. Arborescent Stem-Succulents	tall	–	(succ.)
47. Typical Stem-Succulents	med.-small	–	(succ.)
48. Bush Stem-Succulents	med.-small	–	(succ.)
Graminoids			
49. Arborescent Grasses	tall	large	mal.
50. Tall Cane-Grasses	tall-med.	med.-large	mal.
51. Typical Tall Grasses	medium	medium	mal.
52. Short Sward-Grasses	small	small	mal.
53. Short Bunch-Grasses	small	small	mal.
54. Tall Tussock-Grasses	medium	med.-large	mal.
55. Short Tussock-Grasses	small	small-med.	mal.
56. Sclerophyllous Grasses	small	medium	scler.
57. Desert Grasses	dwarf	small	mal.
Forbs			
58. Tropical Evergreen Forbs	tall-dwarf	large-med.	cor.-scler.
59. Temperate Evergreen Forbs	med.-dwarf	med.-small	scler.
60. Raingreen Forbs	med.-dwarf	large-small	mal.
61. Summergreen Forbs	med.-dwarf	large-small	mal.
62. Succulent Forbs	small-dwarf	med.-small	succ.
Undifferentiated small herbs			
63. Xeric Cushion-Herbs	small-dwarf	small	mal.-cor.
64. Ephemeral Dry-Desert Herbs	small-dwarf	small	mal.
65. Summergreen Cold-Desert Herbs	dwarf	small	mal.
66. Raingreen Cold-Desert Herbs	dwarf	small	mal.
Vines and lianas			
67. Tropical Broad-Evergreen Lianas	tall	large-med.	cor.
68. Broad-Evergreen Vines	med.-small	large-small	cor.-scler.
69. Broad-Raingreen Vines	med.-dwarf	large-small	mal.
70. Broad-Summergreen Vines	med.-dwarf	large-small	mal.
Ferns			
71. Evergreen Ferns	med.-small	med.-small	cor.
72. Summergreen Ferns	med.-small	med.-small	mal.-cor.
Epiphytes (evergreen)			
73. Tropical Broad-Evergreen Epiphytes	large-med.	large-med.	cor.
74. Narrow-Leaved Epiphytes	med.-small	small	cor.
75. Broad-Wintergreen Epiphytes	medium	medium	cor.
Thallophytes (poikilohydrous)			
76. Mat-Forming Thallophytes	med.-large	med.-small	mal.-cor.
77. Xeric Thallophytes	small-dwarf	small-dwarf	cor.-crust.

cor. = coriaceous
scler. = sclerophyllous
mal. = malacophyllous
lign. = ligneous

pub. = pubescent
succ. = succulent
crust. = crustose

Trees

True trees (as opposed to rosette-trees) are tall woody phanerophytes with a single main stem (trunk) and usually well developed lateral branches forming a more or less characteristic crown. Typical branching and crown forms may range from multi-stemmed 'overgrown bushes' (treated separately below) through typically 'dendritic' (see de Laubenfels 1975) to the highly monopodial columnar or umbrella-like growth forms of most conifers. Trees can grow to over 100 m in height, but heights of 10–30 m are typical. Trees shorter than 5–10 m may be called treelets, and certain scrubby but very tree-like forms of 1–5 m may be called dwarf-trees (e.g. 'campos cerrados' of Brazil, trunk-succulent euphorbs of Africa). Some understorey trees taller than 10 m, mainly in the tropics, may also be called treelets in order to distinguish them from much taller canopy trees.

Trees may have broad, narrow (linear), or needle/scale leaves, or may have no leaves at all (e.g. *Haloxylon ammodendron*). Tree leaves may be large to quite small and may be evergreen (always present, without regard initially to actual length of tenure), deciduous (summergreen or raingreen), or semi-evergreen (persistent depending on degree of cold and/or drought).

Trees generally have well-developed underground root systems, but these may be deep in drier environments or quite shallow in environments with high water tables and permanently saturated soil (e.g. not more than 50 cm in many equatorial rainforests, necessitating the characteristic above-ground buttressing). Trees generally have significantly more standing biomass above ground than below. Since trees generally have the largest transpiring surface area of any growth form, they are more important in wetter climates of the world. Because of their size trees require more time to develop but then generally attain dominance by shading shorter growth froms and by more effective water uptake from greater area and depth around them. Trees occur but do not attain dominance only where water is too limited for sufficient canopy development or where tree invasion is prevented in the seedling stage by well developed grass covers.

Stands of trees in which adjacent tree crowns overlap to form a more or less closed canopy are called forests (except for artificial plantations). Formations which do not present at least a seasonally closed canopy are called (open) woodlands and generally have better developed, more diverse understoreys. The term rainforest generally refers to closed, mesic evergreen forests which also have at least one closed understorey.

Broad-leaved trees

Broad-leaved trees have leaves which are generally of at least microphyll size and are produced in order to make maximum use of at least seasonally favorable growing conditions. Where evergreen, these trees are often dominants. Where leaves are produced anew seasonally, broadleaved trees can dominate but may be outcompeted in marginal areas (e.g. boreal mixed forests) by evergreen sclerophyll trees better able to utilize a short growing season. Leaves produced seasonally often attain higher photosynthetic rates but are generally softer (malacophyllous) than evergreen leaves and have less control over water loss. Deciduous trees generally transpire freely during their growing seasons and evade drier conditions by shedding their leaves. Leaves produced for longer tenure (generally one year for semi-evergreens and at least one year for true evergreens) become harder and are described as coriaceous (leathery, i.e. still pliable) or sclerophyllous. Such leaves usually photosynthesize more slowly but control water loss better.

General references: Kozlowski 1962, 1971; Gates & Schmerl 1975; Kramer 1962; Larcher 1976; Levitt 1972; Mooney 1974; Leathart 1977; Lieth 1974; Lieth & Whittaker 1975; Thoday 1931; Tranquillini 1979; Cooper 1975; Elton & Meentemeyer 1979.

1. Tropical Rainforest Trees are the tall, evergreen canopy-trees of equatorial regions with abundant, often daily rainfall, high humidity, and short if any dry seasons. The leaves of these trees are generally fairly large, entire, and most often described as coriaceous. Such leaves are designed to withstand short daily periods of water stress around midday but do not require stronger reinforcing, as against protracted drought or winter cold. Because of the climatic differences at canopy level and within the rainforest, trees typically show distinct morphologic differences between their more xeromorphic canopy leaves and more mesomorphic lower leaves.

Tropical rainforest trees are typically 30–50 m tall. Although the great canopy height is the result of generally favorable year-round growing conditions, the heights of individual trees are controlled also by competition for light. This can be seen easily where terrain undulates but canopy level remains relatively constant. Tropical rainforest tree crowns usually have a characteristic rounded appearance. Brünig (1970) has described a number of crown forms characteristic of different radiation environments. Because tropical rainforest soils are often saturated to within 1–3 decimeters of the surface, the trees root laterally and form dense networks of roots within the top 3 dm of the soil. Above-ground prop-roots and buttressing are common in order to provide additional support. Tropical rainforest trees have been described as lauraceous in appearance, but the actual diversity in species and even families is overwhelming. Many species still have not been identified. The list of characteristic families is very long but includes especially *Lauraceae* (s.l.), *Rubiaceae, Anacardiaceae, Moraceae, Dipterocarpaceae, Sapindaceae, Sapotaceae, Meliaceae, Vochysiaceae, Myrtaceae,* and some *Leguminosae* and *Compositae.*

As one ascends from lowland equatorial rainforests one finds that the adjacent montane rainforests are often even wetter and luxurious. Epiphytes and some other forms increase, but the canopy trees generally decrease gradually in height, and the number of distinct strata generally decreases from as many as 5–6 to 2–3 at mid-mountain. Tropical montane rainforest trees are much like the lowland trees but become less tall and gradually show somewhat smaller leaves. Diversity is often greatest at the base of windward mountain slopes and decreases upward.

References: Whitmore 1975; Richards 1952; Brünig 1970; Aubert de la Ruë et al. 1957; Walter 1973; Hueck 1966; Lind & Morrison 1974; Shreve 1914; Schnell 1970–77; Troll 1959; Hallé et al. 1978; UNESCO 1978; Vareschi 1968; Fosberg 1960; Bünning 1956; Meggers et al. 1973; Hargreaves & Hargreaves 1965; Knapp 1965, 1973; Daubenmire 1978.

2. Tropical Evergreen Microphyll Trees include a wide variety of tropical evergreen trees with either compound leaves consisting of numerous small leaflets (prototypically *Leguminosae*) or small simple leaves. The leaves are typically coriaceous. These trees occur widely in closed forests and in open woodlands throughout the tropics, both as canopy trees and as understorey forms, but they almost always represent admixtures or at best co-dominants. Characteristic taxa include not only *Leguminosae* but also a variety of other families with typically compound leaves, such as *Meliaceae, Rutaceae,* and *Simaroubaceae,* plus small-leaved eucalypts, etc. Specific examples include *Grevillea robusta* (silky oak of Australian rainforests), *Swietenia* (mahogony), *Butea frondosa, Schleichera trijuga* (*Sapindac.*), *Caesalpinia coriaria,* and *Crescentia cujete* (calabash tree). Tropical evergreen microphyll trees can be important rainforest components but become more important toward the drier margins of tropical evergreen forests, which Vareschi (1968) preferred not to call true rainforests. His comparison of leaf sizes (biótipos) in an evergreen 'selva eupluvial' in Borneo and a largely raingreen 'selva alisia' in Venezuela shows the generally smaller size of the evergreen leaves in the drier Venezuelan environment. Tropical evergreen microphyll trees are often shorter than canopy trees and form evergreen understoreys in tropical 'semi-evergreen forests' which have largely deciduous canopies (see, for example, Eyre 1968). Such semi-evergreen tropical forests are well developed on Trinidad (Beard 1946), as belts around the tropical rainforests of South America and southern-southeastern Asia (Eyre 1968; Stamp 1924), in northern Australia, in scattered parts of West Africa ('dry evergreen forest', Eyre 1968), and in Indo-Malaysia (Eyre 1968; Schimper 1898). The tropical evergreen microphyll trees of northern Australia are mainly eucalypts.

References: Eyre 1968; Vareschi 1968; Richards 1952; Whitmore 1975; Beard 1946, 1955, 1967; Stamp 1924; Ogawa et al. 1961; Milton Moore 1970; Brockman 1968.

3. Tropical Evergreen Sclerophyll Trees include prototypically the tall, more mesic, tropical and subtropical eucalypts of Australia plus a variety of similar trees in other areas. These occur in moderately humid to distinctly subhumid environments, extending well into subtropical Australia. Such eucalypts have been successfully introduced into similar tropical and subtropical climates, including highlands, throughout the world. The leaves are classified as broad but are typically elongate, often somewhat curved (*Eucalyptus*), and may be broadly linear in some species. Eucalypt leaf textures range in a continuum from coriaceous in more mesic situations to sclerophyllous in drier situations, with any division being probably arbitrary and not really important. The sclerophylly characteristic of Australian evergreen trees is considered (e.g. Webb 1959) to be an adaptation to erratic rainfall, an attribute which serves well also when these trees are introduced outside Australia. Total leaf area also varies with conditions but is typically low for such tall trees. *Eucalyptus* wood is often lightweight, and the trees can grow quickly. Tropical evergreen sclerophyll trees occur most commonly in the semi-evergreen and sclerophyll forests of northern and eastern Australia and as admixtures in the Australian rainforsts (Milton Moore, 1970; Webb 1959; Eyre 1968). Though extending to 35 °S in mediterranean climates, the leathery-leaved Jarrah (*Eucalyptus marginata*) can be included also. (The more mesic and much taller *E. diversicolor*, Karri, is more problematic.) Equivalent sclerophyllous trees can be found especially in East Africa (Burtt 1942; Lind & Morrison 1974), but the success with which eucalypts have spread after introduction into other tropical and subtropical areas suggests a particular ecological strategy superior to that developed in similar climates elsewhere.

References: Milton Moore 1970; Keast et al. 1959; Beard 1955, 1965, 1967; Eyre 1968; Walter 1968, 1973; Knapp 1973; Burtt 1942; Lind and Morrision 1974; Webb 1959; Gentilli 1960; Aubert de la Ruë et al. 1957; Specht 1972; Min. Agric. Rabat 1960.

4. Temperate Broad-Evergreen Trees include several types of tall and shorter trees which occur in milder, generally maritime temperate-zone climates with a variety of rainfall regimes. The broad leaves of these trees can be large or small but are generally harder than those of tropical trees (i.e. sclerophyllous) since they must withstand winter cold as well as varying degrees of periodic drought. Temperate evergreen trees are often co-dominants but are also often outcompeted in mixed stands by faster-growing, softer-leaved and taller summergreen trees. At least three distinct subtypes occur, namely those of warm-temperate summer-rain, winter-rain and maritime-perhumid climates. The eucalypts of Australia present a special problem because of their height and lack of major physiognomic differences in different climates. The tall, broad-leaved eucalypts of mesic southern Australia and Tasmania can be grouped with the temperate broad-leaved rainforest trees of other areas, while the more xeromorphic majority of the tall eucalypts are classified as Tropical Evergreen Sclerophyll Trees, even in subtropical areas.

4a) Warm-Temperate Evergreen Trees are the evergreen broad-sclerophyll trees of subtropical areas with summer rainfall, mild winters with some frost, and some summer moisture stress due to high potential evaporation rates. Such climates occur primarily near the east sides of land masses in the subtropical belts, namely in southern China and Japan, West Indies and southeastern USA, southern Brazil, southeastern Australia, southeastern South Africa and northern New Zealand. The leaves of these trees are typically mesophyllous to microphyllous in size. Some resist saltwater spray, as in coastal southeastern USA (e.g. *Quercus virginiana*, see Boyce 1954). Typical taxa include *Quercus* (USA and Far East), other *Fagaceae* in China and Japan (e.g. *Lithocarpus, Castanopsis*), *Ilex* (USA and Far East), *Persea* (USA and South America), other *Lauraceae* (USA, Far East, South America), *Nothofagus* (Australia, New Zealand, Chile-Argentina), and mainly tropical taxa in southeastern South Africa. Several of these genera (usually different species) occur also in winter-rain climates, such as *Quercus, Lithocarpus* and *Castanopsis* in California and *Quercus* in the Mediterranean area. The *Citrus* species, though cultivated most commonly in mediterranean climates, are native to summer-rain climates and appear to belong in this type.

References: Daubenmire 1978; Hueck 1966; Knapp 1965, 1973; Wang 1961; Ogawa et al. 1961; Numata 1974; Boyce 1954; Küchler 1964; Zinderen-Bakker 1973; Monk 1966, 1968; Kan et al. 1965; Wells 1928, 1942; Kusumoto 1961; Brockman 1968; Walter 1968; Holloway 1954; Webb 1959; Lindman & Ferri 1974.

4b) Mediterranean Evergreen Trees are the evergreen broad-sclerophyll trees of the warm-temperate to subtropical areas with mild and rainy winters but very dry summers. Such 'mediterranean' climates occur on the west sides of land masses at the same latitudes or slightly higher (30°–44°) than the summer-rain subtropics, namely around the Mediterranean Sea, at the southwest tip of South Africa, along the southwest coasts of Australia, in southern California, and in central Chile. Mediterranean Evergreen Trees differ little in physiognomy from many Warm-Temperate Evergreen Trees, with the exception that mediterranean trees may be shorter in stature with more spreading crowns. The leaves of mediterranean evergreen trees are usually microphyllous in size and typically quercoid. Many of these leaves may be shed during especially dry summers. Leaf tenure is commonly one year, the old being shed in spring as the new are produced. Many taxa resist saltwater spray. The most characteristic genus in the Northern Hemisphere is *Quercus*, including *Q. suber* and *Q. ilex* around the Mediterranean and *Q. chrysolepis* and *Q. wislizenii* in California-Arizona. Other typical taxa include *Olea*, *Ceratonia siliqua*, and *Arbutus unedo* around the Mediterranean and *Arbutus menziesii*, *Castanopsis chrysophylla*, *Umbellularia californica* (*Laurac.*) and *Lithocarpus densiflorus* in California. *Leucadendron argenteum* (*Proteaceae*) appears to be the only important mediterranean tree of South Africa, its relatives all being shrubs (Walter 1968). Australia is problematic, since it has produced tall eucalypts in its mediterranean areas rather than more typical mediterranean tree forms. Mediterranean evergreen trees of central Chile include *Lithraea caustica* (*Anacardiac.*), *Quillaja saponaria* (*Rosac.*) and *Peumus boldus (Monimiac.)*.

References: Walter 1956, 1968, 1975; di Castri & Mooney 1973; Hueck 1966; Cooper 1922; Brockman 1968; Oppenheimer 1932; Larcher 1961, 1970, 1972; Sauvage 1961; Horvát et al. 1974; Zohary 1973; Zohary & Orshan 1966; Pisek & Rehner 1958, 1960; Braun-Blanquet & Walter 1931; Rouschal 1939; Grieve 1955; Lange & Lange 1963; Ern 1966; Lüdi 1935; Schmithüsen 1956; Adamson 1927; Specht 1969, 1972; Knapp 1965, 1973.

4c) Temperate Broad-Leaved Rainforest Trees are the often larger-leaved but still sclerophyllous trees of warm-temperate maritime climates with year-round rainfall. Such 'rainforest' climates can occur on temperate-zone windward west coasts (e.g. southern Chile, southern New Zealand, Atlantic Europe, northwestern USA) and on subtropical east coasts which get summer rainfall from tropical systems (e.g. southern China and Japan, southeastern USA, southeastern Australia, northern New Zealand, southern Brazil). Such areas differ climatically from adjacent mediterranean areas by having no summer drought and from summer-rain areas by having higher annual rainfall and humidity. Temperate broad-leaved rainforest trees can have much larger leaves, especially in Chile and the warmer east-coast areas (e.g. *Magnolia grandiflora*), or can have the microphyll-mesophyll leaves most typical of *Nothofagus* in the Southern Hemisphere (see Webb 1959). Typical taxa include *Magnolia grandiflora* and *Gordonia lasianthus* (*Theaceae*) in southeastern U.S.A.; *Laurelia* (*Monimiaceae*), *Eucryphia* (*Eucryphiac.*), and the shorter *Aextoxicon punctatum* (*Aextoxicac.*) and *Drimys winteri* (*Magnoliac.*) in southern Chile; a variety of *Apocynaceae* (e.g. *Aspidosperma*) and *Lauraceae* (e.g. *Phoebe, Nectandra*) plus many others in southern Brazil (see Hueck 1966); *Nothofagus* in Australia and New Zealand, plus very tall *Eucalyptus* spp. in Australia (e.g. *Eu. regnans, Eu. saligna, Eu. diversicolor*); and various *Lauraceae* (e.g. *Machilus, Actinodaphne*) and *Magnoliaceae* in southern China and Japan.

References: Daubenmire 1978; Hueck 1966; Knapp 1965, 1973; Wang 1961; Ogawa et al. 1961; Numata 1974; Monk 1968; Kan et al. 1965; Radford et al. 1968; Brockman 1968; Walter 1968, 1974; Kusumoto 1961; Webb 1959; Holloway 1954; Lindman & Ferri 1974; Veblen 1979.

5. Raingreen Broad-Leaved Trees are generally tropical trees with larger or smaller, usually malacophyllous broad leaves which are lost in response to a pronounced, regularly occurring dry season. Tropical climates with such dry seasons are found mainly along the tropical margins in each hemisphere, where the seasonal shift of the global atmospheric circulation patterns causes dry, subsiding air masses during the low-sun period of the year. This tropical 'winter' is not cold (except in highlands) and in many areas is in fact the warmest season, due to the lack of a cloud cover. The raingreen leaves may be shed at the beginning of the dry season (suggesting some degree of adaptation to seasonal drought) or may be lost more gradually as the drought progresses. Experiments performed with raingreen trees in various botanical gardens, in which water was provided throughout the dry period, have suggested that leaf-fall in tropical trees (raingreenness) is facultative, as opposed to the more obligatory leaf-fall and winter dormancy in temperate deciduous (summergreen) trees (see Walter 1973). The length of the dry season normally increases as one goes toward the subtropical arid margins of tropical summer-rain climates, causing length of wet season and generally also total precipitation (both annual and wet-season monthly) to decrease correspondingly. As this happens the character of the vegetation changes also, from quite mesomorphic raingreen 'monsoon forests' which resemble rainforests during the wet season (e.g. southern Asia) to much more open, xeromorphic 'dry forests' with less undergrowth (e.g. East Africa). Although the change in physiognomy is primarily a property of the vegetation stand, there are also some changes in the form and physiology of the constituent raingreen trees. As a result, two subtypes are distinguished.

5a) Mesic Raingreen Trees are generally taller and closely resemble rainforest trees except for their deciduousness. Leaves may be either large or small, the latter often occurring as compound leaves in *Leguminosae.* The leaves are designed to produce optimally during the wet season and then be lost. Their size and structure are thus affected only indirectly by the relative lengths of wet and dry seasons. There is a great variety of raingreen trees, but typical taxa include *Tectona* (teak), *Terminalia, Shorea, Xylia* and various *Dipterocarpaceae* in southern and southeastern Asia; various *Anacardiaceae, Bombacaceae, Moraceae* and *Leguminosae* in Central America and northern South America; a great number of *Leguminosae* in Brazil; and a variety of scattered taxa, including many *Leguminosae* in Africa. Mesic raingreen forests and semi-evergreen forests (with many mesic raingreen canopy trees) are best developed in southern and southeastern Asia (including Indonesia) and are also important in Caribbean America and in the mountains of the summer-rain tropics.

5b) Xeric Raingreen Trees include a somewhat greater variety of both tall (enormous, e.g. *Adansonia*) and shorter (e.g. *Acacia*) trees which in one way or another have a more xeromorphic structure. Leaves may be either entire mesophylls (e.g. *Colophospermum mopane*) or the more common soft, often compound (e.g. *Leguminosae*) microphylls. Two important and quite characteristic subtypes are the trunk-succulent 'bottle trees' of Africa (e.g. *Adansonia digitata*) and Brazil (e.g. *Cavanillesia arborea*) and the spreading, flat-topped umbrella acacias of the African savannas. More typical, dendritic taxa include above all the many *Leguminosae* of the open 'miombo' woodlands of Africa (e.g. *Brachystegia, Isoberlinia, Julbernardia*), the 'alisio' and similar forests of Caribbean America, and scattered other woodlands and savannas on all tropical continents. Dry raingreen forests and open woodlands are best developed in eastern (highlands) and south-central Africa, in a sub-Saharan belt across northern Africa, in the Caatingas and 'chaco' regions of Brazil, along the Pacific coast of Mexico and Caribbean coast of Venezuela, through much of central and western India, and in the 'dry belts' of Southeast Asia. Xeric raingreen trees are the main trees in the extensive African and other savannas, with the exception of Australia. Even among the typically evergreen eucalypts of Australia, however, there are several raingreen species (e.g. *Eu. alba, Eu. brachyandra*) which occur in the savannas and open woodlands of the summer-rain northern sections (Walter 1973; Keast et al. 1959), especially in the Kimberley Plateau area.

References: Walter 1939, 1973; Hueck 1966; Knapp 1965, 1973; Eyre 1968; Daubenmire 1972, 1978; Lind & Morrision 1974; Bhatia 1958; Beard 1944a, 1944b, 1953, 1955, 1967; Boaler 1966; Lauer 1952; Burtt 1942; Ernst 1971; Mani 1974; Malaisse 1974; Owen 1974; Stocker 1970, 1971; Keast et al. 1959; Müller 1977; Werger 1977; UNESCO 1978; Aubert de la Ruë et al. 1957; Meggers et al. 1973; Schmidt 1973; Sreedhara 1978; Shrimal & Vyas 1977; Misra & Gopal 1968; Mata et al. 1972; Schnetter 1968; Scholz 1967; Fanshawe 1969; Stamp 1924; Ogawa et al. 1961; Champion & Seth 1968; Ewel & Madriz 1968; Golley & Medina 1975; Lawson et al. 1970.

6. Summergreen Broad-Leaved Trees are generally temperate-zone trees (tropical in some mountains but never equatorial) with large to small, generally malacophyllous leaves which are lost in response to winter cold. Since the leaves are usually soft and designed for high productivity during the warm season, broad-summergreen trees are typical of temperate areas with summer rainfall and extend into dry and summerdry climates only on moist sites, such as along rivers. Summergreen trees are the characteristic dominants of the typical four-season (nemoral) temperate climates of eastern North America, much of Europe, and eastern Asia. For a variety of reasons, summergreen trees in the Southern Hemisphere are restricted to small areas in New Zealand and the southern Andes. Summergreen leaves are generally 'killed' by autumn frost and gradually fall off. Before falling they typically turn bright yellow and/or red (orange, violet) for a 2–4 week period in mid-autumn, the intensity of coloration being greatest in areas with many warm, sunny autumn days but cool, clear nights (i.e. North America and East Asia more than Europe). Refoliation occurs when springtime temperatures are sufficient and frost has subsided. Both leaf-fall and sufficient winter cold (vernalization) are apparently required by summergreen trees. This can be seen from the fact that summergreen species shed their leaves even in the mild winters of south-central Florida (January mean temperatures around 15° C) and by the fact that summergreen trees grown in the winterless tropics suffer from confused flowering cycles and gradually die. Summergreen tree taxa are mostly holarctic and common to the three major Northern Hemisphere formations of summergreen forest, as well as to more local formations. These taxa include *Quercus, Fagus, Castanea* (all *Fagaceae*), *Acer, Tilia, Fraxinus, Ulmus, Carya* and *Carpinus*. As summers become cooler and shorter, summergreen leaves tend to become softer and smaller and do not survive protracted drought (e.g. Neuwirth & Polster 1960). The main taxa from these cool boreal and montane areas are *Betula, Salix, Populus* and *Alnus*, plus *Nothofagus* in the Southern Hemisphere. Whereas the more shade-tolerant, malacophyllous summergreens are generally climatic-climax species in temperate areas, the boreal summergreens are usually successional species, being shaded out by the dense canopies of the taller boreal conifers. Transition zones of mixed forest are well developed in North America, Europe and East Asia, and Eyre (1968) notes that many historical cultural capitals of Europe (e.g. London, Stockholm, Moscow, Berlin) plus Boston and Beijing (Peking) all developed within this narrow belt. Smaller areas of summergreen forest and/or open woodland occur in northwestern U.S.A. (Küchler 1946), in southern Chile (Hueck 1966), in New Zealand, and across Russia and the mountains of Central Asia (Walter 1974). Summergreen forests composed of holarctic species also penetrate into the summer-rain tropics along the mountain ranges of eastern Mexico (Miranda & Sharp 1950; Leopold 1950), Bolivia-Argentina (Hueck 1954, 1966), and Southeast Asia (Ogawa et al. 1961; Walter 1968).

References: Walter 1956, 1968, 1974; Daubenmire 1978; Knapp 1965; Braun 1950; Wang 1961; Ellenberg 1963; Horvát et al. 1974; Eyre 1968; Hueck 1954, 1966; Ogawa et al. 1961; Leopold 1950; Schweinfurth 1957; Küchler 1946, 1964; Brockman 1968; Cockayne 1958; Godley 1960; Grebenshchikov 1974; Hamet-Ahti et al. 1974; Hartmann & Schnelle 1970; Lieth 1974; Miranda & Sharp 1950; Numata 1974; Radford et al. 1968; Axelrod 1966; Jäger 1968; Reichle 1970; Elton & Meentemeyer 1979; Reed 1971; Veblen 1979.

Narrow-leaved trees

Narrow-leaved trees have either flat linear leaves (e.g. *Podocarpus*) or needle-leaves, which may be reduced to even smaller scale-leaves (e.g. *Cupressaceae*). Most narrow-leaved trees are conifers (gymnosperms), which implies further that most narrow-leaved trees have strongly monopodial growth forms and tend to be tall. In general the smaller, harder leaves or needles of conifers provide greater protection during unfavorable periods than do broad leaves. As a result, narrow-leaved trees are mostly evergreen and occur, in one form or another, in all tree climates except the drier tropical summer-rain areas. In addition, the smaller, harder leaves (needles) and tall stature at maturity permit maintenance of greater total leaf areas than those of broad-leaved trees. In some areas, such as boreal and wet mediterranean climates, this provides a definite competitive advantage, and tall needle-leaved trees become the dominant form, despite relatively slow growth rates. Narrow-leaved trees are rarely climatic dominants in the tropics and subtropics but remain important as canopy constituents and often as individually taller emergents. In all areas where evergreen plants can survive unfavorable periods they have two important advantages over deciduous and other seasonal forms:

1. Evergreens have effectively longer growing seasons since their leaves are already deployed.
2. Evergreens survive poorer nutrient supplies since they waste less on seasonally discarded leaves.

General references: Dallimore & Jackson 1966; Krüssmann 1972; Tolmatschew 1954; Tranquillini 1979; de Laubenfels 1953, 1969; Hare 1954; Hustich 1953; Tranquillini & Holzer 1958.

7. Tropical Linear-Leaved Trees are tall (often emergent), monopodial, frost-sensitive evergreen trees with flat, linear leaves and stately, somewhat conical growth. The combination of height, high leaf area and coriaceous to sclerophyllous linear leaves makes these trees well suited to any warm humid climate with only short stress periods. As a result, such trees (primarily species of *Podocarpus, Agathis* and *Araucaria*) are found not only throughout the humid tropics but also throughout the largely frostless maritime climates of the Southern Hemisphere subtropics and temperate zone. The similarity between the Southern Hemisphere maritime climates and the climates of humid tropical mountains has been well recognized (Troll & Lauer 1978), and many genera and forms are common to both (Troll 1961, 1970). Tropical linear-leaved trees occur in moist tropical lowlands primarily as individuals but become more important in adjacent mountains, where they can become dominants in montane and cloud forests. Montane *Podocarpus* forests are best developed, within the tropics, in Africa (Knapp 1973; Lind & Morrison 1974) and in the East Indies (Whitmore 1975; van Steenis 1962; de Laubenfels 1969). In the southern temperate zone podocarps form important forests in New Zealand (Dawson 1962; Robbins 1962) and mix with other, more temperate conifers in mesic forests of various areas. Some podocarps and araucarias have been successfully introduced as ornamentals in northern warm-temperate and subtropical areas.

References: Troll & Lauer 1978; Dallimore & Jackson 1966; Whitmore 1975; Knapp 1973; Hueck 1966; Lind & Morrison 1974; Walter 1960, 1968, 1973; Troll 1959, 1961, 1970; van Steenis 1962; Robbins 1961, 1962; Dawson 1962; de Laubenfels 1969; Costin 1957; Kuschel 1975; UNESCO 1978; Krüssmann 1972; Godley 1975; Webb 1959.

8. Tropical Xeric Needle-Trees include a very small number of needle or scale-leaved but still relatively tall, evergreen trees which occur primarily in the drier montane forests of Africa. Because they are the tallest trees in these drier areas, these trees are generally dominants or co-dominants, but they occur also as individuals in more mesic montane forests. The main taxa are *Juniperus procera* (East Africa and Arabia), certain needle-leaved podocarps (e.g. *Podocarpus gracilior*), and *Widdringtonia* (*Cupressaceae*) in southern Africa. These taxa form dry coniferous montane forests (Knapp 1973: Trocken-Koniferen-Bergwälder) in drier mountain areas from Arabia (von Wissmann 1972) to South Africa. In general the *Cupressaceae* form the driest forests, with needle-leaved podocarps and then linear-leaved podocarps (e.g. *Podocarpus latifolius*) forming the gradient to more mesic broad-leaved montane forests (Knapp 1973). Equivalent forests from other tropical or subtropical areas have not been described, but some subtropical pine forests may be similar.

References: Knapp 1973; von Wissmann 1972; Kerfoot 1960–61; Zohary 1973; Lind & Morrison 1974; Troll 1959.

9. Temperate Rainforest Needle-Trees are tall, magnificent, monopodial trees with a variety of evergreen needle types and sizes which all add up to the same ecological significance, namely the highest leaf area indices of any plant form in the world and many of the world's tallest tree species. The needle-leaves of these trees may be flattened linear needles (as in *Taxus, Tsuga, Cunninghamia, Sciadopitys* and *Abies*), flattened scales (as in *Thuja* and *Chamaecyparis*), small falcate needles (as in *Cryptomeria, Sequoiadendron* and some species of *Araucaria*), or the equivalent of compound needles with generally flattened linear leaflets (as in *Sequoia*). The density of individual needles is always high. Most needle types are flattened for greater surface area, even in otherwise round-needled genera (e.g. *Picea sitchensis* and *P. breweriana*). Leaf area indices of 8–18 are typical (e.g. Tadaki 1977), and values as high as 50 (!) have been reported (Waring et al. 1978). Other taxa include temperate, smaller-leaved podocarps, scale-leaved *Dacrydium,* and *Saxegothaea,* all from the Southern Hemisphere. *Sequoia sempervirens* (to 110 m) and *Sequoiadendron giganteum* (to over

100 m) are the tallest trees in the world, occurring respectively along the north coast and in the Sierra Nevada of California. Temperate needle-leaved rainforest trees occur, in general, in perhumid temperate climates with mild winters and no severe frost. Such situations and the corresponding rainforests are best developed in the windward, often foggy marine west-coast climates of northern California to British Columbia, southern Chile, northwestern Europe, the west coast of New Zealand, and western Tasmania. The greater dominance of needle-leaved trees in the North American rainforest may be due to the greater mediterranean influence there, to which the needle-trees may be better adapted than broad-leaved trees. It is interesting to note that the tallest trees occur in the more mediterranean southern part of the American rainforest area, where potential evapotranspiration rates are higher. The world's tallest broad-leaved tree, *Eucalyptus regnans* of southwestern Australia, occurs in a similar climate. Elton & Meentemeyer (1979) found, on the other hand, that the tallest summergreen trees (in North America) tend to occur near the low-evaporation ends of their species ranges.

References: Daubenmire 1978; Walter 1960, 1968; Tadaki 1977; Waring et al. 1978; Hueck 1966; Leathart 1977; Eyre 1968; de Laubenfels 1953, 1969; Brockman 1968; Küchler 1964, 1977; Franklin & Dyrness 1973; Shelford 1963; Schmithüsen 1956, 1960; Numata 1974; Costin 1957; Davis 1964; Webb 1959; Cockayne 1958; Holloway 1954; Godley 1975; Troll & Lauer 1978; Williams 1974; Krajina 1969; Schuler 1977; Kilgore & Taylor 1979.

10. Temperate Needle-Trees include both generally round-needled and scale-leaved evergreen trees from a variety of temperate-zone climates with varying degrees of winter cold but generally warm, sunny summers or at least long growing seasons, as in more maritime climates. These trees are generally monopodial and of more normal height than the tall rainforest and mesic tropical conifers. The needles are generally long or of medium length (e.g. *Pinus*) or occur as tightly overlapping scales (e.g. *Cupressaceae*) which render the entire branchlet photosynthetic. Though somewhat xeromorphic in appearance, these trees can present moderately large leaf areas. The hardness of the needles, however, makes them well adapted to withstand both moderate freezing temperatures and summer water stress. 'Harder' photosynthetic surfaces generally result in lower photosynthetic and growth rates. In this respect the pines (*Pinus*) represent an anomaly, as many pine species grow quite rapidly and dominate successional vegetation until finally shaded out by more shade-tolerant but not significantly taller broad-leaved trees. Many *Pinus* species have extremely high light-saturation levels. The *Cupressaceae,* on the other hand, generally grow more slowly. Temperate needle-trees can be grouped into warm-temperate (summer rain), mediterranean (winter rain), and cold-winter subtypes, based on physiologic and morphologic adaptations to the climates they inhabit. The fact that they usually occur in corresponding but disjunct regions on the different continents reinforces their status as distinct subtypes.

10a) Heliophilic Long-Needle Trees are, prototypically, the monopodial, somewhat xeromorphic 'southern pines' of the humid subtropical and warm-temperate climates of southeastern U.S.A. and adjacent Caribbean Islands. In the U.S.A. this type includes only 'hard pine' species (Brockman 1968), which generally have fewer needles per bundle, harder needles and harder wood. The climatic limits of a prototypic species, *Pinus taeda* (loblolly pine), have been studied in detail by Hocker (1956), Mather & Yoshioka (1966), and the American commercial forestry industry. Other species from the same region include *P. palustris* (longleaf pine), *P. echinata* (shortleaf pine, with needles of only 8–12 cm in length), and *P. elliottii* (= *P. caribbea,* slash pine or Caribbean pine). Although these trees are mainly successional on sites with better soil and are often considered to be relicts, similar needle-tree forms are important and sometimes dominant in other east-coast subtropical areas with similar climates. These include *Araucaria angustifolia* in southern Brazil (Hueck 1966) and other species of *Pinus* in eastern Mexico and Central America (Leopold 1950; Parsons 1955), in Southeast Asia (Stamp 1924; Ogawa et al. 1961; Wang 1961), and in southernmost Japan (Numata 1974). Ecologically similar species of *Casuarina* (Australian pine, a leafless non-conifer) occur in subtropical-tropical eastern Australia but are frost-sensitive. *Pinus elliottii* has been widely introduced for reforestation in Australia, New Zealand and South Africa. Heliophilic needle-trees are not winter-dormant and are active whenever winter temperatures are high enough. Hocker's (1956) study suggested that *Pinus taeda* is limited both to the north and west (i.e. toward more continental climates) mainly by inability to take up sufficient water from cold or frozen soil during the winter.

References: Hocker 1956; Dallimore & Jackson 1966; Brockman 1968; Hueck 1966; Leopold 1950; Parsons 1955; Küchler 1964; Daubenmire 1978; Shelford 1963; Knapp 1965, 1973; Mather & Yoshioka 1966; Ogawa et al. 1961; Milton Moore 1970; Wells 1928, 1942; Braun 1950.

10b) Mediterranean Needle Trees include both needle and scale-leaved trees which occur in maritime winter-rain climates with greater summer moisture stress and which, as a result, are more variable in size and shape and generally grow more slowly. Included are the true cedars (*Cedrus*) of less severe montane climates, various mediterranean pines of lower and middle elevations (e.g. *Pinus pinea, P. maritima, P. halepensis*), and certain scale-leaved *Cupressaceae* from California (e.g. *Libocedrus decurreus, Cupressus macrocarpa*), Chile (*Libocedrus chilensis*), the Mediterranean region (*Cupressus* and *Juniperus* spp.), and South Africa (*Widdringtonia*). Mediterranean needle-trees are generally restricted to coastal or montane areas which have somewhat moderated summer temperatures, higher humidity, and occasional light summer rainfall and/or runoff inputs. These trees tolerate some frost but remain active in winter whenever temperatures permit. Many of these trees show characteristic flattened crowns with well developed lateral branches, as often occur in windy and water-stressed environments. The needles of *Cedrus* and the mediterranean pines are generally shorter than those of heliophilic needle-trees but longer than those of boreal conifers.

References: Dallimore & Jackson 1966; Walter 1968; Brockman 1968; Leathart 1977; Hueck 1966; Mayer & Sevim 1958; Zohary & Orshan 1966; Zohary 1973; Beals 1965; de Wilde 1961; Emberger 1939; Schmithüsen 1956, 1960; Knapp 1965, 1973; Jenny et al. 1969; Kilgore & Taylor 1979.

10c) Typical Temperate Needle-Trees are the cold-winter needle-leaved trees which occur in a variety of temperate-zone climates with extended periods of sub-freezing temperature. Such climates are probably restricted to the Northern Hemisphere, and the main taxon is the northern genus *Pinus*, including *P. ponderosa* and similar species from the drier western U.S.A., *P. strobus* and *P. rigida* from eastern U.S.A., *P. roxburghii* from the Himalaya, and *P. yunnanensis* from higher elevations in southern China. *Pinus sylvestris*, the most wide-spread tree species in Eurasia, spans the temperate and boreal zones, functioning in Europe largely as a typical temperate-zone needle-tree. These trees are largely successional species on better soils in the more mesic climates of Europe and eastern U.S.A. Despite this and the taxonomic poverty of the form, these trees can be very numerous, dominating with nearly pure stands in various montane belts, successional areas, and on sites with edaphic problems.

References: Walter 1968, 1974; Daubenmire 1943, 1978; Ern 1974; Shelford 1963; Knapp 1965; Schweinfurth 1957; Wang 1961; Numata 1974; Brockman 1968; Küchler 1964; Mather & Yoshioka 1966; Kilgore & Taylor 1979.

11. Boreal/Montane Needle-Trees are the tall, monopodial, conical to columnar evergreen trees of the boreal forest and subarctic forest-tundra, and upper montane belts of temperate-zone mountains. All are relatively short-needled conifers, mainly from the genera *Picea, Abies,* and *Pinus*. The compact evergreen needles have the ability to withstand temperatures down to −60 °C when cold-hardened and to withstand long periods of sub-freezing temperature with almost no respiration (Larcher 1976; Pisek & Rehner 1958, 1960; Polster & Fuchs 1963; Tranquillini 1963, 1964; Tranquillini & Holzer 1958). Yet when springtime or even mid-winter temperatures are high enough (air temperature around the freezing point) the leaves are quickly ready to resume photosynthesis. Boreal and montane needle-trees have dense, thick canopies with high leaf area indices (8–15 in closed forests). They attain dominance in the extensive boreal forests of Canada-Alaska and northern Scandinavia-Russia but become less important in the drier and coldest winters of Siberia, where only deciduous larches survive as important trees. At its northern edge the boreal forest opens into the forest-tundra woodland, with tall, cylindrical trees well spaced to take advantage of the low sun angle. The northernmost limit of boreal needle-trees closely follows the 10 °C isotherm for July mean temperature in most areas (Hare 1950, 1954; Hustich 1953; and many others). Closed forest generally requires 2–3 months with mean temperatures above 10 °C. Boreal/montane needle-trees also dominate characteristic montane and subalpine belts in most temperate-zone mountain ranges of the Northern Hemisphere, and there are some fairly similar montane forests in New Zealand (Walter 1960, 1968). The montane trees are generally tall and cylindrical, but a subalpine equivalent of the open forest-tundra woodland is usually absent due to greater topographic influence in mountains. Walter (1974) and others have listed equivalent North American and Eurasian species in both boreal and montane areas.

References: Walter 1960, 1968, 1974; Hare & Ritchie 1972; Hare 1950, 1954; Hustich 1953; Dallimore & Jackson 1966; Brockman 1968; Daubenmire 1943, 1978; Knapp 1965; Rowe 1959; Jäger 1968; Tranquillini 1963, 1964, 1979; Tranquillini & Holzer 1958; Pisek & Rehner 1958, 1960; Larcher 1976; Reiners & Lang 1979.

12. Hydrophilic Summergreen Needle-Trees include only three species from two genera of *Taxodiaceae* which occur in quite moist temperate environments with cool winters. The species are *Taxodium distichum* (bald cypress) and *T. ascendens* (pond cypress) of swamps and mesic forests (including hammocks) in the southeastern U.S.A. and *Metasequoia glyptostroboides* of China, which was identified from fossils and discovered living only in 1948. These trees have compound leaves consisting of relatively soft needle-leaflets. They grow slowly and require large amounts of water but attain impressive height and age, becoming dominant where sufficient water is available. *Taxodium* grows most commonly in standing water.

References: Penfound 1952; Wharton 1978; Brockman 1968.

13. Boreal Summergreen Needle-Trees are tall, monopodial boreal conifers with summergreen deciduous needles, as found in the genera *Larix* and *Pseudolarix*. Because they must first produce new leaves these trees have a shorter effective growing season than the evergreen boreal conifers. The larches have softer needles with less inactive sclerenchyma and photosynthesize rapidly at first, especially at the beginning of the growing season. After ten years or more, however, the evergreens begin to grow faster than the larches and eventually attain dominance (Walter 1968). Larches survive as individuals, both in the boreal forest and in montane forests, but dominate only in the driest and coldest winters of Siberia where it is too cold and there is too little protective snow cover for evergreen trees. Much of Siberia has relatively low rainfall and dry summers. The trees in such areas probably could not survive without the deeply frozen soil, which thaws slowly and spreads the meltwater more evenly over the course of the summer. Because of dryness and low sun angles, *Larix* stands in Siberia are often quite open. The main *Larix* species are *L. decidua* in Eurasia (including Siberia), *L. dahurica* in Siberia, *L. leptolepis* in Japan, and *L. laricina* and *L. occidentalis* in North America. *Pseudolarix* occurs only in China.

References: Walter 1968, 1974; Tranquillini 1963, 1979; Knapp 1965; Brockman 1968; Numata 1974.

Small and dwarf trees

Small trees (treelets) are most commonly simply smaller versions of larger trees. Smaller stature means less leaf surface and less standing biomass, which (other factors being more or less the same) means reduced water requirements and some reduction of photosynthetic requirements due to reduced respiration. Smaller size can thus be advantageous in forest understoreys with reduced light, in drier climates, on shallow soils with reduced water-holding capacity, and in windy environments with higher potential evapotranspiration. Smaller size can also be an advantage where saturated soil prevents the deeper root systems needed by larger trees. The term 'dwarf tree' is reserved for even smaller treelets (generally not exceeding 5 m) whose stunted growth form results from some form of more pronounced stress. Dwarf-trees and krummholz may not always be mutually exclusive, but dwarf-trees include forms such as the stunted treelets of the 'campos cerrados' of Brazil, where wind is not a factor. Increasing dryness, requiring some reduction in transpiring surface area, can be met by reduction in above-ground vegetation cover (e.g. open woodlands instead of closed forests), by reduction of plant size, or, most commonly, by some combination of these responses and others. As canopies open up and more light is available to lower strata, smaller trees become more important and can attain dominance. In fact, only in Australia (*Eucalyptus*) and in the forest-tundra (low sun angle) is it common for open woodlands still to be dominated by taller trees.

General references: Walter 1968, 1973; Milton Moore 1970.

14. Tropical Broad-Evergreen Small Trees include rainforest-understorey, seasonal forest-understorey, thicket, open-woodland and even savanna species but without obvious boundaries between subtypes. Most forms are coriaceous, with leaf sizes ranging from megaphyllous in the shade of tropical rainforests to the nanophyllous leaflets of more xeric *Leguminosae*. The diversity in rainforest understorey forms is enormous, but it is only in thickets (usually edaphic climaxes) and open woodlands or scrublands that small evergreen trees gain dominance or co-dominance in the tropics. Both usually occur locally or in mosaics. Since they do not dominate large areas, tropical small evergreen trees have been studied very little as a group. Important taxa include *Leguminosae, Eucalyptus, Myrtaceae* and *Sapotaceae*.

References: Richards 1952; Beard 1953, 1955, 1967; Milton Moore 1970; Brockman 1968.

15. Tropical Broad-Evergreen Dwarf-Trees are, prototypically, the stunted, fire-resistant, large-leaved evergreen treelets of the 'campos cerrados,' which occur in a tropical wet-dry climate on the deep soils of central Brazil, south of the Amazon rainforest. The dry season of the 'campos cerrados' is not as extreme as is more common in much of Africa, but the deep soil also appears to be necessary for the development of an evergreen woodland in Brazil. The treelets are typically 2–4 m tall, often with twisted trunks, and make a strikingly xeromorphic impression (see, for example, Hueck 1966) which Walter (1973) and others consider to be 'oligotrophic scleromorphy'. As a well developed formation, the Brazilian 'campos cerrados' appear to be unique, though similarly stunted and xeromorphic treelets occur also in other areas, including the llanos of Venezuela and (azonally) on laterite soils in the Congo basin in central Africa (Walter 1973). The most important taxa in the 'campos cerrados' include *Kielmeyera, Byrsonima, Erythroxylum, Hancornia, Xanthoxylum, Machaerium* (*Leg.*) and many others.

References: Ferri 1973; Rawitscher 1948; Ferri & Coutinho 1959; Walter 1973; Hueck 1966; Cole 1960; Goodland 1971; Eiten 1972; Knapp 1973.

16. Cloud-Forest Small Trees include a variety of both broad-microphyll and narrow-leaved smaller trees growing in and above the cloud-belts of tropical and subtropical mountains and in similarly perhumid climates down even to sea-level along windward coasts in the Southern Hemisphere. The trees may be visibly stunted and gnarled, both near alpine treeline or other exposed sites and along windy coastlines. In the montane cloud-forests and subalpine 'elfin woodlands', however, the trees are often almost obscured by their dense coverings of mosses and other epiphytes. *Podocarpus* species are especially important in the cloud-forests of both the Andes and East Africa (as well as the southern temperate zone), while holarctic taxa such as *Quercus* and *Pinus* are important in both Mexico and Southeast Asia. *Eucalyptus* and *Nothofagus* are important in the Southern Hemisphere. *Ericaceae* are generally important as shrubs, but *Erica arborea* and *Philippia excelsa* are important small trees in African subalpine forests, along with *Hagenia abyssinica* (*Rosac.*), *Olea chrysophylla, Ilex witis* and *Juniperus procera. Ericaceae* and *Melastomaceae* are important as small trees in Central America. Despite the constantly high humidity, large leaves are not an advantage in the cloud forest because of the constantly cool temperatures and possibility of light frost. The leaves are commonly hard.

References: Troll & Lauer 1978; Troll 1959, 1966, 1968, 1970, 1973; Walter 1968, 1973; Richards 1952; Lind & Morrison 1974; Whitmore 1975; Hueck 1966; Knapp 1965, 1973; Vareschi 1973; Costin 1957; Daubenmire 1978; van Steenis 1962; Shreve 1914.

17. Temperate Broad-Evergreen Small Trees include a variety of both understorey and low-canopy trees with generally mesophyllous or slightly smaller (e.g. notophyllous, see Webb 1959) sclerophylls designed to withstand some winter cold. These trees occur most commonly as slow-growing understorey trees (e.g. *Ilex*, some *Quercus* spp., some *Prunus* spp.) but can form the upper stratum in drier forests and open woodlands (e.g. *Quercus*) and in coastal and moist montane forests of the cooler Southern Hemisphere (e.g. *Nothofagus*, montane *Eucalyptus* spp.). These trees are perhaps most important as dominants in the extremely stenothermal, perhumid, windward-coastal climates of southern Chile (*Nothofagus dombeyi, N. betuloides*), subalpine Tasmania and Australia (*N. cunninghamii, Eucalyptus coccifera, Eu. subcrenulata*), and southern New Zealand (*N. menziesi, N. fusca* et al.). In colder areas these trees do not always form closed stands, even in very moist climates.

References: Walter 1960, 1968; Hueck 1966; Schmithüsen 1956; Costin 1957; Cockayne 1958; Godley 1960; Webb 1959; Veblen 1979.

18. Broad-Raingreen Small Trees include many taxa of deciduous forest, woodland and savanna trees but most typically small or compound-leaved forms such as the *Leguminosae*. In more open areas these trees tend to be low, spreading and flat-topped, as are the familiar scattered acacias of the African savannas. In more wooded but scrubby areas these trees may occur as individual emergents or simply as part of the more amorphous thorn-scrub. Small raingreen trees can also form understoreys in raingreen forests or occur as less conspicuous understorey components in semi-evergreen forests. Important taxa include *Acacia, Albizia* and a great number of other *Mimosaceae, Papillionaceae* and *Caesalpinaceae* (legumes), plus such others as *Bulnesia sarmientoi* (a raingreen zygophyll) and *Zizyphus mistol* (*Rhamnac.*) from the South American 'chaco' and *Commiphora* (*Burserac.*) from Africa and India. The leaves (leaflets) of many *Leguminosae* can be extremely small.

References: Walter 1973; Knapp 1973; Hopkins 1962–70; Hueck 1966.

19. Broad-Summergreen Small Trees include mesic understorey and more xeric open-woodland taxa of the temperate zone, both of which most typically have leaves of mesophyll size. Particularly important taxa include *Prunus* and other fruit trees, the colorful *Cornus florida* (dogwood) of eastern North America, *Cercis*, and certain species of *Quercus* in more open, scrubby woodlands and of *Nothofagus* in Chile and New Zealand. This form occurs most commonly in mesic forest understoreys but gains dominance in degraded vegetation (e.g. the šibljak of the Balkans), in drier areas (e.g. the summer-green scrub-oak woodlands of intermontane western U.S.A.), and in cooler, windier forest areas of southern Chile (*Nothofagus pumilio* and, more typically, *N. antarctica*) and New Zealand. A particularly interesting occurrence of smaller summergreen trees as dominants is in the wide-spread open parklands of mountains from the Caucasus and Iran to Mongolia and northwestern China. These parklands are composed mainly of fruit and nutbearing trees, including *Malus, Pyrus, Prunus, Amygdalus, Cerasus, Armeniaca, Crataegus, Pistacia* and the generally larger *Juglans fallax* (Walter 1974; Walter & Box, in press). The trees of the montane parklands occur in both summer-rain and winter-rain climates but receive additional water in summer from runoff. Small summergreen trees are also important in submediterranean transitional climates (e.g. southern Europe) where they form low, mixed woodlands with mediterranean shrubs.

References: Walter 1956, 1968, 1974; Horvát et al. 1974; Zohary 1973; Schweinfurth 1957; Hueck 1966; Küchler 1964; Walter & Box, in press; Reichle 1970; Veblen 1979; Reich & Hinckley 1980.

20. Needle-Leaved Small Trees are small, generally slow-growing, xeromorphic evergreen trees which include primarily certain arborescent junipers, certain smaller pines (e.g. the piñon pines of southwestern U.S.A.) and certain convergent forms in other areas (e.g. *Callitris* and *Actinostrobus* in Australia). These trees can have either true needles (e.g. *Pinus*) or scale-leaves (e.g. *Juniperus*). They occur most commonly in dry climates and generally become dominants only there, but can also occur in quite mesic climates (e.g. the widespread *Juniperus virginiana* of eastern North America and *J. communis* of both Europe and North America). These trees are particularly conspicuous as individuals in heaths and similar vegetation on sandy substrates. In the western U.S.A. small junipers and piñon pines occur together in characteristic low, quite open woodlands covering large areas. *Actinostrobus* and even some *Proteaceae* (e.g. *Banksia ericifolia* and *B. spinulosa*, the latter with long 'needles'), *Leguminosae* (phyllodious *Acacia* spp.), and *Casuarina* spp. are similar in Australia, particularly in the 'sand heath' areas. *Juniperus* and

Pinus are common in the mountains of drier parts of Asia.

References: Walter 1968, 1974; Küchler 1964; Shelford 1963; Knapp 1965, 1973; Milton Moore 1970; Zohary 1973; Specht 1972; Brockman 1968.

Rosette-trees

Rosette-trees are trees which grow from a single terminal bud to produce, typically, a monopodial, unbranched bole carrying a terminal rosette (tuft) of usually large, often compound evergreen leaves. Such 'trees' are also called tuft-trees (e.g. Küchler 1967). The trunk grows not by adding a new layer each year, as true trees do, but rather from the inside by adding new cells within the expanding trunk-like structure formed originally from remnants of earlier parts of the terminal rosette. Trunks are most commonly unbranched but may be clustered, entirely underground (appearing only later in some taxa), missing altogether (rosette-shrubs), or moderately branched. True rosette-trees are monocots. Palms are the most common taxon, but other taxa have also produced rosette-tree or equivalent forms, such as the Joshua tree (*Yucca brevifolia*) of the Sonora-Mojave desert region, the tree-ferns, the 'grass trees' of Australia, other *Liliaceae,* and certain equivalent tropical-alpine dicots (e.g. *Senecio keniodendron*). The leaves, morphologically, may be either simple or compound (pinnately or palmately), but the ecological unit, i.e. the simple leaf, leaf segment, or frond segment, generally has a characteristic linear or broad-graminoid shape and may be hardened or even succulent to reduce water loss. All species are evergreen and discard older leaves only as new ones are produced. Many rosette-tree forms (though not the tree-ferns and not all palms) are associated with high-insolation situations. Rosette-trees are also associated with generally frost-free climates or at least with situations (e.g. tropical alpine areas) where frost does not last longer than overnight.

General references: McCurrach 1960; Langlois 1976; Moore 1973; Hallé et al. 1978; Aubert de la Ruë at al. 1957; Corner 1966.

21. Palmiform Tuft-Trees are the typical tall, monopodial evergreen 'palm trees' of the tropics, most of which are palms. Most are unbranched and carry terminal tufts of strikingly large compound leaves borne on stout, often green branches. The leaves are generally sclerophyllous and may be pinnately or palmately compound. Physiognomically similar nonpalm taxa which may attain tree height include *Phenakospermum, Ravenala* and *Strelitzia* (*Musaceae*), *Carica papaya* and other *Caricaceae,* some arborescent *Xanthorrhoeaceae* and *Bromeliaceae* (more commonly treelets), and some *Pandanaceae* (usually branched and more commonly treelets or non-trees). Palmiform tuft-trees occur in a wide range of generally frost-free tropical climates with sufficient water provided by rainfall and/or groundwater. Because palms are mostly quite tolerant of saturated soil they are often the only tree forms found in such places as tropical coastlines (also salt spray-tolerant), desert oases, and seasonally (or irregularly) flooded floodplains and savannas. The sclerophyllous leaf structure of palms apparently provides quite adequate protection against water loss in even the driest desert climates, provided that groundwater is available. Palms grow rapidly and are most often successional species, in many ways the tropical ecological equivalent of pines. Palmiform trees can dominate in a variety of situations, both dry and mesic, where other trees are excluded. Palmiform trees generally require high light intensity, however, and occur in forests mainly as successional species or where climatic-climax trees are precluded by soil or other conditions (as opposed to many palmiform treelets which are understorey forms). Because of the milder, more maritime winter climates, palmiform trees extend further poleward in the Southern Hemisphere than in the Northern.

References: McCurrach 1960; Langlois 1976; Moore 1973; Walter 1973; Hueck 1966; Knapp 1973; Richards 1952; Lind & Morrison 1974; Whitmore 1975; Hallé et al. 1978; Aubert de la Ruë et al. 1957; Fosberg 1960; Corner 1966.

Rosette-treelets

Rosette-treelets (tuft-treelets) are small rosette-trees which do not attain tree proportions (generally at least 5–8 m) and which generally show a somewhat greater variety of forms and ecological relations. Like rosette-trees, the rosette-treelets are evergreen, most typically monopodial and unbranched (with a variety of exceptions), and grow with a terminal tuft of large leaves. Rosette-treelets, however, also occur as less light-demanding understorey forms, as groundwater-independent xeric forms, and include the ecophysiognomically similar but taxonomically unrelated tree-ferns and tropical-alpine tuft-dicots.

22. Palmiform Rosette-Treelets are smaller equivalents of Palmiform Rosette-Trees but can occur in a greater variety of forms and habitats. Palmiform rosette-treelets can be either light-demanding or understorey forms, xeric sclerophylls or quite mesic forms with softer leaves. Clumped trunks and some degree of branching may be more common, and some forms can take on the appearance of huge palmiform bushes (i.e. with very short trunks but yet with fronds reaching heights of 8 m). Palmiform treelets are much more common in climatic-climax tropical forests than are palmiform trees, and they are also important in certain temperate maritime climates of the Southern Hemisphere. Palmiform treelets tolerate saturated soil well and can be important in periodically flooded or waterlogged areas. The true palms are the main representatives but cycads and some palmiform *Musaceae, Pandanaceae,* etc. also occur as rosette-treelets.

References: McCurrach 1960; Langlois 1976; Moore 1973; Walter 1973; Hueck 1966; Knapp 1973; Richards 1952; Lind & Morrison 1974; Whitmore 1975; Hallé et al. 1978; Aubert de la Ruë et al. 1957; Fosberg 1960.

23. Tree-Ferns are woody, arborescent ferns which are quite similar in growth form to palmiform treelets but more demanding ecologically, requiring cool, very moist, typically mesic-understorey conditions. Tree-ferns are found most commonly in the cloud-belt forests of tropical mountains and in extremely maritime, generally frost-free perhumid climates of the Southern Hemisphere. The trunks of tree-ferns are normally unbranched and carry large, rosette-like terminal crowns of usually compound, moderately soft, evergreen leaves. Tree-ferns can grow to about 10 m. True tree-ferns are found in the family *Cyatheaceae* (plus *Dicksoniaceae*), but the monocotyledonous family *Velloziaceae* contains some equivalent forms. Climatic requirements and the world distribution of tree-ferns are described by Troll (1970).

References: Troll 1959; 1968, 1970, 1972b; Kroener 1968; Christ 1910; Troll & Lauer 1978; Shreve 1914; Richards 1952; Whitmore 1975; Cockayne 1958.

24. Tropical Alpine Tuft-Treelets are the short but erect, xeromorphic, often pubescent non-palm woody tuft-plants of the alpine and subalpine belts of the Andes, the mountains of East Africa and some mountains in Central America and elsewhere. The climates of these equatorial areas are diurnal rather than seasonal, with high insolation and warm surface temperatures during the day and freezing temperatures most nights of the year. The leaves of the plants are generally either sclerophyllous or thickly white-pubescent. The plants rarely exceed about 6 m in height and are reduced to stocky, almost trunkless tuft-scrub as temperatures decrease with altitude. The most common taxa include *Lobelia* and *Senecio* in Africa, *Espeletia* and *Lupinus* in the Andes, *Argyroxiphium sanwicensis* in Hawaii, and *Anaphalis* in Indonesia. Most of these dicot genera are known for erect, often weedy species in the northern temperate zone. Their assumption of a caulirosette (stemmed-rosette) growth form in tropical alpine areas is an often-cited example of convergent evolution, though the step is probably not so great here as in some other cases.

References: Cuatrecasas 1968; Coe 1967; Hedberg 1951; Salt 1954; Weber 1958; Walter 1973; Troll 1959, 1968; Vareschi 1970; Daubenmire 1978; Van Steenis 1962; Sturm 1978; Knapp 1973; Lind & Morrison 1974; Baruch 1979.

25. Xeric Tuft-Treelets include various evergreen, mostly unbranched, monocotyledonous treelets with smaller sclerophyll tufts and generally xeromorphic appearance, as occur in the semi-deserts and other drier areas of the tropics and subtropics. Characteristic forms include the stem-yuccas (*Liliaceae*) in the Western Hemisphere, the tuft-treelet species of *Dracaena* in Africa, and the familiar 'grass trees' of Australia (*Kingia* spp. and *Xanthorrhoea* spp.). A striking but less characteristic form, due to its arborescent branching, is the Joshua Tree (*Yucca brevifolia*) of the Sonora and Mojave deserts, which becomes a significant tree with sufficient age. The leaves of xeric tuft-treelets are always evergreen and generally sclerophyllous.

They may be either stiff and succulent as in *Yucca* or more graminoid (but still large) as in *Kingia* and some dracaenas. Though partly succulent some taxa can extend into quite mesic climates on drier sites (e.g. *Yucca aloifolia* in the sandy coastal plain of the southeastern U.S.A.). Xeric tuft-treelets tolerate some winter frost but suffer whenever temperatures remain below freezing. They prefer high insolation and occur throughout the world's warm, dry climates.

References: Walter 1968, 1973; Shreve & Wiggins 1964; Whittaker & Niering 1965; Knapp 1965, 1973; Keast et al. 1959; Orians & Solbrig 1977; Werger 1977; Richard-Vindard & Battistini 1972; Turage & Hinckley 1938; MacDougal 1912.

Arborescents

The nouns 'large bush' and 'arborescent' (used here almost interchangeably) were suggested by de Laubenfels' (1975) 'overgrown bush form' and are used to refer to woody plants which branch near or even under ground-level but grow to tree height. Arborescents are thus an intermediate form between trees and shrubs, for which the exact branching pattern is not as important as general plant size. Arborescents, as a category, can therefore include a variety of intermediate forms, notably sprawling xeric 'large scrub' (e.g. thorn-scrub), clumped-stem forms associated with underground tubers (e.g. the mallee form), and other rather indeterminate but slowly low-arborescent forms characteristic of drier areas (e.g. the leafless 'trees' of Middle Asian semi-deserts). Because their growth form is not monopodial, arborescents can never become truly tall trees. They are typically open-woodland forms which, for various reasons, assume some intermediate combination of height and leaf area as appropriate for a generally intermediate water supply. Some quite malacophyllous forms also occur, however, in quite wet situations (including standing water), where they co-exist with larger trees but may better tolerate the largely anaerobic soil conditions.

26. Evergreen Arborescents include prototypically the tall, lignotuberous, phreatophytic 'mallee' (i.e. multi-stemmed) species of *Eucalyptus,* which form extensive open woodlands in southern Australia further into semi-arid climates than any other forms of comparable size. These are consistently the tallest of the 'overgrown bushes' and can maintain heights of 10 meters right up to their distributional limits. Heights of mallee eucalypts range from one meter in mallee heaths to 12 meters in woodlands (Barrow & Pearson 1970). The shorter mallee shrubs are called marlocks, have more poorly developed underground woody stocks (Burbidge 1952), and are best represented simply as shrubs. Lignotubers, promoting multi-stemmed growth, are fairly common in drier climates and in various eucalypts. Burbidge (1952) seems to ascribe 'the significance of the mallee habit' entirely to the (genetic) existence of lignotubers and does not offer an ecological explanation. He does suggest that the multi-stemmed habit is a late evolutionary development within *Eucalyptus,* restricted primarily to the sections *Dumosae* and *Subulatae*. He lists the following mallee species: *E. morrisii, E. diversifolia, E. leptophylla, E. gracilis, E. eneorifolia,* and most commonly *E. dumosa* and *E. oleosa.* Pryor (1959) states that all *Eucalyptus* species of the 'highly successful xerophytic Mallee communities' display the multi-stemmed growth form, which suggests a climatic (water supply) role even though the mallees are strongly associated with calcareous soils. Certainly the phreatophyte's connection with a relatively constant groundwater supply would permit taller growth with greater leaf area. And certainly multiple stems from a rootstock with many dormant buds will be favored by any processes (e.g. grazing, fire, cold, drought) which damage the main meristem of existing shoots, as has been noted in connection with *Prosopis* (Simpson 1977). Similar multi-stemmed, phreatophytic growth forms are found in other taxa in other similarly dry climates (e.g. *Prosopis, Acacia*) and in some wet situations (e.g. riparian *Salix*), but these taxa are deciduous. Other multi-stemmed evergreen taxa which can be groundwater-connected and may be ecologically similar include *Myrica cerifera* (sandy coastal areas of southeastern U.S.A.) and various riparian taxa throughout the tropics.

References: Burbidge (1952), Pryor (1959), Barrow & Pearson (1970), Keast et al. 1959; Milton Moore 1970; Walter 1968, 1973; Simpson 1977.

27. Raingreen Thorn-Scrub includes various forms of larger and smaller, non-phreatophytic, often sprawling bushes with usually soft, often quite small raingreen leaves (often compound) and often definite thorns. This form is very widespread in the summer-rain tropics, often dominating large areas which are too dry for forest or taller woodland. It is perhaps best represented by many non-phyllodious species of *Acacia* and many other legumes (e.g. *Mimosa, Caesalpinia, Cassia, Albizia, Brachysema*). Raingreen scrub forms often have evergreen, photosynthetic stems and may show other variations, such as the virgate *Fouquiera splendens* (ocotillo) of Arizona-Sonora and even some dwarfed but 'bottle tree'-like treelets (e.g. *Idria columnaris,* also *Fouquieriaceae,* and certain *Burseraceae* in southwestern Africa, Sonora, and elsewhere). Raingreen thorn-scrub has been well recognized as a formation type for a long time (e.g. von Humboldt 1807, Rübel 1930). It is important on all continents except Europe and Antarctica.

References: Walter 1973; Knapp 1965, 1973; Hueck 1966; Shantz & Marbut 1923; Beard 1944, 1955, 1967; Leopold 1950; Daubenmire 1978; Milton Moore 1970; Lauer 1952; Boaler 1966; Burtt 1942; Lind & Morrison 1974; Mani 1974; Stocker 1970; Orians & Solbrig 1977.

28. Summergreen Arborescents include prototypically the phreatophytic, readily re-sprouting species of *Prosopis, Salix* and other holarctic and tropical genera which arise from woody rootstocks in both wet areas (e.g. carr) and most importantly in dry areas along streambeds and wherever else groundwater contact can be maintained. The ecology of this form appears to be much like that of the mallee eucalypts, at least in dry areas, but the summergreen forms occur in areas with colder winters or involve tropical taxa (e.g. *Prosopis*) which have a deciduous habit. The rootstocks of these plants contain many dormant buds. Whenever the terminal meristem of existing shoots is destroyed, as by fire, winter cold, grazing, or felling, dormant buds are activated and new shoots form. Summergreen *Prosopis* woodlands are especially well developed in Texas but occur in similar areas of the Americas, Africa and Asia where rainfall is low but sufficient groundwater can be reached by the extensive root systems, especially along washes and canyons. *Prosopis* can also function as a non-phreatophyte and assume normal tree form where water is sufficient. The observed decrease (*Prosopis*) of tree forms and increase in bush forms in the southwestern U.S.A. has been related to the increased use of the area for livestock grazing (Simpson 1977). Similar *Prosopis* and other summergreen woodlands are found in much of northern and Pacific Mexico, in Argentina and Uruguay ('algarrobos'), in sub-Saharan Africa (where *Prosopis* is raingreen, i.e. raingreen scrub), and in much of the Middle East, including Iran, the Arabian Peninsula, and Middle Asia. Other taxa which form multi-stemmed summergreen large bushes include various acacias (*Acacia greggii* and *A. constricta* in North America, *A. aroma* and *A. furcatispina* in Argentina), *Celtis spinosa* (Argentina), and various holarctic genera such as *Salix, Betula* and *Populus,* which occur as phreatophytic bushes in both dry and wet climates.

References: Simpson 1977; Orians & Solbrig 1977; Walter 1968, 1973, 1974; Hueck 1966; Zohary 1973; Eyre 1968.

29. Leafless Arborescents include both arborescent and shrubby, generally phreatophytic woody scrub with evergreen photosynthetic stems and either no leaves at all or at most soft, semi-ephemeral nanophylls. These are much like Xeric Dwarf-Shrubs in their ecology but are larger. Leafless arborescents occur in a variety of dry climates but are best developed in the cold-winter, continental-mediterranean semi-deserts of Middle Asia (Kazakhstan, Uzbekistan, Turkmenistan, and adjacent areas), where they can form the dominant vegetation. Summers are hot and dry, and winters can be very cold, which leaves only a very short period in spring during which leaves are advantageous. Rainfall coming in the spring, plus snowmelt, provides a brief spring growing season (up to two months), but growth is often made possible well into the summer and perhaps to autumn by the extensive root systems and the extremely low, resistant transpiring surface. Water-loss rates are comparable to those of cuticular transpiration. Growth is slow, but plants reach ages of 100 years or more. This form grows especially well in the extensive sand-deserts of Middle Asia, where lack of a more extensive vegetation cover permits groundwater to accumulate close to the surface. Typical taxa in Middle Asia and the Middle East include above all the arborescent saksa'ul species *Haloxylon aphyllum* (to 12 m.) and *H. persicum* (to

8 m.), similarly arborescent *Ammodendron* spp. (to 6 m.) and *Calligonum* spp. (to 8 m.), and a larger variety of shrubby taxa (and morphs), e.g. *Salsola, Anabasis, Astragalus, Eremospartum.* Several of these taxa are used extensively in reforestation. Similar forms occur in other semi-desert areas (e.g. *Bulnesia retama,* a nearly leafless zygophyll growing to 8 m. in northwestern Argentina, see Hueck 1966), but with some exceptions they are generally not arborescent.

References: Walter & Box, in press; Walter 1968, 1973, 1974; Zohary 1973; Hueck 1966; Bobrovskaya 1971; Petrov 1966–67; Korovin 1961–62; Nechayeva et al. 1973; Nikitin 1966.

Krummholz

Krummholz ('crooked wood') is wind-stunted woody scrub occurring primarily at treeline and other exposed sites in mountains and along some windy seacoasts. It is the result of constantly strong winds, often from one direction, which kill new growth on the windward side and/or reduce most upward growth. Krummholz forms resulting from these conditions are usually classified as:

1. flagged (upward growth but sheared on windward side)
2. flag-matt combination
3. matt (reptant or pulvinate with no emergent growth).

Along gradients of increasing climatic severity (e.g. increasing elevation on a mountain), one can often observe a sequence of krummholz forms, from flagged only at treeline through flag-matt islands to espalier-like reptant matt forms well above treeline. The height of the matt forms is generally delimited by the thickness of the winter snow cover, since winter, with its ice-blasting just above the snow level represents the most severe environmental limit. Certain other forms can also be called krummholz but are somewhat different and treated separately, notably the cushion-shrubs of more exposed, generally snowless windy areas (see Rauh 1939) and the large 'butt-sweep' forms produced by the alternating seasonal pressures of downward snow-creep and tree orthotropy (see Tranquillini 1979).

30. Needle-Leaved Treeline Krummholz is the most common form of treeline krummholz in mountains and includes all three of the morphs described above, as well as the butt-sweep form further downslope. Of the needle-leaved treeline krummholz taxa only *Larix* is not evergreen, but it is also rather uncommon at treeline. More typical taxa include the familiar boreal-montane forest genera *Picea* and *Abies* plus *Juniperus* in the Northern Hemisphere and *Podocarpus, Dacrydium* et al. in the tropics and Southern Hemisphere. Species from these genera all form tall trees under better conditions downslope but are gradually reduced to reptant forms at treeline and above. Reptant treeline krummholz is uncommon on tropical mountains, being replaced largely by stunted rosette-treelets and by cushion-shrubs. The subalpine 'elfin woodland' within the cloud belt of many tropical mountains is usually wind-sheared on top and stunted but generally maintains a closed forest canopy. The physiological and population ecology of woody treeline species has been studied in detail in many parts of the world, but simple explanations for the cause and location of treelines in general remain elusive.

References: Tranquillini 1959, 1963, 1964, 1967, 1979; Troll 1960, 1968, 1972b, 1973; Walter 1968, 1974; Wardle 1965, 1971; Michaelis 1932–34; Schiechtl 1966; Allg. Forstz. 1966; Sharpe 1970; Ives & Barry 1974; Marchand & Chabot 1978; Hansen (in prep.); Webber 1978; Costin 1957; Gersmehl 1973; Brockmann-Jerosch 1919, 1928; Yoshino 1973; Ellenberg 1966; Holtmeier 1974; Larcher 1957, 1972; Pisek & Rehner 1958, 1960; Pisek & Winkler 1958.

Shrubs

Shrubs are woody plants with multiple stems arising from a common base at or near ground level. Because they lack a single main trunk shrubs do not grow as tall as trees. This reduction in height generally also means a reduction in total leaf area, the area of greatest potential water loss. As a result, shrubs are found both in tree climates and in climates too dry for trees. Although shrubs can form as dense a ground cover as can trees, shrubs are more commonly found, especially when dominant, in more open stands in drier areas. In such situations the shrubs often have greater below-ground 'cover' (i.e. area of lateral rooting and water uptake) than they have above ground, which accounts for their often regular spacing. Many woody plants, unlike monopodial trees, begin as shrub forms but grow with age into small trees (by developing a main trunk) or into what de Laubenfels (1975) has called 'overgrown bushes'. Those forms which sprout profusely but then grow to tree size, such as phreatophytes growing from laterally elongated rootstocks, have been included as arborescents (treated previously). Those forms and taxa with more determinate fruticose growth form, even if they grow later into larger bushes or small treelets, are not so different ecologically and are included as shrubs.

31. Tropical Broad-Evergreen Shrubs include a wide variety of generally large, frost-sensitive shrubs, typically with normal or larger coriaceous leaves. A familiar example might be *Coffea arabica,* which can also grow into a small tree. Since competition for light can be much greater in humid tropical environments, making vertical growth of greater urgency, the distinction between true shrub forms and more ascending forms probably is less clear in the tropics. Shrubs are generally not as common in humid tropical forests as in their temperate counterparts. Broad-leaved evergreen shrubs are often more important as understoreys in transitional and raingreen forests and in more open woodlands. Broad-evergreen shrubs in humid areas come typically from families such as *Rubiaceae, Anonaceae* or *Euphorbiaceae*, while the taxa in drier areas are generally more diverse. Though many broad-evergreen shrub taxa occur in tropical and subtropical Australia, they are perhaps better designated as mediterranean sclerophyll shrubs. Evergreen shrubs become more important on some equatorial mountains, such as *Erica arborea* and *Philippia excelsa* in the subalpine ericaceous belt of East Africa. These species, however, can also grow as trees to 10 meters. Tropical evergreen shrubs have not been studied as a geoecological type, and their ecology remains poorly understood.

References: Walter 1973; Knapp 1973; Lind & Morrison 1974; Whitmore 1975; Richards 1952; Milton Moore 1970.

32. Temperate Broad-Evergreen Shrubs (excluding the xeric types) are normal to large shrubs with somewhat harder, cold-resistant leaves than the typically coriaceous leaves of tropical evergreens. Though smaller evergreen forms can occur under snow even in the tundra, typical broad-leaved evergreen shrubs are restricted to regions where winter temperatures are not so extreme. Three subtypes are readily apparent.

32a) Mediterranean Evergreen Shrubs are typically medium to large shrubs with generally small, sclerophyllous broad or linear leaves, as found in the mediterranean climates of the Mediterranean borderlands, the Cape region of South Africa, south-coastal Australia, southern California, and central Chile. The consistent occurrence of this form, arising in quite unrelated taxonomic groups to form more or less dense, extensive shrublands in the world's five widely separated mediterranean climates, has long been cited as the classic example of convergent evolution in physiognomy. Since temperature stress and water stress fall in different seasons in mediterranean climates, the vegetation must be evergreen (or very nearly so) in order to make use of all favorable periods as best it can to maintain a positive annual production-respiration balance. In order to reduce water loss the shrubs may lose half their leaves during either the winter cold or summer drought, but they remain evergreen and resume more active growth in spring (new growth and flowering) and in autumn (additional net production, with new growth in some taxa). Characteristic taxa include *Quercus coccifera, Arbutus unedo, Myrtus communis, Rhamnus alaternus* and *Nerium oleander* in the maquis and similar shrublands of the Mediterranean area; *Protea grandiflora,* other *Protea* species, *Leucospermum conocarpum,* and *Mimetes* spp. in the fynbosch of South Africa; a variety of often more linear-leaved *Proteaceae, Myrtaceae, Pittosporaceae* and phyllodious legumes (mainly *Acacia*) in southern Australia; *Quercus dumosa, Rhus ovata, Arctostaphylos, Ceanothus,* and *Rhamnus* in southern California and Arizona chaparral; and *Kageneckia (Rosac.), Colliguaja (Euphorbiac.),* and *Escallonia (Saxifragac.)* in the matorral and, more commonly, sclerophyll woodlands of central Chile.

References: Walter 1968; di Castri and Mooney 1973; Cooper 1922; Mooney & Dunn 1970; Mooney et al. 1974; Specht 1969, 1979; Schmithüsen 1956; Adamson 1927; Milton Moore 1970; Grieve 1955;

Larcher 1970, 1972; Oppenheimer 1932; Pisek & Rehner 1958; Daubenmire 1978; Vogl & Schorr 1972; Hellmers et al. 1955; Horton & Kraebel 1955; Hanes 1971; Polunin 1972; Ng & Miller 1980.

32b) Temperate Broad-Evergreen Shrubs include a variety of less xeromorphic, large to small shrubs with cold-resistant but often larger and less sclerophyllous, often shiny leaves, as appear to be more appropriate for summer-rain temperate climates without extremely cold winters. Such climates are found in southern China and Japan, in the southeastern and eastern U.S.A., in southern Brazil and adjacent Uruguay-Argentina, in much of Europe west of the Soviet Union, in New Zealand, and in eastern South Africa, plus smaller mountainous areas. Winter-rain climates with only moderately dry summers may also support such shrubs, and such occur in southern Australia, southern Chile, and the Pacific Northwest of North America. Characteristic taxa include *Ligustrum, Pittosporum* and *Osmanthus* (East Asia but widely introduced elsewhere), *Gardenia* (originally from Africa), *Ilex, Berberis,* certain species of *Rhododendron* and other *Ericaceae, Buxus sempervirens* (submediterranean Europe), *Daphne* spp. (Europe), *Hypericum calycinum* (Europe), and a wide variety of others, many of which are well known and widely planted as ornamentals. New leaves appear in the spring and may replace older leaves, but generally very few leaves are lost during the cold winter.

References: Walter 1968, 1974; Polunin 1972; Knapp 1973; Radford et al. 1968; Horvát et al. 1974.

32c) Perhumid Broad-Ericoid Evergreen Shrubs are an especially mesic subtype of the typical broad-evergreen shrubs, differing mainly by having generally larger leaves and by being restricted typically to the mesic conditions found in humid to perhumid temperate-zone montane and tropical subalpine belts. Broad-ericoid evergreen shrubs typically form dense understoreys, especially along streambanks and other forest edges, and often attain dominance in characteristic subalpine 'heath' or ericaceous belts and/or on exposed sites (e.g. 'heath balds' in the Appalachians). The most characteristic taxa are the evergreen montane species of *Rhododendron* and other *Ericaceae,* which occur in humid montane and subalpine belts in North America, Europe, Africa and Asia. Equivalent forms (largely *Epacridaceae*) occur in the humid mountains of Australia and New Zealand, and mesomorphic *Cunoniaceae* and *Myrtaceae* occur in southern Chile. The world's various ericaceous and ericoid montane belts have been especially well interpreted by Troll (1959, 1966, 1972b) and by Specht (1979).

References: Walter 1968, 1973, 1974; Troll 1959, 1966, 1968, 1972a, 1972b; Hedberg 1951; Schweinfurth 1957; Whittaker 1956, 1963; Salt 1954; Specht 1979; Weberbauer 1911; Knapp 1973; Daubenmire 1978; Van Steenis 1962.

33. Hot-Desert Evergreen Shrubs are moderate-sized, generally virgate shrubs with small hard leaves or phyllodes, as found in the warm semi-deserts of southwestern U.S.A. and adjacent Mexico (e.g. *Larrea tridentata,* = *divaricata*), the shrub-steppe region of northwestern Argentina (also *Larrea* primarily), and large regions of the Middle East, interior Australia, and southern Africa. Typical taxa, other than *Larrea* (*Zygophyllaceae*), include the equivalent mulga (*Acacia aneura*) and other acacias of Australia, various other *Zygophyllaceae* (Middle East, Australia and Africa), and a variety of ecologically similar taxa in Australia such as *Eremophila, Hakea* and *Rhagodia.* The leaves (or phyllodes) are generally small and hard, often grayish-green (e.g. *Larrea*), and cast little shade. Many leaves are shed during extreme periods and replaced after sufficient rainfall. These shrubs generally have large underground uptake areas and thus form quite open, regularly spaced stands.

References: Walter 1968, 1973; Hueck 1966; Zohary 1973; Specht 1972, 1979; Shreve & Wiggins 1964; Shreve 1942; Cannon 1911; Cunningham & Strain 1969; Mabry et al. 1977; Orians & Solbrig 1977; Werger 1977; Knapp 1973; Keast et al. 1959; Milton Moore 1970; Chew & Chew 1965; Cunningham & Reynolds 1978.

34. Leaf-Succulent Shrubs are woody shrubs (or miniature treelets) with broad, succulent, evergreen leaves and woody but sometimes semi-succulent stems. This form is best developed in the cooler, fog-moderated Namib and Karroo arid regions of southern Africa. The most familiar example is perhaps the jade-tree (*Crassula argentea*); other examples from southern Africa include *Aloë* *asperifolia* and *A. littoralis* (*rubrolutea*), and *Sansevieria pearsonii*. Leaf-succulent shrubs occur also in other dry areas, e.g. *Simmondsia chinensis* (= *californica*) in southern California and Sonora-Arizona, but are nowhere else as important as in southwestern Africa. Leaf-succulent shrubs have a somewhat larger transpiring area than stem-succulents and thus usually require moderated desert conditions, as in coastal deserts.

References: Walter 1973; Innes 1977; Al-Ani et al. 1972.

35. Cold-Winter Xeromorphic Shrubs include a variety of usually small, often heterophyllous, xeromorphic shrubs and suffrutescent semi-shrubs which are found mainly in the cold-winter, continental-mediterranean steppes and semi-deserts of interior Eurasia and interior western North America. The most common taxon is *Artemisia* and the prototypic species *A. tridentata* (sagebrush) of the American Great Basin and various *Artemisia* species in the semi-deserts of the area the Russians call Middle Asia (primarily Kazakhstan and the adjacent lowland areas). The plants have woody bases and either completely woody branches or annual, partially lignified branches. Though many species can be called evergreen, the plants die back to varying degrees in the cold winters, grow mainly in the spring, hold on through the hot, dry summers, and may show renewed growth in the autumn (Walter and Box, in press). The leaves are generally quite small and often white-pubescent and/or grayish-green in color. Other taxa characteristic for Middle Asia and adjacent regions include a wide variety of *Chenopodiaceae, Leguminosae,* and *Zygophyllaceae* (Walter 1974). Other taxa from western North America include *Chrysothamnus, Tetradymia, Grayia,* and also various *Leguminosae* and *Chenopodiaceae* (e.g. *Atriplex*). Similar forms are less important in the Middle East, in southern interior Australia, and in southern South America. *Artemisia* generally does not regenerate quickly after fire.

References: Walter 1968, 1974; Walter & Box, in press; Billings 1949; Shreve 1942; Fautin 1946; Daubenmire 1978; Zohary 1973; Bobek 1951; Popov 1940; Korovin 1961–62.

36. Broad-Summergreen Shrubs are the seasonally green, broad-malacophyllous shrubs of temperate climates with cool to cold winter and at least some summer rain. Such shrubs range from generally larger-leaved mesic types to generally smaller-leaved forms found mainly in drier climates. Because of the differences in both form and strategy between plants from favorable and stressful environments, the summergreen shrubs can be divided into two basic types.

36a) Mesic Summergreen Shrubs are the larger-leaved shrubs of typical four-season temperate climates with precipitation in all seasons. Conditions for growth are generally good throughout the warm portion of the year, though potential evapotranspiration may exceed precipitation in summer causing mild stresses as soils dry out. Dry years can cause greater stress, but the vegetation normally recovers by the following year. Climatic requirements are essentially the same as those for summergreen trees since the reduced transpiring surface of shrubs is usually offset by shallower root systems. During dry years shrubs can be affected by drought before trees because of the shallower roots. Mesic summergreen shrubs rarely form the dominant vegetation and occur most commonly in forest understoreys, along forest edges, and in opened agricultural land. Typical taxa include *Rosa, Vaccinium, Ribes, Rubus, Spiraea* (all *Rosaceae*), *Viburnum, Astragalus, Aesculus,* and others from a wider variety of families. Many are planted as ornamentals, and many other taxa include species which can be either shrubs or small trees.

36b) Xeric Summergreen Shrubs are generally smaller-leaved shrubs which sometimes also show other traits of xeromorphy, such as somewhat harder leaves, leaf pubescence, or smaller overall plant size. These shrubs are found especially in cold-winter temperate climates with less total rainfall (generally an annual deficit, meaning that soil moisture is rarely at field capacity or above) and pronounced summer moisture stress, such as in more continental mediterranean and grassland or even semi-desert climates. In mediterranean-type climates the shrubs may lose their leaves before autumn. Xeric summergreen shrubs are often widespread in such climates but attain dominance in generally small areas such as the 'deciduous chaparral' of western Colorado and northeastern Utah, drier parts of the Balkan Peninsula (e.g. Šibljak, actually a degradation form), and some mountainous areas of Middle Asia and the Middle East. Typical taxa from the 'deciduous chaparral' include *Symphoricarpos oreophilus, Amelanchier alnifolia, Cercocarpus montanus, Rhus trilobata* (also from evergreen California chaparral), and *Prunus virginiana* (also a small tree). Taxa from other areas include a large number of *Leguminosae* and some *Compositae* plus *Pistacia terebinthus, Cotinus cogyria, Thymelaea, Cistus,* etc.

References: Daubenmire 1978; Walter 1968, 1974; Radford et al. 1968; Polunin 1972; Buchanan 1974; Walter & Box, in press; Horvát et al. 1974; Zohary 1973; Ellenberg 1963; Küchler 1964.

37. Needle-Leaved Evergreen Shrubs include a relatively small number of generally sclerophyllous needle or linear-leaved shrubs of varying sizes from a variety of environments. Some species, such as *Juniperus communis* have wide environmental tolerance ranges and extremely wide distributions. Others, such as *Adenostoma fasciculatum* from the California chaparral, certain linear-leaved dwarf-shrubs (e.g. *Chenopodiaceae*) from semi-deserts, and various needle-shrubs from Australia (e.g. *Banksia ericifolia, B. spinulosa*) have much more restricted distributions. All, however, have a similar response to generally drier conditions, namely the reduction of plant size and leaf surface area to restrict water loss. Such shrubs (and some small, shrubby trees, e.g. *Tamarix*) occur in drier climates, generally in the temperate and subtropical regions, but can also penetrate well into more mesic areas, as does *Juniperus communis* (though often on sandy soil).

References: Walter 1968, 1973, 1974; Daubenmire 1978; Horton & Kraebel 1955; McPherson & Muller 1969; Brockman 1968; Knapp 1973; Blombery 1967.

Dwarf-shrubs

Dwarf-shrubs are woody, generally small-leaved chamaephytes not taller than 50 cm and more commonly 30 cm. Included are also the suffrutescent semi-shrubs, which have woody bases and largely herbaceous, annual upper branches. The smaller size of drawf-shrubs offers several advantages over larger forms, including reduced transpiring surface and greater bud protection near the ground during the cold season (usually within a litter or snow cover). Dwarf-shrubs are uncommon in the winterless tropics, where they would be outcompeted by larger forms, but become more important in cool and/or drier temperate climates and other temperate areas where trees and larger shrubs are excluded (e.g. by poor or excessive drainage, by anthropogenic factors). Dwarf-shrubs may have broad, narrow (including needle), or no leaves and can be evergreen, summergreen, or have less obvious seasonal habits, as in mediterranean and maritime-heath climates. Because the plants and their leaves are small, differences between broad and even needle leaves are less important.

38. Mediterranean Dwarf-Shrubs include a variety of xeromorphic, generally shallow-rooted and fire-sensitive, typically aromatic semi-shrubs and dwarf-shrubs found mainly in the world's mediterranean climates with insufficient rainfall for more demanding vegetation. These plants are typically suffrutescent semi-shrubs which gradually lose their small, malacophyllous, usually grayish and often pubescent leaves during the dry summer but may produce new ones (though usually fewer) after autumn rains begin. Certain aromatic evergreen sclerophylls (e.g. *Rosmarinus*) could also be included, however, because of their similar size and ecology. Mediterranean dwarf-shrubs occur in maritime and more continental mediterranean climates where some growth is possible throughout the winter. This includes woodland and some steppe areas, but the form attains dominance only in drier areas, such as the 'coastal sage' of southern California (e.g. *Eriogonum fasciculatum, Artemisia californica*), and in degraded areas, such as the *garrigue* of southern France and similar Mediterranean borderlands. Mediterranean dwarf-shrubs are generally ousted by larger, deeper rooting, fire-resistant sclerophyll shrubs where rainfall and soil water storage permit. Typical dwarf-shrub taxa include many familiar mediterranean cooking 'herbs' such as *Thymus* and *Salvia,* mostly *Labiatae;* other aromatics such as *Lavandula;* and similar semi-shrub taxa from other mediterranean areas, such as *Pelargonium* (South Africa and Australia), *Olearia* and *Calocephalus* (Australia), *Stoebe* and *Felicia* (South Africa), and *Encelia farinosa* and *Franseria* spp. (Arizona-Sonora).

References: Walter 1968, 1973; Daubenmire 1978; Zohary 1973; Horvát et al. 1974; Kirkpatrick & Hutchinson 1977; Milton Moore 1970; Küchler 1964; di Castri & Mooney 1973; Oppenheimer 1932; Quézel 1965; Schreve 1936; Polunin 1972; Harrison et al. 1971; Epling & Lewis 1942.

39. Temperate Evergreen Dwarf-Shrubs include a relatively small number of broad and narrow-leaved, generally sclerophyllous and typically ericoid dwarf-shrubs which occur in a variety of mesic cold-winter situations, most notably the extensive heaths of western Europe plus certain polar and alpine areas. Many of the species are *Ericaceae* in the Northern Hemisphere. These plants are much like the typical temperate broad-evergreen shrubs in their ecology, but their smaller size and leaves permit them both to endure much colder winters under a protective snow cover and to thrive, at least in cool-maritime climates, on sandy, well drained, and often degraded heath sites, where they can become the dominant form over large areas. Ericoid dwarf-shrubs also occur with graminoids in peaty, poorly drained moors and other wetlands, on more exposed tundra and alpine sites, and in micro-communities in polar and alpine snowbanks (Schneetälchen). The heaths of Atlantic Europe are dominated most often by *Calluna vulgaris* (heather) and extend from near Gibraltar (*Erica cinerea, Lavandula stoechas, Cistus* spp.) along the Atlantic coast and throughout the British islands to Norway and the Faeroes (Walter 1968). The distributions of both *Calluna vulgaris* and the Atlantic heaths are delimited closely by the 0 °C isotherm for January and the 22 °C isotherm for July (Böcher 1940). Rübel (1930) states that monthly rainfall in heath areas of Europe always varies between 5% and 11% of the annual total. The climate is thus cool and quite maritime. Some species of *Erica* (e.g. *E. herbacea,* Schneeheide) even bloom in January. In polar and alpine tundra many of the dwarf-shrubs are also evergreen or at least semi-evergreen (i.e. with overwintering lower leaves), since this permits the plants to react quickly and make use of the full snow-free period. Evergreen (and semi-evergreen) dwarf-shrubs occur also in similar climates outside Europe, both as dominants in dwarf-shrub heaths and as important components of other treeless formations (Specht 1979). Typical taxa include *Vaccinium, Gaylussacia, Gaultheria* and *Arctostaphylos* in Nova Scotia; *Vaccinium vitis-idaea, V. uliginosum, Cassiope tetragona, Rhododendron lapponicum, Rh. ferrugineum, Phyllodoce, Arctostaphylos uva-ursi,* and *Empetrum* in circumpolar arctic heaths and tundra; *Empetrum rubrum* et al. on Tierra del Fuego and the subantarctic islands; *Erica* and *Philippia* spp. in African alpine belts; and *Epacris* and *Kunzea* spp. in the mountains of New Guinea, eastern Australia, and New Zealand (Gimingham 1972). Many species (including *Calluna vulgaris*) occur in both temperate-heath and arctic-alpine situations.

References: Gimingham 1972, 1975; Walter 1960, 1968, 1974; Böcher 1938, 1940; Lötschert 1962–65; Rübel 1930; Tansley 1949; Stocker 1923; Adamson 1927; Sorenson 1941; Grieve 1955; Specht 1972, 1979; Daubenmire 1978; Hanson 1953; Costin 1957; Tieszen 1978; Ulmer 1937; Tranquillini 1964; Cartellieri 1935; Larcher 1976; Aleksandrova 1960; Bliss 1956; Hansen 1930; Rosswall & Heal 1975; Preusser 1975; Polunin 1926.

40. Summergreen Dwarf-Shrubs are primarily boreal and tundra/alpine species which can be either typical shrub taxa (e.g. *Vaccinium*) or dwarf-shrub forms (morphs or species) from typically tree taxa (e.g. *Salix, Betula*). Most species have both high water requirements (malacophyllous) and low optimum temperatures and, as a result, are easily outcompeted by trees and larger shrubs in warmer areas. In tundra and alpine areas these shrubs are most common in maritime or other mesic situations which reduce winter cold and/or provide sufficient snow for protection against desiccation, cold, and ice-blasting. The ecology of this form is otherwise much like that of the Scandinavian dwarf-birch scrub (treelets or larger shrubs), and indeed the genus *Betula* provides both forms, across North America and Eurosiberia as well as in Scandinavia. The most common arctic species include *B. nana* and various *Salix* species, such as *S. artica* and *S. reptans*. Some *Salix* species in particular also have somewhat more reinforced leaves, e.g. *S. reticulata* and *S. retusa* of European mountains.

References: Walter 1968, 1974; Daubenmire 1978; Böcher 1938; Aleksandrova 1960; Bliss 1956; Wielgolaski et al. 1975; Hansen 1930; Tieszen 1978; Rosswall & Heal 1975; Preusser 1975; Sorenson 1941; Hanson 1953; Polunin 1926.

41. Xeric Dwarf-Shrubs include a variety of leafless and nanophyllous, seasonally leafless dwarf-shrubs and semi-shrubs which generally have green, photosynthetic stems and can remain at least partially active even during unfavorable cold and dry seasons. These shrubs are especially characteristic of the cold-winter semi-deserts of Middle Asia, the Middle East, and the intermontane western U.S.A. The leaves, if present at all, usually are produced in the autumn or spring and remain for a few months at most. The leaves are typically malacophyllous and permit greater productivity during short favorable seasons. The remaining evergreen woody stems are very resistant to both winter cold and desiccation. Zohary (1952, 1973) has outlined the following 'hydro-economic' types from the warmer-winter, dry-summer Middle Eastern deserts:

1. Semi-deciduous dwarf-shrubs, which reduce their leaf surface drastically at the beginning of the dry season (e.g. *Reaumuria palaestina, Salsola villosa, Suaeda palaestina, Artemisia monosperma, Zygophyllum dumosum*).
2. Heterophyllous dwarf-shrubs, which lose their lower, mesomorphic winter leaves and retain smaller xeromorphic leaves (e.g. *Artemisia herba-alba*).
3. Semi-evergreen virgate shrubs, which lose their leaves during winter but retain active, evergreen woody shoots (e.g. *Retama retam, Calligonum comosum*).
4. Bud-evergreen dwarf-shrubs, which lose their green winter shoots but retain evergreen buds located in the axils (e.g. *Reaumuria*).
5. Thorny, ephemeral-leaved dwarf-shrubs, which produce small, early-summer leaves and lose them gradually (e.g. *Noaea*).
6. Leafless semi-shrubs, which lose some evergreen shoots during the summer (e.g. *Ephedra*).
7. Leafless, modular semi-shrubs with green bark, which shed in summer the bark of the previous year's shoots (e.g. *Anabasis articulata, Haloxylon articulatum*).

Other variations occur also in other areas. *Ephedra, Artemisia,* and various *Zygophyllaceae* are also important in the American semi-deserts, while *Zygophyllacae, Casuarina humilis* and others are important in Australia. Some species are definitely halophytes (non-succulent).

References: Walter 1968, 1973, 1974; Walter & Box, in press; Zohary 1952, 1973; Orshan 1953; Korovin 1961–62; Nechayeva et al. 1973; Nikitin 1966; Petrov 1966–67; Milton Moore 1970.

Cushion-shrubs

Cushion-shrubs are small krummholz forms occurring where nearly constant desiccating, often cold winds completely limit vertical branch extension, resulting in an extremely dense mass of short branches forming a flat or rounded, almost hemispheric dwarf-shrub. The definitive treatment of cushion-plants in general remains the exhaustive monograph by W. Rauh (1939), who describes cushion-plants initially, woody or herbaceous, as follows (translated from the German):

> 'The cushion-plants are perennial evergreen plants with barely emerging main (vertical) axis and numerous, radially oriented shoots, deployed in tiers, whose tips, as a result of equal longitudinal extension, produce a common, more or less compact surface with the shape of a hemisphere or a flat mat'.

As a definition, Rauh (1939) later states that cushion-plants are 'perennial herbaceous or woody, mostly evergreen plants (chamaephytes) with well developed allorhizal root systems'. He lists a number of basic growth forms, including 'true cushion-plants', creeping cushions, sward cushions, rosette cushions, succulent cushion-plants, cushion-mosses, hapaxanthous cushion-plants, and pseudo-cushion forms. Most of these forms are included elsewhere in the current classification under Succulent Forbs, Xeric Cushion-Herbs, cold-desert herbs, and Mat-Forming Thallophytes. Cushion-shrubs include primarily the woody 'true cushion-plants' and some woody creeping forms. The 'true cushion-plants' exhibit radial growth and are divided physiognomically into radial spherical cushions (Radialkugelpolster), including treelet, shrub, hollow-cushion, and full-cushion subforms, and radial flat cushions (Radialflachpolster), including hollow and full-cushion subforms. Leaf forms are classified as needle-leaves, spatulate leaves (e.g. *Aretia, Azorella* types), calyciform leaves (Keulenblätter), and rolled leaves (revolute and involute). Leaves are invariably quite small, regardless of form.

Although Rauh did not consider convergent

evolution alone to be the best explanation for cushion-plants, they are generally considered to be an adaptation to extremely windy conditions, as found on high mountains, in the polar regions, and in certain deserts. The world distribution of all 'cushion species' is given by Rauh as follows:

South America (primarily Andine)	50.5%
Central Asia and Asia Minor	16%
Subantarctic and New Zealand	13.6%
Europe (primarily alpine)	11.9%
Arctic	2.7%
Africa (primarily North Africa and alpine)	2.9%
North America (primarily alpine)	2.4%

Rauh also lists a variety of site types for cushion-plants, including:

1. wet sites (e.g. moors, streambanks, snow-valleys, periodically flooded gravel terraces, rainy maritime areas, saline soils);
2. intermediate sites (e.g. rock crevices and surfaces, moraine soils, mats); and
3. dry sites (e.g. snowless windy sites, steppes, dry alpine areas, volcanic soils, dry deserts, wind deserts).

Considering the world distributions of cushion-plants and common site types, typical cushion-plant climates could be considered to include those of the equatorial alpine páramo (wet and windy), the tropical alpine puna (seasonally wet-dry and windy), Patagonia (dry and windy), the Asian cold-winter deserts (dry and windy), subpolar islands (wet and windy), dry deserts, the arctic tundra and cold-desert, and wet alpine areas. Cushion-plants are reported as important in all of these areas (see, for example, Cuatrecasas 1968; Cabrera 1968; Walter 1968, 1973, 1974; Soriano 1956; Wace 1961; Godley 1960; Ozenda 1958; Jani 1969; Quézel 1965; Zohary 1973; Tieszen 1978). The climates of these alpine, lowland-desert, and maritime areas have little in common except wind. In considering ecological types of cushion-shrubs, the first important distinction appears to be between those of wet climates (generally also cool) and those of dry climates.

General references: Rauh 1939; Schröter & Hauri 1914; Weberbauer 1931; Heilborn 1925; Reiche 1893; Troll 1937–38; Walter 1968, 1973, 1974; Hodge 1946.

42. Perhumid Evergreen Cushion-Shrubs are compact rounded and flat cushion-shrubs found primarily in the cool, extremely windy, maritime climates of the subantarctic islands, southern Patagonia and Tierra del Fuego, some subarctic islands, some other cool-perhumid windward coasts, and some similarly perhumid páramo climates of equatorial mountains. Perhaps the most characteristic taxon is *Azorella,* which forms very dense, compact, hemispheric cushions in all of the Southern Hemisphere areas just mentioned. *Azorella selago* forms the dominant vegetation over large portions of the subantarctic islands (Wilhelmy 1963; Werth 1928; Walter 1968). Cushion-shrubs of wet climates are most commonly full-cushion forms.

References: Rauh 1939; Werth 1928; Wilhelmy 1963; Walter 1968; Weberbauer 1931; Cuatrecasas 1968; Schröter & Hauri 1914; Heilborn 1925.

43. Xeric Cushion-Shrubs are small rounded and flat cushion-shrubs found in a variety of windy and at least seasonally dry climates, which may otherwise be quite dissimilar. Such situations are found most commonly in the subtropical and temperate-continental regions, especially on mountains, and are probably best represented by the puna regions of the middle Andes and the rain-shadow of windy Patagonia. Xeric Cushion-Shrubs can differ physiognomically from Perhumid Evergreen Cushion-Shrubs by being generally more xeromorphic in structure (e.g. smaller, fewer and harder leaves) and by involving more flat and hollow-cushion forms. Xeric cushion-shrubs are represented in the moist puna by hard-cushion taxa such as *Azorella compacta, Opuntia atacamensis, Adesmia* ssp., and *Pycnophyllum* spp. (Cabrera 1968), and in Patagonia by often larger and thorny taxa such as *Mulinum spinosum* (Umbellif.) and *Chuquiraga aurea* (Walter 1968). Similar forms in other areas include *Anabasis aretioides* in North Africa (Hauri 1912); *Astragalus* and *Acantholimon* spp. in Anatolia (Walter 1968; Gams 1954); *Alyssum* (= *Ptilotrichum*) *spinosum, Arenaria pungens* and others (from a variety of families) in the Atlas Mountains (Walter 1968) and other drier Mediterranean mountains; *Aizoanthemum dinteri* and *Zygophyllum simplex* in southwestern Africa (Walter 1973); and *Acantholimon* and *Artemisia* spp. in Iran (Walter 1968). Walter (1974), considering the vegetation of Pamir, distinguishes five types of xeric, mostly woody cushion-plants characteristic of the dry Central Asian highlands: hard cushions with taproot (e.g. *Acantholimon hedini, Potentilla pamiroalaica*), firm cushions with rooting lateral branches (e.g. *Sibbaldia tetrandra*), loose cushions without adventive roots (e.g. *Oxytropis* spp., *Astragalus myriophyllus*), loose cushions with adventive roots (e.g. *Artemisia rapestris* f. *alpina*), and air-cushions without adventive roots (e.g. *Acantholimon pamiricum*).

References: Rauh 1939; Walter 1968, 1973, 1974; Cabrera 1968; Hodge 1946; Hauri 1912; Gams 1954; Schröter & Hauri 1914; Weberbauer 1931; Reiche 1893; Heilborn 1925.

Rosette-shrubs

Rosette-shrubs are trunkless analogs of the various forms of rosette-trees and treelets. Included are a variety of ground forms, leaf-succulent rosettes, bunch-palms, etc. Some forms consist of basal rosettes with taller flowering stalks (e.g. certain species of *Yucca* and *Agave*), while others are terminal rosettes (often with underground trunks, e.g. many palms). Although palms and agaves, for example, are quite different in many respects, both are monocots, can have trunkless rosette growth forms, occur in largely frost-free climates with high insolation, can be of similar size, have similar leaf (or pinna) shapes and sizes, and are evergreen.

44. Mesic Rosette-Shrubs include primarily the trunkless palms, which occur as understorey forms in most tropical evergreen forests and elsewhere as often dense ground covers. Both palmate (fan) and pinnate leaf types are included. All taxa are evergreen. These plants may grow to several meters in height in tropical rainforests or can occur as smaller ground forms, such as the familiar *Serenoa repens* and *Sabal minor* of the southeastern U.S.A. and *Chamaerops humilis* of North Africa. The trunkless shrub-palms have been studied very little as an ecological type. Some palms grow for many years as rosettes from underground trunks before developing short trunks above ground. Palms are often early successional species because of their rapid reproduction and growth rates and their tolerance of poorer soil conditions. In this respect they often represent tropical ecological equivalents to temperate-zone pines.

References: McCurrach 1960; Moore 1973; Langlois 1976; Knapp 1973; Kurz 1945; Monk 1968; Daubenmire 1978; Walter 1973; Lind & Morrison 1974; Whitmore 1975.

45. Xeric Rosette-Shrubs include primarily the leaf-succulent, basal-rosette agaves, aloes, and similar forms found in the world's tropical and subtropical dry climates. Because of the dryness these plants usually occur as scattered or well-spaced individuals in open woodland, savanna or semi-desert landscapes. The leaves are evergreen, generally more or less succulent (often with spiny or denticulate edges), and characteristically shaped like large, stiff blades arranged in a basal rosette. Common taxa include *Agave* spp. and stemless yuccas (e.g. *Yucca glauca*) in the Americas, ground *Tillandsia* spp. and *Puya chilensis* (both *Bromeliaceae*) in coastal Chile-Peru, *Aloë* spp. (e.g. *A. ferox*) in Africa and the Mediterranean region, and others. These plants generally do not penetrate into humid climates because of their vulnerabilty to rotting.

References: Walter 1968, 1973; Daubenmire 1978; Knapp 1973; Innes 1977; Shreve & Wiggins 1964; Whittaker & Niering 1965.

Stem-succulents

Stem-succulents are moderately large plants with fleshy stems (trunks) capable of holding large amounts of water over long periods of time. (Shrubs and forbs with fleshy leaves are included elsewhere.) Stem-succulents are evergreen and generally leafless (some exceptions), with moderate branching in most larger forms. Columnar and barrel-shaped forms usually have ribbed surfaces which can expand to store whatever water is available (Spalding 1905). Other forms, such as many species of *Opuntia,* have stems flattened into expandable cladodes. The root-systems of stem-succulents typically are shallow but spread laterally over several meters in order to intercept quickly whatever rain penetrates the soil surface, even from short showers (Cannon 1911).

Most large stem-succulents are restricted to mostly frost-free, high-insolation, semi-arid to arid climates, especially those with two rain seasons per year (e.g. Arizona-Sonora, equatorial East Africa). Only the bushier stem-succulents with more lignified bases (e.g. *Opuntia*) penetrate into more mesic climates, where rotting can be a problem. Some stem-succulents do penetrate, however, to colder climates, such as *Opuntia frigida* in Canada and *Oreocereus celsianus* in the Andine puna (Cabrera 1968).

Stem-succulents include most *Cactaceae* of the Americas, the stem-succulent euphorbs of Africa, plus endemics in southern Africa and elsewhere. Related are the more lignified, dendroid trunk-succulent treelets of southern Africa (e.g. *Commiphora, Aloë dichotoma, Cyphostemma, Pachypodium*), Socotra (e.g. *Adenium socotranum, Dracaena cinnabari*), and Arizona-Sonora (e.g. *Idria*) (Walter 1973; Popov 1957). It appears that these can be included in the same climatic envelope with the Arborescent Stem-Succulents, though the trunk-succulent treelets represent a somewhat different growth form.

46. Arborescent Stem-Succulents are tall, columnar, usually branched forms best represented by the saguaro (*Carnegiea gigantea*) of the Sonoran Desert, by many freely branching species of *Cereus* and related cacti in Mexico and South America (Caatingas, Chaco, Caribbean thorn-scrub), and by the large 'candelabra euphorbs' of East Africa. When branched, these forms may have individual branches (e.g. *Carnegiea*), clustered branches from a main trunk (true tree form, as in *Euphorbia*), or branches clustered at the base (bush form, as in *Lemaireocereus* and *Lophocereus*). As already mentioned, various trunk-succulent treelets seem to be closely related ecologically. Arborescent stem-succulents generally require dry to arid climates in which frost, if it occurs (even nightly as in the puna), does not last more than 24 hours (Turage and Hinckley 1938; Shreve 1911). Arborescent stem-succulents can also penetrate to some extent into woodland areas and occur in the thick thorn-scrub of northern Venezuela, East Africa, and the Caatingas and even in the dry forests of southern Florida. Because of the long time required to reach arborescent size, arborescent stem-succulents are almost always found as individuals.

Being evergreen, large stem-succulents require a certain regularity of precipitation and for this reason are largely replaced in areas with longer dry periods (e.g. Saharan Africa) by deciduous and/or ephemeral forms.

References: Shreve & Wiggins 1964; Shreve 1911; Shelford 1963; Walter 1973; Niering et al. 1963; Hueck 1966; Knapp 1973; Turage & Hinckley 1938; Kausch 1965; MacDougal 1912; Marloth 1908; Spalding 1905; MacDougal et al. 1915; Popov 1957; Innes 1977; Nobel 1980a.

47. Typical Stem-Succulents include the variety of cylindrical, barrel, spherical, and cushion forms best developed in the New World cacti, especially in the semi-deserts of Arizona-Sonora. These are not as tall as other succulents (seldom over 1–2 meters) and thus generally occur more often in formations more open than the dry woodlands of the tropics. Because more of the biomass may rest on the ground, these succulents may also be confined to drier sites where the danger of rotting is reduced. Separation of arborescent and non-arborescent stem-succulent climates, however, is very problematic. Common taxa include *Ferocactus* (short columns and barrels), *Echinocereus* (often cushion-like), *Mammillaria* (cushion-form), *Opuntia* (cylindrical forms), *Trichocaulon* (small spherical forms from rocky sites in the Namib Desert), and *Euphorbia* (small cylinders). These forms generally are not branched. Other small stem-succulents, generally even smaller and with succulent leaves, are included as Succulent Forbs.

References: Shreve & Wiggins 1964; Shelford 1963; Walter 1973; Kausch 1965; MacDougal 1912; MacDougal et al. 1915; Innes 1977; Rawé 1968.

48. Bush Stem-Succulents are branched, more externally reinforced stem-succulents which tolerate both colder and wetter conditions. These include primarily certain species of *Opuntia*, which occur as shrubs in the dry regions of the Americas (and well into wetter climates on dry sites) and as smaller forms in the drier grasslands as far north as southern Saskatchewan. (Branched opuntias can grow into significant treelets in the dry tropics and subtropics, but these are included more properly as Arborescent Stem-Succulents). Similar forms from the Old World are not as widespread, though certain branched species of *Opuntia*, notably *O. stricta* in Australia (Dodd 1959), have expanded rapidly after introduction to become pest species.

Though often reported to have no important native stem-succulents, Australia does have many bushy, semi-succulent, salt-tolerant native *Chenopodiaceae* (e.g. *Atriplex, Kochia*) which appear to occupy the same climatic ranges as certain bush-opuntias.

References: Walter 1973; Shelford 1963; Shreve & Wiggins 1964; Daubenmire 1978.

Graminoids

Graminoids are narrow-leaved herbs (*Graminiae, Cyperaceae,* and various ecological equivalents) growing from generally well developed underground rootstocks which may be either perennial (e.g. rhizomes) or annual. Graminoid growth forms may be classified as bunched (caespitose), spreading (sward-forming), and rooting (Bews 1929), though intermediate forms may be more common. Leaves may be flat-linear (typical-graminoid), revolute, or even spiny-revolute. Ecologically it seems appropriate to distinguish three basic climatically related types:

1. spreading graminoids, which are favored by any process which kills vertical shoots (e.g. cold winter temperatures, herbivory) and in sandy areas with very low vegetation cover but available soil moisture (e.g. rooting psammophilic grasses);
2. typical bunch-graminoids, which may occur almost anywhere but tend to dominate in areas too dry for spreading graminoids; and
3. tussock-graminoids, an extreme case of bunched growth which provides protection in cool (but not extremely cold) climates for new shoots arising from the middle of the tussock.

In general tall grasses (except tall tussock-grasses of the Southern Hemisphere) can be considered to be spreading, since they usually require more mesic habitats with sufficient moisture for a continuous ground cover. Arborescent bamboos, however, though spreading, are restricted to particular localities by peculiarities of their flowering and seed-production patterns. Bunch-grasses (and cold-climate sedges) often maintain a larger below-ground water-uptake area to support a smaller above-ground transpiring surface area, as is the strategy of many plants in drier environments. This contributes to their apparent spacing in drier areas. Tussock-grasses (and some sedges) occur in many maritime climates which appear to be quite humid but which often have high evaporation rates due to constantly strong winds, as in Patagonia and on many isolated islands.

Graminoids (mainly grasses) can occur wherever there is sufficient light and a sufficient growing season. As climates become cooler and/or drier, however, both graminoids and forbs have several advantages over larger woody forms (Daubenmire 1978, p. 59):

1. Herbs can devote a greater proportion of their production to seed formation.
2. Herbs can migrate faster because of shorter time required to begin flowering.
3. Herbs can aestivate or survive as seeds (annuals) during unfavorably dry periods.

One might also add that:

1. Herbs can occur in climates too cold for woody plants, such as polar and alpine tundra areas.
2. Spreading graminoids, once established in a continuous ground cover, can very effectively prevent the survival of woody seedlings because of the speed with which graminoids can produce a taller, effective canopy.

Graminoids occur in some form in almost all types of climates, though natural dominance may be confined to particular subhumid 'grassland climates' (Thornthwaite 1952). Graminoids are usually seasonal (especially in climates where they are important components) but actually respond quickly to almost any occurrence of adequate growing conditions. Their seasonal physiognomy can perhaps best be described as marcescent.

49. Arborescent Grasses are tall (1–8 m), partially lignified or otherwise reinforced grasses which occur in both warm and relatively cool environments with sufficient moisture. The main taxon is the bamboos (*Bambuseae*), though some bamboos are not so tall. Bamboos can form locally dense thickets and even dominant montane belts in the tropics, but otherwise do not dominate large areas because of their dependence on vegetative propagation. Bamboos generally flower only after several years (sometimes decades), but when they do, huge areas of genetically identical plants can flower simultaneously. Arborescent grasses generally tolerate cool, moist winter conditions well into the temperate zones, especially in the more maritime Southern Hemisphere where they reach as far poleward as southern Chile. Arborescent grasses generally have broader leaf blades, greater total leaf area, and require rather mesic conditions.

References: Arber 1934; Bews 1929; Whyte 1968; French 1979; Walter 1973; Numata, in press; Coupland 1979; McClure 1966; Knapp 1973; Hueck 1966; Troll 1959, 1966, 1972b.

50. Tall Cane-Grasses, the grasses of tropical tall-grass savannas, are tall (2–6 m), reinforced (often slightly lignified), generally broader-leaved grasses which occur in climates with at least seasonally coincident warm temperatures and abundant rainfall. These conditions are met throughout much of the tropical and subtropical zones, but such grasses are best developed in the summer-rain tropics and dominate as savannas in climates with long dry seasons and insufficient total moisture for forest or woodland. Such savannas can also occur where sufficient water is available but where tree growth is suppressed by any combination of fire, herbivory, or water loss from a pre-existing grass cover. Tall Cane-Grasses begin growing at the beginning of the favorable season and grow quickly to a height of several meters. At the end of the wet season, however, they die back to ground-level, leaving the standing dead stalks. If fire occurs during the dry season the dead stalks (and any living woody seedlings) are eliminated, but the grasses regenerate quickly at the beginning of the next rainy season and often benefit from the removal of the dead debris. Tall cane-grasses can be either perennial or annual, C_3 or C_4, but are more commonly C_4 perennials. Typical taxa are *Hyparrhenia, Imperata, Pennisetum, Tristachya, Arundinaria* (a bamboo), and tropical species of *Andropogon*. Tall cane-grasses, including *Arundinaria,* also grow in forest understoreys both in the tropics and well into the warm-temperate zone. They are also found commonly as aquatics or semi-aquatics in swamps and marshes, along shorelines, and in periodically flooded areas.

References: Beard 1953, 1967; Eyre 1968; Rattray 1960; Shantz & Marbut 1923; Bews 1929; Arber 1934; Whyte 1954, 1968; Knapp 1973; Coupland 1979; Goodland 1971; Walter 1939, 1973; Hueck 1966; Wilhelmy 1957.

51. Typical Tall Grasses differ from Tall Cane-Grasses in that they are shorter (seldom over 2 meters), generally not reinforced (more pliable), and have generally softer, narrower leaves. These are typically the grasses of the temperate tall grasslands and of montane meadows but can occur also in the tropics. The same genera as in the tropics can be involved (e.g. *Andropogon*), but the particular species are less tall, due to less total warmth and moisture during the (often shorter) growing season and to the fact that more species are C_3, with lower net production rates. Typical tall grasses occur in many areas but are best developed in the temperate continental interiors with early-summer rainfall maxima and cold winters, i.e. the true prairies of North America and the meadow-steppe of southeastern Europe. Common taxa include *Andropogon, Agropyron, Stipa, Festuca,* and many others.

References: Walter 1967, 1968, 1974; Weaver & Albertson 1956; Bews 1929; Coupland 1979; French 1979; Shelford 1963; Daubenmire 1978; Eyre 1968; Knapp 1965, 1973; Küchler 1964.

52. Short Sward-Grasses are spreading grasses usually less than 50 cm in height which produce the smooth, continuous short-grass swards of the northern Great Plains (into Canada) and similar but less homogeneous swards in parts of the Ukraine. Their growth form is the result of quite cold winter, grazing, and insufficent summer warmth and/or moisture for taller growth. The result is a thick, dense sod so firm that early settlers built their houses with it. Rain falls primarily in early summer, but late-summer drought is often less severe because of the shorter warm season, lower transpiration levels, and greater soil moisture (chernozem soils) than in warmer grasslands. Typical taxa include *Buchloë dactyloides* and *Bouteloua gracilis* in the northern Great Plains; *Poa* spp. in Europe, eastern North America, and East Asia; *Kobresia, Poa, Deschampsia,* and sedges in mesic montane swards of western North America; *Festuca* (more normally a bunch-grass) in the Ukraine; the common lawn-grass *Cynodon dactylon,* common throughout the winterless tropics; and the same or similar taxa, plus more sedges, in arctic and alpine mesic tundras.

References: Walter 1968, 1974; Weaver & Albertson 1956; Küchler 1964; Daubenmire 1978; Bews 1929; Eyre 1968.

53. Short Bunch-Grasses (graminoids) are the common, caespitose grasses usually less than 50 cm in height which can occur in most parts of the world and which dominate most short grasslands, from the tropics to the polar regions, in areas too dry (or cold) for taller grasses. Typical taxa are too numerous to mention but include most sedges as well as true grasses. This form occurs wherever there is an adequate growing period. It dominates most typically in tropical dry savannas and temperate steppes, in the tropical alpine puna and similar montane-alpine grasslands, in the dry montane grasslands of temperate mountains (e.g. Asia, southern Africa, Andes), and in drier tundra areas. Though seasonal growth is normally in the spring and early summer, short bunch-grasses also dominate the winter-rain grasslands of the eastern Mediterranean region (Anatolia, Iran) and the western U.S.A. (Central Valley of California, Palouse region).

References: Eyre 1968; Walter 1968, 1973, 1974; Weaver & Albertson 1956; Bews 1929; Beard 1953, 1967; Knapp 1965, 1973; Shanz & Marbut 1923; Troll 1959; Rattray 1960; Rawé 1968; Daubenmire 1978.

54. Tall Tussock-Grasses are the 1–3 meter tall tussocks of the windy, maritime Southern Hemisphere grasslands, including the wetter areas of the Argentine pampas, the montane grasslands of New Zealand, and the sub-antarctic islands, as well as similar climates on islands, in moors, and in tropical mountains in other parts of the world. The tussock growth form is thought to protect the active shoots in the center against near-freezing temperatures. Tall-tussocks are generally found in areas too cool or dry for tree growth, as well as in exposed areas along forest borders. The most important genera are quite familiar from other grasslands but have developed the characteristic tall-tussock growth form only under certain conditions, e.g. *Stipa brachychaeta* in the pampas, *Danthonia* (*Chionochloa*) spp., *Festuca matthewsii*, and *Poa colensoi* in New Zealand.

References: Connor 1961; Zotov 1947; Walter 1968; Cockayne 1958; Drummond & Leatham 1959; Eyre 1968.

55. Short Tussock-Grasses differ from tall tussock-grasses only in size but are much more widespread, both in Southern Hemisphere grasslands and throughout the world. The short-tussocks extend into drier areas than do tall-tussocks, including most of the pampas and Patagonia, the lowland grasslands of New Zealand, alpine and montane grasslands of Australia and New Zealand, and the subantarctic islands. Short tussock-grasses are also a dominant form in the Andine puna, in moors throughout the world, and are important in a variety of cool island or montane grasslands where winters are not too cold or there is a sufficient winter snow cover, as in much of Europe. Typical genera are again familiar from other areas but form short-tussocks only where winter conditions are not too extreme: *Stipa trichotoma* in the pampas, *S. speciosa* and *Festuca pallescens* in Patagonia, *F. novae-zelandiae* and *Poa caespitosa* in New Zealand, and *P. flabellata* on the Falkland Islands.

References: Walter 1968; Zotov 1947; Cockayne 1958; Bews 1929; Wilhelmy 1963; Drummond & Leatham 1959; Coupland 1979; Eyre 1968.

56. Sclerophyllous grasses are xeromorphic grasses with permanently revolute, often spine-like leaves, as best suggested by the Australian 'porcupine grass' (*Triodia* spp.). Sclerophyllous grasses can occur as individual bunch-grasses or as large tussocks, the hummocks of *Triodia* measuring up to several meters across. *Triodia* hummocks dominate large areas of interior Australia. Sclerophyllous grasses are generally evergreen, though the active shoots are usually all but hidden by the mass of brownish old (dead) shoots. Sclerophyllous grasses are especially common in mediterranean climates (e.g. *Stipa tenacissima* in North Africa) and in dry areas with irregular rainfall.

References: Burbidge 1945–46, 1953; Beard 1969; Bews 1929; Walter 1973; Specht 1972.

57. Desert-Grasses generally represent an extreme form of bunch-grass in which a large root network supports a much smaller above-ground bunch of shoots. The spreading psammophilic grasses could perhaps also be included. Desert-grasses grow as well spaced individuals and do not form a dense ground-cover. Above all, desert-grasses represent a growth form necessitated by aridity. Many desert-grass taxa could be short bunch-grasses under better conditions. Desert-grasses occur throughout the world's arid areas and often form the only permanent vegetation in the driest areas, especially if these involve sandy substrates. *Aristida* is a common genus in many dry areas, including the 'desert plains' of Texas-New Mexico and the drier steppes of Middle Asia. The most important dry-puna species is *Stipa jehu.* Both perennial and annual species can be involved.

References: Walter 1968, 1973, 1974; Walter & Box, in press; Weaver & Albertson 1956; Küchler 1964; Shelford 1963; Bews 1929; Eyre 1968; Rattray 1960; Knapp 1965, 1973; Daubenmire 1978; Nobel 1980b.

Forbs

Forbs are herbs which are not graminoids and not ferns, i.e. all herbaceous dicots plus all herbaceous monocots except the graminoid families *Graminiae, Cyperaceae, Juncaceae,* and various ecologically equivalent forms. Physiognomically, forbs are more or less broad-leaved herbs, although erect, linear-leaved forms (e.g. many *Compositae*) are included with the leafier, more mesic forest forms. Forb form and strategy can be classified into many subtypes, most of which are not easily relatable to climate since forbs occur most commonly as understorey components. Ellenberg & Mueller-Dombois (1967b) classify forb types based on the following:

1. life form (Raunkiaer): phanerophytes, chamaephytes, hemicryptophytes, geophytes, therophytes.
2. general structure: caespitose, scapose (with or without basal rosette), reptant, pulvinate.

In addition one can easily make other ecologically useful distinctions based on:

1. leaf shape and size: broad, linear, needle.
2. leaf and stem structure: hard, soft, succulent.
3. plant size (height).
4. seasonality: evergreen, summergreen, raingreen, ephemeral (for annuals: springgreen, summergreen, raingreen, winter annuals).
5. regional affinity: tropical, temperate, polar.
6. other attributes, such as carnivorousness, reproductive strategy (e.g. r or K-strategists), propagation mechanisms.

Herbaceous phanerophytes (e.g. *Musa*) and chamaephytes are generally restricted to the humid tropics; more typically forbs are either perennial hemicryptophytes or geophytes or annual therophytes. Since all three generally occur in the same climates, the most ecologically useful initial distinction seems to be that based on seasonality plus regional affinity.

Outside the favorable climates of the humid tropical and temperate forests, both forbs and graminoids have several advantages over woody plants, as already noted in the section on graminoids. Although the geoecology of graminoids, as a life form and a synusia, has been studied extensively worldwide, the same has not been true for forbs, mainly because forbs do not form the dominant vegetation of large areas as do some graminoids. The worldwide geography and comparative ecology of forbs as a group remains one of the more neglected areas of plant ecology and biogeography.

58. Tropical Evergreen Forbs include an enormous variety of forms and strategies which cannot really be separated climatically and which occur throughout the tropical region. Classification of such forms and subforms, both taxonomically and ecologically, remains a largely untouched field of plant ecology. Certain taxa such as *Musa* (banana, a phanerophyte), *Begonia* and *Coleus* from rainforest understoreys, and the herbaceous dwarf-palms are familiar types, but they do not begin to suggest the variety of forms which actually exists. Forbs are often not as important in tropical forests as in temperate forests because of the greater number of larger woody competitors. To some extent forbs have been replaced by epiphytes, which overcame the reduced light on the rainforest floor by moving upward onto the trees. In other tropical areas evergreen forbs may be more important components of the vegetation, as in high mountains (e.g. the tall, often white-pubescent rosette-forbs of the páramo, plus more familiar-looking members of temperate-zone taxa), in open woodlands, and around rainforest margins. The most common subtypes of Tropical Evergreen Forbs probably include the familiar leafy dicots, leafy monocots, herbaceous palms and other rosette-forbs, and (outside rainforest areas) familiar erect, single-stemmed forbs (e.g. *Compositae*) and succulent forms. Annual but aseasonal forbs can be considered evergreen if they produce evergreen stands.

References: Walter 1973; Cuatrecasas 1968; Richards 1952; Cabrera 1968; Schnell 1970; Löhr & Müller 1968; Everard & Morley 1970.

59. Temperate Evergreen Forbs include a variety of often smaller, sometimes harder-leaved (sclerophyllous) forms which occur most commonly in the understoreys of mesic forests. Temperate Evergreen Forbs are generally hemi-cryptophytes, as geophytes and therophytes are usually seasonal in the temperate zone and temperate herbaceous phanerophytes and chamaephytes, except for graminoids, generally do not occur. Most temperate evergreen forbs are small dicots, such as *Pyrola, Chimaphila, Gaultheria, Hexastylis,* and *Plantago.* Included also are many taxa which remain green in milder temperate winters but would be considered summergreen elsewhere, for example *Taraxacum, Duchesnia, Stellaria,* and other prostrate, often weedy herbs with softer leaves. Many carnivorous temperate forbs are semi-evergreen (e.g. *Drosera*). Temperate succulent forbs are usually evergreen but are included as a separate type elsewhere. Temperate Evergreen Forbs occur both in evergreen forests and woodlands and well into areas with colder winters, where they are often protected in winter by coverings of fallen leaves and snow. These forbs are especially important in temperate coastal rainforests, where they can form continuous evergreen ground-covers.

References: Walter 1968, 1974; Radford et al. 1968; Polunin 1972; Everard & Morley 1970; Reichle 1970; Penfound 1952; Ellenberg 1963.

60. Raingreen Forbs are seasonal forms which sprout anew at the beginning of the tropical rainy season, grow and reproduce, and generally die back to ground-level as the dry season progresses. Of the five seasonal types of forb described in this model, raingreen forbs are probably the least studied and are certainly the least known in the temperate zone. Raingreen forbs may be phanerophytes, chamaephytes, hemi-cryptophytes, geophytes, or therophytes, though the first is generally uncommon. Most raingreen forbs are small, since larger plants in the winterless tropics generally tend to become woody. Many raingreen forbs are familiar-looking broad-leaved dicots (e.g. *Leguminosae, Compositae, Rubiaceae, Labiatae*), and many are considered weeds. Annuals are common. Raingreen forbs are found in both forested and more open tropical summer-rain regions. Except in the extreme case of desert ephemerals, raingreen forbs generally do not attain dominance in formations. They may, however, form significant and even continuous understoreys in open areas where rainfall is concentrated in short periods but where grasses are not totally dominant. In the Sonora Desert and other dry areas with two distinct rain periods both summer-rain and winter-rain forb covers are common, each involving different sets of species. The distinction between Raingreen Forbs and Ephemeral Desert Herbs is not well defined, except that the former are generally seasonal and the latter more irregular. Some taxa which taxonomically are forbs could be better classified ecologically as graminoids, e.g. certain *Eriocaulaceae, Xyridaceae.*

References: Walter 1973; Beard 1953; Eiten 1972; Knapp 1973; Everard & Morley 1970.

61. Summergreen Forbs include a great variety of seasonal forms which sprout anew in spring or early summer, grow and reproduce, and generally die back to ground-level as summer or autumn progresses. The end of the growing period may be brought on either by summer heat and desiccation or by autumn cold. Summergreen forbs involve hemi-cryptophytes, geophytes, and therophytes, which may be leafy or nearly scapose, single-stemmed or highly branched, or of a number of other basic forms. The potential variety in subtypes is suggested by a few well known and typical families such as *Liliaceae, Compositae, Leguminosae, Ranunculaceae, Araceae, Umbelliferae,* and *Violaceae.* Three common ecological subtypes are mesic forest 'wildflowers', spring-ephemeral geophytes, and summer-blooming 'weeds'. Summergreen forbs are well developed in all but the driest temperate-zone and boreal-polar climates (with sufficient summer warmth) and extend along mountain ranges well into the summer-rain tropics. Many individual summergreen species, however, have soft leaves and high transpiration and respiration rates, and are limited toward warmer climates, especially in colder areas, by increased respiration which may be too much for a positive net production. Summergreen forbs generally are not dominants but do constitute important forest synusiae and often subdominant components of mesic meadows and 'meadow-steppes', as in southeastern Europe and the former true prairies of the U.S.A. Summergreen tall-forbs growing to 2–3 meters can be locally dominant on small, mesic, well fertilized pasture sites ('Läger') in the Alps and Caucasus (Rübel 1930, Walter 1968), as well as on more natural wet-montane, warmer-tundra, and floodplain sites. Of the five forb types described herein, the summergreen temperate forbs are by far the best studied.

References: Walter 1968, 1974; Rübel 1930; Anderson 1964; Lieth 1960; Kojić 1966; Bethke et al. 1965; Goryshina 1972; Ellenberg 1963; Daubenmire 1978; Radford et al. 1968; Polunin 1972; Daxer 1934; Mudrack 1935; Walter & Box, in press; Schemske et al. 1978.

62. Succulent Forbs are small, generally evergreen, herbaceous leaf-succulents which may also have succulent stems and/or root-systems. The most common forms are rosulate (e.g. *Sedum, Echeveria*) and prostrate (e.g. *Portulaca*). Many more exotic forms can also be included, however, such as the caudex-succulent 'living stones' (*Lithops*) of south-western Africa, the well known, non-rosette *Kalanchoë* of dry tropical areas, and many *Mesembryanthemaceae* (*Aizoaceae, Ficoidaceae*) of southern Africa. Some quite succulent species are annuals, e.g. *Mesembryanthemum cryptanthum* (South Africa) and many species of *Portulaca.* The variety in forms is quite striking (especially in southern Africa), but all have in common the requirement of moderately wet to dry sites, since they are limited both by extreme aridity and by the danger of rotting in excessively wet areas. Succulent forbs are especially common in mediterranean, warm semi-desert, and warm woodland climates and on other drier and saline sites such as coastal sands.

References: Innes 1977; Walter 1973; Polunin 1972; Shreve & Wiggins 1964; Rawé 1968; Hueck 1966; Zemke 1939.

Small herbs

The term 'small herb' is used to refer to groups of both small forbs and graminoids occurring in extreme environments. In such situations the differences in leaf shape, plant structure, and taxonomy are of much less importance than the fact that the plants are quite small and herbaceous. Generally they are also highly seasonal, which forms the most important ecological basis for further differentiation. Such plants may be either annual or perennial (usually geophytes) but generally must complete seasonal growth and seed production during the short periods in which water and warmth are available.

63. Xeric Cushion-Herbs are usually small, often rosette or mat-forming herbs found in dry and often windy (exposed) areas which may be either cold or warm. The xeromorphy of such plants may be temporal as much as morphologic, in that leaves may be quite soft but life-cycles appropriately short to be completed using the small amount of water and/or warmth seasonally available. The best examples are perhaps the rosette, mat, and cushion-herbs of exposed polar and alpine tundra sites, such as certain species of *Draba, Dryas, Cerastium, Stellaria, Silene* (the last three *Caryophyllaceae*), and above all *Saxifraga.* The ground-hugging form provides protection against desiccation, even on exposed sites, both by reducing the transpiring surface and by remaining within the calmer near-surface air layer. In cold climates the cushion form also permits better use of near-surface warmth during the active period. Larger but otherwise similar cushion forms can be found in most warmer xeric environments. Most Xeric Cushion-Herbs are summergreen (or raingreen in tropical summer-rain areas).

64. Ephemeral Desert Herbs are small, generally malacophyllous raingreen ephemerals which can grow quickly and complete their life cycles using the small amounts of water provided by short rainy periods or even individual showers. As little as 15 mm of rainfall or less may be sufficient (Tevis 1958). Both perennials with underground storage organs and annual therophytes are included, the latter probably requiring less water for survival (Walter 1973). Both types of plants can quickly develop ephemeral carpets and can attain dominance in at least two distinct environmental situations: (1) the warm deserts with no permanent vegetation, and (2) the continental spring-ephemeral semi-deserts of Middle Asia, which have a 4–8 week growing season in March-May provided by some spring rainfall and melting snow. This latter type (see Walter and Box, in press, and Walter 1974) is dominated by mini-geophytes such as *Poa bulbosa, Tulipa biflora,* and *Carex pachystylis,* and occurs on loess, which has a relatively high water-holding capacity in the upper few centimeters. The warm deserts and semi-deserts may support winter and/or summer ephemerals, depending upon the rainfall regime. The ephemeral semi-desert and desert carpets of *Weitzia aurea* in Australia (Walter 1973), of *Nolana* and *Calandrinia* species in the Atacama Desert (Kohler 1967), and of both summer and winter ephemerals in Arizona-Sonora (Shreve and Wiggins 1964, Tevis 1958, Walter 1973, Ehleringer et al. 1979) are especially well known, as well as some spring-ephemeral carpets of less arid regions (e.g. the bluebonnets, *Lupinus subcarnosa*, of Texas woodlands).

65. Summergreen Cold-Desert Herbs include a variety of often quite small graminoids and forbs (usually mini-geophytes with some therophytes) which grow in the shortest, most extreme growing seasons of the polar cold-deserts and temperate-zone subnival belts. Their main distinguishing characteristics are small size and the ability to grow rapidly and complete their life-cycles within favorable periods sometimes as short as two weeks in midsummer (Caldwell et al. 1978; Billings et al. 1978; Bliss 1956). Because soil may be scarce in such environments, and water thus limiting as well as warmth, the plants are often found only in crevices and other more mesic microsites. Each summer's photosynthate must be channelled as efficiently as possible into the production of seeds (therophytes) or into replenishing the underground storage organs (geophytes). The small size of such plants reduces potential water loss from surface areas, reduces the respiratory requirements of larger amounts of standing biomass and greater annual growth, and permits maximum use of near-surface warmth. Summergreen Cold-Desert Herbs are essentially summergreen ephemerals, as opposed to the raingreen ephemerals of arid regions. Important genera include *Eriophorum, Carex, Saxifraga, Geum, Gentiana*, and *Ranunculus*.

66. Raingreen Cold-Desert Herbs are small herbs (forbs and graminoids, annuals and perennials) which occur generally in tropical alpine areas which are both cold and subject to seasonal drought such as the Andine puna. Since mean temperatures are usually above freezing, it is the precipitation regime which delimits the growing season. Low temperatures (and nightly frost), however, limit the degree of activity even during the growing season, probably even more than in the case of the summergreen cold-desert, which receives little nightly frost and up to 24 hours of sunshine during the growing season. (The differences outside the growing season are of less importance to the vegetation). Raingreen Cold-Desert Herbs are, in form and function, much like summergreen cold-desert herbs and, at least in the Andine puna, involve many of the same taxa (e.g. *Plantago, Scirpus, Juncaceae, Poaceae, Cyperaceae*), as well as other temperate-zone taxa (e.g. *Iridaceae, Astragalus, Amaryllidaceae*) (Cabrera 1968).

Additional general references:
summergreens: Böcher 1938; Tieszen 1978; Walter 1968, 1974; Bliss 1956; Walter & Box, in press; Salisbury & Spommer 1964; Tranquillini 1978.
raingreens: Walter 1973, 1974; Went 1948–49; Shreve & Wiggins 1964; Walter & Box, in press; Zohary 1973.

Vines and lianas

Vines are weak-stemmed plants whose long, slender, fast-growing shoots rely on other plants or objects for support (Menninger 1970). Vines may be either woody or herbaceous, perennial or annual. Large woody vines are generally called lianas and occur in warmer humid climates of the tropics and subtropics. In such climates the larger vines and lianas can grow quickly to canopy height, spread out laterally, and almost completely cover the existing vegetation.

Vines germinate in the soil but then elongate rapidly, climbing or sprawling over other plants (or other objects) in order to reach more sunlight quickly and without having to invest large amounts of photosynthate in self-supporting structure. Climbing is accomplished by a variety of mechanisms, including grasping by means of tendrils, clinging by means of hooks and thorns, adventitious rooting, weaving and twining growth, and even sticking by means of adhesives. One can also distinguish photophytic vines (e.g. *Vitis, Pueraria, Ficus, Lonicera*) which grow quickly to fuller sunlight and skiophytic (shade-loving) vines (e.g. *Hedera, Parthenocissus*) which climb more slowly in the understorey, often by rooting along existing trunks. Most vines and lianas have broad leaves (or compound leaves). Hallé et al. (1978) and Cremers (1973) have recently identified a number of basic architectural types.

Because of their rapid rates of growth and need for existing vegetation for support, vines and lianas are best developed in warmer humid climates. Perhumid climates are most favorable, since vines and lianas often must compete for water with deeper-rooting trees as well as with many shallow-rooting forms.

67. Tropical Broad-Evergreen Lianas are the large, woody, perennial lianas of tropical rainforests. These are almost always photophytic vines and generally climb quickly into the existing forest canopy, where they produce their flowers, fruits, and most of their leaves. The ropelike stems can become several decimeters in diameter, and the resulting tangle at canopy level can bind trees together so that they cannot be felled by cutting at ground-level. Included in this group are the pseudo-lianas (or hemi-epiphytes), plants which germinate as epiphytes and then send down roots into the soil (see Walter 1973). Typical tropical liana genera include *Ficus* (including the stranglers), *Landolphia, Combretum, Calycobolus, Anodendron, Entada* (*Leguminosae*), and the climbing palms (e.g. *Calamus*). Included also must be taxa which commonly become self-supporting trees upon reaching the canopy (e.g. *Ceiba, Freycinetia, Bougainvillea, Securidaca*).

68. Broad-Evergreen Vines, as opposed to lianas, are smaller (perennial) vines which may be either woody or herbaceous (more commonly the former) and which may climb or sprawl in a greater variety of ways. These include both tropical taxa (e.g. *Philodendron, Vanilla*) and distinctly temperate taxa (e.g. *Lonicera, Hedera, Smilax*). In the humid tropics such smaller vines are more likely to be root-climbers and skiophytes than twining photophytes, though no sharp distinction can be made. Temperate broad-evergreen vines can be quite aggressive, however, as in the case of *Lonicera japonica* (honeysuckle), which sprawled quickly over large areas of the southeastern U.S.A. after introduction. Included are also the horizontal vines such as *Hedera* and *Pachysandra* and various climbing palms, aroids, and grasses (mainly bamboos).

69. Broad-Raingreen Vines occur in tropical and subtropical areas with pronounced dry seasons. Though little-studied, they can be particularly important in open woodlands and thorn-scrub, where they sprawl at will over existing trees and bushes. Though the leaves are deciduous, stems are often green and photosynthetically active throughout at least part of the dry season. Much of the generally impenetrable character of the tropical thorn-scrub can be due to the tangle formed by raingreen vines binding other plants together. *Leguminosae* and *Euphorbia* are important taxa.

70. Broad-Summergreen Vines occur in temperate summergreen areas with cold winters and summer rain. Included are both native taxa (e.g. *Vitis, Rhus, Parthenocissus*) and tropical invaders which already have the deciduous habit (e.g. *Pueraria lobata, = thunbergiana,* the ubiquitous Kudzu vine of the southeastern U.S.A.). Many summergreen vines are locally important and may be planted as ornamentals because of their flowers (e.g. *Wisteria, Ipomoea*) or their autumn colors (e.g. *Parthenocissus*). *Vitis* is important economically. Herbaceous vines are perhaps best represented among the summergreens (e.g. *Vicia, Ipomoea*) but woody forms are more common.

References Menninger 1970; Hallé et al. 1978; Cremers 1973; Walter 1973; Schnell 1970; Hueck 1966; Richards 1952.

Ferns

Ferns (*Pteridophyta*) are an old group of generally small plants with prostrate, even underground stems and large, erect, often compound leaves called fronds. (The larger, trunk-forming tree-ferns of the humid tropics are included, physiognomically, under rosette-treelets. Epiphytic ferns are included under epiphytes). Some ferns are quite tough and withstand drier conditions. For a variety of reasons (non-vascularized gametophytes, flagellated sperms, no protected seed-stage during reproduction), however, ferns generally grow more successfully in habitats which are at least moderately wet. Ferns are generally best developed in humid forests and, since many prefer the reduced amount of sunlight, can often form closed understoreys in both tropical and temperate rainforests. Ferns are generally poikilohydrous, the degree of hydration depending entirely on ambient moisture conditions. Included in this group are both the ephemeral ferns of rocky arid sites (e.g. *Cheilanthes lindheimeri* of Arizona, *Pellaea calomeanos* of the Namib) and many rainforest forms, both terrestrial and epiphytic (e.g. *Hymenophyllaceae*).

Ferns may be divided into evergreen and seasonal forms for ecological purposes, the latter losing its fronds seasonally but sending out new ones from a permanent rootstock or prostrate stem. Many individual fern species can be evergreen or seasonal depending on local climatic regimes.

71. Evergreen Ferns include both delicate rainforest forms which occur where there is no seasonal dryness and more xeromorphic, generally coriaceous forms whose fronds survive seasonal dryness (or winter cold). Evergreen, in this sense, means that the fronds are persistent and active (hydrated) throughout the year. Evergreen Ferns are best developed in the humid tropics and subtropics but can extend into temperate areas where winter temperatures are not too severe, such as in temperate rainforests. In areas with winter frost the evergreen fronds may become prostrate, at least during the winter. Evergreen Ferns often form luxurious closed ground-covers in cool perhumid rainforests of both tropical mountains and windward temperate coastal areas, especially in the Southern Hemisphere. Though still important in warmer tropical lowland rainforests, their understorey dominance is generally broken by larger competitors. Commonly evergreen genera include *Asplenium, Cheilantes, Pellaea, Polystichum,* and *Polypodium* (although certain species of these genera can be quite ephemeral in desert situations).

72. Summergreen Ferns sprout anew each spring after being killed back to ground-level by winter cold. Summergreen Ferns also prefer cool, perhumid or seasonally perhumid environments, though some taxa with more coriaceous foliage do grow in warmer, drier areas. The more hygrophilic summergreen ferns tolerate quite cool growing seasons and can be found as closed understoreys in cool-temperate forests all the way to alpine and subpolar forest limits. Some seasonal ferns might more properly be called raingreen, including not only the poikilohydrous desert forms but also certain ferns of mediterranean climates. Page (1977), for example, reports that *Asplenium onopteris, Davallia canariensis,* and *Polypodium macaronesicum* occur in the montane laurel forests of the Canary Islands but are summer-deciduous in response to the mediterranean summer dryness. Other species of *Asplenium* (spleenworts, including tropical rainforest nest-ferns) and *Polypodium* (often epiphytic or growing on rocks) are commonly evergreen in other tropical and temperate climates. Commonly deciduous genera (again with exceptions) include *Athyrium* and *Cystopteris* (both *Aspidiaceae*), common in North America, Eurasia, and in the Southern Hemisphere (e.g. *Cystopteris fragilis* of Australian mountains and the subantarctic islands). Different species of *Dryopteris* (also *Aspidiaceae*) tend to be summergreen in central European deciduous forests (Ellenberg 1963; e.g. *D. filix mas, D. Linnaeana, D. austriaca dilatata*) but more commonly evergreen or semi-evergreen in the milder (but more continental) summergreen climates of the southeastern U.S.A. (Radford et al. 1968: e.g. *D. marginalis, D. intermedia, D. cristata,* but summergreen *D. spinulosa* and *D. campyloptera,* both less common).

References: Christ 1910; Jermy, Crabbe, and Thomas 1973; Blombery 1967; Ellenberg 1963; Page 1977; Walter 1968, 1973.

Epiphytes

Epiphytes, strictly speaking, are plants which germinate and grow on other plants, primarily on trees and presumably as a means of obtaining more sunlight. Most have only aerial roots; those which do send roots down from their tree-branch locations into the soil are more properly called hemi-epiphytes (or pseudo-lianas). Only a few epiphytes are found in the temperate zone, most commonly the mistletoes, certain tropical invaders (e.g. *Tillandsia usneoides*), and various mosses and lichens. Epiphytes are quite common and striking in the tropics, however, especially in rainforests where rainfall is frequent and humidity is high, alleviating potential stress on the aerial roots.

Epiphyte variety is quite great, both in taxonomic groups (orchids, aroids, bromeliads, cacti, ferns, etc.) and basic forms (leafy, filiform, rosette, tussock, stoloniferous, thorny, succulent, etc.). True epiphytes are not parasitic and generally do not damage or greatly stress the trees which carry them. Most epiphytes do prefer living trees, however, often of particular taxa, possibly because of bark characteristics and because the bark of dead trees may soon fall off. Epiphytes often settle in the saddles formed by branching, since water is more readily held in these areas.

Tropical-rainforest epiphytes can be classified into sun-loving (heliophilic) photophytes, which live on upper and exposed branches, and shade-loving (ombrophilic) skiophytes, which live on lower, more protected branches. Heliophilic epiphytes are generally somewhat more xeromorphic (e.g. coriaceous, sclerophyllous, succulent) in structure but may also have larger leaves. Photophytic and skiophytic epiphytes generally occur in the same climates.

73. Tropical Broad-Evergreen Epiphytes include the wide range of generally larger, leafier rainforest forms, many of which are sun-epiphytes. The most characteristic form is perhaps the agave-like, leaf-succulent bromeliads, which sit on exposed branches and saddles and catch rainwater in their rosette bases, hence a common name 'water-cups'. Included also, however, are a variety of often mildly leaf-succulent aroids, cacti, and orchids, with leaves of varying sizes. Tropical Broad-Evergreen Epiphytes are both drought and cold-sensitive and are not common in tropical areas with extended dry seasons. They do not extend far into the temperate zone, but this is perhaps due more to seasonal dryness than to extreme frost-sensitivity, since some such epiphytes occur in the highest forested belts (cloud forests) of tropical mountains.

74. Narrow-Leaved Epiphytes involve generally smaller forms with filiform, graminoid, or other narrow leaves. These include a wider range of moisture requirements, from quite moisture-demanding, filamentous or filmy ferns (e.g. *Hymenophyllaceae*) of rainforest understoreys to certain xerophytic species of *Tillandsia,* which look more like clumps of xeric grass and may occur even in distinctly dry deciduous woodlands. The hygrophilic forms are best developed in the cloud forests of tropical mountains, where they can cover and obscure other plants almost completely. They occur also downslope and generally throughout the lowland rainforests. The more xerophytic forms extend into the seasonally-dry subtropics. *Tillandsia usneoides* (Spanish moss) extends into the temperate southeastern U.S.A., where temperatures fall below freezing on many winter nights.

75. Broad-Wintergreen Epiphytes represent the 'mistletoe habit' and include a variety of at least partly parasitic, generally somewhat more xeromorphic taxa which are found both in the tropical and the temperate zones of both the Old World and the Americas. Kuijt (1961, p. 19), in reference to *Dendrophthora,* describes the plants as 'rather small bushes of olive green or yellowish color, glabrous, parasitic on a great variety of trees and shrubs'. Wintergreen Epiphytes are seen most commonly in treetops during the season when the trees are bare. The epiphytes remain green during the foliated period, however, and can occur on evergreen trees, both in rainforests and in surprisingly dry areas (e.g. *Phoradendron juniperii,* which parasitizes *Juniperus* species in the southwestern U.S.A.). The degree of parasitism is generally not extreme and appears to fill primarily a need for nutrients. The most important taxa are *Dendrophthora* (53 spp. in tropical America, see Kuijt 1961), *Phoradendron* (found also throughout tropical America and the southern U.S.A., see Trelease 1916), *Viscum* (the common mistletoe of Europe), certain *Myzodendraceae* (which parasitize *Nothofagus* in southern South America), plus *Psittacanthus, Phacellaria,* and various *Loranthaceae* in Australia.

References: Kuijt 1961; Trelease 1916; Walter 1973; Richards 1952; Hosokawa 1954, 1967; Bünning 1956; Schnell 1970; Stanford 1969; Biebl 1964.

Thallophytes

The term thallophyte is used here in the sense of Ellenberg & Mueller-Dombois (1967b) to refer to non-vascular cryptogams, i.e. mosses, liverworts, algae and lichens. Thallophytes are always small and most are poikilohydrous, i.e. active and conspicuous only when sufficiently wet. The larger chamaephytic mosses and lichens can form significant cushions, hummocks, or mats (carpets) on both ground and biotic surfaces. Smaller thallophytes of all four taxonomic classes can form flatter coverings on a variety of surfaces (e.g. trees, rocks, ground) and are usually less apparent. Some are thero-thallophytes and escape unfavorable conditions as seeds. Epiphytic thallophytes are included here rather than as epiphytes.

76. Mat-Forming Thallophytes include the larger, relatively permanent chamaephytes, primarily pulvinate mosses and liverworts and some fruticose lichens. These can be either evergreen or seasonally active but generally maintain a permanent structure. The two habits are perhaps best represented by the evergreen carpets and hanging mats (e.g. *Selaginella*) in tropical rain and cloud forests and by the ground covering of seasonally active mosses and lichens in tundra areas. In the latter case the poikilohydrous permanent structure permits the thallophytes to compete successfully with larger but more slowly developing vascular herbs. Mat-Forming Thallophytes also occur in most closed forests, wetlands, on sufficiently moist rock and wall surfaces, and in more mesic grasslands and meadows, including lawns. In these cases the thallophytes may be more seasonal, making use of the less favorable season (e.g. humid temperate-zone winters, as in Western Europe) while being quickly outcompeted during the main growing season.

77. Xeric Thallophytes include the smaller, generally less conspicuous thallophytes, primarily flat-matted mosses and liverworts, foliose and crustose lichens, various algae (including endolithic lichens and algae), and many annual thallophytes of all groups. These may occur under quite mesic conditions but become more important as conditions become drier, as in drier forests and woodlands, on more exposed and primitive surfaces, and perhaps most importantly on drier tundra and cold-desert sites. Xeric Thallophytes can be the dominant and often the only vegetation in polar and subnival deserts, including the vegetated portions of Antarctica, and in the driest desert areas. In forests these can form epiphyllous microsystems on leaves.

References: Ellenberg & Mueller-Dombois 1967b; Walter 1968, 1973; Vogel 1955; Folmann 1965; Segal 1969; Ruinen 1961; Renner 1933; Tieszen 1978; Noerr 1974a, 1974b; Steere 1978; Watson 1913; Britton 1967; Billings & Mooney 1968; Bliss 1971; Ahmadjian & Hale 1973; Lange 1965; Wielgolaski 1972; Crum 1972; Gams 1934.

References for world and regional vegetation

World (descriptive)
Eyre (1968), Rübel (1930), Schimper & von Faber (1935), Troll (1959, 1960, 1961), Troll & Paffen (1964), Walter (1968, 1973, 1977).

World (comparative)
Beard (1967), Lauer (1952), Meggers et al. (1973), Troll & Lauer (1978), Vareschi (1968).

Tropics
Aubert de la Rüe et al. (1957), Beard (1944a, 1955), Golley & Medina (1975), Misra & Gopal (1968), Odum & Pigeon (1970), Schnell (1970–77), Troll (1959).

Mountains
Allg. Forst. (1966), Beals (1969), Cabrera (1958, 1968), Coe (1967), Cuatrecasas (1968), Daubenmire (1943), Donita (1965), Ellenberg (1975), Ern (1966), Gersmehl (1973), Gorchakovsky (1966), Grebenshchikov (1963), Hedberg (1951), Hodge (1946), Ives & Barry (1974), Michaelis (1932–34), Schweinfurth (1957), Sturm (1978), Tranquillini (1964, 1979), Troll (1959, 1966, 1968, 1972a, 1972b, 1973), Walter (1960, 1975), Wardle (1965), Webber (1978), Whittaker (1956).

Arctic/Subarctic
Aleksandrova (1960), Billings & Mooney (1968), Bliss (1956, 1971, in press), Bliss et al. (1973), Böcher (1938), Bryson (1966), Hulten (1962), Hustich (1953), Ives & Barry (1974), Rosswall & Heal (1975), Schwartzenbach (1960), Tieszen (1978), Wielgolaski et al. (1975).

North America
Billings (1949), Boyce (1954), Braun (1950), Brockman (1968), Bruner (1931), Damman (1977), Daubenmire (1943, 1978), Franklin & Dyrness (1973), Hare (1950, 1954), Hettinger & Janz (1974), Kirkpatrick & Hutchinson (1977), Knapp (1965), Küchler (1964, 1977), Livingston & Shreve (1921), Merriam (1898), Rowe (1959), Shelford (1963), Shreve (1942, 1951), Weaver & Albertson (1956), Wells (1928, 1942), Westman (1978), Wharton (1978), Whittaker (1956).

Central America and Caribbean
Asprey & Robbins (1953), Beard (1944a, 1944b, 1953, 1955), Coker (1905), Ern (1974), Graham (1973), Hargreaves & Hargreaves (1965), Lauer (1968), Lauer et al. (in press), Leopold (1950), Parsons (1955), Sarmiento (1976), Sawyer and Lindsey (1971), Shreve (1914, 1934), Weber (1958).

South America
Beard (1944a, 1953, 1955), Boelcke (1957), Cabrera (1958), Cole (1960), Cuatrecasas (1968), Eiten (1972, 1978), Ellenberg (1975), Ewel & Madriz (1968), Fittkau et al. (1969), Goodland (1971), Graham (1973), Hueck (1966), Hueck & Seibert (1972), Kummerow et al. (1961), Lescure (1978), Lindman & Ferri (1974), Lewis & Collattes (1975), Müller (1977), Rauh (1958), Rawitscher (1948), Sarmiento (1976), Schmithüsen (1956, 1960), Schnetter (1968), Solbrig (1976), Smith (1971), Soriano (1956), Sturm (1978), Veblen & Ashton (1978), Walter (1967), Walter & Medina (1969), Weinberger et al. (1973).

Atlantic Islands
Hansen (1930), Jonsson (1905), Kunkel (1976), Preusser (1975), Schmid (1954), Sunding (1972), Wace (1961), Wace & Holdgate (1976).

Europe
Acta Phytogeogr. Suecica (1965), Adamović (1933), Ahti et al. (1968), Brockmann-Jerosch (1925), Burnett (1964), Donita (1965), Eijsink et al. (1978), Ellenberg (1963), Ern (1966), Eyre (1968), Feoli-Chiapella & Feoli (1977), Gils et al. (1975), Hartmann & Schnelle (1970), Heal & Perkins (1978), Horvát (1978), Horvát et al. (1974), Lötschert (1962–65), Lüdi (1935), Malmer (1978), Markgraf (1949), Meyer (1978), Polunin (1972), Sissingh (1977), Szafer (1966), Tansley (1949), UNESCO (1969), Walter (1956, 1975), Zohary & Orshan (1966).

Africa
Ayyad and El-Ghonemy (1976), Bouxin (1975), Burtt (1942), Campbell & Moll (1977), Cannon (1934), Coe (1967), Emberger (1939), Fanshawe (1969), Frankenberg (1978), Hamilton (1975), Hedberg (1951), Holland & Hove (1975), Keay (1949, 1959), Knapp (1973), Koechelin et al. (1974), Langdale-Brown et al. (1964), Leistner (1967), Leistner & Werger (1973), Lind & Morrison (1974), Livingstone (1967), McKenzie et al. (1977), Missouri Botanical Garden (1968), Mitchell (1971), Phillips (1928), Quézel (1964), Rattray (1960), Rauh (1973), Rawé (1968), Richards (1963), Richard-Vindard & Battistini (1972), Salt (1954), Schmidt (1973, 1975), Schnell (1976), Scholtz (1966), Shantz & Marbut (1923), UNESCO (1969), Vesey-Fitzgerald (1970), Walter (1976), Werger (1977), Werger et al. (1978), Zinderen-Bakker (1970, 1973).

Indian Ocean Islands
Cadet (1974), Chastain (1958), Legris (1969), Popov (1957).

Soviet Union
Gorchakovsky (1966), Grebenshchikov (1963), Gulizashvili et al. (1975), Karpenko (1964), Keller (1927), Korovin (1961–62), Nechayeva et al. (1973), Nikitin (1966), Petrov (1966–67), Restshikov (1961), Walter (1974, 1977), Walter & Box (in press), Worobjev (1963).

Middle East
Abul-Fatih (1975), Beals (1965), Bobek (1951), Bokhari & Khan (1976), Breckle (1975), Czeczött (1938–39), Danin (1978), Davis et al. (1971), UNESCO (1969), Walter (1956, 1974), Zohary (1973).

South Asia
Bhatia (1958), Champion & Seth (1968), Champion et al. (1965), Fernando (1968), Freitag (1971), Gupta (1978), Inst. Franç. Pondichery (1961–70), Kitamura (1964), Mani (1974, 1979), Meusel & Schubert (1971), Schweinfurth (1957), Shrimal & Vyas (1977), Sreedhara Murthy (1978), Troll (1972a), Whyte et al. (1954).

Japan
Hamet-Ahti et al. (1974), Hara (1959), Lieth et al. (1973), Numata (1974), Numata et al. (1972).

East Asia
Küchler (1948), Walter (1974), Walter & Box (in press), Wang (1961), Yunatov (1950).

Southeast Asia
Ogawa et al. (1961), Stamp (1924), Vidal (1961), Whitmore (1975), Williams (1965).

Australia
Beadle (1966), Beard (1967, 1969, 1976), Blombery (1967), Costin (1957), Davis (1964), Keast et al. (1959), Milton Moore (1970), Rule (1967), Specht (1972), Webb (1959, 1964), Williams (1974).

New Zealand
Cockayne (1958), Connor (1961), Dawson (1962), Godley (1975), Holloway (1954), Kuschel (1975), Robbins (1962), Zotov (1947).

Oceania
Balgooy (1975), Brünig (1970), Fosberg (1960), Robbins (1961), Van Steenis (1962), Whitmore (1975), Wiggins & Porter (1971).

Subantarctic and Antarctica
Brockmann-Jerosch (1928), Godley (1960), Holdgate (1970), Rudolph (1966), Wilhelmy (1963).

APPENDIX B

Predicted vegetation at selected representative and well-known sites

The macroclimatic data and predicted vegetation at 111 selected sites are presented below in computer-generated form. These sites were selected as a geographically representative supplement to the list of validation sites in Appendix C. A number of university cities were included in order to encourage comparison of climatically predictable and actually occurring plant forms and vegetation structures. The format of the computer listing is described at the beginning of Appendix C. For interpretation of the predicted results refer to sections 5.F and 6.A in the main text. Since names of countries are given in representations of local languages (see Appendix E), some may not be readily apparent:

AL 'ARABIYAH AS-SA'UDIYAH = Saudi Arabia
AL LUBNAN = Lebanon
AL MAGHREB = Morocco
BHARAT = India
CHOSON = Korea
GRUZHIJA = Georgian S.S.R. (Russian)
MAGYARORSZAG = Hungary
MISRA = Egypt
MYANMA = Burma
PRATHET THAI = Thailand
YAITOPYA = Ethiopia
ZHONGGUO = China (mainland)

These names are truncated after eight characters on the computer listing.

Appendix B. Predicted vegetation at selected representative and well-known sites.

```
                                                                  1
      LOCATION                                        LAT     LONG
 TMAX  TMIN   PRCP   PMAX   PMIN  PMTMAX  MI             ELEV

 7. SYDPROVEN                    GROENLAN             60.00   -45.00
       7.0  -5.0  977.  150.   30.  137.  2.32            15.
              * 1. SHORT SWARD-GRASSES                TMAX     0.29
                2. SHORT BUNCH-GRASSES                MI       0.30
                3. SUMMERGREEN TUNDRA DWARF-SHRUBS    TMAX     0.14
                4. TEMPERATE EVERGREEN DWARF-SHRUBS   TMAX     0.14
                5. TEMPERATE EVERGREEN FORBS          TMIN     0.0
                6. SEASONAL COLD-DESERT HERBS         MULT.    1.00
                7. MAT-FORMING THALLOPHYTES           MI       0.69

34. EDMONTON, ALBERTA            CANADA               53.55 -113.47
      17.0 -13.0  446.   77.   20.   77.  0.90           676.
              * 1. BOREAL/MONTANE SHORT-NEEDLED TREES MI       0.33
              + 2. BOREAL SUMMERGREEN NEEDLE-TREES    TMAX     0.50
                3. BOREAL BROAD-SUMMERGREEN TREES     MI       0.11
                4. BROAD-SUMMERGREEN SMALL TREES      MI       0.22
                5. TALL GRASSES                       TMIN     0.15
              + 6. SHORT SWARD-GRASSES                MI       0.45
                7. SUMMERGREEN GIANT-SCRUB            TMAX     0.26
                8. BROAD-SUMMERGREEN MESIC SHRUBS     TMAX     0.19
                9. XERIC SUMMERGREEN SHRUBS           MI       0.14
              +10. SHORT BUNCH-GRASSES                TMIN     0.54
               11. NEEDLE-LEAVED EVERGREEN SHRUBS     TMAX     0.37
               12. SUMMERGREEN TUNDRA DWARF-SHRUBS    TMAX     0.08
               13. SUMMERGREEN FORBS                  TMAX     0.37
               14. TEMPERATE EVERGREEN DWARF-SHRUBS   TMIN     0.18
               15. XERIC CUSHION-HERBS                MI       0.13
               16. MAT-FORMING THALLOPHYTES           MI       0.11
               17. XERIC THALLOPHYTES                 MULT.    1.00

54. MONTREAL DORVAL, QUEBEC   CANADA                  45.47   -73.73
      21.0  -9.0  973.   93.   70.   71.  1.69            30.
              * 1. SUMMERGREEN BROAD-LEAVED TREES     TMIN     0.30
              * 2. BOREAL/MONTANE SHORT-NEEDLED TREES TMAX     0.09
                3. BOREAL BROAD-SUMMERGREEN TREES     TMAX     0.33
                4. BROAD-SUMMERGREEN SMALL TREES      TMIN     0.45
                5. TALL GRASSES                       TMIN     0.23
                6. SHORT SWARD-GRASSES                TMIN     0.55
                7. BROAD-SUMMERGREEN MESIC SHRUBS     TMAX     0.44
                8. SUMMERGREEN GIANT-SCRUB            TMAX     0.39
                9. BROAD-ERICOID EVERGREEN SHRUBS     TMIN     0.05
               10. SHORT BUNCH-GRASSES                TMIN     0.57
               11. NEEDLE-LEAVED EVERGREEN SHRUBS     MI       0.19
               12. SUMMERGREEN FORBS                  TMAX     0.48
               13. SUMMERGREEN FERNS                  PMTMAX   0.29
               14. TEMPERATE EVERGREEN DWARF-SHRUBS   TMAX     0.17
               15. MAT-FORMING THALLOPHYTES           TMAX     0.39
               16. XERIC THALLOPHYTES                 MI       0.18

62. BARROW,ALASKA                USA                  71.30 -156.78
       5.0 -27.5  104.   20.    2.   20.  0.68             7.
              * 1. SEASONAL COLD-DESERT HERBS         TMAX     0.33
                2. XERIC CUSHION-HERBS                TMAX     0.0
                3. XERIC THALLOPHYTES                 TMAX     0.17
                4. ICE DESERT                         TMAX     0.00
```

```
                                                                  2
      LOCATION                                        LAT     LONG
 TMAX  TMIN   PRCP   PMAX   PMIN  PMTMAX  MI             ELEV

66. ANCHORAGE,ALASKA             USA                  61.22 -149.88
      14.0 -11.0  360.   68.   11.   30.  0.79           36.
              * 1. BOREAL/MONTANE SHORT-NEEDLED TREES MI       0.25
                2. BOREAL SUMMERGREEN NEEDLE-TREES    PMTMAX   0.16
                3. BROAD-SUMMERGREEN SMALL TREES      MI       0.12
                4. SHORT SWARD-GRASSES                PMTMAX   0.16
                5. SUMMERGREEN GIANT-SCRUB            TMAX     0.13
                6. BROAD-SUMMERGREEN MESIC SHRUBS     TMAX     0.0
                7. XERIC SUMMERGREEN SHRUBS           TMAX     0.0
              + 8. SHORT BUNCH-GRASSES                TMIN     0.55
                9. NEEDLE-LEAVED EVERGREEN SHRUBS     TMAX     0.25
               10. SUMMERGREEN TUNDRA DWARF-SHRUBS    MI       0.12
               11. SUMMERGREEN FORBS                  TMAX     0.25
               12. TEMPERATE EVERGREEN DWARF-SHRUBS   MI       0.12
               13. XERIC CUSHION-HERBS                MI       0.27
               14. SEASONAL COLD-DESERT HERBS         TMAX     0.07
               15. XERIC DWARF-SHRUBS                 MI       0.01
               16. XERIC THALLOPHYTES                 MULT.    1.00

77. PORTLAND, OREGON             USA                  45.55 -122.65
      19.0   4.0 1062.  180.   12.   12.  1.61            9.
              * 1. TEMPERATE NEEDLE-TREES             TMAX     0.44
                2. SUB-MEDITERRANEAN NEEDLE-TREES     TMIN     0.31
                3. TEMP. BROAD-EVERGREEN SMALL TREES  TMIN     0.15
                4. TALL CANE-GRAMINOIDS               TMIN     0.10
                5. DWARF-NEEDLE SMALL TREES           TMAX     0.08
                6. SHORT BUNCH-GRASSES                TMIN     0.71
                7. NEEDLE-LEAVED EVERGREEN SHRUBS     MI       0.24
                8. SUMMERGREEN FORBS                  TMAX     0.43
                9. MAT-FORMING THALLOPHYTES           MI       0.51
               10. BROAD-WINTERGREEN EPIPHYTES        TMIN     0.14
               11. XERIC THALLOPHYTES                 MI       0.24

98. SANTA BARBARA,CALIFORNIA  USA                     34.42 -119.70
      18.8  11.2  471.  120.    0.    5.  0.64           37.
              * 1. MEDITERRANEAN EVERGREEN SHRUBS     TMAX     0.20
              + 2. SHORT BUNCH-GRASSES                MI       0.70
                3. NEEDLE-LEAVED EVERGREEN SHRUBS     MI       0.38
                4. XERIC EVERGREEN TUFT-TREELETS      TMAX     0.05
                5. SUMMERGREEN FORBS                  TMAX     0.42
              + 6. XERIC CUSHION-SHRUBS               MI       0.41
                7. BUSH STEM-SUCCULENTS               TMIN     0.37
                8. MEDITERRANEAN DWARF-SHRUBS         TMAX     0.15
                9. XERIC CUSHION-HERBS                TMAX     0.54
               10. BROAD-WINTERGREEN EPIPHYTES        MI       0.07
               11. SUCCULENT FORBS                    TMAX     0.04
               12. XERIC THALLOPHYTES                 MULT.    1.00
```

```
                                                                3
     LOCATION                                          LAT     LONG
  TMAX  TMIN  PRCP  PMAX  PMIN  PMTMAX  MI               ELEV

110. SAN DIEGO, CALIFORNIA       USA                   32.72 -117.15
     20.0  13.0  259.   53.    3.    4.  0.34                6.
           * 1. SHORT BUNCH-GRASSES                    MI      0.43
           * 2. XERIC EVERGREEN TUFT-TREELETS          TMAX    0.12
             3. EVERGREEN GIANT-SCRUB                  TMAX    0.0
           + 4. XERIC CUSHION-SHRUBS                   TMIN    0.60
           + 5. DESERT-GRASSES                         MI      0.48
             6. BUSH STEM-SUCCULENTS                   TMIN    0.43
             7. SUMMERGREEN FORBS                      MI      0.43
             8. LEAFLESS XEROMORPHIC LARGE-SCRUB       TMAX    0.23
             9. MEDITERRANEAN DWARF-SHRUBS             TMAX    0.21
            10. LEAF-SUCCULENT EVERGREEN SHRUBS        TMIN    0.08
            11. TYPICAL STEM-SUCCULENTS                TMAX    0.0
            12. XERIC ROSETTE-SHRUBS                   TMAX    0.0
            13. EPHEMERAL DESERT HERBS                 MI      0.48
            14. XERIC CUSHION-HERBS                    MI      0.43
            15. XERIC DWARF-SHRUBS                     TMAX    0.36
            16. SUCCULENT FORBS                        TMAX    0.09
            17. XERIC THALLOPHYTES                     TMAX    0.91

130. YELLOWSTONE PARK,WYOMING  USA                     44.96 -110.72
     15.5  -7.0  444.   51.   31.   34.  0.96              1899.
           * 1. BOREAL/MONTANE SHORT-NEEDLED TREES     MI      0.38
             2. BOREAL SUMMERGREEN NEEDLE-TREES        PMTMAX  0.27
             3. BOREAL BROAD-SUMMERGREEN TREES         PMTMAX  0.13
             4. BROAD-SUMMERGREEN SMALL TREES          TMAX    0.26
             5. SHORT SWARD-GRASSES                    PMTMAX  0.27
             6. SUMMERGREEN GIANT-SCRUB                TMAX    0.20
             7. BROAD-SUMMERGREEN MESIC SHRUBS         TMAX    0.09
             8. SHORT BUNCH-GRASSES                    TMIN    0.59
             9. NEEDLE-LEAVED EVERGREEN SHRUBS         TMAX    0.31
            10. SUMMERGREEN TUNDRA DWARF-SHRUBS        TMAX    0.21
            11. SUMMERGREEN FORBS                      TMAX    0.31
            12. TEMPERATE EVERGREEN DWARF-SHRUBS       PMTMAX  0.27
            13. MAT-FORMING THALLOPHYTES               MI      0.17
            14. XERIC CUSHION-HERBS                    MI      0.05
            15. XERIC THALLOPHYTES                     MULT.   1.00

140. DENVER,COLORADO             USA                   39.72 -105.02
     19.0  -1.8  357.   56.   10.   43.  0.64              1592.
             1. BOREAL/MONTANE SHORT-NEEDLED TREES     MI      0.07
             2. BOREAL SUMMERGREEN NEEDLE-TREES        TMAX    0.30
             3. SUMMERGREEN GIANT-SCRUB                TMAX    0.33
             4. XERIC SUMMERGREEN SHRUBS               TMAX    0.24
             5. SHORT SWARD-GRASSES                    MI      0.22
             6. DWARF-NEEDLE SMALL TREES               TMAX    0.08
             7. TEMPERATE BROAD-EVERGREEN SHRUBS       TMIN    0.01
           + 8. SHORT BUNCH-GRASSES                    TMIN    0.65
             9. NEEDLE-LEAVED EVERGREEN SHRUBS         MI      0.38
            10. COLD-WINTER XEROMORPHIC SHRUBS         TMAX    0.24
            11. SUMMERGREEN FORBS                      TMAX    0.43
            12. LEAFLESS XEROMORPHIC LARGE-SCRUB       TMAX    0.19
            13. XERIC CUSHION-HERBS                    TMAX    0.52
            14. XERIC DWARF-SHRUBS                     MI      0.25
            15. XERIC THALLOPHYTES                     MULT.   1.00
```

```
                                                                        4
     LOCATION                                                 LAT     LONG
  TMAX  TMIN  PRCP  PMAX  PMIN  PMTMAX  MI                      ELEV

172. HOUSTON, TEXAS                USA                        29.77  -95.37
     28.0  11.0 1171.  120.   77.  110.  1.12                    125.
           * 1. WARM-TEMPERATE BROAD-EVERGREEN TREES          TMAX    0.20
           * 2. SUMMERGREEN BROAD-LEAVED TREES                TMAX    0.13
           * 3. HELIOPHILIC LONG-NEEDLED TREES                MI      0.11
             4. TEMPERATE BROAD-RAINFOREST TREES              TMAX    0.0
             5. TROPICAL EVERGREEN MICROPHYLL-TREES           TMIN    0.17
             6. TROPICAL EVERGREEN SCLEROPHYLL TREES          TMIN    0.14
             7. SWAMP SUMMERGREEN NEEDLE-TREES                TMAX    0.0
             8. PALMIFORM TUFT-TREES                          TMIN    0.0
             9. TALL GRASSES                                  TMAX    0.28
            10. TALL CANE-GRAMINOIDS                          TMAX    0.28
            11. TEMP. BROAD-EVERGREEN SMALL TREES             TMAX    0.20
            12. TROPICAL BROAD-EVERGREEN SMALL TREES          TMIN    0.15
            13. BROAD-SUMMERGREEN SMALL TREES                 TMAX    0.11
            14. ARBORESCENT GRASSES                           TMAX    0.10
            15. SUMMERGREEN GIANT-SCRUB                       TMAX    0.30
            16. SHORT SWARD-GRASSES                           TMAX    0.23
            17. TROPICAL BROAD-EVERGREEN SHRUBS               TMAX    0.18
            18. DWARF-NEEDLE SMALL TREES                      TMAX    0.17
            19. MEDITERRANEAN EVERGREEN SHRUBS                PMTMAX  0.14
            20. TEMPERATE BROAD-EVERGREEN SHRUBS              TMAX    0.13
            21. BROAD-SUMMERGREEN MESIC SHRUBS                TMAX    0.13
            22. PALMIFORM TUFT-TREELETS                       TMIN    0.05
            23. SHORT BUNCH-GRASSES                           TMAX    0.40
            24. NEEDLE-LEAVED EVERGREEN SHRUBS                TMAX    0.28
            25. EVERGREEN GIANT-SCRUB                         TMIN    0.15
            26. PALMIFORM MESIC ROSETTE-SHRUBS                TMIN    0.14
            27. BUSH STEM-SUCCULENTS                          MI      0.34
            28. SUMMERGREEN FORBS                             TMAX    0.28
            29. TEMPERATE EVERGREEN FORBS                     TMAX    0.08
            30. SUMMERGREEN FERNS                             MI      0.02
            31. BROAD-EVERGREEN VINES                         TMAX    0.18
            32. BROAD-WINTERGREEN EPIPHYTES                   TMAX    0.14
            33. NARROW-LEAVED EPIPHYTES                       MI      0.11
            34. BROAD-SUMMERGREEN VINES                       MI      0.11
            35. MAT-FORMING THALLOPHYTES                      TMAX    0.07
            36. XERIC THALLOPHYTES                            TMAX    0.40
```

Appendix B. Predicted vegetation at selected representative and well-known sites.

5

LOCATION TMAX TMIN PRCP PMAX PMIN PMTMAX MI	LAT ELEV	LONG
178. CHICAGO, ILLINOIS USA	41.88	-87.63
23.0 -4.0 832. 100. 36. 71. 1.26	186.	
* 1. SUMMERGREEN BROAD-LEAVED TREES	MI	0.29
* 2. TEMPERATE NEEDLE-TREES	TMIN	0.10
3. BOREAL BROAD-SUMMERGREEN TREES	TMAX	0.17
4. BROAD-SUMMERGREEN SMALL TREES	TMAX	0.37
5. TALL GRASSES	TMIN	0.33
6. SHORT SWARD-GRASSES	TMAX	0.48
7. SUMMERGREEN GIANT-SCRUB	TMAX	0.48
8. BROAD-SUMMERGREEN MESIC SHRUBS	TMAX	0.44
9. DWARF-NEEDLE SMALL TREES	TMIN	0.04
10. SHORT BUNCH-GRASSES	TMIN	0.62
11. NEEDLE-LEAVED EVERGREEN SHRUBS	TMAX	0.48
12. SUMMERGREEN FORBS	TMAX	0.48
13. SUMMERGREEN FERNS	MI	0.13
14. TEMPERATE EVERGREEN DWARF-SHRUBS	TMAX	0.06
15. TEMPERATE EVERGREEN FORBS	TMIN	0.04
16. MAT-FORMING THALLOPHYTES	TMAX	0.28
17. BROAD-SUMMERGREEN VINES	TMIN	0.06
18. XERIC THALLOPHYTES	MI	0.59
193. MIAMI, FLORIDA USA	25.77	-80.20
27.5 20.0 1199. 200. 44. 145. 0.94	2.	
1. TROPICAL RAINFOREST TREES	MI	0.04
2. WARM-TEMPERATE BROAD-EVERGREEN TREES	MI	0.20
3. TROPICAL XERIC NEEDLE-TREES	TMAX	0.17
+ 4. TROPICAL EVERGREEN SCLEROPHYLL TREES	TMAX	0.38
5. PALMIFORM TUFT-TREES	MI	0.25
6. TROPICAL EVERGREEN MICROPHYLL-TREES	MI	0.20
7. TALL CANE-GRAMINOIDS	MI	0.25
8. TALL GRASSES	MI	0.25
9. ARBORESCENT GRASSES	TMAX	0.13
10. TROPICAL BROAD-EVERGREEN SMALL TREES	MI	0.04
11. PALMIFORM TUFT-TREELETS	MI	0.25
12. SHORT SWARD-GRASSES	TMAX	0.25
13. TROPICAL BROAD-EVERGREEN SHRUBS	TMAX	0.20
14. TEMPERATE BROAD-EVERGREEN SHRUBS	TMAX	0.17
+15. SHORT BUNCH-GRASSES	TMAX	0.42
+16. EVERGREEN GIANT-SCRUB	TMAX	0.38
17. XERIC EVERGREEN TUFT-TREELETS	MI	0.09
18. PALMIFORM MESIC ROSETTE-SHRUBS	MI	0.04
19. BUSH STEM-SUCCULENTS	TMAX	0.42
20. RAINGREEN FORBS	TMAX	0.30
21. XERIC ROSETTE-SHRUBS	MI	0.29
22. TROPICAL EVERGREEN FORBS	MI	0.04
23. SUCCULENT FORBS	MI	0.29
24. BROAD-WINTERGREEN EPIPHYTES	TMAX	0.18
25. MAT-FORMING THALLOPHYTES	TMAX	0.08
26. BROAD-EVERGREEN VINES	MI	0.04
27. XERIC THALLOPHYTES	TMAX	0.42

6

LOCATION TMAX TMIN PRCP PMAX PMIN PMTMAX MI	LAT ELEV	LONG
202. HARRISBURG, PENNSYLVANIA USA	40.27	-76.87
24.0 0.0 914. 92. 57. 87. 1.32	102.	
* 1. SUMMERGREEN BROAD-LEAVED TREES	MI	0.32
2. TEMPERATE NEEDLE-TREES	TMAX	0.0
3. SUB-MEDITERRANEAN NEEDLE-TREES	TMIN	0.06
4. BOREAL BROAD-SUMMERGREEN TREES	TMAX	0.08
5. TALL GRASSES	TMIN	0.42
6. BROAD-SUMMERGREEN SMALL TREES	TMAX	0.32
7. SUMMERGREEN GIANT-SCRUB	TMAX	0.48
8. SHORT SWARD-GRASSES	TMAX	0.42
9. BROAD-SUMMERGREEN MESIC SHRUBS	TMAX	0.38
10. DWARF-NEEDLE SMALL TREES	TMIN	0.22
11. TEMPERATE BROAD-EVERGREEN SHRUBS	TMIN	0.09
12. SHORT BUNCH-GRASSES	TMAX	0.62
13. NEEDLE-LEAVED EVERGREEN SHRUBS	TMAX	0.44
14. SUMMERGREEN FORBS	TMAX	0.44
15. TEMPERATE EVERGREEN FORBS	TMIN	0.22
16. SUMMERGREEN FERNS	MI	0.17
17. TEMPERATE EVERGREEN DWARF-SHRUBS	TMAX	0.0
18. BUSH STEM-SUCCULENTS	TMIN	0.0
19. BROAD-SUMMERGREEN VINES	MI	0.24
20. MAT-FORMING THALLOPHYTES	TMAX	0.23
21. BROAD-WINTERGREEN EPIPHYTES	TMIN	0.0
22. XERIC THALLOPHYTES	MI	0.52
206. MT.WASHINGTON,NEW HAMPSH USA	44.30	-71.40
9.5 -14.3 1882. 178. 132. 170. 5.43	1917.	
* 1. SHORT SWARD-GRASSES	TMAX	0.47
2. SUMMERGREEN TUNDRA DWARF-SHRUBS	TMAX	0.37
3. TEMPERATE EVERGREEN DWARF-SHRUBS	TMIN	0.15
4. MAT-FORMING THALLOPHYTES	MI	0.87
5. SEASONAL COLD-DESERT HERBS	TMAX	0.58
207. HOULTON, MAINE USA	46.13	-67.85
20.0 -10.0 947. 95. 63. 85. 1.73	140.	
* 1. SUMMERGREEN BROAD-LEAVED TREES	TMIN	0.27
* 2. BOREAL/MONTANE SHORT-NEEDLED TREES	TMAX	0.18
3. BOREAL BROAD-SUMMERGREEN TREES	TMAX	0.42
4. BROAD-SUMMERGREEN SMALL TREES	TMIN	0.43
5. TALL GRASSES	TMIN	0.21
6. SHORT SWARD-GRASSES	TMIN	0.53
7. BROAD-SUMMERGREEN MESIC SHRUBS	TMAX	0.38
8. SUMMERGREEN GIANT-SCRUB	TMAX	0.36
9. BROAD-ERICOID EVERGREEN SHRUBS	TMIN	0.0
10. SHORT BUNCH-GRASSES	TMIN	0.56
11. NEEDLE-LEAVED EVERGREEN SHRUBS	MI	0.17
12. SUMMERGREEN FORBS	TMAX	0.45
13. SUMMERGREEN FERNS	TMIN	0.31
14. TEMPERATE EVERGREEN DWARF-SHRUBS	TMAX	0.22
15. MAT-FORMING THALLOPHYTES	TMAX	0.45
16. XERIC THALLOPHYTES	MI	0.16

7

LOCATION							LAT	LONG
TMAX	TMIN	PRCP	PMAX	PMIN	PMTMAX	MI	ELEV	

224. MEXICO — MEXICO — LAT 19.40, LONG -99.15

TMAX	TMIN	PRCP	PMAX	PMIN	PMTMAX	MI	ELEV
17.5	12.0	588.	155.	6.	53.	0.82	2280.

	No.	Vegetation	Factor	Value
*	1.	TEMPERATE NEEDLE-TREES	MI	0.21
*	2.	TROPICAL XERIC NEEDLE-TREES	TMAX	0.17
	3.	MONTANE BROAD-RAINGREEN TREES	MI	0.02
	4.	TROPICAL EVERGREEN SCLEROPHYLL TREES	TMAX	0.13
	5.	TROPICAL EVERGREEN MICROPHYLL-TREES	MI	0.08
	6.	PALMIFORM TUFT-TREES	TMIN	0.05
	7.	TALL CANE-GRAMINOIDS	MI	0.15
	8.	TALL GRASSES	MI	0.15
	9.	BROAD-RAINGREEN SMALL TREES	TMAX	0.13
	10.	TEMP. BROAD-EVERGREEN SMALL TREES	MI	0.08
	11.	ARBORESCENT GRASSES	MI	0.02
+	12.	SHORT SWARD-GRASSES	MI	0.39
	13.	TEMPERATE BROAD-EVERGREEN SHRUBS	TMAX	0.17
	14.	TROPICAL BROAD-EVERGREEN SHRUBS	MI	0.15
	15.	RAINGREEN THORN-SCRUB	TMAX	0.13
	16.	MEDITERRANEAN EVERGREEN SHRUBS	TMAX	0.11
	17.	PALMIFORM TUFT-TREELETS	TMIN	0.10
+	18.	SHORT BUNCH-GRASSES	MI	0.76
	19.	NEEDLE-LEAVED EVERGREEN SHRUBS	TMAX	0.38
	20.	BUSH STEM-SUCCULENTS	TMAX	0.38
	21.	RAINGREEN FORBS	TMIN	0.22
	22.	TEMPERATE EVERGREEN DWARF-SHRUBS	MI	0.15
	23.	XERIC CUSHION-SHRUBS	MI	0.10
	24.	XERIC CUSHION-HERBS	MI	0.23
	25.	BROAD-WINTERGREEN EPIPHYTES	TMAX	0.11
	26.	BROAD-RAINGREEN VINES	TMIN	0.10
	27.	MAT-FORMING THALLOPHYTES	MI	0.02
	28.	XERIC THALLOPHYTES	MULT.	1.00

244. LA CRUZ — COSTA RI — LAT 11.07, LONG -85.65

TMAX	TMIN	PRCP	PMAX	PMIN	PMTMAX	MI	ELEV
28.0	25.0	1450.	310.	1.	88.	0.87	246.

	No.	Vegetation	Factor	Value
*	1.	TROPICAL EVERGREEN SCLEROPHYLL TREES	TMAX	0.35
*	2.	PALMIFORM TUFT-TREES	MI	0.20
*	3.	TROPICAL EVERGREEN MICROPHYLL-TREES	MI	0.14
	4.	XERIC RAINGREEN TREES	MI	0.20
+	5.	BROAD-RAINGREEN SMALL TREES	TMAX	0.35
	6.	TALL CANE-GRAMINOIDS	MI	0.20
	7.	TROPICAL BROAD-EVERGREEN DWARF-TREES	MI	0.08
	8.	PALMIFORM TUFT-TREELETS	MI	0.20
	9.	TROPICAL BROAD-EVERGREEN SHRUBS	TMAX	0.18
	10.	RAINGREEN THORN-SCRUB	MI	0.15
+	11.	SHORT BUNCH-GRASSES	TMAX	0.40
+	12.	EVERGREEN GIANT-SCRUB	TMAX	0.33
	13.	RAINGREEN FORBS	TMAX	0.28
	14.	BROAD-RAINGREEN VINES	TMAX	0.35
	15.	BROAD-WINTERGREEN EPIPHYTES	TMAX	0.14
	16.	MAT-FORMING THALLOPHYTES	TMAX	0.07
	17.	XERIC THALLOPHYTES	TMAX	0.40

8

LOCATION							LAT	LONG
TMAX	TMIN	PRCP	PMAX	PMIN	PMTMAX	MI	ELEV	

249. NASSAU, NEW PROVIDENCE — BAHAMAS — LAT 25.08, LONG -77.35

TMAX	TMIN	PRCP	PMAX	PMIN	PMTMAX	MI	ELEV
28.0	22.0	1181.	180.	33.	130.	0.85	3.

	No.	Vegetation	Factor	Value
*	1.	TROPICAL XERIC NEEDLE-TREES	TMAX	0.13
+	2.	TROPICAL EVERGREEN SCLEROPHYLL TREES	TMAX	0.35
	3.	PALMIFORM TUFT-TREES	MI	0.18
	4.	TROPICAL EVERGREEN MICROPHYLL-TREES	MI	0.12
	5.	TALL CANE-GRAMINOIDS	MI	0.18
	6.	TALL GRASSES	MI	0.18
	7.	TROPICAL BROAD-EVERGREEN DWARF-TREES	MI	0.06
	8.	ARBORESCENT GRASSES	MI	0.06
	9.	SHORT SWARD-GRASSES	TMAX	0.23
	10.	TROPICAL BROAD-EVERGREEN SHRUBS	MI	0.18
	11.	PALMIFORM TUFT-TREELETS	MI	0.18
+	12.	SHORT BUNCH-GRASSES	TMAX	0.40
+	13.	EVERGREEN GIANT-SCRUB	TMAX	0.33
	14.	XERIC EVERGREEN TUFT-TREELETS	MI	0.20
	15.	BUSH STEM-SUCCULENTS	TMAX	0.40
	16.	XERIC ROSETTE-SHRUBS	TMAX	0.40
	17.	RAINGREEN FORBS	TMAX	0.28
	18.	SUCCULENT FORBS	MI	0.41
	19.	BROAD-WINTERGREEN EPIPHYTES	TMAX	0.14
	20.	MAT-FORMING THALLOPHYTES	MI	0.06
	21.	XERIC THALLOPHYTES	TMAX	0.40

252. KINGSTON — JAMAICA — LAT 18.00, LONG -76.83

TMAX	TMIN	PRCP	PMAX	PMIN	PMTMAX	MI	ELEV
27.5	25.0	800.	190.	15.	40.	0.51	34.

	No.	Vegetation	Factor	Value
	1.	TROPICAL EVERGREEN SCLEROPHYLL TREES	MI	0.02
	2.	XERIC RAINGREEN TREES	TMAX	0.22
	3.	BROAD-RAINGREEN SMALL TREES	TMAX	0.38
+	4.	RAINGREEN THORN-SCRUB	TMAX	0.50
	5.	SHORT SWARD-GRASSES	MI	0.02
+	6.	XERIC EVERGREEN TUFT-TREELETS	TMAX	0.44
+	7.	SHORT BUNCH-GRASSES	TMAX	0.42
	8.	EVERGREEN GIANT-SCRUB	TMAX	0.38
	9.	ARBORESCENT STEM-SUCCULENTS	TMAX	0.31
	10.	BUSH STEM-SUCCULENTS	TMAX	0.42
	11.	XERIC ROSETTE-SHRUBS	TMAX	0.38
	12.	RAINGREEN FORBS	TMAX	0.30
	13.	LEAF-SUCCULENT EVERGREEN SHRUBS	TMAX	0.26
	14.	XERIC CUSHION-SHRUBS	TMAX	0.25
	15.	TYPICAL STEM-SUCCULENTS	MI	0.18
	16.	SUCCULENT FORBS	TMAX	0.43
	17.	BROAD-RAINGREEN VINES	TMAX	0.38
	18.	XERIC CUSHION-HERBS	TMAX	0.10
	19.	XERIC THALLOPHYTES	TMAX	0.42

Appendix B. Predicted vegetation at selected representative and well-known sites.

```
                                                                   9
      LOCATION                                               LAT     LONG
  TMAX  TMIN  PRCP  PMAX  PMIN  PMTMAX  MI                      ELEV

271. CARACAS                     VENEZUEL                    10.50  -66.93
    20.8  18.0  809.  100.   10.   78.  0.93                   1042.
          * 1. TROPICAL XERIC NEEDLE-TREES                   TMAX      0.39
          * 2. MONTANE BROAD-RAINGREEN TREES                 MI        0.14
            3. TROPICAL MONTANE RAINFOREST TREES             MI        0.03
            4. TROPICAL EVERGREEN SCLEROPHYLL TREES          TMAX      0.29
            5. PALMIFORM TUFT-TREES                          TMAX      0.21
            6. TROPICAL EVERGREEN MICROPHYLL-TREES           MI        0.19
            7. BROAD-RAINGREEN SMALL TREES                   TMAX      0.29
            8. TALL CANE-GRAMINOIDS                          MI        0.25
            9. TALL GRASSES                                  MI        0.25
           10. ARBORESCENT GRASSES                           MI        0.14
           11. TROPICAL BROAD-EVERGREEN SMALL TREES          MI        0.03
          +12. SHORT SWARD-GRASSES                           MI        0.47
           13. TROPICAL BROAD-EVERGREEN SHRUBS               MI        0.25
           14. PALMIFORM TUFT-TREELETS                       MI        0.25
           15. RAINGREEN THORN-SCRUB                         MI        0.08
          +16. SHORT BUNCH-GRASSES                           MI        0.79
           17. XERIC EVERGREEN TUFT-TREELETS                 MI        0.09
           18. EVERGREEN GIANT-SCRUB                         TMAX      0.07
           19. PALMIFORM MESIC ROSETTE-SHRUBS                MI        0.03
           20. RAINGREEN FORBS                               TMAX      0.47
           21. BUSH STEM-SUCCULENTS                          TMAX      0.47
           22. TEMPERATE EVERGREEN DWARF-SHRUBS              TMAX      0.18
           23. XERIC ROSETTE-SHRUBS                          TMAX      0.04
           24. TROPICAL EVERGREEN FORBS                      MI        0.03
           25. BROAD-WINTERGREEN EPIPHYTES                   TMAX      0.34
           26. BROAD-RAINGREEN VINES                         TMAX      0.29
           27. MAT-FORMING THALLOPHYTES                      MI        0.14
           28. SUCCULENT FORBS                               TMAX      0.13
           29. XERIC CUSHION-HERBS                           MI        0.09
           30. BROAD-EVERGREEN VINES                         MI        0.03
           31. XERIC THALLOPHYTES                            TMAX      0.84

274. PARAMO DE MUCUCHIES         VENEZUEL                     8.70  -70.85
     4.0   2.0  682.  150.    5.   44.  1.32                   4221.
          * 1. SHORT BUNCH-GRASSES                           TMAX      0.33
          * 2. TROPICAL ALPINE TUFT-TREELETS                 TMIN      0.20
            3. MESIC EVERGREEN CUSHION-SHRUBS                MI        0.02
            4. TROPICAL EVERGREEN FORBS                      TMIN      0.0
            5. MAT-FORMING THALLOPHYTES                      MI        0.40
            6. RAINGREEN COLD-DESERT HERBS                   TMIN      0.38
            7. XERIC THALLOPHYTES                            MI        0.52
```

```
                                                                  10
      LOCATION                                               LAT     LONG
  TMAX  TMIN  PRCP  PMAX  PMIN  PMTMAX  MI                      ELEV

283. BOGOTA                       COLOMBIA                    4.60  -74.08
    15.0  14.0 1061.  165.   50.  150.  1.56                   2660.
            1. TROPICAL LINEAR-LEAVED TREES                  TMAX      0.0
            2. TROPICAL MONTANE RAINFOREST TREES             TMAX      0.0
            3. TROPICAL EVERGREEN MICROPHYLL-TREES           TMAX      0.0
            4. TROPICAL EVERGREEN SCLEROPHYLL TREES          TMAX      0.0
            5. TROPICAL CLOUD-FOREST DWARF-TREES             MI        0.36
            6. ARBORESCENT GRASSES                           TMAX      0.29
            7. TALL CANE-GRAMINOIDS                          TMAX      0.29
            8. TALL GRASSES                                  TMAX      0.29
            9. TROPICAL BROAD-EVERGREEN SMALL TREES          TMAX      0.0
           10. SHORT SWARD-GRASSES                           TMAX      0.59
           11. TROPICAL BROAD-EVERGREEN SHRUBS               TMAX      0.29
           12. TREE FERNS                                    MI        0.23
           13. PALMIFORM TUFT-TREELETS                       TMAX      0.06
           14. SHORT BUNCH-GRASSES                           TMAX      0.76
           15. NEEDLE-LEAVED EVERGREEN SHRUBS                MI        0.29
           16. PALMIFORM MESIC ROSETTE-SHRUBS                TMAX      0.0
           17. TEMPERATE EVERGREEN DWARF-SHRUBS              TMAX      0.53
           18. TROPICAL EVERGREEN FORBS                      MI        0.42
           19. RAINGREEN FORBS                               TMAX      0.29
           20. EVERGREEN FERNS                               TMAX      0.29
           21. MAT-FORMING THALLOPHYTES                      MI        0.49
           22. NARROW-LEAVED EPIPHYTES                       TMIN      0.36
           23. BROAD-EVERGREEN VINES                         TMAX      0.29
           24. TROPICAL BROAD-EVERGREEN EPIPHYTES            TMAX      0.18
           25. XERIC THALLOPHYTES                            MI        0.29

295. LIMA                         PERU                      -12.05  -77.05
    24.0  16.0   48.   10.    0.    0.  0.06                    158.
          + 1. DESERT-GRASSES                                TMAX      0.62
            2. EPHEMERAL DESERT HERBS                        TMAX      0.62
            3. SUCCULENT FORBS                               MI        0.11
            4. DRY DESERT                                    TMAX      0.92
            5. XERIC THALLOPHYTES                            TMAX      0.62

297. CERRO DE PASCO               PERU                      -10.68  -76.27
     7.0   5.0  885.  115.   22.  114.  1.69                   4350.
          * 1. SHORT SWARD-GRASSES                           TMAX      0.22
            2. SHORT BUNCH-GRASSES                           TMAX      0.56
            3. TROPICAL ALPINE TUFT-TREELETS                 MI        0.31
            4. MESIC EVERGREEN CUSHION-SHRUBS                MI        0.23
            5. TEMPERATE EVERGREEN DWARF-SHRUBS              TMAX      0.11
            6. TROPICAL EVERGREEN FORBS                      TMIN      0.11
            7. RAINGREEN COLD-DESERT HERBS                   TMIN      0.62
            8. MAT-FORMING THALLOPHYTES                      MI        0.53
            9. XERIC THALLOPHYTES                            MI        0.18
```

LOCATION	TMAX TMIN PRCP PMAX PMIN PMTMAX MI	LAT / ELEV	LONG
313. MANAOS BRASIL		-3.13	-60.02
	28.0 26.0 1771. 245. 34. 55. 1.04	45.	
* 1.	TROPICAL RAINFOREST TREES	MI	0.14
2.	TROPICAL XERIC NEEDLE-TREES	TMAX	0.13
3.	TROPICAL EVERGREEN SCLEROPHYLL TREES	TMAX	0.35
4.	PALMIFORM TUFT-TREES	MI	0.33
5.	TROPICAL EVERGREEN MICROPHYLL-TREES	TMAX	0.24
6.	TALL CANE-GRAMINOIDS	TMAX	0.28
7.	TALL GRASSES	TMAX	0.28
8.	TROPICAL BROAD-EVERGREEN SMALL TREES	MI	0.14
9.	TROPICAL BROAD-EVERGREEN DWARF-TREES	PMTMAX	0.10
10.	TROPICAL BROAD-EVERGREEN LIANAS	MI	0.04
11.	PALMIFORM TUFT-TREELETS	MI	0.33
12.	TROPICAL BROAD-EVERGREEN SHRUBS	TMAX	0.18
13.	SHORT BUNCH-GRASSES	TMAX	0.40
14.	EVERGREEN GIANT-SCRUB	TMAX	0.33
15.	PALMIFORM MESIC ROSETTE-SHRUBS	MI	0.14
16.	BUSH STEM-SUCCULENTS	TMAX	0.40
17.	RAINGREEN FORBS	TMAX	0.28
18.	TROPICAL EVERGREEN FORBS	MI	0.14
19.	BROAD-WINTERGREEN EPIPHYTES	TMAX	0.14
20.	BROAD-EVERGREEN VINES	MI	0.14
21.	MAT-FORMING THALLOPHYTES	TMAX	0.07
22.	NARROW-LEAVED EPIPHYTES	MI	0.04
23.	XERIC THALLOPHYTES	TMAX	0.40
325. IGUATU BRASIL		-6.37	-39.30
	29.0 26.0 786. 195. 3. 30. 0.45	208.	
* 1.	XERIC RAINGREEN TREES	MI	0.22
2.	BROAD-RAINGREEN SMALL TREES	TMAX	0.30
+ 3.	RAINGREEN THORN-SCRUB	TMAX	0.44
+ 4.	ARBORESCENT STEM-SUCCULENTS	TMAX	0.39
+ 5.	SHORT BUNCH-GRASSES	TMAX	0.35
6.	XERIC EVERGREEN TUFT-TREELETS	TMAX	0.35
7.	EVERGREEN GIANT-SCRUB	TMAX	0.25
8.	XERIC ROSETTE-SHRUBS	TMAX	0.45
9.	BUSH STEM-SUCCULENTS	TMAX	0.37
10.	TYPICAL STEM-SUCCULENTS	MI	0.35
11.	RAINGREEN FORBS	TMAX	0.24
12.	DESERT-GRASSES	MI	0.12
13.	SUCCULENT FORBS	TMAX	0.50
14.	BROAD-RAINGREEN VINES	TMAX	0.30
15.	XERIC THALLOPHYTES	TMAX	0.35
365. VALPARAISO CHILE		-33.03	-71.63
	18.0 11.0 490. 150. 0. 1. 0.74	41.	
* 1.	MEDITERRANEAN EVERGREEN SHRUBS	TMAX	0.14
+ 2.	SHORT BUNCH-GRASSES	MI	0.74
3.	NEEDLE-LEAVED EVERGREEN SHRUBS	TMAX	0.40
4.	XERIC EVERGREEN TUFT-TREELETS	TMAX	0.0
5.	SUMMERGREEN FORBS	TMAX	0.40
6.	BUSH STEM-SUCCULENTS	TMIN	0.37
7.	XERIC CUSHION-SHRUBS	MI	0.21
8.	MEDITERRANEAN DWARF-SHRUBS	TMAX	0.11
9.	XERIC CUSHION-HERBS	MI	0.35
10.	SUCCULENT FORBS	TMAX	0.0
11.	XERIC THALLOPHYTES	MULT.	1.00

LOCATION	TMAX TMIN PRCP PMAX PMIN PMTMAX MI	LAT / ELEV	LONG
374. BAHIA DOUGLAS CHILE		-55.90	-67.70
	9.2 1.0 915. 115. 42. 65. 2.03	5.	
* 1.	SHORT SWARD-GRASSES	TMAX	0.46
* 2.	TALL TUSSOCK-GRASSES	TMIN	0.18
3.	SHORT BUNCH-GRASSES	MI	0.53
4.	SUMMERGREEN TUNDRA DWARF-SHRUBS	TMAX	0.35
5.	MESIC EVERGREEN CUSHION-SHRUBS	TMIN	0.0
6.	TEMPERATE EVERGREEN DWARF-SHRUBS	TMAX	0.35
7.	TEMPERATE EVERGREEN FORBS	TMIN	0.26
8.	MAT-FORMING THALLOPHYTES	MI	0.67
390. BUENOS AIRES ARGENTIN		-34.60	-58.45
	23.0 10.0 962. 125. 52. 73. 1.37	25.	
* 1.	SUMMERGREEN BROAD-LEAVED TREES	MI	0.35
* 2.	WARM-TEMPERATE BROAD-EVERGREEN TREES	TMIN	0.23
* 3.	TEMPERATE BROAD-RAINFOREST TREES	PMTMAX	0.18
* 4.	TEMPERATE NEEDLE-TREES	TMAX	0.11
5.	MEDITERRANEAN BROAD-EVERGREEN TREES	PMTMAX	0.02
6.	SUB-MEDITERRANEAN NEEDLE-TREES	TMAX	0.36
7.	TROPICAL EVERGREEN MICROPHYLL-TREES	TMIN	0.13
8.	TROPICAL EVERGREEN SCLEROPHYLL TREES	TMIN	0.09
9.	TALL GRASSES	TMAX	0.48
10.	TEMP. BROAD-EVERGREEN SMALL TREES	TMAX	0.45
11.	BROAD-SUMMERGREEN SMALL TREES	TMAX	0.37
12.	TALL CANE-GRAMINOIDS	TMIN	0.31
13.	ARBORESCENT GRASSES	TMIN	0.29
14.	TROPICAL BROAD-EVERGREEN SMALL TREES	TMIN	0.10
15.	TROPICAL BROAD-EVERGREEN LIANAS	TMIN	0.0
16.	SHORT SWARD-GRASSES	TMAX	0.48
17.	SUMMERGREEN GIANT-SCRUB	TMAX	0.48
18.	TEMPERATE BROAD-EVERGREEN SHRUBS	TMAX	0.47
19.	BROAD-SUMMERGREEN MESIC SHRUBS	TMAX	0.44
20.	DWARF-NEEDLE SMALL TREES	TMAX	0.42
21.	MEDITERRANEAN EVERGREEN SHRUBS	TMIN	0.38
22.	TROPICAL BROAD-EVERGREEN SHRUBS	TMIN	0.17
23.	PALMIFORM TUFT-TREELETS	TMIN	0.0
24.	TALL TUSSOCK-GRASSES	TMAX	0.0
25.	SHORT BUNCH-GRASSES	TMAX	0.68
26.	NEEDLE-LEAVED EVERGREEN SHRUBS	MI	0.46
27.	EVERGREEN GIANT-SCRUB	TMIN	0.10
28.	PALMIFORM MESIC ROSETTE-SHRUBS	TMIN	0.09
29.	SUMMERGREEN FORBS	TMAX	0.48
30.	TROPICAL EVERGREEN FORBS	TMIN	0.29
31.	TEMPERATE EVERGREEN FORBS	TMAX	0.28
32.	EVERGREEN FERNS	MI	0.20
33.	SUMMERGREEN FERNS	MI	0.20
34.	BUSH STEM-SUCCULENTS	MI	0.10
35.	TEMPERATE EVERGREEN DWARF-SHRUBS	TMAX	0.06
36.	MARITIME HEATH DWARF-SHRUBS	TMAX	0.0
37.	BROAD-WINTERGREEN EPIPHYTES	TMIN	0.36
38.	MAT-FORMING THALLOPHYTES	TMAX	0.28
39.	BROAD-SUMMERGREEN VINES	MI	0.27
40.	BROAD-EVERGREEN VINES	TMIN	0.23
41.	NARROW-LEAVED EPIPHYTES	TMIN	0.20
42.	XERIC THALLOPHYTES	MI	0.46

Appendix B. Predicted vegetation at selected representative and well-known sites.

```
                                                                   13
     LOCATION                                            LAT     LONG
  TMAX  TMIN  PRCP  PMAX  PMIN  PMTMAX  MI                  ELEV

397. REYKJAVIK                  ISLAND                   64.15  -21.85
     11.0  -1.0  870.   98.   49.   49.  1.87                 5.
             1. BOREAL/MONTANE SHORT-NEEDLED TREES       TMAX    0.0
             2. TALL GRASSES                             TMAX    0.08
             3. BROAD-SUMMERGREEN SMALL TREES            TMAX    0.0
             4. SHORT SWARD-GRASSES                      TMAX    0.46
             5. TALL TUSSOCK-GRASSES                     TMIN    0.0
             6. SHORT BUNCH-GRASSES                      MI      0.61
             7. SUMMERGREEN TUNDRA DWARF-SHRUBS          TMAX    0.42
             8. NEEDLE-LEAVED TREELINE KRUMMHOLZ         TMAX    0.33
             9. NEEDLE-LEAVED EVERGREEN SHRUBS           TMAX    0.08
            10. TEMPERATE EVERGREEN DWARF-SHRUBS         TMAX    0.38
            11. TEMPERATE EVERGREEN FORBS                TMIN    0.17
            12. SUMMERGREEN FORBS                        TMAX    0.08
            13. MAT-FORMING THALLOPHYTES                 MI      0.58
            14. SEASONAL COLD-DESERT HERBS               TMAX    0.31
            15. XERIC THALLOPHYTES                       MI      0.07

416. VARBERG                    SVERIGE                  57.10   12.25
     16.0  -1.0  571.   88.   28.   67.  1.02                 7.
           * 1. BOREAL/MONTANE SHORT-NEEDLED TREES       MI      0.42
           * 2. TEMPERATE NEEDLE-TREES                   TMAX    0.11
             3. SUMMERGREEN BROAD-LEAVED TREES           TMAX    0.07
             4. BOREAL SUMMERGREEN NEEDLE-TREES          TMAX    0.40
             5. SUB-MEDITERRANEAN NEEDLE-TREES           TMIN    0.0
             6. BOREAL BROAD-SUMMERGREEN TREES           MI      0.22
             7. TALL GRASSES                             MI      0.32
             8. BROAD-SUMMERGREEN SMALL TREES            TMAX    0.28
             9. SHORT SWARD-GRASSES                      MI      0.52
            10. SUMMERGREEN GIANT-SCRUB                  TMAX    0.22
            11. BROAD-SUMMERGREEN MESIC SHRUBS           TMAX    0.13
            12. TEMPERATE BROAD-EVERGREEN SHRUBS         TMIN    0.05
            13. TALL TUSSOCK-GRASSES                     TMIN    0.0
            14. SHORT BUNCH-GRASSES                      TMIN    0.66
            15. NEEDLE-LEAVED EVERGREEN SHRUBS           TMAX    0.33
            16. SUMMERGREEN TUNDRA DWARF-SHRUBS          TMAX    0.17
            17. SHORT TUSSOCK-GRASSES                    TMIN    0.0
            18. SUMMERGREEN FORBS                        TMAX    0.33
            19. TEMPERATE EVERGREEN DWARF-SHRUBS         MI      0.32
            20. TEMPERATE EVERGREEN FORBS                MI      0.12
            21. MARITIME HEATH DWARF-SHRUBS              TMIN    0.0
            22. MAT-FORMING THALLOPHYTES                 MI      0.22
            23. BROAD-SUMMERGREEN VINES                  MI      0.02
            24. XERIC THALLOPHYTES                       MI      0.97
```

```
                                                                   14
     LOCATION                                            LAT     LONG
  TMAX  TMIN  PRCP  PMAX  PMIN  PMTMAX  MI                  ELEV

431. CAMBRIDGE                  ENGLAND                  52.22    0.13
     17.0   4.0  551.   62.   34.   62.  0.88                14.
           * 1. TEMPERATE NEEDLE-TREES                   TMAX    0.22
             2. BOREAL BROAD-SUMMERGREEN TREES           MI      0.10
             3. TALL GRASSES                             MI      0.21
             4. BROAD-SUMMERGREEN SMALL TREES            MI      0.21
             5. TEMP. BROAD-EVERGREEN SMALL TREES        MI      0.15
           + 6. SHORT SWARD-GRASSES                      MI      0.44
             7. SUMMERGREEN GIANT-SCRUB                  TMAX    0.26
             8. BROAD-SUMMERGREEN MESIC SHRUBS           TMAX    0.19
             9. TEMPERATE BROAD-EVERGREEN SHRUBS         TMAX    0.13
            10. TALL TUSSOCK-GRASSES                     MI      0.10
           +11. SHORT BUNCH-GRASSES                      TMIN    0.71
           +12. SHORT TUSSOCK-GRASSES                    TMAX    0.42
            13. NEEDLE-LEAVED EVERGREEN SHRUBS           TMAX    0.37
            14. SUMMERGREEN FORBS                        TMAX    0.37
            15. TEMPERATE EVERGREEN DWARF-SHRUBS         MI      0.21
            16. BUSH STEM-SUCCULENTS                     TMIN    0.13
            17. XERIC CUSHION-HERBS                      MI      0.14
            18. MAT-FORMING THALLOPHYTES                 MI      0.10
            19. BROAD-WINTERGREEN EPIPHYTES              TMAX    0.07
            20. XERIC THALLOPHYTES                       MULT.   1.00

434. AMSTERDAM                  NEDERLAN                 52.35    4.87
     18.0   3.0  648.   70.   33.   54.  1.03                 2.
           * 1. BOREAL/MONTANE SHORT-NEEDLED TREES       TMAX    0.36
           * 2. TEMPERATE NEEDLE-TREES                   TMAX    0.33
           * 3. SUMMERGREEN BROAD-LEAVED TREES           MI      0.12
             4. BOREAL SUMMERGREEN NEEDLE-TREES          TMAX    0.40
             5. BOREAL BROAD-SUMMERGREEN TREES           MI      0.22
             6. TALL GRASSES                             MI      0.32
             7. BROAD-SUMMERGREEN SMALL TREES            MI      0.32
             8. TEMP. BROAD-EVERGREEN SMALL TREES        TMIN    0.08
             9. SHORT SWARD-GRASSES                      MI      0.52
            10. SUMMERGREEN GIANT-SCRUB                  TMAX    0.30
            11. BROAD-SUMMERGREEN MESIC SHRUBS           TMAX    0.25
            12. TALL TUSSOCK-GRASSES                     MI      0.22
            13. TEMPERATE BROAD-EVERGREEN SHRUBS         TMAX    0.20
            14. DWARF-NEEDLE SMALL TREES                 TMAX    0.0
            15. SHORT BUNCH-GRASSES                      TMIN    0.70
            16. NEEDLE-LEAVED EVERGREEN SHRUBS           TMAX    0.40
            17. SHORT TUSSOCK-GRASSES                    MI      0.25
            18. SUMMERGREEN TUNDRA DWARF-SHRUBS          TMAX    0.0
            19. SUMMERGREEN FORBS                        TMAX    0.40
            20. TEMPERATE EVERGREEN DWARF-SHRUBS         MI      0.32
            21. TEMPERATE EVERGREEN FORBS                MI      0.12
            22. BUSH STEM-SUCCULENTS                     TMIN    0.10
            23. MARITIME HEATH DWARF-SHRUBS              MI      0.03
            24. MAT-FORMING THALLOPHYTES                 MI      0.22
            25. BROAD-WINTERGREEN EPIPHYTES              TMIN    0.11
            26. BROAD-SUMMERGREEN VINES                  MI      0.03
            27. XERIC THALLOPHYTES                       MI      0.96
```

15

LOCATION / TMAX TMIN PRCP PMAX PMIN PMTMAX MI / LAT LONG / ELEV

439. STRASBOURG — FRANCE — LAT 48.58, LONG 7.75, ELEV 141.

TMAX 18.8, TMIN 1.0, PRCP 696., PMAX 89., PMIN 34., PMTMAX 89., MI 1.10

* 1.	TEMPERATE NEEDLE-TREES	TMIN	0.33
* 2.	BOREAL/MONTANE SHORT-NEEDLED TREES	TMAX	0.29
* 3.	SUMMERGREEN BROAD-LEAVED TREES	MI	0.18
4.	BOREAL SUMMERGREEN NEEDLE-TREES	TMAX	0.32
5.	SUB-MEDITERRANEAN NEEDLE-TREES	PMTMAX	0.12
6.	BOREAL BROAD-SUMMERGREEN TREES	MI	0.27
7.	BROAD-SUMMERGREEN SMALL TREES	MI	0.36
8.	TALL GRASSES	MI	0.36
9.	SHORT SWARD-GRASSES	MI	0.55
10.	SUMMERGREEN GIANT-SCRUB	TMAX	0.33
11.	BROAD-SUMMERGREEN MESIC SHRUBS	TMAX	0.30
12.	TALL TUSSOCK-GRASSES	TMIN	0.18
13.	TEMPERATE BROAD-EVERGREEN SHRUBS	TMIN	0.14
14.	DWARF-NEEDLE SMALL TREES	TMAX	0.07
15.	SHORT BUNCH-GRASSES	TMIN	0.68
16.	NEEDLE-LEAVED EVERGREEN SHRUBS	TMAX	0.42
17.	SHORT TUSSOCK-GRASSES	MI	0.15
18.	SUMMERGREEN FORBS	TMAX	0.42
19.	TEMPERATE EVERGREEN DWARF-SHRUBS	TMAX	0.29
20.	TEMPERATE EVERGREEN FORBS	MI	0.18
21.	MARITIME HEATH DWARF-SHRUBS	MI	0.09
22.	BUSH STEM-SUCCULENTS	TMIN	0.03
23.	MAT-FORMING THALLOPHYTES	MI	0.27
24.	BROAD-SUMMERGREEN VINES	MI	0.09
25.	BROAD-WINTERGREEN EPIPHYTES	TMIN	0.04
26.	XERIC THALLOPHYTES	MI	0.83

448. MARSEILLE — FRANCE — LAT 43.30, LONG 5.40, ELEV 75.

TMAX 22.5, TMIN 6.8, PRCP 572., PMAX 98., PMIN 18., PMTMAX 18., MI 0.77

* 1.	TEMPERATE NEEDLE-TREES	MI	0.16
2.	TEMP. BROAD-EVERGREEN SMALL TREES	MI	0.03
+ 3.	DWARF-NEEDLE SMALL TREES	TMAX	0.38
4.	XERIC SUMMERGREEN SHRUBS	MI	0.30
5.	MEDITERRANEAN EVERGREEN SHRUBS	TMIN	0.14
6.	TROPICAL BROAD-EVERGREEN SHRUBS	TMIN	0.03
+ 7.	SHORT BUNCH-GRASSES	TMAX	0.71
8.	NEEDLE-LEAVED EVERGREEN SHRUBS	MI	0.49
9.	SUMMERGREEN FORBS	TMAX	0.51
10.	MEDITERRANEAN DWARF-SHRUBS	TMIN	0.32
11.	BUSH STEM-SUCCULENTS	TMIN	0.23
12.	XERIC CUSHION-SHRUBS	MI	0.17
13.	XERIC ROSETTE-SHRUBS	TMAX	0.13
14.	LEAFLESS XEROMORPHIC LARGE-SCRUB	MI	0.04
15.	XERIC CUSHION-HERBS	MI	0.30
16.	BROAD-WINTERGREEN EPIPHYTES	MI	0.23
17.	SUCCULENT FORBS	TMIN	0.07
18.	XERIC DWARF-SHRUBS	MI	0.04
19.	XERIC THALLOPHYTES	TMAX	0.71

16

LOCATION / TMAX TMIN PRCP PMAX PMIN PMTMAX MI / LAT LONG / ELEV

456. PORTO — PORTUGAL — LAT 41.18, LONG -8.60, ELEV 100.

TMAX 20.0, TMIN 9.0, PRCP 1189., PMAX 150., PMIN 21., PMTMAX 21., MI 1.61

* 1.	TEMPERATE NEEDLE-TREES	TMAX	0.44
* 2.	TROPICAL LINEAR-LEAVED TREES	TMIN	0.20
3.	TROPICAL MONTANE RAINFOREST TREES	TMIN	0.07
4.	MONTANE BROAD-RAINGREEN TREES	PMTMAX	0.05
5.	MEDITERRANEAN BROAD-EVERGREEN TREES	TMAX	0.0
6.	SUB-MEDITERRANEAN NEEDLE-TREES	TMAX	0.43
7.	TROPICAL EVERGREEN MICROPHYLL-TREES	TMIN	0.09
8.	TROPICAL EVERGREEN SCLEROPHYLL TREES	TMIN	0.05
9.	TEMP. BROAD-EVERGREEN SMALL TREES	TMAX	0.40
10.	TALL CANE-GRAMINOIDS	TMIN	0.28
11.	TROPICAL BROAD-EVERGREEN SMALL TREES	TMIN	0.05
12.	ARBORESCENT GRASSES	PMTMAX	0.05
13.	BROAD-SUMMERGREEN SMALL TREES	PMTMAX	0.05
14.	BROAD-RAINGREEN SMALL TREES	TMIN	0.05
15.	MEDITERRANEAN EVERGREEN SHRUBS	MI	0.25
16.	DWARF-NEEDLE SMALL TREES	TMAX	0.17
17.	TROPICAL BROAD-EVERGREEN SHRUBS	TMIN	0.13
18.	BROAD-SUMMERGREEN MESIC SHRUBS	PMTMAX	0.05
19.	SUMMERGREEN GIANT-SCRUB	PMTMAX	0.05
20.	SHORT BUNCH-GRASSES	TMIN	0.77
21.	NEEDLE-LEAVED EVERGREEN SHRUBS	MI	0.24
22.	PALMIFORM MESIC ROSETTE-SHRUBS	TMIN	0.05
23.	EVERGREEN GIANT-SCRUB	TMAX	0.0
24.	SUMMERGREEN FORBS	TMAX	0.45
25.	RAINGREEN FORBS	TMIN	0.09
26.	TEMPERATE EVERGREEN FORBS	PMTMAX	0.05
27.	MARITIME HEATH DWARF-SHRUBS	PMTMAX	0.05
28.	MAT-FORMING THALLOPHYTES	TMAX	0.45
29.	BROAD-WINTERGREEN EPIPHYTES	TMAX	0.29
30.	XERIC THALLOPHYTES	MI	0.24

459. ZARAGOZA — ESPANA — LAT 41.63, LONG -0.88, ELEV 205.

TMAX 24.0, TMIN 5.5, PRCP 312., PMAX 46., PMIN 16., PMTMAX 16., MI 0.42

* 1.	XERIC SUMMERGREEN SHRUBS	PMTMAX	0.38
+ 2.	SHORT BUNCH-GRASSES	MI	0.54
+ 3.	COLD-WINTER XEROMORPHIC SHRUBS	TMAX	0.47
4.	NEEDLE-LEAVED EVERGREEN SHRUBS	MI	0.05
5.	SUMMERGREEN FORBS	TMAX	0.44
6.	LEAFLESS XEROMORPHIC LARGE-SCRUB	TMAX	0.36
+ 7.	XERIC CUSHION-SHRUBS	TMIN	0.35
8.	MEDITERRANEAN DWARF-SHRUBS	TMIN	0.21
9.	DESERT-GRASSES	MI	0.19
10.	BUSH STEM-SUCCULENTS	TMIN	0.18
11.	XERIC ROSETTE-SHRUBS	TMIN	0.09
12.	XERIC DWARF-SHRUBS	TMAX	0.46
13.	XERIC CUSHION-HERBS	TMAX	0.24
14.	SUCCULENT FORBS	TMIN	0.02
15.	XERIC THALLOPHYTES	TMAX	0.62

Appendix B. Predicted vegetation at selected representative and well-known sites.

```
                                                                          17
      LOCATION                                              LAT     LONG
   TMAX  TMIN  PRCP  PMAX  PMIN  PMTMAX  MI                     ELEV

465. SEVILLA                       ESPANA                   37.38   -5.98
     30.0  11.0  493.   74.    2.     6.   0.49                 20.
              1. DWARF-NEEDLE SMALL TREES                   TMAX    0.0
              2. MEDITERRANEAN EVERGREEN SHRUBS             TMAX    0.0
            + 3. SHORT BUNCH-GRASSES                        TMAX    0.31
            + 4. HOT-DESERT EVERGREEN SHRUBS                TMIN    0.21
              5. NEEDLE-LEAVED EVERGREEN SHRUBS             MI      0.18
              6. EVERGREEN GIANT-SCRUB                      TMIN    0.15
              7. XERIC EVERGREEN TUFT-TREELETS              TMIN    0.14
              8. LEAFLESS XEROMORPHIC LARGE-SCRUB           TMAX    0.40
              9. BUSH STEM-SUCCULENTS                       TMAX    0.33
             10. XERIC ROSETTE-SHRUBS                       TMIN    0.30
            +11. MEDITERRANEAN DWARF-SHRUBS                 TMAX    0.26
             12. SUMMERGREEN FORBS                          TMAX    0.20
             13. TYPICAL STEM-SUCCULENTS                    TMIN    0.17
             14. XERIC CUSHION-SHRUBS                       TMAX    0.17
             15. DESERT-GRASSES                             MI      0.03
             16. XERIC DWARF-SHRUBS                         TMAX    0.36
             17. SUCCULENT FORBS                            TMIN    0.24
             18. EPHEMERAL DESERT HERBS                     MI      0.03
             19. XERIC CUSHION-HERBS                        TMAX    0.0
             20. XERIC THALLOPHYTES                         TMAX    0.31

470. PALMA, MALLORCA               ESPANA                   39.57    2.65
     24.8  10.2  481.   63.    9.    22.   0.59                 20.
            * 1. TROPICAL EVERGREEN SCLEROPHYLL TREES       TMIN    0.10
              2. BROAD-RAINGREEN SMALL TREES                TMIN    0.10
            + 3. XERIC SUMMERGREEN SHRUBS                   TMAX    0.49
              4. MEDITERRANEAN EVERGREEN SHRUBS             MI      0.24
              5. DWARF-NEEDLE SMALL TREES                   MI      0.24
              6. SUMMERGREEN GIANT-SCRUB                    PMTMAX  0.11
              7. RAINGREEN THORN-SCRUB                      TMIN    0.10
            + 8. SHORT BUNCH-GRASSES                        TMAX    0.57
              9. NEEDLE-LEAVED EVERGREEN SHRUBS             MI      0.33
             10. HOT-DESERT EVERGREEN SHRUBS                TMAX    0.16
             11. EVERGREEN GIANT-SCRUB                      TMIN    0.11
             12. XERIC EVERGREEN TUFT-TREELETS              TMIN    0.10
             13. ARBORESCENT STEM-SUCCULENTS                TMIN    0.10
            +14. MEDITERRANEAN DWARF-SHRUBS                 TMAX    0.46
             15. SUMMERGREEN FORBS                          TMAX    0.41
             16. XERIC CUSHION-SHRUBS                       TMAX    0.38
             17. LEAFLESS XEROMORPHIC LARGE-SCRUB           MI      0.36
             18. BUSH STEM-SUCCULENTS                       TMIN    0.34
             19. XERIC ROSETTE-SHRUBS                       TMAX    0.24
             20. RAINGREEN FORBS                            TMIN    0.14
             21. TYPICAL STEM-SUCCULENTS                    MI      0.02
             22. SCLEROPHYLLOUS GRASSES                     TMIN    0.01
             23. XERIC DWARF-SHRUBS                         MI      0.36
             24. XERIC CUSHION-HERBS                        TMAX    0.21
             25. SUCCULENT FORBS                            TMIN    0.21
             26. BROAD-RAINGREEN VINES                      TMIN    0.01
             27. XERIC THALLOPHYTES                         TMAX    0.57
```

```
                                                                          18
      LOCATION                                              LAT     LONG
   TMAX  TMIN  PRCP  PMAX  PMIN  PMTMAX  MI                     ELEV

478. NAPOLI                        ITALIA                   40.85   14.28
     24.0   8.2  875.  120.   16.    41.   1.08                149.
            * 1. MEDITERRANEAN BROAD-EVERGREEN TREES        MI      0.26
            * 2. SUMMERGREEN BROAD-LEAVED TREES             MI      0.17
              3. MONTANE BROAD-RAINGREEN TREES              TMIN    0.01
              4. TROPICAL XERIC NEEDLE-TREES                TMIN    0.01
              5. TEMPERATE NEEDLE-TREES                     TMAX    0.0
              6. SUB-MEDITERRANEAN NEEDLE-TREES             MI      0.22
              7. TROPICAL EVERGREEN MICROPHYLL-TREES        TMIN    0.05
              8. TROPICAL EVERGREEN SCLEROPHYLL TREES       TMIN    0.01
              9. BROAD-SUMMERGREEN SMALL TREES              TMAX    0.32
             10. TEMP. BROAD-EVERGREEN SMALL TREES          MI      0.31
             11. TALL GRASSES                               PMTMAX  0.26
             12. TALL CANE-GRAMINOIDS                       TMIN    0.25
             13. ARBORESCENT GRASSES                        TMIN    0.20
             14. TROPICAL BROAD-EVERGREEN SMALL TREES       TMIN    0.01
             15. BROAD-RAINGREEN SMALL TREES                TMIN    0.01
             16. DWARF-NEEDLE SMALL TREES                   TMAX    0.50
             17. SUMMERGREEN GIANT-SCRUB                    TMAX    0.48
             18. SHORT SWARD-GRASSES                        PMTMAX  0.39
             19. BROAD-SUMMERGREEN MESIC SHRUBS             MI      0.36
             20. TEMPERATE BROAD-EVERGREEN SHRUBS           PMTMAX  0.26
             21. MEDITERRANEAN EVERGREEN SHRUBS             TMIN    0.25
             22. TROPICAL BROAD-EVERGREEN SHRUBS            TMIN    0.09
             23. SHORT BUNCH-GRASSES                        TMAX    0.62
             24. NEEDLE-LEAVED EVERGREEN SHRUBS             TMAX    0.44
             25. SHORT TUSSOCK-GRASSES                      TMAX    0.04
             26. EVERGREEN GIANT-SCRUB                      TMIN    0.01
             27. PALMIFORM MESIC ROSETTE-SHRUBS             TMIN    0.01
             28. SUMMERGREEN FORBS                          TMAX    0.44
             29. BUSH STEM-SUCCULENTS                       TMIN    0.27
             30. TEMPERATE EVERGREEN FORBS                  MI      0.17
             31. MEDITERRANEAN DWARF-SHRUBS                 MI      0.11
             32. XERIC ROSETTE-SHRUBS                       MI      0.11
             33. RAINGREEN FORBS                            TMIN    0.05
             34. TEMPERATE EVERGREEN DWARF-SHRUBS           TMAX    0.0
             35. BROAD-WINTERGREEN EPIPHYTES                TMIN    0.29
             36. MAT-FORMING THALLOPHYTES                   TMAX    0.23
             37. BROAD-EVERGREEN VINES                      TMIN    0.16
             38. SUCCULENT FORBS                            MI      0.11
             39. NARROW-LEAVED EPIPHYTES                    MI      0.08
             40. BROAD-SUMMERGREEN VINES                    PMTMAX  0.02
             41. XERIC THALLOPHYTES                         TMAX    0.62

482. SAN BERNARDINO, TICINO     SCHWEIZ                     46.47    9.20
      8.0  -7.0 2595.  290.   92.   254.   6.87               2073.
            * 1. SHORT SWARD-GRASSES                        TMAX    0.38
              2. SUMMERGREEN TUNDRA DWARF-SHRUBS            TMAX    0.25
              3. TEMPERATE EVERGREEN DWARF-SHRUBS           TMAX    0.25
              4. MAT-FORMING THALLOPHYTES                   MI      0.90
              5. SEASONAL COLD-DESERT HERBS                 TMAX    0.88
```

LOCATION							LAT	LONG
TMAX	TMIN	PRCP	PMAX	PMIN	PMTMAX	MI	ELEV	

484. ZUERICH — SCHWEIZ — LAT 47.48, LONG 8.53

TMAX	TMIN	PRCP	PMAX	PMIN	PMTMAX	MI	ELEV
17.0	-2.0	1105.	140.	52.	140.	1.94	569.

* 1.	BOREAL/MONTANE SHORT-NEEDLED TREES	TMAX	0.45
* 2.	TEMPERATE NEEDLE-TREES	TMIN	0.19
* 3.	SUMMERGREEN BROAD-LEAVED TREES	TMAX	0.13
4.	BOREAL BROAD-SUMMERGREEN TREES	TMAX	0.33
5.	TALL GRASSES	TMAX	0.37
6.	BROAD-SUMMERGREEN SMALL TREES	TMAX	0.32
7.	SHORT SWARD-GRASSES	TMAX	0.63
8.	BROAD-ERICOID EVERGREEN SHRUBS	MI	0.28
9.	SUMMERGREEN GIANT-SCRUB	TMAX	0.26
10.	BROAD-SUMMERGREEN MESIC SHRUBS	TMAX	0.19
11.	TEMPERATE BROAD-EVERGREEN SHRUBS	TMIN	0.0
12.	SHORT BUNCH-GRASSES	MI	0.55
13.	SUMMERGREEN TUNDRA DWARF-SHRUBS	TMAX	0.08
14.	NEEDLE-LEAVED EVERGREEN SHRUBS	MI	0.04
15.	TEMPERATE EVERGREEN DWARF-SHRUBS	TMAX	0.39
16.	SUMMERGREEN FORBS	TMAX	0.37
17.	SUMMERGREEN FERNS	TMAX	0.37
18.	TEMPERATE EVERGREEN FORBS	TMIN	0.13
19.	MAT-FORMING THALLOPHYTES	MI	0.59
20.	BROAD-SUMMERGREEN VINES	TMAX	0.13
21.	XERIC THALLOPHYTES	MI	0.03

488. OBERGURGL — OESTERRE — LAT 47.00, LONG 10.76

TMAX	TMIN	PRCP	PMAX	PMIN	PMTMAX	MI	ELEV
10.8	-4.0	831.	105.	49.	91.	1.97	1927.

1.	TALL GRASSES	TMAX	0.06
2.	SHORT SWARD-GRASSES	TMAX	0.45
3.	SHORT BUNCH-GRASSES	MI	0.53
4.	SUMMERGREEN TUNDRA DWARF-SHRUBS	TMAX	0.40
5.	NEEDLE-LEAVED TREELINE KRUMMHOLZ	TMAX	0.27
6.	NEEDLE-LEAVED EVERGREEN SHRUBS	MI	0.02
7.	TEMPERATE EVERGREEN DWARF-SHRUBS	TMAX	0.37
8.	SUMMERGREEN FERNS	TMAX	0.06
9.	SUMMERGREEN FORBS	TMAX	0.06
10.	TEMPERATE EVERGREEN FORBS	TMIN	0.04
11.	MAT-FORMING THALLOPHYTES	MI	0.60
12.	SEASONAL COLD-DESERT HERBS	TMAX	0.33
13.	XERIC THALLOPHYTES	MI	0.02

LOCATION							LAT	LONG
TMAX	TMIN	PRCP	PMAX	PMIN	PMTMAX	MI	ELEV	

490. WIEN (HOHE WARTE) — OESTERRE — LAT 48.22, LONG 16.33

TMAX	TMIN	PRCP	PMAX	PMIN	PMTMAX	MI	ELEV
19.5	-1.0	685.	85.	40.	85.	1.10	203.

* 1.	TEMPERATE NEEDLE-TREES	TMIN	0.24
* 2.	BOREAL/MONTANE SHORT-NEEDLED TREES	TMAX	0.23
* 3.	SUMMERGREEN BROAD-LEAVED TREES	MI	0.18
4.	BOREAL SUMMERGREEN NEEDLE-TREES	TMAX	0.25
5.	SUB-MEDITERRANEAN NEEDLE-TREES	TMIN	0.0
6.	BOREAL BROAD-SUMMERGREEN TREES	MI	0.27
7.	BROAD-SUMMERGREEN SMALL TREES	MI	0.37
8.	TALL GRASSES	MI	0.37
9.	SHORT SWARD-GRASSES	MI	0.55
10.	SUMMERGREEN GIANT-SCRUB	TMAX	0.35
11.	BROAD-SUMMERGREEN MESIC SHRUBS	TMAX	0.34
12.	DWARF-NEEDLE SMALL TREES	TMAX	0.13
13.	TEMPERATE BROAD-EVERGREEN SHRUBS	TMIN	0.05
14.	SHORT BUNCH-GRASSES	TMIN	0.66
15.	NEEDLE-LEAVED EVERGREEN SHRUBS	TMAX	0.44
16.	SUMMERGREEN FORBS	TMAX	0.44
17.	TEMPERATE EVERGREEN DWARF-SHRUBS	TMAX	0.25
18.	TEMPERATE EVERGREEN FORBS	TMIN	0.17
19.	MAT-FORMING THALLOPHYTES	MI	0.27
20.	BROAD-SUMMERGREEN VINES	MI	0.09
21.	XERIC THALLOPHYTES	MI	0.83

496. KOELN — DEUTSCHL — LAT 50.93, LONG 6.98

TMAX	TMIN	PRCP	PMAX	PMIN	PMTMAX	MI	ELEV
18.0	2.5	660.	78.	41.	78.	1.03	56.

* 1.	BOREAL/MONTANE SHORT-NEEDLED TREES	TMAX	0.36
* 2.	TEMPERATE NEEDLE-TREES	TMAX	0.33
* 3.	SUMMERGREEN BROAD-LEAVED TREES	MI	0.13
4.	BOREAL SUMMERGREEN NEEDLE-TREES	TMAX	0.40
5.	SUB-MEDITERRANEAN NEEDLE-TREES	MI	0.18
6.	BOREAL BROAD-SUMMERGREEN TREES	MI	0.23
7.	TALL GRASSES	MI	0.33
8.	BROAD-SUMMERGREEN SMALL TREES	MI	0.33
9.	TEMP. BROAD-EVERGREEN SMALL TREES	TMIN	0.04
10.	SHORT SWARD-GRASSES	MI	0.52
11.	SUMMERGREEN GIANT-SCRUB	TMAX	0.30
12.	BROAD-SUMMERGREEN MESIC SHRUBS	TMAX	0.25
13.	TALL TUSSOCK-GRASSES	MI	0.23
14.	TEMPERATE BROAD-EVERGREEN SHRUBS	TMAX	0.20
15.	DWARF-NEEDLE SMALL TREES	TMAX	0.0
16.	SHORT BUNCH-GRASSES	TMIN	0.69
17.	NEEDLE-LEAVED EVERGREEN SHRUBS	TMAX	0.40
18.	SHORT TUSSOCK-GRASSES	MI	0.24
19.	SUMMERGREEN TUNDRA DWARF-SHRUBS	TMAX	0.0
20.	SUMMERGREEN FORBS	TMAX	0.40
21.	TEMPERATE EVERGREEN DWARF-SHRUBS	MI	0.33
22.	TEMPERATE EVERGREEN FORBS	MI	0.13
23.	BUSH STEM-SUCCULENTS	TMIN	0.08
24.	MARITIME HEATH DWARF-SHRUBS	MI	0.03
25.	MAT-FORMING THALLOPHYTES	MI	0.23
26.	BROAD-WINTERGREEN EPIPHYTES	TMIN	0.09
27.	BROAD-SUMMERGREEN VINES	MI	0.03
28.	XERIC THALLOPHYTES	MI	0.94

Appendix B. Predicted vegetation at selected representative and well-known sites.

LOCATION: TMAX TMIN PRCP PMAX PMIN PMTMAX MI — LAT LONG ELEV — 21

504. BERLIN	DDR	52.45	13.43
17.5 -0.5 527. 65. 30. 65. 0.87		42.	
* 1.	BOREAL/MONTANE SHORT-NEEDLED TREES	MI	0.32
* 2.	TEMPERATE NEEDLE-TREES	MI	0.26
+ 3.	BOREAL SUMMERGREEN NEEDLE-TREES	TMAX	0.45
4.	BOREAL BROAD-SUMMERGREEN TREES	MI	0.09
5.	BROAD-SUMMERGREEN SMALL TREES	MI	0.20
6.	TALL GRASSES	MI	0.20
+ 7.	SHORT SWARD-GRASSES	MI	0.43
8.	SUMMERGREEN GIANT-SCRUB	TMAX	0.28
9.	BROAD-SUMMERGREEN MESIC SHRUBS	MI	0.20
10.	XERIC SUMMERGREEN SHRUBS	MI	0.17
11.	TEMPERATE BROAD-EVERGREEN SHRUBS	TMIN	0.07
12.	TALL TUSSOCK-GRASSES	TMIN	0.05
+13.	SHORT BUNCH-GRASSES	TMIN	0.66
14.	NEEDLE-LEAVED EVERGREEN SHRUBS	TMAX	0.38
15.	SHORT TUSSOCK-GRASSES	TMIN	0.05
16.	SUMMERGREEN TUNDRA DWARF-SHRUBS	TMAX	0.04
17.	SUMMERGREEN FORBS	TMAX	0.38
18.	TEMPERATE EVERGREEN DWARF-SHRUBS	MI	0.20
19.	XERIC CUSHION-SHRUBS	MI	0.03
20.	XERIC CUSHION-HERBS	MI	0.16
21.	MAT-FORMING THALLOPHYTES	MI	0.09
22.	XERIC THALLOPHYTES	MULT.	1.00

506. WARSZAWA	POLSKA	52.25	21.00
18.0 -3.0 545. 79. 30. 79. 0.94		121.	
* 1.	BOREAL/MONTANE SHORT-NEEDLED TREES	TMAX	0.36
* 2.	TEMPERATE NEEDLE-TREES	TMIN	0.14
3.	SUMMERGREEN BROAD-LEAVED TREES	MI	0.04
+ 4.	BOREAL SUMMERGREEN NEEDLE-TREES	TMAX	0.40
5.	BOREAL BROAD-SUMMERGREEN TREES	MI	0.15
6.	TALL GRASSES	MI	0.26
7.	BROAD-SUMMERGREEN SMALL TREES	MI	0.26
+ 8.	SHORT SWARD-GRASSES	MI	0.47
9.	SUMMERGREEN GIANT-SCRUB	TMAX	0.30
10.	BROAD-SUMMERGREEN MESIC SHRUBS	TMAX	0.25
11.	XERIC SUMMERGREEN SHRUBS	MI	0.08
12.	DWARF-NEEDLE SMALL TREES	TMAX	0.0
+13.	SHORT BUNCH-GRASSES	TMIN	0.63
14.	NEEDLE-LEAVED EVERGREEN SHRUBS	TMAX	0.40
15.	SUMMERGREEN TUNDRA DWARF-SHRUBS	TMAX	0.0
16.	SUMMERGREEN FORBS	TMAX	0.40
17.	TEMPERATE EVERGREEN DWARF-SHRUBS	MI	0.26
18.	TEMPERATE EVERGREEN FORBS	MI	0.04
19.	MAT-FORMING THALLOPHYTES	MI	0.15
20.	XERIC CUSHION-HERBS	MI	0.07
21.	XERIC THALLOPHYTES	MULT.	1.00

LOCATION: TMAX TMIN PRCP PMAX PMIN PMTMAX MI — LAT LONG ELEV — 22

509. DEBRECEN-PALLAG	MAGYAROR	47.53	21.63
20.8 -2.0 583. 70. 32. 58. 0.90		135.	
* 1.	TEMPERATE NEEDLE-TREES	TMIN	0.19
* 2.	BOREAL/MONTANE SHORT-NEEDLED TREES	TMAX	0.11
3.	SUMMERGREEN BROAD-LEAVED TREES	MI	0.00
4.	BOREAL SUMMERGREEN NEEDLE-TREES	TMAX	0.12
5.	BOREAL BROAD-SUMMERGREEN TREES	MI	0.11
6.	BROAD-SUMMERGREEN SMALL TREES	MI	0.22
7.	TALL GRASSES	MI	0.22
+ 8.	SHORT SWARD-GRASSES	MI	0.45
9.	SUMMERGREEN GIANT-SCRUB	TMAX	0.39
10.	BROAD-SUMMERGREEN MESIC SHRUBS	MI	0.22
11.	DWARF-NEEDLE SMALL TREES	TMIN	0.13
12.	TEMPERATE BROAD-EVERGREEN SHRUBS	TMIN	0.0
+13.	SHORT BUNCH-GRASSES	TMIN	0.64
14.	NEEDLE-LEAVED EVERGREEN SHRUBS	TMAX	0.47
15.	SUMMERGREEN FORBS	TMAX	0.47
16.	TEMPERATE EVERGREEN DWARF-SHRUBS	TMAX	0.18
17.	TEMPERATE EVERGREEN FORBS	MI	0.00
18.	XERIC CUSHION-HERBS	MI	0.12
19.	MAT-FORMING THALLOPHYTES	MI	0.11
20.	XERIC THALLOPHYTES	TMAX	0.84

510. LJUBLJANA	JUGOSLAV	46.07	14.48
19.4 -1.8 1415. 155. 71. 134. 2.29		306.	
* 1.	SUMMERGREEN BROAD-LEAVED TREES	TMAX	0.29
* 2.	BOREAL/MONTANE SHORT-NEEDLED TREES	TMAX	0.24
* 3.	TEMPERATE NEEDLE-TREES	TMIN	0.20
4.	BOREAL BROAD-SUMMERGREEN TREES	TMAX	0.47
5.	BROAD-SUMMERGREEN SMALL TREES	TMAX	0.44
6.	TALL GRASSES	TMIN	0.38
7.	SHORT SWARD-GRASSES	TMIN	0.64
8.	BROAD-ERICOID EVERGREEN SHRUBS	TMIN	0.41
9.	SUMMERGREEN GIANT-SCRUB	TMAX	0.38
10.	BROAD-SUMMERGREEN MESIC SHRUBS	TMAX	0.34
11.	DWARF-NEEDLE SMALL TREES	TMAX	0.12
12.	TEMPERATE BROAD-EVERGREEN SHRUBS	TMIN	0.01
13.	SHORT BUNCH-GRASSES	MI	0.34
14.	SUMMERGREEN FERNS	TMAX	0.48
15.	SUMMERGREEN FORBS	TMAX	0.48
16.	TEMPERATE EVERGREEN DWARF-SHRUBS	TMAX	0.26
17.	TEMPERATE EVERGREEN FORBS	TMIN	0.14
18.	MAT-FORMING THALLOPHYTES	TMAX	0.55
19.	BROAD-SUMMERGREEN VINES	TMIN	0.19

```
                                                                  23
      LOCATION                                          LAT     LONG
  TMAX  TMIN   PRCP   PMAX   PMIN   PMTMAX   MI            ELEV

516. IRAKLION, KRITI                HELLAS              35.33    25.15
     26.0  11.8  533.    98.     1.      1.  0.59             27.
         * 1. MEDITERRANEAN EVERGREEN SHRUBS            MI       0.25
         + 2. SHORT BUNCH-GRASSES                       TMAX     0.50
           3. NEEDLE-LEAVED EVERGREEN SHRUBS            MI       0.33
           4. HOT-DESERT EVERGREEN SHRUBS               TMAX     0.22
           5. EVERGREEN GIANT-SCRUB                     TMIN     0.19
           6. XERIC EVERGREEN TUFT-TREELETS             TMIN     0.17
         + 7. MEDITERRANEAN DWARF-SHRUBS                TMAX     0.47
           8. BUSH STEM-SUCCULENTS                      TMIN     0.39
           9. SUMMERGREEN FORBS                         TMAX     0.36
          10. LEAFLESS XEROMORPHIC LARGE-SCRUB          MI       0.35
         +11. XERIC CUSHION-SHRUBS                      TMAX     0.32
          12. TYPICAL STEM-SUCCULENTS                   MI       0.01
          13. XERIC DWARF-SHRUBS                        MI       0.35
          14. SUCCULENT FORBS                           TMIN     0.27
          15. XERIC CUSHION-HERBS                       TMAX     0.16
          16. XERIC THALLOPHYTES                        TMAX     0.50

524. SOFIJA                         BALGARIJ            42.68    23.32
     20.0  -1.0  630.    83.    28.     57.  1.02            550.
         * 1. TEMPERATE NEEDLE-TREES                    TMIN     0.24
         * 2. BOREAL/MONTANE SHORT-NEEDLED TREES        TMAX     0.18
         * 3. SUMMERGREEN BROAD-LEAVED TREES            MI       0.12
           4. BOREAL SUMMERGREEN NEEDLE-TREES           TMAX     0.20
           5. SUB-MEDITERRANEAN NEEDLE-TREES            TMIN     0.0
           6. BOREAL BROAD-SUMMERGREEN TREES            MI       0.22
           7. BROAD-SUMMERGREEN SMALL TREES             MI       0.32
           8. TALL GRASSES                              MI       0.32
           9. SHORT SWARD-GRASSES                       MI       0.52
          10. SUMMERGREEN GIANT-SCRUB                   TMAX     0.36
          11. BROAD-SUMMERGREEN MESIC SHRUBS            MI       0.32
          12. DWARF-NEEDLE SMALL TREES                  TMAX     0.17
          13. TEMPERATE BROAD-EVERGREEN SHRUBS          TMIN     0.05
          14. SHORT BUNCH-GRASSES                       TMIN     0.66
          15. NEEDLE-LEAVED EVERGREEN SHRUBS            TMAX     0.45
          16. SUMMERGREEN FORBS                         TMAX     0.45
          17. TEMPERATE EVERGREEN DWARF-SHRUBS          TMAX     0.22
          18. TEMPERATE EVERGREEN FORBS                 MI       0.12
          19. MAT-FORMING THALLOPHYTES                  MI       0.22
          20. BROAD-SUMMERGREEN VINES                   MI       0.02
          21. XERIC THALLOPHYTES                        TMAX     0.91

530. IZANA,TENERIFA                 ISLAS CA            28.31   -16.40
     17.0   3.5  369.    75.     2.      2.  0.64           2367.
         * 1. SHORT BUNCH-GRASSES                       MI       0.70
         * 2. NEEDLE-LEAVED EVERGREEN SHRUBS            TMAX     0.37
         * 3. COLD-WINTER XEROMORPHIC SHRUBS            TMAX     0.12
           4. SUMMERGREEN FORBS                         TMAX     0.37
           5. XERIC CUSHION-SHRUBS                      TMIN     0.28
           6. BUSH STEM-SUCCULENTS                      TMIN     0.12
           7. LEAFLESS XEROMORPHIC LARGE-SCRUB          TMAX     0.11
           8. MEDITERRANEAN DWARF-SHRUBS                TMIN     0.04
           9. XERIC CUSHION-HERBS                       MI       0.56
          10. XERIC DWARF-SHRUBS                        MI       0.25
          11. XERIC THALLOPHYTES                        MULT.    1.00
```

```
                                                                  24
      LOCATION                                          LAT     LONG
  TMAX  TMIN   PRCP   PMAX   PMIN   PMTMAX   MI            ELEV

544. DAR-EL-BEIDA (CASABLANCA  AL-MAGHR                 33.65    -7.58
     23.0  12.0  406.    77.     0.      2.  0.51             55.
         * 1. MEDITERRANEAN EVERGREEN SHRUBS            MI       0.11
         + 2. SHORT BUNCH-GRASSES                       MI       0.62
           3. NEEDLE-LEAVED EVERGREEN SHRUBS            MI       0.21
           4. EVERGREEN GIANT-SCRUB                     TMIN     0.20
           5. XERIC EVERGREEN TUFT-TREELETS             TMIN     0.18
           6. HOT-DESERT EVERGREEN SHRUBS               TMAX     0.06
           7. SUMMERGREEN FORBS                         TMAX     0.48
         + 8. XERIC CUSHION-SHRUBS                      TMAX     0.48
           9. BUSH STEM-SUCCULENTS                      TMIN     0.40
         +10. MEDITERRANEAN DWARF-SHRUBS                TMAX     0.37
          11. LEAFLESS XEROMORPHIC LARGE-SCRUB          TMAX     0.32
          12. TYPICAL STEM-SUCCULENTS                   TMAX     0.15
          13. XERIC ROSETTE-SHRUBS                      PMTMAX   0.05
          14. LEAF-SUCCULENT EVERGREEN SHRUBS           TMIN     0.0
          15. XERIC DWARF-SHRUBS                        TMAX     0.44
          16. XERIC CUSHION-HERBS                       TMAX     0.28
          17. SUCCULENT FORBS                           TMAX     0.23
          18. XERIC THALLOPHYTES                        TMAX     0.68

563. TAMANRASSET                    ALGERIE             22.93     5.50
     29.0  12.5   45.    14.     0.      2.  0.04           1463.
         + 1. DESERT-GRASSES                            TMAX     0.35
           2. EPHEMERAL DESERT HERBS                    TMAX     0.35
           3. DRY DESERT                                TMAX     0.94
           4. XERIC THALLOPHYTES                        TMAX     0.35

571. AWJILAH                        LIBIYA              29.15    21.25
     29.0  13.0   13.     6.     0.      0.  0.01             30.
           1. DRY DESERT                                TMAX     0.94
           2. XERIC THALLOPHYTES                        MI       0.10

600. TOMBOUKTOU                     MALI                16.77    -3.02
     34.0  22.0  285.    98.     0.     15.  0.16            250.
           1. RAINGREEN THORN-SCRUB                     TMAX     0.24
         + 2. HOT-DESERT EVERGREEN SHRUBS               TMAX     0.33
         + 3. ARBORESCENT STEM-SUCCULENTS               TMAX     0.33
           4. XERIC ROSETTE-SHRUBS                      TMAX     0.30
           5. TYPICAL STEM-SUCCULENTS                   TMAX     0.30
         + 6. SCLEROPHYLLOUS GRASSES                    TMAX     0.30
           7. LEAFLESS XEROMORPHIC LARGE-SCRUB          MULT.    0.24
           8. BUSH STEM-SUCCULENTS                      TMAX     0.20
           9. DESERT-GRASSES                            TMAX     0.17
          10. XERIC CUSHION-SHRUBS                      TMAX     0.03
          11. SUCCULENT FORBS                           TMAX     0.27
          12. XERIC DWARF-SHRUBS                        TMAX     0.21
          13. EPHEMERAL DESERT HERBS                    TMAX     0.17
          14. XERIC THALLOPHYTES                        TMAX     0.17
```

Appendix B. Predicted vegetation at selected representative and well-known sites.

```
                                                                    25
      LOCATION                                          LAT    LONG
   TMAX  TMIN  PRCP  PMAX  PMIN  PMTMAX  MI               ELEV

618. JOS                         NIGERIA                9.92     8.88
     26.0  21.0 1444.  300.    1.  103.  1.31            1047.
             * 1. MONSOON BROAD-RAINGREEN TREES          TMAX     0.25
               2. TROPICAL EVERGREEN SCLEROPHYLL TREES   TMAX     0.45
               3. PALMIFORM TUFT-TREES                   TMAX     0.42
               4. TROPICAL EVERGREEN MICROPHYLL-TREES    MULT.    0.35
               5. BROAD-RAINGREEN SMALL TREES            TMAX     0.45
               6. TROPICAL BROAD-EVERGREEN DWARF-TREES   TMAX     0.40
               7. TALL CANE-GRAMINOIDS                   TMAX     0.36
               8. PALMIFORM TUFT-TREELETS                TMAX     0.43
               9. TROPICAL BROAD-EVERGREEN SHRUBS        TMAX     0.27
              10. EVERGREEN GIANT-SCRUB                  TMAX     0.50
              11. SHORT BUNCH-GRASSES                    TMAX     0.50
              12. RAINGREEN FORBS                        TMAX     0.36
              13. BUSH STEM-SUCCULENTS                   MI       0.14
              14. BROAD-RAINGREEN VINES                  TMAX     0.45
              15. BROAD-WINTERGREEN EPIPHYTES            TMAX     0.29
              16. BROAD-EVERGREEN VINES                  TMAX     0.27
              17. MAT-FORMING THALLOPHYTES               TMAX     0.14
              18. XERIC THALLOPHYTES                     TMAX     0.50

637. WADI HALFA                  AS-SUDAN              21.93    31.33
     31.5  15.0    5.     2.    0.     2.  0.00           155.
               1. DRY DESERT                             MULT.    1.00

641. ADDIS-ABEBA                 YAITOPYA               9.00    38.33
     17.5  14.8 1302.  300.    6.   90.  1.80            2440.
             * 1. TROPICAL MONTANE RAINFOREST TREES      TMAX     0.28
             * 2. MONTANE BROAD-RAINGREEN TREES          TMAX     0.25
             * 3. TROPICAL LINEAR-LEAVED TREES           TMAX     0.19
               4. TROPICAL EVERGREEN MICROPHYLL-TREES    TMAX     0.15
               5. TROPICAL EVERGREEN SCLEROPHYLL TREES   TMAX     0.13
               6. PALMIFORM TUFT-TREES                   TMAX     0.08
               7. TROPICAL CLOUD-FOREST DWARF-TREES      TMAX     0.25
               8. TALL CANE-GRAMINOIDS                   TMAX     0.38
               9. TALL GRASSES                           TMAX     0.38
              10. ARBORESCENT GRASSES                    TMAX     0.38
              11. TROPICAL BROAD-EVERGREEN SMALL TREES   TMAX     0.15
              12. BROAD-RAINGREEN SMALL TREES            TMAX     0.13
              13. SHORT SWARD-GRASSES                    TMAX     0.64
              14. TROPICAL BROAD-EVERGREEN SHRUBS        TMIN     0.37
              15. PALMIFORM TUFT-TREELETS                TMAX     0.18
              16. SHORT BUNCH-GRASSES                    MI       0.67
              17. PALMIFORM MESIC ROSETTE-SHRUBS         TMAX     0.13
              18. NEEDLE-LEAVED EVERGREEN SHRUBS         MI       0.12
              19. TROPICAL EVERGREEN FORBS               TMIN     0.46
              20. TEMPERATE EVERGREEN DWARF-SHRUBS       TMAX     0.36
              21. RAINGREEN FORBS                        TMIN     0.34
              22. MAT-FORMING THALLOPHYTES               MI       0.56
              23. BROAD-EVERGREEN VINES                  TMAX     0.38
              24. BROAD-RAINGREEN VINES                  TMAX     0.13
              25. BROAD-WINTERGREEN EPIPHYTES            TMAX     0.11
              26. XERIC THALLOPHYTES                     MI       0.11
```

```
                                                                    26
      LOCATION                                          LAT    LONG
   TMAX  TMIN  PRCP  PMAX  PMIN  PMTMAX  MI               ELEV

657. SUQUTRA                     SUQUTRA               12.50    54.00
     30.8  25.0  171.   91.    0.    1.  0.11              10.
             + 1. HOT-DESERT EVERGREEN SHRUBS            MI       0.27
               2. TYPICAL STEM-SUCCULENTS                TMAX     0.46
               3. LEAFLESS XEROMORPHIC LARGE-SCRUB       TMAX     0.37
               4. BUSH STEM-SUCCULENTS                   TMAX     0.31
             + 5. DESERT-GRASSES                         TMAX     0.28
               6. XERIC CUSHION-SHRUBS                   MI       0.06
               7. SUCCULENT FORBS                        TMAX     0.42
               8. XERIC DWARF-SHRUBS                     TMAX     0.33
               9. EPHEMERAL DESERT HERBS                 TMAX     0.28
              10. XERIC THALLOPHYTES                     TMAX     0.28

658. LAMU                        KENYA                 -2.27    40.90
     27.0  24.0  929.  350.    3.  128.  0.61               9.
             * 1. TROPICAL EVERGREEN SCLEROPHYLL TREES   MI       0.19
               2. XERIC RAINGREEN TREES                  TMAX     0.19
               3. BROAD-RAINGREEN SMALL TREES            TMAX     0.40
             + 4. RAINGREEN THORN-SCRUB                  TMAX     0.48
               5. SHORT SWARD-GRASSES                    MI       0.19
             + 6. SHORT BUNCH-GRASSES                    TMAX     0.45
             + 7. EVERGREEN GIANT-SCRUB                  TMAX     0.42
               8. RAINGREEN FORBS                        TMAX     0.32
               9. BROAD-RAINGREEN VINES                  TMAX     0.40
              10. BROAD-WINTERGREEN EPIPHYTES            MI       0.02
              11. XERIC THALLOPHYTES                     TMAX     0.45

671. LULUABOURG                  ZAIRE                 -5.90    22.42
     25.0  24.0 1411.  280.    6.   69.  1.12             660.
             * 1. TROPICAL RAINFOREST TREES              MI       0.20
               2. MONSOON BROAD-RAINGREEN TREES          TMAX     0.17
               3. TROPICAL LINEAR-LEAVED TREES           MI       0.11
               4. TROPICAL EVERGREEN SCLEROPHYLL TREES   TMAX     0.50
               5. PALMIFORM TUFT-TREES                   TMAX     0.38
               6. TROPICAL EVERGREEN MICROPHYLL-TREES    MI       0.34
               7. BROAD-RAINGREEN SMALL TREES            TMAX     0.50
               8. TALL CANE-GRAMINOIDS                   MI       0.38
               9. TALL GRASSES                           MI       0.38
              10. TROPICAL BROAD-EVERGREEN DWARF-TREES   PMTMAX   0.27
              11. ARBORESCENT GRASSES                    TMAX     0.25
              12. TROPICAL BROAD-EVERGREEN SMALL TREES   MI       0.20
              13. TROPICAL BROAD-EVERGREEN LIANAS        MI       0.11
              14. PALMIFORM TUFT-TREELETS                MI       0.38
              15. SHORT SWARD-GRASSES                    TMAX     0.37
              16. TROPICAL BROAD-EVERGREEN SHRUBS        TMAX     0.32
              17. SHORT BUNCH-GRASSES                    TMAX     0.56
              18. EVERGREEN GIANT-SCRUB                  TMAX     0.42
              19. PALMIFORM MESIC ROSETTE-SHRUBS         MI       0.20
              20. RAINGREEN FORBS                        TMAX     0.40
              21. BUSH STEM-SUCCULENTS                   MI       0.34
              22. TROPICAL EVERGREEN FORBS               MI       0.20
              23. BROAD-RAINGREEN VINES                  TMAX     0.50
              24. BROAD-WINTERGREEN EPIPHYTES            TMAX     0.36
              25. BROAD-EVERGREEN VINES                  MI       0.20
              26. MAT-FORMING THALLOPHYTES               TMAX     0.19
              27. XERIC THALLOPHYTES                     TMAX     0.56
```

```
                                                                       27
       LOCATION                                             LAT      LONG
TMAX  TMIN  PRCP  PMAX  PMIN  PMTMAX  MI                        ELEV

700. LOURENCO MARQUES          MOCAMBIQ                   -25.97    32.58
       26.0  18.0  716.  135.   12.  110.  0.68                  59.
          * 1. TROPICAL EVERGREEN SCLEROPHYLL TREES       MI        0.27
            2. XERIC RAINGREEN TREES                      TMAX      0.13
            3. BROAD-RAINGREEN SMALL TREES                TMAX      0.45
            4. RAINGREEN THORN-SCRUB                      TMAX      0.44
            5. SHORT SWARD-GRASSES                        MULT.     0.27
            6. MEDITERRANEAN EVERGREEN SHRUBS             PMTMAX    0.14
            7. TEMPERATE BROAD-EVERGREEN SHRUBS           MI        0.12
          + 8. EVERGREEN GIANT-SCRUB                      TMAX      0.50
          + 9. SHORT BUNCH-GRASSES                        TMAX      0.50
          +10. XERIC EVERGREEN TUFT-TREELETS              TMIN      0.45
           11. ARBORESCENT STEM-SUCCULENTS                MI        0.17
           12. HOT-DESERT EVERGREEN SHRUBS                MI        0.10
           13. BUSH STEM-SUCCULENTS                       TMAX      0.50
           14. RAINGREEN FORBS                            TMAX      0.36
           15. XERIC CUSHION-SHRUBS                       MI        0.32
           16. XERIC ROSETTE-SHRUBS                       TMAX      0.30
           17. LEAF-SUCCULENT EVERGREEN SHRUBS            MI        0.18
           18. BROAD-RAINGREEN VINES                      TMIN      0.40
           19. SUCCULENT FORBS                            TMAX      0.36
           20. XERIC CUSHION-HERBS                        TMAX      0.16
           21. BROAD-WINTERGREEN EPIPHYTES                MI        0.12
           22. XERIC THALLOPHYTES                         TMAX      0.50

718. TABLE MOUNTAIN,CAPE PROV  SOUTH AF                   -33.95    18.42
       17.0   9.5 1780.  280.   55.   55.  2.84                 761.
          * 1. TEMPERATE RAINFOREST NEEDLE-TREES          TMAX      0.38
          * 2. TROPICAL LINEAR-LEAVED TREES               TMAX      0.15
          * 3. TROPICAL MONTANE RAINFOREST TREES          TMIN      0.10
            4. TEMPERATE NEEDLE-TREES                     MI        0.07
            5. TROPICAL EVERGREEN MICROPHYLL-TREES        TMIN      0.11
            6. SUB-MEDITERRANEAN NEEDLE-TREES             MI        0.08
            7. TALL GRASSES                               TMAX      0.41
            8. TEMP. BROAD-EVERGREEN SMALL TREES          TMAX      0.29
            9. ARBORESCENT GRASSES                        TMIN      0.26
           10. TROPICAL BROAD-EVERGREEN SMALL TREES       TMIN      0.07
           11. TALL CANE-GRAMINOIDS                       MI        0.07
           12. SHORT SWARD-GRASSES                        PMTMAX    0.55
           13. TALL TUSSOCK-GRASSES                       TMAX      0.35
           14. BROAD-ERICOID EVERGREEN SHRUBS             TMAX      0.33
           15. TROPICAL BROAD-EVERGREEN SHRUBS            TMIN      0.15
           16. TEMPERATE BROAD-EVERGREEN SHRUBS           TMAX      0.13
           17. PALMIFORM MESIC ROSETTE-SHRUBS             TMIN      0.07
           18. MESIC EVERGREEN CUSHION-SHRUBS             TMAX      0.07
           19. SHORT BUNCH-GRASSES                        MI        0.06
           20. TEMPERATE EVERGREEN FORBS                  TMIN      0.63
           21. MARITIME HEATH DWARF-SHRUBS                TMAX      0.45
           22. SUMMERGREEN FORBS                          TMAX      0.41
           23. TEMPERATE EVERGREEN DWARF-SHRUBS           TMAX      0.41
           24. TROPICAL EVERGREEN FORBS                   TMIN      0.27
           25. EVERGREEN FERNS                            PMTMAX    0.09
           26. MAT-FORMING THALLOPHYTES                   TMAX      0.76
           27. BROAD-EVERGREEN VINES                      TMIN      0.21
           28. NARROW-LEAVED EPIPHYTES                    TMIN      0.18
           29. BROAD-WINTERGREEN EPIPHYTES                TMAX      0.07
```

```
                                                                       28
       LOCATION                                             LAT      LONG
TMAX  TMIN  PRCP  PMAX  PMIN  PMTMAX  MI                        ELEV

746. KERGUELEN                 KERGUELE                   -49.25    69.17
        6.8   1.0  925.  150.   40.   82.  2.10                  16.
          * 1. SHORT SWARD-GRASSES                        TMAX      0.26
          * 2. TALL TUSSOCK-GRASSES                       TMIN      0.18
            3. SHORT BUNCH-GRASSES                        MI        0.47
            4. SUMMERGREEN TUNDRA DWARF-SHRUBS            TMAX      0.12
            5. MESIC EVERGREEN CUSHION-SHRUBS             TMIN      0.0
            6. TEMPERATE EVERGREEN FORBS                  TMIN      0.26
            7. TEMPERATE EVERGREEN DWARF-SHRUBS           TMAX      0.12
            8. MAT-FORMING THALLOPHYTES                   MI        0.68

759. JEVPATORIJA,UKRAINA       SSSR                        45.20    33.37
       25.0   0.0  349.   36.   24.   26.  0.51                   6.
          * 1. XERIC SUMMERGREEN SHRUBS                   TMAX      0.48
          * 2. DWARF-NEEDLE SMALL TREES                   MULT.     0.12
          + 3. SHORT BUNCH-GRASSES                        TMAX      0.56
          + 4. COLD-WINTER XEROMORPHIC SHRUBS             TMAX      0.41
            5. LEAFLESS XEROMORPHIC LARGE-SCRUB           TMAX      0.40
            6. SUMMERGREEN FORBS                          TMAX      0.40
            7. BUSH STEM-SUCCULENTS                       TMIN      0.0
            8. XERIC DWARF-SHRUBS                         TMIN      0.40
            9. XERIC CUSHION-HERBS                        TMAX      0.20
           10. XERIC THALLOPHYTES                         TMAX      0.56

761. LENINGRAD,ROSSIJA         SSSR                        59.92    30.25
       17.0  -8.0  470.   65.   22.   59.  0.96                   6.
          * 1. BOREAL/MONTANE SHORT-NEEDLED TREES         MI        0.38
            2. BOREAL SUMMERGREEN NEEDLE-TREES            TMAX      0.50
            3. BOREAL BROAD-SUMMERGREEN TREES             MI        0.17
            4. BROAD-SUMMERGREEN SMALL TREES              MI        0.28
            5. TALL GRASSES                               TMIN      0.25
            6. SHORT SWARD-GRASSES                        MI        0.49
            7. SUMMERGREEN GIANT-SCRUB                    TMAX      0.26
            8. BROAD-SUMMERGREEN MESIC SHRUBS             TMAX      0.19
            9. XERIC SUMMERGREEN SHRUBS                   MI        0.05
           10. SHORT BUNCH-GRASSES                        TMIN      0.58
           11. NEEDLE-LEAVED EVERGREEN SHRUBS             TMAX      0.37
           12. SUMMERGREEN TUNDRA DWARF-SHRUBS            TMAX      0.08
           13. SUMMERGREEN FORBS                          TMAX      0.37
           14. TEMPERATE EVERGREEN DWARF-SHRUBS           MI        0.28
           15. MAT-FORMING THALLOPHYTES                   MI        0.17
           16. XERIC CUSHION-HERBS                        MI        0.04
           17. XERIC THALLOPHYTES                         MULT.     1.00

764. ASTRACHAN',ROSSIJA        SSSR                        46.35    48.05
       25.0  -7.0  156.   18.    9.   14.  0.23                 -14.
          + 1. COLD-WINTER XEROMORPHIC SHRUBS             TMAX      0.41
          + 2. DESERT-GRASSES                             TMAX      0.56
            3. LEAFLESS XEROMORPHIC LARGE-SCRUB           TMIN      0.26
            4. EPHEMERAL DESERT HERBS                     TMIN      0.27
            5. XERIC DWARF-SHRUBS                         TMIN      0.26
            6. XERIC CUSHION-HERBS                        MI        0.12
            7. XERIC THALLOPHYTES                         TMAX      0.56
```

Appendix B. Predicted vegetation at selected representative and well-known sites.

```
                                                            29
      LOCATION                                     LAT     LONG
 TMAX  TMIN  PRCP  PMAX  PMIN  PMTMAX  MI            ELEV

797. KOLGUJEV , O. KOLGUJEV     SSSR               69.35   49.00
   8.0 -12.5  224.   40.    3.   25.  0.73                7.
          1. SHORT SWARD-GRASSES                   PMTMAX  0.0
        + 2. SHORT BUNCH-GRASSES                   TMIN    0.54
          3. SEASONAL COLD-DESERT HERBS            TMAX    0.70
          4. XERIC CUSHION-HERBS                   TMAX    0.30
          5. XERIC THALLOPHYTES                    MULT.   1.00

802. BATUMI, GRUZHIJA           SSSR               41.63   41.63
  23.0    7.0 2404.  305.   84.  204.  3.28               3.
        * 1. SUMMERGREEN BROAD-LEAVED TREES        TMAX    0.47
        * 2. TEMPERATE BROAD-RAINFOREST TREES      TMIN    0.15
          3. WARM-TEMPERATE BROAD-EVERGREEN TREES  TMIN    0.0
          4. SWAMP SUMMERGREEN NEEDLE-TREES        TMAX    0.17
          5. TROPICAL EVERGREEN MICROPHYLL-TREES   TMIN    0.0
          6. TALL GRASSES                          TMAX    0.52
          7. TEMP. BROAD-EVERGREEN SMALL TREES     TMIN    0.38
          8. BROAD-SUMMERGREEN SMALL TREES         TMAX    0.37
          9. ARBORESCENT GRASSES                   TMIN    0.14
         10. SHORT SWARD-GRASSES                   TMAX    0.52
         11. SUMMERGREEN GIANT-SCRUB               TMAX    0.48
         12. BROAD-SUMMERGREEN MESIC SHRUBS        TMAX    0.44
         13. TEMPERATE BROAD-EVERGREEN SHRUBS      TMIN    0.41
         14. BROAD-ERICOID EVERGREEN SHRUBS        TMAX    0.17
         15. TROPICAL BROAD-EVERGREEN SHRUBS       TMIN    0.04
         16. TALL TUSSOCK-GRASSES                  TMAX    0.0
         17. SUMMERGREEN FORBS                     TMAX    0.52
         18. SUMMERGREEN FERNS                     TMAX    0.35
         19. TEMPERATE EVERGREEN FORBS             TMAX    0.30
         20. TEMPERATE EVERGREEN DWARF-SHRUBS      TMAX    0.06
         21. MARITIME HEATH DWARF-SHRUBS           TMAX    0.0
         22. BROAD-SUMMERGREEN VINES               TMAX    0.47
         23. MAT-FORMING THALLOPHYTES              TMAX    0.30
         24. BROAD-WINTERGREEN EPIPHYTES           TMIN    0.25
         25. BROAD-EVERGREEN VINES                 TMIN    0.12
         26. NARROW-LEAVED EPIPHYTES               TMIN    0.08

804. PEREVAL KRESTOVYJ,GRUZHI   SSSR               42.53   44.47
  11.0  -11.5 1502.  220.   65.  188.  3.97            2380.
          1. BOREAL/MONTANE SHORT-NEEDLED TREES    TMAX    0.0
          2. TALL GRASSES                          TMAX    0.09
          3. BROAD-SUMMERGREEN SMALL TREES         TMAX    0.0
          4. SHORT SWARD-GRASSES                   TMIN    0.51
          5. SUMMERGREEN TUNDRA DWARF-SHRUBS       TMAX    0.45
          6. NEEDLE-LEAVED TREELINE KRUMMHOLZ      TMAX    0.33
          7. TEMPERATE EVERGREEN DWARF-SHRUBS      TMIN    0.22
          8. SUMMERGREEN FERNS                     TMAX    0.09
          9. SUMMERGREEN FORBS                     TMAX    0.09
         10. MAT-FORMING THALLOPHYTES              MI      0.82
         11. SEASONAL COLD-DESERT HERBS            TMAX    0.36
```

```
                                                            30
      LOCATION                                     LAT     LONG
 TMAX  TMIN  PRCP  PMAX  PMIN  PMTMAX  MI            ELEV

830. FRUNZE, KIRGHIZ            SSSR               42.90   74.60
  23.0   -4.5  382.   70.   16.   27.  0.59             750.
        * 1. XERIC SUMMERGREEN SHRUBS              TMAX    0.43
        * 2. SUMMERGREEN GIANT-SCRUB               PMTMAX  0.25
          3. SHORT SWARD-GRASSES                   PMTMAX  0.06
          4. DWARF-NEEDLE SMALL TREES              TMIN    0.02
        + 5. SHORT BUNCH-GRASSES                   TMIN    0.62
        + 6. COLD-WINTER XEROMORPHIC SHRUBS        TMAX    0.47
          7. NEEDLE-LEAVED EVERGREEN SHRUBS        MI      0.33
          8. SUMMERGREEN FORBS                     TMAX    0.48
          9. LEAFLESS XEROMORPHIC LARGE-SCRUB      TMIN    0.31
         10. XERIC DWARF-SHRUBS                    TMIN    0.31
         11. XERIC CUSHION-HERBS                   TMAX    0.28
         12. XERIC THALLOPHYTES                    TMAX    0.68

833. PAMIRSKI POST, TADZHIKIS   SSSR               37.80   73.25
  14.0  -17.0   59.   12.    1.   12.  0.15            3640.
        + 1. DESERT-GRASSES                        TMIN    0.48
          2. SEASONAL COLD-DESERT HERBS            TMAX    0.07
          3. EPHEMERAL DESERT HERBS                TMIN    0.06
          4. XERIC DWARF-SHRUBS                    TMIN    0.06
          5. DRY DESERT                            MI      0.65
          6. XERIC THALLOPHYTES                    TMIN    0.64

856. JENISEJSK, SIBIR'          SSSR               58.45   92.17
  19.0  -22.0  426.   65.   13.   60.  0.93              78.
        * 1. BOREAL/MONTANE SHORT-NEEDLED TREES    TMIN    0.24
        + 2. BOREAL SUMMERGREEN NEEDLE-TREES       TMAX    0.30
          3. BOREAL BROAD-SUMMERGREEN TREES        MI      0.14
          4. BROAD-SUMMERGREEN SMALL TREES         TMIN    0.17
        + 5. SHORT SWARD-GRASSES                   TMIN    0.37
          6. BROAD-SUMMERGREEN MESIC SHRUBS        TMIN    0.17
          7. SUMMERGREEN GIANT-SCRUB               TMIN    0.17
          8. XERIC SUMMERGREEN SHRUBS              MI      0.09
        + 9. SHORT BUNCH-GRASSES                   TMIN    0.49
         10. NEEDLE-LEAVED EVERGREEN SHRUBS        TMIN    0.18
         11. SUMMERGREEN FORBS                     TMAX    0.43
         12. MAT-FORMING THALLOPHYTES              MI      0.14
         13. XERIC CUSHION-HERBS                   MI      0.09
         14. XERIC THALLOPHYTES                    MULT.   1.00

897. JAKUTSK, SIBIR'            SSSR               62.22  129.82
  19.0  -44.0  187.   40.    4.   25.  0.44             102.
          1. BOREAL SUMMERGREEN NEEDLE-TREES       PMTMAX  0.0
        + 2. SHORT BUNCH-GRASSES                   TMIN    0.29
          3. DESERT-GRASSES                        MI      0.13
          4. SUMMERGREEN FORBS                     TMIN    0.11
          5. XERIC CUSHION-HERBS                   TMIN    0.29
          6. XERIC THALLOPHYTES                    MULT.   1.00
```

```
       LOCATION                                                LAT     LONG
  TMAX  TMIN  PRCP  PMAX  PMIN  PMTMAX  MI                        ELEV

931. LEVCOSIA                      KIPROS                      35.18   33.35
  28.4    9.8  379.   80.    1.      3.  0.39                     159.
        * 1. SHORT BUNCH-GRASSES                               TMAX    0.38
        * 2. HOT-DESERT EVERGREEN SHRUBS                       TMIN    0.16
        * 3. EVERGREEN GIANT-SCRUB                             TMIN    0.09
        * 4. XERIC EVERGREEN TUFT-TREELETS                     TMIN    0.08
          5. LEAFLESS XEROMORPHIC LARGE-SCRUB                  TMAX    0.46
        + 6. MEDITERRANEAN DWARF-SHRUBS                        TMAX    0.35
          7. BUSH STEM-SUCCULENTS                              TMIN    0.33
        + 8. DESERT-GRASSES                                    MI      0.28
          9. SUMMERGREEN FORBS                                 TMAX    0.26
         10. XERIC ROSETTE-SHRUBS                              TMIN    0.25
         11. XERIC CUSHION-SHRUBS                              TMAX    0.22
         12. TYPICAL STEM-SUCCULENTS                           TMIN    0.12
         13. XERIC DWARF-SHRUBS                                TMAX    0.41
         14. EPHEMERAL DESERT HERBS                            MI      0.28
         15. SUCCULENT FORBS                                   TMIN    0.19
         16. XERIC CUSHION-HERBS                               TMAX    0.06
         17. XERIC THALLOPHYTES                                TMAX    0.38

941. CEDRES                        AL-LUBNA                    34.35   36.10
  17.5    0.5  768.  190.    0.      7.  1.35                    1930.
        * 1. SUB-MEDITERRANEAN NEEDLE-TREES                    TMIN    0.09
          2. SHORT BUNCH-GRASSES                               TMIN    0.67
          3. NEEDLE-LEAVED EVERGREEN SHRUBS                    TMAX    0.38
          4. SUMMERGREEN FORBS                                 TMAX    0.38
          5. BUSH STEM-SUCCULENTS                              TMIN    0.02
          6. BROAD-WINTERGREEN EPIPHYTES                       TMIN    0.02
          7. XERIC THALLOPHYTES                                MI      0.48

944. TEL AVIV-YAFO                 YISRA'EL                    32.05   34.77
  25.0   13.0  476.  135.    0.      3.  0.53                      3.
        * 1. MEDITERRANEAN EVERGREEN SHRUBS                    MI      0.15
        + 2. SHORT BUNCH-GRASSES                               TMAX    0.56
          3. EVERGREEN GIANT-SCRUB                             TMIN    0.25
          4. NEEDLE-LEAVED EVERGREEN SHRUBS                    MI      0.25
          5. XERIC EVERGREEN TUFT-TREELETS                     TMIN    0.23
          6. HOT-DESERT EVERGREEN SHRUBS                       TMAX    0.17
        + 7. MEDITERRANEAN DWARF-SHRUBS                        TMAX    0.47
          8. BUSH STEM-SUCCULENTS                              TMIN    0.43
          9. SUMMERGREEN FORBS                                 TMAX    0.40
        +10. XERIC CUSHION-SHRUBS                              TMAX    0.37
         11. XERIC ROSETTE-SHRUBS                              PMTMAX  0.23
         12. TYPICAL STEM-SUCCULENTS                           MI      0.14
         13. LEAF-SUCCULENT EVERGREEN SHRUBS                   TMIN    0.08
         14. SUCCULENT FORBS                                   TMAX    0.32
         15. XERIC CUSHION-HERBS                               TMAX    0.20
         16. XERIC THALLOPHYTES                                TMAX    0.56

948. ADH-DHARAN                    AL-'ARAB                    26.30   50.13
  36.0   16.0   54.   17.    0.      0.  0.04                     150.
        + 1. DESERT-GRASSES                                    TMAX    0.11
          2. EPHEMERAL DESERT HERBS                            TMAX    0.11
          3. DRY DESERT                                        TMAX    0.95
          4. XERIC THALLOPHYTES                                TMAX    0.11
```



```
       LOCATION                                                LAT     LONG
  TMAX  TMIN  PRCP  PMAX  PMIN  PMTMAX  MI                        ELEV

950. SANA'A                        AL-YAMAN                    15.38   44.20
  21.0   14.0  390.  100.    0.      2.  0.49                    2360.
        * 1. MEDITERRANEAN EVERGREEN SHRUBS                    MI      0.09
        + 2. SHORT BUNCH-GRASSES                               MI      0.61
          3. NEEDLE-LEAVED EVERGREEN SHRUBS                    MI      0.19
          4. XERIC EVERGREEN TUFT-TREELETS                     TMAX    0.18
          5. EVERGREEN GIANT-SCRUB                             TMAX    0.08
        + 6. XERIC CUSHION-SHRUBS                              TMAX    0.61
          7. SUMMERGREEN FORBS                                 TMAX    0.48
          8. BUSH STEM-SUCCULENTS                              TMIN    0.47
          9. MEDITERRANEAN DWARF-SHRUBS                        TMAX    0.26
         10. LEAFLESS XEROMORPHIC LARGE-SCRUB                  TMAX    0.26
         11. LEAF-SUCCULENT EVERGREEN SHRUBS                   TMIN    0.15
         12. TYPICAL STEM-SUCCULENTS                           TMAX    0.05
         13. DESERT-GRASSES                                    MI      0.01
         14. XERIC ROSETTE-SHRUBS                              PMTMAX  0.0
         15. XERIC CUSHION-HERBS                               TMAX    0.39
         16. XERIC DWARF-SHRUBS                                TMAX    0.39
         17. SUCCULENT FORBS                                   TMAX    0.14
         18. EPHEMERAL DESERT HERBS                            MI      0.01
         19. XERIC THALLOPHYTES                                TMAX    0.83

968. TEHERAN                       IRAN                        35.67   51.43
  29.5    2.5  246.   50.    1.      3.  0.26                    1220.
        + 1. SHORT BUNCH-GRASSES                               MI      0.25
          2. COLD-WINTER XEROMORPHIC SHRUBS                    TMAX    0.15
          3. LEAFLESS XEROMORPHIC LARGE-SCRUB                  TMAX    0.42
        + 4. DESERT-GRASSES                                    TMAX    0.33
          5. SUMMERGREEN FORBS                                 TMAX    0.22
          6. BUSH STEM-SUCCULENTS                              TMIN    0.08
          7. XERIC DWARF-SHRUBS                                TMAX    0.38
          8. EPHEMERAL DESERT HERBS                            TMAX    0.33
          9. XERIC CUSHION-HERBS                               TMAX    0.02
         10. XERIC THALLOPHYTES                                TMAX    0.33

974. KABUL                         AFGHANES                    34.50   69.18
  24.0   -2.0  316.   90.    2.     11.  0.44                    1815.
          1. XERIC SUMMERGREEN SHRUBS                          MULT.   0.06
        + 2. SHORT BUNCH-GRASSES                               MI      0.56
        + 3. COLD-WINTER XEROMORPHIC SHRUBS                    TMAX    0.47
          4. NEEDLE-LEAVED EVERGREEN SHRUBS                    MI      0.10
          5. SUMMERGREEN FORBS                                 TMAX    0.44
          6. LEAFLESS XEROMORPHIC LARGE-SCRUB                  TMAX    0.36
          7. DESERT-GRASSES                                    MI      0.13
          8. XERIC DWARF-SHRUBS                                TMIN    0.36
          9. XERIC CUSHION-HERBS                               TMAX    0.24
         10. EPHEMERAL DESERT HERBS                            MI      0.13
         11. XERIC THALLOPHYTES                                TMAX    0.62
```

Appendix B. Predicted vegetation at selected representative and well-known sites.

```
                                                                          33
     LOCATION                                               LAT      LONG
 TMAX  TMIN  PRCP  PMAX  PMIN  PMTMAX  MI                       ELEV

983. PUNA                     BHARAT                        18.52    73.90
   29.0  20.0  672.  160.    2.   38.  0.51                    556.
            1. TROPICAL EVERGREEN SCLEROPHYLL TREES         MI       0.02
            2. XERIC RAINGREEN TREES                        TMAX     0.31
            3. BROAD-RAINGREEN SMALL TREES                  TMAX     0.30
          + 4. RAINGREEN THORN-SCRUB                        TMAX     0.44
            5. SHORT SWARD-GRASSES                          MI       0.02
          + 6. ARBORESCENT STEM-SUCCULENTS                  TMAX     0.39
          + 7. SHORT BUNCH-GRASSES                          TMAX     0.35
            8. XERIC EVERGREEN TUFT-TREELETS                TMAX     0.35
            9. EVERGREEN GIANT-SCRUB                        TMAX     0.25
           10. XERIC ROSETTE-SHRUBS                         TMAX     0.45
           11. BUSH STEM-SUCCULENTS                         TMAX     0.37
           12. RAINGREEN FORBS                              TMAX     0.24
           13. XERIC CUSHION-SHRUBS                         TMAX     0.20
           14. TYPICAL STEM-SUCCULENTS                      MI       0.19
           15. LEAF-SUCCULENT EVERGREEN SHRUBS              TMAX     0.18
           16. SUCCULENT FORBS                              TMAX     0.50
           17. BROAD-RAINGREEN VINES                        TMAX     0.30
           18. XERIC CUSHION-HERBS                          TMAX     0.04
           19. XERIC THALLOPHYTES                           TMAX     0.35

995. SIMLA                    BHARAT                        31.15    77.25
   19.8   5.0 1558.  435.    7.  144.  2.29                   2200.
          * 1. SUMMERGREEN BROAD-LEAVED TREES               TMAX     0.32
            2. TALL GRASSES                                 TMAX     0.49
            3. BROAD-SUMMERGREEN SMALL TREES                TMAX     0.46
            4. TEMP. BROAD-EVERGREEN SMALL TREES            TMIN     0.23
            5. TALL CANE-GRAMINOIDS                         TMIN     0.14
            6. ARBORESCENT GRASSES                          TMIN     0.05
            7. SHORT SWARD-GRASSES                          TMIN     0.73
            8. BROAD-ERICOID EVERGREEN SHRUBS               MI       0.42
            9. SUMMERGREEN GIANT-SCRUB                      TMAX     0.39
           10. BROAD-SUMMERGREEN MESIC SHRUBS               TMAX     0.36
           11. TEMPERATE BROAD-EVERGREEN SHRUBS             TMIN     0.32
           12. SHORT BUNCH-GRASSES                          MI       0.34
           13. SUMMERGREEN FORBS                            TMAX     0.49
           14. SUMMERGREEN FERNS                            TMAX     0.49
           15. TEMPERATE EVERGREEN DWARF-SHRUBS             TMAX     0.23
           16. TROPICAL EVERGREEN FORBS                     TMIN     0.11
           17. MAT-FORMING THALLOPHYTES                     TMAX     0.52
           18. BROAD-SUMMERGREEN VINES                      TMAX     0.32
           19. BROAD-WINTERGREEN EPIPHYTES                  TMIN     0.18
           20. BROAD-EVERGREEN VINES                        TMIN     0.04
```

```
                                                                          34
     LOCATION                                               LAT      LONG
 TMAX  TMIN  PRCP  PMAX  PMIN  PMTMAX  MI                       ELEV

1005. COX'S BAZAAR            BANGLADE                      21.43    91.98
   28.0  20.0 3560.  900.    4.  356.  2.57                     12.
          * 1. MONSOON BROAD-RAINGREEN TREES                MI       0.21
            2. PALMIFORM TUFT-TREES                         TMIN     0.47
            3. TROPICAL EVERGREEN MICROPHYLL-TREES          TMAX     0.24
            4. BROAD-RAINGREEN SMALL TREES                  TMAX     0.35
            5. TALL CANE-GRAMINOIDS                         MI       0.19
            6. TROPICAL BROAD-EVERGREEN LIANAS              TMAX     0.17
            7. PALMIFORM TUFT-TREELETS                      TMAX     0.33
            8. SHORT SWARD-GRASSES                          TMAX     0.25
            9. TROPICAL BROAD-EVERGREEN SHRUBS              TMAX     0.18
           10. TEMPERATE BROAD-EVERGREEN SHRUBS             TMAX     0.13
           11. SHORT BUNCH-GRASSES                          MI       0.18
           12. RAINGREEN FORBS                              TMAX     0.28
           13. BROAD-RAINGREEN VINES                        TMAX     0.35
           14. BROAD-EVERGREEN VINES                        TMAX     0.18
           15. BROAD-WINTERGREEN EPIPHYTES                  TMAX     0.14
           16. MAT-FORMING THALLOPHYTES                     TMAX     0.07

1007. GALLE                   SRI LANK                       6.03    80.22
   28.0  25.0 2737.  300.   78.  300.  1.83                    100.
          * 1. TROPICAL RAINFOREST TREES                    TMAX     0.20
            2. HELIOPHILIC LONG-NEEDLED TREES               TMAX     0.38
            3. TROPICAL LINEAR-LEAVED TREES                 TMAX     0.0
            4. PALMIFORM TUFT-TREES                         TMAX     0.50
            5. TROPICAL EVERGREEN SCLEROPHYLL TREES         TMAX     0.35
            6. TROPICAL EVERGREEN MICROPHYLL-TREES          TMAX     0.24
            7. TALL GRASSES                                 TMAX     0.28
            8. TALL CANE-GRAMINOIDS                         TMAX     0.28
            9. TROPICAL BROAD-EVERGREEN SMALL TREES         TMAX     0.24
           10. TROPICAL BROAD-EVERGREEN LIANAS              TMAX     0.17
           11. ARBORESCENT GRASSES                          TMAX     0.10
           12. PALMIFORM TUFT-TREELETS                      TMAX     0.33
           13. SHORT SWARD-GRASSES                          TMAX     0.23
           14. TROPICAL BROAD-EVERGREEN SHRUBS              TMAX     0.18
           15. SHORT BUNCH-GRASSES                          TMAX     0.40
           16. PALMIFORM MESIC ROSETTE-SHRUBS               TMAX     0.35
           17. EVERGREEN GIANT-SCRUB                        MI       0.10
           18. TROPICAL EVERGREEN FORBS                     TMAX     0.23
           19. EVERGREEN FERNS                              TMAX     0.10
           20. BROAD-EVERGREEN VINES                        TMAX     0.18
           21. NARROW-LEAVED EPIPHYTES                      TMAX     0.17
           22. BROAD-WINTERGREEN EPIPHYTES                  TMAX     0.14
           23. TROPICAL BROAD-EVERGREEN EPIPHYTES           TMAX     0.11
           24. MAT-FORMING THALLOPHYTES                     TMAX     0.07
           25. XERIC THALLOPHYTES                           MI       0.09
```

LOCATION TMAX TMIN PRCP PMAX PMIN PMTMAX MI		LAT LONG ELEV	
1013. DALANDZADGAD MONGOL A		43.58	104.50
21.5 -16.0 119. 30. 1. 11. 0.22		1466.	
1.	COLD-WINTER XEROMORPHIC SHRUBS	TMIN	0.15
2.	SHORT BUNCH-GRASSES	MI	0.08
+ 3.	DESERT-GRASSES	TMIN	0.52
4.	LEAFLESS XEROMORPHIC LARGE-SCRUB	TMIN	0.08
5.	SUMMERGREEN FORBS	MI	0.08
6.	EPHEMERAL DESERT HERBS	TMIN	0.08
7.	XERIC DWARF-SHRUBS	TMIN	0.08
8.	XERIC CUSHION-HERBS	MI	0.08
9.	XERIC THALLOPHYTES	TMAX	0.79
1041. NANNING, GUANGXI ZHUANG ZHONGGUO		22.80	108.33
29.0 15.0 1382. 240. 30. 210. 1.17		120.	
* 1.	HELIOPHILIC LONG-NEEDLED TREES	MI	0.15
* 2.	WARM-TEMPERATE BROAD-EVERGREEN TREES	TMAX	0.10
3.	MONSOON BROAD-RAINGREEN TREES	TMIN	0.07
4.	SUMMERGREEN BROAD-LEAVED TREES	TMAX	0.07
5.	TROPICAL EVERGREEN SCLEROPHYLL TREES	TMAX	0.30
6.	PALMIFORM TUFT-TREES	TMIN	0.21
7.	TROPICAL EVERGREEN MICROPHYLL-TREES	TMAX	0.18
8.	BROAD-RAINGREEN SMALL TREES	TMAX	0.30
9.	TALL CANE-GRAMINOIDS	TMAX	0.24
10.	TALL GRASSES	TMAX	0.24
11.	TROPICAL BROAD-EVERGREEN SMALL TREES	TMAX	0.18
12.	TEMP. BROAD-EVERGREEN SMALL TREES	TMAX	0.15
13.	TROPICAL BROAD-EVERGREEN LIANAS	TMAX	0.08
14.	BROAD-SUMMERGREEN SMALL TREES	TMAX	0.05
15.	ARBORESCENT GRASSES	TMAX	0.05
16.	TROPICAL BROAD-EVERGREEN DWARF-TREES	TMIN	0.0
17.	SUMMERGREEN GIANT-SCRUB	TMAX	0.26
18.	PALMIFORM TUFT-TREELETS	TMIN	0.25
19.	SHORT SWARD-GRASSES	TMAX	0.20
20.	TROPICAL BROAD-EVERGREEN SHRUBS	TMAX	0.14
21.	TEMPERATE BROAD-EVERGREEN SHRUBS	TMAX	0.07
22.	BROAD-SUMMERGREEN MESIC SHRUBS	TMAX	0.06
23.	SHORT BUNCH-GRASSES	TMAX	0.35
24.	EVERGREEN GIANT-SCRUB	TMAX	0.25
25.	NEEDLE-LEAVED EVERGREEN SHRUBS	TMAX	0.24
26.	PALMIFORM MESIC ROSETTE-SHRUBS	MI	0.24
27.	BUSH STEM-SUCCULENTS	MI	0.28
28.	RAINGREEN FORBS	TMAX	0.24
29.	SUMMERGREEN FORBS	TMAX	0.24
30.	TROPICAL EVERGREEN FORBS	TMAX	0.19
31.	EVERGREEN FERNS	TMAX	0.05
32.	TEMPERATE EVERGREEN FORBS	TMAX	0.04
33.	BROAD-RAINGREEN VINES	TMIN	0.25
34.	BROAD-EVERGREEN VINES	TMAX	0.14
35.	NARROW-LEAVED EPIPHYTES	TMAX	0.13
36.	BROAD-WINTERGREEN EPIPHYTES	TMAX	0.07
37.	MAT-FORMING THALLOPHYTES	TMAX	0.03
38.	XERIC THALLOPHYTES	TMAX	0.35

LOCATION TMAX TMIN PRCP PMAX PMIN PMTMAX MI		LAT LONG ELEV	
1047. SHANGHAI, SHANGHAI SHI ZHONGGUO		31.22	121.47
27.0 4.0 1120. 200. 34. 102. 1.37		12.	
* 1.	SUMMERGREEN BROAD-LEAVED TREES	TMAX	0.20
* 2.	HELIOPHILIC LONG-NEEDLED TREES	TMIN	0.09
3.	SWAMP SUMMERGREEN NEEDLE-TREES	PMTMAX	0.02
4.	TALL GRASSES	TMAX	0.32
5.	BROAD-SUMMERGREEN SMALL TREES	TMAX	0.16
6.	TEMP. BROAD-EVERGREEN SMALL TREES	TMIN	0.15
7.	TALL CANE-GRAMINOIDS	TMIN	0.10
8.	ARBORESCENT GRASSES	TMIN	0.0
9.	SUMMERGREEN GIANT-SCRUB	TMAX	0.35
10.	SHORT SWARD-GRASSES	TMAX	0.28
11.	DWARF-NEEDLE SMALL TREES	TMAX	0.25
12.	TEMPERATE BROAD-EVERGREEN SHRUBS	TMAX	0.20
13.	BROAD-SUMMERGREEN MESIC SHRUBS	TMAX	0.19
14.	SHORT BUNCH-GRASSES	TMAX	0.45
15.	NEEDLE-LEAVED EVERGREEN SHRUBS	TMAX	0.32
16.	SUMMERGREEN FORBS	TMAX	0.32
17.	SUMMERGREEN FERNS	TMAX	0.15
18.	TEMPERATE EVERGREEN FORBS	TMAX	0.12
19.	BUSH STEM-SUCCULENTS	MI	0.09
20.	BROAD-SUMMERGREEN VINES	TMAX	0.20
21.	BROAD-WINTERGREEN EPIPHYTES	TMIN	0.14
22.	MAT-FORMING THALLOPHYTES	TMAX	0.10
23.	BROAD-EVERGREEN VINES	TMIN	0.0
24.	XERIC THALLOPHYTES	TMAX	0.45
1064. KAGOSHIMA, KYUSHU NIHON		31.60	130.55
26.0 7.5 2244. 435. 88. 207. 2.74		5.	
* 1.	SUMMERGREEN BROAD-LEAVED TREES	TMAX	0.27
* 2.	TEMPERATE BROAD-RAINFOREST TREES	TMAX	0.13
3.	WARM-TEMPERATE BROAD-EVERGREEN TREES	TMIN	0.04
4.	SWAMP SUMMERGREEN NEEDLE-TREES	TMIN	0.26
5.	TROPICAL EVERGREEN MICROPHYLL-TREES	TMIN	0.02
6.	TALL GRASSES	TMAX	0.36
7.	TEMP. BROAD-EVERGREEN SMALL TREES	TMAX	0.30
8.	BROAD-SUMMERGREEN SMALL TREES	TMAX	0.21
9.	ARBORESCENT GRASSES	TMIN	0.17
10.	TALL CANE-GRAMINOIDS	MI	0.11
11.	SUMMERGREEN GIANT-SCRUB	TMAX	0.39
12.	SHORT SWARD-GRASSES	TMAX	0.35
13.	TEMPERATE BROAD-EVERGREEN SHRUBS	TMAX	0.27
14.	BROAD-SUMMERGREEN MESIC SHRUBS	TMAX	0.25
15.	TROPICAL BROAD-EVERGREEN SHRUBS	TMIN	0.06
16.	SHORT BUNCH-GRASSES	MI	0.10
17.	SUMMERGREEN FORBS	TMAX	0.36
18.	SUMMERGREEN FERNS	TMAX	0.20
19.	TEMPERATE EVERGREEN FORBS	TMAX	0.16
20.	BROAD-WINTERGREEN EPIPHYTES	TMIN	0.27
21.	BROAD-SUMMERGREEN VINES	TMAX	0.27
22.	MAT-FORMING THALLOPHYTES	TMAX	0.15
23.	BROAD-EVERGREEN VINES	TMIN	0.13

Appendix B. Predicted vegetation at selected representative and well-known sites.

37

LOCATION TMAX TMIN PRCP PMAX PMIN PMTMAX MI		LAT ELEV	LONG
1069. TAEGU	CHOSON	35.88	128.58
26.0 -1.0 980. 235. 19. 132. 1.37		52.	
* 1. SUMMERGREEN BROAD-LEAVED TREES		TMAX	0.27
2. TALL GRASSES		TMAX	0.36
3. BROAD-SUMMERGREEN SMALL TREES		TMAX	0.21
4. SUMMERGREEN GIANT-SCRUB		TMAX	0.39
5. SHORT SWARD-GRASSES		TMAX	0.32
6. BROAD-SUMMERGREEN MESIC SHRUBS		TMAX	0.25
7. DWARF-NEEDLE SMALL TREES		TMIN	0.17
8. TEMPERATE BROAD-EVERGREEN SHRUBS		TMIN	0.05
9. SHORT BUNCH-GRASSES		TMAX	0.50
10. NEEDLE-LEAVED EVERGREEN SHRUBS		TMAX	0.36
11. SUMMERGREEN FORBS		TMAX	0.36
12. SUMMERGREEN FERNS		MI	0.20
13. TEMPERATE EVERGREEN FORBS		TMAX	0.16
14. BROAD-SUMMERGREEN VINES		TMIN	0.24
15. MAT-FORMING THALLOPHYTES		TMAX	0.14
16. XERIC THALLOPHYTES		MI	0.47
1070. T'AINAN	T'AI-WAN	23.00	120.18
28.0 17.0 1679. 410. 15. 310. 1.42		14.	
* 1. MONSOON BROAD-RAINGREEN TREES		TMIN	0.21
* 2. WARM-TEMPERATE BROAD-EVERGREEN TREES		TMAX	0.20
* 3. SUMMERGREEN BROAD-LEAVED TREES		TMAX	0.13
4. TROPICAL LINEAR-LEAVED TREES		TMAX	0.0
5. TROPICAL EVERGREEN SCLEROPHYLL TREES		TMAX	0.35
6. PALMIFORM TUFT-TREES		TMIN	0.32
7. TROPICAL EVERGREEN MICROPHYLL-TREES		TMAX	0.24
8. BROAD-RAINGREEN SMALL TREES		TMAX	0.35
9. TALL GRASSES		TMAX	0.28
10. TALL CANE-GRAMINOIDS		TMAX	0.28
11. TROPICAL BROAD-EVERGREEN SMALL TREES		TMAX	0.24
12. TROPICAL BROAD-EVERGREEN LIANAS		TMAX	0.17
13. TROPICAL BROAD-EVERGREEN DWARF-TREES		TMIN	0.13
14. BROAD-SUMMERGREEN SMALL TREES		TMAX	0.11
15. ARBORESCENT GRASSES		TMAX	0.10
16. PALMIFORM TUFT-TREELETS		TMAX	0.33
17. SUMMERGREEN GIANT-SCRUB		TMAX	0.30
18. SHORT SWARD-GRASSES		TMAX	0.23
19. TROPICAL BROAD-EVERGREEN SHRUBS		TMAX	0.18
20. TEMPERATE BROAD-EVERGREEN SHRUBS		TMAX	0.13
21. BROAD-SUMMERGREEN MESIC SHRUBS		TMAX	0.13
22. SHORT BUNCH-GRASSES		TMAX	0.40
23. PALMIFORM MESIC ROSETTE-SHRUBS		TMAX	0.35
24. EVERGREEN GIANT-SCRUB		TMAX	0.33
25. RAINGREEN FORBS		TMAX	0.28
26. TROPICAL EVERGREEN FORBS		TMAX	0.23
27. TEMPERATE EVERGREEN FORBS		TMAX	0.08
28. BROAD-RAINGREEN VINES		TMAX	0.35
29. BROAD-EVERGREEN VINES		MULT.	0.18
30. NARROW-LEAVED EPIPHYTES		TMAX	0.17
31. BROAD-WINTERGREEN EPIPHYTES		TMAX	0.14
32. TROPICAL BROAD-EVERGREEN EPIPHYTES		TMAX	0.11
33. MAT-FORMING THALLOPHYTES		TMAX	0.07
34. XERIC THALLOPHYTES		TMAX	0.40

38

LOCATION TMAX TMIN PRCP PMAX PMIN PMTMAX MI		LAT ELEV	LONG
1075. RANGCON	MYANMA	16.78	96.17
30.0 25.0 2633. 600. 6. 173. 1.58		5.	
1. TROPICAL RAINFOREST TREES		TMAX	0.0
2. MONSCON BROAD-RAINGREEN TREES		TMAX	0.42
3. PALMIFORM TUFT-TREES		TMAX	0.42
4. TROPICAL EVERGREEN SCLEROPHYLL TREES		TMAX	0.25
5. TROPICAL EVERGREEN MICROPHYLL-TREES		TMAX	0.12
6. TROPICAL BROAD-EVERGREEN DWARF-TREES		TMAX	0.33
7. BROAD-RAINGREEN SMALL TREES		TMAX	0.25
8. TALL GRASSES		TMAX	0.20
9. TALL CANE-GRAMINOIDS		TMAX	0.20
10. TROPICAL BROAD-EVERGREEN SMALL TREES		TMAX	0.12
11. ARBORESCENT GRASSES		TMAX	0.0
12. TROPICAL BROAD-EVERGREEN LIANAS		TMAX	0.0
13. PALMIFORM TUFT-TREELETS		TMAX	0.24
14. SHORT SWARD-GRASSES		TMAX	0.17
15. TROPICAL BROAD-EVERGREEN SHRUBS		TMAX	0.09
16. SHORT BUNCH-GRASSES		TMAX	0.31
17. PALMIFORM MESIC ROSETTE-SHRUBS		TMAX	0.25
18. EVERGREEN GIANT-SCRUB		TMAX	0.17
19. RAINGREEN FORBS		TMAX	0.20
20. TROPICAL EVERGREEN FORBS		TMAX	0.16
21. BROAD-RAINGREEN VINES		TMAX	0.25
22. BROAD-EVERGREEN VINES		TMAX	0.09
23. BROAD-WINTERGREEN EPIPHYTES		TMAX	0.0
24. MAT-FORMING THALLOPHYTES		TMAX	0.0
25. XERIC THALLOPHYTES		MI	0.27
1087. SAIGON	VIET-NAM	10.75	106.67
28.0 25.0 1985. 340. 4. 108. 1.21		8.	
* 1. MONSOON BROAD-RAINGREEN TREES		MI	0.26
2. PALMIFORM TUFT-TREES		MI	0.42
3. TROPICAL EVERGREEN SCLEROPHYLL TREES		TMAX	0.35
4. TROPICAL EVERGREEN MICROPHYLL-TREES		TMAX	0.24
5. BROAD-RAINGREEN SMALL TREES		TMAX	0.35
6. TROPICAL BROAD-EVERGREEN DWARF-TREES		MI	0.34
7. TALL CANE-GRAMINOIDS		TMAX	0.28
8. TROPICAL BROAD-EVERGREEN LIANAS		TMAX	0.17
9. PALMIFORM TUFT-TREELETS		TMAX	0.33
10. SHORT SWARD-GRASSES		TMAX	0.23
11. TROPICAL BROAD-EVERGREEN SHRUBS		TMAX	0.18
12. SHORT BUNCH-GRASSES		TMAX	0.40
13. EVERGREEN GIANT-SCRUB		TMAX	0.33
14. RAINGREEN FORBS		TMAX	0.28
15. BROAD-RAINGREEN VINES		TMAX	0.35
16. BROAD-EVERGREEN VINES		TMAX	0.18
17. BROAD-WINTERGREEN EPIPHYTES		TMAX	0.14
18. MAT-FORMING THALLOPHYTES		TMAX	0.07
19. XERIC THALLOPHYTES		TMAX	0.40

LOCATION / LAT LONG
TMAX TMIN PRCP PMAX PMIN PMTMAX MI / ELEV

1093. MANILA, LUZON PILIPINA 14.58 120.98
28.0 25.0 2068. 420. 11. 171. 1.28 19.

* 1.	TROPICAL RAINFOREST TREES	TMAX	0.20
2.	MONSOON BROAD-RAINGREEN TREES	MI	0.30
3.	TROPICAL LINEAR-LEAVED TREES	TMAX	0.0
4.	PALMIFORM TUFT-TREES	MI	0.46
5.	TROPICAL EVERGREEN SCLEROPHYLL TREES	TMAX	0.35
6.	TROPICAL EVERGREEN MICROPHYLL-TREES	TMAX	0.24
7.	TROPICAL BROAD-EVERGREEN DWARF-TREES	MI	0.40
8.	BROAD-RAINGREEN SMALL TREES	TMAX	0.35
9.	TALL GRASSES	TMAX	0.28
10.	TALL CANE-GRAMINOIDS	TMAX	0.28
11.	TROPICAL BROAD-EVERGREEN SMALL TREES	TMAX	0.24
12.	TROPICAL BROAD-EVERGREEN LIANAS	TMAX	0.17
13.	ARBORESCENT GRASSES	TMAX	0.10
14.	PALMIFORM TUFT-TREELETS	TMAX	0.33
15.	SHORT SWARD-GRASSES	TMAX	0.23
16.	TROPICAL BROAD-EVERGREEN SHRUBS	TMAX	0.18
17.	SHORT BUNCH-GRASSES	TMAX	0.40
18.	EVERGREEN GIANT-SCRUB	TMAX	0.33
19.	PALMIFORM MESIC ROSETTE-SHRUBS	MI	0.30
20.	RAINGREEN FORBS	TMAX	0.28
21.	TROPICAL EVERGREEN FORBS	TMAX	0.23
22.	BROAD-RAINGREEN VINES	TMAX	0.35
23.	BROAD-EVERGREEN VINES	TMAX	0.18
24.	NARROW-LEAVED EPIPHYTES	TMAX	0.17
25.	BROAD-WINTERGREEN EPIPHYTES	TMAX	0.14
26.	TROPICAL BROAD-EVERGREEN EPIPHYTES	TMAX	0.11
27.	MAT-FORMING THALLOPHYTES	TMAX	0.07
28.	XERIC THALLOPHYTES	TMAX	0.40

1100. DJAKARTA, DJAWA INDONESI -6.17 106.80
26.5 25.5 1793. 290. 42. 101. 1.16 8.

* 1.	TROPICAL RAINFOREST TREES	MI	0.23
2.	TROPICAL EVERGREEN SCLEROPHYLL TREES	TMAX	0.42
3.	PALMIFORM TUFT-TREES	MI	0.40
4.	TROPICAL EVERGREEN MICROPHYLL-TREES	TMAX	0.32
5.	TALL CANE-GRAMINOIDS	TMAX	0.34
6.	TALL GRASSES	TMAX	0.34
7.	TROPICAL BROAD-EVERGREEN SMALL TREES	MI	0.23
8.	TROPICAL BROAD-EVERGREEN LIANAS	MI	0.14
9.	PALMIFORM TUFT-TREELETS	MI	0.40
10.	TROPICAL BROAD-EVERGREEN SHRUBS	TMAX	0.25
11.	SHORT BUNCH-GRASSES	TMAX	0.47
12.	EVERGREEN GIANT-SCRUB	TMAX	0.46
13.	PALMIFORM MESIC ROSETTE-SHRUBS	MI	0.23
14.	RAINGREEN FORBS	TMAX	0.34
15.	BUSH STEM-SUCCULENTS	MI	0.29
16.	TROPICAL EVERGREEN FORBS	MI	0.23
17.	EVERGREEN FERNS	MI	0.05
18.	BROAD-WINTERGREEN EPIPHYTES	TMAX	0.25
19.	BROAD-EVERGREEN VINES	MI	0.23
20.	NARROW-LEAVED EPIPHYTES	MI	0.14
21.	MAT-FORMING THALLOPHYTES	TMAX	0.12
22.	TROPICAL BROAD-EVERGREEN EPIPHYTES	MI	0.05
23.	XERIC THALLOPHYTES	TMAX	0.47

LOCATION / LAT LONG
TMAX TMIN PRCP PMAX PMIN PMTMAX MI / ELEV

1127. CHARLOTTE WATERS,N.TERRI AUSTRALI -26.00 134.83
25.5 11.0 236. 28. 12. 18. 0.32 166.

* 1.	HOT-DESERT EVERGREEN SHRUBS	TMAX	0.19
* 2.	EVERGREEN GIANT-SCRUB	TMIN	0.15
* 3.	ARBORESCENT STEM-SUCCULENTS	TMIN	0.14
+ 4.	DESERT-GRASSES	TMAX	0.53
5.	LEAFLESS XEROMORPHIC LARGE-SCRUB	TMAX	0.42
6.	BUSH STEM-SUCCULENTS	TMIN	0.37
+ 7.	XERIC CUSHION-SHRUBS	TMAX	0.35
8.	XERIC ROSETTE-SHRUBS	TMAX	0.27
9.	TYPICAL STEM-SUCCULENTS	TMIN	0.17
10.	SCLEROPHYLLOUS GRASSES	TMIN	0.05
11.	EPHEMERAL DESERT HERBS	TMAX	0.53
12.	XERIC DWARF-SHRUBS	TMAX	0.49
13.	SUCCULENT FORBS	TMIN	0.24
14.	XERIC CUSHION-HERBS	TMAX	0.18
15.	XERIC THALLOPHYTES	TMAX	0.53

1152. GAYNDAH, QUEENSLAND AUSTRALI -25.62 151.60
25.5 14.5 775. 125. 31. 112. 0.85 112.

* 1.	TROPICAL XERIC NEEDLE-TREES	TMAX	0.30
* 2.	WARM-TEMPERATE BROAD-EVERGREEN TREES	MI	0.12
3.	TROPICAL EVERGREEN SCLEROPHYLL TREES	TMIN	0.30
4.	PALMIFORM TUFT-TREES	MI	0.18
5.	TROPICAL EVERGREEN MICROPHYLL-TREES	MI	0.12
6.	TALL CANE-GRAMINOIDS	MI	0.18
7.	TALL GRASSES	MI	0.18
8.	BROAD-SUMMERGREEN SMALL TREES	MI	0.18
9.	TEMP. BROAD-EVERGREEN SMALL TREES	MI	0.12
10.	ARBORESCENT GRASSES	MI	0.06
11.	SUMMERGREEN GIANT-SCRUB	TMAX	0.41
+12.	DWARF-NEEDLE SMALL TREES	TMAX	0.38
+13.	SHORT SWARD-GRASSES	TMAX	0.35
14.	TEMPERATE BROAD-EVERGREEN SHRUBS	MI	0.30
15.	BROAD-SUMMERGREEN MESIC SHRUBS	MI	0.18
16.	PALMIFORM TUFT-TREELETS	MI	0.18
17.	TROPICAL BROAD-EVERGREEN SHRUBS	MI	0.18
18.	MEDITERRANEAN EVERGREEN SHRUBS	PMTMAX	0.12
+19.	SHORT BUNCH-GRASSES	TMAX	0.53
20.	NEEDLE-LEAVED EVERGREEN SHRUBS	TMAX	0.38
21.	EVERGREEN GIANT-SCRUB	TMIN	0.32
22.	XERIC EVERGREEN TUFT-TREELETS	MI	0.20
23.	BUSH STEM-SUCCULENTS	TMIN	0.48
24.	SUMMERGREEN FORBS	TMAX	0.38
25.	RAINGREEN FORBS	TMIN	0.33
26.	XERIC ROSETTE-SHRUBS	TMAX	0.27
27.	SUCCULENT FORBS	TMAX	0.34
28.	BROAD-WINTERGREEN EPIPHYTES	MI	0.30
29.	XERIC CUSHION-HERBS	TMAX	0.18
30.	MAT-FORMING THALLOPHYTES	MI	0.06
31.	XERIC THALLOPHYTES	TMAX	0.53

Appendix B. Predicted vegetation at selected representative and well-known sites.

41

LOCATION TMAX TMIN PRCP PMAX PMIN PMTMAX MI	LAT / ELEV	LONG
1162. COWRA, NEW SOUTH WALES AUSTRALI	-33.83	148.68
24.0 8.0 591. 63. 39. 51. 0.83	320.	
1. MEDITERRANEAN BROAD-EVERGREEN TREES	MI	0.04
2. TEMPERATE NEEDLE-TREES	TMAX	0.0
3. TROPICAL EVERGREEN SCLEROPHYLL TREES	TMIN	0.0
4. BROAD-SUMMERGREEN SMALL TREES	MI	0.16
5. TALL GRASSES	MI	0.16
6. TEMP. BROAD-EVERGREEN SMALL TREES	MI	0.10
7. SUMMERGREEN GIANT-SCRUB	TMAX	0.48
+ 8. DWARF-NEEDLE SMALL TREES	MI	0.46
+ 9. SHORT SWARD-GRASSES	MI	0.40
10. TEMPERATE BROAD-EVERGREEN SHRUBS	MI	0.28
11. MEDITERRANEAN EVERGREEN SHRUBS	TMIN	0.23
12. BROAD-SUMMERGREEN MESIC SHRUBS	MI	0.16
+13. SHORT BUNCH-GRASSES	TMAX	0.62
14. NEEDLE-LEAVED EVERGREEN SHRUBS	TMAX	0.44
15. SHORT TUSSOCK-GRASSES	TMAX	0.04
16. XERIC EVERGREEN TUFT-TREELETS	TMIN	0.0
17. EVERGREEN GIANT-SCRUB	TMIN	0.0
18. SUMMERGREEN FORBS	TMAX	0.44
19. BUSH STEM-SUCCULENTS	TMIN	0.27
20. XERIC ROSETTE-SHRUBS	TMIN	0.19
21. RAINGREEN FORBS	TMIN	0.04
22. TEMPERATE EVERGREEN DWARF-SHRUBS	TMAX	0.0
23. BROAD-WINTERGREEN EPIPHYTES	MI	0.28
24. XERIC CUSHION-HERBS	MI	0.21
25. SUCCULENT FORBS	TMIN	0.12
26. MAT-FORMING THALLOPHYTES	MI	0.04
27. XERIC THALLOPHYTES	TMAX	0.62
1170. MIENA,TASMANIA AUSTRALI	-42.40	146.50
12.0 1.5 832. 89. 48. 48. 1.82	1087.	
* 1. BOREAL/MONTANE SHORT-NEEDLED TREES	TMAX	0.09
2. TEMPERATE RAINFOREST NEEDLE-TREES	TMAX	0.0
3. TALL GRASSES	TMAX	0.14
4. BROAD-SUMMERGREEN SMALL TREES	TMAX	0.07
5. SHORT SWARD-GRASSES	PMTMAX	0.48
6. TALL TUSSOCK-GRASSES	TMIN	0.23
7. SUMMERGREEN GIANT-SCRUB	TMAX	0.0
8. SHORT BUNCH-GRASSES	MI	0.65
9. SUMMERGREEN TUNDRA DWARF-SHRUBS	TMAX	0.50
10. NEEDLE-LEAVED TREELINE KRUMMHOLZ	TMAX	0.33
11. NEEDLE-LEAVED EVERGREEN SHRUBS	MI	0.11
12. MESIC EVERGREEN CUSHION-SHRUBS	TMIN	0.06
13. TEMPERATE EVERGREEN DWARF-SHRUBS	TMAX	0.43
14. TEMPERATE EVERGREEN FORBS	TMIN	0.28
15. SUMMERGREEN FORBS	TMAX	0.14
16. MARITIME HEATH DWARF-SHRUBS	TMAX	0.0
17. MAT-FORMING THALLOPHYTES	MI	0.56
18. XERIC THALLOPHYTES	MI	0.10

42

LOCATION TMAX TMIN PRCP PMAX PMIN PMTMAX MI	LAT / ELEV	LONG
1179. GLENBERVIE, NORTH ISLAND NEW ZEAL	-35.72	174.38
17.0 9.0 1680. 210. 93. 98. 2.63	250.	
* 1. TEMPERATE RAINFOREST NEEDLE-TREES	TMAX	0.38
* 2. TEMPERATE BROAD-RAINFOREST TREES	TMAX	0.27
* 3. TEMPERATE NEEDLE-TREES	MI	0.16
* 4. TROPICAL LINEAR-LEAVED TREES	TMAX	0.15
* 5. SUMMERGREEN BROAD-LEAVED TREES	TMAX	0.13
6. TROPICAL MONTANE RAINFOREST TREES	TMIN	0.07
7. TROPICAL EVERGREEN MICROPHYLL-TREES	TMIN	0.09
8. SUB-MEDITERRANEAN NEEDLE-TREES	PMTMAX	0.02
9. TALL GRASSES	TMAX	0.41
10. BROAD-SUMMERGREEN SMALL TREES	TMAX	0.35
11. TEMP. BROAD-EVERGREEN SMALL TREES	TMAX	0.29
12. ARBORESCENT GRASSES	TMIN	0.24
13. TALL CANE-GRAMINOIDS	MI	0.16
14. TROPICAL BROAD-EVERGREEN SMALL TREES	TMIN	0.05
15. SHORT SWARD-GRASSES	TMAX	0.71
16. TALL TUSSOCK-GRASSES	TMAX	0.35
17. BROAD-ERICOID EVERGREEN SHRUBS	TMAX	0.33
18. SUMMERGREEN GIANT-SCRUB	TMAX	0.29
19. BROAD-SUMMERGREEN MESIC SHRUBS	TMAX	0.19
20. TEMPERATE BROAD-EVERGREEN SHRUBS	TMAX	0.13
21. TROPICAL BROAD-EVERGREEN SHRUBS	TMIN	0.13
22. SHORT BUNCH-GRASSES	MI	0.15
23. MESIC EVERGREEN CUSHION-SHRUBS	TMAX	0.07
24. PALMIFORM MESIC ROSETTE-SHRUBS	TMIN	0.05
25. TEMPERATE EVERGREEN FORBS	TMIN	0.61
26. MARITIME HEATH DWARF-SHRUBS	TMAX	0.45
27. SUMMERGREEN FORBS	TMAX	0.41
28. SUMMERGREEN FERNS	TMAX	0.41
29. TEMPERATE EVERGREEN DWARF-SHRUBS	TMAX	0.41
30. TROPICAL EVERGREEN FORBS	TMIN	0.25
31. EVERGREEN FERNS	TMIN	0.17
32. MAT-FORMING THALLOPHYTES	MI	0.75
33. BROAD-EVERGREEN VINES	TMIN	0.19
34. NARROW-LEAVED EPIPHYTES	TMIN	0.16
35. BROAD-SUMMERGREEN VINES	TMAX	0.13
36. BROAD-WINTERGREEN EPIPHYTES	TMAX	0.07

```
                                                              43
        LOCATION                                       LAT    LONG
  TMAX  TMIN  PRCP  PMAX  PMIN  PMTMAX  MI               ELEV

1185. NORFOLK ISLAND             NORFOLK              -29.03  167.95
      22.5  16.2 1357.  150.   70.   85.  1.61            15.
         * 1. TROPICAL LINEAR-LEAVED TREES             MI      0.38
         * 2. TEMPERATE BROAD-RAINFOREST TREES         PMTMAX  0.29
         * 3. TROPICAL MONTANE RAINFOREST TREES        TMAX    0.17
         * 4. HELIOPHILIC LONG-NEEDLED TREES           PMTMAX  0.12
           5. TROPICAL EVERGREEN MICROPHYLL-TREES      TMIN    0.40
           6. TROPICAL EVERGREEN SCLEROPHYLL TREES     TMIN    0.37
           7. PALMIFORM TUFT-TREES                     TMAX    0.27
           8. TALL GRASSES                             TMAX    0.51
           9. TALL CANE-GRAMINOIDS                     TMAX    0.51
          10. TROPICAL BROAD-EVERGREEN SMALL TREES     TMIN    0.41
          11. ARBORESCENT GRASSES                      TMAX    0.38
          12. TROPICAL BROAD-EVERGREEN LIANAS          TMIN    0.31
          13. SHORT SWARD-GRASSES                      TMAX    0.51
          14. TEMPERATE BROAD-EVERGREEN SHRUBS         TMAX    0.50
          15. TROPICAL BROAD-EVERGREEN SHRUBS          TMIN    0.42
          16. PALMIFORM TUFT-TREELETS                  TMIN    0.31
          17. MEDITERRANEAN EVERGREEN SHRUBS           MI      0.25
          18. SHORT BUNCH-GRASSES                      TMAX    0.71
          19. PALMIFORM MESIC ROSETTE-SHRUBS           TMIN    0.37
          20. EVERGREEN GIANT-SCRUB                    TMAX    0.21
          21. TROPICAL EVERGREEN FORBS                 MI      0.44
          22. EVERGREEN FERNS                          MI      0.32
          23. TEMPERATE EVERGREEN FORBS                TMAX    0.31
          24. TEMPERATE EVERGREEN DWARF-SHRUBS         TMAX    0.08
          25. BROAD-WINTERGREEN EPIPHYTES              TMAX    0.46
          26. BROAD-EVERGREEN VINES                    TMAX    0.43
          27. NARROW-LEAVED EPIPHYTES                  MI      0.38
          28. TROPICAL BROAD-EVERGREEN EPIPHYTES       TMIN    0.31
          29. MAT-FORMING THALLOPHYTES                 TMAX    0.31
          30. XERIC THALLOPHYTES                       MI      0.24

1187. SUVA                         FIJI               -18.13  178.42
      26.3  23.0 2926.  365.  130.  323.  2.18             6.
         * 1. TROPICAL RAINFOREST TREES                TMAX    0.37
           2. TROPICAL LINEAR-LEAVED TREES             TMAX    0.13
           3. PALMIFORM TUFT-TREES                     TMAX    0.43
           4. TROPICAL EVERGREEN MICROPHYLL-TREES      TMAX    0.34
           5. TALL GRASSES                             TMAX    0.35
           6. TALL CANE-GRAMINOIDS                     TMAX    0.35
           7. TROPICAL BROAD-EVERGREEN SMALL TREES     TMAX    0.34
           8. TROPICAL BROAD-EVERGREEN LIANAS          TMAX    0.31
           9. ARBORESCENT GRASSES                      TMAX    0.19
          10. PALMIFORM TUFT-TREELETS                  TMAX    0.41
          11. SHORT SWARD-GRASSES                      TMAX    0.33
          12. TROPICAL BROAD-EVERGREEN SHRUBS          TMAX    0.26
          13. PALMIFORM MESIC ROSETTE-SHRUBS           TMAX    0.44
          14. SHORT BUNCH-GRASSES                      MI      0.41
          15. TROPICAL EVERGREEN FORBS                 TMAX    0.33
          16. EVERGREEN FERNS                          TMAX    0.19
          17. BROAD-WINTERGREEN EPIPHYTES              TMAX    0.26
          18. BROAD-EVERGREEN VINES                    TMAX    0.26
          19. NARROW-LEAVED EPIPHYTES                  TMAX    0.24
          20. TROPICAL BROAD-EVERGREEN EPIPHYTES       TMAX    0.21
          21. MAT-FORMING THALLOPHYTES                 TMAX    0.14
```

```
                                                              44
        LOCATION                                       LAT    LONG
  TMAX  TMIN  PRCP  PMAX  PMIN  PMTMAX  MI               ELEV

1197. PAPEETE,TAHITI             POLYNESI             -17.53 -149.57
      27.5  25.0 1889.  335.   47.  212.  1.21            90.
         * 1. TROPICAL RAINFOREST TREES                TMAX    0.25
           2. HELIOPHILIC LONG-NEEDLED TREES           MI      0.17
           3. TROPICAL LINEAR-LEAVED TREES             TMAX    0.04
           4. PALMIFORM TUFT-TREES                     MI      0.42
           5. TROPICAL EVERGREEN SCLEROPHYLL TREES     TMAX    0.38
           6. TROPICAL EVERGREEN MICROPHYLL-TREES      TMAX    0.26
           7. TALL CANE-GRAMINOIDS                     TMAX    0.30
           8. TALL GRASSES                             TMAX    0.30
           9. TROPICAL BROAD-EVERGREEN SMALL TREES     MI      0.26
          10. TROPICAL BROAD-EVERGREEN LIANAS          MI      0.17
          11. ARBORESCENT GRASSES                      TMAX    0.13
          12. PALMIFORM TUFT-TREELETS                  TMAX    0.36
          13. SHORT SWARD-GRASSES                      TMAX    0.25
          14. TROPICAL BROAD-EVERGREEN SHRUBS          TMAX    0.20
          15. SHORT BUNCH-GRASSES                      TMAX    0.42
          16. EVERGREEN GIANT-SCRUB                    TMAX    0.38
          17. PALMIFORM MESIC ROSETTE-SHRUBS           MI      0.26
          18. RAINGREEN FORBS                          TMAX    0.30
          19. TROPICAL EVERGREEN FORBS                 TMAX    0.25
          20. EVERGREEN FERNS                          MI      0.09
          21. BROAD-EVERGREEN VINES                    TMAX    0.20
          22. BROAD-WINTERGREEN EPIPHYTES              TMAX    0.18
          23. NARROW-LEAVED EPIPHYTES                  MI      0.17
          24. TROPICAL BROAD-EVERGREEN EPIPHYTES       MI      0.09
          25. MAT-FORMING THALLOPHYTES                 TMAX    0.08
          26. XERIC THALLOPHYTES                       TMAX    0.42

1204. HONOLULU,OAHO              HAWAIIAN              21.32 -157.87
      25.0  21.0  703.  100.   21.   45.  0.61            17.
         * 1. TROPICAL EVERGREEN SCLEROPHYLL TREES     MI      0.18
           2. XERIC RAINGREEN TREES                    TMAX    0.06
         + 3. BROAD-RAINGREEN SMALL TREES              TMAX    0.50
           4. SHORT SWARD-GRASSES                      MI      0.18
         + 5. SHORT BUNCH-GRASSES                      TMAX    0.56
         + 6. EVERGREEN GIANT-SCRUB                    TMAX    0.42
           7. XERIC EVERGREEN TUFT-TREELETS            TMAX    0.41
           8. BUSH STEM-SUCCULENTS                     TMAX    0.56
           9. RAINGREEN FORBS                          TMAX    0.40
          10. XERIC CUSHION-SHRUBS                     TMAX    0.37
          11. LEAF-SUCCULENT EVERGREEN SHRUBS          MI      0.32
          12. XERIC ROSETTE-SHRUBS                     TMAX    0.25
          13. SCLEROPHYLLOUS GRASSES                   TMAX    0.25
          14. BROAD-RAINGREEN VINES                    TMAX    0.50
          15. SUCCULENT FORBS                          TMAX    0.32
          16. XERIC DWARF-SHRUBS                       MI      0.32
          17. XERIC CUSHION-HERBS                      TMAX    0.20
          18. BROAD-WINTERGREEN EPIPHYTES              MI      0.02
          19. XERIC THALLOPHYTES                       TMAX    0.56
```

```
                                                          45
   LOCATION                                          LAT     LONG
TMAX  TMIN  PRCP  PMAX  PMIN  PMTMAX  MI                  ELEV

1208. VOLCANO OBSERVATORY,HAWA  HAWAIIAN              19.43 -155.23
      17.5  14.5 2626.  300.  135.  181.  3.61           1200.
         * 1. TROPICAL MONTANE RAINFOREST TREES       TMAX      0.28
         * 2. TROPICAL LINEAR-LEAVED TREES            TMAX      0.19
           3. TROPICAL EVERGREEN MICROPHYLL-TREES     TMAX      0.15
           4. PALMIFORM TUFT-TREES                    TMAX      0.09
           5. TROPICAL CLOUD-FOREST DWARF-TREES       TMAX      0.25
           6. TALL GRASSES                            TMAX      0.43
           7. ARBORESCENT GRASSES                     TMAX      0.43
           8. TROPICAL BROAD-EVERGREEN SMALL TREES    MULT.     0.15
           9. SHORT SWARD-GRASSES                     TMAX      0.71
          10. TREE FERNS                              TMAX      0.38
          11. TROPICAL BROAD-EVERGREEN SHRUBS         TMIN      0.35
          12. PALMIFORM TUFT-TREELETS                 TMAX      0.20
          13. PALMIFORM MESIC ROSETTE-SHRUBS          TMAX      0.14
          14. TROPICAL EVERGREEN FORBS                TMIN      0.45
          15. EVERGREEN FERNS                         TMIN      0.41
          16. TEMPERATE EVERGREEN DWARF-SHRUBS        TMAX      0.37
          17. MAT-FORMING THALLOPHYTES                TMAX      0.71
          18. BROAD-EVERGREEN VINES                   TMIN      0.40
          19. NARROW-LEAVED EPIPHYTES                 TMIN      0.38
          20. TROPICAL BROAD-EVERGREEN EPIPHYTES      TMIN      0.22
          21. BROAD-WINTERGREEN EPIPHYTES             TMAX      0.11

1211. CAMPBELL ISLAND           CAMPBELL             -52.50  169.00
      9.0   4.5 1382.  150.   65.   65.   2.71              23.
         * 1. TALL TUSSOCK-GRASSES                    TMAX      0.44
         * 2. SHORT SWARD-GRASSES                     TMAX      0.44
           3. MESIC EVERGREEN CUSHION-SHRUBS          PMTMAX    0.38
           4. SHORT BUNCH-GRASSES                     MI        0.11
           5. TEMPERATE EVERGREEN DWARF-SHRUBS        TMAX      0.33
           6. TROPICAL EVERGREEN FORBS                TMIN      0.09
           7. MAT-FORMING THALLOPHYTES                MI        0.76
```

```
                                                          46
   LOCATION                                          LAT     LONG
TMAX  TMIN  PRCP  PMAX  PMIN  PMTMAX  MI                  ELEV

1212. CUMBERLAND BAY                 S.GEORGI        -54.25  -36.75
      5.3  -2.0 1313.  145.   68.  121.   4.22                2.
           1. SHORT SWARD-GRASSES                     TMAX      0.01
           2. TEMPERATE EVERGREEN FORBS               TMAX      0.01
           3. SEASONAL COLD-DESERT HERBS              TMAX      0.35
           4. MAT-FORMING THALLOPHYTES                TMAX      0.18
           5. ICE DESERT                              TMAX      0.00

1215. DECEPTION ISLAND, PENIN.  ANTARCTI             -62.95  -60.63
      1.0  -8.5  563.  100.    7.   43. 15.03                 8.
           1. SEASONAL COLD-DESERT HERBS              TMAX      0.07
           2. MAT-FORMING THALLOPHYTES                TMAX      0.03
           3. ICE DESERT                              TMAX      0.00

1225. SOUTH POLE                     ANTARCTI        -90.00     0.0
    -28.1 -59.2   70.   15.    2.   6.700.00              2800.
           1. ICE DESERT                              TMAX      0.03
```

APPENDIX C

Predicted vegetation at the validation sites

The macroclimatic data and predicted vegetation at each of the 74 validation sites are presented below in computer-generated form. The listing contains all the information provided on the primary ECOSIEVE printout but has been reformatted by ECOSIEVE in order to fit a narrower page. Station name, country (truncated after eight characters), and geographic coordinates are listed on the first line for each site, with climatic data and elevation on the second line in the format suggested by the heading on each page. Temperatures are in °C, precipitation amounts in millimeters, and elevations in meters. The two columns after the predicted forms represent closest environmental limits and the respective distances to them, expressed as standardized fractions of the largest applicable range (see section 5.E). The annual moisture index MI (last climatic value in line two) is understood as an estimate of total vegetation cover in percent (100% when MI > 1.0). Asterisks indicate potentially dominant forms in the highest dominance level present, especially in closed formations (MI > 0.9). Potential dominants in drier, more open formations (MI < 0.9) are indicated by plus signs, with asterisks retained to indicate potentially larger but widely spaced forms (e.g. trees in a savanna). Asterisks and plusses are deleted when the corresponding forms are near environmental limits. For interpretation of the predicted results see sections 5.F and 6.A in the main text.

Appendix C. Predicted vegetation at the validation sites.

1

LOCATION: TMAX TMIN PRCP PMAX PMIN PMTMAX MI — LAT LONG ELEV

1. FAIRBANKS/ALASKA USA — 64.85 -147.72
15.4 -23.0 303. 56. 9. 46. 0.72 — 134.

*	1.	BOREAL/MONTANE SHORT-NEEDLED TREES	MI	0.17
+	2.	BOREAL SUMMERGREEN NEEDLE-TREES	TMAX	0.34
	3.	BROAD-SUMMERGREEN SMALL TREES	MI	0.03
+	4.	SHORT SWARD-GRASSES	MI	0.31
	5.	SUMMERGREEN GIANT-SCRUB	TMIN	0.15
	6.	XERIC SUMMERGREEN SHRUBS	TMAX	0.08
	7.	BROAD-SUMMERGREEN MESIC SHRUBS	MI	0.03
+	8.	SHORT BUNCH-GRASSES	TMIN	0.48
	9.	NEEDLE-LEAVED EVERGREEN SHRUBS	TMIN	0.16
	10.	SUMMERGREEN FORBS	TMAX	0.31
	11.	XERIC CUSHION-HERBS	MI	0.39
	12.	XERIC THALLOPHYTES	MULT.	1.00

2. VANCOUVER(PMO)/BRIT.COLU CANADA — 49.27 -123.12
18.0 3.0 1438. 235. 34. 34. 2.25 — 39.

*	1.	BOREAL/MONTANE SHORT-NEEDLED TREES	TMAX	0.36
*	2.	TEMPERATE NEEDLE-TREES	TMAX	0.33
*	3.	TEMPERATE RAINFOREST NEEDLE-TREES	PMTMAX	0.27
*	4.	SUMMERGREEN BROAD-LEAVED TREES	PMTMAX	0.12
	5.	BOREAL SUMMERGREEN NEEDLE-TREES	PMTMAX	0.27
	6.	SUB-MEDITERRANEAN NEEDLE-TREES	TMIN	0.25
	7.	BOREAL BROAD-SUMMERGREEN TREES	PMTMAX	0.12
	8.	BROAD-SUMMERGREEN SMALL TREES	TMAX	0.39
	9.	TALL GRASSES	PMTMAX	0.12
	10.	TEMP. BROAD-EVERGREEN SMALL TREES	TMIN	0.08
	11.	TALL CANE-GRAMINOIDS	TMIN	0.07
	12.	SUMMERGREEN GIANT-SCRUB	TMAX	0.33
	13.	SHORT SWARD-GRASSES	PMTMAX	0.27
	14.	BROAD-SUMMERGREEN MESIC SHRUBS	TMAX	0.25
	15.	TEMPERATE BROAD-EVERGREEN SHRUBS	PMTMAX	0.12
	16.	TALL TUSSOCK-GRASSES	PMTMAX	0.12
	17.	BROAD-ERICOID EVERGREEN SHRUBS	PMTMAX	0.12
	18.	DWARF-NEEDLE SMALL TREES	TMAX	0.0
	19.	SHORT BUNCH-GRASSES	MI	0.36
	20.	SUMMERGREEN TUNDRA DWARF-SHRUBS	TMAX	0.0
	21.	SUMMERGREEN FORBS	TMAX	0.44
	22.	MARITIME HEATH DWARF-SHRUBS	TMIN	0.36
	23.	TEMPERATE EVERGREEN FORBS	TMIN	0.35
	24.	TEMPERATE EVERGREEN DWARF-SHRUBS	PMTMAX	0.27
	25.	MAT-FORMING THALLOPHYTES	TMAX	0.67
	26.	BROAD-WINTERGREEN EPIPHYTES	TMIN	0.11

2

LOCATION: TMAX TMIN PRCP PMAX PMIN PMTMAX MI — LAT LONG ELEV

3. BANFF/ALBERTA CANADA — 51.17 -115.57
14.5 -10.0 454. 64. 26. 40. 0.98 — 1397.

*	1.	BOREAL/MONTANE SHORT-NEEDLED TREES	TMAX	0.32
	2.	BOREAL SUMMERGREEN NEEDLE-TREES	TMAX	0.25
	3.	BOREAL BROAD-SUMMERGREEN TREES	TMAX	0.13
	4.	BROAD-SUMMERGREEN SMALL TREES	TMAX	0.21
	5.	TALL GRASSES	TMIN	0.21
	6.	SHORT SWARD-GRASSES	PMTMAX	0.38
	7.	SUMMERGREEN GIANT-SCRUB	TMAX	0.15
	8.	BROAD-SUMMERGREEN MESIC SHRUBS	TMAX	0.03
	9.	XERIC SUMMERGREEN SHRUBS	MI	0.03
	10.	SHORT BUNCH-GRASSES	TMIN	0.56
	11.	SUMMERGREEN TUNDRA DWARF-SHRUBS	MI	0.29
	12.	NEEDLE-LEAVED EVERGREEN SHRUBS	TMAX	0.27
	13.	SUMMERGREEN FORBS	TMAX	0.27
	14.	TEMPERATE EVERGREEN DWARF-SHRUBS	TMIN	0.26
	15.	MAT-FORMING THALLOPHYTES	MI	0.19
	16.	SEASONAL COLD-DESERT HERBS	TMAX	0.03
	17.	XERIC CUSHION-HERBS	MI	0.03
	18.	XERIC THALLOPHYTES	MULT.	1.00

4. SASKATOON/SASKATCHEWAN CANADA — 52.12 -106.63
18.9 -17.0 358. 68. 14. 58. 0.70 — 501.

*	1.	BOREAL/MONTANE SHORT-NEEDLED TREES	MI	0.15
+	2.	BOREAL SUMMERGREEN NEEDLE-TREES	TMAX	0.31
	3.	BROAD-SUMMERGREEN SMALL TREES	MI	0.00
	4.	TALL GRASSES	MI	0.00
+	5.	SHORT SWARD-GRASSES	MI	0.29
	6.	SUMMERGREEN GIANT-SCRUB	TMIN	0.28
	7.	XERIC SUMMERGREEN SHRUBS	TMAX	0.23
	8.	BROAD-SUMMERGREEN MESIC SHRUBS	MI	0.00
+	9.	SHORT BUNCH-GRASSES	TMIN	0.52
	10.	NEEDLE-LEAVED EVERGREEN SHRUBS	TMIN	0.29
	11.	COLD-WINTER XEROMORPHIC SHRUBS	PMTMAX	0.03
	12.	SUMMERGREEN FORBS	TMAX	0.43
	13.	LEAFLESS XEROMORPHIC LARGE-SCRUB	TMIN	0.06
	14.	TEMPERATE EVERGREEN DWARF-SHRUBS	MI	0.00
	15.	XERIC CUSHION-HERBS	MI	0.43
	16.	XERIC DWARF-SHRUBS	TMIN	0.06
	17.	XERIC THALLOPHYTES	MULT.	1.00

3

LOCATION / TMAX TMIN PRCP PMAX PMIN PMTMAX MI	LAT / ELEV	LONG
5. SWIFT CURRENT/SASKATCHEW CANADA	50.28	-107.83
19.0 -12.0 377. 74. 15. 52. 0.71	744.	
* 1. BOREAL/MONTANE SHORT-NEEDLED TREES	MI	0.15
+ 2. BOREAL SUMMERGREEN NEEDLE-TREES	TMAX	0.30
3. BROAD-SUMMERGREEN SMALL TREES	MI	0.01
4. TALL GRASSES	MI	0.01
5. SUMMERGREEN GIANT-SCRUB	TMAX	0.33
+ 6. SHORT SWARD-GRASSES	MI	0.30
7. XERIC SUMMERGREEN SHRUBS	TMAX	0.24
8. BROAD-SUMMERGREEN MESIC SHRUBS	MI	0.01
+ 9. SHORT BUNCH-GRASSES	TMIN	0.55
10. NEEDLE-LEAVED EVERGREEN SHRUBS	TMIN	0.40
11. COLD-WINTER XEROMORPHIC SHRUBS	PMTMAX	0.15
12. SUMMERGREEN FORBS	TMAX	0.43
13. LEAFLESS XEROMORPHIC LARGE-SCRUB	MI	0.13
14. TEMPERATE EVERGREEN DWARF-SHRUBS	MI	0.01
15. XERIC CUSHION-HERBS	MI	0.42
16. XERIC DWARF-SHRUBS	MI	0.13
17. XERIC THALLOPHYTES	MULT.	1.00
6. SIOUX LOOKOUT/ONTARIO CANADA	50.10	-91.92
18.0 -18.0 694. 95. 32. 81. 1.44	373.	
* 1. BOREAL/MONTANE SHORT-NEEDLED TREES	TMAX	0.36
2. SUMMERGREEN BROAD-LEAVED TREES	TMIN	0.05
3. BOREAL SUMMERGREEN NEEDLE-TREES	MULT.	0.40
4. BOREAL BROAD-SUMMERGREEN TREES	TMIN	0.34
5. BROAD-SUMMERGREEN SMALL TREES	TMIN	0.26
6. TALL GRASSES	TMIN	0.04
7. SHORT SWARD-GRASSES	TMIN	0.43
8. SUMMERGREEN GIANT-SCRUB	TMIN	0.26
9. BROAD-SUMMERGREEN MESIC SHRUBS	TMAX	0.25
10. SHORT BUNCH-GRASSES	TMIN	0.51
11. NEEDLE-LEAVED EVERGREEN SHRUBS	TMIN	0.27
12. SUMMERGREEN TUNDRA DWARF-SHRUBS	TMAX	0.0
13. SUMMERGREEN FORBS	TMAX	0.40
14. SUMMERGREEN FERNS	TMIN	0.06
15. TEMPERATE EVERGREEN DWARF-SHRUBS	TMIN	0.05
16. MAT-FORMING THALLOPHYTES	MI	0.45
17. XERIC THALLOPHYTES	MI	0.39

4

LOCATION / TMAX TMIN PRCP PMAX PMIN PMTMAX MI	LAT / ELEV	LONG
7. ALBANEL (LAC)/QUEBEC CANADA	51.15	-73.00
17.5 -17.2 775. 100. 32. 97. 1.60	127.	
* 1. BOREAL/MONTANE SHORT-NEEDLED TREES	TMIN	0.39
2. SUMMERGREEN BROAD-LEAVED TREES	TMIN	0.08
3. BOREAL SUMMERGREEN NEEDLE-TREES	TMAX	0.45
4. BOREAL BROAD-SUMMERGREEN TREES	TMIN	0.37
5. BROAD-SUMMERGREEN SMALL TREES	TMIN	0.27
6. TALL GRASSES	TMIN	0.06
7. SHORT SWARD-GRASSES	TMIN	0.44
8. SUMMERGREEN GIANT-SCRUB	TMIN	0.27
9. BROAD-SUMMERGREEN MESIC SHRUBS	TMAX	0.22
10. SHORT BUNCH-GRASSES	TMIN	0.52
11. NEEDLE-LEAVED EVERGREEN SHRUBS	MI	0.25
12. SUMMERGREEN TUNDRA DWARF-SHRUBS	TMAX	0.04
13. SUMMERGREEN FORBS	TMAX	0.38
14. SUMMERGREEN FERNS	TMIN	0.09
15. TEMPERATE EVERGREEN DWARF-SHRUBS	TMIN	0.07
16. MAT-FORMING THALLOPHYTES	MI	0.50
17. XERIC THALLOPHYTES	MI	0.25
8. OLYMPIA/WASHINGTON USA	47.05	-122.88
17.9 4.0 1301. 215. 15. 16. 2.05	21.	
* 1. TEMPERATE NEEDLE-TREES	TMAX	0.32
2. SUB-MEDITERRANEAN NEEDLE-TREES	TMAX	0.28
3. TEMP. BROAD-EVERGREEN SMALL TREES	TMIN	0.15
4. TALL CANE-GRAMINOIDS	TMIN	0.10
5. SHORT BUNCH-GRASSES	MI	0.51
6. SUMMERGREEN FORBS	TMAX	0.44
7. MAT-FORMING THALLOPHYTES	TMAX	0.68
8. BROAD-WINTERGREEN EPIPHYTES	TMAX	0.14
9. DAVIS/CALIFORNIA USA	38.55	-121.73
24.6 7.8 410. 91. 0. 1. 0.49	16.	
* 1. MEDITERRANEAN EVERGREEN SHRUBS	MI	0.09
+ 2. SHORT BUNCH-GRASSES	TMAX	0.58
3. NEEDLE-LEAVED EVERGREEN SHRUBS	MI	0.19
4. HOT-DESERT EVERGREEN SHRUBS	TMIN	0.07
5. SUMMERGREEN FORBS	TMAX	0.42
+ 6. MEDITERRANEAN DWARF-SHRUBS	TMIN	0.40
+ 7. XERIC CUSHION-SHRUBS	TMAX	0.39
8. LEAFLESS XEROMORPHIC LARGE-SCRUB	TMAX	0.38
9. BUSH STEM-SUCCULENTS	TMIN	0.26
10. TYPICAL STEM-SUCCULENTS	TMIN	0.03
11. DESERT-GRASSES	MI	0.01
12. XERIC DWARF-SHRUBS	TMAX	0.47
13. XERIC CUSHION-HERBS	TMAX	0.22
14. SUCCULENT FORBS	TMIN	0.11
15. EPHEMERAL DESERT HERBS	MI	0.01
16. XERIC THALLOPHYTES	TMAX	0.58

Appendix C. Predicted vegetation at the validation sites.

```
                                                                    5
    LOCATION                                          LAT     LONG
 TMAX  TMIN  PRCP  PMAX  PMIN  PMTMAX  MI                  ELEV

10. TANBARK FLAT/CALIFORNIA    U.S.A.                 34.20 -117.77
    23.0   8.6  642.  140.    0.     6.  0.92              853.
          * 1. MEDITERRANEAN BROAD-EVERGREEN TREES     MI        0.13
            2. SUB-MEDITERRANEAN NEEDLE-TREES          MI        0.07
            3. TALL CANE-GRAMINOIDS                    MI        0.24
            4. MEDITERRANEAN EVERGREEN SHRUBS          TMIN      0.28
            5. DWARF-NEEDLE SMALL TREES                PMTMAX    0.18
            6. TROPICAL BROAD-EVERGREEN SHRUBS         TMIN      0.11
          + 7. SHORT BUNCH-GRASSES                     TMAX      0.68
            8. NEEDLE-LEAVED EVERGREEN SHRUBS          TMAX      0.48
            9. EVERGREEN GIANT-SCRUB                   TMIN      0.03
           10. XERIC EVERGREEN TUFT-TREELETS           TMIN      0.03
           11. SUMMERGREEN FORBS                       TMAX      0.48
           12. MEDITERRANEAN DWARF-SHRUBS              MI        0.31
           13. BUSH STEM-SUCCULENTS                    TMIN      0.29
           14. XERIC ROSETTE-SHRUBS                    TMAX      0.15
           15. BROAD-WINTERGREEN EPIPHYTES             TMIN      0.31
           16. SUCCULENT FORBS                         TMIN      0.14
           17. XERIC CUSHION-HERBS                     MI        0.10
           18. XERIC THALLOPHYTES                      TMAX      0.68

11. WEST YELLOWSTONE/MONTANA  USA                     44.50 -111.08
    15.2 -11.5  570.   72.   25.    39.  1.29              2035.
          * 1. BOREAL/MONTANE SHORT-NEEDLED TREES      TMAX      0.38
            2. SUMMERGREEN BROAD-LEAVED TREES          TMAX      0.01
            3. BOREAL SUMMERGREEN NEEDLE-TREES         TMAX      0.32
            4. BOREAL BROAD-SUMMERGREEN TREES          TMAX      0.18
            5. BROAD-SUMMERGREEN SMALL TREES           TMAX      0.24
            6. TALL GRASSES                            TMIN      0.18
            7. SHORT SWARD-GRASSES                     PMTMAX    0.35
            8. SUMMERGREEN GIANT-SCRUB                 TMAX      0.19
            9. BROAD-SUMMERGREEN MESIC SHRUBS          TMAX      0.07
           10. SHORT BUNCH-GRASSES                     TMIN      0.55
           11. NEEDLE-LEAVED EVERGREEN SHRUBS          TMAX      0.30
           12. SUMMERGREEN TUNDRA DWARF-SHRUBS         TMAX      0.23
           13. SUMMERGREEN FORBS                       TMAX      0.30
           14. TEMPERATE EVERGREEN DWARF-SHRUBS        TMIN      0.22
           15. MAT-FORMING THALLOPHYTES                MI        0.38
           16. XERIC THALLOPHYTES                      MI        0.56

12. LOGAN/UTAH                 U.S.A.                 41.73 -111.82
    22.8  -4.4  418.   50.   13.    14.  0.67              1455.
            1. TEMPERATE NEEDLE-TREES                  MI        0.03
            2. XERIC SUMMERGREEN SHRUBS                PMTMAX    0.26
            3. DWARF-NEEDLE SMALL TREES                TMIN      0.03
          + 4. SHORT BUNCH-GRASSES                     TMIN      0.62
            5. NEEDLE-LEAVED EVERGREEN SHRUBS          MI        0.41
          + 6. COLD-WINTER XEROMORPHIC SHRUBS          MI        0.34
            7. SUMMERGREEN FORBS                       TMAX      0.49
            8. LEAFLESS XEROMORPHIC LARGE-SCRUB        MI        0.19
            9. XERIC CUSHION-HERBS                     TMAX      0.29
           10. XERIC DWARF-SHRUBS                      MI        0.19
           11. XERIC THALLOPHYTES                      TMAX      0.69
```

```
                                                                    6
    LOCATION                                          LAT     LONG
 TMAX  TMIN  PRCP  PMAX  PMIN  PMTMAX  MI                  ELEV

13. CODY/WYOMING                 U.S.A.               44.53 -109.07
    20.8  -4.5  235.   39.    9.    21.  0.44              1519.
          * 1. XERIC SUMMERGREEN SHRUBS                TMAX      0.32
          + 2. SHORT BUNCH-GRASSES                     MI        0.56
          + 3. COLD-WINTER XEROMORPHIC SHRUBS          TMAX      0.34
            4. SUMMERGREEN FORBS                       TMAX      0.47
            5. LEAFLESS XEROMORPHIC LARGE-SCRUB        TMAX      0.25
            6. DESERT-GRASSES                          MI        0.14
            7. XERIC CUSHION-HERBS                     TMAX      0.40
            8. XERIC DWARF-SHRUBS                      TMIN      0.31
            9. EPHEMERAL DESERT HERBS                  MI        0.14
           10. XERIC THALLOPHYTES                      TMAX      0.84

14. FLAGSTAFF/ARIZONA            USA                  35.20 -111.65
    18.0  -2.0  554.   80.   10.    70.  1.04              2070.
          * 1. BOREAL/MONTANE SHORT-NEEDLED TREES      TMAX      0.36
          * 2. TEMPERATE NEEDLE-TREES                  TMIN      0.19
          * 3. SUMMERGREEN BROAD-LEAVED TREES          MI        0.13
            4. BOREAL SUMMERGREEN NEEDLE-TREES         TMAX      0.40
            5. BOREAL BROAD-SUMMERGREEN TREES          MI        0.23
            6. TALL GRASSES                            MI        0.33
            7. BROAD-SUMMERGREEN SMALL TREES           MI        0.33
            8. SHORT SWARD-GRASSES                     MI        0.52
            9. SUMMERGREEN GIANT-SCRUB                 TMAX      0.30
           10. BROAD-SUMMERGREEN MESIC SHRUBS          TMAX      0.25
           11. TEMPERATE BROAD-EVERGREEN SHRUBS        TMIN      0.0
           12. DWARF-NEEDLE SMALL TREES                TMAX      0.0
           13. SHORT BUNCH-GRASSES                     TMIN      0.64
           14. NEEDLE-LEAVED EVERGREEN SHRUBS          TMAX      0.40
           15. SUMMERGREEN TUNDRA DWARF-SHRUBS         TMAX      0.0
           16. SUMMERGREEN FORBS                       TMAX      0.40
           17. TEMPERATE EVERGREEN DWARF-SHRUBS        MI        0.33
           18. TEMPERATE EVERGREEN FORBS               TMIN      0.13
           19. MAT-FORMING THALLOPHYTES                MI        0.23
           20. BROAD-SUMMERGREEN VINES                 MI        0.04
           21. XERIC THALLOPHYTES                      MI        0.93

15. PHOENIX(AIRPORT)/ARIZONA  USA                     33.50 -112.00
    31.8   9.8  197.   27.    1.    27.  0.18               339.
            1. HOT-DESERT EVERGREEN SHRUBS             TMIN      0.16
            2. ARBORESCENT STEM-SUCCULENTS             TMIN      0.08
            3. LEAFLESS XEROMORPHIC LARGE-SCRUB        TMAX      0.33
            4. BUSH STEM-SUCCULENTS                    TMAX      0.27
            5. XERIC ROSETTE-SHRUBS                    TMIN      0.25
          + 6. DESERT-GRASSES                          TMAX      0.24
            7. TYPICAL STEM-SUCCULENTS                 TMIN      0.12
            8. XERIC DWARF-SHRUBS                      TMAX      0.29
            9. EPHEMERAL DESERT HERBS                  TMAX      0.24
           10. SUCCULENT FORBS                         TMIN      0.19
           11. XERIC THALLOPHYTES                      TMAX      0.24
```

LOCATION TMAX TMIN PRCP PMAX PMIN PMTMAX MI	LAT ELEV	LONG
		7
16. TUCSON/ARIZONA USA	32.22	-110.97
30.0 10.0 293. 62. 5. 42. 0.29	739.	
1. XERIC SUMMERGREEN SHRUBS	MI	0.14
2. RAINGREEN THORN-SCRUB	TMIN	0.09
+ 3. SHORT BUNCH-GRASSES	TMAX	0.31
4. HOT-DESERT EVERGREEN SHRUBS	TMIN	0.17
5. EVERGREEN GIANT-SCRUB	TMIN	0.10
6. XERIC EVERGREEN TUFT-TREELETS	TMIN	0.09
7. ARBORESCENT STEM-SUCCULENTS	TMIN	0.09
8. LEAFLESS XEROMORPHIC LARGE-SCRUB	TMAX	0.40
9. BUSH STEM-SUCCULENTS	TMAX	0.33
+10. DESERT-GRASSES	TMAX	0.31
11. XERIC ROSETTE-SHRUBS	TMIN	0.26
12. SUMMERGREEN FORBS	TMAX	0.20
13. MEDITERRANEAN DWARF-SHRUBS	MULT.	0.18
14. XERIC CUSHION-SHRUBS	TMAX	0.17
15. TYPICAL STEM-SUCCULENTS	TMIN	0.13
16. RAINGREEN FORBS	TMIN	0.13
17. SCLEROPHYLLOUS GRASSES	TMIN	0.0
18. XERIC DWARF-SHRUBS	TMAX	0.36
19. EPHEMERAL DESERT HERBS	TMAX	0.31
20. SUCCULENT FORBS	TMIN	0.20
21. BROAD-RAINGREEN VINES	TMIN	0.0
22. XERIC CUSHION-HERBS	TMAX	0.0
23. XERIC THALLOPHYTES	TMAX	0.31
17. SANTA FE/NEW MEXICO USA	35.70	-106.95
20.0 -5.0 365. 63. 18. 56. 0.60	1929.	
* 1. BOREAL SUMMERGREEN NEEDLE-TREES	TMAX	0.20
2. SUMMERGREEN GIANT-SCRUB	MI	0.34
3. XERIC SUMMERGREEN SHRUBS	TMAX	0.29
4. SHORT SWARD-GRASSES	MI	0.17
5. DWARF-NEEDLE SMALL TREES	TMIN	0.0
+ 6. SHORT BUNCH-GRASSES	TMIN	0.61
7. NEEDLE-LEAVED EVERGREEN SHRUBS	MI	0.34
8. COLD-WINTER XEROMORPHIC SHRUBS	PMTMAX	0.07
9. SUMMERGREEN FORBS	TMAX	0.45
10. LEAFLESS XEROMORPHIC LARGE-SCRUB	TMAX	0.23
11. XERIC CUSHION-HERBS	TMAX	0.45
12. XERIC DWARF-SHRUBS	TMIN	0.30
13. XERIC THALLOPHYTES	TMAX	0.91

LOCATION TMAX TMIN PRCP PMAX PMIN PMTMAX MI	LAT ELEV	LONG
		8
18. COLORADO SPRINGS/COLORAD USA	38.83	-104.82
20.0 -1.0 363. 70. 6. 57. 0.64	1855.	
1. BOREAL/MONTANE SHORT-NEEDLED TREES	MI	0.06
2. BOREAL SUMMERGREEN NEEDLE-TREES	TMAX	0.20
3. SUMMERGREEN GIANT-SCRUB	TMAX	0.36
4. XERIC SUMMERGREEN SHRUBS	TMAX	0.29
5. SHORT SWARD-GRASSES	MI	0.22
6. DWARF-NEEDLE SMALL TREES	TMAX	0.17
7. TEMPERATE BROAD-EVERGREEN SHRUBS	TMIN	0.05
+ 8. SHORT BUNCH-GRASSES	TMIN	0.66
9. NEEDLE-LEAVED EVERGREEN SHRUBS	MI	0.38
10. COLD-WINTER XEROMORPHIC SHRUBS	PMTMAX	0.06
11. SUMMERGREEN FORBS	TMAX	0.45
12. LEAFLESS XEROMORPHIC LARGE-SCRUB	TMAX	0.23
13. XERIC CUSHION-HERBS	TMAX	0.45
14. XERIC DWARF-SHRUBS	MI	0.26
15. XERIC THALLOPHYTES	TMAX	0.91
19. CUSTER/SOUTH DAKOTA U.S.A.	43.77	-103.60
17.9 -6.1 455. 79. 9. 69. 0.94	1622.	
* 1. BOREAL/MONTANE SHORT-NEEDLED TREES	MI	0.36
+ 2. BOREAL SUMMERGREEN NEEDLE-TREES	TMAX	0.41
3. TALL GRASSES	MI	0.25
4. BROAD-SUMMERGREEN SMALL TREES	MI	0.25
+ 5. SHORT SWARD-GRASSES	MI	0.47
6. SUMMERGREEN GIANT-SCRUB	TMAX	0.30
7. BROAD-SUMMERGREEN MESIC SHRUBS	TMAX	0.24
8. XERIC SUMMERGREEN SHRUBS	MI	0.09
+ 9. SHORT BUNCH-GRASSES	TMIN	0.60
10. NEEDLE-LEAVED EVERGREEN SHRUBS	TMAX	0.40
11. SUMMERGREEN FORBS	TMAX	0.40
12. TEMPERATE EVERGREEN DWARF-SHRUBS	MI	0.25
13. MAT-FORMING THALLOPHYTES	MI	0.15
14. XERIC CUSHION-HERBS	MI	0.08
15. XERIC THALLOPHYTES	MULT.	1.00
20. OELRICHS/SOUTH DAKOTA U.S.A.	43.18	-103.23
23.3 -5.5 454. 80. 14. 64. 0.74	1017.	
1. TEMPERATE NEEDLE-TREES	TMIN	0.02
2. TALL GRASSES	MI	0.05
3. BROAD-SUMMERGREEN SMALL TREES	MI	0.05
4. SUMMERGREEN GIANT-SCRUB	MI	0.46
+ 5. XERIC SUMMERGREEN SHRUBS	MI	0.36
+ 6. SHORT SWARD-GRASSES	MI	0.33
7. BROAD-SUMMERGREEN MESIC SHRUBS	MI	0.05
+ 8. SHORT BUNCH-GRASSES	TMIN	0.61
9. NEEDLE-LEAVED EVERGREEN SHRUBS	MI	0.46
10. SUMMERGREEN FORBS	TMAX	0.47
11. LEAFLESS XEROMORPHIC LARGE-SCRUB	MI	0.08
12. TEMPERATE EVERGREEN DWARF-SHRUBS	TMAX	0.04
13. XERIC CUSHION-HERBS	TMAX	0.27
14. XERIC DWARF-SHRUBS	MI	0.08
15. XERIC THALLOPHYTES	TMAX	0.66

Appendix C. Predicted vegetation at the validation sites.

LOCATION TMAX TMIN PRCP PMAX PMIN PMTMAX MI	LAT LONG ELEV	9
21. OKLAHOMA CITY/OKLAHOMA USA	35.47	-97.53
27.3 2.6 802. 125. 27. 74. 0.93	383.	
1. SUMMERGREEN BROAD-LEAVED TREES	MI	0.03
2. SUB-MEDITERRANEAN NEEDLE-TREES	TMAX	0.05
+ 3. TALL GRASSES	MI	0.25
4. BROAD-SUMMERGREEN SMALL TREES	TMAX	0.14
5. TALL CANE-GRAMINOIDS	TMIN	0.06
6. TEMP. BROAD-EVERGREEN SMALL TREES	TMIN	0.05
7. SUMMERGREEN GIANT-SCRUB	TMAX	0.33
+ 8. SHORT SWARD-GRASSES	TMAX	0.26
9. DWARF-NEEDLE SMALL TREES	TMAX	0.23
10. TEMPERATE BROAD-EVERGREEN SHRUBS	TMAX	0.18
11. BROAD-SUMMERGREEN MESIC SHRUBS	TMAX	0.17
+12. SHORT BUNCH-GRASSES	TMAX	0.43
13. NEEDLE-LEAVED EVERGREEN SHRUBS	TMAX	0.31
14. SUMMERGREEN FORBS	TMAX	0.31
15. BUSH STEM-SUCCULENTS	TMIN	0.09
16. TEMPERATE EVERGREEN FORBS	MI	0.03
17. BROAD-WINTERGREEN EPIPHYTES	TMIN	0.09
18. MAT-FORMING THALLOPHYTES	TMAX	0.09
19. XERIC CUSHION-HERBS	MI	0.09
20. XERIC THALLOPHYTES	TMAX	0.43
22. BARTLESVILLE/OKLAHOMA U.S.A.	36.75	-96.00
27.8 2.2 921. 127. 39. 75. 1.08	218.	
* 1. SUMMERGREEN BROAD-LEAVED TREES	TMAX	0.15
2. TALL GRASSES	TMAX	0.29
3. BROAD-SUMMERGREEN SMALL TREES	TMAX	0.12
4. TALL CANE-GRAMINOIDS	TMIN	0.04
5. SUMMERGREEN GIANT-SCRUB	TMAX	0.31
6. SHORT SWARD-GRASSES	TMAX	0.24
7. DWARF-NEEDLE SMALL TREES	TMAX	0.18
8. TEMPERATE BROAD-EVERGREEN SHRUBS	TMAX	0.15
9. BROAD-SUMMERGREEN MESIC SHRUBS	TMAX	0.14
10. SHORT BUNCH-GRASSES	TMAX	0.41
11. NEEDLE-LEAVED EVERGREEN SHRUBS	TMAX	0.29
12. SUMMERGREEN FORBS	TMAX	0.29
13. TEMPERATE EVERGREEN FORBS	TMAX	0.09
14. BUSH STEM-SUCCULENTS	TMIN	0.07
15. BROAD-WINTERGREEN EPIPHYTES	TMIN	0.08
16. MAT-FORMING THALLOPHYTES	TMAX	0.07
17. BROAD-SUMMERGREEN VINES	MI	0.07
18. XERIC THALLOPHYTES	TMAX	0.41

LOCATION TMAX TMIN PRCP PMAX PMIN PMTMAX MI	LAT LONG ELEV	10
23. SOUTH BEND/INDIANA USA	41.68	-86.25
23.0 -3.5 904. 99. 40. 78. 1.42	221.	
* 1. SUMMERGREEN BROAD-LEAVED TREES	MI	0.37
* 2. TEMPERATE NEEDLE-TREES	TMAX	0.11
3. BOREAL BROAD-SUMMERGREEN TREES	TMAX	0.17
4. BROAD-SUMMERGREEN SMALL TREES	TMAX	0.37
5. TALL GRASSES	TMIN	0.34
6. SHORT SWARD-GRASSES	TMAX	0.48
7. SUMMERGREEN GIANT-SCRUB	TMAX	0.48
8. BROAD-SUMMERGREEN MESIC SHRUBS	TMAX	0.44
9. DWARF-NEEDLE SMALL TREES	TMIN	0.07
10. BROAD-ERICOID EVERGREEN SHRUBS	MI	0.01
11. SHORT BUNCH-GRASSES	TMIN	0.63
12. NEEDLE-LEAVED EVERGREEN SHRUBS	MI	0.41
13. SUMMERGREEN FORBS	TMAX	0.48
14. SUMMERGREEN FERNS	MI	0.23
15. TEMPERATE EVERGREEN FORBS	TMIN	0.07
16. TEMPERATE EVERGREEN DWARF-SHRUBS	TMAX	0.06
17. MAT-FORMING THALLOPHYTES	TMAX	0.28
18. BROAD-SUMMERGREEN VINES	TMIN	0.09
19. XERIC THALLOPHYTES	MI	0.41
24. COLUMBUS/OHIO USA	39.95	-83.00
23.0 -1.5 963. 102. 57. 95. 1.43	243.	
* 1. SUMMERGREEN BROAD-LEAVED TREES	MI	0.37
* 2. TEMPERATE NEEDLE-TREES	TMAX	0.11
3. BOREAL BROAD-SUMMERGREEN TREES	TMAX	0.17
4. TALL GRASSES	TMIN	0.39
5. BROAD-SUMMERGREEN SMALL TREES	TMAX	0.37
6. SHORT SWARD-GRASSES	TMAX	0.48
7. SUMMERGREEN GIANT-SCRUB	TMAX	0.48
8. BROAD-SUMMERGREEN MESIC SHRUBS	TMAX	0.44
9. DWARF-NEEDLE SMALL TREES	TMIN	0.15
10. TEMPERATE BROAD-EVERGREEN SHRUBS	TMIN	0.02
11. BROAD-ERICOID EVERGREEN SHRUBS	MI	0.02
12. SHORT BUNCH-GRASSES	TMIN	0.65
13. NEEDLE-LEAVED EVERGREEN SHRUBS	MI	0.40
14. SUMMERGREEN FORBS	TMAX	0.48
15. SUMMERGREEN FERNS	MI	0.23
16. TEMPERATE EVERGREEN FORBS	TMIN	0.15
17. TEMPERATE EVERGREEN DWARF-SHRUBS	TMAX	0.06
18. MAT-FORMING THALLOPHYTES	TMAX	0.28
19. BROAD-SUMMERGREEN VINES	TMIN	0.21
20. XERIC THALLOPHYTES	MI	0.40

```
                                                                    11
       LOCATION                                          LAT    LONG
    TMAX  TMIN  PRCP  PMAX  PMIN  PMTMAX  MI                ELEV

25. BAR HARBOR/MAINE            U.S.A.                  44.38  -68.20
     19.2  -5.0 1219.  125.   82.   84.  2.20                  9.
              * 1. SUMMERGREEN BROAD-LEAVED TREES        TMAX     0.28
              * 2. BOREAL/MONTANE SHORT-NEEDLED TREES    TMAX     0.25
                3. TEMPERATE NEEDLE-TREES                TMIN     0.05
                4. BOREAL BROAD-SUMMERGREEN TREES        TMAX     0.48
                5. BROAD-SUMMERGREEN SMALL TREES         TMAX     0.43
                6. TALL GRASSES                          TMIN     0.31
                7. SHORT SWARD-GRASSES                   TMIN     0.60
                8. SUMMERGREEN GIANT-SCRUB               TMAX     0.37
                9. BROAD-SUMMERGREEN MESIC SHRUBS        TMAX     0.32
               10. BROAD-ERICOID EVERGREEN SHRUBS        TMIN     0.25
               11. DWARF-NEEDLE SMALL TREES              TMIN     0.0
               12. SHORT BUNCH-GRASSES                   MI       0.40
               13. SUMMERGREEN FORBS                     TMAX     0.48
               14. SUMMERGREEN FERNS                     PMTMAX   0.40
               15. TEMPERATE EVERGREEN DWARF-SHRUBS      TMAX     0.27
               16. TEMPERATE EVERGREEN FORBS             TMIN     0.0
               17. MAT-FORMING THALLOPHYTES              TMAX     0.56
               18. BROAD-SUMMERGREEN VINES               TMIN     0.0

26. TRENTON/NEW JERSEY          USA                     40.22  -74.75
     24.0   0.0 1013.  110.   68.  107.  1.46                 17.
              * 1. SUMMERGREEN BROAD-LEAVED TREES        MI       0.39
                2. TEMPERATE NEEDLE-TREES                TMAX     0.0
                3. BOREAL BROAD-SUMMERGREEN TREES        TMAX     0.08
                4. TALL GRASSES                          TMIN     0.42
                5. BROAD-SUMMERGREEN SMALL TREES         TMAX     0.32
                6. SUMMERGREEN GIANT-SCRUB               TMAX     0.48
                7. SHORT SWARD-GRASSES                   TMAX     0.42
                8. BROAD-SUMMERGREEN MESIC SHRUBS        TMAX     0.38
                9. DWARF-NEEDLE SMALL TREES              TMIN     0.22
               10. TEMPERATE BROAD-EVERGREEN SHRUBS      TMIN     0.09
               11. BROAD-ERICOID EVERGREEN SHRUBS        MI       0.04
               12. SHORT BUNCH-GRASSES                   TMAX     0.62
               13. NEEDLE-LEAVED EVERGREEN SHRUBS        MI       0.37
               14. SUMMERGREEN FORBS                     TMAX     0.44
               15. SUMMERGREEN FERNS                     MI       0.25
               16. TEMPERATE EVERGREEN FORBS             TMIN     0.22
               17. TEMPERATE EVERGREEN DWARF-SHRUBS      TMAX     0.0
               18. BUSH STEM-SUCCULENTS                  TMIN     0.0
               19. BROAD-SUMMERGREEN VINES               TMIN     0.29
               20. MAT-FORMING THALLOPHYTES              TMAX     0.23
               21. BROAD-WINTERGREEN EPIPHYTES           TMIN     0.0
               22. XERIC THALLOPHYTES                    MI       0.37

                                                                    12
       LOCATION                                          LAT    LONG
    TMAX  TMIN  PRCP  PMAX  PMIN  PMTMAX  MI                ELEV

27. MT.MITCHELL (2 SSW)/N.CA  U.S.A.                    35.73  -82.28
     15.2  -4.0 1736.  173.   92.  156.  3.56               1989.
              * 1. BOREAL/MONTANE SHORT-NEEDLED TREES    TMAX     0.38
                2. SUMMERGREEN BROAD-LEAVED TREES        TMAX     0.01
                3. BOREAL BROAD-SUMMERGREEN TREES        TMAX     0.18
                4. TALL GRASSES                          TMIN     0.33
                5. BROAD-SUMMERGREEN SMALL TREES         TMAX     0.28
                6. SHORT SWARD-GRASSES                   TMIN     0.61
                7. SUMMERGREEN GIANT-SCRUB               TMAX     0.21
                8. BROAD-ERICOID EVERGREEN SHRUBS        TMAX     0.18
                9. BROAD-SUMMERGREEN MESIC SHRUBS        TMAX     0.08
               10. SUMMERGREEN TUNDRA DWARF-SHRUBS       TMAX     0.23
               11. TEMPERATE EVERGREEN DWARF-SHRUBS      TMIN     0.42
               12. SUMMERGREEN FERNS                     TMAX     0.34
               13. SUMMERGREEN FORBS                     TMAX     0.34
               14. TEMPERATE EVERGREEN FORBS             TMIN     0.04
               15. MAT-FORMING THALLOPHYTES              MI       0.82
               16. BROAD-SUMMERGREEN VINES               TMAX     0.01

28. CHARLESTON/SOUTH CAROLIN  USA                       32.80  -79.95
     27.5  10.6 1168.  200.   51.  200.  1.20                  3.
              * 1. WARM-TEMPERATE BROAD-EVERGREEN TREES  TMAX     0.25
              * 2. SUMMERGREEN BROAD-LEAVED TREES        TMAX     0.17
              * 3. HELIOPHILIC LONG-NEEDLED TREES        MI       0.17
                4. TEMPERATE BROAD-RAINFOREST TREES      TMAX     0.03
                5. TROPICAL EVERGREEN MICROPHYLL-TREES   TMIN     0.16
                6. TROPICAL EVERGREEN SCLEROPHYLL TREES  TMIN     0.12
                7. SWAMP SUMMERGREEN NEEDLE-TREES        TMAX     0.08
                8. TALL GRASSES                          TMAX     0.30
                9. TALL CANE-GRAMINOIDS                  TMAX     0.30
               10. TEMP. BROAD-EVERGREEN SMALL TREES     TMAX     0.22
               11. BROAD-SUMMERGREEN SMALL TREES         TMAX     0.13
               12. TROPICAL BROAD-EVERGREEN SMALL TREES  TMIN     0.13
               13. ARBORESCENT GRASSES                   TMAX     0.13
               14. SUMMERGREEN GIANT-SCRUB               TMAX     0.33
               15. SHORT SWARD-GRASSES                   TMAX     0.25
               16. TROPICAL BROAD-EVERGREEN SHRUBS       TMIN     0.19
               17. TEMPERATE BROAD-EVERGREEN SHRUBS      TMAX     0.17
               18. BROAD-SUMMERGREEN MESIC SHRUBS        TMAX     0.16
               19. PALMIFORM TUFT-TREELETS               TMIN     0.03
               20. DWARF-NEEDLE SMALL TREES              PMTMAX   0.0
               21. SHORT BUNCH-GRASSES                   TMAX     0.42
               22. NEEDLE-LEAVED EVERGREEN SHRUBS        TMAX     0.30
               23. EVERGREEN GIANT-SCRUB                 TMIN     0.13
               24. PALMIFORM MESIC ROSETTE-SHRUBS        TMIN     0.12
               25. SUMMERGREEN FORBS                     TMAX     0.30
               26. BUSH STEM-SUCCULENTS                  MI       0.25
               27. TEMPERATE EVERGREEN FORBS             TMAX     0.10
               28. SUMMERGREEN FERNS                     MI       0.08
               29. BROAD-EVERGREEN VINES                 TMAX     0.20
               30. BROAD-WINTERGREEN EPIPHYTES           TMAX     0.18
               31. NARROW-LEAVED EPIPHYTES               MI       0.17
               32. BROAD-SUMMERGREEN VINES               MI       0.17
               33. MAT-FORMING THALLOPHYTES              TMAX     0.08
               34. XERIC THALLOPHYTES                    TMAX     0.42
```

Appendix C. Predicted vegetation at the validation sites.

LOCATION TMAX TMIN PRCP PMAX PMIN PMTMAX MI		LAT LONG ELEV	13
29. SAVANNAH/GEORGIA USA		32.07	-81.08
27.5 10.4 1162. 180. 40. 180. 1.20		15.	
* 1.	WARM-TEMPERATE BROAD-EVERGREEN TREES	TMAX	0.25
* 2.	SUMMERGREEN BROAD-LEAVED TREES	TMAX	0.17
* 3.	HELIOPHILIC LONG-NEEDLED TREES	MI	0.17
4.	TROPICAL EVERGREEN MICROPHYLL-TREES	TMIN	0.15
5.	TROPICAL EVERGREEN SCLEROPHYLL TREES	TMIN	0.11
6.	SWAMP SUMMERGREEN NEEDLE-TREES	TMAX	0.08
7.	TALL GRASSES	TMAX	0.30
8.	TALL CANE-GRAMINOIDS	TMAX	0.30
9.	TEMP. BROAD-EVERGREEN SMALL TREES	TMAX	0.22
10.	BROAD-SUMMERGREEN SMALL TREES	TMAX	0.13
11.	ARBORESCENT GRASSES	TMAX	0.13
12.	TROPICAL BROAD-EVERGREEN SMALL TREES	TMIN	0.12
13.	SUMMERGREEN GIANT-SCRUB	TMAX	0.33
14.	SHORT SWARD-GRASSES	TMAX	0.25
15.	TROPICAL BROAD-EVERGREEN SHRUBS	TMIN	0.18
16.	TEMPERATE BROAD-EVERGREEN SHRUBS	TMAX	0.17
17.	BROAD-SUMMERGREEN MESIC SHRUBS	TMAX	0.16
18.	DWARF-NEEDLE SMALL TREES	PMTMAX	0.11
19.	PALMIFORM TUFT-TREELETS	TMIN	0.02
20.	SHORT BUNCH-GRASSES	TMAX	0.42
21.	NEEDLE-LEAVED EVERGREEN SHRUBS	TMAX	0.30
22.	EVERGREEN GIANT-SCRUB	TMIN	0.12
23.	PALMIFORM MESIC ROSETTE-SHRUBS	TMIN	0.11
24.	SUMMERGREEN FORBS	TMAX	0.30
25.	BUSH STEM-SUCCULENTS	MI	0.25
26.	RAINGREEN FORBS	TMIN	0.15
27.	TEMPERATE EVERGREEN FORBS	TMAX	0.10
28.	SUMMERGREEN FERNS	MI	0.08
29.	XERIC ROSETTE-SHRUBS	MI	0.00
30.	BROAD-EVERGREEN VINES	TMAX	0.20
31.	BROAD-WINTERGREEN EPIPHYTES	TMAX	0.18
32.	NARROW-LEAVED EPIPHYTES	MI	0.17
33.	BROAD-SUMMERGREEN VINES	MI	0.17
34.	MAT-FORMING THALLOPHYTES	TMAX	0.08
35.	SUCCULENT FORBS	MI	0.00
36.	XERIC THALLOPHYTES	TMAX	0.42

LOCATION TMAX TMIN PRCP PMAX PMIN PMTMAX MI		LAT LONG ELEV	14
30. BIRMINGHAM/ALABAMA USA		33.52	-86.82
26.5 7.5 1353. 155. 69. 145. 1.57		182.	
* 1.	HELIOPHILIC LONG-NEEDLED TREES	TMIN	0.24
* 2.	SUMMERGREEN BROAD-LEAVED TREES	TMAX	0.23
* 3.	TEMPERATE BROAD-RAINFOREST TREES	TMAX	0.10
4.	WARM-TEMPERATE BROAD-EVERGREEN TREES	TMIN	0.04
5.	SWAMP SUMMERGREEN NEEDLE-TREES	TMAX	0.25
6.	TROPICAL EVERGREEN MICROPHYLL-TREES	TMIN	0.02
7.	TALL GRASSES	TMAX	0.34
8.	TEMP. BROAD-EVERGREEN SMALL TREES	TMAX	0.27
9.	TALL CANE-GRAMINOIDS	TMIN	0.22
10.	BROAD-SUMMERGREEN SMALL TREES	TMAX	0.18
11.	ARBORESCENT GRASSES	TMIN	0.17
12.	SUMMERGREEN GIANT-SCRUB	TMAX	0.37
13.	SHORT SWARD-GRASSES	TMAX	0.30
14.	DWARF-NEEDLE SMALL TREES	TMAX	0.29
15.	TEMPERATE BROAD-EVERGREEN SHRUBS	TMAX	0.23
16.	BROAD-SUMMERGREEN MESIC SHRUBS	TMAX	0.22
17.	TROPICAL BROAD-EVERGREEN SHRUBS	TMIN	0.06
18.	SHORT BUNCH-GRASSES	TMAX	0.47
19.	NEEDLE-LEAVED EVERGREEN SHRUBS	MI	0.27
20.	SUMMERGREEN FORBS	TMAX	0.34
21.	SUMMERGREEN FERNS	TMAX	0.17
22.	TEMPERATE EVERGREEN FORBS	TMAX	0.14
23.	BROAD-WINTERGREEN EPIPHYTES	TMAX	0.25
24.	BROAD-SUMMERGREEN VINES	TMAX	0.23
25.	BROAD-EVERGREEN VINES	TMIN	0.13
26.	MAT-FORMING THALLOPHYTES	TMAX	0.12
27.	XERIC THALLOPHYTES	MI	0.27

```
                                                                   15
     LOCATION                                              LAT     LONG
  TMAX  TMIN  PRCP  PMAX  PMIN  PMTMAX  MI                     ELEV

31. XILITLA                      MEXICO                    21.33  -98.57
     24.0  15.3 2537.  760.   38.  218.  2.58                1035.
         * 1. WARM-TEMPERATE BROAD-EVERGREEN TREES        TMAX     0.40
         * 2. SUMMERGREEN BROAD-LEAVED TREES               TMAX     0.40
         * 3. TROPICAL LINEAR-LEAVED TREES                 TMAX     0.31
           4. TROPICAL MONTANE RAINFOREST TREES            TMAX     0.0
           5. TROPICAL EVERGREEN MICROPHYLL-TREES          TMIN     0.36
           6. SWAMP SUMMERGREEN NEEDLE-TREES               TMAX     0.33
           7. PALMIFORM TUFT-TREES                         TMIN     0.23
           8. TALL GRASSES                                 TMAX     0.46
           9. TROPICAL BROAD-EVERGREEN SMALL TREES         TMIN     0.36
          10. BROAD-SUMMERGREEN SMALL TREES                TMAX     0.32
          11. ARBORESCENT GRASSES                          TMAX     0.30
          12. TROPICAL BROAD-EVERGREEN LIANAS              TMIN     0.26
          13. TALL CANE-GRAMINOIDS                         MI       0.18
          14. SUMMERGREEN GIANT-SCRUB                      TMAX     0.48
          15. SHORT SWARD-GRASSES                          TMAX     0.46
          16. TEMPERATE BROAD-EVERGREEN SHRUBS             TMAX     0.40
          17. BROAD-SUMMERGREEN MESIC SHRUBS               TMAX     0.38
          18. TROPICAL BROAD-EVERGREEN SHRUBS              TMAX     0.36
          19. PALMIFORM TUFT-TREELETS                      TMIN     0.26
          20. PALMIFORM MESIC ROSETTE-SHRUBS               TMIN     0.33
          21. SHORT BUNCH-GRASSES                          MI       0.18
          22. TROPICAL EVERGREEN FORBS                     TMAX     0.46
          23. RAINGREEN FORBS                              TMIN     0.36
          24. EVERGREEN FERNS                              TMAX     0.30
          25. TEMPERATE EVERGREEN FORBS                    TMAX     0.25
          26. TEMPERATE EVERGREEN DWARF-SHRUBS             TMAX     0.0
          27. BROAD-WINTERGREEN EPIPHYTES                  TMAX     0.43
          28. BROAD-EVERGREEN VINES                        TMAX     0.36
          29. NARROW-LEAVED EPIPHYTES                      TMAX     0.33
          30. TROPICAL BROAD-EVERGREEN EPIPHYTES           TMIN     0.26
          31. MAT-FORMING THALLOPHYTES                     TMAX     0.25

32. MAZATLAN                     MEXICO                    23.22 -106.42
     27.5  19.8   850.  270.    0.  141.  0.70                 78.
         * 1. TROPICAL EVERGREEN SCLEROPHYLL TREES         MI       0.29
           2. XERIC RAINGREEN TREES                        TMAX     0.22
           3. BROAD-RAINGREEN SMALL TREES                  TMAX     0.38
         + 4. RAINGREEN THORN-SCRUB                        MI       0.44
         + 5. XERIC EVERGREEN TUFT-TREELETS                MI       0.44
         + 6. SHORT BUNCH-GRASSES                          TMAX     0.42
           7. EVERGREEN GIANT-SCRUB                        TMAX     0.38
           8. BUSH STEM-SUCCULENTS                         TMAX     0.42
           9. RAINGREEN FORBS                              TMAX     0.30
          10. LEAF-SUCCULENT EVERGREEN SHRUBS              MI       0.16
          11. BROAD-RAINGREEN VINES                        TMAX     0.38
          12. BROAD-WINTERGREEN EPIPHYTES                  MI       0.14
          13. XERIC CUSHION-HERBS                          TMAX     0.10
          14. XERIC THALLOPHYTES                           TMAX     0.42
```

```
                                                                   16
     LOCATION                                              LAT     LONG
  TMAX  TMIN  PRCP  PMAX  PMIN  PMTMAX  MI                     ELEV

33. QUIBDO                       COLOMBIA                   5.70  -76.67
     27.5  27.010545. 1115.  540.  950.  6.10                  43.
         * 1. TROPICAL RAINFOREST TREES                    TMAX     0.25
           2. PALMIFORM TUFT-TREES                         TMAX     0.48
           3. TROPICAL EVERGREEN MICROPHYLL-TREES          TMAX     0.26
           4. TROPICAL BROAD-EVERGREEN SMALL TREES         TMAX     0.26
           5. TROPICAL BROAD-EVERGREEN LIANAS              TMAX     0.21
           6. PALMIFORM TUFT-TREELETS                      TMAX     0.36
           7. TROPICAL BROAD-EVERGREEN SHRUBS              TMAX     0.20
           8. PALMIFORM MESIC ROSETTE-SHRUBS               TMAX     0.38
           9. TROPICAL EVERGREEN FORBS                     TMAX     0.27
          10. EVERGREEN FERNS                              TMAX     0.13
          11. BROAD-EVERGREEN VINES                        TMAX     0.20
          12. NARROW-LEAVED EPIPHYTES                      TMAX     0.19
          13. BROAD-WINTERGREEN EPIPHYTES                  TMAX     0.18
          14. TROPICAL BROAD-EVERGREEN EPIPHYTES           TMAX     0.14
          15. MAT-FORMING THALLOPHYTES                     TMAX     0.09

34. BARINAS                      VENEZUEL                   8.63  -70.20
     27.0  26.0 1863.  350.   11.  305.  1.14                 180.
         * 1. TROPICAL RAINFOREST TREES                    MI       0.21
           2. MONSOON BROAD-RAINGREEN TREES                MI       0.21
           3. TROPICAL EVERGREEN SCLEROPHYLL TREES         TMAX     0.40
           4. PALMIFORM TUFT-TREES                         MI       0.39
           5. TROPICAL EVERGREEN MICROPHYLL-TREES          TMAX     0.29
           6. BROAD-RAINGREEN SMALL TREES                  TMAX     0.40
           7. TALL GRASSES                                 TMAX     0.32
           8. TALL CANE-GRAMINOIDS                         TMAX     0.32
           9. TROPICAL BROAD-EVERGREEN DWARF-TREES         MI       0.30
          10. TROPICAL BROAD-EVERGREEN SMALL TREES         MI       0.21
          11. TROPICAL BROAD-EVERGREEN LIANAS              MI       0.13
          12. PALMIFORM TUFT-TREELETS                      TMAX     0.38
          13. TROPICAL BROAD-EVERGREEN SHRUBS              TMAX     0.23
          14. SHORT BUNCH-GRASSES                          TMAX     0.45
          15. EVERGREEN GIANT-SCRUB                        TMAX     0.42
          16. PALMIFORM MESIC ROSETTE-SHRUBS               MI       0.21
          17. RAINGREEN FORBS                              TMAX     0.32
          18. TROPICAL EVERGREEN FORBS                     MI       0.21
          19. BROAD-RAINGREEN VINES                        TMAX     0.40
          20. BROAD-EVERGREEN VINES                        MI       0.21
          21. BROAD-WINTERGREEN EPIPHYTES                  TMAX     0.21
          22. NARROW-LEAVED EPIPHYTES                      MI       0.13
          23. MAT-FORMING THALLOPHYTES                     TMAX     0.10
          24. TROPICAL BROAD-EVERGREEN EPIPHYTES           MI       0.04
          25. XERIC THALLOPHYTES                           TMAX     0.45
```

17

LOCATION TMAX TMIN PRCP PMAX PMIN PMTMAX MI		LAT LONG ELEV	
35. OBIDOS BRASIL		-1.92	-55.52
27.0 25.7 1680. 315. 20. 60. 1.12		24.	
* 1.	TROPICAL RAINFOREST TREES	MI	0.20
2.	MONSOON BROAD-RAINGREEN TREES	MI	0.20
3.	TROPICAL EVERGREEN SCLEROPHYLL TREES	TMAX	0.40
4.	PALMIFORM TUFT-TREES	MI	0.38
5.	TROPICAL EVERGREEN MICROPHYLL-TREES	TMAX	0.29
6.	BROAD-RAINGREEN SMALL TREES	TMAX	0.40
7.	TALL GRASSES	TMAX	0.32
8.	TALL CANE-GRAMINOIDS	TMAX	0.32
9.	TROPICAL BROAD-EVERGREEN SMALL TREES	MI	0.20
10.	TROPICAL BROAD-EVERGREEN DWARF-TREES	PMTMAX	0.17
11.	TROPICAL BROAD-EVERGREEN LIANAS	MI	0.11
12.	PALMIFORM TUFT-TREELETS	MI	0.38
13.	TROPICAL BROAD-EVERGREEN SHRUBS	TMAX	0.23
14.	SHORT BUNCH-GRASSES	TMAX	0.45
15.	EVERGREEN GIANT-SCRUB	TMAX	0.42
16.	PALMIFORM MESIC ROSETTE-SHRUBS	MI	0.20
17.	RAINGREEN FORBS	TMAX	0.32
18.	TROPICAL EVERGREEN FORBS	MI	0.20
19.	BROAD-RAINGREEN VINES	TMAX	0.40
20.	BROAD-WINTERGREEN EPIPHYTES	TMAX	0.21
21.	BROAD-EVERGREEN VINES	MI	0.20
22.	NARROW-LEAVED EPIPHYTES	MI	0.11
23.	MAT-FORMING THALLOPHYTES	TMAX	0.10
24.	TROPICAL BROAD-EVERGREEN EPIPHYTES	MI	0.02
25.	XERIC THALLOPHYTES	TMAX	0.45
36. QUIXERAMOBIM BRASIL		-5.20	-39.28
28.7 27.0 747. 192. 0. 70. 0.43		199.	
* 1.	XERIC RAINGREEN TREES	MI	0.19
2.	BROAD-RAINGREEN SMALL TREES	TMAX	0.32
+ 3.	RAINGREEN THORN-SCRUB	TMAX	0.45
+ 4.	ARBORESCENT STEM-SUCCULENTS	TMAX	0.37
+ 5.	XERIC EVERGREEN TUFT-TREELETS	TMAX	0.37
6.	SHORT BUNCH-GRASSES	TMAX	0.37
7.	EVERGREEN GIANT-SCRUB	TMAX	0.28
8.	XERIC ROSETTE-SHRUBS	TMAX	0.43
9.	TYPICAL STEM-SUCCULENTS	MI	0.41
10.	BUSH STEM-SUCCULENTS	TMAX	0.38
11.	RAINGREEN FORBS	TMAX	0.25
12.	DESERT-GRASSES	MI	0.17
13.	SUCCULENT FORBS	TMAX	0.49
14.	BROAD-RAINGREEN VINES	TMAX	0.32
15.	XERIC THALLOPHYTES	TMAX	0.37

18

LOCATION TMAX TMIN PRCP PMAX PMIN PMTMAX MI		LAT LONG ELEV	
37. BELO HORIZONTE BRASIL		-19.92	-43.93
22.4 16.0 1534. 330. 9. 303. 1.72		857.	
* 1.	TROPICAL LINEAR-LEAVED TREES	MI	0.42
* 2.	MONTANE BROAD-RAINGREEN TREES	TMAX	0.26
* 3.	TROPICAL MONTANE RAINFOREST TREES	TMAX	0.18
4.	TROPICAL EVERGREEN MICROPHYLL-TREES	TMIN	0.39
5.	TROPICAL EVERGREEN SCLEROPHYLL TREES	TMIN	0.36
6.	PALMIFORM TUFT-TREES	TMIN	0.26
7.	TALL CANE-GRAMINOIDS	TMAX	0.51
8.	TALL GRASSES	TMAX	0.51
9.	TROPICAL BROAD-EVERGREEN SMALL TREES	TMIN	0.40
10.	ARBORESCENT GRASSES	TMAX	0.38
11.	BROAD-RAINGREEN SMALL TREES	TMIN	0.36
12.	TROPICAL BROAD-EVERGREEN LIANAS	TMIN	0.30
13.	TROPICAL BROAD-EVERGREEN DWARF-TREES	TMIN	0.07
14.	SHORT SWARD-GRASSES	TMAX	0.52
15.	TEMPERATE BROAD-EVERGREEN SHRUBS	TMAX	0.49
16.	TROPICAL BROAD-EVERGREEN SHRUBS	TMIN	0.42
17.	PALMIFORM TUFT-TREELETS	TMIN	0.30
18.	SHORT BUNCH-GRASSES	TMAX	0.72
19.	PALMIFORM MESIC ROSETTE-SHRUBS	TMIN	0.36
20.	EVERGREEN GIANT-SCRUB	MI	0.16
21.	TROPICAL EVERGREEN FORBS	MI	0.48
22.	RAINGREEN FORBS	TMIN	0.39
23.	TEMPERATE EVERGREEN DWARF-SHRUBS	TMAX	0.09
24.	BROAD-WINTERGREEN EPIPHYTES	TMAX	0.46
25.	BROAD-EVERGREEN VINES	TMAX	0.44
26.	MAT-FORMING THALLOPHYTES	TMAX	0.31
27.	BROAD-RAINGREEN VINES	TMIN	0.30
28.	XERIC THALLOPHYTES	MI	0.16

LOCATION TMAX TMIN PRCP PMAX PMIN PMTMAX MI		LAT LONG ELEV	
38. SAO PAOLO	BRASIL	-23.53	-46.62
21.0 14.0 1270. 228. 20. 186. 1.63		740.	
* 1.	MONTANE BROAD-RAINGREEN TREES	TMAX	0.40
* 2.	TROPICAL LINEAR-LEAVED TREES	MI	0.39
* 3.	TROPICAL MONTANE RAINFOREST TREES	TMAX	0.33
* 4.	TEMPERATE NEEDLE-TREES	TMAX	0.33
* 5.	TEMPERATE RAINFOREST NEEDLE-TREES	MI	0.14
* 6.	WARM-TEMPERATE BROAD-EVERGREEN TREES	TMAX	0.10
7.	HELIOPHILIC LONG-NEEDLED TREES	TMAX	0.08
8.	TROPICAL EVERGREEN MICROPHYLL-TREES	TMIN	0.30
9.	TROPICAL EVERGREEN SCLEROPHYLL TREES	TMIN	0.27
10.	PALMIFORM TUFT-TREES	TMIN	0.16
11.	TALL GRASSES	TMAX	0.48
12.	TEMP. BROAD-EVERGREEN SMALL TREES	TMAX	0.45
13.	ARBORESCENT GRASSES	TMAX	0.45
14.	TALL CANE-GRAMINOIDS	TMIN	0.45
15.	TROPICAL BROAD-EVERGREEN SMALL TREES	TMIN	0.30
16.	BROAD-RAINGREEN SMALL TREES	TMIN	0.27
17.	TROPICAL BROAD-EVERGREEN LIANAS	TMIN	0.20
18.	SHORT SWARD-GRASSES	TMAX	0.61
19.	TEMPERATE BROAD-EVERGREEN SHRUBS	TMAX	0.40
20.	TROPICAL BROAD-EVERGREEN SHRUBS	TMIN	0.33
21.	PALMIFORM TUFT-TREELETS	TMIN	0.20
22.	SHORT BUNCH-GRASSES	TMIN	0.82
23.	PALMIFORM MESIC ROSETTE-SHRUBS	TMIN	0.27
24.	NEEDLE-LEAVED EVERGREEN SHRUBS	MI	0.23
25.	EVERGREEN GIANT-SCRUB	TMAX	0.08
26.	SUMMERGREEN FORBS	TMAX	0.48
27.	TROPICAL EVERGREEN FORBS	TMIN	0.43
28.	TEMPERATE EVERGREEN FORBS	TMAX	0.39
29.	RAINGREEN FORBS	TMIN	0.30
30.	TEMPERATE EVERGREEN DWARF-SHRUBS	TMAX	0.17
31.	MAT-FORMING THALLOPHYTES	TMAX	0.39
32.	BROAD-EVERGREEN VINES	TMIN	0.38
33.	NARROW-LEAVED EPIPHYTES	TMIN	0.36
34.	BROAD-WINTERGREEN EPIPHYTES	TMAX	0.36
35.	BROAD-RAINGREEN VINES	TMIN	0.20
36.	TROPICAL BROAD-EVERGREEN EPIPHYTES	TMIN	0.20
37.	XERIC THALLOPHYTES	MI	0.23

LOCATION TMAX TMIN PRCP PMAX PMIN PMTMAX MI		LAT LONG ELEV	
39. VILLA NOUGUES/TUCUMAN	ARGENTIN	-27.00	-65.70
18.7 9.2 1437. 250. 22. 250. 2.22		1388.	
* 1.	TEMPERATE RAINFOREST NEEDLE-TREES	MI	0.41
* 2.	SUMMERGREEN BROAD-LEAVED TREES	TMAX	0.25
* 3.	TROPICAL LINEAR-LEAVED TREES	TMIN	0.21
* 4.	TEMPERATE NEEDLE-TREES	PMTMAX	0.20
* 5.	MONTANE BROAD-RAINGREEN TREES	TMIN	0.09
* 6.	TROPICAL MONTANE RAINFOREST TREES	TMIN	0.08
7.	TROPICAL EVERGREEN MICROPHYLL-TREES	TMIN	0.10
8.	TROPICAL EVERGREEN SCLEROPHYLL TREES	TMIN	0.05
9.	TALL GRASSES	TMAX	0.47
10.	BROAD-SUMMERGREEN SMALL TREES	TMAX	0.41
11.	TEMP. BROAD-EVERGREEN SMALL TREES	TMAX	0.36
12.	TALL CANE-GRAMINOIDS	TMIN	0.28
13.	ARBORESCENT GRASSES	TMIN	0.25
14.	TROPICAL BROAD-EVERGREEN SMALL TREES	TMIN	0.06
15.	BROAD-RAINGREEN SMALL TREES	TMIN	0.05
16.	SHORT SWARD-GRASSES	TMAX	0.73
17.	BROAD-ERICOID EVERGREEN SHRUBS	MI	0.41
18.	SUMMERGREEN GIANT-SCRUB	TMAX	0.36
19.	BROAD-SUMMERGREEN MESIC SHRUBS	TMAX	0.29
20.	TEMPERATE BROAD-EVERGREEN SHRUBS	TMAX	0.25
21.	TROPICAL BROAD-EVERGREEN SHRUBS	TMIN	0.13
22.	SHORT BUNCH-GRASSES	MI	0.39
23.	PALMIFORM MESIC ROSETTE-SHRUBS	TMIN	0.05
24.	TEMPERATE EVERGREEN FORBS	TMAX	0.60
25.	SUMMERGREEN FORBS	TMAX	0.47
26.	SUMMERGREEN FERNS	TMAX	0.47
27.	MARITIME HEATH DWARF-SHRUBS	TMAX	0.39
28.	TEMPERATE EVERGREEN DWARF-SHRUBS	TMAX	0.29
29.	TROPICAL EVERGREEN FORBS	TMIN	0.26
30.	RAINGREEN FORBS	TMIN	0.10
31.	MAT-FORMING THALLOPHYTES	TMAX	0.60
32.	BROAD-SUMMERGREEN VINES	TMAX	0.25
33.	BROAD-EVERGREEN VINES	TMIN	0.20
34.	BROAD-WINTERGREEN EPIPHYTES	TMAX	0.19
35.	NARROW-LEAVED EPIPHYTES	TMIN	0.17
40. ISLA VICTORIA/LAGO NAHUE	ARGENTIN	-41.00	-71.58
14.7 3.4 1644. 265. 35. 35. 3.22		900.	
* 1.	TEMPERATE RAINFOREST NEEDLE-TREES	TMAX	0.21
2.	BOREAL BROAD-SUMMERGREEN TREES	TMAX	0.14
3.	BROAD-SUMMERGREEN SMALL TREES	TMAX	0.25
4.	TALL GRASSES	PMTMAX	0.14
5.	TEMP. BROAD-EVERGREEN SMALL TREES	TMIN	0.11
6.	SHORT SWARD-GRASSES	PMTMAX	0.29
7.	SUMMERGREEN GIANT-SCRUB	TMAX	0.18
8.	TALL TUSSOCK-GRASSES	PMTMAX	0.14
9.	BROAD-ERICOID EVERGREEN SHRUBS	TMAX	0.14
10.	BROAD-SUMMERGREEN MESIC SHRUBS	TMAX	0.05
11.	TEMPERATE EVERGREEN FORBS	TMIN	0.37
12.	SUMMERGREEN FORBS	TMAX	0.32
13.	TEMPERATE EVERGREEN DWARF-SHRUBS	PMTMAX	0.29
14.	MARITIME HEATH DWARF-SHRUBS	TMAX	0.25
15.	MAT-FORMING THALLOPHYTES	MI	0.80

Appendix C. Predicted vegetation at the validation sites.

LOCATION TMAX TMIN PRCP PMAX PMIN PMTMAX MI	LAT LONG ELEV	
41. BARILOCHE, SAN CARLOS DE ARGENTIN	-41.13	-71.25
14.9 3.0 1081. 181. 27. 27. 2.16	825.	
* 1. BOREAL/MONTANE SHORT-NEEDLED TREES	TMAX	0.35
2. TEMPERATE RAINFOREST NEEDLE-TREES	PMTMAX	0.07
3. SUB-MEDITERRANEAN NEEDLE-TREES	TMAX	0.06
4. BROAD-SUMMERGREEN SMALL TREES	PMTMAX	0.26
5. TEMP. BROAD-EVERGREEN SMALL TREES	TMIN	0.08
6. TALL CANE-GRAMINOIDS	TMIN	0.07
7. SUMMERGREEN GIANT-SCRUB	TMAX	0.19
8. SHORT SWARD-GRASSES	PMTMAX	0.07
9. BROAD-SUMMERGREEN MESIC SHRUBS	TMAX	0.06
10. SHORT BUNCH-GRASSES	MI	0.43
11. SUMMERGREEN TUNDRA DWARF-SHRUBS	TMAX	0.26
12. SUMMERGREEN FORBS	TMAX	0.33
13. TEMPERATE EVERGREEN FORBS	PMTMAX	0.26
14. MARITIME HEATH DWARF-SHRUBS	PMTMAX	0.26
15. TEMPERATE EVERGREEN DWARF-SHRUBS	PMTMAX	0.07
16. MAT-FORMING THALLOPHYTES	MI	0.69
17. RAINGREEN COLD-DESERT HERBS	TMAX	0.01
42. VALDIVIA CHILE	-39.80	-73.23
15.5 8.2 2489. 420. 62. 66. 4.18	5.	
* 1. TEMPERATE RAINFOREST NEEDLE-TREES	TMAX	0.27
* 2. TEMPERATE BROAD-RAINFOREST TREES	PMTMAX	0.10
3. TROPICAL LINEAR-LEAVED TREES	TMAX	0.04
4. TROPICAL MONTANE RAINFOREST TREES	TMIN	0.01
5. TROPICAL EVERGREEN MICROPHYLL-TREES	TMAX	0.03
6. TALL GRASSES	TMAX	0.35
7. TEMP. BROAD-EVERGREEN SMALL TREES	TMAX	0.23
8. ARBORESCENT GRASSES	TMIN	0.20
9. TROPICAL BROAD-EVERGREEN SMALL TREES	TMIN	0.01
10. SHORT SWARD-GRASSES	PMTMAX	0.62
11. TALL TUSSOCK-GRASSES	TMAX	0.48
12. BROAD-ERICOID EVERGREEN SHRUBS	TMAX	0.21
13. TROPICAL BROAD-EVERGREEN SHRUBS	TMIN	0.09
14. TEMPERATE BROAD-EVERGREEN SHRUBS	TMAX	0.03
15. MESIC EVERGREEN CUSHION-SHRUBS	TMAX	0.17
16. PALMIFORM MESIC ROSETTE-SHRUBS	TMIN	0.01
17. TEMPERATE EVERGREEN FORBS	TMIN	0.57
18. TEMPERATE EVERGREEN DWARF-SHRUBS	TMAX	0.55
19. SUMMERGREEN FORBS	TMAX	0.35
20. MARITIME HEATH DWARF-SHRUBS	TMAX	0.32
21. TROPICAL EVERGREEN FORBS	TMIN	0.22
22. EVERGREEN FERNS	TMIN	0.14
23. MAT-FORMING THALLOPHYTES	MI	0.85
24. BROAD-EVERGREEN VINES	TMIN	0.16
25. NARROW-LEAVED EPIPHYTES	TMIN	0.13

LOCATION TMAX TMIN PRCP PMAX PMIN PMTMAX MI	LAT LONG ELEV	
43. USHUAIA CHILE	-54.78	-68.33
10.0 1.6 556. 65. 40. 42. 1.23	34.	
1. TALL GRASSES	TMAX	0.0
2. SHORT SWARD-GRASSES	PMTMAX	0.41
3. TALL TUSSOCK-GRASSES	TMIN	0.24
4. SHORT BUNCH-GRASSES	TMAX	0.67
5. SUMMERGREEN TUNDRA DWARF-SHRUBS	TMAX	0.33
6. NEEDLE-LEAVED TREELINE KRUMMHOLZ	TMAX	0.0
7. NEEDLE-LEAVED EVERGREEN SHRUBS	TMAX	0.0
8. TEMPERATE EVERGREEN DWARF-SHRUBS	TMAX	0.33
9. TEMPERATE EVERGREEN FORBS	MI	0.27
10. BUSH STEM-SUCCULENTS	TMAX	0.0
11. SUMMERGREEN FORBS	TMAX	0.0
12. MAT-FORMING THALLOPHYTES	MI	0.35
13. XERIC THALLOPHYTES	MI	0.63
44. THINGVELLIR ISLAND	64.25	-21.12
11.5 -2.4 1340. 162. 71. 84. 2.97	103.	
1. BOREAL/MONTANE SHORT-NEEDLED TREES	TMAX	0.05
2. TALL GRASSES	TMAX	0.13
3. BROAD-SUMMERGREEN SMALL TREES	TMAX	0.04
4. SHORT SWARD-GRASSES	TMAX	0.57
5. NEEDLE-LEAVED TREELINE KRUMMHOLZ	TMAX	0.50
6. SUMMERGREEN TUNDRA DWARF-SHRUBS	TMAX	0.48
7. SHORT BUNCH-GRASSES	MI	0.01
8. TEMPERATE EVERGREEN DWARF-SHRUBS	TMIN	0.46
9. SUMMERGREEN FERNS	TMAX	0.13
10. SUMMERGREEN FORBS	TMAX	0.13
11. TEMPERATE EVERGREEN FORBS	TMIN	0.11
12. MAT-FORMING THALLOPHYTES	MI	0.76
13. SEASONAL COLD-DESERT HERBS	TMAX	0.30
45. ABISKO SVERIGE	68.33	18.85
11.0 -11.0 267. 42. 11. 42. 0.74	388.	
1. BOREAL/MONTANE SHORT-NEEDLED TREES	TMAX	0.0
2. BROAD-SUMMERGREEN SMALL TREES	TMAX	0.0
+ 3. SHORT SWARD-GRASSES	MI	0.32
+ 4. SHORT BUNCH-GRASSES	TMIN	0.55
5. NEEDLE-LEAVED TREELINE KRUMMHOLZ	MI	0.19
6. NEEDLE-LEAVED EVERGREEN SHRUBS	TMAX	0.08
7. SUMMERGREEN TUNDRA DWARF-SHRUBS	MI	0.05
8. SUMMERGREEN FORBS	TMAX	0.08
9. TEMPERATE EVERGREEN DWARF-SHRUBS	MI	0.05
10. XERIC CUSHION-HERBS	MI	0.36
11. SEASONAL COLD-DESERT HERBS	TMAX	0.31
12. XERIC THALLOPHYTES	MULT.	1.00

```
       LOCATION                                         LAT     LONG
   TMAX  TMIN  PRCP  PMAX  PMIN  PMTMAX  MI                ELEV

46. LUND(DALBY SOEDERSKOG)/S   SVERIGE                  55.70   13.18
      16.0  -0.7  616.   75.   32.   67.  1.12                65.
            * 1. BOREAL/MONTANE SHORT-NEEDLED TREES     TMAX    0.45
            * 2. TEMPERATE NEEDLE-TREES                 TMAX    0.11
              3. SUMMERGREEN BROAD-LEAVED TREES         TMAX    0.07
              4. BOREAL SUMMERGREEN NEEDLE-TREES        TMAX    0.40
              5. SUB-MEDITERRANEAN NEEDLE-TREES         TMIN    0.02
              6. BOREAL BROAD-SUMMERGREEN TREES         TMAX    0.25
              7. TALL GRASSES                           TMAX    0.33
              8. BROAD-SUMMERGREEN SMALL TREES          TMAX    0.28
              9. SHORT SWARD-GRASSES                    MI      0.56
             10. SUMMERGREEN GIANT-SCRUB                TMAX    0.22
             11. BROAD-SUMMERGREEN MESIC SHRUBS         TMAX    0.13
             12. TEMPERATE BROAD-EVERGREEN SHRUBS       TMIN    0.06
             13. TALL TUSSOCK-GRASSES                   TMIN    0.03
             14. SHORT BUNCH-GRASSES                    TMIN    0.66
             15. NEEDLE-LEAVED EVERGREEN SHRUBS         TMAX    0.33
             16. SUMMERGREEN TUNDRA DWARF-SHRUBS        TMAX    0.17
             17. SHORT TUSSOCK-GRASSES                  TMIN    0.03
             18. TEMPERATE EVERGREEN DWARF-SHRUBS       MI      0.38
             19. SUMMERGREEN FORBS                      TMAX    0.33
             20. TEMPERATE EVERGREEN FORBS              TMIN    0.19
             21. MARITIME HEATH DWARF-SHRUBS            TMIN    0.03
             22. SUMMERGREEN FERNS                      MI      0.02
             23. MAT-FORMING THALLOPHYTES               MI      0.29
             24. BROAD-SUMMERGREEN VINES                TMAX    0.07
             25. XERIC THALLOPHYTES                     MI      0.79

47. MIKULOV/WEINVIERTEL        OESTERRE                  0.0     0.0
      20.5  -1.0  511.   80.   19.   53.  0.95               236.
            * 1. TEMPERATE NEEDLE-TREES                 TMIN    0.24
            * 2. BOREAL/MONTANE SHORT-NEEDLED TREES     TMAX    0.14
              3. SUMMERGREEN BROAD-LEAVED TREES         MI      0.06
              4. BOREAL SUMMERGREEN NEEDLE-TREES        TMAX    0.15
              5. SUB-MEDITERRANEAN NEEDLE-TREES         TMIN    0.0
              6. BOREAL BROAD-SUMMERGREEN TREES         MI      0.16
              7. BROAD-SUMMERGREEN SMALL TREES          MI      0.27
              8. TALL GRASSES                           MI      0.27
              9. SHORT SWARD-GRASSES                    MI      0.48
             10. SUMMERGREEN GIANT-SCRUB                TMAX    0.38
             11. BROAD-SUMMERGREEN MESIC SHRUBS         MI      0.27
             12. DWARF-NEEDLE SMALL TREES               TMIN    0.17
             13. XERIC SUMMERGREEN SHRUBS               MI      0.06
             14. TEMPERATE BROAD-EVERGREEN SHRUBS       TMIN    0.05
             15. SHORT BUNCH-GRASSES                    TMIN    0.66
             16. NEEDLE-LEAVED EVERGREEN SHRUBS         TMAX    0.47
             17. SUMMERGREEN FORBS                      TMAX    0.47
             18. TEMPERATE EVERGREEN DWARF-SHRUBS       TMAX    0.19
             19. TEMPERATE EVERGREEN FORBS              MI      0.06
             20. MAT-FORMING THALLOPHYTES               MI      0.16
             21. XERIC CUSHION-HERBS                    MI      0.06
             22. XERIC THALLOPHYTES                     TMAX    0.87
```



```
       LOCATION                                         LAT     LONG
   TMAX  TMIN  PRCP  PMAX  PMIN  PMTMAX  MI                ELEV

48. AIN-DRAHAM                   TUNISIE                36.78    8.70
      25.0   6.7 1534.  190.    4.   58.  2.01               739.
            * 1. SUMMERGREEN BROAD-LEAVED TREES         TMAX    0.33
            * 2. MEDITERRANEAN BROAD-EVERGREEN TREES    TMIN    0.17
              3. SUB-MEDITERRANEAN NEEDLE-TREES         TMAX    0.21
              4. BROAD-SUMMERGREEN SMALL TREES          TMAX    0.26
              5. TALL CANE-GRAMINOIDS                   TMIN    0.20
              6. SUMMERGREEN GIANT-SCRUB                TMAX    0.43
              7. DWARF-NEEDLE SMALL TREES               TMAX    0.42
              8. SHORT SWARD-GRASSES                    TMAX    0.40
              9. TEMPERATE BROAD-EVERGREEN SHRUBS       TMAX    0.33
             10. BROAD-SUMMERGREEN MESIC SHRUBS         TMAX    0.31
             11. TROPICAL BROAD-EVERGREEN SHRUBS        TMIN    0.03
             12. BROAD-ERICOID EVERGREEN SHRUBS         TMAX    0.0
             13. SHORT BUNCH-GRASSES                    MI      0.55
             14. SUMMERGREEN FORBS                      TMAX    0.40
             15. SUMMERGREEN FERNS                      PMTMAX  0.13
             16. BROAD-SUMMERGREEN VINES                PMTMAX  0.31
             17. BROAD-WINTERGREEN EPIPHYTES            TMIN    0.24
             18. MAT-FORMING THALLOPHYTES               TMAX    0.20
             19. BROAD-EVERGREEN VINES                  TMIN    0.10

49. AZROU                        AL-MAGHR               33.45   -5.23
      26.0   7.0  837.  110.    7.   16.  1.11              1250.
            * 1. MEDITERRANEAN BROAD-EVERGREEN TREES    TMIN    0.20
              2. SUB-MEDITERRANEAN NEEDLE-TREES         TMAX    0.14
              3. TROPICAL EVERGREEN MICROPHYLL-TREES    TMIN    0.0
              4. TEMP. BROAD-EVERGREEN SMALL TREES      TMAX    0.30
              5. TALL CANE-GRAMINOIDS                   TMIN    0.21
              6. DWARF-NEEDLE SMALL TREES               TMAX    0.33
              7. MEDITERRANEAN EVERGREEN SHRUBS         TMIN    0.15
              8. TROPICAL BROAD-EVERGREEN SHRUBS        TMIN    0.04
              9. SHORT BUNCH-GRASSES                    TMAX    0.50
             10. NEEDLE-LEAVED EVERGREEN SHRUBS         TMAX    0.36
             11. SUMMERGREEN FORBS                      TMAX    0.36
             12. BUSH STEM-SUCCULENTS                   TMIN    0.23
             13. MEDITERRANEAN DWARF-SHRUBS             MI      0.08
             14. XERIC ROSETTE-SHRUBS                   MI      0.08
             15. RAINGREEN FORBS                        TMIN    0.0
             16. BROAD-WINTERGREEN EPIPHYTES            TMIN    0.25
             17. MAT-FORMING THALLOPHYTES               TMAX    0.14
             18. SUCCULENT FORBS                        MI      0.08
             19. XERIC THALLOPHYTES                     TMAX    0.50

50. KSAR-ES-SOUQ                 AL-MAGHR               31.93    4.43
      31.8   9.1  134.   24.    3.    3.  0.13              1060.
              1. HOT-DESERT EVERGREEN SHRUBS            TMIN    0.13
              2. LEAFLESS XEROMORPHIC LARGE-SCRUB       TMAX    0.33
              3. BUSH STEM-SUCCULENTS                   TMAX    0.27
            + 4. DESERT-GRASSES                         TMAX    0.24
              5. XERIC ROSETTE-SHRUBS                   TMIN    0.23
              6. TYPICAL STEM-SUCCULENTS                TMIN    0.09
              7. XERIC DWARF-SHRUBS                     TMAX    0.29
              8. EPHEMERAL DESERT HERBS                 TMAX    0.24
              9. SUCCULENT FORBS                        TMIN    0.16
             10. DRY DESERT                             MI      0.89
             11. XERIC THALLOPHYTES                     TMAX    0.24
```

25

LOCATION TMAX TMIN PRCP PMAX PMIN PMTMAX MI	LAT ELEV	LONG
51. DEMANHUR MISRA	0.0	0.0
26.2 12.7 90. 23. 0. 3. 0.09	20.	
1. HOT-DESERT EVERGREEN SHRUBS	MI	0.14
+ 2. DESERT-GRASSES	TMAX	0.49
3. LEAFLESS XEROMORPHIC LARGE-SCRUB	TMAX	0.45
4. BUSH STEM-SUCCULENTS	TMIN	0.42
5. TYPICAL STEM-SUCCULENTS	TMIN	0.25
6. EPHEMERAL DESERT HERBS	TMAX	0.49
7. XERIC DWARF-SHRUBS	MI	0.46
8. SUCCULENT FORBS	TMIN	0.31
9. DRY DESERT	TMAX	0.93
10. XERIC THALLOPHYTES	TMAX	0.49
52. MAHADDAY-WEYNE SOMALIYA	3.00	45.53
29.0 25.0 477. 105. 2. 105. 0.28	115.	
1. BROAD-RAINGREEN SMALL TREES	MI	0.12
+ 2. RAINGREEN THORN-SCRUB	TMAX	0.44
+ 3. ARBORESCENT STEM-SUCCULENTS	TMAX	0.39
4. SHORT BUNCH-GRASSES	MI	0.30
5. XERIC EVERGREEN TUFT-TREELETS	MI	0.12
6. EVERGREEN GIANT-SCRUB	MI	0.12
7. TYPICAL STEM-SUCCULENTS	TMAX	0.45
8. XERIC ROSETTE-SHRUBS	TMAX	0.45
9. BUSH STEM-SUCCULENTS	TMAX	0.37
+10. DESERT-GRASSES	TMAX	0.35
11. RAINGREEN FORBS	TMAX	0.24
12. XERIC CUSHION-SHRUBS	TMAX	0.20
13. LEAF-SUCCULENT EVERGREEN SHRUBS	TMAX	0.18
14. SUCCULENT FORBS	TMAX	0.50
15. EPHEMERAL DESERT HERBS	TMAX	0.35
16. BROAD-RAINGREEN VINES	MI	0.12
17. XERIC CUSHION-HERBS	TMAX	0.04
18. XERIC THALLOPHYTES	TMAX	0.35
53. ASMERA/ERITREA YAITOPYA	15.33	38.92
19.7 15.8 469. 138. 0. 47. 0.59	2372.	
* 1. TROPICAL EVERGREEN SCLEROPHYLL TREES	MI	0.16
2. BROAD-RAINGREEN SMALL TREES	TMAX	0.23
3. RAINGREEN THORN-SCRUB	TMAX	0.22
+ 4. SHORT BUNCH-GRASSES	MI	0.67
5. XERIC EVERGREEN TUFT-TREELETS	TMAX	0.10
+ 6. XERIC CUSHION-SHRUBS	MI	0.53
7. BUSH STEM-SUCCULENTS	TMAX	0.45
8. RAINGREEN FORBS	TMIN	0.38
9. LEAF-SUCCULENT EVERGREEN SHRUBS	TMAX	0.28
10. XERIC CUSHION-HERBS	TMAX	0.47
11. BROAD-RAINGREEN VINES	TMAX	0.23
12. SUCCULENT FORBS	TMAX	0.08
13. XERIC THALLOPHYTES	TMAX	0.94

26

LOCATION TMAX TMIN PRCP PMAX PMIN PMTMAX MI	LAT ELEV	LONG
54. KERICHO KENYA	-0.37	35.28
18.8 17.0 1816. 190. 68. 151. 2.31	2042.	
* 1. TROPICAL MONTANE RAINFOREST TREES	TMAX	0.42
* 2. TROPICAL LINEAR-LEAVED TREES	TMAX	0.29
3. TROPICAL EVERGREEN MICROPHYLL-TREES	TMAX	0.22
4. PALMIFORM TUFT-TREES	TMAX	0.15
5. TROPICAL EVERGREEN SCLEROPHYLL TREES	MI	0.09
6. TROPICAL CLOUD-FOREST DWARF-TREES	TMAX	0.12
7. ARBORESCENT GRASSES	TMAX	0.47
8. TALL GRASSES	TMAX	0.47
9. TALL CANE-GRAMINOIDS	MI	0.33
10. TROPICAL BROAD-EVERGREEN SMALL TREES	TMAX	0.22
11. TROPICAL BROAD-EVERGREEN LIANAS	TMAX	0.07
12. SHORT SWARD-GRASSES	TMAX	0.73
13. TROPICAL BROAD-EVERGREEN SHRUBS	TMIN	0.46
14. TREE FERNS	TMAX	0.27
15. PALMIFORM TUFT-TREELETS	TMAX	0.26
16. SHORT BUNCH-GRASSES	MI	0.33
17. PALMIFORM MESIC ROSETTE-SHRUBS	TMAX	0.20
18. TROPICAL EVERGREEN FORBS	TMIN	0.54
19. EVERGREEN FERNS	TMAX	0.47
20. TEMPERATE EVERGREEN DWARF-SHRUBS	TMAX	0.29
21. MAT-FORMING THALLOPHYTES	TMAX	0.60
22. NARROW-LEAVED EPIPHYTES	TMIN	0.48
23. BROAD-EVERGREEN VINES	TMAX	0.47
24. TROPICAL BROAD-EVERGREEN EPIPHYTES	TMIN	0.35
25. BROAD-WINTERGREEN EPIPHYTES	TMAX	0.20
55. NANYUKI KENYA	0.02	37.07
17.8 15.8 679. 100. 11. 69. 0.93	1947.	
* 1. TROPICAL XERIC NEEDLE-TREES	TMAX	0.19
* 2. MONTANE BROAD-RAINGREEN TREES	MI	0.14
3. TROPICAL EVERGREEN MICROPHYLL-TREES	TMAX	0.16
4. TROPICAL EVERGREEN SCLEROPHYLL TREES	TMAX	0.14
5. PALMIFORM TUFT-TREES	TMAX	0.09
6. TALL CANE-GRAMINOIDS	MI	0.25
7. TALL GRASSES	MI	0.25
8. BROAD-RAINGREEN SMALL TREES	TMAX	0.14
9. ARBORESCENT GRASSES	MI	0.14
+10. SHORT SWARD-GRASSES	MI	0.47
11. TROPICAL BROAD-EVERGREEN SHRUBS	MI	0.25
12. PALMIFORM TUFT-TREELETS	TMAX	0.19
13. RAINGREEN THORN-SCRUB	MI	0.08
+14. SHORT BUNCH-GRASSES	MI	0.79
15. PALMIFORM MESIC ROSETTE-SHRUBS	MI	0.03
16. BUSH STEM-SUCCULENTS	TMAX	0.39
17. RAINGREEN FORBS	TMIN	0.38
18. TEMPERATE EVERGREEN DWARF-SHRUBS	MI	0.25
19. TROPICAL EVERGREEN FORBS	MI	0.03
20. BROAD-RAINGREEN VINES	TMAX	0.14
21. MAT-FORMING THALLOPHYTES	MI	0.14
22. BROAD-WINTERGREEN EPIPHYTES	TMAX	0.13
23. XERIC CUSHION-HERBS	MI	0.09
24. BROAD-EVERGREEN VINES	MI	0.03
25. XERIC THALLOPHYTES	MULT.	1.00

LOCATION							LAT	LONG	27
TMAX	TMIN	PRCP	PMAX	PMIN	PMTMAX	MI	ELEV		
56. MAKURDI			NIGERIA				7.75	8.53	
30.0	25.0	1377.	175.	0.	109.	0.83	111.		

	1.	TROPICAL XERIC NEEDLE-TREES	TMAX 0.0
	2.	TROPICAL EVERGREEN SCLEROPHYLL TREES	TMAX 0.25
	3.	PALMIFORM TUFT-TREES	MI 0.16
	4.	TROPICAL EVERGREEN MICROPHYLL-TREES	MI 0.09
+	5.	XERIC RAINGREEN TREES	MI 0.27
+	6.	BROAD-RAINGREEN SMALL TREES	TMAX 0.25
	7.	TALL CANE-GRAMINOIDS	MI 0.16
	8.	RAINGREEN THORN-SCRUB	MI 0.21
	9.	PALMIFORM TUFT-TREELETS	MI 0.16
	10.	TROPICAL BROAD-EVERGREEN SHRUBS	TMAX 0.09
+	11.	SHORT BUNCH-GRASSES	TMAX 0.31
	12.	XERIC EVERGREEN TUFT-TREELETS	MI 0.23
	13.	EVERGREEN GIANT-SCRUB	TMAX 0.17
	14.	XERIC ROSETTE-SHRUBS	MI 0.46
	15.	BUSH STEM-SUCCULENTS	TMAX 0.33
	16.	RAINGREEN FORBS	TMAX 0.20
	17.	SUCCULENT FORBS	TMAX 0.45
	18.	BROAD-RAINGREEN VINES	TMAX 0.25
	19.	BROAD-WINTERGREEN EPIPHYTES	TMAX 0.0
	20.	XERIC THALLOPHYTES	TMAX 0.31

57. LUSAKA			ZAMBIA				-15.42	28.28
25.0	16.0	835.	165.	0.	62.	0.89	1278.	

*	1.	TROPICAL XERIC NEEDLE-TREES	TMAX 0.33
	2.	MONTANE BROAD-RAINGREEN TREES	TMAX 0.0
	3.	TROPICAL EVERGREEN SCLEROPHYLL TREES	TMIN 0.36
	4.	PALMIFORM TUFT-TREES	MI 0.22
	5.	TROPICAL EVERGREEN MICROPHYLL-TREES	MI 0.16
	6.	XERIC RAINGREEN TREES	TMAX 0.06
+	7.	BROAD-RAINGREEN SMALL TREES	TMIN 0.36
	8.	TALL CANE-GRAMINOIDS	MI 0.22
	9.	MEDITERRANEAN EVERGREEN SHRUBS	TMAX 0.36
	10.	TROPICAL BROAD-EVERGREEN SHRUBS	MI 0.22
	11.	PALMIFORM TUFT-TREELETS	MI 0.22
	12.	RAINGREEN THORN-SCRUB	MI 0.12
+	13.	SHORT BUNCH-GRASSES	TMAX 0.56
+	14.	EVERGREEN GIANT-SCRUB	TMIN 0.40
	15.	XERIC EVERGREEN TUFT-TREELETS	MI 0.14
	16.	BUSH STEM-SUCCULENTS	TMIN 0.53
	17.	RAINGREEN FORBS	TMIN 0.39
	18.	XERIC ROSETTE-SHRUBS	TMAX 0.25
	19.	XERIC CUSHION-SHRUBS	MI 0.01
	20.	BROAD-WINTERGREEN EPIPHYTES	MI 0.33
	21.	SUCCULENT FORBS	TMAX 0.32
	22.	BROAD-RAINGREEN VINES	TMIN 0.30
	23.	XERIC CUSHION-HERBS	MI 0.13
	24.	XERIC THALLOPHYTES	TMAX 0.56

LOCATION							LAT	LONG	28
TMAX	TMIN	PRCP	PMAX	PMIN	PMTMAX	MI	ELEV		
58. MAUN			BOTSWANA				-20.00	23.42	
26.5	15.0	457.	100.	0.	26.	0.43	942.		

*	1.	XERIC RAINGREEN TREES	TMAX 0.16
	2.	BROAD-RAINGREEN SMALL TREES	TMIN 0.32
	3.	RAINGREEN THORN-SCRUB	TMIN 0.32
+	4.	SHORT BUNCH-GRASSES	TMAX 0.47
+	5.	EVERGREEN GIANT-SCRUB	TMIN 0.35
	6.	XERIC EVERGREEN TUFT-TREELETS	TMIN 0.32
	7.	HOT-DESERT EVERGREEN SHRUBS	TMAX 0.25
	8.	ARBORESCENT STEM-SUCCULENTS	TMAX 0.25
	9.	NEEDLE-LEAVED EVERGREEN SHRUBS	MI 0.08
	10.	BUSH STEM-SUCCULENTS	TMAX 0.47
	11.	LEAFLESS XEROMORPHIC LARGE-SCRUB	TMAX 0.46
+	12.	MEDITERRANEAN DWARF-SHRUBS	TMAX 0.45
	13.	RAINGREEN FORBS	TMAX 0.34
	14.	SUMMERGREEN FORBS	TMAX 0.34
	15.	XERIC ROSETTE-SHRUBS	TMAX 0.32
	16.	TYPICAL STEM-SUCCULENTS	TMAX 0.32
	17.	XERIC CUSHION-SHRUBS	TMAX 0.30
	18.	SCLEROPHYLLOUS GRASSES	TMIN 0.25
	19.	LEAF-SUCCULENT EVERGREEN SHRUBS	TMIN 0.23
	20.	DESERT-GRASSES	MI 0.16
	21.	XERIC DWARF-SHRUBS	TMAX 0.48
	22.	SUCCULENT FORBS	TMAX 0.39
	23.	BROAD-RAINGREEN VINES	TMIN 0.25
	24.	EPHEMERAL DESERT HERBS	MI 0.16
	25.	XERIC CUSHION-HERBS	TMAX 0.14
	26.	XERIC THALLOPHYTES	TMAX 0.47

LOCATION TMAX TMIN PRCP PMAX PMIN PMTMAX MI	LAT ELEV	LONG 29
59. KROONSTAD/ORANJE FRYSTAA SOUTH AF	-27.77	27.20
22.5 8.5 606. 95. 8. 64. 0.84	1348.	
* 1. TEMPERATE NEEDLE-TREES	TMAX	0.17
2. MEDITERRANEAN BROAD-EVERGREEN TREES	MI	0.05
3. MONTANE BROAD-RAINGREEN TREES	TMIN	0.04
4. TROPICAL XERIC NEEDLE-TREES	TMIN	0.02
5. TROPICAL EVERGREEN MICROPHYLL-TREES	TMIN	0.07
6. TROPICAL EVERGREEN SCLEROPHYLL TREES	TMIN	0.02
7. BROAD-SUMMERGREEN SMALL TREES	MI	0.17
8. TALL GRASSES	MI	0.17
9. TEMP. BROAD-EVERGREEN SMALL TREES	MI	0.11
10. ARBORESCENT GRASSES	MI	0.05
11. BROAD-RAINGREEN SMALL TREES	TMIN	0.02
12. SUMMERGREEN GIANT-SCRUB	TMAX	0.46
+13. SHORT SWARD-GRASSES	MI	0.41
+14. DWARF-NEEDLE SMALL TREES	TMAX	0.38
15. TEMPERATE BROAD-EVERGREEN SHRUBS	MI	0.29
16. MEDITERRANEAN EVERGREEN SHRUBS	TMIN	0.27
17. XERIC SUMMERGREEN SHRUBS	MI	0.21
18. BROAD-SUMMERGREEN MESIC SHRUBS	MI	0.17
19. TROPICAL BROAD-EVERGREEN SHRUBS	TMIN	0.10
20. RAINGREEN THORN-SCRUB	TMIN	0.02
+21. SHORT BUNCH-GRASSES	TMAX	0.71
22. NEEDLE-LEAVED EVERGREEN SHRUBS	TMAX	0.51
23. EVERGREEN GIANT-SCRUB	TMIN	0.02
24. XERIC EVERGREEN TUFT-TREELETS	TMIN	0.02
25. SUMMERGREEN FORBS	TMAX	0.51
26. BUSH STEM-SUCCULENTS	TMIN	0.28
27. XERIC ROSETTE-SHRUBS	TMAX	0.13
28. TEMPERATE EVERGREEN DWARF-SHRUBS	TMAX	0.08
29. XERIC CUSHION-SHRUBS	MI	0.08
30. RAINGREEN FORBS	TMIN	0.07
31. BROAD-WINTERGREEN EPIPHYTES	MI	0.29
32. XERIC CUSHION-HERBS	MI	0.20
33. SUCCULENT FORBS	TMIN	0.14
34. MAT-FORMING THALLOPHYTES	MI	0.05
35. XERIC THALLOPHYTES	TMAX	0.71

LOCATION TMAX TMIN PRCP PMAX PMIN PMTMAX MI	LAT ELEV	LONG 30
60. CATHEDRAL PEAK(LITTLE BE SOUTH AF	-28.98	29.23
17.0 10.0 1418. 289. 12. 197. 2.21	1860.	
* 1. TEMPERATE NEEDLE-TREES	TMAX	0.22
* 2. TROPICAL LINEAR-LEAVED TREES	TMAX	0.15
* 3. MONTANE BROAD-RAINGREEN TREES	TMIN	0.14
* 4. TROPICAL MONTANE RAINFOREST TREES	TMIN	0.13
5. TROPICAL EVERGREEN MICROPHYLL-TREES	TMAX	0.12
6. TROPICAL EVERGREEN SCLEROPHYLL TREES	TMIN	0.09
7. TALL GRASSES	TMAX	0.41
8. TALL CANE-GRAMINOIDS	TMIN	0.31
9. TEMP. BROAD-EVERGREEN SMALL TREES	TMAX	0.29
10. ARBORESCENT GRASSES	TMIN	0.29
11. TROPICAL BROAD-EVERGREEN SMALL TREES	TMIN	0.10
12. BROAD-RAINGREEN SMALL TREES	TMIN	0.09
13. SHORT SWARD-GRASSES	TMAX	0.71
14. BROAD-ERICOID EVERGREEN SHRUBS	TMAX	0.33
15. TROPICAL BROAD-EVERGREEN SHRUBS	TMIN	0.17
16. TEMPERATE BROAD-EVERGREEN SHRUBS	TMAX	0.13
17. PALMIFORM TUFT-TREELETS	TMIN	0.0
18. SHORT BUNCH-GRASSES	MI	0.39
19. PALMIFORM MESIC ROSETTE-SHRUBS	TMIN	0.09
20. MESIC EVERGREEN CUSHION-SHRUBS	TMAX	0.07
21. TEMPERATE EVERGREEN FORBS	MI	0.65
22. SUMMERGREEN FORBS	TMAX	0.41
23. TEMPERATE EVERGREEN DWARF-SHRUBS	TMAX	0.41
24. TROPICAL EVERGREEN FORBS	TMIN	0.29
25. RAINGREEN FORBS	TMIN	0.13
26. MAT-FORMING THALLOPHYTES	MI	0.70
27. BROAD-EVERGREEN VINES	TMIN	0.23
28. NARROW-LEAVED EPIPHYTES	TMIN	0.20
29. BROAD-WINTERGREEN EPIPHYTES	TMAX	0.07
30. BROAD-RAINGREEN VINES	TMIN	0.0
31. TROPICAL BROAD-EVERGREEN EPIPHYTES	TMIN	0.0

LOCATION / TMAX TMIN PRCP PMAX PMIN PMTMAX MI — LAT LONG ELEV

61. HELL-BOURG — REUNION — -21.10 55.60

19.7 14.0 2444. 500. 20. 150. 3.31 — 935.

* 1. MONTANE BROAD-RAINGREEN TREES	TMIN	0.43
* 2. TEMPERATE RAINFOREST NEEDLE-TREES	TMAX	0.41
* 3. TROPICAL MONTANE RAINFOREST TREES	TMIN	0.40
* 4. TROPICAL LINEAR-LEAVED TREES	TMAX	0.36
5. TROPICAL EVERGREEN MICROPHYLL-TREES	TMAX	0.28
6. PALMIFORM TUFT-TREES	TMIN	0.16
7. TROPICAL CLOUD-FOREST DWARF-TREES	TMAX	0.03
8. TALL GRASSES	TMAX	0.49
9. ARBORESCENT GRASSES	TMIN	0.48
10. TEMP. BROAD-EVERGREEN SMALL TREES	TMAX	0.39
11. TROPICAL BROAD-EVERGREEN SMALL TREES	TMAX	0.28
12. BROAD-RAINGREEN SMALL TREES	TMAX	0.24
13. TROPICAL BROAD-EVERGREEN LIANAS	TMAX	0.14
14. SHORT SWARD-GRASSES	TMAX	0.75
15. TROPICAL BROAD-EVERGREEN SHRUBS	TMIN	0.33
16. TEMPERATE BROAD-EVERGREEN SHRUBS	TMAX	0.31
17. PALMIFORM TUFT-TREELETS	TMIN	0.20
18. PALMIFORM MESIC ROSETTE-SHRUBS	TMAX	0.24
19. TEMPERATE EVERGREEN FORBS	TMAX	0.52
20. TROPICAL EVERGREEN FORBS	TMIN	0.43
21. RAINGREEN FORBS	TMIN	0.30
22. TEMPERATE EVERGREEN DWARF-SHRUBS	TMAX	0.24
23. MAT-FORMING THALLOPHYTES	TMAX	0.52
24. BROAD-EVERGREEN VINES	TMIN	0.38
25. NARROW-LEAVED EPIPHYTES	TMIN	0.36
26. BROAD-WINTERGREEN EPIPHYTES	TMAX	0.26
27. BROAD-RAINGREEN VINES	TMIN	0.20
28. TROPICAL BROAD-EVERGREEN EPIPHYTES	TMIN	0.20

62. BORISOVKA/ROSSIJA — SSSR — 50.63 35.97

19.7 -9.0 537. 74. 28. 64. 0.94 — 200.

* 1. BOREAL/MONTANE SHORT-NEEDLED TREES	TMAX	0.21
2. SUMMERGREEN BROAD-LEAVED TREES	MI	0.05
3. BOREAL SUMMERGREEN NEEDLE-TREES	TMAX	0.23
4. BOREAL BROAD-SUMMERGREEN TREES	MI	0.15
5. BROAD-SUMMERGREEN SMALL TREES	MI	0.26
6. TALL GRASSES	TMIN	0.23
+ 7. SHORT SWARD-GRASSES	MI	0.48
8. SUMMERGREEN GIANT-SCRUB	TMAX	0.35
9. BROAD-SUMMERGREEN MESIC SHRUBS	MI	0.26
10. XERIC SUMMERGREEN SHRUBS	MI	0.07
+11. SHORT BUNCH-GRASSES	TMIN	0.57
12. NEEDLE-LEAVED EVERGREEN SHRUBS	TMAX	0.45
13. SUMMERGREEN FORBS	TMAX	0.45
14. TEMPERATE EVERGREEN DWARF-SHRUBS	TMAX	0.24
15. MAT-FORMING THALLOPHYTES	MI	0.15
16. XERIC CUSHION-HERBS	MI	0.07
17. XERIC THALLOPHYTES	TMAX	0.94

LOCATION / TMAX TMIN PRCP PMAX PMIN PMTMAX MI — LAT LONG ELEV

63. KERKI/TURKMENISTAN — SSSR — 37.83 65.20

29.5 2.0 161. 34. 0. 2. 0.17 — 242.

1. COLD-WINTER XEROMORPHIC SHRUBS	TMAX	0.15
2. LEAFLESS XEROMORPHIC LARGE-SCRUB	TMAX	0.42
+ 3. DESERT-GRASSES	TMAX	0.33
4. BUSH STEM-SUCCULENTS	TMIN	0.07
5. XERIC DWARF-SHRUBS	TMAX	0.38
6. EPHEMERAL DESERT HERBS	TMAX	0.33
7. XERIC THALLOPHYTES	TMAX	0.33

64. CHATKAL MTNS./KIRGHIZ — SSSR — 41.60 70.70

20.6 -3.4 995. 203. 0. 1. 1.63 — 1433.

* 1. SHORT BUNCH-GRASSES	TMIN	0.63
* 2. NEEDLE-LEAVED EVERGREEN SHRUBS	MI	0.23
3. SUMMERGREEN FORBS	TMAX	0.47
4. XERIC THALLOPHYTES	MI	0.23

65. NAJAF — AL-'IRAQ — 31.95 44.30

38.0 10.0 83. 18. 0. 0. 0.06 — 40.

1. TYPICAL STEM-SUCCULENTS	TMAX	0.10
2. LEAFLESS XEROMORPHIC LARGE-SCRUB	TMAX	0.08
3. BUSH STEM-SUCCULENTS	TMAX	0.07
4. DESERT-GRASSES	TMAX	0.05
5. SUCCULENT FORBS	TMAX	0.09
6. XERIC DWARF-SHRUBS	TMAX	0.07
7. EPHEMERAL DESERT HERBS	TMAX	0.05
8. DRY DESERT	TMAX	0.95
9. XERIC THALLOPHYTES	TMAX	0.05

66. MAIMANA — AFGHANES — 35.92 64.63

23.5 3.0 400. 90. 0. 11. 0.55 — 615.

* 1. DWARF-NEEDLE SMALL TREES	MI	0.18
+ 2. SHORT BUNCH-GRASSES	MI	0.65
+ 3. COLD-WINTER XEROMORPHIC SHRUBS	TMAX	0.50
4. NEEDLE-LEAVED EVERGREEN SHRUBS	MI	0.27
5. SUMMERGREEN FORBS	TMAX	0.46
6. LEAFLESS XEROMORPHIC LARGE-SCRUB	TMAX	0.34
7. BUSH STEM-SUCCULENTS	TMIN	0.10
8. XERIC ROSETTE-SHRUBS	TMIN	0.0
9. MEDITERRANEAN DWARF-SHRUBS	TMIN	0.0
10. XERIC DWARF-SHRUBS	TMAX	0.45
11. XERIC CUSHION-HERBS	TMAX	0.26
12. XERIC THALLOPHYTES	TMAX	0.65

67. KOTGAI — AFGHANES — 33.95 69.90

19.0 -5.0 519. 105. 7. 18. 0.94 — 2450.

* 1. BOREAL/MONTANE SHORT-NEEDLED TREES	TMAX	0.27
2. TEMPERATE NEEDLE-TREES	TMIN	0.05
3. DWARF-NEEDLE SMALL TREES	TMIN	0.0
+ 4. SHORT BUNCH-GRASSES	TMIN	0.61
5. NEEDLE-LEAVED EVERGREEN SHRUBS	TMAX	0.43
6. SUMMERGREEN FORBS	TMAX	0.43
7. MAT-FORMING THALLOPHYTES	MI	0.15
8. XERIC CUSHION-HERBS	MI	0.08
9. XERIC THALLOPHYTES	MULT.	1.00

```
                                                                   33
      LOCATION                                              LAT     LONG
   TMAX  TMIN  PRCP  PMAX  PMIN  PMTMAX  MI                    ELEV

68. MUSSOORIE                  BHARAT                     30.45    78.25
    20.2   6.4 2368.  785.    7.     8.   3.32              1500.
             1. TROPICAL BROAD-EVERGREEN SHRUBS           TMIN     0.02
             2. SUMMERGREEN FORBS                         TMAX     0.50
             3. MAT-FORMING THALLOPHYTES                  TMAX     0.49
             4. BROAD-WINTERGREEN EPIPHYTES               TMIN     0.23

69. YUELIN/ORDOS               ZHONGGUO                    0.0      0.0
    23.0  -8.8  405.  120.    4.    83.   0.72              1121.
             1. BROAD-SUMMERGREEN SMALL TREES             MI       0.03
             2. SUMMERGREEN GIANT-SCRUB                   TMIN     0.45
           + 3. SHORT SWARD-GRASSES                       MI       0.31
             4. BROAD-SUMMERGREEN MESIC SHRUBS            MI       0.03
           + 5. SHORT BUNCH-GRASSES                       TMIN     0.57
             6. NEEDLE-LEAVED EVERGREEN SHRUBS            MI       0.45
             7. SUMMERGREEN FORBS                         TMAX     0.48
             8. XERIC CUSHION-HERBS                       TMAX     0.28
             9. XERIC THALLOPHYTES                        TMAX     0.68

70. MITCHELL PLATEAU CAMP/W.  AUSTRALI                   -14.82   125.83
    28.8  19.5 1611.  369.    0.    90.   1.15              268.
           * 1. MONSCON BROAD-RAINGREEN TREES             MI       0.22
             2. PALMIFORM TUFT-TREES                      MI       0.40
             3. TROPICAL EVERGREEN SCLEROPHYLL TREES      TMAX     0.31
             4. TROPICAL EVERGREEN MICROPHYLL-TREES       TMAX     0.19
             5. BROAD-RAINGREEN SMALL TREES               TMAX     0.31
             6. TALL CANE-GRAMINOIDS                      TMAX     0.25
             7. PALMIFORM TUFT-TREELETS                   TMAX     0.30
             8. TROPICAL BROAD-EVERGREEN SHRUBS           TMAX     0.15
             9. SHORT BUNCH-GRASSES                       TMAX     0.36
            10. EVERGREEN GIANT-SCRUB                     TMAX     0.27
            11. RAINGREEN FORBS                           TMAX     0.25
            12. BROAD-RAINGREEN VINES                     TMAX     0.31
            13. BROAD-WINTERGREEN EPIPHYTES               TMAX     0.09
            14. XERIC THALLOPHYTES                        TMAX     0.36

71. WILUNA/W.AUSTRALIA         AUSTRALI                  -26.60   120.22
    30.0  12.0  247.   34.    5.    33.   0.25              556.
             1. RAINGREEN THORN-SCRUB                     TMIN     0.18
           + 2. HOT-DESERT EVERGREEN SHRUBS               TMIN     0.25
             3. SHORT BUNCH-GRASSES                       MI       0.19
             4. ARBORESCENT STEM-SUCCULENTS               TMIN     0.18
             5. LEAFLESS XEROMORPHIC LARGE-SCRUB          TMAX     0.40
             6. XERIC ROSETTE-SHRUBS                      TMIN     0.33
             7. BUSH STEM-SUCCULENTS                      TMAX     0.33
           + 8. DESERT-GRASSES                            TMAX     0.31
             9. TYPICAL STEM-SUCCULENTS                   TMIN     0.22
            10. RAINGREEN FORBS                           MI       0.19
            11. SUMMERGREEN FORBS                         MI       0.19
            12. XERIC CUSHION-SHRUBS                      TMAX     0.17
            13. SCLEROPHYLLOUS GRASSES                    TMIN     0.10
            14. LEAF-SUCCULENT EVERGREEN SHRUBS           TMIN     0.0
            15. XERIC DWARF-SHRUBS                        TMAX     0.36
            16. EPHEMERAL DESERT HERBS                    TMAX     0.31
            17. SUCCULENT FORBS                           TMIN     0.28
            18. XERIC CUSHION-HERBS                       TMAX     0.0
            19. XERIC THALLOPHYTES                        TMAX     0.31
```

```
                                                                   34
      LOCATION                                              LAT     LONG
   TMAX  TMIN  PRCP  PMAX  PMIN  PMTMAX  MI                    ELEV

72. PERTH/WESTERN AUSTRALIA   AUSTRALI                   -31.93   115.83
    24.0  13.0  883.  150.    8.    19.   1.14               65.
           * 1. MEDITERRANEAN BROAD-EVERGREEN TREES       MI       0.30
           * 2. TROPICAL LINEAR-LEAVED TREES              MI       0.12
             3. TEMPERATE NEEDLE-TREES                    TMAX     0.0
             4. TROPICAL MONTANE RAINFOREST TREES         TMAX     0.0
             5. TROPICAL EVERGREEN MICROPHYLL-TREES       TMIN     0.26
             6. SUB-MEDITERRANEAN NEEDLE-TREES            MI       0.25
             7. TROPICAL EVERGREEN SCLEROPHYLL TREES      TMIN     0.23
             8. PALMIFORM TUFT-TREES                      TMIN     0.11
             9. TALL CANE-GRAMINOIDS                      MI       0.39
            10. TEMP. BROAD-EVERGREEN SMALL TREES         MI       0.34
            11. BROAD-RAINGREEN SMALL TREES               TMIN     0.23
            12. TROPICAL BROAD-EVERGREEN SMALL TREES      MI       0.21
            13. DWARF-NEEDLE SMALL TREES                  TMAX     0.50
            14. MEDITERRANEAN EVERGREEN SHRUBS            TMAX     0.43
            15. TROPICAL BROAD-EVERGREEN SHRUBS           TMIN     0.29
            16. PALMIFORM TUFT-TREELETS                   TMIN     0.15
            17. SHORT BUNCH-GRASSES                       TMAX     0.62
            18. NEEDLE-LEAVED EVERGREEN SHRUBS            TMAX     0.44
            19. EVERGREEN GIANT-SCRUB                     TMIN     0.25
            20. SUMMERGREEN FORBS                         TMAX     0.44
            21. BUSH STEM-SUCCULENTS                      MI       0.32
            22. RAINGREEN FORBS                           TMIN     0.26
            23. MEDITERRANEAN DWARF-SHRUBS                MI       0.06
            24. XERIC ROSETTE-SHRUBS                      MI       0.06
            25. SCLEROPHYLLOUS GRASSES                    MI       0.06
            26. BROAD-WINTERGREEN EPIPHYTES               TMAX     0.43
            27. MAT-FORMING THALLOPHYTES                  TMAX     0.23
            28. BROAD-RAINGREEN VINES                     TMIN     0.15
            29. SUCCULENT FORBS                           MI       0.06
            30. XERIC THALLOPHYTES                        TMAX     0.62
```

Appendix C. Predicted vegetation at the validation sites.

LOCATION TMAX TMIN PRCP PMAX PMIN PMTMAX MI	LAT ELEV	LONG 35
73. KIETH(DARK ISLAND)/S.AUS AUSTRALI	-36.10	140.52
22.0 9.1 457. 60. 20. 20. 0.68	0.	
1. TEMPERATE NEEDLE-TREES	MI	0.05
2. TROPICAL EVERGREEN SCLEROPHYLL TREES	TMIN	0.05
3. BROAD-RAINGREEN SMALL TREES	TMIN	0.05
+ 4. XERIC SUMMERGREEN SHRUBS	TMAX	0.38
5. DWARF-NEEDLE SMALL TREES	TMAX	0.33
6. MEDITERRANEAN EVERGREEN SHRUBS	TMIN	0.32
7. RAINGREEN THORN-SCRUB	TMIN	0.05
8. SUMMERGREEN GIANT-SCRUB	PMTMAX	0.0
+ 9. SHORT BUNCH-GRASSES	MI	0.72
10. NEEDLE-LEAVED EVERGREEN SHRUBS	MI	0.42
11. EVERGREEN GIANT-SCRUB	TMIN	0.05
12. XERIC EVERGREEN TUFT-TREELETS	TMIN	0.05
13. ARBORESCENT STEM-SUCCULENTS	TMAX	0.0
14. HOT-DESERT EVERGREEN SHRUBS	TMAX	0.0
15. SUMMERGREEN FORBS	TMAX	0.50
16. XERIC CUSHION-SHRUBS	MI	0.32
17. MEDITERRANEAN DWARF-SHRUBS	TMAX	0.32
18. BUSH STEM-SUCCULENTS	TMIN	0.30
19. LEAFLESS XEROMORPHIC LARGE-SCRUB	MI	0.17
20. XERIC ROSETTE-SHRUBS	TMAX	0.10
21. RAINGREEN FORBS	TMIN	0.09
22. XERIC CUSHION-HERBS	TMAX	0.33
23. XERIC DWARF-SHRUBS	MI	0.17
24. SUCCULENT FORBS	TMIN	0.16
25. BROAD-WINTERGREEN EPIPHYTES	MI	0.13
26. XERIC THALLOPHYTES	TMAX	0.75

LOCATION TMAX TMIN PRCP PMAX PMIN PMTMAX MI	LAT ELEV	LONG 36
74. ALEXANDRA/CTAGO NEW ZEAL	-45.33	169.38
16.6 2.8 330. 49. 15. 33. 0.63	158.	
1. BOREAL/MONTANE SHORT-NEEDLED TREES	MI	0.04
2. SUMMERGREEN GIANT-SCRUB	TMAX	0.25
3. SHORT SWARD-GRASSES	MI	0.20
4. XERIC SUMMERGREEN SHRUBS	TMAX	0.14
5. TEMPERATE BROAD-EVERGREEN SHRUBS	MI	0.04
+ 6. SHORT BUNCH-GRASSES	MI	0.69
7. NEEDLE-LEAVED EVERGREEN SHRUBS	TMAX	0.35
8. SHORT TUSSOCK-GRASSES	MI	0.20
9. COLD-WINTER XEROMORPHIC SHRUBS	TMAX	0.09
10. SUMMERGREEN FORBS	TMAX	0.35
11. XERIC CUSHION-SHRUBS	TMIN	0.26
12. BUSH STEM-SUCCULENTS	TMIN	0.09
13. LEAFLESS XEROMORPHIC LARGE-SCRUB	TMAX	0.09
14. XERIC CUSHION-HERBS	MI	0.61
15. XERIC DWARF-SHRUBS	TMAX	0.25
16. XERIC THALLOPHYTES	MULT.	1.00

APPENDIX D

Actual vegetation at the validation sites

Descriptions of the actual vegetation at the validation sites include a formation name and lists of the most important taxa in each structural class present. In presenting this information a number of abbreviations are necessary, as shown below. The vegetation descriptions and species lists were constructed from various sources, as indicated for each site. Dominant taxa are denoted by asterisks. (Note that, through the author's oversight, taxonomic names in this appendix are not italicized, since type-setting was done from a computer-printed original.)

Abbreviations

Growth forms/life forms:

B - large bushes
Ch - chamaephytes
DS - dwarf-shrubs
DT - dwarf trees
E - epiphytes
Eph - ephemerals
F - forbs
Fn - ferns
G - graminoids
Gp - geophytes
H - herbs
Hc - hemi-cryptophytes
OB - overgrown bushes
RS - rosette-shrubs
RT - rosette-trees
S - shrubs
SS - stem-succulents
ST - small trees
T - trees
TG - tussock-grasses
Th - therophytes
UB - underbrush
US - understorey
UT - understorey trees
V - vines and lianas

Seasonal habits:

EG - evergreen
SG - summergreen
RG - raingreen
E - ephemeral
Sf - suffrutescent

Leaf sizes and shapes:

mic - microphyllous
nan - nanophyllous
lep - leptophyllous
n - narrow
b - broad

Others:

a - arborescent
d - dwarf
b - bunch (grass)
s - sward-forming (grass)
w - white-pubescent

This appendix was largely completed in 1978 and thus omits some more recent studies. No systematic attempt has been made to resolve synonyms, check the accuracy of the original sources, or obtain complete species lists for lower synusiae, since such information is often available only locally and seeking it would greatly delay publication. The author invites readers with (better) local vegetation descriptions (including species lists) and applicable climatic data to send them for comparison with predicted vegetation profiles.

1. FAIRBANKS/ALASKA U.S.A.

Continental, closed Boreal Spruce-Hardwood Forest, with scattered bogs and with lower, open forests on north slopes:

T: *Picea glauca, Larix alaskana, Betula resinifera (= B. alaskana, B. neoalaskana), B. papyrifera, Populus balsamifera, et al.

S: Arctostaphylos rubra, A. uva-ursi, Empetrum nigrum, Ledum decumbens, Ribes triste, Rosa acicularis, Salix alaxensis, S. arbusculoides, S. bebbiana et al., Vaccinium vitis-idaea, V. uliginosum et al., Viburnum edule, Amelanchier alnifolia, et al.

H: Microthermal tundra and meadow herbs (e.g. Carex, Scirpus, Eriophorum, Ranunculus, Oxytropis, Epilobium, Castilleja pallida, Artemisia frigida).

References: Jordal 1951; Knapp 1965; Viereck & Little 1972.

2. VANCOUVER (PMO)/BRIT. COLUMBIA CANADA

Tall, dense Douglas Fir Forest, with Hemlock and Red Cedar:

T: *Pseudotsuga menziesii var. menziesii, Tsuga heterophylla, Thuja plicata, Pinus monticola, P. contorta, Abies grandis.

UT: Alnus rubra, Acer spp., Prunus emarginata, Populus balsamifera ssp. trichocarpa, Arbutus menziesii (EG), Corylus cornuta var. californica, Berberis nervosa (EG), et al.

S: Rhododendron californicum, Vaccinium spp., Arctostaphylos uva-ursi, Rubus spectabilis, et al.

H: Gaultheria shallon, Chimaphila umbellata var. occidentalis, Trientalis europa var. latifolia, Viola sempervirens, Goodyera decipiens, Pyrola spp., Pteridium aquilinum, et al.

References: Rowe 1959; Knapp 1965; Krajina 1969.

3. BANFF/ALBERTA CANADA

Boreal-Montane Transitional Needle Forest:

T: *Picea engelmannii, *Abies lasiocarpa, *Pinus contorta var. latifolia, P. banksiana, Picea glauca var. albertiana, P. mariana, Thuja occidentalis, Larix laricina, L. lyallii, Betula papyrifera, Populus tremuloides, P. balsamifera.

US: Vaccinium ssp., with tundra herbs and dwarf-shrubs.

References: Rowe 1959; Knapp 1965; Shelford 1963.

4. SASKATOON/SASKATCHEWAN CANADA

Northern Grove Belt (Fescue Prairie):

T: Populus tremuloides

S: Rosa arkansasa (Sf)

G: *Festuca scabrella (b), Koeleria cristata, Danthonia spp., Agropyron spp., Stipa spp., Carex spp., with Bromus pumpellianus, Calamagrostis montanensis, Muhlenbergia spp., and many others.

F: prairie forbs, e.g. Solidago, Aster, Achillea, Artemisia, Potentilla, Astragalus, Antennaria, and many others.

References: Coupland & Brayshaw 1953; Weaver & Albertson 1956; Knapp 1965.

5. SWIFT CURRENT/SASKATCHEWAN CANADA

Northern Short and Mixed-Grass Prairie (northern limit):

G: *Bouteloua gracilis et al., *Stipa comata et al., Buchloë dactyloides, Sporobolus cryptandrus, Agropyron smithii et al., Aristida spp., Koeleria cristata, Sitanion hystrix, Carex spp., and many others.

S: Eurotia lanata, Atriplex nuttallii, Artemisia cana.

SS: Opuntia polyacantha.

F: prairie forbs, e.g. Artemisia frigida, Phlox hoodii, Asclepias spp., Astragalus mollissimus, Castilleja sessilifolia, Oenothera spp., Oxytropis lamberti, Plantago purshii, and many others.

References: Weaver & Albertson 1956; Küchler 1964; Knapp 1965.

6. SIOUX LOOKOUT/ONTARIO CANADA

Central Boreal Forest:

T: *Picea glauca, *Abies balsamea, Picea mariana, Betula papyrifera, Populus tremuloides, with Larix laricina, Pinus banksiana, Thuja occidentalis, Betula lutea, Populus grandidentata, et al.

US: Vaccinium spp., Lycopodium spp., Trientalis borealis, Cornus canadensis, Gaultheria procumbens, Pyrola spp., et al.

References: Rowe 1959; Knapp 1965; Shelford 1963.

7. LAC ALBANEL/QUEBEC CANADA

Eastern Boreal Forest:

T: *Picea mariana, P. glauca, Abies balsamea, with Larix laricina, Pinus banksiana, Picea rubens, Thuja occidentalis, et al. and with groves of Betula papyrifera, Populus tremuloides, P. balsamifera, et al.

US: Vaccinium spp., Lycopodium spp., Trientalis borealis, Cornus canadensis, Gaultheria procumbens, Pyrola spp., et al.

References: Rowe 1959; Shelford 1963; Knapp 1965.

8. OLYMPIA/WASHINGTON U.S.A.

Tall, dense Douglas Fir Forest with Red Cedar and Hemlock:

T: *Pseudotsuga menziesii var. viridis, *Tsuga heterophylla, *Thuja plicata, Pinus monticola, Abies grandis.

UT: Acer spp., Alnus rubra, Berberis nervosa, Corylus cornuta var. californica, Prunus virginiana demissa, et al.

S: Rhododendron californicum, Vaccinium spp., Rubus spectabilis, Arctostaphylos uva-ursi, et al.

H: Gaultheria shallon, Chimaphila umbellata var. occidentalis, Trientalis europea var. latifolia, Viola sempervirens, Goodyera decipiens, Pyrola spp., Pteridium aquilinum, et al.

References: Andrews & Cowlin 1937; Küchler 1964; Knapp 1965; Franklin & Dyrness 1973; personal observation.

9. DAVIS/CALIFORNIA U.S.A.

California Bunch-Grass Prairie with many forbs:

G: *Stipa pulchra, *S. cernua, S. lepida, Aristida divaricata, Koeleria cristata, Poa scabrella, Elymus spp., Festuca spp., Hordeum brachyantherum, et al., plus ruderals.

F: spring ephemerals (e.g. Brodiaea, Calochortus, Allium) and prairie forbs (e.g. Lupinus spp., Trifolium spp., Eschscholtzia californica, Gilia spp. (= Ipomopsis)), et al.

DS: white-pubescent dwarf-shrubs, e.g. Salvia, Eriogonum, Artemisia californica.

SS: scattered Opuntia, especially where overgrazed.

References: Knapp 1965; Shelford 1963; Küchler 1964, 1977.

10. TANBARK FLAT/CALIFORNIA U.S.A.

Chamise Chaparral:

S: *Adenostoma fasciculatum, Arctostaphylos spp., Ceanothus spp., Rhamnus spp., Heteromeles arbutifolia, Quercus dumosa, Cercocarpus betuloides, Garrya veatchii, Prunus ilicifolia, Rhus spp., et al.

DS: Eriogonum fasciculatum (Sf), Salvia spp. (Sf. w), et al.

H: various grasses and forbs.

References: Mooney & Parsons 1973; Küchler 1964.

11. WEST YELLOWSTONE/MONTANA U.S.A.

Montane Spruce-Fir Forest with Douglas Fir and Pine:

T: *Picea engelmannii, *Abies lasiocarpa, Pseudotsuga menziesii, Pinus contorta var. latifolia s. str., with Larix spp., Populus tremuloides et al., Alnus tenuifolia, Salix spp., et al.

S: Vaccinium spp., Cornus spp., Spiraea spp., et al.

F: Gaultheria humifusa, Ranunculus spp., Aquilegia spp., Viola spp., Castilleja spp., Trifolium pratense, Gentiana spp., et al.

G: Carex spp., et al.

References: Küchler 1964; Knapp 1965; Shaw 1974; Shelford 1963.

12. LOGAN/UTAH U.S.A.

Summergreen and Semi-Deciduous Scrub ('Deciduous Chaparral'):

S: *Cercocarpus ledifolius (EG), *Quercus gambelii, Qu. utahensis et al., Arctostaphylos spp., Ceanothus velutinus (EG), Purshia tridentata, Cowania stansburiana, Amelanchier utahensis, Acer grandidentatum, Pachystima myrsinites, Physocarpus malvacens, Rhus trilobata, Symphoricarpos spp., Peraphyllum ramosissimum, et al.

H: various grasses and forbs.

References: Küchler 1964; Welsh & Moore 1973; personal observation.

13. CODY/WYOMING U.S.A.

Sagebrush-Short Grass Steppe:

S: *Artemisia tridentata, et al.

G: *Agropyron spicatum et al., Festuca idahoensis, Stipa spp., Poa spp., et al.

F: scattered Phlox spp., Purshia tridentata, Lithospermum ruderale, Lupinus sericens, Atriplex spp., et al.

SS: Opuntia spp.

References: Weaver & Albertson 1956; Küchler 1964; personal observation.

14. FLAGSTAFF/ARIZONA U.S.A.

Southern Rocky Mountain Yellow-Pine Forest:

T: *Pinus ponderosa var. scopulorum, other Pinus spp., Populus tremuloides, et al.

S: Ceanothus fendleri, Quercus spp., Holodiscus dumosus, et al.

G: *Muhlenbergia spp., Festuca spp., Aristida spp., et al.

F: Antennaria spp., Lupinus spp., et al.

References: Küchler 1964; Knapp 1965; Shelford 1963.

15. PHOENIX/ARIZONA U.S.A.

Colorado Valley Larrea-Franseria (90–95%) semi-desert:

S: *Larrea tridentata (EG), *Franseria dumosa (SG), et al.

G: desert grasses, e.g. Bouteloua, Aristida.

SS: occasional Opuntia, Echinocactus, et al.

F: Encelia, et al.

Eph: abundant summer and winter ephemerals.

Small, mostly deciduous trees and bushes (e.g. Prosopis juliflora, Torreyana, Cercidium ssp., Olneya tesota, Dalea) may be found on upper bajadas and in drainage-ways.

References: Shreve & Wiggins 1964; Knapp 1965; Küchler 1964; Shelford 1963; personal observation.

16. TUCSON/ARIZONA U.S.A.

Upland Larrea-Cercidium-Opuntia Scrub, with tall cacti:

T: *Cercidium microphyllum (RG) et al., Prosopis juliflora (RG), Olneya tesota (SG), et al.

S: *Larrea tridentata, *Franseria dumosa (SG), Fouquiera splendens (RG), Ephedra, et al.

SS: Carnegiea gigantea (a), Opuntia spp., Echinocereus, et al.

G: desert-grasses, e.g. Bouteloua, Aristida (locally abundant),

F: Encelia farinosa, et al.

Eph: abundant summer and fewer winter ephemerals.

References: Shreve & Wiggins 1964; Walter 1973; Küchler 1964; Knapp 1965; Shelford 1963; personal observation.

17. SANTA FE/NEW MEXICO U.S.A.

Open Juniper-Pinyon Woodland, with grasses but few forbs:

DT: *Juniperus monosperma, *J. scopulorum, J. osteosperma (= J. utahensis), *Pinus edulis, P. monophylla.

S: Cercocarpus spp., Ceanothus spp., Quercus spp., Arenaria filiorum, Chrysothamnus spp., Artemisia tridentata, et al.

H: turf and bunch grasses (Agropyron smithii, Bouteloua gracilis, Buchloe dactyloides, Sporobolus).

References: Weaver & Albertson 1965; Knapp 1965; Küchler 1964; personal observation.

18. COLORADO SPRINGS/COLORADO U.S.A.

Southern Short and Mixed-Grass Prairie:

G: *Bouteloua gracilis, B. hirsuta, Muhlenbergia gracillima, Aristida longiseta; with Stipa comata et al., Andropogon scoparius, Schedonnardus paniculatus, Koeleria cristata, Sitanion elymoides, Sporobolus cryptandrus, Hilaria jamesii, Bouteloua curtipendula, and other grasses.

DT: scattered Juniperus spp.

RS: Yucca glauca.

SS: occasional Opuntia spp.

F: steppe forbs, e.g. Plantago, Senecio, Haplopappus,

References: Weaver & Albertson 1956; Daubenmire 1943; Shantz 1906; Küchler 1964; Knapp 1965.

19. CUSTER/SOUTH DAKOTA U.S.A.

Black Hills Pine Forest:

T: *Pinus ponderosa.

UT: Prunus virginiana, et al.

S: Arctostaphylos uva-ursi, Juniperus communis, Symphoricarpus spp., Artemisia tridentata, Cercocarpus parviflorus, Physocarpus intermedia, Rhus trilobata, et al.

G: Elymus spp., Poa pratense, Agropyron spp., Koeleria cristata, Bouteloua gracilis(s), Stipa comata, Buchloë dactyloides(s), et al., plus Carex spp.

F: Chrysothamnus graveolens, Gutierrezia sarothrae, et al.

References: Hayward 1928; Weaver & Albertson 1956; Van Bruggen 1976; Küchler 1964; personal observation.

20. OELRICHS/SOUTH DAKOTA U.S.A.

Short Bunch-Grass (Agropyron-Stipa) Steppe:

G(mid-height): *Koeleria cristata, *Agropyron smithii, A. spicatum, *Stipa comata, et al.

G(short): *Buchloë dactyloides(s), *Bouteloua gracilis, other Agropyron and Stipa spp., et al.

F: many steppe forbs, e.g. Liatris, Aster, Penstemon, Mertensia, Solidago, Artemisia frigida(w).

References: Hayward 1928; Weaver & Albertson 1956; Küchler 1964; personal observation.

21. OKLAHOMA CITY/OKLAHOMA U.S.A.

Southern Tall-Grass Prairie (red-clay Stipa-Koeleria association):

T: occasional Quercus stellata, Qu. marilandica, Carya texana.

G: *Andropogon scoparius et al., *Bouteloua racemosa, *Agropyron smithii, *Sporobolus asper, with other tall and mid-height grasses.

F: numerous prairie forbs, e.g. Claytonia virginica, Houstonia minima, Bursa, Draba, Anemone caroliniana, Baptisia spp., Allium spp., Achillea spp., Asclepias, Croton, Psoralea, Solidago, Helianthus, Veronica, and many more.

S: numerous steppe semi-shrubs, e.g. Rhus, Amorpha, Clematis.

References: Knapp 1965; Bruner 1931; Küchler 1964; personal observation.

22. BARTLESVILLE/OKLAHOMA U.S.A.

Oak Woodland-Tall (subclimax) Grassland Mosaic, with cove-like closed oak forests:

T: *Quercus marilandica, *Qu. stellata, other Quercus spp., Carya texana, et al.

UT: Cercis canadensis, Celtis spp., Prunus serotina, et al.

G: *Andropogon scoparius, *A. gerardi, Bouteloua spp., Elymus canadensis, Panicum virgatum, Stipa leucotricha, Sporobolus spp., Eragrostis spp., Sorghastrum nutans, et al.

F: numerous steppe forbs, especially Compositae (e.g. Solidago, Aster, Erigeron, Helianthus, Liatris) and legumes (e.g. Baptisia, Amorpha, Psoralea).

Soil: fine-textured (shale) soils, favoring grasses over trees.

References: Bruner 1931; Ray 1959; Küchler 1964; personal observation.

23. SOUTH BEND/INDIANA U.S.A.

Summergreen Oak-Hickory Forest (western limit):

T: *Quercus macrocarpa, *Qu. velutina, *Qu. alba, *Qu. rubra, *Carya ovata, *C. tomentosa, other Quercus and Carya spp., Acer rubrum, Fraxinus americana, Juglans nigra, Tilia americana, Ulmus americana, et al.

UT: Prunus serotina, Cercis canadensis, Sassafras albidum, Hamamelis virginiana, Malus spp., et al.

H: many grasses and mesic-submesic forbs.

References: Braun 1950; Knapp 1965; Küchler 1964.

24. COLUMBUS/OHIO U.S.A.

Summergreen Beech-Maple Forest:

T: *Fagus grandifolia, *Acer saccharum, Fraxinus americana, Quercus rubra, Qu. alba, Carya spp., Ulmus spp., Betula lutea, Carpinus caroliniana, Liriodendron tulipfera, Juglans cinerea, Tilia americana, et al.

UT: Prunus serotina, Ostrya virginiana, Aesculus spp., Cercis canadensis, et al.

UB: Lonicera canadensis, et al.

F: Hepatica americana, Hydrophyllum macrophyllum, and many others.

References: Braun 1950; Knapp 1965; Küchler 1964.

25. BAR HARBOR/MAINE U.S.A.

Spruce-Northern Hardwood Forest and open Woodlands with bogs:

T: *Picea rubra, *P. canadensis, Abies balsamea, Pinus strobus, Thuja occidentalis, Tsuga canadensis, Betula papyrifera, B. populifolia, B. lutea, Populus tremuloides, P. grandidentata, Acer rubrum, Quercus rubra, Fagus grandifolia, et al.

S: Vaccinium spp., Alnus spp., Viburnum cassinoides, Gaylussacia baccata, Kalmia angustifolia, Acer spp., Rubus spp., et al.

H: Agrostis spp., Festuca spp., Carex spp., et al.

Soil: glacial till, with sand and areas of bare rock.

References: Moore & Taylor 1927; Küchler 1964; Davis 1966; Damman 1977; personal observation.

26. TRENTON/NEW JERSEY U.S.A.

Appalachian Summergreen Oak-Chestnut Forest:

T: *Quercus rubra, Qu, alba, *Castanea dentata (destroyed by Chestnut Blight), Quercus coccinea et al., Liriodendron tulipfera, Acer saccharum, A. rubrum, Fagus grandifolia, Liquidambar styraciflua, Carya spp., Nyssa sylvatica, Pinus rigida, P. strobus, Fraxinus americana, Betula lenta, Carpinus caroliniana, Tsuga canadensis, et al.

UT: Cornus florida, Cercis canadensis, Prunus spp., et al.

S: Vaccinium spp., Kalmia spp., et al.

H: many summergreen mesic forbs.

References: Braun 1950; Knapp 1965; Küchler 1964.

27. MT. MITCHELL/N. CAROLINA U.S.A.

Appalachian Fir-Spruce Forest:

T: *Abies fraseri, Picea rubens.

UT: Sorbus americana, Prunus pennsylvanica, Betula lutea, Acer spicatum, et al.

S: Rhododendron catawbiense, R. maximum, Rubus spp., Vaccinium erythrocarpum, Viburnum alnifolium, Menziesia pilosa, et al.

H: Houstonia serpyllifolia, Oxalis acetosella, Clintonia borealis, Aster acuminatus, Saxifraga michauxii, Chelone lyonii, Carex debilis, Cinna latifolia, et al.

References: Whittaker 1956; Davis 1930; Radford 1976; personal observation.

28. CHARLESTON/SOUTH CAROLINA U.S.A.

Southern Bottomland Mixed Forest (with local swamps):

T: Quercus virginiana, Qu. laurifolia, Qu. alba (SG), Qu. stellata (SG), Qu. phellos (SG), Magnolia grandiflora, M. virginiana, Liquidambar styraciflua (SG), Carya spp. (SG), Pinus taeda, P. echinata, Nyssa sylvatica (SG), et al., with Taxodium distichum (SG, Conif.) in swamps.

UT: Ilex vomitoria, Persea borbonia, Cornus florida (SG), et al.

RS: Sabal palmetto.

S: Vaccinium, Bumelia, Baccharis, Myrica cerifera, et al.

G: Arundinaria gigantea, et al.

Also: Tillandsia usneoides (Ep), with many vines.

References: Braun 1950; Küchler 1964; Radford, Ahles & Bell 1968; personal observation.

29. SAVANNAH/GEORGIA U.S.A.

Coastal Plain Pine-Hardwood Forest (with more pines and with local swamps):

T: Quercus virginiana, Qu. phellos (SG), Pinus taeda, P. flexilis, P. palustris, Liquidambar styraciflua (SG), Nyssa sylvatica (SG), Magnolia grandiflora, et al., with Taxodium distichum (SG, Conif.) in swamps.

UT: Persea borbonia, Ilex vomitoria, Cornus florida (SG), et al.

RS: Sabal palmetto.

S: Vaccinium spp., Baccharis spp., Myrica cerifera, et al.

G: Arundinaria gigantea, Panicum amarum, et al., with Spartina and Juncus prairies (brackish).

Also: Tillandsia usneoides (Ep), with many vines.

References: Braun 1950; Küchler 1964; Shelford 1963; personal observation.

30. BIRMINGHAM/ALABAMA U.S.A.

Southern Pine-Summergreen Oak Forest:

T: *Quercus stellata, *Qu. alba, *Pinus taeda, *P. echinata, P. australis, P. palustris, P. caribaea, Quercus margaretta, Qu. marilandica, Qu. velutina, Carya spp., Nyssa sylvatica, Oxydendron arboreum, et al.

UT: Cornus florida, Cercis canadensis.

S: Vaccinium spp., Chrysobalanus, Eriogonum, Gaylussacia, Croton, Leiophyllum, et al.

G: Aristida stricta, Panicum, Sporobolus, Andropogon, et al.

Also: many vines and spring forbs.

References: Braun 1950; Knapp 1965; Küchler 1963; personal observation.

31. XILITLA MEXICO

Tropical Montane Rainforest with Holarctic Species (lower altitudinal limit):

T: Quercus candolleana, Qu. excelsa, Qu. trinitatis, Qu. xalapensis et al., Carpinus caroliniana, Carya mexicana, Clethra spp., Cornus spp., Liquidambar styraciflua, Pinus patula, Podocarpus spp., et al.

US: Turpinia spp., Vaccinium spp., Leucothoe mexicana, Eugenia, Garrya laurifolia, Ilex spp., et al.

V: Rhus radicans, Parthenocissus quinquefolia, et al.

H: Gaultheria spp., Asplenium spp., et al.

References: Miranda & Sharp 1950; Knapp 1965.

32. MAZATLAN MEXICO

Lowland Dry Raingreen Thorn Forest:

T: *Ipomoea arborescens, *Acacia cymbispina, *Zizyphus sonorensis (EG), Bauhinia longiflora, Crescentia alata, Bunchosia palmeri, Acacia farnesiana, A. pennatula, Bursera, Piscidia piscipula, Alvaradoa amorphoides, Ceiba, Haematoxylon brasiletto, Pedilanthus macrocarpus, Lysiloma, Prosopis juliflora, Pithecellobium sonorae, Lonchocarpus, Caesalpinia, Guaiacum (Zygophyll.), Jatropha cinerea, Ficus petiolaris, Cordia, Mimosa, et al.

S: Croton alamosanus, Fouquiera macdougalii, Cassia spp., et al.

SS: Cereus thurberi, Pachycereus (both arborescent).

H: Bouteloua curtipendula, B. rothrockii, Hilaria semplei, Cathestecum spp., et al.

References: Leopold 1950; Shreve 1934; Pesman 1962; Mata et al. 1972; Miranda & Hernandez 1963; Knapp 1965.

33. QUIBDO COLOMBIA

Tropical coastal-lowland Rainforest (tall, multistoreyed, with numerous lianas and epiphytes, and with palm forests in swamps):

T: Cespedetia apathulata, Brosimum utile, Terminalia amazonica, Ficus, Dialyanthera, Hura crepitans, Aniba perutilis, Hevea, Nectandra, Cedrela, Tabebuia, Castilla, Cariniana, Prioria, Myroxylon, Ochroma, Cecropia, Schizolobium, and many others.

UT: numerous palms (e.g. Jessenia polycarpa, Welfia regia, Phytelephas spp.), and many other trees.

Also: many shrubs, lianas, epiphytes, and ferns.

References: Verdoorn 1945; Hueck 1966; West 1958.

34. BARINAS VENEZUELA

Tradewind Raingreen (Monsoon) Forest:

T: Spondias mombin, Pterocarpus vernalis et al., Bombacopsis sepium, Sapium, Hura crepitans, Astronium graveolens, Brosimum, Tabebuia spp., Cochlospermum vitifolium, Cordia spp., Pouteria, Terminalia, Cedrela mexicana, Swietenia macrophylla, Samanea saman, et al.

US: palms (e.g. Attalea, Orbignya, Oenocarpus, Syagrus, Oreodoxa), et al.

S: Psychotria, Miconia, Urera, Croton, et al.

H: many grasses and forbs, including Scitaminaceae.

E: Ficus spp., Coussapoa (Morac.) spp., Phoradendron spp.

References: Hueck 1966; Veillon 1955.

35. OBIDOS BRAZIL

Dry Tropical Rainforest with Campos Cobertos (islands of drier open forest):

T: Curatella americana, Anacardium occidentale, Bowdichia, Aeschynomene, Qualea, Vochysia, Byrsonima, et al., with Cassia, Calliandra, Crotalaria, Eugenia, et al.

H: tall and short tropical grasses and many other species.

Soil: terra roxa

References: Hueck 1966; Ule 1915.

36. QUIXERAMOBIM BRAZIL

True Caatinga (Raingreen Thorn-Woodland with palms and arborescent stem-succulents):

T: Zizyphus joazeiro (EG), Mimosa, Cassia, Caesalpinia, Acacia, Piptadenia (Leg.), Amburana, Cavanillesia arborea (Bombac.), Chorisia ventricosa (Bombac.), Melanoxylon, Aspidosperma, Platymiscium, Machaerium, Spondias, Pithecolobium, et al.

RT: Cocos schizophylla, Copernicia cerifera, et al.

RS: Ground-bromeliads, e.g. Aechmoa, Ananas, Billbergia, Dyckia, Pitcairnia, Neoglaziovia.

SS: Cereus jamacaru, C. squamosus, Opuntia, Pilocereus gounellei, Melanocactus (most arborescent).

S: Capparis yco (EG), Euphorbia, Croton, Caesalpinia, Jatropha, Anonaceae, Coccoloba, Cnidoscolus, et al.

H: short grasses but few forbs.

References: Hueck 1966; Corrêa 1926; Hueck & Seibert 1972.

37. BELO HORIZONTE BRAZIL

Campos Cerrados (dwarf evergreen woodland of 2–4 m, with thick grass cover; near transition to mixed forests):

T: Kielmeyera, Byrsonima, Dimorphandra, Machaerium (EG Leg.), Stryphnodendron, Dalbergia (EG Leg.), Copaifera (EG Leg.), Caryocar, Aspidosperma tomentosum (RG) et al., Bombax, Curatella americana, Quassia versicolor, Luehea paniculata, Caryocar brasilense, Palicourea rigida, Terminalia argentea, Shinus, Lithraea, Astronium, et al.

RS: dwarf-palms, e.g. Attalea exigua, Cocos petraea, C. acaulis, Asterocaryum avenarium, Diplothemium campestre, Acrocomia sclerocarpa.

S: Erythroxylon, Byrsonima, Cassia, et al.

H: many tall and mid-height grasses, with numerous forbs.

References: Hueck 1966; Ferri 1969; Hueck & Seibert 1972.

38. SAO PAOLO BRAZIL

Subtropical Species-Rich Semi-Evergreen Forest:

T: Cedrela fissilis, C. glaziovii, Balfourodendron riedelianum, Hymenaea stilbocarpa, Inga edulis, Centrolobium robustum, Myroxylon peruiferum, Dalbergia nigra, Aspidosperma polyneutron, Myrocarpus frondosus, Machaerium spp., Piptadenia spp., Holocalyx glariovii, Luehea divaricata, Gallesia guararema, Cabralea cangarena, Phoebe porosa, Tabebuia spp., Cariniana estrellensis, Melanoxylon brauna, Plathymenia foliosa, Ficus spp., Vochysia spp., et al.

UT: palms, e.g. Arecastrum romanzoffianum, Acrocomia totai.

UB: thick brush, with Euphorbiaceae, lianas, epiphytes.

Current use: fields and plantations, especially coffee.

References: Hueck 1956, 1966; Hueck & Seibert 1972.

39. VILLA NOUGUES/TUCUMAN ARGENTINA

Southern Andes Subalpine Alder Belt (near lower limit):

T: *Alnus jorullensis (SG), with species from the montane belt, e.g. Juglans australis (SG), Podocarpus parlatorei (EG), Prunus spp.

US: Sambucus, Ilex, Berberis, et al.

F: Geum, Ranunculus, Anemone, Vicia, Urtica, Lathyrus, Thalictrum, Geranium, Stachys, Salvia, Veronica, Cerastium, et al.

References: Hueck 1954, 1966; Hieronymus 1945.

40. ISLA VICTORIA/LAGO NAHUEL HUAPI ARGENTINA

Patagonian Lower-Montane Nothofagus Rainforest (30–40 m):

T: *Nothofagus dombeyi, with N. betuloides, Eucryphia cordifolia.

UT: Drimys winteri, Nothofagus pumilio (SG), scattered Libocedrus tetragona (= Pilgerodendron uviferum), Podocarpus (e.g. P. andinus, P. poepp. ex Endl., P. nubigenus Lindl.), Saxegothea, et al.

UB: Weinmannia trichosperma, Caldcluvia paniculata, Laurelia serrata, Myrtaceae, et al;

Also: many ferns, epiphytes, and often bamboo.

Reference: Hueck 1966.

41. BARILOCHE, SAN CARLOS DE ARGENTINA

Patagonian Lower-Montane Libocedrus Forest (20–25 m)

T: *Libocedrus chilensis (up to 90%), with Lomatia hirsuta, Diostea juncea, Schinus crenatus, Maytenus boaria, and isolated Araucaria araucana.

S: Berberis buxifolia, B. darwinii, Azara microphylla, Fabiana imbricata, Pernettya poeppigii.

Also: Patagonian steppe-grasses and dwarf-shrubs.

References: Hueck 1966; Schmithüsen 1960.

42. VALDIVIA CHILE

Temperate Species-Rich Broad-Macrophyll Rainforest (40–50 m) (near transition to summergreen mesophytic forest):

T: Aextoxicon punctatum, Eucryphia cordifolia, Laurelia aromatica, L. serrata, Drimys winteri, et al. (all with large, lauraceous EG leaves); Myrceugenella apiculata, et al. (with small, myricaceous EG leaves); plus Flotovia diacanthoides, Guerina avellana, Lomatia ferruginea, L. hirsuta, Maytenus boaria, Nothofagus obliqua (SG), N. procera (SG), scattered Fitzroya patagonica (Coniferae), et al.

S: Weinmannia trichosperma, Caldcluvia paniculata, Rhaphithamnus cyanocarpus, Ugni molinae, Coriaria ruscifolia, et al.

Also: many lianas (e.g. Boquila trifoliata, Mitraria coccinea, Griselinia ruscifolia, Dioscorea brachybotrya), epiphytes, climbing and rigid bamboo-grasses, and a thick ground cover of forbs and mosses.

References: Hueck 1966; Veblen & Ashton 1978; Schmithüsen 1956, 1960.

43. USHUAIA CHILE

Subantarctic Summergreen Microphyllous Forest (10–20 m), near southern limit:

T: *Nothofagus pumilio, with N. antarctica and scattered N. betuloides (EG) and N. dombeyi (EG).

S(few): Ribes magellanicum, R. cucullatum, Berberis spp., Pernettya mucronata, Escallonia spp.

G: Carex, Festuca, Trisetum, Elymus, plus the bamboo Chusquea.

F: Viola, Rubus, Ranunculus, Senecio, Sisyrinchium, Anemone, Vicia, et al.

Reference: Hueck 1966.

44. THINGVELLIR ICELAND

Icelandic Maritime 'Mo' Tundra (herbs, dwarf-shrubs, and mosses):

G: *Festuca rubra et al., *Carex spp., Agrostis canina, Luzula spicata, Juncus trifidus, Deschampsia flexuosa, Poa, Trisetum, et al.

DS: Salix herbacea, Empetrum nigrum, Calluna vulgaris, Arctostaphylos uva-ursi, et al.

F: Polygonum viviparum, Galium spp., Selaginella selaginoides, Equisetum spp., et al.

Also: many mosses and lichens (locally dominant).

References: Hansen 1930; Jonsson 1905; Ostenfeld & Grontved 1934; personal observation.

45. ABISKO SWEDEN

Summergreen Fell-Birch Forest-Tundra:

T: *Betula tortuosa (= B. pubescens ssp. tortuosa), B. callosa, Populus tremula, with Pinus silvestris on moors and on serpentine soils.

S: Betula nana, Empetrum hermaphroditum, Vaccinium myrtillus, with Arctostaphylos alpina, A. uva-ursi, Loiseleuria procumbens, et al.

Also: Carex ssp., grasses, mosses, lichens (Cladonia, Nephroma).

References: Blüthgen 1960; Acta Phytogeogr. Suecica 1965; personal observation.

46. DALBY SÖDERSKOG (LUND)/SKÅNE — SWEDEN

Summergreen Elm-Oak Forest, with low-tree and shrub layers (undisturbed since 1916):

T: *Ulmus glabra, *Quercus robur, Fagus sylvatica, Fraxinus excelsior
UT: mainly Ulmus saplings, Corylus avellana, Crataegus laevigata, Acer platanoides, Euonymus europaeus
S: Prunus padus, Crataegus, saplings
H: Anemone nemorosa, Ranunculus ficaria, et al., Oxalis acetosella, Corydalis, Gagea, Geum, Veronica, Carex, Poa, Viola, Campanula, Equisetum, etc.
V: Lonicera xylosetum.

Reference: Malmer et al. 1978.

47. MIKULOV/WEINVIERTEL — AUSTRIA

Summergreen Woodland and Scrub with dry grassland mosaic:

T: *Quercus pubescens, Carpinus, Pinus sylvestris, P. nigra, Robinia pseudoacacia, Betula, Corylus
UB: *Quercus, Carpinus, Corylus, Betula
G: *Festuca (Festucetalia valesiacae), Agropyron repens, A. intermedium, Poa angustifolia, P. compressa, Bromus inermis, Stipa, Danthonia, et al.
V: Convolvulus arvensis
F: Cardaria draba, Equisetum arvense, Kochia prostrata, and many others

Reference: Eijsink et al. 1978.

48. AIN-DRAHAM — TUNISIA

Semi-Evergreen Mesic Oak Forest (10–20 m):

T: *Quercus faginea ssp. baetica (SG), Qu. suber (EG), Sorbus torminalis, Prunus avium, Acer campestre, Ilex aquifolium.
S: Crataegus ssp., Erica arborea, Cytisus triflorus, et al.
V: Rubus discolor, Tamus communis, Hedera helix s.l., et al.
F: Cyclamen africanum, Geranium spp., Scilla aristidis, Moehringia pentandra, Asperula laevigata, and many others.
G: Festuca drymeja var. grandis, Melica minuta, Carex, et al.

References: Knapp 1973; UNESCO 1969.

49. AZROU — MOROCCO

Mediterranean Montane Oak Forest:

T: *Quercus ilex (EG).
US: Sarothamnus baeticus, Cytisus triflorus, Crataegus monogyna, Cistus salvifolius, Jasminum fruticans, Thymus zygis, Viburnum tinus, Rosa spp., and individually Buxus sempervirens, Phillyrea spp., Pistacia spp., et al.
V: Lonicera, Smilax, Clematis, Rubia, Tamus.
Ch: Ononis arborescens, Psoralea bituminosa, Ruta montana, Ajuga pseudo-Iva, Cephalaria leucantha, Artemisia sp., Ilex aquifolium, Daphne laureola, Rhamnus, et al.
Hc: Festuca, Carex, Rumex, Ranunculus, Eryngium, Haynaldia hordacea, Salvia, Geum, et al.

References: Knapp 1973; Walter 1968, 1977b; UNESCO 1969; Braun-Blanquet & Maine 1924; Emberger 1938; personal observation.

50. KSAR-ES-SOUQ — MOROCCO

Sub-Saharan High-Plateau Shrub-Bunch Grass Steppe:

G: *Stipa tenacissima(b), Aristida pungens, A. plumosa, Panicum turgidum, et al.
S: Artemisia herba-alba, A. campestris, Ephedra alata, Anabasis spp., Balanites aegyptiaca, Salvadora persica, Zizyphus lotus, Boscia senegalensis, Calotropis procera, Genista raetum, Lygeum spartum, et al.
F: yellow-flowered Compositae, et al.

References: Shantz & Marbut 1923; Zohary 1973; UNESCO 1969; Emberger 1938; personal observation.

51. DAMANHUR AREA EGYPT

(west, along Cairo-Alexandria road)

Sparse Artemisia-Thymelaea Desert Scrub (10% cover):

S: *Artemisia monosperma, Zygophyllum coccineum, *Thymelaea hirsuta, Anabasis articulata, Salsola, Lithospermum callosum

G: Aristida plumosa, A. pungens, Panicum turgidum et al.

F: Allium, Muscari, Plantago albicans, Asphodelus, Helianthemum lippii

Eph: Mesembryanthemum forskalei, Trigonella, Ifloga, Erodium, Matthiola, Salvia lanigera, Malva, Chrysanthemum coronarium, et al.

V: Convolvulus lanatus

References: Ayyad & El-Ghonemy 1976; Walter 1973.

52. MAKURDI NIGERIA

Raingreen Wooded Savanna (Wet-Savanna Zone):

T: *Daniellia oliveri (Caesalp.), *Lophira alata, *L. lanceolata, *Parinari polyandra, Terminalia spp., Afzelia africana, Bridelia ferruginea, Butyrospermum parkii, Cussonia barteri, Hymenocardia acida, Vitex doniana, Annona, Detarium senegalense, Gardenia, Lannea, Prosopis africana, et al.

G: *Hyparrhenia spp., *Andropogon spp., *Pennisetum spp., Panicum, Schizachyrium, Tephrosia, Eriosema, et al.

F: Eulophia spp., Indigofera, Curculigo pilosa, Amorphophallus spp., Kaempferia, Costus spectabilis.

References: Knapp 1973; Keay 1949.

53. MAHADDAY-WEYNE SOMALIA

Dry Raingreen (Commiphora) Thorn-Scrub:

T/B: *Commiphora spp., *Acacia spp., Capparidaceae (e.g. Boscia, Cadaba, Maerua, Capparis, Courbonia), Grewia spp., Delonix elata, Sterculia, Heeria (Anacard.), Lannea (Anacard.), Sesamothamnus (Pedaliac.), Salvadora persica, et al.

G: Chrysopogon aucheri, Aristida stipoides, Cenchrus ciliaris, Sporobolus variegatus, Schoenfeldia gracilis, et al. (all short), with tall Beckeropsis and Hyparrhenia in wide river valleys.

References: Knapp 1973; Keay 1959.

54. ASMERA/ERITREA ETHIOPIA

Region of Dry Thorn-Scrub, Dwarf-Shrub, and Montane (short) Grassland Mosaic:

T: *Acacia spp., Juniperus procera, Olea africana, Balanites aegyptica, Sanseveria longiflora, Barbeya oleoides (Ulmac.), Celtis africana, Sideroxylon oxyacantha, Tarchonanthus camphoratus.

S: *Acokanthera schimperi, *Buxus hildebrandtii, et al.

SS: Euphorbia spp.

RT: Dracaena.

G: *Festuca abyssinica, *Pentaschistis mannii, *Agrostis isopholis, et al.

References: Knapp 1973; Keay 1959; Shantz & Marbut 1923.

55. KERICHO KENYA

Region of Equatorial Montane (Evergreen) Rainforest:

T_2: *Ocotea usambarensis, *Podocarpus spp., with Aningeria, Casearia, Chrysophyllum, Ekebergia, Entandrophrgama, Ficus, Macaranga, Polyscias, Pygeum, Strombosia, Schefflera, Vitex, and many others.

T_1: Afrocrania, Allophylus, Canthium, Cassine, Cassipourea, Croton, Cyclicomorpha, Conopharyngia, Cola, Dracaena, Enneastemon, Ficalhoa, Galiniera, Grumelia, Lasianthus, Oxyanthus, Peddiea, Premna, Trichilia, Trichocladus, Uvariodendron, Xymalos, et al.

References: Knapp 1973; Lind & Morrison 1974.

56. NANYUKI KENYA

Region of potential Dry Montane Conifer Forest (open-forest mosaic with thorn-scrub and wooded steppe):

T: *Juniperus procera, *Olea africana, *O. hochstetteri, Podocarpus spp., Croton megalocarpus, Cussonia spicata, Tarchonanthus camphoratus, et al.

UB: *Acacia spp., *Commiphora spp., Acanthus emiens, et al.

G: Chrysopogon, Aristida, et al.

References: Knapp 1973; Keay 1959; Lind & Morrison 1974.

57. LUSAKA ZAMBIA

Miombo Forest (somewhat open raingreen forest dominated by legumes with mid-size, compound leaves):

T: *Brachystegia (Caesalp.) spp., Julbernardia (Leg.), Isoberlinia (Leg.), Berlinia (Caesalp.), Guibourtia (Caesalp.), Marquesia (Caesalp.), Afrormosia (Leg.), Parinari (Ros.), Pterocarpus (Leg.) et al.

UT: Uapaca (Euphorb.) spp., Diospyros spp., Monotes (Dipterocarp.) spp., Pseudolachnostylis (Euphorb.), Faurea (Proteac.), et al.

S: Hexalobus, Hymnocardia, Maprounea, Allophylus, et al.

H: many grasses and forbs, especially active during the dry season.

References: Knapp 1973; Keay 1959; Shantz & Marbut 1923.

58. MAUN BOTSWANA

Mopane Dry Forest (open raingreen forest with sparse undergrowth):

T: *Colophospermum mopane (Caesalp.: entire mesophylls), with Acacia nigrescens, Kirkia acuminata (Simaroub.), Terminalia prunioides, et al.

S: Grewia, Combretum, Capparidaceae (e.g. Courbonia), Ximenia, Boscia, et al.

H: relatively few grasses and forbs.

References: Knapp 1973; Keay 1959.

59. KROONSTAD/ORANJE FRYSTAAT SOUTH AFRICA

Region of Tall, Forb-Poor Themeda Grassland (High Veld):

G: *Themeda triandra, Heteropogon contortus et al., Setaria flabellata, Elyonurus argenteus, Eragrostis, Cymbopogon, Microchloa caffra, Tristachya spp., Helichrysum rugulosum, Vernonia kraussii, Andropogon spp., Hyparrhenia hirta, et al.

F: Elephantorrhiza burchellii, Hypoxis costata, Aster serrulatus, Clerodendron triphyllum, Indigofera hedyantha, et al.

References: Shantz & Marbut 1923; Knapp 1973; Acocks 1953; Zinderen Bakker 1973.

60. CATHEDRAL PEAK (LITTLE BERG)/DRAKENSBERG SOUTH AFRICA

Montane Evergreen Heathland (Fynbos):

S: *Passerina filiformis, *Philippia evansii, *Widdringtonia nodiflora (n, Coniferae), Passerina montana, Protea, Erica, Myrica, Cliffortia, Macowania, Buchenroedera, Anthospermum, Rhus discolor, R. dentata, Buddleja, Syncolostemon, Calpurnia, Melianthus, Asparagus, Senecio, with Stoebe, Euphorbia, Myrsine, Artemisia, Psoralea, Polygala and many more.

RT: Encephalartos ghellinekii (cycad)

V: Riocreuxia torulosa var. tomentosa, Dioscorea sylvatica, Clematis brachiata.

H: Restio spp., Polystichum, Cymbopogon validus, Berkheya macrocephala, Festuca costata, Helichrysum spp., Indigofera, and many others.

Reference: Killick 1979.

61. HELL-BOURG REUNION

Region of Montane Evergreen Rainforest:

T: Cassine (= Elaeodendron), Diospyros, Dodonaea,Doratoxylon, Evodia, Linociera, Mimusops, Nuxia, Ocotea, Olea, Psiloxylon, Sideroxylon, Tabernaemontana, Terminalia, Weinmannia, Xylopia, Danais, et al.

References: Knapp 1973; Cadet 1974.

62. BORISOVKA/RUSSIA U.S.S.R.

Summergreen Oak-Basswood Forest:

T: *Quercus robur, *Tilia cordata, Acer platanoides, A. campestre, Ulmus montana, Malus sylvestris, Crataegus curvipendula, et al.

S: Euonymus europea, E. verrucosa, et al.

F: spring geophytes, e.g. Scilla sibirica,Corydalis halleri, Anemone ranunculoides, Gagea lutea, Ficaria verna; summer forbs, e.g. Aegopodium podagraria, Galium (Asperula) odorata, Stellaria holostea, Glechoma hirsuta, Pulmonaria obscura, Polygonatum multiflorum, Geum urbanum, Viola suavis, Asarum europaeum.

G: Carex pilosa, Poa nemoralis, Festuca gigantea, et al.

Reference: Walter 1976a.

63. KERKI/TURKMENISTAN U.S.S.R.

Kara-Kum Semi-Desert (with leafless and nanophyllous trees):

T: *Ammodendron conollyi (Leg.), *Haloxylon aphyllum (= ammodendron) (Chenopod.).

S: *Haloxylon persicum, Calligonum spp. (Polygon.), Salsola richteri, S. arbuscula, Astragalus paucijugus, Aellenia subaphylla, Ephedra strobilacea.

S(Sf): Smirnovia turkestana (Leg.), Artemisia spp., Astragalus spp., Salsola spp.

H: Carex physodes, C. subphysodes, C. pachystylis, Poa bulbosa, Tournefortia sogdiana, Heliotropium arguzioides, Aristida karelinii, Allium spp., Gagea spp., Tulipa sogdiana, Iris spp., Rhinopetalum arianum, Eremurus inderiensis (Lil.), Eminium lehmannii (Arac.), Rheum turkestanicum, Dorema sabulosa, et al.; and many annual herbs.

References: Walter 1976a; Walter & Box, in press.

64. CHATKAL MOUNTAINS/KIRGHIZ U.S.S.R.

Montane Summergreen Walnut Belt:

T: *Juglans fallax (= J. regia ssp. fallax), Prunus divaricata, Acer spp., Malus spp., Crataegus spp., et al.

G: Melica altissima, Festuca gigantea, Agropyron caninum, Poa nemoralis, Bromus spp., Carex polyphylla, et al.

F: Impatiens parviflora, Aegopodium podagraria, Lamium album, Cerastium, Senecio, Allium, Polygonatum, Arum, et al.

Also: Lonicera spp., Rosa spp., Galanthus, Brachypodium, et al.

Reference: Walter 1974.

65. NAJAF DEPRESSION IRAQ

Salsola-Zygophyllum Shrub Steppe (5–15% cover) with saline marshes:

S: *Salsola crassa, *Zygophyllum coccineum, Nitraria retusa, Seidlitzia rosmarinus, Halocnemum strobilaceum

H: Limonium carnosum, Cleome arabica, Bienertia cycloptera, Frankenia pulverulenta, Schangenia aegyptiaca, Atriplex leucoclada, Aeluropus logopoides

Reference: Abul-Fatih 1975.

66. MAIMANA AFGHANISTAN

Open Summergreen Pistacia Woodland (5–40% cover, 2–5 m., with shrubs, grasses, and forbs):

T: *Pistacia vera, Amygdalus (Prunus) spp., Cercis griffithii.

S: Amygdalus spinosissima, Cerasus (Prunus) bifrons, Colutea gracilis, Ephedra spp., Rosa spp.

DS: Artemisia tenuiseta, Acantholimon sp.

Hc: Cousinia spp., Phlomis bucharia, Chrysanthemum umbelliferum, Helichrysum plicatum, Delphinium zalil, Poa bulbosa, Chaetolimon sogdianum, Salvia pterocalyx, Solenanthus turkestanicus, et al.

Gp: Carex stenophylla, Eremurus olgae, Ungernia trisphaera, Bellevalia atroviolacea, Bongardia chrysogonum, Eranthis longistipata, Anemone sp., et al.

Th: Aegilops crassa et al., Bromus spp., Taeniantherum crinitum, Astragalus severtzowii, and many others.

References: Walter 1974; Freitag 1971; Kitamura 1960.

67. KOTGAI AFGHANISTAN

Montane Cedrus deodara Forest:

T: *Cedrus deodara, Juniperus seravschanica, Pinus gerardiana (dominates belt immediately lower), P. wallachiana, Quercus semecarpifolia, Qu. baloot (EG).

S: Sophora griffithii, Rosa ecae, Cotoneaster sp., Berberis sp., Lonicera quinquelocularis, et al.

H: Salvia nubicola, Phlomis cashmeriana, Ph. spectabilis, Nepeta pinetorum, Astragalus (Myobroma) erythrosemius, Carex cardiolepis, Poa sp., Oryzopsis gracilis, Isatis koelzii, Fragaria nubicola, Saussurea amplifolia, Taraxacum stenolepum, Thlaspi griffithianum, et al.

References: Freitag 1971; Breckle 1975.

68. MUSSOORIE INDIA

Montane Evergreen Quercus incana Forest, with Rhododendron, Lyonia, Viburnum and Berberis:

T: *Quercus incana, with Cedrela serrata (locally) and Cedrus deodara and Pinus roxburghii on dry exposures.

UT: Rhododendron arboreum, Lyonia ovalifolia, Euonymus tingens, Populus ciliata, Lonicera, Cornus macrophylla, Persea odoratissima, Rhus cotinus, Ilex dipyrena, Pyrus pashia

S: Rhododendron, Leptodermis lanceolata, Rhamnus virgatus, Viburnum cylindricum, Berberis, Indigofera

V: Hedera nepalensis, Smilax aspera, Rubus, Parthenocissus

G: Arundinaria, Cympogon, Themeda, Chrysopogon

F: Bergenia ciliata, Myriactis wallichii, Impatiens, Rubus, Desmodium, Plectranthus, Artemisia roxburghiana, Sarcococca saligna, Commelina oblique, Polygonum

Fn: Polystichum aculeatum, Dryopteris, Pteris

Reference: Saxena & Srivastava 1973.

69. YÜJLIN/ORDOS REGION CHINA

Cold-Winter Xeric Shrub Steppe (15–40% cover) on aeolian sand:

OB: Caragana korshinskii (only on mobile dunes)

S: *Artemisia ordosica, Caragana microphylla, C. stenophylla

G: Carex duriuscula, Stipa glareosa et al.

F: Artemisia frigida, Allium mongolicum, Pycnostelma lateriflorum, Peganum nigellastrum

References: Walter 1968, 1974; Walter & Box, in press.

70. WILUNA/W. AUSTRALIA AUSTRALIA

Mulga Scrub (large, evergreen shrubs with polymorphic phyllodes):

S: *Acacia aneura, with other Acacia spp., Eremophila (Myorpor.), Cassia (Caesalp.), Hakea (Proteac.), smaller Eucalyptus spp., et al.

G: patches of Triodia (sclerophyllous hummock-grasses), et al.

H: ephemeral carpets of Weitzia aurea and Helipterum spp. (both Compositae), et al.

Also: Chenopodiaceae on saline sites.

References: Walter 1973, 1977b; Beard 1965; Milton Moore 1970.

71. KEITH (DARK ISLAND)/S. AUSTRALIA AUSTRALIA

Open Dry Evergreen Heath (1–2 m.) with scattered mallee:

T: Eucalyptus spp.

S: *Banksia ornata (b, mic), *Casuarina pusilla (n), *Leptospermum myrsinoides (b, lep), B.marginata (b, nan), Phyllota (n), Adenanthos terminalis (n), Spyridium subochreatum var. laxiusculum (n)

RS: *Xanthorrhoea australis

DS: Hibbertia spp. (n), Leucopogon spp. (b, lep), Baeckea ericaea (n), Calytrix alpestris (n), Boronia caerulescens (n)

G: Hypolaena fastigiata (n), Lepidosperma spp., Schoenus, Lepidobolus, Amphipogon, Lomandra juncea, Danthonia, Stipa, Triodia.

F: Thysanotus and other Liliaceae, Orchidaceae, Proteaceae, Epacridaceae, Asteraceae, and various others.

Reference: Specht 1979.

72. MITCHELL PLATEAU CAMP/W. AUSTRALIA AUSTRALIA

Open Eucalyptus Woodland (to 18 m) on the plateau, with patches of Savanna Woodland and Monsoon Forest on laterite rubble below plateau level.

Plateau woodland:

T: *Eucalyptus tetrodonta, *Eu. miniata, Eu. nesophila
RT: Livistona eastonii (3–5 m)
ST: Erythrophloeum chlorostachys, Terminalia circumalata
S: Cochlospermum heteroneurum, Grevillea spp., Petalostigma quadriloculare, Persoonia falcata
G: *Heteropogon contortus (?), Sorghum plumosum, Themeda australis, Chrysopogon latifolius, Plectrachne pungens (rare)
F: Gomphrena canescens

In open grassy glades:

ST: Eucalyptus latifolia
S: Calytrix achaeta, C. exstipulata, Grevillea heliosperma
G: *Eriachne (?)

Monsoon forest patches (1–25 ha thickets with scattered emergent trees):

T: *Zizyphus quadrilocularis, *Albizia lebbek, *Atalaya variifolia, *Cochlospermum fraseri, *Pouteria sericea, *Wrightia pubescens plus Aglaia sp., Bauhinia cunninghamii, Mimusops elengi, Terminalia platyphylla, et al.
ST/S: Litsea, Randia, Cassia, Acacia, Murraya, Cassine, Euphorbia (all rather site-specific), plus a common but unidentified 'prickly shrub'
V: various climbers
H: various forbs, grasses, some ferns (mostly unidentified)

Savanna woodland matrix:

T: *Eucalyptus tectifica, *Eu. grandifolia, *Erythrophloeum chlorostachys, Terminalia circumalata, Melaleuca viridiflora, Adansonia gregorii
ST: Acacia pachyphloea, Hakea arborescens
RT: Cycas basaltica
G: Sorghum et al.

Reference: Beard 1976b.

73. PERTH/W. AUSTRALIA AUSTRALIA

Evergreen Jarrah (Eucalyptus marginata) Forest:

T: *Eucalyptus marginata, with scattered other Euc. spp.
US: Casuarina priesii, Banksia (Proteac.), et al.
RS: the 'grass-trees' Kingia australia and Xanthorrhoea priesii (both Liliaceae).
S: many Proteaceae, Myrtaceae, Leguminosae, Epacridaceae, et al.
H: Goodeniaceae, et al., with scattered orchids and insectivorous Drosera.

References: Walter 1968; Beard 1965; Milton Moore 1970.

74. ALEXANDRA/SOUTH ISLAND NEW ZEALAND

Low-Tussock Grassland (valley site):

TG(short): *Festuca novae-zelandiae, *Poa caespitosa, P. intermedia, Agropyron scabrum, et al.
S: Discaria toumatoa (Rhamn.), Carmichaelia (Leg.), Aciphylla (yucca-like Umbellif.), with Raoulia spp. on highly degraded sites.
DS: Leucopogon (Epacrid.).
G: Carex spp., Danthonia spp., Poa, Agrostis, et al.
F: Luzula, Chrysobactron (Lilac.), Muehlenbeckia (Polygonac.), Scleranthus (Caryoph.), Ranunculus, Acaena (Ros.), Oxalis, Gnaphalium, Epilobium (Onagrac.), Plantago, et al.
Also: numerous ruderal European grasses and weeds, e.g. Agrostis tenuis, Anthoxanthum odoratum, Holcus lanatus, Rumex acetosella, Hypochoeris radicata, et al.

References: Walter 1968; Cockayne 1958.

APPENDIX E

The macroclimatic data-base

The data-base for generating the world climatic and vegetation distributions consists of 1225 meteorological stations for which basic temperature and precipitation data are available. The data were read from climate diagrams in the *Climate-Diagram World Atlas* (Walter & Lieth 1960–67) and, to a much lesser extent, from other sources as needed especially in the polar areas. The climate-diagram atlas contains data for about 8000 stations, more than four times as many as in any other known compilation available at the time the data-gathering began (1973). Various compendia of digital monthly data were available (e.g. the *World Weather Records* compiled by the Environmental Sciences Services Administration, 1966), but these cover too few stations and could be used only for individual stations not available in the climate-diagram atlas. A computerized file of monthly data for about 2000 sites (Spangler & Jenne 1979) has recently been acquired and is being adapted for use with all future world models.

The data-base consisted initially of 988 sites used by Box, Lieth, & Wolaver (1971). These sites were spread evenly over the earth such that there was at least one and were generally three of four sites in each 10° by 10° quadrat on the land areas. This data-set contained geographic coordinates which were accurate to only 1°, as well as mean annual temperature and average annual precipitation only. The accuracy of the geographic coordinates was increased to two decimal digits, and other climatic data were added over the period 1973–1974. In 1975–76 the data were scrutinized for errors and another 237 data-sites were added, primarily in regions of high topographic complexity and in poorly represented regions such as polar areas and Central Asia.

The data items in the primary data-base (CLIM177, 177 = January 1977) are shown below. Because of the large number of values to be gathered, it was necessary to reduce the number of values actually read. Thus, only the annual values and necessary local maxima and minima of the temperature and precipitation curves were read from hard data. The other monthly values were interpolated by the routine CLIMFIT (see Appendix F). The hard data for each data-site require from two to five punchcards, depending on the complexity of the annual temperature and precipitation curves. The number of data items gathered for each site varies from a minimum of 17 to a maximum of over 40.

Data-sites are arranged by political or physiographic unit beginning with northern North America (Greenland) and proceeding south through the Americas, through Europe, Africa, the Soviet Union, Asia, Australia, and Oceania, and ending in Antarctica. In order to avoid problems of translation, both station names and names of appropriate political or physiographic units are rendered in the most appropriate local language or in a romanization of that language based on the International Phonetic Alphabet, which corresponds closely to English consonants and Latin vowels. In most cases the spellings are those of the Rand-McNally *International Atlas* (Rand-McNally and Co. 1969), from which the geographic coordinates were also obtained.

CLIM177 contains the climatic data necessary for estimating annual curves for mean temperature and average precipitation. Each CLIM177 record consists of two card-images in a standard format plus one to three additional card-images containing additional extrema of mean temperature (first card) or average precipitation (second and third cards). The records have the following format.

Card 1. Basic site and climatic data
column 1: level of data provided.
0 = two card-images with monthly data only for the two extreme months (temperature and precipitation);
1 = basic data (level 0) plus additional extrema.
columns 2–9: data source, assumed to be the climate-diagram atlas unless otherwise indicated. Abbreviations include the author, year, and page or table number.
columns 10–15: site latitude, coded with two decimal digits. Northern latitudes are coded as positive numbers, southern latitudes as negative numbers.
columns 16–22: site longitude, coded with two decimal digits. Eastern longitudes are coded as positive numbers, western longitudes as negative numbers.
columns 23–28: site elevation, in meters.
columns 29–33: mean annual temperature, in °C.
columns 34–38: highest monthly mean temperature, in °C.
columns 39–43: lowest monthly mean temperature, in °C.
columns 44–49: average annual precipitation, in mm.
columns 50–55: highest average monthly precipitation, in mm.
columns 56–61: lowest average monthly precipitation, in mm.
columns 62–65: month of highest monthly precipitation.
columns 66–69: month of lowest monthly precipitation.
columns 70–73: month of highest mean monthly temperature.
(The last three items are cloded with one decimal digit. Integer values represent monthly mid-points.)
columns 77–80: site sequence number.

Card 2. Site description and other data
columns 1–32: name of data-site (additional data as desired).
columns 33–48: name of appropriate political or physiographic unit (e.g. U.S.A., England, but Hawaiian Islands).
columns 63–34: number of temperature extrema per year.
columns 67–68: number of precipitation extrema per year.
columns 77–80: site sequence number.

Card 3. Additional temperature reference values
Card 3 is included only if the number of annual extrema exceeds two. Values are coded as above; integers represent monthly mid-points. Extrema must be coded in annual sequence.
columns 3–6: month of lowest mean monthly temperature.
columns 9–12, 21–24, 33–36, etc.: months of additional extrema in the annual curve of mean temperature.
columns 15–18, 27–30, 39–42, etc.: corresponding extrema of mean temperature.
columns 77–80: site sequence number.

Cards 4 and 5. Additional precipitation reference values
Cards 4 and 5 are included only if the number of annual extrema exceeds two. (Card 5 only if needed for overflow from card 4). Extrema must be coded in annual sequence.
columns 3–6, 15–18, 27–30, etc.: months of additional extrema in the annual curve of average precipitation.
columns 9–12, 21–24, 33–36, etc.: corresponding extrema of average precipitation.
columns 77–80: site sequence number.

APPENDIX F

The processing and mapping programs

Development of the vegetation model and its application to large data-sets, with output as computer maps, required the development of a number of computer programs and modular sub-programs. Most of these are described generally or at least mentioned in the main text. Except for the mapping program SYMAP (Harvard University), all programming components were developed entirely by the author. Most have been revised and improved several times, as indicated by version numbers attached to the programs. SYMAP has been described in section 6.B and is documented by Dougenik and Sheehan (1975). Some SYMAP adjunct routines and the use of SYMAP to make world maps are described by Box (1979a). The other programs are described below. All are in FORTRAN.

ECOSIEVE (Version 2.6)

ECOSIEVE is the main processing program for applying the vegetation model (or other models based on environmental limitation) to particular environmental situations, which can represent either actual data sites or hypothetical situations. This is accomplished by series of screening operations based on the environmental limits of the objects to be predicted. Predictions generated by ECOSIEVE can be useful both during the development of ecological hypotheses and models and to check model accuracy by geographic simulation and extrapolation.

The structure of ECOSIEVE is illustrated by the flow-chart in this section. ECOSIEVE consists of a main program, a subroutine SIEVE which performs the screening operations, and three user-programmed subroutines for definition of generated secondary variables (subroutine GENER), expression of non-constant environmental limits (subroutine FUNCTN), and application of additional operations (subroutine APPLY) to be performed after the basic screening. The main program provides for definition of the screening structure, which consists of a set of environmental variables defining an environmental hyperspace, a set of objects to be predicted (called ecoforms), and sets of upper and lower environmental limits which partition the hyperspace into environmental envelopes for the individual ecoforms. Variables and ecoforms are identified by names. Values for variables may be read from an external source or generated (in GENER) from other variables. ECOSIEVE currently permits up to 24 variables and 100 ecoforms. The main program also provides for user-specifiable formats and headings, access to the screening and adjunct subroutines, and various processing and input/output options. These options may involve additional printed or written output, processing of only selected sites and/or ecoforms, cross-tabulation of results, and/or identification of rejecting variables or closest environmental limits.

Processing consists of set-up and execution phases. After reading all instructions and constructing the screening structure, ECOSIEVE applies it to the data sites in the order in which they are presented. Each site is independent and is completed before the next is begun. For each site,

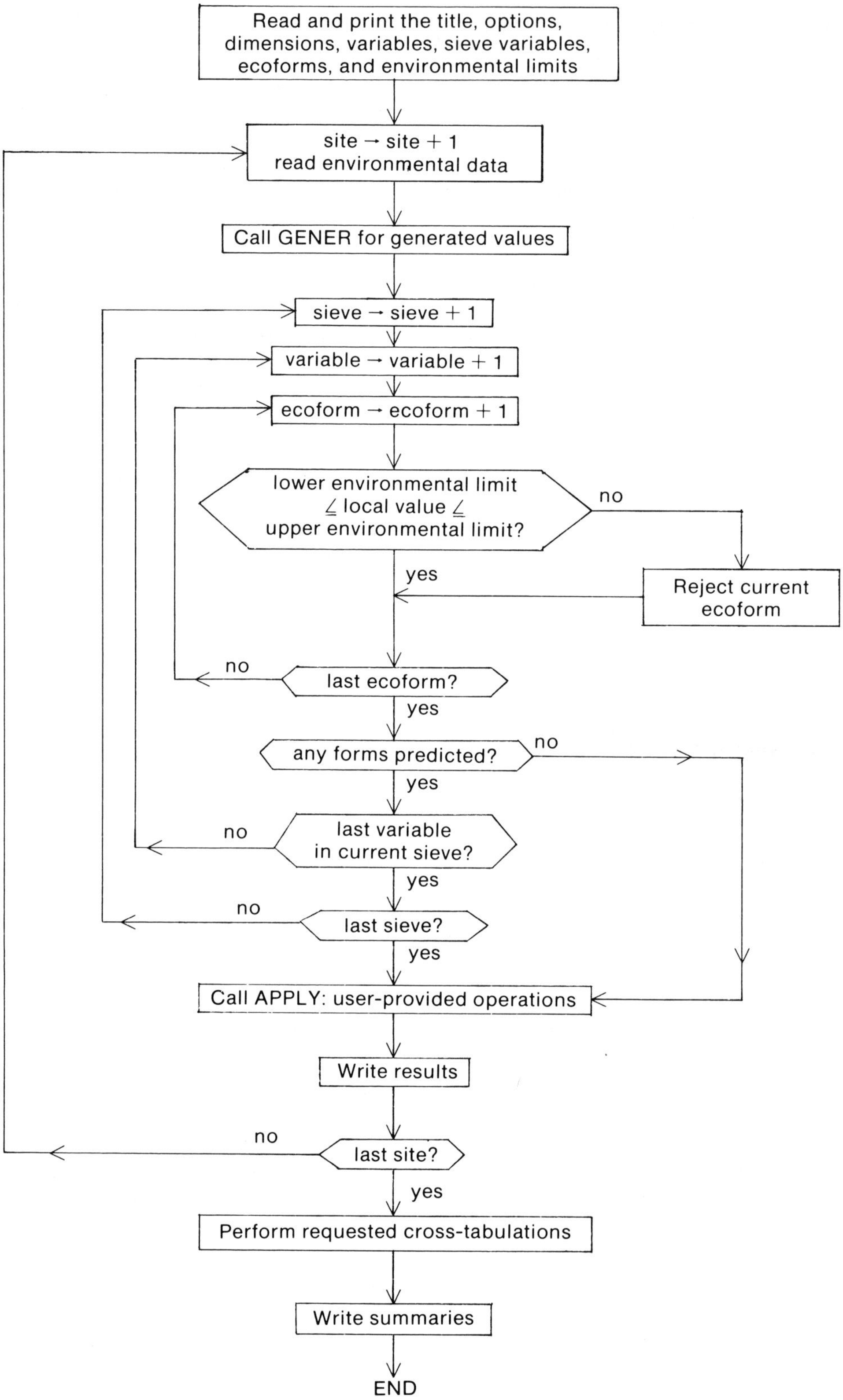
Read and print the title, options, dimensions, variables, sieve variables, ecoforms, and environmental limits
site → site + 1
read environmental data
Call GENER for generated values
sieve → sieve + 1
variable → variable + 1
ecoform → ecoform + 1
lower environmental limit ≤ local value ≤ upper environmental limit?
no
yes
Reject current ecoform
no
last ecoform?
yes
any forms predicted?
no
yes
no
last variable in current sieve?
yes
no
last sieve?
yes
Call APPLY: user-provided operations
Write results
no
last site?
yes
Perform requested cross-tabulations
Write summaries
END

every ecoform is considered by comparing its environmental limits with the corresponding local values. Ecoforms are rejected immediately whenever the following relation does not hold:

$$L \leq v \leq U$$

where L and U are corresponding lower and upper environmental limits respectively, and v is the corresponding local value. Limits may be constant-valued (resulting in rectangular envelopes) or functional, in order to express factor interactions (resulting in curved boundaries). Factor interactions may also be expressed, however, through integrative predictive variables, which may greatly simplify the mathematical form of the envelopes. Envelopes may overlap, be contiguous, or be disjunct, and are entirely independent of each other. (A utility program to check for 'holes' in a predictive design is being developed).

When all screening operations have been performed, the successful ecoforms are available for adjunct operations (e.g. dominance considerations) provided through the user-programmed subroutine APPLY. After these operations the successful ecoforms are listed on the primary printout (see Appendix B or C), written on auxiliary files if requested, and tabulated for the final summary. The summary includes prediction frequencies for all ecoforms and optional, cross-tabulated listings of successful sites for each ecoform.

The ECOSIEVE configuration for the world vegetation model presented herein also involves the computation of secondary variables through GENER and a dominance-based general successional algorithm CLIMAX, which is called through APPLY. CLIMAX eliminates all ecoforms from lower dominance levels (see Table 9) according to user-specifiable criteria. APPLY interprets these results and assigns actual dominance (+ or * on the listings in Appendices B and C), depending also on potential total vegetation cover and on proximity of forms to environmental limits, as described in sections 5.C–5.F. The computation of the Thornthwaite estimate of annual potential evapotranspiration is performed in a module PETCWT which is called from GENER. GENER divides this value into annual precipitation to produce the annual moisture index (MI) and computes annual temperature range (DTY), precipitation of the warmest month (PMTMAX), and various other climatic values not used in the final model.

ECOSIEVE was originally written in 1973 at the Nuclear Research Center in Jülich, Germany. In addition to its use in this book, it has also been used to generate world vegetation distributions used as the basis for world terrestrial maps of estimated average annual photosynthetic energy fixation (Box 1976) and photosynthetic efficiency (Box 1977). ECOSIEVE is not limited to vegetation models but can be used to predict any object for which environmental limits are appropriate (e.g. species, ecosystem types, management options). The structure of the program is particularly suited to overlay approaches. Version 2.0 of ECOSIEVE was documented by Box (1979a). A user's manual for the current version 2.6 is available (Box, 1981b).

CLIMFIT (Version 2.0)

CLIMFIT is a program which converts the minimal sets of hard climatic data of Appendix E into sequences of interpolated monthly values, based on the assumption of at least piecewise sinusoidal temporal patterns. Climatic phenomena which appear to meet the sinusoidal criterion include long-term averages of mean temperature, precipitation, and most other important macroclimatic factors. CLIMFIT is used herein to estimate the precipitation of the warmest month (PMTMAX), when it is not a measured extremum, and to estimate the annual curve of mean temperature used as the basis for estimating potential evapotranspiration.

Required input data for each curve to be estimated are all local maxima and minima during the year, the yeartimes at which they occur, and the annual total (e.g. precipitation) or mean value (e.g. temperature), whichever is appropriate. The annual curve is generated by interpolation of half-cycle cosine curve segments between adjacent extrema, coupled with an iterative correction procedure based on the annual total or mean. The correction is equivalent to deforming each curve segment after initial fitting toward one extremum or the other (i.e. up or down) until the corresponding annual total or mean matches that provided in the hard data. Mathematical details of the procedure and its validation are given in Box (1978a). The method is an interpolation, not a curve-fitting procedure, so the estimated curve is constrained to pass through each reference-point (extremum) provided. CLIMFIT estimates only the values between the measured extrema.

There are a number of sources of potential error in the CLIMFIT procedure, especially in certain, fairly well-defined cases in which CLIMFIT assumptions are violated. In order to test the accuracy of CLIMFIT results, annual curves of mean temperature and average precipitation estimated by CLIMFIT were compared with monthly values obtained from the *World Weather Records* (Environmental Sciences Services Administration 1966). The comparisons were made for 38 sites which were selected from a larger set of 113 geographically representative climate-diagram sites by the availability of the corresponding digital data in the *World Weather Records*. Since different data sources may provide different values based on different periods of measurement and/or different sets of records, the comparisons were standardized by basing the CLIMFIT estimates on the data in the *World Weather Records* rather than on the data in the computerized data-base (Appendix E). The comparisons were made by a program called FITCHECK, with results summarized in Box (1978a).

The only model variables which can be affected by CLIMFIT errors are the annual moisture index (MI) and the precipitation of the warmest month (PMTMAX). These effects are minimized since only values for intermediate months are estimated by CLIMFIT. Systematic (i.e. unbalanced) errors in the temperature curves causing possible shifts in evaporation estimates could not be seen. Errors significantly affecting PMTMAX may occur in a very few cases where PMTMAX is very low but not the yearly minimum. Such errors usually permit eco-forms to occur rather than to be precluded erroneously, and the effects of such errors are minimized by consideration of proximity to environmental limits.

CLIMFIT estimates appear to be acceptable for most applications in which the focus is on annual levels and extremes rather than on all monthly values. Although discrepancies may be significant in a few cases, the interannual variation in actual values is also large, and the discrepancies are certainly within the year-to-year variation. Most significant discrepancies occur in situations where they are not of utmost ecological importance (e.g. during the middle part of a monsoonal upswing in the precipitation curve). The recent provision of a computerized world climatic data-base (with monthly values for individual years) by Spangler & Jenne (1979) will make CLIMFIT unnecesary for those sites in their data-base. CLIMFIT remains useful for reconstructing interesting but insufficiently described sites. It also contains a routine for more drastic reconstruction by triangulation using adjacent sites. A computer-printable user's manual is available.

FLEXVEG (Version 3.3)

FLEXVEG is the routine for converting ECOSIEVE-generated model predictions into a form which can be used to produce SYMAP maps. FLEXVEG is built into SYMAP through the latter's subroutine FLEXIN (FLEXible INput) and provides general mechanisms for:

1. Specifying which plant types are to appear on which maps.
2. Grouping similar plant forms into single mapping units.
3. Applying the dominance hierarchy.
4. Identifying basic physiognomic classes and the life forms belonging to each.
5. Producing maps of life-form occurrence, importance and dominance.
6. Producing maps of life-form ecotones.
7. Producing maps of life-form diversity, both total and within specifiable sub-groupings (e.g. synusiae).
8. Producing complementary partial maps for subsequent overlaying.

The operations to be performed, as well as necessary parameters, are communicated to FLEXVEG by instruction cards placed in the SYMAP instruction packages. FLEXVEG also includes logic for converting site geographic coordinates to SYMAP coordinates at execution-time, based on the coordinate-conversion routine FLEXPROJ (see section 6.B or Box 1979a).

FLEXVEG is mainly an interface routine and performs mostly identification, grouping and input/output operations. Where it does involve cover, importance or dominance considerations its logic follows that of the ECOSIEVE system. An earlier version of FLEXVEG was described in Box (1978a). Documentation for the current expanded version is in preparation.

Other programs

A variety of other programs were involved peripherally in the project. MAPCOUNT (Box 1979a) was used to check the areal accuracy of the mapping basis (Box 1975) but has been used more in studies involving world and regional budgets for plant productivity and other functional processes (e.g. Box 1978b). MAPCOUNT could be used to quantify predicted vegetation areas on computer maps (e.g. under hypothetical past or future world climate), but site density and consequent confidence in predicted boundaries should first be increased. MAPMERGE (Box 1979a) was used during model development to overprint complementary partial maps to produce more complex world maps corresponding to the Rübel (1930) and Holdridge (1947) models. SOLWAT (Box, 1981a), began as an improved programming of the Thornthwaite-Mather (1957) soil water budgeting procedure and was then generalized to be applicable also to a variety of more natural situations (e.g. saturated soil, permafrost). SOLWAT was not used directly in the final model but has been used for a variety of other applications, including model development (see section IV.B), estimates of actual evapotranspiration used as the basis for world models of phytomass accumulation (Box, in preparation), litter production and decomposition (Meentemeyer et al., in preparation), and soil chemistry (Folkoff, in preparation).

References

Abul-Fatih, H. 1975. 'Vegetation and soil of saline depressions near Najaf, central Iraq'. *Vegetatio* 30: 107–115.

Acocks, J. P. H. 1953. 'Veld types of South Africa'. *Bot. Survey of S. Africa, Memoirs* 28: 1–192.

Acta Phytogeographica Suecica 1965. *The Plant Cover of Sweden*. Uppsala: Almqvist and Wiksells Boktryckeri. 314 pp.

Adamović, L. 1933. *Die Pflanzengeographische Stellung und Gliederung Italiens*. Jena: Gustav-Fischer-Verlag. 260 pp + 31 maps.

Adamson, R. S. 1927. 'The plant communities of Table Mountain'. *J. Ecol.* 15: 278–309.

Ahmadjian, V., and M. E. Hale (eds.) 1973. *The Lichens*. New York: Academic Press.

Ahti, T., L. Hamet-Ahti, and J. Jalas 1968. 'Vegetation zones and their sections in northwestern Europe'. *Ann. Bot. Fenn.* 5: 169–211.

Al-Ani, H. A., B. R. Strain, and H. A. Mooney 1972. 'The physiological ecology of diverse populations of the desert shrub *Simmondsia chinensis*'. *J. Ecol.* 60: 41–57.

Aleksandrova, V. D. 1960. 'Some regularities in the distribution of the vegetation in the arctic tundra'. *Arctic* 13: 147–162.

Allgemeine Forst- und Jagdzeitung 1966. Special issue on timberline. Vol. 77, pp. 105–137.

Anderson, M. C. 1964. 'Studies of the woodland light climate'. *J. Ecol.* 52: 27–41, 643–663.

Andrews, H. J., and R. W. Cowlin 1940. *Forest Resources of the Douglas-Fir Region*. Pacific Northwest Forest and Range Experimental Station. Washington: U.S. Dept. Agric., Forest Service, Misc. publ. 389.

Arber, A. 1934. *The Gramineae: A study of Cereal, Bamboo and Grass*. Cambridge: Univ. Press. 480 pp.

Asprey, G. F., and R. G. Robbins 1953. 'The Vegetation of Jamaica'. *Ecol. Monogr.* 23: 353–412.

Aubert de le Ruë, E., F. Bourlière, and J.-P. Harroy 1957. *The Tropics*. New York: A. A. Knopf, Inc.

Axelrod, D. I. 1966. 'Origin of deciduous and evergreen habits in temperate forests'. *Evolution* 20: 1–15.

Ayyad, M. A., and A. A. El-Ghonemy 1976. 'Phytosociological and environmental gradients in a sector of the Western Desert of Egypt'. *Vegetatio* 31: 93–102.

Balgooy, M. M. J. van, 1975. *Pacific Plant Areas*. Vol. 3. Leiden: J. J. Groen en Zoon. 386 pp.

Barrow, P. M., and F. B. Pearson 1970. The Mallee and Mallee Heaths. In: *Australian Grasslands* (R. Milton Moore, ed.).

Baruch, Z. 1979. 'Elevational differentiation in *Espeletia schultzii* (Compositae), a giant rosette plant of the Venezuelan Páramos'. *Ecology* 60: 85–98.

Beadle, N. C. W. 1966. 'Soil phosphate and its role in molding segments of the Australian flora and vegetation, with special reference to xeromorphy and sclerophylly'. *Ecology* 47: 992–1007.

Beals, E. W. 1965. 'The remnant cedar forests of Lebanon'. *J. Ecol.* 53: 679–694.

Beals, E. W. 1969. 'Vegetational change along altitudinal gradients'. *Science* 165: 981–985.

Beard, J. S. 1944a. 'Climax vegetation in tropical America'. *Ecology* 25: 127–158.

Beard, J. S. 1944b. 'The natural vegetation of the island of Tobago, British West Indies'. *Ecol. Monogr.* 14: 135–163.

Beard, J. S. 1946. *The Natural Vegetation of Trinidad*. Oxford Forestry Memoirs, no. 20.

Beard, J. S. 1953. 'The savanna vegetation of northern tropical America'. *Ecol. Monogr.* 23: 149–215.

Beard, J. S. 1955. 'The classification of tropical American vegetation types'. *Ecology* 36: 89–100.

Beard, J. S. (ed.) 1965. *Descriptive Catalogue of West Australian Plants*. Perth (Australia): King's Park Board. 122 pp.

Beard, J. S. 1967. 'Some vegetation types of tropical Australia in relation to those of Africa and America'. J. Ecol., 55: 247–299.

Beard, J. S. 1969. 'The natural regions of the deserts of Western Australia'. *J. Ecol.* 57: 677–711.

Beard, J. S. 1973. The Physiognomic Approach. In: *Ordination and Classification of Communities* (R. H. Whittaker, ed.). Handbk. Veg. Sci. 5. The Hague: Dr W. Junk Publishers.

Beard, J. S. 1976a. The Evolution of Australian Desert Plants. In: *Evolution of Desert Biota* (D. W. Goodall, ed.). Austin: Univ. of Texas Press.

Beard, J. S. 1976b. 'The monsoon forests of the Admiralty Gulf, Western Australia'. *Vegetatio* 31: 177–192.

Bethke, H., H. Haas, and O. Stocker 1965. 'Über den Wasser- und Photosynthesehaushalt einiger Frühjahrsgeophyten'. *Flora* Sect. B, 156: 8–49.

Bews, J. W. 1929. *The World's Grasses: Their Differentiation, Distribution, Economics and Ecology*. London, New York, Toronto: Longmans, Green and Co. 408 pp.

Bhatia, K. K. 1958. 'A mixed teak forest of central India'. *J. Ecol.* 46: 43–63.

Biebl, R. 1964. 'Zum Wasserhaushalt von *Tillandsia recurvata* L. und *Tillandsia usneoides* L. auf Porto Rico'. *Protoplasma* 58: 345–368.

Billings, W. D. 1949. 'The shadscale vegetation zone of Nevada and eastern California in relation to climate and soils'. *Amer. Midl. Nat.* 41: 87–109.

Billings, W. D., and H. A. Mooney 1968. 'The ecology of arctic and alpine plants'. *Biol. Rev.* 43: 481–529.

Billings, W. D., K. M. Peterson, and G. R. Shaver 1978. Growth, Turnover, and Respiration of Roots and Tillers in Tundra Graminoids. In: *Vegetation and Production Ecology of an Alaskan Arctic Tundra* (L. L. Tieszen, ed.). Springer-Verlag.

Bliss, L. C. 1956. 'A comparison of plant development in microenvironments of arctic and alpine tundras'. *Ecol. Monogr.* 26: 303–337.

Bliss, L. C. 1971. 'Arctic and alpine plant life cycles'. *Ann. Rev. Ecol. Syst.* 2: 405–438.

Bliss, L. C. (ed.) (in press). *Tundra Ecosystems*. Intl. Biolog. Pgm. Series, vol. 25. Cambridge Univ. Press.

Bliss, L. C., et al. 1973. Arctic Tundra Ecosystems. *Ann. Rev. Ecol. Syst.* 4: 359–399.

Bliss, L. C., and F. E. Wielgolaski (eds.) 1974. *Primary Production and Production Processes, Tundra Biome*. Edmonton and Oslo: IBP Tundra Biome Steering Committee and Univ. of Alberta. 256 pp.

Blombery, A. M. 1967. *A Guide to Native Australian Plants*. Sydney: Halstead Press. 482 pp.

Blüthgen, J. 1960. 'Der skandinavische Fjällbirkenwald als Landschaftsformation'. *Petermanns Geogr. Mitteil.* 1960: 119–144.

Boaler, S. B. 1966. 'Ecology of a miombo site, Lupa North Forest Preserve, Tanzania. II. Plant communities and seasonal variation in the vegetation'. *J. Ecol.* 54: 447–479.

Bobek, H. 1951. 'Die natürlichen Wälder und Gehölzfluren Irans'. *Bonner Geogr. Abh.* 8: 1–60.

Bobrovskaya, N. I. 1971. 'Concerning the water balance of the tree and shrub species of the Kara-Kum sand-desert'. *Bot. Zh.* 56: 361–368 (in Russian).

Böcher, T. W. 1938. *Biological Distributional Types in the Flora of Greenland*. København: C. A. Reitzels Forlag.

Böcher, T. W. 1940. 'Studies on the plant geography of the North-Atlantic heath formation'. *Kgl. Danske Vidensk. Selskab., Biol. Medd.* 15: 1–64.

Bodenheimer, F. S. 1938. *Problems of Animal Ecology*. London: Oxford Univ. Press.

Boelcke, O. 1957. 'Communidades herbáceas del Norte de Patagonia y sus relaciones con la ganaderia'. *Rev. Investig. Agric.* 11: 1–98.

Bokhari, M. H., and R. A. Khan 1976. 'Preliminary observations on plant-rock type relationships in the Shiraz-Dashte Arjan area, southwest Iran'. *Vegetatio* 31: 193–199.

Bouxin, G. 1975. 'Action des feux saissoniers sur la strate ligneuse dans le Parc National de l'Akagera (Rwanda, Afrique Centrale)'. *Vegetatio* 30: 189–196.

Box, E. O. 1975. Quantitative Evaluation of Global Productivity Models Generated by Computers. In: *Primary Productivity of the Biosphere* (H. Lieth and R. H. Whittaker, eds.). Ecol. Studies 14. New York: Springer-Verlag.

Box, E. O. 1976. Average Annual Energy Fixation of Natural Vegetation ('Berlin Model' world map). In: *Applications of Calorimetry in the Life Sciences* (I. Lamprecht and B. Schaarschmidt, eds.), p. 328. Berlin: Walter de Gruyter.

Box, E. O. 1977. *Annual Photosynthetic Efficiency of Terrestrial Vegetation*. World map presented at European Seminar on Biolog. Solar Energy Conversion Systems, Grenoble, May 1977 (reproduced in Box 1979c).

Box, E. O. 1978a. *Ecoclimatic Determination of Terrestrial Vegetation Physiognomy*. Chapel Hill: Ph.D. dissertation, University of North Carolina. 108 pp + 11 tables, 22 maps, 7 appendices. (Ann Arbor/Michigan: University Microfilms International, 1979.).

Box, E. O. 1978b. 'Geographic dimensions of terrestrial net and gross primary productivity'. *Radiation and Envl. Biophysics* 15: 305–322.

Box, E. O. 1979a. *Quantitative Cartographic Analysis: A Summary (with Geoenvironmental Applications) of SYMAP Auxiliary Programs Developed at the Jülich Nuclear Research Center*. Jülich (Germany): Kernforschungsanlage Jülich GmbH, Jül-Bericht, no. 1582.

Box, E. O. 1979b. The MAPCOUNT Series of Programs for Analysis and Manipulation of SYMAP Maps. In: *Harvard Library of Computer Mapping*. Vol. 2: Mapping Software and Cartographic Data-Bases. Cambridge (Massachusetts): Harvard University.

Box, E. O. 1979c. Use of Synagraphic Computer Mapping in Geoecology. In: *Harvard Library of Computer Mapping*. Vol. 5: Computer Mapping in Education, Research, and Medicine. Cambridge (Massachusetts): Harvard University.

Box, E. O. 1981a. SOLWAT: A Minimal-Input Soil Water Budgeting System Applicable to a Full Range of Natural Conditions. User's Manual for Version 2.2. Computer-printed manual plus program.

Box, E. O. 1981b. ECOSIEVE: An Environmental Limitation-Based Generating Program for Geographical and Management Applications. User's Manual for Version 2.6. Computer-printed manual plus program.

Box, E. O. 1981c. 'Predicting Physiognomic Vegetation Types with Climate Variables'. *Vegetatio*.

Box, E. O. (in preparation). How Accurate Worldwide Are Climatic Estimates and Maps of Evapotranspiration and the Water Balance?

Box, E. O. (in preparation). 'Life forms of Carolina vegetation'.

Box, E. O. (in preparation). 'What determines the accumulation of phytomass in climatic-climax terrestrial vegetation?'.

Box, E. O., H. Lieth, and T. Wolaver 1971. Miami Model: Primary Productivity Predicted from Temperature and Precipitation Averages (world map). In: *Publ. in Climatol.* 25: 37–46. Elmer (New Jersey).

Boyce, S. G. 1954. 'The Salt Spray Community'. *Ecol. Monogr.* 24: 29–67.

Braun, E. L. 1950. *Deciduous Forests of Eastern North America.* Philadelphia: The Blakiston Co. 596 pp.
Braun-Blanquet, J. 1928. *Pflanzensoziologie.* Berlin: Springer-Verlag.
Braun-Blanquet, J., and R. Maine 1924. Etudes sur la végétation et la flore Marocaines. *Mémoires de la Société des Sciences Naturelles du Maroc* 8(1).
Braun-Blanquet, J., and H. Walter 1931. 'Zur Ökologie der Mediterranpflanzen'. *Jb. wiss. Bot.* 74: 697–748.
Breckle, S.-W. 1975. 'Ökologische Beobachtungen oberhalb der Waldgrenze des Safed Koh (Ost-Afghanistan)'. *Vegetatio* 30: 89–97.
Britton, M. E. 1967. Vegetation of the Arctic Tundra. In: *Arctic Biology: 18th Biology Colloquium* (H. P. Hansen, ed.), 2nd ed. Corvallis (Oregon): Oregon State Univ. Press.
Brockman, C. F. 1968. *Trees of North America: A Field Guide to the Major Native and Introduced Species North of Mexico.* New York: Golden Press. 280 pp.
Brockmann-Jerosch, H. 1919. *Baumgrenze and Klimacharakter.* Zürich: Beiträge zur Geobotanischen Landesaufnahme, vol. 6. 256 pp.
Brockmann-Jerosch, H. 1925. *Die Vegetation der Schweiz.* Leipzig and Stuttgart: Verlag von Rascher & Co. 500 pp.
Brockmann-Jerosch, H. 1928. Die südpolare Baumgrenze. *Vierteljahresschr. d. Naturforsch. Ges. Zürich* 73: 705–719.
Brünig, E. F. 1970. 'Stand structure, physiognomy, and environmental factors in some lowland forests in Sarawak'. *Trop. Ecol.* 11: 26–43.
Bruner, W. E. 1931. 'The vegetation of Oklahoma'. *Ecol. Monogr.* 1: 101–188.
Bryson, R. 1966. 'Air masses, streamlines, and the boreal forest'. *Geogr. Bull. (Canada)* 8: 228–269.
Buchanan, H. 1974. *Living Color: Wildflower Communities of Bryce Canyon and Cedar Breaks.* Bryce Canyon (Utah): Bryce Canyon Nat. Hist. Assn. 70 pp.
Budyko, M. I. 1974. *Climate and Life.* New York: Academic Press.
Bünning, E. 1956. 'Der tropische Regenwald'. *Verständl. Wissensch.* 56.
Burbidge, N. T. 1945–46. 'Morphology and anatomy of the Western Australian species of *Triodia* R. Br.' I and II. *Trans. Roy. Soc. S. Australia* 69: 303–308; 70: 221–229.
Burbidge, N. T. 1952. 'The significance of the Mallee Habit in *Eucalyptus*'. *Proc. Roy. Soc. Qld.* 62: 73–78.
Burbidge, N. T. 1953. 'The genus *Triodia* R. Br. (Gramineae)'. *Aust. J. Bot.* 1: 121.
Burnett, J. H. 1964. *The Vegetation of Scotland.* Edinburgh and London: Oliver and Boyd. 614 pp.
Burt, B. D. 1942. 'Some East African vegetation communities'. *J. Ecol.* 30: 65–146.
Cabrera, A. L. 1958. 'La vegetación de la Puna Argentina'. *Revista Invest. Agric.* 11(4).
Cabrera, A. L. 1968. 'Ecología vegetal de la puna'. *Colloq. Geogr. (Bonn)* 9: 91–116.
Cadet, Th. 1974. 'Etude sur la végétation des hautes altitudes de l'île de la Réunion (Océan Indien)'. *Vegetatio* 29: 121–130.
Cain, S. A. 1950. 'Life forms and phytoclimates'. *Bot. Rev.* 16: 1–329.
Caldwell, M. M., D. A. Johnson, and M. Fareed 1978. Constraints on Tundra Productivity. In: *Vegetation and Production Ecology of an Alaskan Arctic Tundra* (L. L. Tieszen, ed.), Springer-Verlag.
Campbell, B. M., and E. J. Moll 1977. 'The forest communities of Table Mountain, South Africa'. *Vegetatio* 34: 105–115.
Cannon, W. A. 1911. *The Root Habits of Desert Plants.* Washington: Carnegie Inst. publ. 131.
Cannon, W. A. 1934. *General and Physiological Features of the Vegetation of the More Arid Portions of Southern Africa.* Washington: Carnegie Inst. publ. 354.
Cartellieri, E. 1935. 'Jahresgang des osmotischen Wertes, der Transpiration und Assimilation einiger Ericaceen der alpinen Zwergstrauchheide usw.'. *Jb. wiss. Bot.* 82: 460–506.
Carter, D. B., and J. R. Mather 1966. 'Climatic Classification for Environmental Biology'. *Publ. in Climatol.* 19(4): 305–395. Elmer (New Jersey).
Caswell, H. 1976. The Validation Problem. In: *Systems Analysis and Simulation in Ecology,* Vol. 4 (B. C. Patten, ed.). New York: Academic Press.
Champion, H. G. 1936. *A Preliminary Survey of Forest Types of India and Burma.* Indian Forest Records, N. S. Silvicult., 1. New Delhi.
Champion, H. G., S. K. Seth, and G. M. Khattak 1965. *Forest Types of Pakistan.* Sukkur: Forest Research Inst.
Champion, H. G., and S. K. Seth 1968. *A Revised Survey of the Forest Types of India.* Delhi: Govt. Printer.
Chastain, A. 1958. La Flore et la Végétation des îles de Kerguélen. *Mém. du Museum National d'Histoire Naturelle* New Series B, XI(1). Paris. 136 pp + 36 plates.
Chew, R. M., and A. E. Chew 1965. 'The primary productivity of a desert-shrub (*Larrea tridentata*) community'. *Ecol. Monogr.* 35: 355–375.
Christ, H. 1910. *Die Geographie der Erde.* Jena. 358 pp.
Cockayne, L. 1958. *The Vegetation of New Zealand.* 3rd ed. (reprint by H. R. Engelmann). Meisenheim (Germany): Anton Hain, K. G. 456 pp.
Coe, M. J. 1967. *The Ecology of the Alpine Zone of Mount Kenya.* Monogr. Biol. 17. The Hague: W. Junk. 136 pp.
Coker, W. C. 1905. *Vegetation of the Bahama Islands.* Baltimore: Geogr. Soc. Baltimore. 270 pp + 47 plates.
Cole, M. M. 1960. 'Cerrado, Caatinga and Pantanal: the distribution and origin of the savanna vegetation of Brazil', *Geogr. J.* 74: 168–179.
Connor, H. E. 1961. 'A tall-tussock grassland community in New Zealand'. *N. Z. J. Sci. Technol.* 43A: 825–835.
Cooper, W. S. 1922. *The Broad-Sclerophyll Vegetation of California.* Washington: Carnegie Inst. publ. 319.
Cooper, J. P. (ed.) 1975. *Photosynthesis and Productivity in Different Environments.* Intl. Biolog. Pgm. Series, vol. 3. Cambridge: Univ. Press. 715 pp.
Corner, E. J. H. 1966. *The Natural History of Palms.* Berkeley: University of California Press. 394 pp.
Corrêa, M. P. 1926. *Diccionário das Plantas Úteis do Brasil e das Exóticas Cultivadas.* Rio de Janeiro: Imprensa Nacional. 747 pp.
Costin, A. B. 1957. 'The High-Mountain Vegetation of Australia'. *Austr. J. Bot.* 5: 173–189.
Coupland, R. T. (ed.) 1979. *Grassland Ecosystems of the World: Analysis of Grasslands and their Uses.* Intl. Biolog.

Pgm. Series, vol. 18. Cambridge: Univ. Press.

Coupland, R. T., and T. C. Brayshaw 1953. 'The fescue grassland in Saskatchewan'. *Ecology* 34: 386–405.

Cremers, G. 1973–1974. 'Architecture de quelques lianes d'Afrique tropicale'. *Candollea* 28: 249–280; 29: 57–110.

Crum, H. A. 1972. 'The geographic origins of the mosses of North America's eastern deciduous forest'. *J. Hattori Bot. Lab.* 35: 269–298.

Cuatrecasas, J. 1968. 'Páramo Vegetation and its Life Forms'. *Colloq. Geogr. (Bonn)* 9: 91–116.

Cunningham, G. L., and J. F. Reynolds 1978. 'A simulation model of primary production and carbon allocation in the creosote bush (*Larrea tridentata* DC Cov.)'. *Ecology* 59: 37–52.

Cunningham, G. L., and B. R. Strain 1969. 'Ecological significance of seasonal leaf variability in a desert shrub'. *Ecology* 50: 400–408.

Czeczött, H. 1938–39. *A Contribution to the Knowledge of the Flora and Vegetation of Turkey.* Dahlem bei Berlin: Verlag des Repertoriums. 281 pp + 39 plates.

Dallimore, W., and A. B. Jackson 1966. *A Handbook of Coniferae and Gingkoaceae.* 4th ed. (revised by S. G. Harrison). London: Edward Arnold Ltd. 729 pp.

Damman, A. W. H. 1977. 'Geographical changes in the vegetation pattern of raised bogs in the Bay of Fundy region of Maine and New Brunswick'. *Vegetatio* 35: 137–151.

Danin, A. 1978. 'Plant species diversity and ecological districts of the Sinai Desert'. *Vegetatio* 36: 83–93.

Daubenmire, R. F. 1943. 'Vegetational zonation in the Rocky Mountains'. *Bot. Rev.* 9: 325–393.

Daubenmire, R. F. 1972. 'Phenology and other characteristics of tropical semi-deciduous forest in northwestern Costa Rica'. *J. Ecol.* 60: 147–170.

Daubenmire, R. F. 1974. *Plants and Environment: A Textbook of Plant Autecology.* New York: Wiley. 422 pp.

Daubenmire, R. F. 1978. *Plant Geography with Special Reference to North America.* New York: Academic Press. 338 pp.

Davis, J. H. 1930. 'Vegetation of the Black Mountains of North Carolina: an ecological study'. *J. Elisha Mitchell Sci. Soc.* 45: 291–318.

Davis, J. L. 1964. 'A vegetational map of Tasmania'. *Geogr. Rev.* 54: 249–253.

Davis, P. H., P. C. Harper, and I. C. Hedge 1971. *Plant Life of South-West Asia.* Edinburgh: Botanical Society of Edinburgh. 335 pp.

Davis, R. B. 1966. 'Spruce-fir forests of the coast of Maine'. *Ecol. Monogr.* 36: 79–94.

Dawson, J. W. 1962. 'The New Zealand lowland podocarp forest. Is it subtropical?' *Tuatara* 9: 98–116.

Daxer, H. 1934. 'Über die Assimilationsökologie der Waldbodenflora'. *Jb. wiss. Bot.* 80: 363–420.

de Laubenfels, D. J. 1953. 'The external morphology of conifer leaves'. *Phytomorphology* 3: 1–20.

de Laubenfels, D. J. 1969. 'A revision of the Malesian and Pacific rainforest conifers'. *J. Arnold Arboretum* 50: 274–369.

de Laubenfels, D. J. 1975. *Mapping the World's Vegetation: Regionalization of Formations and Floras.* Syracuse (New York): Syracuse University Press. 246 pp.

de Wilde, J. J. F. E. 1961. '*Cedrus atlantica* Manetti in Marokko'. *Medel. Bot. Tuin, Belmonte Arbor. (Wageningen)* 5: 93–103.

di Castri, F., and H. A. Mooney (eds.) 1973. *Mediterranean Type Ecosystems.* Ecol. Studies 7. New York: Springer-Verlag. 406 pp.

Dobson, J. R. (in press). 'Theoretical aspects of environmental envelope formation in N-dimensions'. (Presented at 74th annual meeting of Assn. of Amer. Geographers, New Orleans, April 1978).

Dodd, A. P. 1959. The Biological Control of Prickly Pear in Australia. In: *Biogeography and Ecology in Australia* (A. Keast, R. L. Crocker and C. S. Christian, eds.). Monogr. Biolog., 8: 565–577. The Hague: Dr W. Junk Publishers.

Donita, N. 1965. 'Vegetationsstufen in den Karpaten Rumäniens'. *Rev. Roum. de Biol., Ser. Bot.* 10: 455–467.

Dougenik, J. A., and D. E. Sheehan 1975. SYMAP User's Reference Manual. Cambridge (Massachusetts): Harvard University, Laboratory for Computer Graphics and Spatial Analysis.

Drummond, H. M., and E. H. Leatham 1959. *Bibliography of New Zealand Tussock Grasslands.* Wellington: DSIR Res. Bull. 132.

Duck, L. G., and J. B. Fletcher 1943. *A Game Type Map of Oklahoma.* Oklahoma City: Division of Wildlife Restoration.

Duke, J. A. 1978. The Quest for Tolerant Germplasm. In: *Crop Tolerance to Suboptimal Land Conditions* (G. A. Jung, ed.), Amer. Soc. Agron. spec. publ. 32. Madison (Wisconsin): Amer. Soc. Agron., Crop Sci. Soc. Amer., and Soil Sci. Soc. Amer.

Ehleringer, J., H. A. Mooney, and J. A. Berry 1979. 'Photosynthesis and microclimate of *Camissonia claviformis,* a desert winter annual'. *Ecology* 60: 280–286.

Eijsink, J., G. Ellenbroek, W. Holzner, and M. J. A. Werger 1978. 'Dry and semi-dry grasslands in the Weinviertel, Lower Austria'. *Vegetatio* 36: 129–148.

Eiten, G. 1972. 'The Cerrado vegetation of Brazil'. *Bot. Rev.* 38: 201–341.

Eiten, G. 1978. 'Delimitation of the 'cerrado' concept'. *Vegetatio* 36: 169–178.

Ellenberg, H. 1963. Die Vegetation Mitteleuropas mit den Alpen. *Einführung in die Phytologie* (H. Walter, ed.), Vol. IV/2. Stuttgart: Verlag Eugen Ulmer.

Ellenberg, H. 1966. 'Leben und Kampf an den Baumgrenzen der Erde'. *Naturwiss. Rundschau* 19: 133–139.

Ellenberg, H. 1975. 'Vegetationsstufen in perhumiden bis perariden Bereichen der tropischen Anden'. *Phytocoenologia* 2: 368–387.

Ellenberg, H., and D. Mueller-Dombois 1967a. 'Tentative physiognomic-ecological classification of plant formations of the earth'. *Ber. Geobot. Inst. Rübel* 37: 21–55. (Finalized 1973 for UNESCO and reprinted as Appendix B in Mueller-Dombois and Ellenberg 1974.).

Ellenberg, H., and D. Mueller-Dombois 1967b. 'A key to Raunkiaer plant life forms with revised subdivisions'. *Ber. Geobot. Inst. Rübel* 37: 56–73. (Reprinted as Appendix A in Mueller-Dombois and Ellenberg 1974.).

Elton, W. M., and V. Meentemeyer 1979. 'Environmental Locations of the Largest Trees of Eastern Deciduous Species'. *Amer. Midl. Nat.* 101: 182–190.
Emberger, L. 1938. Carte Phytogéographique du Maroc. *Veröff. Geobot. Inst. Rübel* 14 (supplement).
Enquist, F. 1924. *Sambandet mellan klimat och växtgränser.* Geol. Fören. i Stockholm Förhandl., vol. 46. pp. 202.
Environmental Sciences Services Administration 1966. *World Weather Records 1951*–60. 6 vols. Washington: U.S. Dept. Commerce (Govt. Printing Ofc.).
Epling, C., and H. Lewis 1942. 'The centers of distribution of the chaparral and coastal sage associations'. *Amer. Midl. Nat.* 27: 445–462.
Ern, H. 1966. 'Die dreidimensionale Anordnung der Gebirgs-vegetation auf der Iberischen Halbinsel'. *Bonner Geogr. Abhandl.* 37.
Ern, H. 1974. 'Zur Ökologie und Verbreitung der Kiefern im östlichen Zentralmexiko'. *Dt. Dendrol. Ges.* 67: 164–198.
Ernst, W. 1971. 'Zur Ökologie der Miombo-Wälder'. *Flora* 160: 317–331.
Everard, B., and B. D. Morley 1970. *Wild Flowers of the World.* New York: G. P. Putnam's Sons. 432 pp, 192 plates.
Ewel, J. J., and A. Madriz 1968. *Zonas de Vida de Venezuela. Memoria Explicativa sobre el Mapa Ecológico.* Caracas: Min. de Agric. y Cría. 264 pp + map.
Eyre, S. R. 1968. *Vegetation and Soils, a World Picture.* 2nd ed. Chicago: Aldine Publ. Co. 328 pp.
Fanshawe, D. B. 1969. *The Vegetation of Zambia.* Forest Research Bulletin 17. Lusaka: Govt. Printer.
Fautin, R. W. 1946. 'Biotic communities of the northern desert scrub biome in western Utah'. *Ecol. Monogr.* 16: 251–310.
Feoli-Chiapella, L., and E. Feoli 1977. 'A numerical phyto-sociological study of the summits of the Majella Massive (Italy)'. *Vegetatio* 34: 21–39.
Fernando, S. N. U. 1968. *The Natural Vegetation of Ceylon.* Colombo: Swabhasha Printers. 86 pp + 4 maps.
Ferri, M. G. 1969. *Plantas do Brasil: Espécies do Cerrado.* São Paolo: Editora Edgard Blucher, 240 pp.
Ferri, M. G. 1973. *A Vegetação de Cerrados Brasileiros.* Belo Horizonte. 362 pp.
Ferri, M. G., and L. M. Coutinho 1959. *Contribução ao conhecimento da ecologia do Cerrado.* São Paolo: Univers. do São Paolo, Bol. 224: 103–150.
Fittkau, E. J., J. Illies, H. Klinge, G. H. Schwabe, and H. Siolo (eds.) 1969. *Biogeography and Ecology in South America.* Monogr. Biolog. 18–19. The Hague: Dr W. Junk Publishers.
Folkoff, M., V. Meentemeyer, and E. O. Box (in preparation). 'Environmental control of the geography of soil pH'.
Folmann, G. 1965. 'Fensterflechten in der Atacamawüste'. *Die Naturwiss.* 52: 434–435.
Fosberg, F. R. 1960. *The Vegetation of Micronesia.*
Fosberg, F. R. 1967. A Classification of Vegetation for General Purposes. Appendix I in: *Guide to the Check Sheet for IBP Areas* (G. F. Peterken, ed.). IBP Handbook no. 4. Oxford: Blackwell Sci. Publications.
Fosberg, F. R. 1976. 'Geography, ecology and biogeography'. *Annals, Assn. Amer. Geogr.* 66: 117–128.
Foster, M. B. 1945. 'Blueprint of the Jungle, As Depicted by the Altitude of Growth of the Bromeliads – with Notes on the Culture of Certain Tropical Epiphytes'. *J. N. Y. Bot. Gard.* 46: 9–16.
Frankenberg, P. 1978. 'Lebensformen und Florenelemente im nordafrikanischen Trockenraum'. *Vegetatio* 37: 91–100.
Franklin, J. F., and C. T. Dyrness 1973. *Natural Vegetation of Oregon and Washington.* Portland (Oregon): Pacific Northwest Forest and Range Experimental Station. Washington: U.S. Dept. Agric., Forest Service Genl. Techn. Rept. PNW-8. 417 pp.
Freitag, H. 1971. 'Die natürliche Vegetation Afghanistans'. *Vegetatio* 22: 285–344.
French, N. R. (ed.) 1979. *Perspectives in Grassland Ecology.* Ecolog. Studies, vol. 32. New York: Springer-Verlag. 204 pp.
Fry, F. E. J. 1947. 'Effects of the environment on animal activity'. *Univ. Toronto Studies, Biol. Ser.* 55: 1–62.
Gams, H. 1934. 'Beiträge zur Kenntnis der Steppenmoose'. *Ann. Bryol.* 7: 37–56.
Gams, H. 1954. 'Die *Tragacantha*-Igelheiden der Gebirge um das Kaspische, Schwarze und Mittelländische Meer'. *Veröff. Geobot. Inst. Rübel* 31: 217–243.
Gates, D. M. 1968. 'Transpiration and leaf temperature'. *Ann. Rev. Plant Physiol.* 19: 211–238.
Gates, D. M. 1969. Climate and Stability. In: *Diversity and Stability in Ecological Systems* (G. M. Woodwell and H. H. Smith, eds.). Brookhaven Symp. Biol., 22. Upton (New York): Brookh. Natl. Lab.
Gates, D. M., and R. Janke 1966. 'The energy environment of the alpine tundra'. *Oecol. Plant.* 1: 39–61.
Gates, D. M., and R. B. Schmerl (eds.) 1975. *Perspectives of Biophysical Ecology.* Ecol. Studies 12. New York: Springer-Verlag.
Gentilli, J. 1960. 'Il fattore termico nell'ecologia degli eucalitti'. *Centro Sperimet. Agric. e Forest. (Roma), Publ.* 4: 53–119.
Gersmehl, P. J. 1973. 'Pseudo-Timberline: the Southern Appalachian Grassy Balds'. *Arctic Alpine Res.* 5: A137–A138.
Gersmehl, P. J. 1976. 'An Alternative Biogeography'. *Annals Assn. Amer. Geogr.* 66: 223.
Gimingham, C. H. 1972. *Ecology of Heathlands.* London: Chapman and Hall. 266 pp.
Gimingham, C. H. 1975. *An Introduction to Heathland Ecology.* Edinburgh: Oliver and Boyd. 124 pp.
Givnish, T. J. 1975. The Contribution of Seasonality to Plant Diversity. Cited by R. M. May in: *Unifying Concepts in Ecology* (W. H. van Dobben and R. H. Lowe-McConnell, eds.), pp. 166–167. The Hague: Dr W. Junk Publishers.
Givnish, T. J., and G. J. Vermeij 1976. 'Sizes and shapes of liane leaves'. *Amer. Nat.* 110: 743–778.
Godley, E. J. 1960. 'The botany of southern Chile in relation to New Zealand and the Subantarctic'. *Proc. Roy. Soc., Ser. B* 152: 457–475.
Godley, E. J. 1975. Flora and Vegetation. In: *Biogeography and Ecology in New Zealand* (G. Kuschel, ed.), Monogr. Biolog., 27. The Hague: Dr W. Junk Publishers.
Golley, F. B. 1972. Energy Flux in Ecosystems. In: *Ecosystem Structure and Function* (J. A. Wiens, ed.). Corvallis: Oregon State University, Ann. Biol. Colloq., 31: 69–90.
Golley, F. B., and E. Medina (eds.) 1975. *Tropical Ecological Systems.* Ecol. Studies 11. New York: Springer-Verlag. 398 pp.

Goodall, D. W. (ed.) 1978. *Arid Land Ecosystems*, vol. 1. Intl. Biolog. Pgm. Series, vol. 16. Cambridge: Univ. Press.

Goodland, R. 1971. 'A physiognomic analysis of the 'Cerrado' vegetation of central Brazil'. *J. Ecol.* 59: 411–419.

Gorchakovsky, P. L. 1966. *The High-Mountain Flora and Vegetation of the Ural.* Sverdlovsk. 270 pp (in Russian).

Goryshina, T. K. 1972. 'Recherches écophysiologiques sur les plantes éphéméroides printanières dans les chênaies de la zone forêt-steppe de la Russie centrale'. *Ecol. Plant.* 7: 241–258.

Graham, A. (ed.) 1973. *Vegetation and Vegetational History of Northern Latin America.* Amsterdam: Elsevier. 394 pp.

Grebenshchikov, O. S. 1974. 'An essay on climate characteristics of the main vegetation formations of the Caucasus'. *Bot. Zh.* 59: 161–174 (in Russian).

Grieve, B. J. 1955. 'The physiology of sclerophyll plants'. *J. Roy. Soc. W. Australia* 39: 31–45.

Grime, P. J. 1979. *Plant Strategies and Vegetation Processes.* New York: Wiley. 222 pp.

Gulizashvili, V. S., L. B. Makhatadze, and L. I. Prilipko 1975. *The Vegetation of the Caucasus.* Moskva. 232 pp (in Russian).

Gupta, R. K. 1978. 'Impact of human influences on the vegetation of the western Himalaya'. *Vegetatio* 37: 111–118.

Haefner, J. W. 1978. 'Ecosystem Assembly Grammars: Generative Capacity and Empirical Adequacy'. *J. Theor. Biol.* 73: 293–318.

Hallé, F., R. A. A. Oldeman, and P. B. Tomlinson 1978. *Tropical Trees and Forests.* New York: Springer-Verlag.

Hamet-Ahti, L., T. Ahti, and T. Konen 1974. 'A scheme of vegetation zones for Japan and adjacent regions'. *Ann. Bot. Fenn.* 11: 59–88.

Hamilton, A. C. 1975. 'A quantitative analysis of altitudinal zonation in Uganda forests'. *Vegetatio* 30: 99–106.

Hanes, T. L. 1971. 'Succession after fire in the chaparral of southern California'. *Ecol. Monogr.* 41: 27–52.

Hansen, H. M. 1930. *Studies on the Vegetation of Iceland.* København: J. Frimodt. 186 pp.

Hansen, K. J. (in prep.). Environmental Controls on Growth Variability of the Coniferous Species at Treeline, Front Range, Colorado. Ph.D. Dissertation (University of Colorado).

Hanson, H. C. 1953. 'Vegetation types in northwestern Alaska und comparison with communities in other Arctic regions'. *Ecology* 34: 111–140.

Hara, H. (ed.) 1959. *Distribution Maps of Flowering Plants in Japan.* 2 vols. (with 200 maps by H. Hara and H. Kanai). Tokyo: Inoue Book Co.

Hare, F. K. 1950. 'Climate and zonal divisions of the boreal forest formation in eastern Canada'. *Geogr. Rev.* 40: 615–635.

Hare, F. K. 1954. 'The boreal conifer zone'. *Geogr. Studies* 1: 4–18.

Hare, F. K., and J. C. Ritchie 1972. 'The boreal bioclimates'. *Geogr. Rev.* 62: 333–365.

Harrison, A. T., E. Small, and H. A. Mooney 1971. 'Drought relationships and distribution of two mediterranean-climate California plant communities'. *Ecology* 52: 869–875.

Hargreaves, D., and B. Hargreaves 1965. *Tropical Trees found in the Caribbean, South America, Central America, and Mexico.* Kailua (Hawaii): Hargreaves Co. Inc. 64 pp. (incl. 125 color plates).

Hart, J. S. 1952. 'Lethal temperatures of fish from different latitudes'. *Publ. Ontario Fish. Res. Lab.* 72: 1–79.

Hartmann, F. K., and F. Schnelle 1970. *Klimagrundlagen natürlicher Waldstufen und ihrer Waldgesellschaften in deutschen Mittelgebirgen.* Stuttgart: Gustav-Fischer-Verlag. 175 pp.

Hauri, H. 1912. '*Anabasis aretoides* Moq. et Coss, eine Polsterpflanze der algerischen Sahara'. *Beih. Bot. Centralbl.* 28 (Abt. I).

Hayward, H. E. 1928. 'Studies of plants in the Black Hills of South Dakota'. *Bot. Gaz.* 85: 353–412.

Heal, O. W., and D. F. Perkins 1978. *Production Ecology of British Moors and Montane Grasslands.* Ecological Studies, 27. New York: Springer-Verlag. 426 pp.

Hedberg, O. 1951. 'Vegetation belts of the East African mountains'. *Svensk Bot. Tidskr.* 45: 140–202.

Heilborn, O. 1925. 'Contributions to the ecology of the Ecuadorian páramos with special reference to cushion plants and osmotic pressure'. *Svensk Bot. Tidskrift* 19.

Hellmers, H., et al. 1955. 'Root systems of some chaparral plants in southern California'. *Ecology* 36: 667–678.

Hettinger, L. R., and A. J. Janz 1974. *Vegetation and Soils of Northeastern Alaska.* Canadian Arctic Gas Study Ltd and Alaskan Arctic Gas Study Co. 206 pp.

Hieronymous, J. 1945. *Observaciones sobre la Vegetación de la Provincia de Tucumán.* Tucumán (Argentina): Univ. Nac. de Tucumán, Monografías Inst. Estudios Geográficos, 5.

Hocker, H. W. 1956. 'Certain aspects of climate as related to the distribution of Loblolly Pine (*Pinus taeda* L.)'. *Ecology* 37: 824–834.

Hodge, W. H. 1946. 'Cushion-plants of the Peruvian puna'. *J. N. Y. Bot. Gard.* 47: 133–141.

Holdgate, M. W. (ed.) 1970. *Antarctic Ecology.* 2 vols. London and New York: Academic Press. 998 pp.

Holdridge, L. R. 1947. 'Determination of world plant formations from simple climatic data'. *Science* 105: 267–268.

Holdridge, L. R. 1959. 'A simple method for determining potential evapotranspiration from temperature data'. *Science* 130: 572.

Holland, P. G., and A. R. T. Hove 1975. 'The distribution of *Euphorbia candelabrum* in the southern Rift Valley, Kenya'. *Vegetatio* 30: 49–54.

Holloway, J. T. 1954. 'Forests and climate on the South Island of New Zealand'. *Trans. Royal Soc. N. Z.* 82: 392–410.

Holtmeier, F. K. 1974. *Geoökologische Beobachtungen und Studien an der subarktischen und alpinen Waldgrenze in Vergleichender Sicht.* Wiesbaden: Steiner-Verlag.

Holzner, H., and M. Numata (eds., in preparation). *Biology and Ecology of Weeds.* Geobotany Series. The Hague: Dr W. Junk Publishers.

Hopkins, B. 1962–70. 'Vegetation of the Olokemeji Forest Reserve, Nigeria'. (I-VII). *J. Ecol.* 50: 559–598; 53: 109–138; 54: 687–703; 56: 97–115; 58: 765–825.

Horton, J. S., and C. J. Kraebel 1955. 'Development of vegetation after fire in the chamise chaparral of southern California'. *Ecology* 36: 244–262.

Horvát, I., V. Glavać, and H. Ellenberg 1974. Vegetation

Südosteuropas. *Geobotanica Selecta* (R. Tüxen, ed.), vol. IV. Stuttgart: Gustav-Fischer-Verlag. 768 pp.
Hosokawa, T. 1954. 'On the *Campnosperma* forests of Kusaie in Micronesia, with special reference to the community units of epiphytes'. *Vegetatio* 5/6: 351–360.
Hosokawa, T. 1967. Ecological Studies of Tropical Epiphytes in Forest Ecosystems. In: *Procs. Symp. Intern. Soc. Trop. Ecology, Varanasi (India)*, pp. 16–21.
Hueck, K. 1954. 'Der Anden-Erlenwald (Die *Alnus jorullensis*-Assoziation) in der Provinz Tucumán'. *Angew. Pflanzensoz.* 1 (Klagenfurt).
Hueck, K. 1956. 'Mapa fitogeográfico do estado de São Paolo'. *Bol. Paulista de Geografia* 22: 19–25.
Hueck, K. 1966. *Die Wälder Südamerikas: Ökologie, Zusammensetzung und wirtschaftliche Bedeutung. Vegetationsmonographien der einzelnen Großräume*, vol. II. Stuttgart: Gustav-Fischer-Verlag. 422 pp.
Hueck, K., and P. Seibert 1972. *Vegetationskarte von Südamerika. Vegetationsmonographien der einzelnen Großräume*, Vol. IIa. Stuttgart: Gustav-Fischer-Verlag. 71 pp + map.
Hulten, E. 1962. 'The circumpolar plants'. *Kungl. Svenska Vetensk. Akad. Handl.* Ser. 4, 8(5). Stockholm.
Humboldt, A. von 1807. *Ideen zu einer Geographie der Pflanzen.* Tübingen: F. G. Cotta (reprinted 1963 by Wissenschaftl. Buchges. Darmstadt).
Hustich, J. 1953. 'The boreal limits of conifers'. *J. Arctic Inst. N. America*, 6(2).
Hylckama, T. E. A. van 1956. *The Water Balance of the Earth.* Publ. in Climatol., 9(2). Elmer (New Jersey).
Innes, C. 1977. *The Complete Handbook of Cacti and Succulents.* New York: Van Nostrand Reinhold Co. 224 pp.
Innis, G. S. 1975. 'One Direction for Improving Ecosystems Modeling'. *Behavioral Science* 20: 68–74.
Inst. Français de Pondichery 1961–70. *Cartes de la végétation de l'Inde.* Pondichery: Travaux Sect. Scient. Techn. de l'Inst. 10 maps.
Ives, J. D., and R. G. Barry (eds.) 1974. *Arctic and Alpine Environments.* London: Methuen and Co. Ltd.
Jacobsen, H. 1970. *Das Sukkulentenlexikon.* Jena and Stuttgart: Fischer-Verlag. 590 pp.
Jäger, E. 1968. 'Die klimatischen Bedingungen des Areals der dunklen Taiga und der sommergrünen Breitlaubwälder'. *Ber. Dt. Bot. Ges.* 81: 397–408.
Jani, E. 1969. Pflanzen aus der Libyschen Sahara. *Willdenowia (Berlin)* 5: 295–341.
Jenny, H., et al. 1969. 'The pygmy forest-podzol ecosystem and its dune associates of the Mendocino coast'. *Madroño* 20: 60–74.
Jermy, A. C., J. A. Crabbe, and B. A. Thomas (eds.) 1973. The Phylogeny and Classification of the Ferns. *Bot. J. Linnean Soc.* 67: suppl. 1. London: Academic Press. 284 pp.
Jonsson, H. 1905. 'Vegetationen i Syd-Island'. *Bot. Tidskrift* 27: 1–82.
Jordal, L. H. 1951. 'Plants from the vicinity of Fairbanks, Alaska'. *Rhodora* 53: 156–159.
Kan, M., H. Saito, and T. Shidei 1965. 'Studies on the productivity of evergreen broad-leaved forests'. *Bull. Kyoto Univ. For.* 37: 55–75 (in Japanese with English summary).
Karpenko, A. S. 1964. 'The geography of the plant cover of the Lower Amur Basin'. *Bot. Zhurn.* 49: 1408–1420 (Russian).
Kausch, W. 1965. 'Beziehungen zwischen Wurzelwachstum, Transpiration und CO_2-Gaswechsel bei einigen Kakteen'. *Planta* 66: 229–238.
Keast, A., R. L. Crocker, and C. S. Christian (eds.) 1959. *Biogeography and Ecology in Australia.* Monogr. Biolog., 8. The Hague: Dr W. Junk Publishers. 640 pp.
Keay, R. W. J. 1949. *An Outline of Nigerian Vegetation.* Lagos: Government Printer. 52 pp + map.
Keay, R. W. J. 1959. *Vegetation Map of Africa South of the Tropic of Cancer.* Oxford University Press. 24 pp + map.
Kerfoot, O. 1960–61. '*Juniperus procera* Ende (the African Pencil Cedar) in Africa and Arabia'. *E. Afr. Agric. Forest. J.* 26: 170–177.
Kilgore, B. M., and D. Taylor 1979. 'Fire history of a *Sequoia*-mixed conifer forest'. *Ecology* 60: 129–142.
Killick, D. J. B. 1979. African Montane Heathlands. In: *Heathlands and Related Shrublands* (R. L. Specht, ed.), Ecosystems of the world, vol. 9A. Amsterdam: Elsevier.
Kirkpatrick, J. B., and C. F. Hutchinson 1977. 'The community composition of Californian coastal sage scrub'. *Vegetatio* 35: 21–33.
Kitamura, S. 1960. *Flora of Afghanistan.* Kyoto: Committee of Kyoto Univ. Sci. Exped. to Karakoram and Hindukush. 486 pp.
Knapp, R. 1965. *Die Vegetation von Nord- und Mittelamerika und der Hawaii-Inseln.* Vegetationsmonographien der einzelnen Großräume, vol. I. Stuttgart: Gustav-Fischer-Verlag. 373 pp.
Knapp, R. 1973. *Die Vegetation von Afrika unter Berücksichtigung von Umwelt, Entwicklung, Wirtschaft, Agrar- und Forstgeographie.* Vegetationsmonographien der einzelnen Großräume, vol. III. Stuttgart: Gustav-Fischer-Verlag. 626 pp.
Koechlin, J., J.-L. Guillaumet, and P. Morat 1974. *Flore et Végétation de Madagascar.* Flora et Vegetatio Mundi (R. Tüxen, ed.), vol. 5. Vaduz: J. Cramer. 687 pp.
Kohler, A. 1967. 'Beiträge zur Kenntnis der ephemeren Vegetation am Südrand der Atacama-Wüste (Chile)'. *Ber. Dt. Bot. Ges.* 80: 563–572.
Kojić, M. 1966. 'Über die Wasserbilanz und Stoffproduktion bei Frühjahrsgeophyten'. *Ber. Dt. Bot. Ges.* 79: 188–197.
Köppen, W. 1931. *Grundriß der Klimakunde.* Berlin: W. de Gruyter.
Korovin, E. P. 1961–62. *The Vegetation of Middle Asia and Southern Kazakhstan.* 2nd ed. Tashkent. 452 pp + 578 pp (2 vols, in Russian).
Kozlowski, T. T. 1962. Photosynthesis, Climate and Growth of Trees. In: *Tree Growth* (Kozlowski, ed.). New York: Ronald Press.
Kozlowski, T. T. 1971. *Growth and Development of Trees.* 2 vols. New York: Academic Press. 443 pp + 514 pp.
Kramer, P. J. 1962. The Role of Water in Tree Growth. In: *Tree Growth* (Kozlowski, ed.). New York: Ronald Press.
Krajina, V. J. 1969. *Ecology of Forest Trees in British Columbia.* Ecology of Western North America, 2(1): 1–146. Vancouver: Univ. Brit. Colombia, Dept. Botany.
Krebs, C. J. 1972. *Ecology: The Experimental Analysis of*

Distribution and Abundance. New York: Harper and Row. 694 pp.

Kroener, H. E. 1968. Die Verbreitung der echten Baumfarne (Cyatheaceen) und ihre klimaökologischen Voraussetzungen. Bonn: Ph.D. Dissertation, Univ. Bonn, Math-Nat. Sci. Faculty. 269 pp.

Krüssman, G. 1972. *Handbuch der Nadelgehölze*. Berlin and Hamburg: V. Parcy., 366 pp. + 779 figs.

Küchler, A. W. 1946. 'The broadleaf deciduous forests of the Pacific Northwest'. *Annals, Assn. Amer. Geogr.* 36: 122–147.

Küchler, A. W. 1948. 'A new vegetation map of Manchuria'. *Ecology* 29: 513–516.

Küchler, A. W. 1951. 'The relation between classifying and mapping vegetation'. *Ecology* 32: 275–283.

Küchler, A. W. 1964. *The Potential Natural Vegetation of the Conterminous United States*. New York: Amer. Geogr. Soc., Special Research Publ. 36.

Küchler, A. W. 1967. *Vegetation Mapping*. New York: Ronald Press. 472 pp.

Küchler, A. W. (ed.) 1965–68. *International Bibliography of Vegetation Maps*. 4 vols. Lawrence (Kansas): Univ. of Kansas Publications, Library Series, 21.

Küchler, A. W. 1977. Natural Vegetation of California (map and explanatory notes). In: *Terrestrial Vegetation of California* (M. G. Barbour and J. Major). New York: Wiley.

Kuijt, J. 1961. A Revision of *Dendrophthora* (Loranthaceae). *Wentia* 6: 1–145.

Kummerow, J., V. Matte, and F. Schlegel 1961. 'Zum Problem der Nebelwälder an der zentralchilenischen Küste'. *Ber. Dt. Bot. Ges.* 74: 135–145.

Kunkel, G. G. (ed.) 1976. Biogeography and Ecology in the Canary Islands. Monogr. Biolog. Vol. 30. The Hague: Dr W. Junk Publishers.

Kurz, H. 1945. 'Secondary forest succession in the Tallahassee Red Hills'. *Proc. Fla. Acad. Sci.* 7: 59–100.

Kuschel, G. (ed.) 1975. *Biogeography and Ecology in New Zealand*. Monogr. Biolog. vol. 27. The Hague: Dr W. Junk Publishers.

Kusumoto, T. 1961. 'An ecological analysis of the distribution of broad-leaved evergreen trees, based on dry-matter production'. *Jap. J. Botany* 17: 307–331.

Langdale-Brown, I., H. A. Osmaston, and J. G. Wilson 1964. *The Vegetation of Uganda and its Bearing on Land-Use*. Entebbe: Government Printer. 159 pp + 8 maps.

Lange, O. L. 1965. 'Der CO_2-Gaswechsel von Flechten bei tiefen Temperaturen'. *Planta* 64: 1–49.

Lange, O. L., and R. Lange 1963. 'Untersuchungen über Blattemperaturen, Transpiration und Hitzeresistenz an Pflanzen mediterraner Standorte (Costa Brava, Spanien)'. *Flora* 153: 387–425.

Langlois, A. C. 1976. *Supplement to 'Palms of the World'*. Gainesville: Univ. Presses of Florida. 252 pp.

Larcher, W. 1957. 'Frosttrocknis an der Waldgrenze und in der alpinen Zwergstrauchheide'. *Veröff. Mus. Fernandeum Innsbruck* 37: 49–81.

Larcher, W. 1961. 'Jahresgang des Assimilations- und Respirationsvermögens von *Olea europea* L. ssp. *sativa* Hoff. et Link., *Quercus ilex* L. und *Quercus pubescens* Willd. aus dem nördlichen Gardaseegebiet'. *Planta* 56: 575–606.

Larcher, W. 1970. 'Kälteresistenz und Überwinterungsvermögen mediterraner Holzpflanzen'. *Oecol. Plant.* 5: 267–286.

Larcher, W. 1972. 'Der Wasserhaushalt immergrüner Holzpflanzen im Winter'. *Ber. Dt. Bot. Ges.* 85: 315–327.

Larcher, W. 1976. *Ökologie der Pflanzen*. 2nd ed. Uni-Taschenbücher no. 232. Stuttgart: Verlag Eugen Ulmer. 320 pp.

Lauer, W. 1952. 'Humide und aride Jahreszeiten in Afrika und Südamerika und ihre Beziehung zu den Vegetationsgürteln'. *Bonner Geogr. Abhandl.* 9.

Lauer, W. 1968. 'Problemas de la División Fitogeográfica en América Central'. *Colloq. Geogr. (Bonn)* 9: 139–156.

Lauer, W., Klink, and Kalus (in press). *Die Vegetation und ihre Umwelt im Hochland von Mexiko*. Wiesbaden: Steiner.

Lawson, G. W., K. O. Armstrong-Mensah, and J. B. Hall 1970. 'A catena in tropical moist semi-deciduous forest near Kade, Ghana'. *J. Ecol.* 58: 371–398.

Leathart, S. 1977. *Trees of the World*. New York: A & W Publishers. 224 pp.

Legris, P. 1969. 'La Grande Comore, climats et végétation'. *Inst. Franç. Pondicherie, Trav. Sect.Scient. Techn.* III(5): 28.

Leistner, O. A. 1967. *The Plant Ecology of the Southern Kalahari*. Pretoria: Govt. Printer. 172 pp + map and 51 plates.

Leistner, O. A., and M. J. A. Werger 1973. 'Southern Kalahari Phytosociology'. *Vegetatio* 28: 353–399.

Leopold, A. S. 1950. 'Vegetation zones of Mexico'. *Ecology* 31: 507–518.

Lescure, J. P. 1978. 'An architectural study of the vegetation's regeneration in French Guiana'. *Vegetatio* 37: 53–60.

Levins, R. 1966. 'The strategy of model building in population biology'. *Amer. Scientist* 54: 421–431.

Levitt, J. 1969. 'Growth and survival of plants at extremes of temperature – a unified concept'. *Symp. Soc. Exptl. Biol.* 23: 395–448.

Levitt, J. 1972. *Responses of Plants to Environmental Stresses*. New York: Academic Press. 697 pp.

Lewis, J. P., and M. B. Collates 1975. 'La vegetación de la Provincia de Santa Fe'. *Bol. Soc. Argent. de Bot.* 16: 151–179.

Lieth, H. 1956. 'Ein Beitrag zur Frage der Korrelation zwischen mittleren Klimawerten und Vegetationsformationen'. *Ber. Dt. Bot. Ges.* 69: 169–176.

Lieth, H. 1960. 'Über den Lichtkompensationspunkt der Landpflanzen'. *Planta* 54: 530–576.

Lieth, H. (ed.) 1974. *Phenology and Seasonality Modeling*. Ecol. Studies 8. New York: Springer-Verlag. 444 pp.

Lieth, H. 1975. Primary Production of the Major Vegetation Units of the World. In: *Primary Productivity of the Biosphere* (H. Lieth and R. H. Whittaker, eds.), Ecol. Studies 14. New York: Springer-Verlag.

Lieth, H., and E. O. Box 1972. Evapotranspiration and Primary Productivity; C. W. Thornthwaite Memorial Model. *Publ. in Climatol.* 25: 37–46. Elmer (New Jersey).

Lieth, H., and E. O. Box 1977. 'The Gross Primary Productivity Pattern of the Land Vegetation: A First Attempt'. *Trop. Ecol.* 18: 109–115.

Lieth, H., M. Numata, and T. Suganuma 1973. 'Studies of the

grassland vegetation in the Kawatabi Special Research Area of the Japanese IBP'. *Vegetatio* 28: 41–56.
Lieth, H., and R. H. Whittaker (eds.) 1975. *Primary Productivity of the Biosphere.* Ecological Studies, vol. 14. New York: Springer-Verlag. 339 pp.
Lind, E. M., and M. E. S. Morrison 1974. *East African Vegetation.* London: Longman Group Ltd. 257 pp.
Lindman, C. A. M., and M. G. Ferri 1974. *A Vegetação no Rio Grande do Sul.* Coleção Reconquista do Brasil (vol. 2). Belo Horizonte: Livraria Itatiaia Editora Ltda. 378 pp.
Livingston, B. E., and F. Shreve 1921. *The Distribution of Vegetation in the United States, as Related to Climatic Conditions.* Washington: Carnegie Inst. publ. 284. 590 pp.
Livingstone, D. A. 1967. 'Postglacial vegetation of the Ruwenzori Mountains in equatorial Africa'. *Ecol. Monogr.* 37: 25–52.
Löhr, E., and D. Müller 1968. 'Blatt-Atmung der höheren Bodenpflanzen im tropischen Regenwald'. *Physiol. Plant.* 21: 673–675.
Lötschert, W. 1962–65. 'Beiträge zur Ökologie der subatlantischen Zwergstrauchheide NW-Deutschlands', I-III. *Beitr. Biol. Pfl.* 37: 331–410; 41: 359–366.
Lötschert, W. 1969. *Pflanzen an Grenzstandorten.* Stuttgart.
Lüdi, W. 1935. 'Beitrag zur regionalen Vegetationsgliederung der Apenninenhalbinsel'. *Veröff. Geobot. Inst. Rübel* 12: 212–239.
Mabry, T. J., J. H. Hunziker, and D. R. DiFeo jr. (eds.) 1977. *Creosote Bush: Biology and Chemistry of* Larrea *in New World Deserts.* US/IBP Synthesis Series, vol. 6. Stroudsburg (Pa.): Dowden, Hutchinson, and Ross, Inc. 304 pp.
MacDougal, D. T. 1912. 'The water balance of desert plants'. *Ann. Bot.* 26: 71.
MacDougal, D. T., E. R. Long, and I. G. Brown 1915. 'End results of desiccation and respiration in succulent plants'. *Physiol. Res.* 1–289.
Major, J. 1963. 'A climatic index to vascular plant activity'. *Ecology* 44: 485–498.
Major, J. 1967. Potential Evapotranspiration and Plant Distribution in Western States with Emphasis on California. In: *Ground Level Climatology* (R. H. Shaw, ed.). Washington: Amer. Assn. Adv. Sci.
Malaisse, Fr. P. 1974. Phenology of the Zambezian Woodland Area with Emphasis on the Miombo System. In: *Phenology and Seasonality Modeling* (H. Lieth, ed.). Ecol. Studies 8. New York: Springer-Verlag.
Malmer, N., L. Lindgren, and S. Persson 1978. 'Vegetational succession in a south Swedish deciduous wood'. *Vegetatio* 36: 17–29.
Mani, M. S. (ed.) 1974. *Ecology and Biogeography in India.* Monogr. Biolog. vol. 23. The Hague: Dr W. Junk Publishers.
Mani, M. S. 1979. *Ecology and Phytogeography of High-Altitude Plants of the Himalaya.* New York: Halstead. 300 pp.
Marchand, P. J., and B. F. Chabot 1978. 'Winter water relations of tree-line plant species on Mt. Washington, New Hampshire'. *Arc. Alp. Res.* 10: 105–116.
Markgraf, F. 1949. 'Eine neue Höhenstufenkarte der Vegetation Albaniens'. *Ber. Geobot. Inst. Rübel* 1948: 109–118.
Marloth, R. 1908. *Das Kapland.* Wiss. Ergebn. Dt. Tiefsee-Exp., vol. II. Jena.
Mata, G. F., et al. 1972. *Mapa y Descripción de los Tipos de Vegetación de la República Mexicana.* México: Talleres Gráficos de la Nación. 60 pp + map.
Mather, J. R. 1974. *Climatology: Fundamentals and Applications.* New York: McGraw-Hill Book Co. 412 pp.
Mather, J. R., and G. A. Yoshioka 1966. The Role of Climate in the Distribution of Vegetation. In: *Climate Classification for Environmental Biology* (D. B. Carter and J. R. Mather, eds.), Publ. in Climatol., 19(4): 372–384. Elmer (New Jersey).
Mayer, H., and M. Sevim 1958. 'Die Libanonzeder'. *Jb. Ver. Schütz Alpenpfl.* (München), 1958: 1–20.
McClure, F. A. 1966. *The Bamboos, a Fresh Perspective.* Cambridge (Massachusetts): Harvard University Press. 347 pp.
McCurrach, J. C. 1960. *Palms of the World.* New York: Harper and Brothers (Reprinted 1976 by Horticultural Books Inc., Stuart/Florida). 290 pp.
McKenzie, B., E. J. Moll, and B. M. Campbell 1977. 'A phytosociological study of Orange Kloof, Table Mountain, South Africa'. *Vegetatio* 34: 41–53.
McPherson, J. K., and C. H. Muller 1969. 'Allelopathic effects of *Adenostoma fasciculatum*, 'chamise', in the California chaparral'. *Ecol. Monogr.* 39: 177–198.
Meentemeyer, V., and W. Elton 1977. 'The potential implementation of biogeochemical cycles in biogeography'. *Prof. Geographer* 29: 266–271.
Meentemeyer, V., E. O. Box, and R. Thompson (in press). 'World patterns and correlates of plant litter production'.
Meentemeyer, V., R. Thompson, and E. O. Box (in preparation). 'World geography of terrestrial plant litter decomposition and accumulation'.
Meggers, B. J., E. S. Ayensu, and W. D. Duckworth (eds.) 1973. *Tropical Forest Ecosystems in Africa and South America: A Comparative Review.* Washington: Smithsonian Inst. Press. 350 pp.
Menaut, J. C., and J. Cesar 1980. 'Structure and primary productivity of Lamto savannas, Ivory Coast'. *Ecology* 60: 1197–1210.
Menninger, E. A. 1970. *Flowering Vines of the World: An Encyclopedia of Climbing Plants.* New York: Hearthside Press. 410 pp.
Merriam, C. H. 1898. *Life Zones and Crop Zones.* Washington: U. S. Dept. Agric., Biological Survey Bull. 10.
Meusel, H., and R. Schubert 1971. 'Beiträge zur Pflanzengeographie des Westhimalajas'. *Flora* 160: 137–194, 370–432, 573–606.
Meyer, M. 1978. 'Vergleich verschiedener *Chrysopogon gryllus*-reicher Trockwiesen des insubrischen Klimabereichs und angrenzender Gebiete'. *Vegetatio* 35: 107–114.
Michaelis, P. 1932–34. 'Ökologische Studien an der Baumgrenze'. I: *Ber. Dt. Bot. Ges.*, 50: 31–42; II and III: *Beih. Bot. Zbl.* 52B: 310–332, 333–377; IV and V: *Jb. wiss. Bot.* 80: 169–243, 337–362.
Milton Moore, R. (ed.) 1970. *Australian Grasslands.* Canberra: Austral. Natl. Univ. Press. 455 pp.
Ministère de l'Agriculture 1960. *L'Eucalyptus au Maroc.* Rabat.
Miranda, F., and E. Hernandez 1963. 'Los Tipos de Vegetación de México y su Clasificación'. *Bol. Soc. Bot. de México* 28: 29–179.

Miranda, F., and A. J. Sharp 1950. 'Characteristics of the vegetation in certain temperate regions of eastern Mexico'. *Ecology* 31: 313–333.

Misra, R., and B. Gopal (eds.) 1968. *Proceedings of the Symposium on Recent Advances in Tropical Ecology*. 2 vols. Varanasi (India): Intl. Soc. Trop. Ecology.

Missouri Botanical Garden 1978. *Phytogeography of Africa.* Papers from 24th Systematics Symposium. St. Louis. 222 pp.

Mitchell, J. (ed.) 1971. *Guide Book to Mount Kenya and Kilimanjaro.* Nairobi: Mountain Club of Kenya (Kenya Litho Ltd.).

Monk, C. D. 1966. 'An ecological significance of evergreenness'. *Ecology* 47: 504–505.

Monk, C. D. 1968. 'Successional and environmental relationships of the forest vegetation of north central Florida'. *Amer. Midl. Nat.* 79: 441–457.

Mooney, H. A. 1974. Plant Forms in Relation to Environment. In: *Vegetation and Environment* (B. R. Strain and W. D. Billings, eds.), Handbk. Veg. Sci. vol. 6. The Hague: Dr W. Junk Publishers.

Mooney, H. A., and E. L. Dunn 1970. 'Convergent evolution in mediterranean-climate evergreen sclerophyll shrubs'. *Evolution* 24: 292–303.

Mooney, H. A., and D. J. Parsons 1973. Structure and Function of the California Chaparral – an Example from San Dimas. In: *Mediterranean Type Ecosystems, Origin and Structure* (di Castri and Mooney, eds.), Ecol. Studies 7. New York: Springer-Verlag.

Mooney, H. A., D. J. Parsons, and J. Kummerow 1974. Plant Development in Mediterranean Climates. In: *Phenology and Seasonality Modeling* (H. Lieth, ed.), Ecol. Studies 8. New York: Springer-Verlag.

Moore, H. E. 1973. 'The major groups of palms and their distribution'. *Gentes Herbarium* 11: 27–141.

Moore, B., and N. Taylor 1927. 'Vegetation of Mt. Desert Island, Maine, and its environment'. *Mem. Brooklyn Bot. Gard.* 3: 1–151.

Mudræck, F. 1935. 'Über die Assimilationstätigkeit und das Wachstum von *Ficaria verna*'. *Planta* 23: 71–104.

Mueller-Dombois, D., and H. Ellenberg 1974. *Aims and Methods of Vegetation Ecology*. New York: John Wiley and Sons. 547 pp.

Muller, P. (ed.) 1977. Ecosystem Research in South America. Biogeographica, vol. 8. The Hague: Dr W. Junk Publishers. 158 pp.

Naylor, E. E. 1945. 'Air Plants and their Problems of Survival'. *J. N. Y. Bot. Gard.* 46: 55–59.

Nechayeva, N. T., V. K. Vassilevskaya, and K. G. Antonova 1973. *Life Forms of the Plants in the Kara-Kum Desert.* Moskva. 243 pp (in Russian).

Neuwirth, G., and H. Polster 1960. 'Wasserverbrauch und Stoffproduktion der Schwarzpappel und Espe unter Dürrebelastung'. *Arch. Forstw.* 9: 789–810.

Ng, E., and P. C. Miller 1980. 'Soil moisture relations in southern California chaparral'. *Ecology* 61: 98–107.

Niering, W. A., et al. 1963. 'The saguaro: a population in relation to environment'. *Science* 142: 15–23.

Nikitin, S. A. 1966. *The Tree and Shrub Vegetation of the Deserts of the USSR.* Moskva. 256 pp (in Russian).

Nobel, P. S. 1980a. 'Morphology, surface temperatures, and northern limits of columnar cacti in the Sonoran Desert'. *Ecology* 61: 1–7.

Nobel, P. S. 1980b. 'Water vapor conductance and CO_2 uptake for leaves of a C_4 desert grass, *Hilaria rigida*'. *Ecology* 61: 252–258.

Noerr, M. von 1974a. 'Trockenresistenz bei Moosen'. *Flora* 163: 371–387.

Noerr, M. von 1974b. 'Hitzeresistenz bei Moosen'. *Flora* 163: 388–397.

Numata, M. 1974. *The Flora and Vegetation of Japan.* Tokyo: Kodansha and Amsterdam: Elsevier. 294 pp.

Numata, M. (ed., in prep.). *Ecology of Grasslands and Bamboolands in the World.* The Hague: Dr W. Junk Publishers.

Numata, M., A. Miyawaki, and D. Itow 1972. 'Natural and semi-natural vegetation in Japan'. *Blumea* 20: 435–481 (with 26 photos, veg. maps, and altitudinal profile).

Odum, E. P. 1971. *Fundamentals of Ecology.* 3rd ed. Philadelphia: W. B. Saunders and Co. 574 pp.

Odum, H. T., and R. F. Pigeon (eds.) 1970. *A Tropical Rain Forest: A Study of Irradiation and Ecology at El Verde, Puerto Rico.* Oak Ridge (Tennessee): U. S. Atomic Energy Commission.

Ogawa, H., K. Yoda, and T. Kira 1961. 'A preliminary survey on the vegetation of Thailand'. *Nature Life SE Asia* 1: 21–157.

Oppenheimer, H. R. 1932. 'Zur Kenntnis der hochsommerlichen Wasserbilanz mediterraner Gehölze'. *Ber. Dt. Bot. Ges.* 50a: 185–245.

Orians, G. H., and O. T. Solbrig (eds.) 1977. *Convergent Evolution in Warm Deserts: An Examination of Strategies and Patterns in Deserts of Argentina and the U.S.* US/IBP Synthesis Series, vol. 3. Stroudsburg (Pennsylvania): Dowden, Hutchinson, and Ross, Inc. 352 pp.

Orshan, G. 1953. 'Note on the application of Raunkiaer's system of life forms in arid regions'. *Pal. J. Bot., Jerus.* Ser. 6: 120–122.

Ostenfeld, C. H., and J. Grontved 1934. *The Flora of Iceland and the Faeroes.* København: Levin and Munksgaard. 196 pp.

Owen, D. F. 1974. 'A contribution to the ecology of the African baobab'. *Savanna* 3: 1–14.

Ozenda, P. 1958. *Flore du Sahara septentrional et central.* Paris: Centre Nat. Rech. Sci.

Page, C. N. 1977. 'An ecological survey of the ferns of the Canary Islands'. *Fern Gaz.* 11: 297–312.

Parkhurst, D. F., and O. L. Loucks 1972. 'Optimal leaf size in relation to environment'. *J. Ecol.* 60: 505–537.

Parsons J. J. 1955. 'The Miskito pine savanna of Nicaragua and Honduras'. *Annals Assn. Amer. Geogr.* 45: 36–63.

Parsons, R. F., and D. G. Cameron 1974. 'Maximum plant species diversity in terrestrial communities'. *Biotropica* 6: 202–203.

Pelton, W. L., K. M. King, and C. B. Tanner 1960. 'An evaluation of the Thornthwaite and mean temperature methods for determining potential evapotranspiration'. *Agron. J.* 52: 387–395.

Penfound, W. T. 1952. 'Southern swamps and marshes'. *Bot. Rev.* 18: 413–446.

Pesman, M. W. 1962. *Meet Flora Mexicana.* Globe (Arizona): Dale S. King. 278 pp.

Petrov, M. P. 1966–67. *The Deserts of Central Asia.* 2 vols. Moskva-Leningrad. 274 pp + 286 pp (in Russian).
Phillips, J. F. V. 1928. 'The principal forest types in the Knysna region'. *S. Afr. J. Sci.* 25: 188.
Phillips, W. S. 1963. 'Depth of roots in soils'. *Ecology* 44: 424.
Pisek, A., and G. Rehner 1958. 'Temperaturminima der Netto-assimilation von mediterranen und nordisch-alpinen Immergrünen'. *Ber. Dt. Bot. Ges.* 71: 188–193.
Pisek, A., and G. Rehner 1960. 'Immergrüne Pflanzen (einschließlich Coniferen)'. *Handb. Pfl.-physiol.* 5: 415–459.
Pisek, A., and W. Tranquillini 1954. 'Assimilation und Kohlenstoffhaushalt in der Krone von Fichten (*Picea excelsa* Link) und Rotbuchenbäumen (*Fagus silvatica* L.)'. *Flora* 141: 237–270.
Pisek, A., and E. Winkler 1958. 'Assimilationsvermögen und Respiration der Fichte (*Picea excelsa* Link) in verschiedener Höhenlage und der Zirbe (*Pinus cembra* L.) an der alpinen Waldgrenze'. *Planta* 51: 518–543.
Polster, H., and S. Fuchs 1963. 'Winterassimilation und -Atmung der Kiefer (*Pinus silvestris*) im mitteldeutschen Binnenlandklima'. *Arch. Forstwesen* 12: 1011–1024.
Polunin, N. 1926. 'Plant succession in Norwegian Lapland'. *J. Ecol.* 24: 372–391.
Polunin, N. 1948. *Botany of the Canadian Eastern Arctic.* Part III: Vegetation and Ecology. Natl. Mus. Canada bull. 104. 304 pp.
Polunin, 01. 1972. *Pflanzen Europas.* (Translated from English by T. Schauer.) München: BLV Verlagsgesellschaft. 336 pp. (incl. 1088 color plates).
Popov, M. G. 1940. *The Plant Cover of Kazakhstan.* Moskva. 216 pp. (in Russian).
Popov, G. B. 1957. 'The vegetation of Socotra'. *J. Linnean Soc. London Bot.* 55: 706–720.
Praeger, L. R. 1921. *An Account of the Genus* Sedum *as found in Cultivation.* Plant Monograph Reprints (J. Cramer and H. K. Swann, eds.), vol. 2 (1967). Lehre (Germany): Verlag J. Cramer. 314 pp.
Precht, H., J. Christophersen, H. Hensel, and W. Larcher (eds.) 1973. *Temperature and Life.* 2nd ed. Berlin: Springer-Verlag.
Preusser, H. 1975. *The Landscapes of Iceland, Types and Regions.* The Hague: Dr W. Junk Publishers.
Pryor, L. D. 1959. Species Distribution and Association in *Eucalyptus.* In: *Biogeography and Ecology in Australia* (Keast et al., eds.), Monogr. Biol., vol. 8. The Hague: Dr. W. Junk Publishers.
Quarterman, E. 1950. 'Major plant communities of Tennessee cedar glades'. *Ecology* 31: 234–255.
Quézel, P. 1965. *La Végétation du Sahara.* Geobotanica Selecta (R. Tüxen, ed.), vol. II. Stuttgart: Gustav-Fisher-Verlag. 333 pp.
Radford, A. E. (ed.) 1976. *Vegetation-Habitats-Floras. Natural Areas in the Southeastern United States: Field Data and Information.* Chapel Hill (North Carolina): Univ. of North Carolina Student Stores. 290 pp.
Radford, A. E., H. E. Ahles, C. R. Bell 1968. *Manual of the Vascular Flora of the Carolinas.* Chapel Hill (North Carolina): University of North Carolina Press. 1184 pp.
Rand-McNally and Co. 1969. *The International Atlas.* Chicago: Rand-McNally. 558 pp.
Rand-McNally and Co. (without date). *Standard Outline Map of the World* (Robinson projection, 1' = 500 miles equatorial scale). Chicago: Rand-McNally.
Rase, W.-D. 1976. *SYMAP-Handbuch. Version 5.20.* Bonn-Bad Godesberg: Bundesforschungsanstalt für Landeskunde und Raumordnung, EDV-Report No. 2/76. 96 pp.
Rattray, J. M. 1960. *The Grass Cover of Africa.* Roma: FAO Agric. Studies, 49.
Rauh, W. 1939. 'Über polsterförmigen Wuchs'. *Nova Acta Leopoldina* 7: 1–239.
Rauh, W. 1958. 'Beitrag zur Kenntnis der peruanischen Kakteenvegetation'. *Sitz.-ber. Heidelberg Akad. Wiss., Math.-Naturw. Kl., 1. Abhandl.*
Rauh, W. 1973. 'Über Zonierung und Differenzierung der Vegetation Madagaskars'. *Akad. Wiss. Mainz, Math.-Naturw. Kl., 1. Wiesbaden.*
Raunkiaer, O. 1934. *The Life Forms of Plants and Statistical Plant Geography.* Oxford: Clarendon Press. 632 pp.
Rawé, R. 1968. *Succulents in the Veld.* Cape Town: Howard Timmins. 104 pp.
Rawitscher, F. 1948. 'The water economy of the 'Campos Cerrados' of southern Brazil'. *J. Ecol.* 36: 237–268.
Ray, R. J. 1959. 'A phytosociological analysis of the tallgrass prairie in northeastern Oklahoma'. *Ecology* 40: 255–261.
Reader, R., and UNC Applied Programming Group 1972. SYMAP (version 5.16a) *Instruction Manual.* Document no. LSR-139-0. Research Triangle Park (North Carolina): Triangle Universities Computation Center. 66 pp.
Reed, R. M. 1971. 'Aspen forests of the Wind River Mountains, Wyoming'. *Amer. Midl. Nat.* 86: 327–343.
Reich, P. B., and T. M. Hinckley 1980. 'Water relations, soil fertility, and plant nutrient composition of a pygmy oak ecosystem'. *Ecology* 61: 400–416.
Reiche, K. 1893. 'Über polster- und deckenförmig wachsende Pflanzen'. *Verh. wiss. Ver. zu Santiago* 2.
Reichle, D. E. (ed.) 1970. *Analysis of Temperate Forest Ecosystems.* Ecolog. Studies, vol. 1. New York: Springer-Verlag. 340 pp.
Reiners, W. A., and G. E. Lang 1979. 'Vegetational patterns and processes in the balsam fir zone, White Mountains, New Hampshire'. *Ecology* 60: 403–417.
Renner, 01. 1933. 'Zur Kenntnis des Wasserhaushalts javanischer Kleinepiphyten'. *Planta* 18: 215–287.
Restshikov, M. A. 1961. *The Steppes of the Trans-Baikal Region.* Moskva: Akad. Nauk, Ot'd. Sib., Biol. 34 (in Russian).
Richards, P. W. 1952. *The Tropical Rainforest: an Ecological Study.* Cambridge University Press. 450 pp.
Richards, P. W. 1963. 'The upland forests of Cameroon Mountain'. *J. Ecol.* 51: 529–554.
Richard-Vindard, G., and R. Battistini (eds.) 1972. *Biogeography and Ecology of Madagascar.* Monogr. Biolog., vol. 21. The Hague: Dr W. Junk Publishers. 780 pp.
Robbins, R. G. 1961. 'The montane vegetation of New Guinea'. *Tuatara* 8: 121–133.
Robbins, R. G. 1962. 'The podocarp broadleaf forest of New Zealand'. *Trans. Royal Soc. N. Z., Bot.* I., 33–75.
Robinson, A. H. 1974. 'A new map projection: its development and characteristics'. *Intl. Yrbk. Cartogr.* 14: 145–155.

Robinson, A. H., and R. D. Sale 1969. *Elements of Cartography*. 3rd ed. New York: John Wiley and Sons. 415 pp.
Ronning, O. I. 1960. 'The vegetation and flora north of the arctic circle'. *Tromsö Museums Skr.* 8: 50–72.
Rosenzweig, L. M. 1968. 'Net primary productivity of terrestrial communities: prediction from climatological data'. *Amer. Naturalist* 102: 67–74.
Rosswall, T., and O. W. Heal (eds.) 1975. *Structure and Function of Tundra Ecosystems*. Ecological Bulletins, 20. Stockholm: NFR.
Rouschal, E. 1939. 'Zur Ökologie der Macchien: I. Der sommerliche Wasserhaushalt der Macchienpflanzen'. *Jb. wiss. Bot.* 87: 436–523.
Rowe, J. S. 1959. *Forest Regions of Canada*. Ottawa: Dept. of Northern Affairs and Natural Resources, Forestry Branch. 71 pp (with vegetation map).
Rubinstein, E. 1924. 'Beziehungen zwischen dem Klima und dem Pflanzenreich'. *Meteorol. Ztschr.* 41.
Rudolph, E. D. 1963. 'Vegetation of Hallett Station area, Victoria Land, Antarctica'. *Ecology* 44: 585–586.
Rudolph, E. D. 1966. 'Terrestrial vegetation of Antarctica'. *Antarc. Res.* Ser. 8: 109–124.
Rübel, E. F. 1930. *Pflanzengesellschaften der Erde*. Bern: Verlag Hans Huber. 464 pp.
Ruinen, J. 1961. 'The Phyllosphere I: An Ecologically Neglected Milieu'. *Plant and Soil* 15: 81–109.
Salisbury, F. B., and G. G. Spommer 1964. 'Leaf temperature of alpine plants in the field'. *Planta* 60: 497–505.
Salt, G. 1954. 'A contribution to the ecology of the upper Kilimanjaro'. *J. Ecol.* 42: 375–423.
Sarmiento, G. 1976. Evolution of Arid Vegetation in Tropical America. In: *Evolution of Desert Biota* (D. W. Goodall, ed.). Austin: Univ. of Texas Press.
Sauvage, Ch. 1961. *Recherches géobotaniques sur les subéraies maroccaines*. Trav. Inst. Sc. Chérifien, Série Botan., no. 21. Rabat.
Sawyer, J. O., and A. A. Lindsey 1971. *Vegetation of the Life Zones in Costa Rica*. Indianapolis: Indiana Acad. Sci. 214 pp.
Saxena, H. O., and P. B. L. Srivastava 1973. 'Forest communities of Mussoorie'. *Trop. Ecol.* 14: 197–218.
Scheichtl, H. M. 1966. 'Physiognomie der Waldgrenze im Gebirge'. *Allg. Forstz. (Wien)* 77: 105–111.
Schemske, D. W., et al. 1978. 'Flowering ecology of some spring woodland herbs'. *Ecology* 59: 351–366.
Schimper, A. F. W. 1898. *Pflanzengeographie auf physiologischer Grundlage*. Jena: Fischer-Verlag. 876 pp.
Schimper, A. F. W., and F. C. von Faber 1935. *Pflanzengeographie auf physiologischer Grundlage*. 3rd ed. 2 vols. Jena: Fischer. 1612 pp.
Schmid, E. 1954. 'Beiträge zur Flora und Vegetation der Kanarischen Inseln'. *Ber. Geobot. Inst. Rübel* 1953: 28–53.
Schmidt, W. 1973. 'Vegetationskundliche Untersuchungen im Savannenreservat Lamto (Elfenbeinküste)'. *Vegetatio* 28: 145–200.
Schmidt, W. 1975. 'Plant communities on permanent plots of the Serengeti Plains'. *Vegetatio* 30: 133–145.
Schmithüsen, J. 1956. 'Die räumliche Ordnung der chilenischen Vegetation'. *Bonner Geogr. Abhandl.* 17: 1–86.
Schmithüsen, J. 1960. 'Die Nadelhölzer in den Waldgesellschaften der südlichen Anden'. *Vegetatio* 9: 313–327.
Schnell, R. 1970–77. *Introduction à la Phytogéographie des Pays Tropicaux*. Vols. 1–4. Paris: Gauthier-Villars.
Schnetter, R. 1968. 'Die Vegetation des Cerro San Fernando und des Cerro La Llorona im Trockengebiet bei Santa Marta, Kolumbien'. *Ber. Dt. Bot. Ges.* 81: 289–302.
Scholz, H. 1966. 'Beitrag zur Flora des Tibesti-Gebirges (Tschad)'. *Willdenowia (Berlin)* 4: 183–202.
Scholz, H. 1967. 'Baumbestand, Vegetationsgliederung und Klima des Tibesti-Gebirges'. *Berliner Geogr. Abh.* 5: 11–16.
Schröter, E., and H. Hauri 1914. Versuch einer Übersicht der angiospermen Polsterpflanzen. *Engl. Bot. Jahrb.* vol. 50 (supplement: commemorative volume for A. Engler).
Schuler, S. 1977. *Simon and Schuster's Guide to Trees*. (Translated from Italian by H. Young.) New York: Simon and Schuster. Unpaginated, 300 color plates.
Schwarzenbach, F. H. 1960. 'Die arktische Steppe in den Trockengebieten Ost- und Nordgrønlands'. *Ber. Geobot. Inst. Rübel* 1959: 42–64.
Schweinfurth, U. 1957. 'Die horizontale und vertikale Verbreitung der Vegetation im Himalaya'. *Bonner Geogr. Abhandl.* 20.
Segal, S. 1969. *Ecological Notes on Wall Vegetation*. The Hague: Dr. W. Junk Publishers. 326 pp.
Shantz, H. L. 1906. 'A study of the vegetation of the mesa region east of Pike's Peak: the *Bouteloua* formation'. *Bot. Gaz.* 42: 16–47.
Shantz, H. L., and C. F. Marbut 1923. *The Vegetation and Soils of Africa*. New York: National Research Council and American Geographical Society. 264 pp.
Sharpe, D. M. 1970. 'The Effective Climate in the Dynamics of Alpine Timberline Ecosystems in Colorado'. *Publ. Climatol.* 23(1): 1–82.
Shaw, R. S. 1974. *Plants of Yellowstone and Grand Teton National Parks*. Salt Lake City: Wheelwright Press. 160 pp.
Shelford, V. E. 1963. *The Ecology of North America*. Urbana (Illinois): University of Illinois Press. 610 pp.
Shepard, D. S. 1970. SYMAP Interpolation Characteristics. In: *Computer Mapping as an Aid in Air Pollution Studies*, Vol. II: Individual Reports, Grant No. 68A-2405D. Washington: Natl. Air Poll. Control Adm., Pub. Health Serv., U.S. Dept. Health, Educ., Welfare. 70 pp.
Shimwell, D. W. 1971. *The Description and Classification of Vegetation*. London: Sidgwick and Jackson (USA: Univ. of Washington). 322 pp.
Shreve, F. 1911. 'The influence of low temperatures on the distribution of the Giant Cactus'. *Plant World* 14: 136–146.
Shreve, F. 1914. *A Montane Rain Forest: A Contribution to the Physiological Plant Geography of Jamaica*. Washington: Carnegie Inst. publ. 199.
Shreve, F. 1934. 'Vegetation of the northwestern coast of Mexico'. *Torrey Bot. Club Bull.* 61: 373–380.
Shreve, F. 1936. 'The transition from desert to chaparral in Baja California'. *Madroño* 3: 257–264.
Shreve, F. 1942. 'The desert vegetation of North America'. *Bot. Rev.* 8: 195–246.
Shreve, F., and I. L. Wiggins 1964. *Vegetation and Flora of the Sonoran Desert*. 2 vols. Stanford (California): Stanford Univ. Press.

Shrimal, R. L., and L. N. Vyas 1977. 'Community structure of the Motamagra forest block (Udaipur)'. *Vegetatio* 34: 127–132.

Simonis, W. 1954. 'Beobachtungen zur Ökologie einiger Therophyten in zwei mediterranen Pflanzengesellschaften'. *Vegetatio* 5/6: 553–561.

Simpson, B. B. (ed.) 1977. *Mesquite: Its Biology in Two Desert-Scrub Ecosystems.* US/IBP Synthesis Series, vol. 4. Stroudsburg (Pa.): Dowden, Hutchinson, and Ross, Inc. 272 pp.

Sissingh, G. 1977. 'Optional woodland development on sandy soils in the Netherlands'. *Vegetatio* 35: 187–191.

Smith, A. 1971. *Mato Grosso.*

Solbrig, O. T. 1976. The Origin and Floristic Affinities of the South American Temperate Desert and Semidesert Regions. In: *Evolution of Desert Biota* (D. W. Goodall, ed.). Austin: Univ. Texas Press.

Sonesson, M. 1970. *Studies on the Mire Vegetation in the Torneträsk Area, Northern Sweden.* III. Communities of the Poor Mires. Lund: Opera Botanica, no. 26.

Sorenson, T. 1941. 'Temperature relations and phenology of the northeast Greenland flowering plants'. *Medd. Groenl.* 125(9): 1–302.

Soriano, A. 1956. 'Los distritos floristicos de la Provincia Patagonia'. *Rev. Investig. Agric.* 10: 323–347.

Spalding, E. S. 1905. 'Mechanical adjustments of the Sahuaro (*Cereus giganteus*) to varying quantities of stored water'. *Bull. Torrey Bot. Club* 32: 57.

Spangler, W. M. L., and R. L. Jenne 1979. *World Monthly Surface Station Climatology.* Boulder (Colorado): National Center for Atmospheric Research. 16 pp. + 2 computer tapes.

Specht, R. L. 1969. 'A comparison of sclerophyllous vegetation characteristic of mediterranean type climates in France, California, and southern Australia'. *Austral. J. Bot.* 17: 277–292.

Specht, R. L. 1972. *The Vegetation of South Australia. Handbook of the Flora and Fauna of South Australia.* Adelaide: Govt. Printer. 328 pp.

Specht, R. L. (ed.) 1979. *Heathlands and Related Shrublands.* Ecosystems of the World, vol. 9A. Amsterdam: Elsevier Sci. Publ. Co. 498 pp.

Sreedhara Murthy, M. 1978. 'Relationships between vegetation and climate in the upper catchment area of the Narmada River, central India'. *Vegetatio* 36: 53–60.

Stamp, L. D. 1924. *The Vegetation of Burma from an Ecological Standpoint.* Rangoon: Univ. of Rangoon.

Stanford, W. W. 1969. 'The distribution of epiphytic orchids in relation to each other and to the geographic location and climate'. *Biol. J. Linn. Soc. London* 1: 247–285.

Steere, W. C. 1978. *The Mosses of Arctic Alaska.* Lehre (Germany): J. Cramer. 508 pp, 48 maps.

Stocker, O. 1923. 'Die Transpiration und Wasserökologie nordwestdeutscher Heide- und Moorpflanzen am Standort'. *Z. Bot.* 15: 1–41.

Stocker, O. 1970. 'Der Wasser- und Photosynthese-Haushalt von Wüstenpflanzen der mauretanischen Sahara: I. Regengrüne und immergrüne Bäume'. *Flora* 159: 539–572.

Stocker, O. 1971. 'Der Wasser- und Photosynthese-Haushalt von Wüstenpflanzen der mauretanischen Sahara. II. Wechselgrüne, Rutenzweig- und stammsukkulente Bäume'. *Flora* 160: 445–494.

Strain, B. R., and W. D. Billings (eds.) 1974. *Vegetation and Environment.* Handbk. Veg. Sci. Vol. 6. The Hague: Dr W. Junk Publishers.

Sturm, H. 1978. *Zur Ökologie der andinen Páramoregion.* Biogeographica, vol. 14. The Hague: Dr W. Junk Publishers. 130 pp.

Sunding, P. 1972. *The Vegetation of Gran Canaria.* Oslo: Universitetsforlaget. 186 pp + map and tables.

Szafer, W. (ed.) 1966. *The Vegetation of Poland.* (Transl. by W. H. Paryski). Oxford: Pergamon Press (from Polish Scientific Publishers, Warsaw). 738 pp.

Tadaki, Y. 1977. Leaf Biomass. In: *Primary Productivity of Japanese Forests* (Shidei and Kira, eds.). JIBP Synthesis, vol. 16. Tokyo: Univ. Tokyo Press.

Tansley, A. G. 1949. *The British Islands and their Vegetation.* 2 vols. Cambridge: University Press. 930 pp.

Taylor, S. E. 1975. Optimal Leaf Form. In: *Perspectives of Biophysical Ecology* (D. M. Gates and R. B. Schmerl, eds.), Ecol. Studies 12. New York: Springer-Verlag.

Tevis, L. 1958. 'A population of desert ephemerals germinated by less than one inch of rain'. *Ecology* 39: 688–695.

Thoday, D. 1931. 'The significance of reduction in the size of leaves'. *J. Ecol.* 19: 297–303.

Thornthwaite, C. W. 1948. 'An approach toward a rational classification of climate'. *Geogr. Rev.* 38: 55–94.

Thornthwaite, C. W. 1952. *Grassland Climates.* Publ. in Climatol, 5(6). Elmer (New Jersey).

Thornthwaite, C. W., and J. R. Mather 1957. *Instructions and Tables for Computing Potential Evapotranspiration and the Water Balance.* Publ. in Climatol., 10(3): 185–311. Elmer (New Jersey).

Tieszen, L. L. (ed.) 1978. *Vegetation and Production Ecology of an Alaskan Arctic Tundra.* Ecological Studies, 29. New York: Springer-Verlag. 736 pp.

Tolmatschew, A. I. 1954. *On the Origin and Developmental History of the Dark Needle-Leaved Taiga.* Moskva-Leningrad (in Russian).

Tomlinson, P. B. 1979. 'Systematics and Ecology of the *Palmae*'. *Ann. Rev. Ecol. Syst.* 10: 85–107.

Tranquillini, W. 1959. 'Die Stoffproduktion der Zirbe (*Pinus cembra* L.) an der Waldgrenze während eines Jahres'. *Planta* 54: 107–151.

Tranquillini, W. 1963. 'Die Abhängigkeit der CO_2-Assimilation junger Lärchen, Fichten und Zirben von der Luft- und Bodenfeuchte'. *Planta* 60: 70–94.

Tranquillini, W. 1964. 'The physiology of plants at high altitudes'. *Ann. Rev. Plant Physiol.* 15: 345–362.

Tranquillini, W. 1967. 'Über die physiologischen Ursachen der Wald- und Baumgrenze'. *Mitt. Forstl. Bundesversuchsanst. Wien* 75: 457–487.

Tranquillini, W. 1979. *Physiological Ecology of the Alpine Timberline.* Ecological Studies, 31. New York: Springer-Verlag. 137 pp.

Tranquillini, W., and K. Holzer 1958. 'Über das Gefrieren und Auftauen von Coniferennadeln'. *Ber. Dt. Bot. Ges.* 71: 143–156.

Trelease, W. 1916. *The Genus* Phoradendron: *A Monographic Revision.* Urbana: Univ. Illinois Press. 224 pp.

Troll, C. 1959. 'Die tropischen Gebirge'. *Bonner Geogr. Abhandl.* 25.

Troll, C. 1960. Die Physiognomik der Gewächse als Ausdruck der ökologischen Lebensbedingungen. In: *Tagesber. wiss. Abhandl. Dt. Geographentages 1959.* Wiesbaden.

Troll, C. 1961. 'Klima und Pflanzenkleid der Erde in dreidimensionaler Sicht'. *Die Naturwiss.* 48: 332–348.

Troll, C. 1966. *Ökologische Landschaftsforschung und vergleichende Hochgebirgsforschung* (Collected works, vol. 1). Erdkundliches Wissen, II. Wiesbaden: Franz-Steiner-Verlag.

Troll, C. (ed.) 1968. *Geo-ecology of the Mountainous Regions of the Tropical Americas.* Colloquium Geographicum, vol. 9 Bonn: Ferd. Dummlers Verlag. 224 pp.

Troll, C. 1970. 'Das 'Baumfarnklima' und die Verbreitung der Baumfarne auf der Erde'. *Tübinger Geogr. Studien* 34: 179–189.

Troll, C. 1972a. 'The three-dimensional zonation of the Himalayan system'. *Erdwiss. Forschungen* 4: 264–275.

Troll, C. (ed.) 1972b. *Geoecology of the High-Mountain Regions of Eurasia.* Erdwissenschaftl. Forschung, 4. Wiesbaden: Franz-Steiner-Verlag.

Troll, C. 1973. 'The upper timberlines in different climate zones'. *Arc. Alp. Res.* 5: A3–A18.

Troll, C. and W. Lauer (eds.) 1978. *Geoecological Relations between the South Temperate Zone and Tropical Mountains.* Erdwiss.-Forschung, II. Akad. d. Wiss. und Lit. Mainz. Wiesbaden: Steiner.

Troll, C., and K. H. Paffen 1964. *Karte der Jahreszeiten-Klimate der Erde.* Erdkunde, Arch. Wiss. Geogr., 18: 5–28.

Troll, W. 1937–38. *Vergleichende Morphologie der höheren Pflanzen.* Vol. 1, parts 1 and 2. Berlin.

Turage, W. V., and A. L. Kinckley 1938. 'Freezing weather in relation to plant distribution in the Sonoran Desert'. *Evol. Monogr.* 8: 529–550.

Twomey, A. C. 1936. 'Climograph studies of certain introduced and migratory birds'. *Ecology* 17: 122–132.

Udvardy, M. D. F. 1975. *A Classification of the Biogeographical Provinces of the World.* Morges (Switzerland): Internatl. Union for Conserv. of Nature. 48 pp.

Ule, E. 1915. *Biologische Beobachtungen im Amazonasgebiet.* Berlin.

Ulmer, W. 1937. 'Über den Jahresgang der Frosthärte einiger immergrüner Arten der alpinen Stufe, sowie der Zirbe und Fichte'. *Jb. wiss. Bot.* 84: 553–592.

UNESCO 1969. *Vegetation Map of the Mediterranean Zone* (with Explanatory Notes). Arid Zone Research, 30. Paris.

UNESCO 1978. *Tropical Forest Ecosystems.* Natural Resources Research, vol. 14. Paris.

Van Bruggen, T. 1976. *The Vascular Plants of South Dakota.* Ames (Iowa): The Iowa State Univ. Press. 538 pp.

Van Dyne, G. M., F. M. Smith, R. L. Czaplewski, and R. G. Woodmansee 1978. Analyses and Syntheses of Grassland Ecosystem Dynamics. In: *Glimpses of Ecology* (J. S. Singh and B. Gopal, eds.). Jaipur (India): Internatl. Scientific Publications.

Van Steenis, C. G. G. J. 1962. 'Die Gebirgsflora der malesischen Tropen'. *Endeavour* 21: 183–194.

Vareschi, V. 1968. 'Comparación entre selvas Neotropicales y Paleotropicales en base a su espectro de biotipas'. *Acta Bot. Venez.* 3: 239–263.

Vareschi, V. 1970. *Flora de los páramos de Venezuela.* Mérida: Univ. de los Andes. 429 pp (126 figs.).

Vareschi, V. 1973. El Problema de la Vegetación Óptima. In: I⁰ Congreso Latinoamericano V⁰ Mexicano de Botánica, Memorias de Symposia. México: Soc. Bot. de México.

Veblen, T. T. 1979. 'Structure and dynamics of *Nothofagus* forests near timberline in south-central Chile'. *Ecology* 60: 937–945.

Veblen, T. T., and D. H. Ashton 1978. 'Catastrophic influences on the vegetation of the Valdivian Andes, Chile'. *Vegetatio* 36: 149–167.

Veillon, J. P. 1955. *Bosques andinos de Venezuela.* Mérida.

Verdoorn, F. (ed.) 1945. *Plants and Plant Science in Latin America.* Waltham (Massachusetts): Chronica Botanica. 381 pp.

Vesey-Fitzgerald, D. F. 1970. 'The origin and distribution of valley grasslands in East Africa'. *J. Ecol.* 58: 51–75.

Vidal, J. 1961. *La Végétation du Laos.* Travaux du Laboratoire Forestier de Toulouse, 5(1), article III.

Viereck, L. A., and E. L. Little jr. 1972. *Alaska Trees and Shrubs.* Washington: U.S. Dept. Agric., Forest Serv., Agriculture Handbook 410. 265 pp (with vegetation map).

Vogel, S. 1955. 'Niedere Fensterpflanzen in der südafrikanischen Wüste'. *Beitr. Biol. Pfl.* 31: 45–135.

Vogl, R. J., and P. K. Schorr 1972. 'Fire and manzanita chaparral in the San Jacinto Mountains, California'. *Ecology* 53: 1179–1188.

Volk, O. H. 1954. 'Klima und Pflanzenverbreitung in Afghanistan'. *Vegetatio* 5/6: 422–433.

Wace, N. M. 1961. 'The vegetation of Gough Island'. *Ecol. Monogr.* 31: 337–367.

Wace, N. M., and M. W. Holdgate 1976. *Man and Nature in the Tristan da Cunha Islands.* Morges: Internatl. Union for Conserv. of Nature. 114 pp.

Walter, H. 1931. *Die Hydratur der Pflanzen und ihre physiologisch-ökologische Bedeutung.* Jena.

Walter, H. 1939. 'Grassland, Savanne und Busch der ariden Teile Afrikas in ihrer ökologischen Bedingtheit'. *Jb. wiss. Bot.* 87: 750–860.

Walter, H. 1955. 'Die Klimadiagramme als Mittel zur Beurteilung der Klimaverhältnisse für ökologische, vegetationskundliche und landwirtschaftliche Zwecke'. *Ber. Dt. Bot. Ges.* 68: 331–334.

Walter, H. 1956a. 'Die Vegetationsgliederung Anatoliens'. *Flora* 143: 295–326.

Walter, H. 1956b. 'Die heutige ökologische Problemstellung und der Wettbewerb zwischen der mediterranen Hartlaubvegetation und den sommergrünen Laubwäldern'. *Ber. Dt. Bot. Ges.* 69: 263–273.

Walter, H. 1960. 'Höhenstufen und alpine Vegetation in Australien, auf Tasmanien, und auf Neuseeland'. *Ber. Geobot. Inst. Rübel* 31: 67–71.

Walter, H. 1967. 'Das Pampaproblem in vergleichend ökologischer Betrachtung'. *Erdkunde (Bonn)* 21: 181–203.

Walter, H. 1968. *Die Vegetation der Erde in öko-physiologischer*

Betrachtung. Vol. II: Die gemäßigten und arktischen Zonen. Jena: VEB Gustav-Fischer-Verlag. 1002 pp.

Walter, H. 1973. *Die Vegetation der Erde in öko-physiologischer Betrachtung*. Vol. I: Die tropischen und subtropischen Zonen. 3rd ed. Stuttgart: Gustav-Fischer-Verlag. 744 pp.

Walter, H. 1974. *Die Vegetation Osteuropas, Nord- und Zentralasiens*. Vegetationsmonographien der einzelnen Großräume, vol. VII. Stuttgart: Gustav-Fischer-Verlag. 452 pp.

Walter, H. 1975. 'Betrachtungen zur Höhenstufenfolge im Mediterrangebiet (insbesondere in Griechenland) in Verbindung mit dem Wettbewerbsfaktor'. *Veröff. Geobot. Inst. Rübel.* 55: 72–83.

Walter, H. 1976a. *Die ökologischen Systeme der Kontinente (Biogeosphäre): Prinzipien ihrer Gliederung mit Beispielen.* Stuttgart: Gustav-Fisher-Verlag. 132 pp.

Walter, H. 1976b. 'Gibt es in der Namib Nebelpflanzen?'. *Namib und Meer* 7: 5–13.

Walter, H. 1977a. 'The oligotrophic peatlands of western Siberia – the largest peino-helobiome in the world'. *Vegetatio* 34: 167–178.

Walter, H. 1977b. *Vegetationszonen und Klima: Die ökologische Gliederung der Biogeosphäre.* 3rd ed. Uni-Taschenbücher, 14. Stuttgart: Verlag Eugen Ulmer. 310 pp.

Walter, H., and E. O. Box 1976. 'Global classification of natural terrestrial ecosystems'. *Vegetatio* 32: 75–81.

Walter, H., and E. O. Box (in press). Continental Deserts and Semi-Deserts of Eurasia. Chapters 2–9 in: *Temperate Deserts and Semi-Deserts* (N. E. West, ed.), Ecosystems of the World 5. Amsterdam: Elsevier.

Walter, H., and H. Lieth 1960–67. *Klimadiagramm-Weltatlas.* Stuttgart: Gustav-Fischer-Verlag. 8000 stations.

Walter, H. and E. Medina 1969. 'Die Bodentemperatur als ausschlaggebender Faktor für die Gliederung der subalpinen und alpinen Stufe in den Anden Venezuelas'. *Ber. Dt. Bot. Ges.* 82: 275–281.

Wang, T. 1941. 'Die Dauer der ariden, humiden und nivalen Zeiten des Jahres in China'. *Tübinger Geogr. Abh.* 2(7).

Wang, Chi-Wu 1961. *The Forests of China, with a Survey of Grassland and Desert Vegetation.* Cambridge (Massachusetts): Cabot Found. publ. 5. 313 pp.

Wardle, E. P. 1965. 'A comparison of alpine timber lines in New Zealand and North America'. *N. Z. J. Bot.* 3: 113–135.

Wardle, E. P. 1971. 'An explanation for alpine timberline'. *N. Z. J. Bot.* 9: 371–402.

Waring, R. H., W. H. Emmingham, H. L. Gholz, and C. C. Grier 1978. 'Variation in Maximum Leaf Area of Coniferous Forests in Oregon and its Ecological Significance'. *Forest Science* 24: 131–140.

Warming, E. 1895. *Plantesamfund: Grundtrak af den økologiske Plantgeografi.* København.

Warming, E. 1909. *Oecology of Plants: An Introduction to the Study of Plant-Communities.* (Revised English edition). London: Humphrey Milford and Oxford Univ. Press. 422 pp.

Watson, W. 1913. 'Xerophytic adaptations of bryophytes in relation to habitat'. *New Phytol.* 13: 149–169, 181–190.

Weaver, J. E., and F. W. Albertson 1956. *Grasslands of the Great Plains.* Lincoln (Nebraska): Johnsen Publ. Co. 395 pp.

Webb, L. J. 1959. 'A physiognomic classification of Australian rainforests'. *J. Ecol.* 551–570.

Webb, L. J. 1964. 'An historical interpretation of the grass balds of the Bunya Mountains, southern Queensland'. *Ecology* 45: 159–162.

Webber, P. J. (ed.) 1978. *High-Altitude Geoecology.* Washington: Amer. Assn. Adv. Science, Selected-Symposia Series.

Weber, H. 1958. *Die Páramos von Costa Rica und ihre pflanzengeographische Verkettung mit den Hochanden Südamerikas.* Wiesbaden: Verlag Akad. Wiss. Literatur Mainz. 194 pp + 90 photos.

Weberbauer, A. 1911. *Die Pflanzenwelt der peruanischen Anden.* Leipzig.

Weberbauer, A. 1931. 'Über die Polsterpflanze *Pycnophyllum aristatum* und die Polsterpflanzen im allgemeinen'. *Ber. Dt. Bot. Ges.* 49: 29–34.

Weinberger, P., M. Romero, and M. Oliva 1973. 'Untersuchungen über die Dürreresistenz patagonischer immergrüner Gehölze'. *Vegetatio* 28: 75–98.

Wells, B. W. 1928. 'Plant Communities of the Coastal Plain of North Carolina and their Successional Relations'. *Ecology* 9: 230–242.

Wells, B. W. 1942. 'Ecological Problems of the Southeastern United States Coastal Plain'. *Bot. Rev.* 8: 533–561.

Welsh, S. I., and G. Moore 1973. *Utah Plants, Tracheophyta.* 3rd ed. Provo (Utah): Brigham Young Univ. Press. 474 pp.

Wendelberger, G. 1959. 'Die Waldsteppen des pannonischen Raumes'. *Veröff. Geobot. Inst. Rübel* 35: 77–113.

Went, F. W. 1940. 'Soziologie der Epiphyten eines tropischen Urwaldes'. *Ann. Jard. Bot. Buitenz.* 50: 1–98.

Went, F. W. 1948–49. 'Ecology of desert plants'. I and II. *Ecology* 29: 242–353; 30: 1–3, 26–38.

Went, F. W. 1957. *The Experimental Control of Plant Growth.* Waltham (Massachusetts): Chronica Botanica. 343 pp.

Werger, M. J. A. (ed.) 1977. *Biogeography and Ecology of Southern Africa.* Monogr. Biolog., vol. 31. The Hague: Dr W. Junk Publishers. 1436 pp.

Wernstedt, F. L. 1972. *World Climatic Data.* Lemont (Pennsylvania): Climatic Data Press. 554 pp.

Werth, E. 1928. Die Vegetation der subantarktischen Inseln Kerguelen, Possession und Heard-Eiland. In: *Dtsch. Südpolar Exp. 1901/03* (E. Drygalski, ed.), vol. 8, Berlin-Leipzig.

West, R. C. 1958. *The Pacific Lowlands of Colombia.* Baton Rouge: Louisiana State University Press. 278 pp.

Westman, W. E. 1978. 'Patterns of nutrient flow in the pygmy forest region of northern California'. *Vegetatio* 36: 1–15.

Wharton, C. H. 1978. *The Natural Environments of Georgia.* Atlanta: Georgia Dept. of Natural Resources. 228 pp.

Whitmore, T. C. 1975. *Tropical Rainforests of the Far East.* Oxford: Clarendon Press. 282 pp.

Whittaker, R. H. 1956. 'Vegetation of the Great Smoky Mountains'. *Ecol. Monogr.* 26: 1–80.

Whittaker, R. H. 1962. 'Classification of natural communities'. *Bot. Rev.* 28: 1–239.

Whittaker, R. H. 1963. 'Net production of heath balds and forest heaths in the Great Smoky Mountains'. *Ecology* 46: 176–182.

Whittaker, R. H. 1975. *Communities and Ecosystems.* 2nd ed. New York: Macmillan Publ. Co. 387 pp.

Whittaker, R. H., and W. A. Niering 1965. 'Vegetation of the Santa Catalina Mountains, Arizona: a gradient analysis of the south slope'. *Ecology* 46: 429–452.

Whyte, R. O. 1968. *Grasslands of the Monsoon.* New York and Washington: Frederick A. Praeger. 326 pp.

Whyte, R. O., et al. 1954. 'The grassland of India'. *8. Congr. Intern. Bot., Comptes Rendues* Sect. 7–8: 46–53. Paris.

Wielgolaski, F. E. 1972. 'Vegetation types and plant biomass in tundra'. *Arc. Alp. Res.* 4: 291–305.

Wielgolaski, F. E., P. Kallio, and T. Rosswall 1975. *Fennoscandian Tundra Ecosystems.* Part 1: Plants and Microorganisms. Ecol. Studies 16. New York: Springer-Verlag. 366 pp.

Wiggins, I. L., and D. M. Porter 1971. *Flora of the Galapagos Islands.* Stanford Univ. Press.

Wilhelmy, H. 1957. Das große Pantanal im Mato Grosso. *Dtsch. Geographentag* (*Würzburg*), 1957: 45–71.

Wilhelmy, H. 1963. 'Die Falkland-Inseln (Islas Malvinas)'. *Geogr. Ztschr.* 51: 53–62.

Williams, L. 1965. *Vegetation of Southeast Asia: Studies of Forest Types 1963*–1965. Washington: U.S. Dept. Agric. 302 pp.

Williams, W. D. (ed.) 1974. *Biogeography and Ecology in Tasmania.* Monogr. Biolog., vol. 25. The Hague: Dr W. Junk Publishers. pp.

Wissmann, H. von 1972. 'Die *Juniperus* Gebirgswälder in Arabien'. *Erdwiss. Forschung (Wiesbaden)* 4: 157–176.

Worobjew, D. P. 1963. *Die Pflanzenwelt der Kurilischen Inseln.* Moskva-Leningrad: Akad. Nauk.

Yoshino, M. M. 1973. 'Wind-shaped trees in the subalpine zone in Japan'. *Arc. Alp. Res.* 5(2): A115–A126.

Yunatov, A. A. 1950. *The Main Features of the Plant Cover of the Mongolian People's Republic.* Moskva-Leningrad: Publ. Mongol. Comm., 39. 223 pp (in Russian).

Zemke, E. 1939. 'Anatomische Untersuchungen an Pflanzen der Namibwüste'. *Flora, N. F.* 33: 365–416.

Zinderen-Bakker, E. M. van, 1970. 'Observations on the distribution of *Ericaceae* in Africa'. *Colloq. Geogr. (Bonn)* 12: 89–97.

Zinderen-Bakker, E. M. van 1973. 'Ecological investigations of forest communities in the eastern Orange Free State and the adjacent Natal Drakensberg'. *Vegetatio* 28: 299–334.

Zohary, M. 1952. Hydro-Economical Types in the Vegetation of Near East Deserts. *Symp. Biol. Product. of Hot and Cold Deserts (London).* pp. 56–67.

Zohary, M. 1973. *Geobotanical Foundations of the Middle East.* 2 vols. Stuttgart: Gustav-Fischer-Verlag. 738 pp.

Zohary, M., and G. Orshan 1966. 'An outline of the geobotany of Crete'. *Israel J. Bot.* 14 (supplement).

Zotov, V. D. 1947. 'Survey of the tussock-grasslands of the South Island'. *N. Z. J. Sci. Techn.* Ser. A, 20: 212–244.

AF854994

Carnegie Endowment for International Peace
DIVISION OF INTERNATIONAL LAW
James Brown Scott, Director

THE PROBLEM OF AN INTERNATIONAL COURT OF JUSTICE

BY

DR. HANS WEHBERG
GERICHTSREFERENDAR IN DÜSSELDORF

TRANSLATED FROM THE GERMAN BY

CHARLES G. FENWICK, PH.D.
ASSOCIATE PROFESSOR OF POLITICAL SCIENCE
BRYN MAWR COLLEGE

THE LAWBOOK EXCHANGE, LTD.
Clark, New Jersey

ISBN-13: 978-1-58477-915-5
ISBN-10: 1-58477-915-2

Lawbook Exchange edition 2008

Second printing 2010

The quality of this reprint is equivalent to the quality of the original work.

THE LAWBOOK EXCHANGE, LTD.
33 Terminal Avenue
Clark, New Jersey 07066-1321

Please see our website for a selection of our other publications and fine facsimile reprints of classic works of legal history:
www.lawbookexchange.com

Library of Congress Cataloging-in-Publication Data

Wehberg, Hans, 1885-
[Problem eines internationalen Staatengerichtshofes. English]
The problem of an international court of justice / by Hans Wehberg ; translated from the German by Charles G. Fenwick.
p. cm.
Originally published: Oxford : Clarendon Press, 1918.
Includes bibliographical references and index.
ISBN-13: 978-1-58477-915-5 (cloth : alk. paper)
ISBN-10: 1-58477-915-2 (cloth : alk. paper)
1. International courts. 2. International Court of Justice. 3. Arbitration, International. I. Title.
KZ6250.W5513 2008
341.5'52--dc22 2008021125

Printed in the United States of America on acid-free paper

Carnegie Endowment for International Peace
DIVISION OF INTERNATIONAL LAW
James Brown Scott, Director

THE PROBLEM OF AN INTERNATIONAL COURT OF JUSTICE

BY

DR. HANS WEHBERG

GERICHTSREFERENDAR IN DÜSSELDORF

TRANSLATED FROM THE GERMAN BY

CHARLES G. FENWICK, PH.D.

ASSOCIATE PROFESSOR OF POLITICAL SCIENCE
BRYN MAWR COLLEGE

OXFORD: AT THE CLARENDON PRESS
London, Edinburgh, New York, Toronto, Melbourne and Bombay
HUMPHREY MILFORD
1918

PRINTED IN ENGLAND
AT THE OXFORD UNIVERSITY PRESS

TO

JAMES BROWN SCOTT

UNITED STATES DELEGATE TO THE SECOND HAGUE CONFERENCE

IN RESPECT AND GRATITUDE

'May I say that while, fortunately for some of us, justice is tardy, I think it proper in a meeting such as this to say that you have in this room the man who more than all others devoted his energy, thought, time and wonderful capacity for work in the endeavour made by the United States delegation to work out a permanent judicial court at The Hague. The failure to carry the plan through was certainly the greatest possible success, because the project for the Court, which was his work, was unanimously recommended by the Conference. I suppose that in some way when the world demands this Court, some person will evolve a scheme by which its judges can be designated. I refer with great pleasure to one of my warmest friends, Dr. James Brown Scott, a member of the American delegation at the Second Peace Conference at the Hague.'

(Address by the Honorable William I. Buchanan, Delegate of the United States to the Second Hague Peace Conference, at the fifteenth Lake Mohonk Conference, 1909. *Report*, p. 53.)

INTRODUCTORY NOTE BY THE DIRECTOR

In the spring of 1912 there was issued, in the German language, a book entitled *Der Staatenverband der Haager Konferenzen,* or as it may be rendered in English, *The Union created by the Hague Conferences.* This work is the first of a series of volumes under the general title *Das Werk vom Haag,* or *The Work of The Hague,* published by Messrs. Duncker and Humblot of Munich and Leipzig, and due to the enterprise and devotion of Dr. Walther Schücking, professor in the University of Marburg, Germany. The series, as a whole, was intended to make known the results of the First and Second Hague Conferences by publishing monographs dealing with the declarations, conventions, and recommendations, as well as the problems of the Conferences, so that the students and scholars of Germany might be furnished with all information essential to a correct understanding of the Conferences.

In the conduct of this enterprise Professor Schücking has secured the co-operation of the following well-known publicists: von Bar, Fleischmann, Kohler, Lammasch, von Liszt, Meurer, Niemeyer, Nippold, von Ullmann, and Wehberg. Thus far two volumes of the series have been issued, viz., the volume by Professor Schücking referred to above, and one by Dr. Hans Wehberg, entitled *Das Problem eines internationalen Staatengerichtshofes,* or *The Problem of an International Court of Justice.* A third work in three volumes, entitled *Judicial Decisions of the Permanent Court at The Hague,* is in process of publication, being a careful and accurate account of the cases which have been tried by temporary tribunals formed from the Permanent Court since its establishment in 1902.

The present volume, which is Dr. Wehberg's monograph on the problem of an international court of justice, cannot be read by any one without admiration for the skilful manner in which he has discussed the various problems connected with an international court and without a feeling of gratitude for the balanced judgment and the spirit of fairness which he has displayed in the attempt to solve the problems connected with the proposed court, which are both many and difficult.

Dr. Wehberg approached the question of an international court of justice with considerable misgivings, and, in a previous publication, he declared himself to be a partisan of the so-called Permanent Court of Arbitration and opposed to the establishment of a truly permanent court composed of professional judges acting under a sense of judicial responsibility. Study and reflection, however, have convinced him of the necessity of the latter court, and the volume under review is calculated to strengthen the faith of those who believe in the proposed court and to persuade many doubters and waverers who are open to argument.

Dr. Wehberg believes that international law must be developed in order more adequately to meet the world's needs ; that the conflicts inevitably arising between nations, in so far as they are of a legal character, can best be decided by a court of justice composed of permanent and professional judges whose duty it will be both to find and to apply the principle of justice decisive of the conflicts submitted to the court; and that such a tribunal can safely be entrusted to develop the system of law, as is the wont of courts, which it applies to the decision of concrete cases. He believes, and rightly, that the nations appearing before the court should not be represented by judges upon the bench, as the presence of national judges is not only embarrassing in itself, but tends to question the impartiality of the decision. He is inclined to think that the framers of the so-called Permanent Court of Arbitration devised by the First Hague Conference did not intend to create a judicial tribunal; that their purpose was to provide machinery for the adjustment of a dispute rather than for its decision according to principles of law; and he cites on this point the views not only of delegates of the First but also of the Second Conference in support of his contention. The question is one of no little difficulty, but even if the partisans of the so-called Permanent Court of Arbitration meant it to be a judicial tribunal, Dr. Wehberg is very clear in his mind that it is not and that an international court of justice in the technical sense of the word should be created. The learned and industrious author approves the project of the Court of Arbitral Justice drafted by the Second Hague Conference, and his ninth chapter is a brief and serviceable comment upon it.

The undersigned has devoted much thought and attention to the question of a permanent international court of justice, and he commends this volume to the interested public as the best statement of the problems connected with the necessity and organization of an international court which it has been his fortune to read, and as calculated to advance the cause of judicial and therefore of peaceable settlement of justiciable disputes between and among the nations.

James Brown Scott,
Director of the Division of International Law.

Washington, D.C.,
June 29, 1917.

CONTENTS

BIBLIOGRAPHY

(Additional works are referred to in the text.)

Abbreviations

American Journal—*American Journal of International Law*, New York.

Proceedings—*Proceedings of International Conference under the auspices of American Society for Judicial Settlement of International Disputes, December 15–17, 1910*, Baltimore.

Proceedings of the Second Hague Peace Conference are quoted in the text without the title being given. If no Roman numerals are given in the citation, the reference is to the second volume.

Report—*Report of the Annual Meeting of the Lake Mohonk Conference on International Arbitration*, Lake Mohonk.

Revue de Droit international—*Revue de Droit international et de Législation comparée*, Brussels.

Revue générale—*Revue générale de Droit international public*, Paris.

Adler: *Der Krieg, die Kongressidee und die allgemeine Wehrpflicht.* Prague, 1868.

Anderson: 'The Relations between International Tribunals of Arbitration and the Jurisdiction of National Courts,' in *Proceedings of the American Society of International Law*, 1909, vol. iii, pp. 35 et seq.

Annuaire de l'Institut de Droit international. Brussels, 1877 et seq.

—— *de l'Union interparlementaire.* Brussels, 1911 and 1912.

Asser: 'De Tweede Vredes-Conferentie, Een Terugblik,' in the *Tijdspiegel*, 1908, p. 1.

Aylesworth: 'The North Atlantic Fisheries Arbitration,' in *Canadian Law Times*, 1911, pp. 144 et seq.

Aymar: 'The Relations between International Tribunals of Arbitration and the Jurisdiction of National Courts,' in *Proceedings of the American Society of International Law*, vol. iii, 1909, pp. 52 et seq.

Balch: *International Courts of Arbitration.* Philadelphia, 1896.

—— 'L'Adjudication de la question de la frontière entre l'Alaska et le Canada,' in *Revue de Droit international*, 1904, pp. 38 et seq.

—— 'L'Évolution de l'arbitrage international,' in *Revue de Droit international*, 1908, pp. 261 et seq., and pp. 362 et seq.

—— 'La Décision de la Cour permanente d'arbitrage au sujet des pêcheries de l'Atlantique dans le différend entre les États-Unis et l'Empire Britannique,' in *Revue de Droit international*, 1911, p. 5.

Baldwin: 'Respect for international law and The Hague Tribunal,' in *Report*, 1907, pp. 40 et seq.

—— 'The Proposed International Court of Arbitral Justice,' in *Report*, 1910, pp. 78 et seq.

—— *The New Era of International Courts.* Baltimore, 1910.

—— Evolution of the International Court, in *Proceedings*, pp. 222 et seq.

v. Bar: 'Die Umgestaltung des Haager internationalen Schiedsgerichtshofs,' in the *Berliner Tageblatt*, 1907, vol. 22, p. viii.

Bara: *La Science de la paix.* Paris, 1872.

Barclay: 'La Question des pêcheries dans la mer de Behring,' in *Revue de Droit international*, 1893, pp. 417 et seq.

—— Problems of International Practice and Diplomacy. London, 1907.

de la Barra: 'The Juridical Character of an International Permanent Court,' in *Proceedings*, pp. 5 et seq.

Bartholdt: The Interparliamentary Plan, pp. 332 et seq., of the *Proceedings of the National Arbitration and Peace Congress.* New York, 1907.

Baty: *International Law.* London, 1909.

—— 'L'Arbitrage anglo-américain,' in *La Paix par le Droit,* 1911, pp. 192 et seq.

Bonfils: *Manuel de droit international public.* Paris, 1911.

Borchardt: 'The North Atlantic Coast Fisheries Arbitration,' in *Columbia Law Review,* 1911, p. 1.

—— Review of Ralston's book, *International Arbitral Law and Procedure,* in *American Journal,* 1911, p. 534.

Bressonnet: 'L'Arbitrage franco-anglais dans l'affaire des boutres de Mascate,' in *Revue générale,* 1906, p. 145.

Brewer: The Mission of the United States of America in the Court of Peace, in *Advocate of Peace,* 1909, No. 10.

Brown: 'Interstate Controversies in the Supreme Court of the United States,' in *Proceedings,* pp. 76 et seq.

—— 'International Courts,' in *Yale Law Journal,* New Haven, Conn., 1911, No. 1.

Bry: *Précis élémentaire de Droit international public.* Paris, 1910.

Buchanan: The Importance of Conservative Work for Arbitration, in *Report,* 1909, pp. 50 et seq.

——'The Application of the Principle of Arbitration,' p. 242 of the *Proceedings of the Second National Peace Congress,* Boston, 1909.

Bustamante y Sirven: *La seconde Conférence de la Paix,* translated from the Spanish by Scelle. Paris, 1909.

Butler: Opening Address at the Lake Mohonk Conference on International Arbitration. Lake Mohonk, N.Y., 1911.

Carnegie: 'The Moral Issue in War,' in *Proceedings,* pp. 68 et seq.

Chavet: *Justice sociale et justice internationale.* Paris, 1900.

Chester: 'A Permanent Court,' in *Report*, 1899, pp. 46 et seq.

Choate: 'The Court of Arbitral Justice as Recommended by the Second Hague Conference,' in *Proceedings*, pp. 195 et seq.

Clarke: 'A Permanent Tribunal of International Arbitration; Its Necessity and Value,' in *American Journal*, vol. i, No. 2.

Courrier de la Conférence de la Paix. The Hague, 1907.

Crapsey: 'The Permanent Court at The Hague,' in *Case and Comment*, Rochester, 1911, pp. 59 et seq.

Daehne van Varick: *Actes et Documents relatifs au programme de la Conférence de la Paix de La Haye.* The Hague, 1899.

Darby: *International Tribunals.* London, 1904.

Davis: *The Elements of International Law.* New York, 1908.

Dennis: 'The Orinoco Steamship Company Case before The Hague Tribunal,' in *American Journal*, vol. v, pp. 35 et seq.

—— 'The Great Defect of Arbitration,' in *Columbia Law Review*, 1911, pp. 493 et seq.

Descamps: *Essai sur l'organisation de l'arbitrage international.* Brussels, 1896 (quoted from Fried's German translation).

Despagnet: *Cours de Droit international public.* Paris, 1910.

Dietz: *Franz Wirth und der Frankfurter Friedensverein.* Frankfort-on-the-Main, 1911.

Diomedes: *The Second Hague Conference*, vol. i. Athens, 1908 (in Greek).

Dreyfus: *L'Arbitrage international.* Paris, 1892.

Dumas: *Les Sanctions de l'arbitrage international.* Paris, 1905.

—— 'Pourrait-on donner aux États un droit d'action directe devant la Cour de La Haye?' in *La Paix par le Droit*, 1907, No. 6.

Duplessix: *L'Organisation internationale.* Paris, 1909.

Earl: 'An International Court,' in *Report*, 1898, pp. 36 et seq.

Eickhoff: *Die internationale Schiedsgerichtsbewegung*. Berlin, 1910.

—— 'Die Ergebnisse der 16. interparlamentarischen Konferenz,' in the *Tag*, September 24, 1910.

Eliot: 'Defects of Arbitration as a Means of Settling International Disputes,' in *Proceedings*, pp. 275 et seq.

d'Estournelles de Constant: 'Dangers of Haste in Establishing the International Court,' in *Report*, 1911, pp. 113 et seq.

Fauchille: 'Le Conflit de limites entre le Brésil et la Grande-Bretagne et la sentence arbitrale du roi d'Italie,' in *Revue générale*, 1905, p. 25.

Ferrer: *L'Ère nouvelle ; nécessité d'un code international et d'un tribunal arbitral des nations*. Paris, 1863.

Fiore: 'La Sentence arbitrale du Président de la République Argentine,' in *Revue générale*, 1910, No. 3.

—— *Le Droit international codifié*. Paris, 1911.

Fischer: *Die Verfolgung vermögensrechtlicher Ansprüche gegen fremde Schuldnerstaaten*. 1911.

Foster: *International Arbitration and the Hague Court*. New York, 1901.

—— 'Were the Questions involved in the Foreign Wars of the United States of such a nature that they could have been submitted to Arbitration or settled without recourse to War ? ' in *Proceedings*, pp. 44 et seq.

—— 'The Relations between International Tribunals of Arbitration and the Jurisdiction of National Courts,' in *Proceedings of the American Society of International Law*, vol. iii, 1909, pp. 25 et seq.

Foulke: 'Concentration of Effort upon Judicial Settlement of International Disputes,' in *Proceedings*, pp. 282 et seq.

Foulke: 'Arbitration the Shortest Road to International Justice,' in *Report*, 1911, pp. 69 et seq.

Fried: 'Eine ständige Schiedsgerichtskammer im Haag,' in *Friedenswarte*, 1903, p. 170.

—— 'Rasches und billiges Recht im Haag,' in *Friedenswarte*, 1906, p. 13.

—— 'Der ständige Schiedshof im Haag,' in *Pester Lloyd*, July 13 and August 8, 1907.

—— *Die zweite Haager Konferenz.* Leipzig, 1908.

—— 'Die Reichsregierung und die Friedensbestrebungen,' in *Friedenswarte*, 1909, p. 52.

—— 'Der wirklich ständige Schiedsgerichtshof,' in *Friedenswarte*, 1910, p. 13.

—— 'Der wirklich ständige Schiedshof im Haag,' in *Friedenswarte*, 1910, p. 48.

—— 'Die bevorstehende Errichtung des wirklich ständigen Schiedshofes im Haag,' in *Friedenswarte*, 1910, p. 107.

—— 'Zur XVI. interparlamentarischen Konferenz,' in *Friedenswarte*, 1910, p. 186.

—— 'Der Knoxsche Vorschlag,' in *Friedenswarte*, 1911, p. 55.

—— 'Deutschland gegen den Knoxschen Vorschlag,' in *Friedenswarte*, 1911, p. 120.

—— 'Oberprisengerichtshof und Schiedsgerichtshof,' in *Kölnische Volkszeitung*, February 7, 1910.

—— *Pan-Amerika.* Berlin, 1910.

—— *Der Kaiser und der Weltfrieden.* Berlin, 1910.

—— *Handbuch der Friedensbewegung*, vol. i. Leipzig, 1911.

Friedmann: 'Schiedsvertrag und Schiedsspruch über die peruanisch-bolivianischen Grenzstreitigkeiten,' in *Zeitschrift für internationales Recht*, 1910, pts. 2 and 3.

de Fuente: *La segunda Conferencia internacional de la Paz.* Havana, 1908.

Gidel: 'L'Arbitrage de Casablanca,' in *Revue générale*, 1910, No. 4.

Ginn: 'Some of the Things that must be done in order to secure a Permanent Judicial Court,' in *Proceedings*, pp. 313 et seq.

Gobat: 'L'Arbitrage des phoques à fourrure,' in *La Conférence interparlementaire*, 1893, pp. 54 et seq.

——'La cinquième Conférence interparlementaire,' in *La Conférence interparlementaire*, 1894, pp. 213 et seq.

—— 'Procès-verbaux des séances de la Conférence interparlementaire à La Haye,' in *La Conférence interparlementaire*, 1894, p. 230.

—— *Projet d'organisation d'une Cour permanente internationale d'arbitrage.* 1895 (printed in manuscript form).

—— 'La sixième Conférence interparlementaire,' in *La Conférence interparlementaire*, 1895, pp. 343 et seq.

Goldschmidt: 'Projet de règlement pour tribunaux arbitraux internationaux,' in *Revue de Droit international*, 1874, p. 421.

Gray: 'International Public Opinion,' pp. 349 et seq. of the *Proceedings of the National Arbitration and Peace Congress.* New York, 1907.

Gregory: 'Aviation as affecting the Judicial Settlement of International Disputes,' in *Proceedings*, pp. 168 et seq.

Hall: *International Law.* Oxford, 1909.

van Hamel: 'Les Principes du droit d'extradition et leur application dans l'affaire Savarkar,' in *Revue de Droit international*, 1911, p. 370.

Harmening: *Das Weltparlament.* Geneva, 1910.

Haussmann: 'Savarkar und das Schiedsgericht,' *März*, March 14, 1911.

Higgins: *The Hague Peace Conferences.* Cambridge, 1909.

Hill: *Völkerorganisation und der moderne Staat.* Berlin, 1911.

Hirst: 'The Arbiter in Council,' in *Proceedings*, pp. 202 et seq.

Hodgson Pratt: *Établissement de tribunaux d'arbitrage.* 1892 (printed in manuscript form).

Hogg: 'A Note on the Hague Award in the Atlantic Fisheries Arbitration,' in *The Law Quarterly Review,* 1910, No. 104.

Hold v. Ferneck: 'Eine Lanze für den Prisenhof,' in *Zeitschrift für Völkerrecht und Bundesstaatsrecht,* 1912, pp. 1 et seq.

Holls: *The Peace Conference at The Hague.* New York and London, 1900.

Hornby: *Constitution d'un Tribunal international.* Berne, 1893.

—— 'Report on the Necessity of a Permanent Court of Arbitration,' in *Bulletin du sixième Congrès de la Paix,* 1894, pp. 135 et seq.

Houzeau de Lehaie: *Convention pour l'établissement d'une juridiction internationale.* 1895 (printed in manuscript form).

—— 'Rapport présenté au nom de la Commission chargée de formuler un Projet d'organisation de la Cour arbitrale,' pp. 59 et seq. of the Memorial of Descamps to the Powers.

Hubbard: *La Justice internationale.* Paris, 1903–5.

Huber: 'Die Fortbildung des Völkerrechts auf dem Gebiete des Prozess- und Landkriegsrechts,' in *Jahrbuch des öffentlichen Rechts,* Tübingen, 1908.

—— 'Die Gleichheit der Staaten,' pp. 88 et seq. of the anniversary papers presented by foreign jurists on the occasion of Josef Kohler's 60th Birthday. Stuttgart, 1909.

Hull: *The Two Hague Conferences.* Boston, 1908.

Humphrey: 'President Taft and the International Court,' in *Report,* 1910, pp. 94 et seq.

Hyde: 'Legal Problems capable of Settlement by Arbitration,' pp. 221 et seq. of the *Proceedings of the Second National Peace Congress,* Boston, 1909.

International Law Association: Reports of the 17th, 18th, 19th, and 20th Conferences, Brussels, 1895; Buffalo, 1899; Rouen, 1900; Glasgow, 1901; and London, 1896, 1900, 1901.

Interparliamentary Union: Report of the 15th Conference at Berlin, 1908. (In German.)

—— Report of the 16th Conference at Brussels, 1910. (In French.)

Jordan: 'The Waste of War,' in *Proceedings*, pp. 234 et seq.

Judson, F. N.: 'The Jurisdiction of the Supreme Court of the United States over Controversies,' in *Proceedings*, pp. 258 et seq.

—— H. P.: 'Between Diplomacy and War,' in *Proceedings*, pp. 300 et seq.

Kamarovski: *Le Tribunal international.* Paris, 1887.

Karnebeek: *Discours du Président du Comité des Directeurs de la Fondation Carnegie*, July 30, 1907 (printed in manuscript form).

Kaufmann: *Das Wesen des Völkerrechts und die clausula rebus sic stantibus.* Tübingen, 1911.

Kirchwey: 'The International Court a wonderful Achievement,' in *Report*, 1910, pp. 95 et seq.

Kohler: 'Der Savarkarfall,' in *Zeitschrift für Völkerrecht und Bundesstaatsrecht*, 1911, pp. 302 et seq.

Laband: 'Die völkerrechtlichen Schiedsgerichte,' in *Blätter für vergleichende Rechtswissenschaft*, 1905, No. 1.

La Fontaine: *Rapport de la Commission juridique du Bureau et Projet voté par le sixième Congrès universel de la Paix à Anvers en 1894* (printed in manuscript form).

—— *Pasicrisie internationale.* Berne, 1902.

Lammasch: 'Die Fortbildung des internationalen Schiedsgerichts seit der Haager Konferenz und die Kolonialfrage,' in *Deutsche Revue*, November, 1905, pp. 146 et seq.

Lammasch: 'Internationale Schiedsgerichtsbarkeit,' in *Herders Staatslexikon*, 3rd ed., vol. ii, Freiburg im Breisgau, 1909.

—— 'Der internationale Schiedsgerichtshof im Haag,' No. 1 of the *Allgemeine Österreichische Gerichtszeitung*, Vienna, 1911.

—— 'Die Ausgestaltung des Haager Schiedsgerichtshofes,' in *Das Recht*, 1911, p. 148.

—— 'Der Englisch-Amerikanische Schiedsgerichtsvertrag,' in pt. i, vol. xxix of the *Österreichische Rundschau*, Vienna, 1911.

Lansing: 'The North Atlantic Coast Fisheries Arbitration,' in *American Journal*, vol. v, p. 1, and *University of Pennsylvania Law Review*, 1911, p. 119.

de Lapradelle: 'La Conférence de la Paix,' in *Revue générale*, 1899.

—— 'Les transformations de l'arbitrage international,' in *Almanach de la Paix*, Paris, 1905, p. 56.

de Lapradelle et Politis: *Recueil des arbitrages internationaux*, vol. i, 1798–1855. Paris, 1905.

—— 'La deuxième Conférence de la Paix,' in *Revue générale*, 1909.

Lawrence: *International Problems and Hague Conferences*. London, 1908.

—— *A Handbook of Public International Law*. London, 1911.

—— *The Principles of International Law*. London, 1911.

Lémonon: *La seconde Conférence de la Paix*. Paris, 1908.

Lepert: *Projet d'Organisation de la justice internationale*. Monaco, 1907.

v. Liszt: *Das Völkerrecht*. Berlin, 1911.

—— *Das Wesen des völkerrechtlichen Staatenverbandes und der internationale Prisenhof*. Breslau, 1910.

Loewenthal: *Obligatorische Friedensjustiz, nicht Schiedsgericht*. Berlin, 1897.

—— *Geschichte der Friedensbewegung*. Berlin, 1907.

Logan: 'A Permanent Tribunal,' in *Report*, 1896, pp. 59 et seq.; 1898, pp. 39 et seq.

Loomis: 'The Price of Peace,' in *Proceedings*, pp. 228 et seq.

Lorimer: *Principes de droit international.* Paris, 1885.

Loudon: 'Holland as an International Host,' in *Proceedings*, pp. 186 et seq.

de Louter: *Het stellig volkenrecht.* 2 vols. The Hague, 1910.

—— 'L'Arbitrage dans le conflit anglo-américain concernant les pêcheries de l'Atlantique,' in *Revue de Droit international*, 1911, pp. 131 et seq.

—— 'De Toekomst van het volkenrecht,' in *De Gids*, Amsterdam, 1912, No. 2.

Macfarland: 'The Hopefulness of the Knox Proposition,' in *Report*, 1910, pp. 76 et seq.

—— 'The Difficulties as to International Arbitral Courts,' in *Proceedings*, pp. 162 et seq.

Mac Veagh: 'The Relations between International Tribunals of Arbitration and the Jurisdiction of National Tribunals,' in *Proceedings of the American Society of International Law*, vol. iii, 1909, pp. 48 et seq.

Mallarmé: 'L'Arbitrage vénézuélien devant la Cour de La Haye,' in *Revue générale*, 1906, pp. 423 et seq.

van der Mandere: 'Eene reorganisatie van het permanente hof van arbitrage,' in *Vragen des Tijds*, September, 1910.

—— 'Het permanente Hof van arbitrage en zijne werking,' in *Eigen Haard*, November 19, 1910, et seq.

—— *De haagsche overeenkomst betreffende de vreedzame beslechting van internationale geschillen en het permanente hof van arbitrage*, pt. i, Rhenen, 1907 (under the nom-de-plume of Melville).

Marburg: 'The Significance of the International Court of Arbitral Justice,' in *Report*, 1910, pp. 85 et seq.

Marburg: 'The Washington Meeting of the American Society for the Judicial Settlement of International Disputes,' in *American Political Science Review*, 1911, pp. 181 et seq.

Martens: 'Le tribunal d'arbitrage de Paris et la mer territoriale,' in *Revue générale*, 1894, pp. 32 et seq.

—— *La Conférence de la Paix à La Haye*. Paris, 1900.

McGrath: 'North Atlantic Fisheries Arbitration,' in *University Magazine*, Toronto, 1910, pp. 542 et seq.

McKenney: 'Objections to the present Hague Court,' in *Proceedings*, pp. 94 et seq.

Mead: *Dr. Hale's Services in the Cause of International Justice*. Boston, 1909.

—— 'The Proposal of Secretary Knox consistent with the Hague Conventions,' in *Report*, 1910, pp. 92 et seq.

Mérignhac: *Traité théorique et pratique de l'arbitrage international*. Paris, 1895.

—— *La Conférence internationale de la Paix*. Paris, 1900.

—— *Traité de droit public international*. Paris, 1895.

—— *La deuxième Conférence de la Paix*. Paris, 1908.

Meurer: *Die Haager Friedenskonferenz*, vol. i, 'Das Friedensrecht der Haager Konferenz', Munich, 1905.

Moch: *Autour de la Conférence interparlementaire*. Paris, 1895.

—— *Vers la Fédération d'Occident: Désarmons les Alpes*. Paris, 1905.

—— *Histoire sommaire de l'arbitrage permanent*. Monaco, 1910.

Montague: 'The Supreme Court as a Prototype of an International Court,' in *Proceedings*, pp. 210 et seq.

Moore: *History and Digest of the International Arbitrations to which the United States has been a Party*. Washington, 1898, 6 vols.

Morrow: 'The Judiciary and Arbitration,' in *Proceedings of the National Arbitration and Peace Congress*, New York, 1907, p. 338.

Mougins de Roquefort: *De la solution juridique des conflits internationaux.* Paris, 1892.

Nippold: *Die Fortbildung des Verfahrens in völkerrechtlichen Streitigkeiten.* Leipzig, 1907.

—— *Die zweite Haager Konferenz.* Vol. i, Leipzig, 1908; vol. ii, Leipzig, 1911.

Nowakowitsch: *Die zweite Haager Friedenskonferenz.* Belgrade, 1911.

Nys: 'L'arbitrage,' in *Revue de Droit international*, 1906, pp. 5 et seq.

—— *La révision de la sentence arbitrale.* Brussels, 1910.

—— *The Necessity of a Permanent Tribunal.* Baltimore, 1910.

Official Report of the Pennsylvania Arbitration and Peace Conference. Philadelphia, 1908.

Oppenheim: *International Law*, pt. i, 'Peace'. London, 1906.

—— *Die Zukunft des Völkerrechts.* Leipzig, 1911.

Parnes: *Tribunal d'arbitrage armé.* Vienna, 1907.

Penfield: 'International Courts of Justice,' pp. 59 et seq. of the *Official Report of the Proceedings of the Pennsylvania Arbitration and Peace Conference*, Philadelphia, 1908.

Phillipson: *Two Studies on International Law.* New York, 1908.

Pohl: *Deutsche Prisengerichtsbarkeit.* Tübingen, 1911.

Poignand de Fontenioux: *De l'évolution de l'idée de tribunal permanent d'arbitrage à travers les âges et de son avenir.* Poitiers, 1904.

Den Beer Poortugael: 'De Beteekenis van de Tweede Vredesconferentie,' in *Onze Eeuw*, March, April, May, 1908.

Pradier Fodéré: *Cours de droit diplomatique*. Paris, 1881.

Proceedings of the American Society of International Law. Washington, vol. i et seq.

Proceedings of the National Arbitration and Peace Congress. New York, 1907.

Proceedings of the Second National Peace Congress. Boston, 1909.

Ralston: *Venezuela Arbitrations of 1903*. Carácas, 1903.

—— 'Some Suggestions as to the Permanent Court of Arbitration,' in *American Journal*, vol. i, No. 2.

—— *International Arbitral Law and Procedure*. Boston, 1910.

—— 'Some Considerations as to International Arbitral Courts,' in *Proceedings*, pp. 151 et seq.

Regelsperger: 'L'Affaire du Costa Rica Packet et la sentence arbitrale de M. de Martens,' in *Revue générale*, 1897, pp. 735 et seq.

Reinsch: 'The Concept of Legality in International Arbitration,' in *American Journal*, vol. v, pp. 604 et seq.

Renault: 'Une nouvelle mission donnée aux arbitres dans les litiges internationaux à propos de l'arbitrage de Behring,' in *Revue générale*, 1894, pp. 44 et seq.

—— *L'œuvre de La Haye en 1899 et 1907*. Paris, 1908.

Reuter: 'Über den Weg zum allgemeinen ständigen internationalen Schiedsgericht,' in *Die Waffen nieder*, 1897, pp. 361 et seq.

Revon: *L'Arbitrage international*. Paris, 1892.

Richard: 'Treaty Obligations of the United States as related to the International Court,' in *Report*, 1910, p. 90.

Richards: 'The North Atlantic Coast Fisheries Arbitration,' in *Journal of the Society of Comparative Legislation*, 1911, p. 18.

Richardson: 'A World Court,' in *Advocate of Peace*, 1911, No. 5.

Richet: *Die Vergangenheit des Krieges und die Zukunft des Friedens.* Vienna, 1909.

Riddell: 'The International Relations between the United States and Canada,' in *Proceedings*, pp. 14 et seq.

Robin: 'Un Différend franco anglais devant la Cour d'arbitrage de La Haye—Affaire de l'Hindou Savarkar,' in *Revue générale*, 1911, pp. 303 et seq.

Rolin-Jaequemyns: 'Chronique des arbitrages internationaux,' in *Revue de Droit international*, 1890, p. 349; 1891, pp. 76, 201.

Root: 'The Importance of Judicial Settlement,' in *Proceedings*, pp. 9 et seq.

—— 'The American Sentiment of Humanity,' pp. 34 et seq. of the *Proceedings of the National Arbitration and Peace Congress*, New York, 1907.

Rouard de Card: *Les destinées de l'arbitrage international depuis la sentence rendue par le tribunal de Genève.* Paris, 1892.

Saint-Georges d'Armstrong: *De l'utilité de l'arbitrage.* Paris, 1890.

de Saint Maurice: *La deuxième Conférence de la Paix.* Paris, 1908.

Scelle: 'Une instance en révision devant la Cour de La Haye—l'affaire de la Orinoco Steamship Company,' in *Revue générale*, 1911, pp. 164 et seq.

Schlief: *Der Friede in Europa.* Leipzig, 1892.

—— 'Völkerrechtliche Organisationen und Schiedsgerichte,' in *Friedenswarte*, 1906, p. 45.

Schücking: *Die Organisation der Welt.* Leipzig, 1909.

Schücking: 'Droit international, Traités, Conférences et Cour de La Haye,' pp. 424 et seq. of the *Mémoires sur le contact des races*, London, 1911.

—— *Der Staatenverband der Haager Konferenzen.* Leipzig, 1912.

Scott: 'William Ladd's Project of a Congress and Court of Nations,' in *Advocate of Peace*, 1908, Nos. 8–9.

—— 'The Proposed Court of Arbitral Justice,' in *American Journal*, vol. ii, No. 4.

—— 'The Need of an International Court of Justice,' pp. 98 et seq. of the *Official Report of the Pennsylvania Arbitration and Peace Conference.* Philadelphia, 1908.

—— *The Hague Peace Conferences of 1899 and 1907*, vol. i, 'The Conferences'; vol. ii, 'Documents'. Baltimore, 1909.

—— 'The Proposed International Court of Arbitral Justice,' in *Report*, 1909, pp. 54 et seq.

—— *American Addresses at the Second Hague Peace Conference.* Boston, 1910.

—— *Judicial Proceedings as a Substitute for War or International Self-redress.* Baltimore, 1910.

—— 'Progress toward an International Court of Arbitral Justice,' in *Report*, 1910, pp. 67 et seq.

—— 'A Court of the World,' in *The Outlook*, June 18, 1910.

—— *The American Society for Judicial Settlement of International Disputes, Its Scope and Work.* Baltimore, 1910.

—— 'Address of the President,' in *Proceedings*, pp. 2 et seq.

—— 'A Permanent International Court of Justice,' in *The Editorial Review*, April, 1911.

—— *Some Phases of the Peace Movement.* New York, 1911.

—— 'One Aspect of the Peace Movement,' in *The Alumni Quarterly*, July, 1911.

—— 'The Evolution of a Permanent International Judiciary,' in *American Journal*, 1912, pp. 316 et seq.

Seth Low: 'International Arbitration,' pp. 329 et seq. of the *Proceedings of the National Arbitration and Peace Congress,* New York, 1907.

Sigaud: *Confédération européenne.* Paris, 1859.

de Sillac: 'Les Conférences périodiques de la Paix,' pp. 448 et seq. of the *Mémoires sur le contact des races,* London, 1911.

Smith: 'Arbitration at the Second Hague Conference,' in *Proceedings of the American Society of International Law,* vol. ii, 1908, pp. 90 et seq.

—— F. E.: *International Law.* London, 1911.

Snow: 'The Development of the American Doctrine of Jurisdiction of Courts over States,' in *Proceedings,* pp. 100 et seq.

Stanhope: 'Rapport sur l'organisation d'une Cour internationale permanente d'arbitrage,' in *La Conférence interparlementaire,* 1894, pp. 224 et seq.

Stead: *Chronique de la Conférence de La Haye.* The Hague, 1900.

v. Stengel: *Weltstaat und Friedensproblem.* Berlin, 1909.

Stipriaan Luiscius: *L'Avenir de l'arbitrage international.* Brussels, 1907.

Stockton: 'Some Reasons why Judicial Methods for Settlement of International Disputes are superior to other Methods,' in *Proceedings,* pp. 294 et seq.

Straus: Address in the *Proceedings,* pp. 183 et seq.

Strupp: 'Zwei praktische Fälle aus dem Völkerrecht,' pt. i, vol. v of the *Zeitschrift für Völkerrecht und Bundesstaatsrecht,* 1911.

Sturm: *Die psychologische Grundlage des Rechts.* Hanover, 1910.

—— *Das Recht auf Frieden.* Leipzig, 1912.

v. Suttner: *Memoiren.* Stuttgart and Leipzig, 1909.

Suyling: *The Hague Peace Conferences.* The Hague, 1910.

Tettenborn: *Das Haager Schiedsgericht, eine völkerrechtliche Studie.* Bonn, 1911.

Touchet: 'La justice internationale dans la guerre et la paix d'après l'Église,' in *Annales franc-comtoises,* July–August, 1893.

Tripp: 'A Permanent International Tribunal,' in *Report,* 1907, pp. 46 et seq.

Tryon: 'The Proposed High Court of Nations,' in *Advocate of Peace,* 1910, No. 1.

v. Ullmann: 'Die Haager Konferenz und die Weiterbildung des Völkerrechts,' in *Jahrbuch des öffentlichen Rechts,* Tübingen, 1907.

—— *Völkerrecht.* Tübingen, 1908.

Union Interparlementaire: *Résolutions des Conférences et Décisions principales du Conseil.* Brussels, 1911.

Vavasseur: *L'Organisation d'une juridiction arbitrale internationale.* Monaco, 1907.

Wambaugh: 'Why the Growth of Law is aided by Courts more than by Commissions,' in *Proceedings,* pp. 139 et seq.

Wehberg: *Kommentar zu dem Haager Abkommen betreffend die friedliche Erledigung internationaler Streitigkeiten.* Tübingen, 1911.

Weiss: 'L'Arbitrage de 1909 entre la Bolivie et le Pérou,' in *Revue générale,* 1910, No. 2.

Westlake: *International Law,* pt. i, 'Peace'. Cambridge, 1904.

Wheeler: 'The Constitution and Scope of the International Court,' in *Report,* 1910, pp. 90 et seq.

—— 'The Advance from Force to Reason,' in *Proceedings,* pp. 40 et seq.

White, Andrew D.: 'Some Hints as to the Future Work of the Hague Conference,' in *Report*, 1907, pp. 29 et seq.

—— Th. R.: 'Underlying Principles which should govern Appointment of Judges of the International Court,' in *Advocate of Peace*, 1911, No. 7.

Whittuck: *International Documents.* London, 1909.

Wilson-Tucker: *International Law.* London, 1909.

Wood: 'The Juridical Equality of States,' pp. 117 et seq. of the *Bulletin du sixième Congrès de la Paix*, Antwerp, 1894.

Zorn: 'Die völkerrechtlichen Ergebnisse der Haager Konferenz,' in *Im neuen Reich*, Bonn, 1902, pp. 319–94.

—— 'Kodifikation des Völkerrechts,' in *Das Recht*, January 10, 1907.

—— 'Die zweite Haager Friedenskonferenz,' in *Marine-Rundschau*, November, 1907.

—— 'Das völkerrechtliche Werk der beiden Haager Konferenzen,' in *Zeitschrift für Politik*, vol. ii, pt. 3, 1909.

—— *Das Deutsche Reich und die internationale Schiedsgerichtsbarkeit.* Berlin, 1911.

MISCELLANEOUS

'The Central American Court of Justice,' in *Advocate of Peace*, 1908, No. 1.

'Conflit entre Grande-Bretagne et Vénézuéla,' in *Revue générale*, 1901, pp. 71 et seq.

'The Constitution and Powers which an International Court of Arbitral Justice should possess,' by Burton, Montague, Ion and Harris, in *Proceedings of the American Society of International Law*, vol. iii, 1909, pp. 221 et seq.

'Debate relativo al ingreso en la Corte del Nuevo Magistrado por Nicaragua,' Doctor Guttierrez Navas, in *Anales de la Corte de Justicia centro-americana,* 1911, No. 1.

'De quaestie der Noord-Atlantische Kust-Visscherijen,' in *Vrede door Recht,* The Hague, 1910, No. 8.

'De uitspraek in de Savarkar-Zaak,' in *Vrede door Recht,* 1911, No. 3.

'The Development of International Law by Judicial Decisions in the United States,' by Brown and Peelle, in *Proceedings of the American Society of International Law,* vol. iii, 1909, pp. 166 et seq.

'Een permanent hof van arbitrage,' in *Vrede door Recht,* 1910, No. 1.

'The Equality of Nations,' in *Advocate of Peace,* 1907, No. 11.

'The Equality of Nations,' by Hicks, Evans, Harris, Aymar, Ion, Stowell, Montague and Abbott, in *Proceedings of the American Society of International Law,* vol. iii, 1909, pp. 238 et seq.

'États-Unis d'Amérique et Grande-Bretagne,' in *Revue générale,* 1904, pp. 210 et seq.

'La Cour permanente de La Haye et le droit international privé et pénal,' in *Revue de Droit international privé et de Droit pénal international,* 1911, Nos. 3 et seq.

'The Newfoundland Fisheries Settlement,' in *Concord,* London, 1910, No. 10.

'The Newfoundland Fisheries Award,' in *Advocate of Peace,* 1910, No. 10.

'North Atlantic Coast Fisheries Arbitration,' in *Canada Law Journal,* 1910, p. 593.

'The Orinoco Claims Award,' in *Advocate of Peace,* 1910, No. 11.

'Projet de constitution d'un tribunal permanent,' in *Revue générale,* 1895, pp. 334 et seq.

'Proposal to modify the International Prize Court and to invest it as modified with the Jurisdiction and Functions of a Court of Arbitral Justice,' in *American Journal*, vol. iv, No. 1.

Rapport du Conseil Administratif de la Cour permanente d'Arbitrage, 1900–1910.

'The Savarkar Case,' in *American Journal*, vol. v, p. 208.

'Secretary Knox and International Unity,' in *American Journal*, vol. iv, No. 1.

'Secretary Knox's Note to the Powers,' in *Advocate of Peace*, 1910, No. 2.

Statement by the President of the Tribunal that the North Atlantic Fisheries Award was a Compromise, in *American Journal*, 1911, No. 3.

'Una Corte arbitrale permanente,' in *La Vita Internazionale*, 1908, No. 19.

'The United States at the Hague Court of Arbitration,' in *American Journal*, vol. iv, No. 3.

CHAPTER I

THE SOVEREIGNTY OF STATES AND THE INTERNATIONAL COMMUNITY

WITH the great development in the economic relations of states there has come about an increasing change in the conception of the position of the individual state within the international community. In previous centuries, when a state regulated its commercial relations with other states, it could proceed according to its own unhampered will. It might perhaps inflict an injury upon other states and thus bring upon itself the risk of war; nevertheless it was not on the whole dependent upon the political economy of other countries to such an extent as to be obliged to regulate its commercial relations with that object chiefly in mind. It could enter into and promulgate treaties after its own free will, as its sovereignty might dictate. In consequence, political scientists considered sovereignty as the highest attribute of the state, dependent upon no other. Sovereignty was, to the scholars of the old school, the first and most important characteristic of states.

But the situation of the individual states has entirely changed since world economy has created countless bonds between the hitherto isolated states, since commerce, industry, and the system of insurance, as well as numerous other institutions, have been built up upon international foundations. This economic interdependence of the individual states has continued to grow stronger and greater with each day. In his book, *The Great Illusion*,[1] Norman

[1] Leipzig, 1910, p. 44.

Angell has shown us that a failure of the Bank of England during a war between Germany and England would completely ruin German commerce. In place of the former isolated political economy of states, we have to-day a world economy upon which each state is in the highest degree dependent.

This daily increasing interdependence of individual countries, which brings about numerous points of contact between them, has induced states to enter into conventions with one another of an international character, regulating the legal relations between them and between their subjects. When we recall that before the year 1800 no international congress worthy of the name had met with the object of regulating the legal relations [1] of groups of states, and that, on the contrary, since 1815 about two hundred congresses of states with an ever-increasing number of participants have taken place, we recognize clearly the line of development, that an ever-stronger union of states upon a legal basis is coming about, and that the day is no longer far off when the most important fields of international interest will be ruled by law.

Can there be any doubt that the states which are still ruled by old political ideas agree, to some extent unwillingly, to bind themselves towards other states by international conventions ? Will not sovereignty be more and more limited through the increase of international unions ? Will not those conventions in many cases interfere with the administrative and legislative departments, and even with the judicial department, of the domestic government of states ? If the legislative, administrative, and judicial activities of the individual states can no longer be regulated as each state chooses, will not the glory of sovereignty, which formerly was the chief characteristic of states, have

[1] Taken in contrast to political relations.

in large part disappeared? But while states enter into all those conventions with apparent willingness, there can be no doubt that at the present time the community of interests so far forces them to enter into such conventions that it is now a matter of necessity, not of choice. Which is to-day the stronger of the two, the community of interests or the selfish interest of separate states? Must not the latter yield more and more to the former? With irresistible force world economy is developing, and day by day the solidarity of interests is growing. The individual state is to-day no longer free to determine whether it will enter into a particular convention. As with all other states, its hand is forced by the community of interests. Von Liszt[1] says pertinently: 'The community of international law is the necessary consequence of a community of interests, which points out to the sovereign will of the individual state the way which it must go, if it is to survive.' It is true that this is not yet evident in all fields, because in many respects the interdependence of states upon one another has not yet become great enough; but to a greater or less extent it is true with respect to the regulation of postal and railway systems, the prevention of contagious diseases, the regulation of commercial relations, and numerous other affairs. Likewise, there can be no doubt that the community of interests, with each day that it becomes greater, is coming more and more to dominate states. In view of this power standing over it, every modern government must, paraphrasing the discerning words of Aristippus, say with resignation: ἔχομαι, οὐκ ἔχω (others rule me, it is not I that rule).[2]

[1] *Völkerrecht*, p. 10.

[2] Taken from an address of my former teacher, Paul Cauer ('Woher? und Wohin?', Düsseldorf, 1902, pp. 39 et seq.).

But it would be a serious mistake to conclude from the preceding statements that states have in any way lost their former significance. Though they are no longer sovereign in the old sense, their individuality as subjects of law (and this is of prime importance) has not been narrowed with respect to other states. When Bavaria and the other federal states renounced their sovereignty in favour of a united German state, they disappeared from the circle of independent states and retained only certain reserved rights. Henceforth Bavaria no longer stood upon the same footing as France, England, or North America. Granting that the establishment of the German Empire was on the whole for the good of Bavaria, she nevertheless lost by that glorious act her position in the world of states. But the diminution of the so-called sovereignty of states, in consequence of the influence of world economy, is of an entirely different character. Although the community of interests affects all states alike, still no state loses anything of its significance as a subject of law with respect to the other states; for the same bonds which bind it bind at the same time all other states of the international community. It will not be long before states, having put aside their old ideas, will recognize clearly how greatly their bonds are to their benefit; and in consequence they will be led to conclude further common conventions. But if it is inevitable that the law between states shall exercise an ever greater control, and if in this respect greater restrictions are placed upon the special characteristic of states, governments will come to a clearer recognition of the fact and will themselves take in hand the development and no longer let themselves be driven along by the current of events.

Thus far governments have concerned themselves much too little with the development of international law and have been led to enter into negotiations only through important events which positively demanded a codification of principles. They have often failed to recognize the necessity of regulating anew their international relations and have only with difficulty been brought to adopt an advantageous convention, as in the case of the German Empire, in 1899, with regard to the establishment of the Permanent Court of Arbitration at The Hague. As a result of this attitude it has often happened that the initiative in important reforms has often come merely from private effort. I merely refer to the project of an arbitral court for disputes between individual persons and debtor states, as also to the proposal of a world bureau after the model of the Pan American Bureau.

To be thus driven on by circumstances is not in keeping with the high character of modern governments which, in regard to their own domestic affairs, have already entered upon elaborate schemes for the improvement of social conditions. They must become systematic in dealing with foreign affairs as well, and before hard necessity forces them they must ponder long over questions and develop plans for the future. Only thus can they put an end to the present unnatural position in which they are ruled by their ever-increasing common interests, and with proud consciousness say: *ἔχω—καὶ ἔχομαι*! (I am not only ruled by circumstances, but I likewise rule them!).

In his epoch-making book *Der Staatenverband der Haager Konferenzen*,[1] Schücking has pointed out that the idea of international organization is a marked feature of modern international law. In this connexion I may observe that this organization of states must be systematically and scientifically developed. Had this been done to a sufficient

[1] p. 18.

extent at the time, there would have been less obscurity in the discussions at the Hague Peace Conferences concerning the legal nature of the *Cour de la justice arbitrale*, and numerous other points.

The task of the science of international law is, by its preliminary labours, to assist and to guide governments in the making of international conventions. One of the most important problems of international law at the present day is the question of a truly permanent court of justice for disputes between states. No continental writer has thus far treated the question as carefully as it deserves to be treated, and in consequence it is a timely undertaking, before the next great Peace Conference, to attempt to throw light upon this problem which affects in the highest degree the welfare of states.

CHAPTER II

THE NECESSITY OF THE DEVELOPMENT OF INTERNATIONAL LAW THROUGH INTERNATIONAL JUDICIAL DECISIONS

IN his noteworthy work *Die Grundlagen des revolutionären Pacifismus*[1] (Tübingen, 1908), A. H. Fried has drawn the distinction between the reforming and the revolutionary pacifists. By the former he understands those who merely combat the symptoms of the disease, and by the latter those who, in addition, seek to overthrow the present false system of disorganization in the community of states. Similarly, the advocates of the judicial settlement of international disputes are taking two distinct attitudes. The one party looks no further than to settle in a peaceful way the existing disputes by means of arbitration, without considering whether, at the same time, the causes of the present differences are thereby removed. The other party is more far-sighted and seeks not only to settle present conflicts but also, through the gradual development of international law by means of international judicial decisions, to establish more securely the reign of law and thus prevent future disputes.

The pacific settlement of international disputes becomes easier in proportion as there exist legal principles upon which a decision may be based. International disputes frequently become acute because, in default of rules of international law, the parties have no basis of agreement

[1] Schücking aptly proposes (op. cit., p. 4) that the term '*organisatorischer Pacifismus*' [constructive pacifism] be substituted for the expression '*revolutionärer Pacifismus*' [revolutionary pacifism].

and their views are in complete opposition. Such disputes, persisting year after year, are indeed not without danger to the parties, even though they have for their object questions in themselves insignificant. Lammasch justly observes:

> When a dispute or when, it may be, several disputes between two states drag on for a long time, it can easily happen that from a comparatively trivial cause the breach between the two powers will become so wide that the continuance of their friendly relations will be seriously endangered.[1]

In fact, it is quite self-evident that international disputes must, so far as possible, not only be peacefully adjusted but must be decided upon a purely legal basis. If a legal question is decided in the form of a compromise by an arbitral court, and if a similar question later arises in a new dispute between two other parties, the latter have no basis upon which the dispute may be peacefully settled and mutual claims adjusted. For in the second case it may be that both parties, or one of them, are dissatisfied with a decision by compromise, because each state claims neither more nor less than is its due. Moreover, the first arbitral decision has done nothing for the solution of future differences, having merely settled the existing dispute. A decision according to equity, as Hartzfeld has recently suggested in his excellent book *Der Streit der Parteien*,[2] is only of value in private life where the legal principles to be applied are well established. But in international life, where much obscurity prevails concerning the rights and duties of states, a definitive decision according to principles of equity must lead to arbitrariness and confusion. How much more important it is then, in a definite case, to develop in a precise way the principles of inter-

[1] *Herders Staatslexikon*, vol. ii, 1909, p. 1408.

[2] Berlin, 1911.

national law so that in all similar future disputes nations may be able to know what the law is, than to render an equitable decision, which may indeed settle the dispute in hand, but which leaves the legal problem involved in it in an obscure condition for the future.

But, one may object, should international disputes after all be settled upon a purely legal basis, so that in many instances one of the parties loses its case entirely? Is not the important thing the settlement of disputes, and is it not much better to compromise? Will not the state which loses regard the affair with bitterness and thenceforth make no further use of international institutions for the peaceful settlement of disputes? These objections appear to me unfounded. It must be accepted as a principle that through purely legal decisions the right and wrong of a dispute will be brought into clearer light. In that case, if a state is actually in the wrong it will more easily perceive it than in the case of arbitral decisions by compromise. But if, on the other hand, a state which is wholly in the right obtains only half its due, it will have all the more reason for dissatisfaction. A state which loses its case when international disputes are settled upon a purely legal basis may trust that at some future time, when another dispute arises in which it is itself in the right, it will win its case, and then it will not be satisfied with a decision in the form of a compromise. In addition to this it must be remembered that a state frequently loses its case completely even when international disputes are settled by friendly adjustment. In the decision of the Hague Court in the Savarkar case, France completely lost her suit, although the judges were less concerned to give justice than merely to settle the dispute in as friendly a way as possible. Only in strictly political disputes will a settlement by compromise be resorted to. So far, we are in accord with the opinion of Hold von Ferneck in the *Zeitschrift für Völkerrecht und Bundesstaatsrecht.*[1]

On June 17, 1897, the Norwegian Prime Minister, Hagerup, declared in the Storthing that there were still many difficulties to be overcome in the matter of arbitration, and that the rules of law according to which decisions were rendered were still too uncertain, and that there was yet lacking a permanent international arbitral court.

In many cases it has appeared as a fact that the two parties were unwilling to turn over their dispute to an arbitral court, because they feared that an award in the form of a compromise would be rendered, instead of a legal decision. This point has been emphasized in particular by Root at the New York Peace Congress of 1907, and by Oppenheim in his work *Die Zukunft des Völkerrechts.*[2] In this connexion Marburg[3] and Th. R. White[4] make the pertinent observation that many disputes would be settled in a friendly way if the parties were certain that a purely legal decision would be rendered.

Kaufmann[5] is entirely in accord with us when he says that the future progress of the idea of arbitration depends less upon juristic questions and rules than merely upon the possibility of finding judges of such excellence that they can also be entrusted with questions which are not of secondary importance.

But, it will be asked, is international law as yet sufficiently developed to make a purely legal decision of the majority of questions possible? This question may, with the support of the majority of writers—in particular, Descamps,[6]

[1] 1912, p. 14.
[2] 1911, p. 49.
[3] *Proceedings*, p. xvi.
[4] *Report*, 1911, p. 105.
[5] *Das Wesen des Völkerrechts*, p. 4.
[6] *Denkschrift*, pp. 63 et seq

Mérignhac,[1] Lammasch,[2] and Nippold[3]—without hesitation be answered in the affirmative. Were this not the case, the two Hague Peace Conferences would not have laid it down that differences should be decided 'on the basis of respect for law'.

In particular, Professor Hyde[4] has convincingly shown that all the cases which thus far have been decided between the United States and other nations by means of arbitration could have been decided in a strictly legal manner.

While it would seem that a comprehensive codification of international law is not at present possible, its greatly increasing scope makes a gradual development of that science through legal decisions absolutely necessary.

What we also need is an international institution which will give purely legal decisions when the parties wish them. At the same time, decisions according to equity may continue to be given when the parties prefer to have recourse to them.

The plan does not naturally commend itself that the parties shall, as in the *Alabama* case, determine beforehand in the arbitral agreement what the arbitrators are to regard as the rules of international law in the dispute in question, as Saint-Georges d'Armstrong[5] has advocated. Not only is an understanding on this point exceedingly difficult, but such a procedure, which is similar to a compromise, will prevent the development of international law.

The solution, therefore, is: A further development of international law through international decisions! That this demand is justified there is a general consensus of opinion. In his article *Supériorité des arbitrages juridiques*[6]

[1] *Conférence*, p. 363. [2] *Deutsche Revue*, 1905.
[3] *Fortbildung des Verfahrens*, pp. 562 et seq.
[4] *Proceedings of the Second National Peace Congress*, 1909, pp. 221 et seq
[5] p. xvii. [6] *La Justice internationale*, 1903, pp. 209 et seq.

Manoel da Junqueira has definitely condemned compromises. Even Nippold, the German, who in his extreme caution is opposed to the question here at issue of a truly permanent court of justice, is forced to admit [1] that 'if substantive international law, unlike adjective international law, does not lend itself to codification, there is none the less need of its further development, and this, so we jurists hope, will be done not through codification, but rather through the judicial decisions of arbitral courts. That is the proper basis upon which, by means of rules determined by conventions and conferences, by means of partial codification, international law can best receive its further development. In this way the Hague Court of Arbitration, in particular, can be of special significance for international law, in that it can serve not only to settle disputes between nations, but also to develop international law.' Of the same opinion are Balch,[2] von Bar,[3] Descamps,[4] Lammasch,[5] de Louter,[6] Oppenheim,[7] Stanhope,[8] Zorn,[9] and others. In his opening address in the Orinoco case on September 28, 1910, Lammasch pointed out that national courts can only follow the laws in force in the individual states, whereas international courts have the task of creating the rules of international law.

This question was far too little discussed at the Second Hague Peace Conference. D'Oliveira [10] alone made the following pointed remark: 'Could we not confide to the Court the gradual codification of international jurisprudence ?'

[1] Op. cit., p. 564. [2] *Revue de droit international*, 1908, p. 398.
[3] *Berliner Tageblatt*, 1907, vol. 22, p. viii. [4] *Denkschrift*, p. 53.
[5] Op. cit. [6] *Revue de droit international*, 1911, p. 157.
[7] Op. cit., p. 48. [8] *Union*, 1894, p 226.
[9] *Das Deutsche Reich und die internationale Schiedsgerichtsbarkeit*, 1911, p. 40. [10] p. 616.

CHAPTER III

THE IMPOSSIBILITY OF A DEVELOPMENT OF INTERNATIONAL LAW THROUGH THE PRESENT ARBITRAL PROCEDURE

1. The Legal Nature of Arbitral Decisions according to the Text of the Hague Convention

We must accordingly insist upon this point, that it is possible for international law to be developed in the future through legal decisions. How can this best be done ?

In order to answer this question we must direct our attention to the arbitral processes of the present day and endeavour to determine whether they do not already answer the desired purpose, or, at least, are able to answer it, so that there is no need of a new institution for the attainment of our aim.

In his *International Law* [1] Baty has rightly pointed out that under the term ' arbitration ' in international law we understand to-day much more than the word originally and in its technical meaning properly conveys. As will be later pointed out, the mistake has often been made of confusing arbitration (a decision according to equity) with judicial settlement (a decision according to strict law), as practical institutions of present international law, and of joining them together under the common name of ' international arbitration '. There is then set up in contrast to these two forms of international procedure a method of procedure under the name of ' international judicial settlement ', which is of a Utopian character, and which is enforced by a supreme authority such as we

[1] 1909, p. 8.

possess in national law. In consequence great confusion has arisen. By arbitration in the strict sense we understand in general a decision according to equity by freely chosen judges. Pohl [1] speaks as follows :

> The arbitrator desires to settle the dispute and to satisfy the parties through a decision which will ensure peace. He decides the matter not according to law, but disposes of it *ex aequo et bono*.

There is no reason why this definition should be abandoned in international law. In my opinion, Baty [2] is wrong when he says that the chief characteristic of arbitration is merely the free choice of judges who possess the confidence of the disputants. If one compares arbitration with judicial settlement in international law—in national law there are even greater differences—the comparison can best be made by taking as a basis the objects of the two institutions and not merely the external form of the composition of the tribunal, which is a means to the end. The parties desire from the arbitrator a decision satisfactory to both sides, not one based upon the greater right, since their object is to dispose of the dispute. In consequence, they attach no importance to the point that the judges be jurists in the highest sense of the word, but choose any person at all in whom they have confidence. On the contrary, in judicial settlement, as we shall show later, there must be a permanent tribunal with permanent judges, because only such an institution can guarantee a decision according to strict law. We shall later mark out the distinctions between judicial settlement and arbitration, and the great differences they present. Here we must be content to call attention to the fact that the object of arbitration in international law is a decision according to equity, whereas the object

[1] *Deutsche Prisengerichtsbarkeit*, 1911, p. 208.

[2] p. 3.

of judicial settlement in international law is a decision according to strict law.

Accordingly, when it is asked whether international arbitration is suited to the development of international law, the question must be answered in the negative if the arbitration we now have corresponds to the idea of arbitration in international law. Whether this is the case can best be determined if we take the Hague Convention for the pacific settlement of international disputes of October 18, 1907, as determining the present character of international arbitration. Article 37 of the said Convention reads:

> International arbitration has for its object the settlement of disputes between states by judges of their own choice and on the basis of respect for law.

In Article 38, paragraph 1, it is further enjoined:

> In questions of a legal nature, and especially in the interpretation or application of international conventions, arbitration is recognized by the contracting powers as the most effective, and, at the same time, the most equitable means of settling disputes which diplomacy has failed to settle.

These words make it sufficiently evident that the arbitral decisions of the present day are to be legal decisions. That is the general understanding, although Dr. Tettenborn, in her study on *Das Haager Schiedsgericht*,[1] speaks of 'the insecure basis of respect for law'. The statement of Descamps throws no light upon the question; but it is certain that the Hague Conference of 1899 intended to place the words 'on the basis of respect for law' in apposition with a decision according to equity, as Beernaert also pointed out at the Second Hague Conference,[2] in defence of the arbitral court set up by the First Peace Conference.

[1] 1911, p. 18. [2] p. 332.

Moreover, there can be no doubt that by the terms of that convention we have at hand an international organ, the object of which is to give strictly legal decisions, and which in consequence seems to be well suited to the development of international law. Accordingly, it seems at first sight that we already possess in international law an actual system by which disputes can be judicially settled, and arbitration is in its essence abandoned. The name, however, of the present authoritative institution would in consequence be incorrect. But when Scott, in his excellent remarks concerning the Court of Arbitral Justice,[1] observes that 'it may be held that the foundations of a true court in the juridical sense of the word have been laid, except that in place of judges there would be arbitrators appointed by the free choice of the parties', we are taken aback, and we must set ourselves to examine whether in fact the present system is in its method that of judicial settlement. It may possibly be that what was considered as arbitration before the First Hague Conference was an institution by means of which disputes between states were settled upon a purely legal basis. In that case the Hague Conference would only have given expression to something already in existence, without changing it in any way. It is our duty, therefore, to examine whether the numerous arbitral awards rendered up to the present time have been in truth legal decisions.

2. The Element of Compromise in the Decisions thus far rendered by Special Arbitral Courts

In his excellent work, *Pasicrisie internationale*,[2] La Fontaine is obliged to pass, on p. 17, a severe sentence upon one of the oldest arbitral awards of the latter-day

[1] Vol. i, p. 348.

[2] Berne, 1901.

practice of arbitration. With respect to the arbitral award which was rendered by the Czar of Russia in the dispute between Great Britain and the United States regarding claims for indemnity on the basis of the *compromis* of October 20, 1818, La Fontaine observes:

> The award rendered by the Czar of Russia did no more than lay down a principle all too vague and indecisive. The two nations concerned decided to conclude a new convention and entrusted the dispute to two commissioners and two arbitrators.

With regard to the decision rendered about a year earlier by the commissioners appointed, in accordance with Article 4 of the Treaty of Ghent, to settle the dispute between Great Britain and the United States, de Lapradelle and Politis [1] observe:

> Although in form this decision appears to be a judicial one, it is at bottom only a compromise. . . . The commissioners were not as impartial as might have been desired. Each of them worked as far as he could for his own government, after the manner of a skilful lawyer, not that of a judge.

The arbitral award of the King of Holland, rendered in accordance with the *compromis* of September 29, 1827, in the dispute between Great Britain and the United States concerning the north-east boundary, is unfortunately too well known. 'In place of determining, in his award, the true point at issue,' remarks Calvo,[2] 'this sovereign left the question of law in suspense and confined himself to suggesting a basis of settlement of an entirely new and contingent character.' A similar opinion upon this decision is expressed by Kamarovski [3] in particular, and by Dennis.[4]

[1] p. 304. [2] *Le Droit international*, 3rd ed., 1880, vol. ii, p. 575.

[3] *Le Tribunal international*, pp. 202 et seq.

[4] *Columbia Law Review*, 1911, p. 496.

The parties mutually refused to accept the decision of the King of Holland. The American Minister, Gallatin, spoke as follows with regard to this and other disputes:

> An arbitrator, whether he be king or farmer, rarely decides on strict principle of law; he has always a bias to try if possible to split the difference.

The reasoning upon which arbitral awards were rendered, particularly in the first half of the last century, was regarded as of merely secondary importance. The important point was to obtain an adjustment of the dispute. This idea was carried so far that in La Fontaine alone there are twenty decisions of the last century referred to which were not published at all.[1] A decision in a dispute between the United States and Paraguay, in the sixties of the nineteenth century, was brought to light for the first time forty years later by La Fontaine. That decisions of that kind are of no value for the development of international law is self-evident. Nippold [2] and Wambaugh [3] rightly point out that the publication of these decisions is one of the essential conditions for the advantageous working of international courts, since only in this way can they exercise the requisite influence upon the development of international law. In recognition of this fact, all decisions of the Hague Permanent Court, together with the protocols, are bound together and sent gratis to interested parties. With respect to decisions of special arbitration courts, that has unfortunately thus far not been done. It would be greatly desirable that those decisions also which, according to Article 43, paragraph 3, of the Hague Convention for the pacific settlement of international disputes, must be communicated by the powers to the International Bureau of

[1] pp. 33, 34, 38, 40, 43, 56, 68, 69, 75, 232, 612, 615, 616, 617, 618.

[2] *Fortbildung des Verfahrens*, p. 566.

[3] *Proceedings*, p. 142.

the Arbitration Court, should in like manner be published just as those of the Hague Court. The Division of International Law of the Carnegie Endowment has recently undertaken this commendable task, and has entrusted to Professor John Bassett Moore the task of editing all arbitral decisions. Moreover, the Interparliamentary Union, in its *Annuaire* for 1911, has taken the first step towards the publication of the decisions of special arbitral courts.

So few are the really well-reasoned decisions in the field of international arbitration in the past century, that La Fontaine feels called upon to draw special attention to six cases in which the decision is supported by sufficient grounds.[1] For example, in presenting the detailed decision which was rendered, in accordance with the convention of July, 1863,[2] by the commission in the dispute between England and Peru, he speaks as follows:

> This decision is of particular interest because of the care with which it was drawn up, and the detailed reasons which it contains, in contrast with the majority of the decisions which we have thus far presented.

If out of 170 decisions in the nineteenth century only six can be singled out as based upon good grounds, the facts speak for themselves. It is accordingly not surprising that the arbitrators appointed in accordance with the *compromis* of July 11, 1839, in the dispute between Mexico and the United States, were not agreed whether they were to act as a legal tribunal or as a diplomatic body.[3]

De Lapradelle and Politis,[4] whose knowledge of the arbitrations of the period from 1798 to 1855 is so profound, deny positively that arbitration has done anything for the

[1] pp. 27, 46, 47, 62, 245, 385.

[2] p. 47.

[3] Mérignhac, *L'Arbitrage*, p. 53.

[4] pp. xliii et seq.

development of international law. In their opinion, the arbitrators of that period oscillated between compromise and mediation. Neither of the two principal forms of arbitration from 1798 to 1855—mixed commissions of inquiry or the arbitral awards of sovereigns—exercised any perceptible influence in the development of law. The sovereigns supported their decisions with brief and unsatisfactory arguments, and rendered awards more in the form of a compromise. It is quite characteristic that de Lapradelle and Politis, after having reached that conclusion in all the six cases in which sovereigns had decided between 1798 and 1855, instead of stopping to discuss this exceedingly interesting problem of international law, continue as follows:

> But these questions of law, important as they were, were thus handled: the first was set aside, the second misunderstood, the third not considered closely, the fourth and fifth misconstrued, and the sixth given a diametrically wrong application.[1]

In contrast with the decisions of sovereigns, the members of mixed commissions gave detailed reasons for their decisions. But since they were rarely jurists they lacked the necessary ability, and their decisions are incomplete. In consequence, de Lapradelle and Politis express the following opinion upon the period up to 1855:

> The authority which arbitration might have exercised upon the law was greatly lessened. The lack of ability on the part of those who drew up the decisions discredited arbitration in the eyes of jurists, which was a serious matter at a time when text-books had, for two centuries at least, exercised absolute sway over international law. Authors do not cite the decisions of commissioners, and commissioners never fail, on the other hand, to hunt up and cite the authors. . . . Arbitration was at that time from a legal point of view much more the pupil than the master.

[1] p. xlv.

Likewise in the later period of modern arbitration, which began with the celebrated *Alabama* case, the situation remained unchanged. Even the *Alabama* case was not settled in a precisely legal manner, as Marburg [1] points out. England agreed that the tribunal should take as the basis for its decision the three so-called rules of the Treaty of Washington, declaring at the same time that those rules were not at all a statement of principles of international law in force at the time, but that she accepted those rules in order to give a proof of her desire to strengthen the friendly relations between the two countries. Granting that this was done because England did not wish to admit publicly that she was in the wrong, and that the rules of the Treaty of Washington were in truth principles of international law, so that the *Alabama* arbitral court gave a fair decision upon the basis of existing international law, and not upon the basis of a fictitious law, that fact proves nevertheless that the parties were not at all concerned to determine their legal rights, but that they were desirous to dispose of the dispute exclusively by means of an equitable decision. That the arbitrators confined themselves to the limits imposed is self-evident. Shortly after, the celebrated *Springbok* case between the same parties was settled by arbitration, in accordance with the above-mentioned Treaty of Washington. The decision in this case, given clearly upon political grounds, in recognizing the doctrine of continuous voyage, ran counter to established principles of international law, and called forth a protest from all jurists, in particular from the Institute of International Law, which, at its meeting at Wiesbaden, expressly protested against the decision. The decision in the case of the *Circassian*, likewise rendered in accordance with the

[1] *Proceedings*, p. xiv.

Treaty of Washington, was not accompanied by any extensive opinion of the court.

In this manner were the majority of cases decided. In many cases also, the legal basis of the dispute was so weak that the arbitrators were expressly authorized to compromise. This was particularly the case in the dispute between France and Holland concerning the boundary of Guiana-Surinam.[1] In like manner, the numerous cases in which sovereigns acted as arbitrators show that the parties were exclusively concerned with the adjustment of the dispute, not with a strictly legal decision. If, as in the *Phare* case, settled by the French Court of Cassation, and in the Butterfield case, settled by the English Ambassador Monson as sole arbitrator, the legal questions were, though in too little detail, relatively well decided—in the first case the question of *res iudicata*, and in the second case the question of prescription in international law,—yet these facts prove nothing to the point. The conclusion is rather that even an arbitrator may decide according to rules of law. But the great distinction between the strictly legal settlement of disputes and the settlement by arbitration, or rather mediation, consists in the fact that a judge must apply the rules of strict law, whereas an arbitrator is left free to apply them or not, as Baldwin [2] also points out. An arbitrator may never be forbidden to decide according to the rules of equity. Were he so forbidden, there would be nothing left of the purely arbitral settlement of disputes.

Moreover, the publication of the decision in the *Alabama* case was not followed by a general publication of decisions. As late as 1892, Rouard de Card [3] emphasized

[1] As also more recently in the dispute between Peru and Bolivia. See *Zeitschrift für internationales Recht*, 1910, pp. 205 et seq.

[2] *The New Era of International Courts*, 1910, p. 7.

[3] p. 124.

the fact that he could find no trace of the arbitral award rendered in 1881 in the dispute between Holland and San Domingo relative to the Dutch ship *Havana Packer*, although it was rendered in France by the President of the French Republic. He made the same observation with respect to the decision in the dispute between the United States and Haiti with regard to the Lazare case: 'We have not been able to procure the arbitral decision.'[1] The decision pronounced by Baron Lambermont on August 17, 1889, in the dispute between Germany and Great Britain over Lamu Island, was first published a year later, namely, on August 28, 1890.

What conditions were still possible in recent times may be gathered from the fact that in the arbitral court established in accordance with the *compromis* of January 4, 1883, between Chile and England, one of the three arbitrators, a national of Chile, from the very start protested against every decision unfavourable to his country. This ideal judge made public the secret deliberations of the arbitral court without further ado. The subsequent president of the arbitral court, Lafayette, a Brazilian, in contrast with the former president of the arbitral court, the Brazilian Netto, took sides with his Chilean colleague and decided that offences committed by soldiers in violation of the laws of war without the command of their officers gave no ground for indemnity, although the contrary was undoubtedly the case. The result was that the arbitral agreements concluded by Chile with numerous other states were never carried out, and the great powers of Europe who claimed indemnities of Chile for injuries suffered during the war with Peru preferred to settle the disputes through diplomatic negotiations.

[1] p. 133.

The arbitral award rendered in 1885 in the dispute between the United States and Haiti, relative to the American ship *William*, furnishes a further characteristic example. The United States won the case, but Haiti protested. The American Secretary of State recognized that the decision was unfair, and made a declaration to the effect that the American citizen Pelletier, whose rights were involved in the case, had no claim against Haiti; and accordingly the affair gave rise to no judicial or diplomatic proceedings. The fact that in this case the successful party regarded the decision as unfair and repudiated all rights resulting from it, shows clearly how little concern the judges in that case had to render a decision unassailable from a legal point of view.

That even in the important decisions of recent times the principles remain the same is shown by the fact that at the First Hague Peace Conference, Martens, who at that time frequently acted as arbitrator, forcibly stated as his opinion that arbitration merely sought to dispose of disputes, and that accordingly the legal basis was secondary and superfluous.[1]

The famous arbitral court which dealt with the Bering Sea fisheries rendered, on August 15, 1893, merely an award by compromise, as Balch[2] has clearly put it: 'The decision, taken as a whole; was a compromise.' But at that time this was regarded as entirely natural, and it is not to be wondered that Renault,[3] and also Barclay[4] speak of the decision as an excellent one because of the fact that it

[1] See my remarks concerning 'Friedrich v. Martens und die Haager Friedenskonferenzen' in *Zeitschrift für internationales Recht*, 1910, p. 349.

[2] *Revue de droit international*, 1908, p. 389.

[3] *Revue générale*, 1894, p. 44.

[4] *Revue de droit international*, 1893, p. 444.

was a compromise and declared neither party wholly in the wrong. Renault speaks as follows:

> The decision rendered on August 15, last, by the arbitral tribunal contains only the answers made by the whole body or the majority of the arbitrators to the questions set forth in the *compromis* of February 29, 1892, but the reasons are not given for these answers,—the provisions of the award are not fully conformable to the claims of either of the parties, whose interests the arbitrators naturally sought to reconcile as far as possible.

In like manner Barclay speaks:

> In following the example of King Solomon and in rendering an award which met half the claim of each of the parties, the arbitrators conformed to a precedent of ancient wisdom.

But shortly after the publication of the decision in the Bering Sea case, the American counsel, Carter, wrote as follows to the Secretary of State of the United States with respect to that decision:

> Compromise of some sort seems to have been the necessity of the situation; and when this is said, it means that the tribunal was no court at all, but a body of men aiming to reach a solution which would either equally please, or equally displease, the contending parties. Our friend Judge Hoar of Massachusetts is said to have observed that the circumstance which he felt to be the most disagreeable to him in his official life was that he was never able to decide against both the parties. A place on the present tribunal would have suited him to a charm.

In his report[1] to the Universal Peace Congress held at Antwerp in 1894, Hornby enumerates not less than eight important questions of law which the arbitral court in the Bering Sea case had undertaken to answer.

[1] *Official Report*, p. 141.

Under these circumstances it is a fact worthy of special note that Baron Lambermont, in transmitting to Germany and England his decision relative to Lamu Island, rendered on August 17, 1889, expressly emphasized the point that he had decided the case as a judge and not as a mediator.

The award which Martens rendered in 1897 in the dispute between England and Holland, relative to the ship *Costa Rica*, was anything but a strictly legal decision. As was his wont, Martens allowed political reasons to figure largely in his arbitral decisions. Even at this day, after almost fifteen years, the decision of Martens relative to the ship *Costa Rica* is the object of general indignation in Holland.

The far more important decision, on October 3, 1899, of the Venezuelan arbitration court, which sat at Paris under the presidency of Martens, was in its essence only a compromise. Although five celebrated jurists occupied themselves for several months with the boundary dispute between England and Venezuela, and although numerous and complicated legal questions were involved in the dispute, the award covers only two small printed pages. That the award was a compromise is generally recognized. The *Revue générale* [1] speaks as follows :

> The arbitral tribunal appears to have satisfied the two parties. Venezuela, in fact, saw the British claims cut down, and Great Britain obtained the acknowledgment of her rights over a vast territory.

General Harrison, Foster,[2] and Dennis [3] hold the same view.

As Martens was regarded as the most distinguished

[1] 1901, p. 79.

[2] *Proceedings of the American Society of International Law*, 1909, p. 30.

[3] *Columbia Law Review*, 1911, p. 485.

representative of international arbitration in the whole world—a fact pointed out in particular by Holls at the First Hague Conference—it is quite natural that the other arbitrators of that period adopted his attitude. Accordingly we see that even in recent times a strong emphasis is still placed upon the element of compromise in international arbitration. The decision of the King of England, rendered on November 20, 1902, in the dispute between Chile and Argentine is accompanied by no opinion at all. Again, Riddell, for example, a member of the Canadian High Court of Justice, speaks as follows of the decision rendered on January 24, 1903, between Great Britain and the United States in the Alaskan dispute: 'It was felt that the decision was not a judicial decision on the merits, but that the result was based upon diplomatic grounds.' [1] The observation was made at the time that 'not many more cases of that kind are needed to bring arbitration into discredit'.

The distinguished French jurist, Fauchille, in 1906, criticized in the same way [2] the decision rendered by the King of Italy in the boundary dispute between Brazil and Great Britain. He declared at the time:

> In acting as he did the King of Italy assumed the position of friendly compositor, although the arbitral agreement did not actually give him the right to do so. It was only according to the rights recognized as possessed by one or the other party that the arbitrator should have determined the boundary line of Guiana.

How little such decisions meet with the wishes of the parties is shown by the consequences of the arbitral award rendered in June, 1911, in the Chamizal case between

[1] *Proceedings*, p. 34; see also Balch in *Revue de droit intern.*, 1904, p. 39.

[2] *Revue générale*, p. 139.

Mexico and the United States. As the national judges were unable to come to any agreement, the decision of the arbitrators awarded to each of the parties by compromise half of the territory, the ownership of which, by reason of a change in the bed of the Rio Grande, was disputed. Neither government was satisfied with this decision, so that the parties intend either to submit the case to the Hague Court or to settle it by diplomacy. Fried[1] rightly remarks in this connexion:

> The whole case speaks against the method of mixed commissions and the arbitral settlement of such disputes. It makes the necessity of an international and impartial court of justice all the clearer.

At the Second Hague Conference, Barbosa,[2] the great advocate of a freely chosen arbitral court and the opponent of a permanent court of justice, expressed himself as follows:

> I refer you to the case of the *Costa Rica Packet* in 1897, to the boundary dispute between British Guiana and Venezuela in 1899, and to the Venezuelan dispute of 1904. The decisions in these three cases have been sharply criticized in the international reviews and elsewhere by the most eminent authorities upon the subject; and the second case, which seriously menaced the territorial integrity of my country, gave rise to an energetic protest from its government.

In his treatise, *International Arbitral Law and Procedure*, which appeared in 1910, Ralston has arranged according to the subject-matter the legal principles which international arbitral courts have thus far laid down. The contents of the book show how contradictory the decisions of arbitral awards have been and how meagre the judicial

[1] *Friedenswarte*, 1911, p. 254. [2] p. 342.

opinions accompanying them. In his review of Ralston's work in the *American Journal* [1] Borchard justly observes: 'On many subjects, it is practically impossible to reconcile the conflicting decisions of arbitral tribunals.'

3. The Element of Compromise in the Decisions of the Hague Permanent Court of Arbitration

If in the following pages I shall sharply criticize the decisions of the Hague Court of Arbitration, it becomes me to lay special stress in advance upon the conspicuous ability of those men, such as Lammasch, Renault, de Savornin Lohman, Hammarskjöld, Fry, Fusinato, Beernaert, Asser, Gram, and others, who rendered those decisions. What they have done by their skill as arbitrators to make the Hague Court better appreciated is a matter of history. Accordingly, I direct my remarks exclusively against the prevailing system of compromise according to which those decisions were admirably, and in some cases masterfully, decided. I do not advocate better arbitrators—for where can we find more able men—but merely a better system for the settlement of international disputes.

As regards the first decisions of the Hague Court down to the time of the Second Hague Conference, in the dispute between Mexico and the United States there were only the questions of *res iudicata* and of prescription to be settled. The arbitral court settled both summarily. In particular the question of prescription in international law was not studied with sufficient care, although the arbitrators felt called upon to express themselves in detail upon it. The arbitral court acted as if there were no literature at all upon this subject. If one compares a decision of the

[1] 1911, p. 536.

German Imperial Court with this decision we are impressed most favourably with the consideration shown by the national courts for judicial decisions and for the literature of the subject. But in the decisions of the international courts there is rarely any reference to earlier decisions. The question of prescription had been passed upon already in numerous arbitral awards, and in the field of literature that interesting question had been treated in particular by Rolin-Jaequemyns [1] and by Mérignhac,[2] in connexion with the Butterfield case.[3] In the Hague decision it is merely stated that national laws as to prescription can have no application in international law. But, with the exception of the Butterfield case, the existence of the rule of prescription in international law had been constantly affirmed by the eight international arbitral courts, which before 1902 had passed upon the question of prescription in international law.[4] It was by that principle that the American-Venezuelan Commission of 1889 justified its award in the Williams case in a fairly detailed opinion. But the Hague arbitration court refused to take a stand upon the point. It should also have at least been proved why there was an international dispute involved in the case, since the United States had merely undertaken to support the rights of private persons. The searching criticism of this decision in *La Justice internationale* [5] regards it as very doubtfully correct, and declares that the arbitral court abandoned the strictly legal point of view in this case and took its stand upon the ' ground of arbitral equity '. In thus ignoring previous arbitral awards, and in

[1] *Revue de droit international*, vol. xxii, p. 361.

[2] *L'Arbitrage*, 1895, p. 123.

[3] See also Rouard de Card, p. 129 ; Dreyfus, p. 184.

[4] Moore, pp. 1205, 3130, 3139, 3210, 4181, 4199 et seq.

[5] Vol. i, p. 31.

disdaining to consult the literature of the subject, it is impossible to build up an international judiciary with fixed traditions and authority. Theory and practice must work together if anything useful is to result. But how can this be the case if in practice theories are thrown aside as worthless? As a matter of fact, the decision of the Hague court, rendered by five prominent international jurists, because of its lack of proof had so little moral force that in the following year, 1903, the Venezuelan commission in the Gentini case declared that the principle of prescription was 'well recognized in international law'.[1] How can theory develop and carry on the rules laid down by practice, when it never occurs to international judges to state their reasons, and when, on the contrary, they act as legislators? Under such conditions will not a breach hurtful to the development of international law be made from the start between theory and practice? Do not the numerous attacks to which international arbitral courts have with few exceptions been constantly subjected, now become intelligible? We are unfortunately obliged to criticize in almost all recent international decisions the disregard of theoretical works. Thus far the Hague Court is without a library, which, as the highest international arbitral court, it undoubtedly should have.

In fact, we may to-day hold up before the judges of the Hague Court as a standard what, in 1890, Saint-Georges d'Armstrong wished to see enjoined upon the members of the Academy of International Jurists advocated by him as a duty in their legal decisions: 'Before pronouncing upon a question the arbitrators should investigate and take into account the numerous precedents which we have mentioned, such as the diplomatic books, referred to

[1] Ralston, *Venezuelan Arbitrations of 1903*, p. 720.

in the several countries by different colours, which contain all the official notes exchanged between two or more governments upon a single subject; such as the character of positive laws, the resolutions of international congresses and conferences, arbitral awards of which there is knowledge, and the decisions of prize courts; such as, moreover, the principles admitted by the majority of nations and embodied in their treaties, as also the principles laid down by writers of highest authority in matters of international jurisprudence, and those, more general in character, of customary law.' [1]

The decision of the Venezuelan arbitral court has likewise contributed nothing at all to the development of international law. Mallarmé [2] has passed the following judgment upon it:

> It was a question of law that was above all at issue in the case. The court should have kept to it, and not descended to the rôle of a commission of inquiry interpreting the particular facts of diplomatic negotiations. The decision lacks force and authority, because the grounds upon which it rests are merely considerations of fact which may easily be contested.[3]

The decision of the Hague court in the dispute between Japan and the European states was likewise drawn up in the relatively brief form of previous arbitral awards. That this tribunal also regarded itself more as a body of arbitrators than as a real court, is shown by the fact that one of the three judges, Motono, recorded his dissent in signing the award.

[1] Op. cit., p. xv.

[2] *Revue générale*, 1906, pp. 496 et seq.; see also the criticism expressed in *La Justice internationale*, 1904, pp. 9 et seq.

[3] See also Baldwin, *Report*, 1907, p. 44, and Thayer, *Official Report of the Fourteenth Universal Peace Congress*, p. 50.

Had the decision of the Hague tribunal in the Muscat case, which was a particularly good one under the present system of arbitration, been accompanied with as detailed an opinion as in the case of a German Imperial Court decision, it would have been less sharply criticized—Bressonnet [1] said: 'It is difficult to deny that it has been handled badly'—and would have contributed more to the development of international law.

In none of these cases, however, were there involved the most important juristic principles which could be presented to a court. Such questions were only passed upon by two later arbitrations before the Hague Court, namely, the Orinoco and the Savarkar cases.

In the Orinoco case the arbitral court had to determine whether the decision rendered by Barge in the dispute between Venezuela and the United States, and not accepted by the latter, should hold good or not. It had also to decide the celebrated question whether or not the decision of an international arbitration court is subject to revision. As a ground of nullity there was the express recognition by both parties in the *compromis* that there had been exercise of jurisdiction not granted and essential errors. Now the decision gave no clear information upon the point as to what was to be understood by an essential error, but confounded, as Scelle [2] has pointed out—and indeed intentionally—the terms 'exercise of power not granted' and 'essential error'. Indeed, I myself, in my commentary upon the Hague Convention for the pacific settlement of international disputes,[3] made the statement that the court was willing to admit essential error as a ground of nullity in a narrow sense. But I was only led to this

[1] *Revue générale*, 1906, p. 164.
[2] Ibid., 1911, p. 193.
[3] p. 148.

conclusion because I wished, at all events, to put some sense into what seemed to me an oracular decision. For I could not believe that the court, contrary to the terms of the *compromis,* was unwilling to accept essential error as a ground of revision.

Scelle very justly supposes that the court did not wish to declare either of the parties wholly in the wrong, and accordingly avoided taking a definite position upon the question of essential error. On the same ground it laid down the principle that if one part of an arbitral award should be held void, the other parts need not necessarily be held so. Neither of the parties had in their pleadings reckoned with such partial validity or invalidity, and yet the court decided in that manner. The decision is only intelligible if we keep before our eyes that the arbitral court desired to act to some extent as mediator. Only thus can it be understood why no definite position was taken in the decision upon the question of revision and why, in consequence of the obscurity resulting, Scelle interpreted the decision to be against the question of revision, while I in my commentary defended the decision as *in favour of* it.

If we take into account the fact that the court of arbitration has acted as mediator, we shall see that after all it rendered a good decision. But had the decision been rendered from a strictly legal point of view, it would have been much more useful for the development of international law, and Scelle as well as de Melville,[1] regrets that the arbitral court did not decide the question of revision as the principal one at issue. Neither of the parties would then have been rebuffed, for each reckoned with a rejection of the proposal as a whole, and the Ameri-

[1] *Eigen Haard,* Amsterdam, 1911, p. 96.

cans were greatly surprised to find that while they had won in principle, they only obtained the partial annulment of the decision of Barge, in consequence of the fact that the court had compromised. Scelle is quite right when he says that the arbitrators of the Orinoco case regarded themselves rather as 'friendly compositors' than as real judges, and that diplomatic arbitral courts are a danger to the development of international law. Naturally public opinion, as well in the United States as in Venezuela, recognized the element of compromise in the arbitral award, and expressed its displeasure with the decision.[1]

But if the Orinoco arbitral court, acting, as has been said, under the system of arbitral jurisdiction, rendered nevertheless a good decision, yet the same cannot be said of the decision of the Savarkar arbitration court which met some months later. The question at issue in this case was whether the Indian revolutionist, Savarkar, who had been brought back from French to British territory by the authorities, should be returned by England to France. The French and British cases, counter-cases and replies in this case take up well over three hundred pages. The counsel on both sides discussed the question at issue in the fullest detail. But the decision is accompanied by no legal opinion at all; it merely enumerates the facts. The legal issue was settled with the following words: 'There is no principle of international law according to which, under the circumstances of the present case, a fugitive criminal must be given back.' The decision met with general disapproval.[2]

[1] See in particular Dennis in *Columbia Law Review*, 1911, p. 500, and *El Universal*, Carácas, November 24, 1910.

[2] See Kohler, pt. i of the *Zeitschrift für Völkerrecht und Bundesstaatsrecht*, 1911, pp. 202 et seq.; van Hamel, in the *Revue de droit international*, 1911, pp. 376 et seq.; Strupp, pt. i of the *Zeitschrift für Völkerrecht und Bundes-*

The decision simply passed lightly over the main point; as Kohler remarks, it did not decide it at all. It is, as van Hamel says, 'rather a verdict than a judicial decision.' By a peculiar irony of fate this decision, which exhibits a clear failure of the Hague Court, was rendered under the presidency of Beernaert, the same man who in 1907 pointed out with so much feeling that disputes would be settled by the Hague Court in a strictly legal way.

That the Casablanca decision sought to declare neither party wholly in the right nor wholly in the wrong may be stated as a generally recognized fact. This has been brought out in particular by Gidel [1] and by Dennis.[2] This fact may be readily approved in a dispute in which both parties were greatly wrought up.

The celebrated decision of the Newfoundland arbitration court of 1910 is also in the nature of a compromise. In particular was this clearly the case in the answer to the fifth of the seven questions before the arbitral court. In this instance the arbitral court had to determine upon the extension of territorial waters in the so-called 'bays'. The arbitral court decided that in such cases the three-mile belt should not follow the line of the Canadian coast, but a line stretching from one headland to another. That this decision finds no support in existing international law is recognized in particular by Balch,[3] Lansing,[4] de Louter,[5]

staatsrecht, 1911; Robin, in *Revue générale*, 1911, p. 350; Wehberg, in *Friedenswarte*, 1911, p. 120; and Dennis, in *Columbia Law Review*, 1911, p. 497, who is, however, mistaken.

[1] *Revue générale*, 1910, p. 407.

[2] *Columbia Law Review*, 1911, p. 498.

[3] *Revue de droit international*, 1911, p. 22.

[4] *American Journal*, 1911, p. 31.

[5] *Revue de droit international*, 1911, pp. 150 et seq.

and Dennis.[1] Balch [2] was the first to declare the decision a compromise. Soon after the honourable president of the arbitration court, Lammasch,[3] himself declared that the Newfoundland decision contained elements of a compromise.

Apart from the decisions of special arbitration courts there are no further decisions of the Hague Court. The composition of the special Russo-Turkish arbitration court, in which four or five judges were nationals of the parties, well expresses the spirit of the later tendency.

If we survey impartially what the Hague Court has done up to the present time, we must conclude, indeed, that it has settled numerous disputes, and has justified itself as a means of settling disputes. But, on the other hand, it must be pointed out that all of these decisions have done very little for the development of international law, and have denied and repudiated the most important principles. Robin feels compelled to declare in the *Revue générale* [4] that 'when we examine not only the decision of February 24, but certain others which preceded it, we are forced to admit that the Hague Court of Arbitration regards its functions as those of a mediator not less than as those of a judge'.

When Privy Councillor Fischer (Breslau), at the general convention of the Central European Economic Association at Munich on October 14, 1911, proposed the erection of a permanent international arbitration court for private claims against debtor states in connexion with the Hague Arbitration Court, Professor Harburger declared that he had some doubts upon the subject, because the Hague Arbitration Court did not decide in a legal way, but merely compromised. We have here the clearest recognition of the fact asserted above.

[1] *Columbia Law Review*, 1911, p. 498.

[2] Op. cit.

[3] *Das Recht*, 1911, p. 148.

[4] 1911, p. 351.

It is also evident, from what has been said, that in the above cases very valuable decisions might have been given, and thereby international law might have been developed. Von Plener [1] seems to me exceedingly bold when he makes the assertion that 'the great value which the creators of the new institution attach to a certain continuity of jurisprudence does not play so great a part in the case of arbitration, which has more often a character *sui generis* which requires special treatment in each individual instance.' That is unquestionably wrong. Von Plener has here in mind strictly political disputes only, which, in my opinion also, can still be settled in the future according to equity. That the legal questions are not always of an unique character is best proven by past cases. One need only examine how often the same questions have turned up in the 250 decisions thus far rendered. How often have similar questions of law arisen concerning claims for indemnity brought by foreigners injured in a war, concerning seizures of ships, fishing rights, the principle of prescription and of *res iudicata.* The argument of von Plener is, moreover, not a new one. At the seventeenth meeting of the International Law Association, in 1895, at Brussels, an Englishman, Mr. Webster, declared that the disputes arising in international life were of so diverse a character that there was no need of a permanent court; and that it would be much better to do justice to the individuality of each case by means of a court specially constituted on each occasion.[2]

We see, therefore, that in the prevailing system theory and practice are in complete opposition. Arbitral awards

[1] *Union,* 1911, p. 98.

[2] Report of the Seventeenth Conference held at Brussels, October, 1895, p. 39.

are so unanimously recognized as a compromise—as by Marburg and Wambaugh [1]—that the words of the Hague Convention which speaks of a decision 'on the basis of respect for law' must be regarded as a delusion. From this point of view Fischer,[2] in advocating the erection in connexion with the Hague Court of an international court of arbitration for claims against debtor states, emphasizes the fact that the connexion with the Hague Court should not lead to the mistaken opinion that the new court is to decide primarily according to equity, and not according to law.

When we realize in addition that international arbitration by its very nature calls for a decision according to equity,[3] we see immediately that it is a mistake to introduce into arbitration the judicial element of a strictly legal decision. In the exercise of their office international arbitrators cannot get rid of the thought that arbitration calls for a decision *ex aequo et bono*, and they are afraid that by a strictly judicial—that is to say, legal—settlement of the case, which declares one of the parties wholly in the wrong, they will give offence. Barbosa [4] is wrong when he says that arbitrators and judges are alike in that they both render decisions. In reality, only the judge decides while the arbitrator adjusts. Marburg [5] has justly remarked: 'The aim of a court of arbitration is to compose differences, and the spirit of compromise which prevails as a result thereof can hardly yield lasting principles of law or justice.' Dennis also points out [6] that the reason for the compromising character of arbitral awards is to be found in the very nature of arbitration and, accordingly, that the fundamental

[1] *Proceedings*, pp. xiv, 145. [2] p. 15.

[3] See Hold v. Ferneck in *Zeitschrift für Völkerrecht und Bundesstaatsrecht*, 1912, p. 14. [4] p. 659.

[5] *Proceedings*, p. xiv. [6] *Columbia Law Review*, 1911, p. 502.

remedy for the present condition of things is to be sought in the substitution of judicial settlement for arbitration.

Accordingly, it was a great mistake of the First Hague Conference to introduce into the field of international arbitration the demand for a strictly legal settlement, and such an attempt was necessarily destined to fail because of the incompatibility of international arbitration and judicial settlement. It was, indeed, permissible in 1899 to make a compromise between the two institutions by substituting certain characteristics of a legal decision for those of arbitration, and thereby creating a kind of transitional institution in the direction of a judicial court; but it was not permissible to substitute the most essential feature of a legal judgement, namely, a decision according to strict law, for the most characteristic feature of arbitration, namely, a decision according to equity. For the result was the establishment of an international judicial system in its essential nature; only there were lacking the external forms of such a system. If the intention in 1899 was to establish a real international judiciary, it should have been kept in mind that in so doing arbitration was being abandoned, and the necessary foundation of an international judiciary should have been laid. The whole mistake of the First Hague Conference arose from the fact that upon the most important principles the delegates were groping in the dark. Had the question been studied, another definition of arbitration would have been chosen. For the Conference desired in the last instance merely to develop the method of settling international disputes, not to establish an international judiciary. But in its noble zeal it thought that it could serve two masters, and the result was a mongrel institution which is neither wholly judicial settlement nor yet international arbitration.

What we need, accordingly, for the development of international law is real judges who are commissioned to decide in a strictly legal way and to avoid all awards in the nature of a compromise.

It follows, therefore, that the principle upon which Scott proposed the Judicial Arbitration Court to the Second Hague Conference was unique in character and very serviceable for the development of international law. It is Scott's chief service to have been the first to lay down clearly at a great gathering of states that we have on the one hand international arbitration for the equitable settlement of disputes which the parties regard as strictly political in character, and that on the other hand we have international judicial settlement for the strictly legal decision of disputes regarded by the parties as legal in character. Scott's views had already been expressed with the greatest clearness thirteen years earlier by the American, Hornby, in a report to the Universal Peace Congress at Antwerp. He showed that international judicial settlement must develop international law, and that this could only be done by a truly permanent court made up of professional and impartial judges, and that the existing system of arbitration was unsuited to that purpose. The same views were pointedly expressed by the president of the Peace Palace Endowment, Jonkheer van Karnebeek, on July 30, 1907, in his address at the laying of the corner-stone of the palace, when he declared that 'it is possible, it is perhaps desirable, to take a step from the system of arbitration towards the system of judicial settlement properly so called. That will depend upon the progress which the idea of international justice shall make in the minds of governments and of peoples.'

4. The Impossibility of Developing International Law Except by a Complete Reform of the Present System of Arbitration

But the loyal advocates of international arbitration are not satisfied with what has thus far been accomplished, and assert that if they reform the present system of international arbitration it will be quite possible to develop international law through that means.

In my opinion the failure of international arbitration in the past has been so fully proved that we can expect nothing from it in the future for the development of international law. International arbitration courts and, in particular, the Hague Arbitration Court, are composed not merely of jurists pure and simple, but in part of diplomats and, what is worse, also of nationals of the contending parties. Besides, it is certain that for the future decisions according to equity will be commended in the case of strictly political disputes. Under these circumstances, arbitrators will on every occasion be obliged to ask themselves whether a settlement of the dispute according to equity would not be better than a strictly legal decision. They will regard themselves as arbitrators even in cases where they should give a strictly legal decision. This will remain the case until a distinction is made between arbitration for the equitable settlement of disputes regarded by the parties as strictly political, and a judicial system for the legal settlement of disputes regarded by the parties as legal in character.

De Louter [1] in particular emphasizes the fact that the Hague Arbitration Court can give no guarantee for strictly legal decisions.

[1] *De Gids*, 1912, no. 2

Dennis[1] also is of the opinion that the strictly legal decision of disputes can be attained only through the 'fundamental and ultimate remedy', namely, the establishment of an international judiciary. Nevertheless, he recommends various means to facilitate as far as possible the strictly legal settlement of international disputes until the time comes when we shall have a permanent international court of justice. Accordingly, he proposes five distinct improvements in the present system of arbitration:

1. A legal decision must be provided for in the arbitral agreement and a compromise distinctly forbidden. Dennis compares the Casablanca case with the arbitral agreement in the San Juan boundary arbitration of 1869. In the first of these cases the terms of the arbitral agreement were apparently general in character, so that the intention of the parties that a compromise, rather than a judicial decision, should be reached, was well understood. In the second case the British commissioner had proposed that the dispute should be disposed of by a sort of compromise; but the American commissioners were opposed to this and demanded a legal settlement, and were annoyed at the fact that the arbitral agreement contained a contrary instruction. The decision rendered by the German Kaiser was also in accord with this stipulation. It may be that when there is an express instruction in the arbitral agreement to that effect, the arbitrators will feel compelled to render a strictly legal decision of the dispute. But whenever this is not the case a relapse is to be feared. The arbitrators will not be able to get rid of the idea that they are arbitrators and not real judges.

It is certain, however, that states will rarely include such a provision in their arbitral agreements. In the

[1] *Columbia Law Review*, 1911, pp. 502 et seq.

numerous cases in which the right clearly lies more on one side than the other, such an agreement will not be reached. For in such cases it will be the very party that believes that it will lose its case which will give with difficulty its assent to such terms. For it naturally hopes to obtain a more desirable decision by a compromise.

Hence this remedy is entirely inadequate in character.

2. The body of judges should, as Dennis further proposes, be selected by common agreement of the parties, as has already at times been done. Henceforth neither of the parties are to appoint their judges, whether one or two, independently. For in such case they are generally more or less influenced in favour of the state which has appointed them.

But not even in this way can a compromise be prevented. The history of numerous cases of arbitration during the last century shows how difficult it is for the parties to agree upon several judges. In the only two recent cases in which this has happened, the Savarkar and Newfoundland cases, the decision was clearly a compromise.

3. A detailed 'Code of International Procedure' should be drawn up. In particular a more precise distinction should be made between the case and the counter-case.[1] Moreover, the argument must be regulated.[2] This change would certainly be very useful. But I doubt whether it would help to do away with the element of compromise in arbitral awards.

4. Less value is to be attached to the point of unanimity in the decisions rendered. For unanimity, as a rule, results from the fact that the judges agree upon a compromise.

[1] See *American Journal*, 1911, pp. 55 et seq.

[2] See my commentary upon the Hague Convention for the pacific settlement of international disputes, pp. 110 et seq.

This was particularly the case in the Venezuelan dispute, in which the president, in announcing the decision of the court, called particular attention to the fact that the decision was a unanimous one. Dennis thinks that Drago's dissenting opinion in the Newfoundland decision is a very hopeful incident in the direction of more correct views. He says that the provision in the earlier Convention of 1899, that each judge must sign the decision, should not have been eliminated in 1907.

On this point I cannot entirely agree with Dennis; I mean, I think his praise of Drago is scarcely justifiable.[1] Dennis undoubtedly errs in supposing that compromise has been introduced because it was desired to make the decision unanimous. In the Muscat case the fact of unanimity in the decision was not expressly announced, so that in later cases in which the decision was not unanimous this fact need not have been specially pointed out. According to the provisions of Article 79 of the recent Hague Convention, the decision is signed only by the president, and a dissenting opinion may no longer be separately placed on record. Precisely for that reason it has resulted that henceforth there is no longer any outward indication whether a decision is unanimous or not. I am entirely in accord with the present rule and consider a change unnecessary. Moreover, decisions as important as those in the *Alabama*, Bering Sea, and Newfoundland cases, which were undoubtedly compromises, were not unanimous, but were reached by a majority of votes.

5. In conclusion Dennis proposes that the parties shall have the right to a revision and to an annulment of the decision. On this point I am, as I shall point out again later on, in agreement with Dennis. But how can the

[1] See my commentary, p. 139.

whole system of arbitration be improved thereby ? If the court of first instance renders a compromise instead of a strictly legal decision, we may well suppose that the same principles will be applied in the decision on appeal. If a strictly legal decision can possibly be given, it should be given by the court of first instance.

But even if the arbitral judges of the Hague Court could, by means of positive instructions, be induced to render strictly legal decisions with a detailed judicial opinion, nevertheless that would be no guarantee for the development of international law. Just as the decision of a dispute according to equity entirely excludes a development of international law, so the legal settlement alone of international disputes offers little guarantee of the development of international law, unless in addition the further conditions be fulfilled :

1. The individual decisions must be consistent with one another.
2. The strictly legal settlement of disputes as a body must be made possible. For if individual cases only are decided in a legal manner, the decisions will not be numerous enough to secure the development of international law.

So long as international courts meet without any dependence one upon the other, it is impossible to build up judicial traditions and precedents. An arbitral court does not consider itself bound by the decisions of previous tribunals.

The present system gives us decisions which are not connected with one another and often contradictory. Mallarmé[1] has already pointed this out : 'The manner in which the Hague Court is constituted makes it impossible

[1] *Revue générale*, 1906, p. 500.

for a system of jurisprudence to be formed.'[1] Among many others the American delegate, Mr. Choate,[2] expressed the same view at the Second Hague Conference: 'The absence of continuity in the Permanent Court has greatly lessened its power and its influence. Each of its meetings has been without any connexion with the others, and its decrees, rendered at different intervals and upon diverse subjects, have done little to promote the progress of the science of international law; nor have they built up a system of international jurisprudence which we may rightly expect from a tribunal resting upon the consent of all nations.'[3]

All of these facts will be fully discussed later when we treat of the character of an international court of justice. Here we can only show that under the present system of international arbitration, and in particular under its culmination—the Permanent Court of Arbitration—the legal settlement of disputes as a body is impossible.

For if we look more closely at the Hague Permanent Court of Arbitration we see that in spite of the terms of Article 42 of the Convention: 'The Permanent Court of Arbitration is competent for all arbitration cases,' it has been created only for the settlement of such disputes as are either political in character or of important economic significance.

In addition to the Administrative Council and International Bureau, the Hague Permanent Court of Arbitration consists of a list of judges. Each state selects at most four judges who are inscribed upon this list. If then two parties desire to refer a dispute to the Hague Court, they choose from this list the judges who are to act in the

[1] Of the same opinion is Duplessix (*Official Report of the Seventeenth Universal Peace Congress*, 1909, p. 419).

[2] pp. 311, 594.

[3] Of the same opinion is Leger (p. 338).

individual case. The two parties must themselves pay the honoraria of these judges. Now a number of international disputes arise in which a difficult legal problem may indeed be at issue, but in which the object involved is of trifling importance. Now if a dispute of this kind has no political bearing, it does not pay to call together for its settlement a number of international jurists who demand for pay a thousand marks as an honorarium. For such smaller cases we must have a much cheaper way of settling the affair upon a legal basis. For even the summary procedure provided for in the Hague Convention is not adequate, as von Plener [1] mistakenly thinks. This procedure also demands the calling together of a special tribunal, and is still too expensive for the smaller disputes. Accordingly, a cheaper procedure would be of importance for the development of international law and for the peace of nations. For it is clear that the same dispute, which arises over an object not essentially of a political character and of no great material consequence, may later appear in the form of a difficult political question. Supposing that the unimportant question had first been settled in a strictly legal way, it would be clear in the second case what each of the parties could rightly claim. There would then be more secure foundations upon which to proceed to the settlement of a dispute which might perhaps endanger the peace.

It is quite unintelligible how, in the face of this well-known fact, the over-zealous friends of the Hague Court of Arbitration are able to assert that it is well suited to the development of international law. No, the contrary is the case. Of the endless number of international disputes a decreasingly small number will be settled by means of

[1] *Union*, 1910, p. 98.

arbitration, because the cost of this method of settling disputes is ridiculously high in comparison with the trifling objects at issue. As early as 1904 Renault[1] declared: 'The working of the Hague Court is not without certain complications, and it must also be said, not without a certain expense; the object at issue should be worth the trouble. In many cases an arbitration could be carried out much more simply.' Is it any wonder that in the last one hundred and twenty-five years not three hundred international arbitral decisions have been rendered, whereas the supreme court of every nation renders considerably more decisions in a single year? Does one really believe that the small number of international cases offered for settlement is merely due to the dislike of the powers for international arbitration, or to the simple fact that there have been so few cases offered? In all the disputes which thus far have been submitted to arbitration, there has always been involved either an important political question or a question involving valuable material interests. In the arbitral decisions involving questions of money, millions have always been at issue, and the few exceptions in which trifling objects were in dispute took place forty and more years back. The disputes before the Hague Court, in so far as a money value can be assigned to the object at issue, have always involved large sums. The Mexican-American dispute involved 10½ million marks, the Venezuelan dispute 80 million marks, the Newfoundland case 104 million marks, the Orinoco case 6 million marks, the Italian-Peruvian dispute one million marks, and the Russo-Turkish dispute over 25 million marks. Of the disputes settled by special arbitration courts in recent times there are two which relate merely to money claims. In the dispute between Brazil and

[1] *Recueil des arbitrages*, p. ix.

Peru 55 million marks were involved, and in the Alsop dispute between the United States and Chile 4 million marks.

Disputes involving such large sums naturally do not arise every day, and it is precisely the questions of serious import which states are unfortunately still very reluctant to submit to international arbitration. But when arbitral decisions are so infrequent, how can there be any development of international law? Is it not urgent that smaller disputes also, in so far as they involve questions of interest, should be settled by means of an international process? For that purpose is there not need of an entirely different international organ from that of the Hague Arbitration Court, which is indeed not too expensive for disputes of a political character, the friendly settlement of which saves nations from the demon of war, but which in the case of strictly legal disputes, where the object at issue is not important, is exceedingly costly, and, therefore, not available?

Accordingly, Houzeau de Lehaie rightly declared at the Twelfth Universal Peace Congress at The Hague[1] that the cost of the Hague Arbitration Court was very heavy, and that some relief should be afforded in this connexion. He proposed, therefore, the establishment of an international fund to defray the costs of arbitration; this fund was to be obtained first of all from voluntary contributions. The congress passed a resolution to that effect.

A single example from my experience will show that governments, as was moreover demonstrated at the Hague Conference of 1907 (only not pointedly enough), regard the Hague Permanent Court of Arbitration as entirely unsuited for the smaller disputes. In the year 1910 the Commercial Treaty Union approached the German Foreign Office with the request that it submit to the Hague Court

[1] *Official Report*, p. 130.

a legal dispute of great significance. But as only a few thousand marks were involved, the German Government refused to take up the case. Yet a decision by an international tribunal of the highly significant questions raised in this case would have been of the greatest value for the development of international law. It may be worth while if I reproduce here what the Commercial Treaty Union wrote me, on December 16, 1910, relative to its communications with the Foreign Office:

> In addition, you will perhaps be interested to know that we brought before the Law Department of the Foreign Office here the possibility of submitting questions of this kind at present to the Hague Court of Arbitration. But we were told that under the present circumstances this was in principle only possible in cases involving a public quasi-political question, and that, moreover, the enormous cost made this method of settlement seem in advance impracticable in all cases involving only a few thousand marks.

Of the many official declarations in this connexion, the following are selected. Of the five arbitral agreements which were entered into in September, 1905, between Sweden and Norway, only one provided that jurisdiction should be given to the Hague Court of Arbitration. In explanation of this, Löwland declared in the Storthing on October 8, 1905, that the Hague Court had been abandoned because a decision could be obtained more quickly and more cheaply with other arbitrators. Further, the German Secretary of Foreign Affairs, Baron von Schoen, as a member of the Budget Commission of the German Reichstag, declared on January 26, 1909, that he was greatly in sympathy with the practice of arbitration, but that there were cases in which the heavy costs were out of all proportion to the importance of the object at issue. And the Prussian

Minister of Commerce declared in a decree of September, 1911, that 'in view of the peculiar character of arbitral procedure as an exceptional measure, and in view of the expenses connected with it, only those disputes will be submitted to arbitral decisions which involve large sums of money or questions which are of fundamental or great economic significance'. Needless to say that by questions of fundamental significance only the most important questions of principle are to be understood.

Owing to the absence of a permanent court of arbitration, the arbitral clause contained in the numerous (nine) recent treaties of arbitration concluded by Germany is almost of no practical value. The cost of a special arbitration court is so great that although the arbitral clause has been included, for the most part, in treaties for the past six or seven years, thus far only one arbitral court has been constituted.[1]

We need not wonder at the fact that the Hague Court of Arbitration and the special arbitration courts settle only the more important disputes, namely, those of a political nature, and especially those the decision of which can contribute least to the development of international law. For the Hague Court of Arbitration and all special arbitral courts are a part of the system of international arbitration, which in its essential character demands an equitable settlement of disputes which cannot be peacefully settled by the parties and which endanger the peace of nations. Such disputes arise principally from grave political sources. Arbitration merely seeks to dispose of disputes. It has little concern to decide the law between the parties. That, and that alone, is the aim of an international judicial system.

[1] See Borgius, p. 5, no. 23 of the *Mitteilungen des Handelsvertragsvereins*, 1911.

CHAPTER IV

THE IDEAL OF AN INTERNATIONAL COURT OF JUSTICE

1. The Exclusion of Diplomats

If then the purely legal settlement of international disputes is desirable in the interest of developing international law, the question arises how this international judicial system must be constituted.

The chief element of a judicial system is that the court of justice shall be composed of jurists and not of diplomats. Accordingly Vavasseur [1] is entirely wrong in proposing that the new court of justice shall be composed: (1) Of sovereigns; (2) of the ambassadors accredited to The Hague; and (3) of the Dutch Minister for Foreign Affairs. The Fifteenth Universal Peace Congress, in session at Milan in 1906, unfortunately supported this proposal with certain changes. Diplomats are far too closely connected with the individual groups of powers to be able to guarantee an impartial decision. The settlement of disputes by diplomats can, moreover, at the present time contribute nothing to the progress of international law, as Mougins de Roquefort [2] has already shown. In principle only jurists can render a judicial decision. This has been pointed out in particular by Salisbury in his telegram [3] of March 5, 1896, to Pauncefote, and by Root [4] and Scott.[5] There is no objection to the

[1] *L'Organisation d'une juridiction arbitrale internationale*, p. 23.

[2] pp. 77 et seq.

[3] To be found on pp. 21 et seq. of the third part of the excellent collection, *Actes et documents relatifs au programme de la Conférence de la paix* (Hague, 1899), by Jonkheer van Daehne van Varick.

[4] *The Importance of Judicial Settlement*, p. 10.

[5] p. 315.

appointment of diplomats upon international arbitration tribunals. On the other hand, it would be wrong to maintain that mere jurists are not competent to decide practical cases, which have, to a small extent at most, a political character. There is a great number of distinguished jurists who possess a very keen and practical judgment. I need only name men such as Asser, Fromageot, Lammasch, Renault, Scott, and Zorn. It may also be pointed out that the proposals made thus far by the Institute of International Law, which is a body composed only of international jurists, have been based on very practical foundations. A confidence in the ability of such jurists, who possess a practical point of view, has already induced certain persons to propose a tribunal composed strictly of jurists. In consequence various projects have been offered for the establishment of a sort of international academy for the decision of international disputes. I refer in particular to the proposal made by Kaufmann, a university professor at Bonn, in a publication which appeared in 1855, entitled *The Idea and the Practical Uses of a World Academy of International Law.* In 1890 Saint-Georges d'Armstrong [1] proposed a mixed board of diplomats and jurists. Eight years later, in an article in the *Nation* [2] von Bar advanced this same idea, but rightly limited the members of the academy to jurists only; and scholars such as Brusa, Bajer, and von Ullmann are in accord with him. These plans were, to be sure, not practicable, for an international judicial system can only be established by means of an independent tribunal, not by means of a board whose views would first have to be laid before the nations for their approval, as is the case to-day with the reports of

[1] pp. cxv et seq.

[2] October 15, 1898; see also his article in the *Courrier européen* of April 21, 1905.

commissions of inquiry. It was accordingly a happy thought of Nippold's to develop out of the plan for a world academy for legal opinions the idea of an academy of international law, which is to be opened at The Hague in the summer of 1913.[1]

In his address delivered before the First National Peace Congress at New York, in 1907, Root[2] rightly said: 'What we need for the further development of arbitration is the substitution of judicial action for diplomatic action, the substitution of judicial sense of responsibility for diplomatic sense of responsibility.'

2. The Exclusion of Nationals of the Contending Parties

The principle that no one can be a judge in his own case must be strictly applied not only in national but also in international law as well. If the nations really desire to develop international law through the decisions of international judges, they must offer some guarantee that the judges in a given case will not in any way sympathize with either of the parties. The authority of the highest court of the world must suffer if in its very composition a fundamental principle of law is violated. At the present day a national cannot easily be impartial towards his own country; he will always feel sympathy for its case. Accordingly the United States and Great Britain at the Second Hague Peace Conference rightly desired to see national judges excluded, in order to offer to states the most complete guarantee of the impartiality of the court of justice. It was only

[1] See in this connexion the *Friedenswarte*, 1911, p. 50, and the literature cited there; also the first Year Book of the Carnegie Endowment for International Peace, 1911, pp. 109 et seq. Owing to the European War, the Hague Academy has not as yet been opened.—Translator.

[2] p. 45.

because of the opposition of Baron Marschall,[1] with whom Karandjouloff[2] among others was in accord, that those countries yielded on this question. Within the last few years Tripp,[3] Baldwin,[4] McKenney,[5] Wambaugh,[6] and Ralston[7] have in particular demanded the exclusion of national judges. No better argument for this demand can be found than the fact that the overwhelming majority of writers would like to see nationals excluded even from the so-called 'arbitration courts'.[8]

The presence of two national judges upon a great board of judges of from nine to fifteen persons can to be sure not do much harm. For this reason if the number of judges in the new court of justice should be great, the contending states might well be allowed a national judge upon the tribunal, in order to preserve the principle of the Permanent Court. But in that case the ideal of an international judicial system would not be attained. Moreover, I disapprove of the appointment of a tribunal of from nine to fifteen persons, because in that case the responsibility of the individual members would be too light. I would prefer a court of justice composed of five or seven members. A tribunal of smaller composition has been advocated by von Bar,[9] who proposed five judges, and Lammasch,[10] who proposed seven judges. Barbosa[11] also insisted, in 1907, that in a tribunal composed of too large a number of members, the majority might be too easily made up of few real jurists.

[1] p. 603. [2] p. 345. [3] *Report*, 1907, p. 47.
[4] *The New Era of International Courts*, p. 1.
[5] *Proceedings*, p. 97. [6] Op. cit., p. 145.
[7] Op. cit., pp. 154 et seq.
[8] See my commentary upon the Convention for the pacific settlement of international disputes, pp. 75 et seq.
[9] *Berliner Tageblatt*, August 22, 1907.
[10] *Das Recht*, p. 151. [11] p. 620.

But it is self-evident that in the composition of every tribunal there must be as many men as possible who are well versed in the legal systems in force in the countries of the contending parties. In this way the other judges will receive valuable enlightenment from the deliberations, and the arguments of the agents and counsel will be supplemented. Thus, a Swiss national could very well enlighten his colleagues concerning Germany's legal system, and an American judge in like manner concerning the English legal system, and a Spanish judge concerning the South American legal system. But too great value must not be placed upon the presence of judges who are versed in the prevailing legal systems, because the principal point will always be the fact that the judges shall, during the course of a trial, make themselves perfectly acquainted with the legal systems, in the first place through the arguments of the agents and counsel, and then through the statutes and appropriate books. The decision will, however, always be rendered by the whole body of judges, and accordingly the best security against wrong decisions will always be the fact that the judges as a body have become perfectly acquainted with the given legal systems, and not the fact of the presence on the tribunal of a number of persons who are well versed in the particular legal system in question and can protect their colleagues from serious errors. It needs no proof that the counsel on both sides will throw sufficient light upon the point at issue, and thereby assist the tribunal to an understanding of the case just as much as would be possible if there were specialists upon the tribunal. The judges can, however, by means of the right of interrogation, have any obscurity cleared up by the attorneys.

If the presence of experts in the respective legal systems were to be too strongly emphasized, it would lead finally to the demand that only representatives of the legal systems, or distinguished specialists in this particular field, could be appointed to the tribunal. But that would be a mistake, because it must be the task of the international judiciary to reconcile national characteristics with international principles. Scott [1] aptly remarks: 'Domestic law should be internationalized, if it is to attain the object of permanent arbitration.' But that can only be done by persons whose attitude towards the respective legal systems is an objective one, not by those who are bound up with them, and who accordingly are easily tempted to introduce into international law the peculiarities of their legal system.

Moreover, the necessity of having representatives of the individual legal systems present upon the international tribunal is greatly over-estimated. It should be remembered that international private law disputes would not be heard by the international court of justice, because they would be turned over to a special court, as has been repeatedly proposed in recent times.[2] It is clear that, so far as possible, disputes involving questions of international public law must be sharply distinguished from those involving questions of international private law, as Kamarovski has already advocated. For these two branches of legal science have come to be so comprehensive that the number of those who are well versed in both of them is becoming constantly smaller. In the case of disputes in the domain of strict international law experts in the respective legal systems are not so often met with as in international private law. It would seem that at the Second Hague

[1] p. 319.

[2] See my article *Ein internationaler Gerichtshof für Privatklagen* in the commercial-political pamphlets of the Commercial Treaty Union, 1911; also Daehne van Varick, *Le Droit financier international*, Hague, 1907.

Peace Conference such great importance was attached to the representation of distinct legal systems merely because persons were possessed with the idea that international private law disputes were also to be assigned to the new court of justice.

What I have thus far said is based upon the assumption that an international court of justice for the claims of private persons is not to be erected simultaneously with one for the settlement of disputes between states.

But in case both projects were to be realized simultaneously, it would not be necessary to establish several courts, but merely several divisions of one and the same court, provided, however, that the division for international disputes be organized according to the principles set forth above. On the subject of the close connexion of the two projects, I refer to the excellent remarks of Schücking.[1]

3. The Exclusion of Judges appointed by the Parties

But we must go still farther and assert that in general judges appointed by the parties, even when they are not nationals of the state appointing them, offer no guarantee of an impartial decision. Barbosa [2] declared, in 1907, that the judges of the new court of justice would after all be only men and would not renounce their nationality in accepting their new office. Indeed, Ordoñez [3] even thinks that it is quite impossible to constitute an international court of justice, whose members as a body would possess the necessary impartiality, not merely in one, but in all cases. One need only consider how few states there are, and upon what numerous grounds of race, of geographical

[1] Op. cit., pp. 158 et seq. [2] p. 344. [3] p. 156.

situation, and of interests they stand in relations of friendship with or opposition to one another. This is all the more reason why, in my opinion, every precaution must be taken in order to obtain judges as impartial as possible. It is, however, evident that the individual states will only too readily name persons who will represent the views of the state appointing them, and will accordingly neglect the interests of the whole body. No reproach is to be cast upon states for so doing.

On the other hand, the judges who have been appointed by one party will be reluctant to vote against the state which has chosen them. For in choosing them the state places a certain trust in them, that they will represent its interests as far as possible. Thus Cockburn in the *Alabama* case and Drago in the Newfoundland case were unable to vote against the country by whose government they had been appointed. In earlier cases it had been taken in bad part when judges did not vote in favour of the government by whom they were appointed.

It frequently happens that international judges have already expressed their views upon a given question in their writings, so that states have in many disputes the opportunity of informing themselves of the views of the judges whom they appoint.

McKenney, the American counsel in the Orinoco case,[1] rightly observes that ' the non-nationals, arbitrarily selected by one or the other of the disputants, readily take on the colour, the attitude of mind of the disputant to whom they owe their selection '. McKenney thinks that this difficulty could be removed if each party presented to the other party a list from which the latter should select the judges, so that the choice of either side would be limited

[1] *Proceedings*, p. 97.

to persons who would be acceptable also to the other side. Prescinding entirely from the fact that this procedure would be much too troublesome, the result in this case would be that the judges presented by state *A* to state *B*, and selected by state *B*, would only too readily look upon themselves as representatives of state *A*, which first placed confidence in them, whereas state *B*, as it were, only appointed them under compulsion. It is impossible to obtain an impartial tribunal so long as the parties themselves appoint the judges.

Moreover, since the international court has for its object not only the settlement of individual disputes, but also the development of international law, it is necessarily the agent of the family of nations, and not merely the agent of the parties who make use of it in a particular case. Schlief [1] rightly observes: 'Every judicial decision which results from the arbitration of a case must be not only a mere settlement of the case which prevents war between the parties . . . but must of necessity represent the demands of higher interests of mankind.' Accordingly, the international community has by right a voice in the selection of judges, and may demand that impartial judges be appointed, so that international law may be developed through their legal decisions. The disputants must, accordingly, be forbidden to constitute the court themselves with a view to the particular dispute.

Moreover, the court will often have to interpret world treaties. In such cases especially, not only the parties but the entire body of the states signatory to the particular treaty are interested in an impartial decision, so that the parties should not be given an independent voice in the selection of judges.

[1] *Der Friede in Europa*, pp. 274 et seq.

If we examine now the reasons which are alleged for the appointment of the tribunal by the parties in the individual case, we shall find that they are unsound. It is said in general that the judges would possess in a higher degree the confidence of the parties if they were appointed by them. But that could only be the case if the judges as a body were appointed by both parties together, as was advocated in the plan of the first Pan American Congress, and in that of the International Law Association at its meeting at Brussels in 1895. But, in fact, such an agreement can only very rarely be reached, as is pointed out on p. 67 of the report of the International Law Association. For this reason Article 45 of the Hague Convention for the pacific settlement of international disputes, which was based upon a project of the Institute of International Law, provides a method for the composition of the arbitral court by which each party first appoints one or two arbitrators, who then together choose an umpire. In by far the larger majority of cases the arbitral court is composed in this manner, and of the twelve disputes thus far submitted to the Hague Court only two can be named in which the disputants have together agreed upon a number of arbitrators. This was the case in the Newfoundland dispute between Great Britain and the United States, and in the Savarkar dispute between Great Britain and France. In all other cases the tribunal did not possess the confidence of both parties. If, when there are three judges, judge *X* has been appointed by state *A*, and judge *Y* by state *B*, it is clear that judge *X* only possesses the special confidence of state *A*, not that of state *B*, and that judge *Y* only possesses the special confidence of state *B*, not that of state *A*. Hence, it is incorrect to say that the body of judges possesses the confidence of both states, since this can at most be the case with respect to the umpire. But the choice of the umpire is in almost all cases left to the judges appointed directly by the states. In the Casablanca case the arbitrators appointed by France certainly did not possess the special confidence of Germany; rather Germany was obliged to assume that the arbitrator appointed by France would vote for France. It may be assumed without further ado, that in a dispute between states *A* and *B*, the arbitrator appointed by state *A* will have the confidence of state *B* in a far lesser degree than a judge appointed by an entirely neutral power. The difficulty of appointing judges who are satisfactory to both parties is emphasized also in the programme of the American Society for the Judicial Settlement of International Disputes. If it is certain that recent theory and practice favour a method by which the majority of the arbitrators possess only the confidence of one of the parties, it is unintelligible how the contrary assertion can be made that the free choice of the judges by the parties can alone secure the composition of a tribunal whose members as a body possess the confidence of both parties. Barbosa [1] said in 1907 that a French judge was a guarantor for France, if France herself had appointed him. But he forgot to add that in that case the opposite party would also have a guarantor upon the arbitral tribunal, and that the two guarantors would thus offset each other, and the decision would rest with the umpire who possessed the confidence of the parties in a lesser degree than a judge appointed directly by them. When men such as Barbosa,[2] Baty,[3] Beernaert,[4] Carlin,[5] Bustamante,[6] Guillaume,[7] Lammasch,[8] de Louter,[9] and

[1] p. 696. [2] pp. 340, 659, 696. [3] *International Law*, p. 5 et seq.
[4] pp. 333, 334. [5] p. 145. [6] *La Seconde Conférence*, p. 203.
[7] p. 160. [8] p. 653. [9] Vol. ii, p. 159.

von Plener[1] assert that international arbitration presupposes the free choice of the judges, it is undoubtedly true. What we desire, and what Scott sought to obtain in 1907, is not at all international arbitration, but international judicial settlement. While in international arbitration, as a rule, only a part of the judges enjoy the confidence of one party, in international judicial settlement, as we shall see later, the composition of the court is such as to secure the interests of both sides in the highest degree. Root in particular has expressed that idea in his article 'The Importance of Judicial Settlement'.[2]

The ideal of an international judiciary can only be a court of justice whose members are not directly appointed by the parties.

4. The Exclusion of Judges appointed for Particular Disputes—Permanence of the Court

It is further to be observed, as Hornby pointed out as early as 1894, that international judicial settlement requires judges who are such by profession, and whose tenure of office is for life. Only judges by profession can give such clear and correct decisions as the parties to an international dispute have a right to expect. It is the prevailing opinion at the present day[3] that the duties of a judge are not merely abstract in character, depending for their fulfilment upon a knowledge of the law, but that a satisfactory decision can only be given by judges who are thoroughly conversant with the affairs upon which they are to pass judgment. Such knowledge Kaufmann insists can only be

[1] *Union*, 1908, p. 46; 1910, p. 98. [2] p. 7.

[3] See in particular Danz, *Auslegung der Rechtsgeschäfte*, 3rd ed., 1911, p. 77, and Kaufmann, *Wesen des Völkerrechts*, p. 5.

had through constant practice based upon settled tradition. What a difference there is between a judge appointed for a given case, who has to decide in a special instance upon legal relations with which he has never before in his life come in touch, and a judge by profession, who has studied the foundations of international life during years of activity as an international judge!

It is only upon the hypothesis that the court shall be composed of judges who are such by profession, that an international court can be erected whose members do not at the same time occupy high offices of state in their own countries. 'The fact that an international judge holds a high office in his own country,' says d'Estournelles,[1] 'creates a suspicion of partiality.' For the same reason Bustamante[1] advocates members appointed for life, who hold no national office.

This principle is so decisive that the contrary proposal made by Wheeler at the Lake Mohonk Conference of 1910[2] cannot be seriously considered. Wheeler thinks that it would not be well for the judges of the new court to devote themselves exclusively to the affairs of the Hague Court; that in the beginning there would perhaps not be much to do, and that if the judges were at the same time to sit in national courts they could gain greater experience. In other countries as well as in Great Britain and the United States, the judges are not always exclusively occupied upon a single court. But to say this is to forget that national judges, however varied their activities at a given moment, are exclusively occupied with the courts of one and the same country. Wheeler wishes the judges to be occupied at the same time with the affairs of national and international courts. In this way their impartiality would be

[1] p. 195. [2] p. 91.

seriously affected. The case would be quite different if Wheeler had merely proposed that the judges should be engaged upon various international courts. If, as international law develops, a greater number of these courts were to be established, there would be no objection to carrying out the proposal in that form.

The commission appointed by the Interparliamentary Union in 1894 also arrived at the conviction that the judges should not be forbidden to hold another office; that there was with men a time of life when they would only give up their former profession with reluctance; that in the case of members of the international arbitration court it would be a question for the most part of persons who had won special merit in their former profession; and that in consequence they could not be abruptly torn away from their old pursuits. There is indeed some truth in these assertions. But a man who does not feel that he has such a vocation for the office of international judge, that he will devote for the future his whole life and energy to this new office, is simply not suited to be a judge of the community of states.

But is it not necessary on this supposition to adopt Beernaert's fears [1] that enough judges would not be found for the new court of justice? It seems to me that this question cannot be seriously entertained. Those who put it have no correct conception of the idealism which animates so many men in this world to undertake the great tasks of international law. As there are at the present day jurists constantly seeking a position upon the International Bureau of the Hague Arbitration Court, though no consideration is paid to their request, so in the case of the new court of justice the number of able men will be greater than the need.

[1] p. 333.

Professional judges, whose duty it must be to reside permanently at The Hague, are only possible in the case of a truly permanent court. Now such a court is necessary not only for the sake of the impartiality of the judges, but above all because only a permanent court of justice can guarantee the development of international law. This has been strikingly pointed out in particular by Hale in a paper read before the Fifth Universal Peace Congress at Chicago, in 1893,[1] by Schücking,[2] and by Wambaugh, professor at Harvard University.[3] If a court decides a legal question, and if after a certain time a similar case is presented to the same court, the court will feel called upon by reason of its traditions to decide according to the earlier precedent, unless errors shall have been discovered in the first decision. But if the two disputes should be decided by two distinct courts, each being set up for a single dispute, the second arbitration court, being composed of entirely new members and not connected by any tradition with the first court, would not attribute any authoritative weight to the earlier arbitral award and would readily decide the case in another way. The same thing would happen in every case, and varying decisions would constantly be given. If the judges have familiarized themselves with a difficult question, and if a similar question comes later before the same court, this permanent court, by reason of the experience it has obtained, will be in a much better position to decide the new case than a tribunal composed of other persons and not connected with the past by traditions. For the fact is that the decisions thus far rendered, as has already been pointed out, have not developed international law because none of them possessed the tradition and the authority of the decisions of a world court. Thus far every

[1] *Report*, p. 167. [2] Op. cit., p. 91. [3] *Proceedings*, pp. 140 et seq.

question has been treated from an individual point of view, and no great general principles have been developed. The decisions of international courts of arbitration up to the present time cannot at all be classified according to principles of law.[1] Foulke[2] very aptly remarks: 'The umpire in an arbitration need not decide any case in accordance with the decisions of previous tribunals. He can decide according to his own sweet will. But if a court is established, the precedents of that court go to the making of law, and pretty soon you will find out what that law is.' That this is the case is also conceded by Nys[3] and by Renault.[4] The latter asserts: 'What one [court] has decided has no binding force upon the other, a situation which is not without its objections.' When, then, Renault further says that, on the other hand, the fact that the decisions are rendered without any connexion with one another has also its advantages, he falls into a striking error. He thinks that if the same dispute should arise later between two other states, one of them need not fear in advance that it will lose its case, should the court render the same decision. On the contrary, it is to be observed that if there is a permanent court of justice, states will, in a case such as the above, generally dispense with a trial, because the first case has already been decided according to strictly legal principles, and it is foreseen that a new decision of the court will be to the same effect. Which is to be preferred, that the parties settle their disputes outside of court, because they have a standard by which to test their rights in the case, or that they should once more turn over the question to a court?

[1] See p. 29. [2] *Proceedings*, p. 288.
[3] *Revue de Droit intern.*, 1906, p. 18.
[4] *Recueil des arbitrages*, by Lapradelle and Politis, p. vii.

Wambaugh [1] rightly points out that he searched in vain in all the commentaries upon international treaties for any instance of an interpretation based upon the decision of an international arbitration court. In his well-known essay [2] Descamps has already called attention to the great advantages of a permanent court of justice, which in each particular case would be subject to the influence of a responsibility both towards the past and towards the future.

Years ago Rössler,[3] in advocating a state court, expressed himself in words which are applicable to a court of nations: 'The beneficial result to be expected from a state court is the development of a fixed and uniform tradition, resting upon broad principles, for the settlement of cases of conflict in public law. How can such a tradition arise from temporary courts ?'

5. The Question of the Erection and Composition of the Permanent Court

At the Second Hague Peace Conference very heated discussions took place upon the question whether every state should have an equal voice in the appointment of the judges. The American draft of the Court of Arbitral Justice had in view a court consisting of fifteen members.[4] It was clear from the start that if the judges were to be appointed directly by the states, an equal right of appointment could not be conceded to every state. The next

[1] Op. cit. [2] p. 51.

[3] *Studien zur Fortentwicklung der preussischen Verfassung*, pt. 2, 1864, pp. 87–8.

[4] The original form of the American proposition is not printed in the acts of the Conference, and is to be found only in the work, *American Addresses at the Second Hague Peace Conference*, pp. 206 et seq.

proposal sought to form a system of electoral circuits in which individual countries were grouped together, and then as a group were to appoint a judge, while other states were to appoint each their own judge. Two plans of a similar character were presented. According to one, the American states as a unit were to choose four judges, and the other nations together thirteen judges. A second plan fixed electoral circuits solely according to population. Bustamante [1] in particular pointed out the impracticability of this system, because the neighbour states within a given circuit, who frequently, as in South America, have boundary disputes with one another, would never agree upon the choice of a judge. According to another proposal eight of the great powers and nine other states, which together represented the entire rest of the world, were each to be entrusted by the Conference with the selection of an arbitrator. These proposals were not given detailed consideration. The chief battle at the Conference was fought over the so-called 'rotation system'. According to that plan each state was to appoint a judge for a period of twelve years, but the individual judges were to sit for unequal periods. Only the great powers were to be represented upon the tribunal for the full twelve years; while the other states were to be represented for only ten years, four years, two years, and one year. A classification of states was also drawn up according to which population, territorial extent, political position, and wealth of the individual countries, as well as commercial and industrial interests, number of colonies, and varieties of language and of legal systems (Roman law, common law—in its English and in its modified American form,—Spanish law—with its European and American variations,—&c., were to be

[1] p. 185.

properly represented upon the tribunal) were taken into consideration. According to this classification the states represented upon the tribunal each by a judge and deputy judge were as follows:

Germany, United States, Austria-Hungary, France, Great Britain, Italy, Japan, and Russia, for	12 years.
Spain, Netherlands, Turkey, for . . .	10 years.
Argentine, Belgium, Brazil, Chile, China, Denmark, Greece, Mexico, Norway, Portugal, Roumania, Sweden, and Switzerland, for	4 years.
Bulgaria, Persia, Serbia, and Siam, for . .	2 years.
Bolivia, Columbia, Costa Rica, Cuba, San Domingo, Ecuador, Guatemala, Haiti, Honduras, Luxemburg, Montenegro, Nicaragua, Panama, Paraguay, Peru, Salvador, Uruguay, and Venezuela, for .	1 year.

Certain states, in particular Brazil, promptly declared that this plan was unacceptable. The Brazilian spokesman, Barbosa, on August 20, 1907, at the third session of the *Comité d'examen B*, brought forward another proposal. This proposal, like that of a draft laid before the Conference by Russia, aimed to perfect the Hague Arbitration Court of 1899. The Russian draft had provided that the members of the arbitration court were to assemble every year in full session. At this session the members of the arbitration court were to become acquainted with one another, and, above all, were to elect by ballot a commission of three or more members who were to hold themselves ready at all times to decide disputes. It was hoped that in this way a single and undivided court would be rendered possible in the future, instead of adding a new court to the old Arbitral Court of 1899. For the same reason it was later advocated that the members of the new court should only be chosen from among the members of the old one.

The Brazilian proposal provided that each state should appoint a member for the arbitral court. Brazil hoped that every state would not appoint a judge, or that a number of states would agree together upon the appointment of a single judge, so that in this way the number of judges would be as small as possible. Each judge was to remain in office for nine years. After the erection of the new court the old one was to be done away with. Each state was to pay the judge appointed by it. The members of the court appointed by the individual states were to be divided into three groups according to alphabetical order. Each group was to constitute a tribunal for three years, and its members were to reside near enough to The Hague to be able to arrive there within twenty-four hours. The parties were to be free to submit their dispute to certain persons selected from the list of the Hague Court or to the whole body of the Hague Court then in session. This proposal likewise found no favour. It was thought that the grouping in alphabetical order would leave too much room to chance, and the majority of the states could not reconcile themselves to the fact that they would be represented upon the tribunal for only three years out of nine.

The following additional projects were thereupon laid before the Conference, but they also found no favour :

1. The judges were to be chosen by the Conference, and were to exercise their functions temporarily until the next Conference ;

2. Seventeen states were to be selected by the Conference, which were each to appoint a judge ;

3. Each state was to have an equal right to propose a judge, and those judges were to be regarded as elected who received the most votes ;

4. Each state was to name a person, and the persons so named were together to appoint the judges. A particular form of this proposal provided that the judges of the old Permanent Court should appoint fifteen judges from among themselves.

In order to pass upon all these plans properly we must divide them into two classes, into those which have in view the equality of states, and those which do not have it in view. To the last class belongs in particular the so-called rotation system, which received the most serious consideration of the Conference. In endeavouring to answer the question whether it was right that the great powers should claim for themselves a stronger representation upon the Permanent Court we are led to the following result.

The task of an international judge should properly be to look after the interests of the community of states, not after the interests of the individual state which has appointed him. The individual judge must decide not as a German, a Frenchman or an Englishman, but as an international judge. The idea that the great powers must be more strongly represented upon an international tribunal than the smaller states is only intelligible under the belief that the judges must represent the interests of the states by whom they were appointed. But this belief is a mistaken one, and it follows therefore that the rotation system is not justified.

It may now be objected that international judges are not only to expound the law and to render decisions, but they must also create new law where there are gaps in international law, and are to that extent also law-makers. Granted, it is said, that an impartial settlement of the affair would be the result of a strictly legal decision, yet where the court is acting as an international legislature it must naturally give weight to the interests of each state, and arrange a compromise between the conflicting interests.

On the contrary, it must be pointed out first of all, and here I am in full accord with Schücking,[1] that the decisions of the highest international court have never the binding force of laws when they have contributed to the making of new law. International judgments have no greater binding force than, for example, the decisions of the German Imperial Court. If a decision of the highest court of the world appears incorrect to the other states, they are all free to make a new law by international treaty, and thus settle conclusively a matter hitherto resting only upon the authority of the court. Moreover, each individual state may, in particularly important cases, declare that it is not willing to recognize a special decision of the world court as binding upon itself (though this may not of course be done if the state is itself a party to the dispute). But this declaration will not in general be necessary. Article 84 of the Hague Convention for the pacific settlement of international disputes, already provides that 'the award is only binding on the parties to the proceedings'. Moreover, a decision rendered by the world court has only a moral force, which states can set aside in exceptional cases.

But even if international decisions had the force of laws it would not follow that the individual states should have an unequal share in the making of these laws by the supreme court. At the present day the only approved international legislative body, as Oppenheim,[2] Huber,[3] and Zorn [4] have in particular pointed out, is one in the form of an inter-

[1] Op. cit., p. 139.

[2] *Die Zukunft des Völkerrechts*, p. 33.

[3] *Die Gleichheit der Staaten*, p. 108 et seq.

[4] *Festgabe für Güterbock*, p. 207 et seq.

national conference at which each state has an equal vote. Moreover, if at the present day the complete equality of states in the international legislative body has been recognized we may indeed ask with surprise, as Matte [1] has done, why this principle has been departed from in the case of the supreme international court ?

Matte argues thus : 'If each state at the Conference votes as a unit for the adoption of each and all the resolutions of the Conference, why should there be a different representation in the judicial organization entrusted with putting those resolutions into execution ? '

If later on a legislative body should be introduced into international law, in which resolutions would be passed by a majority of votes, in such a case there might be a question of giving more votes to the larger states than to the smaller ones.[2] An international legislative body acts always by compromise, as our experience of the London Naval Conference well shows. The larger states have in this instance more interests to be represented than the smaller states. But in the case of the international court the chief concern consists in a fair adjustment of individual interests, not in a struggle of those interests in which might makes right. This has been emphasized, for example, by Moch in his work *Autour de la Conférence.*[3] Scott [4] also pointed out in 1907 that 'a court is neither a legislative nor a deliberative body ', but he failed to draw the necessary conclusions. The distinguished Swiss jurist, Huber, has given expression to this idea in the following striking words:

> In the case of a judicial body it must be assumed that the influence of power is entirely eliminated. The only

[1] p. 180.

[2] Quidde, *Zur Organisation der Interparlamentarischen Union*, 1911, p. 23 [3] p. 26 [4] p. 316.

interest which is really entitled to be decisive in the constitution of permanent courts concerns the excellence of the personnel of the court. If the lesser powers, by reason of their smaller share in the appointment of judges, cannot have complete confidence in the court, the objection must be met by directing attention to a suitable method of selecting the judges and to a comprehensive regulation of their qualifications, not by taking into account in the composition of the court an influence which, in the last instance, rests only upon political power entirely foreign to the idea of justice.[1]

The only influence which at first sight could be decisive in favour of allotting a larger number of judges to the great powers would perhaps be the fact that the great powers resort to the court more frequently.[2] Accordingly Scott[3] thinks that it is an argument for the system of rotation when he says that 'the interests of a great state with a large population extend very far, are in fact world-wide, and it is, above all, in connexion with these interests that disputes apparently arise'. But if the states have really the intention of creating an impartial tribunal, this very fact would argue with considerably greater force that the great powers should not be too strongly represented upon the tribunal. For since the great powers resort more than others to the court, if they possessed a particularly large number of judges, there would always be national judges upon the tribunal, and the principle that 'no man can be a judge in his own case' would be completely set aside. The population of individual states should, as Scott proposed, only be considered as a basis for representation of the states upon the tribunal if the greater representation

[1] Smith, *Proceedings of the American Society of International Law*, vol. ii, pp. 95 et seq., holds the opposite view.

[2] See p. 156 of my commentary upon the Convention for the pacific settlement of international disputes.

[3] p. 318.

upon the tribunal of the interests of the individual parties would secure a better decision. But, unquestionably, precisely the opposite is the case. This fact was strikingly recognized by the commission appointed by the Interparliamentary Union at its meeting at The Hague in 1894. It feared that if the great powers were to have a larger number of members of the arbitration court, in disputes between these same powers there would frequently not be enough judges.[1]

From whatever point of view the apportionment of judges according to the rotation system be regarded, the conclusion in every case is that the plan must be unconditionally rejected. It is impossible, as Huber also says,[2] to find any scheme of gradation for a system of rotation. The smaller and the medium-sized states were entirely right in having nothing to do with such a system.

Schücking,[3] who also rejects the system of rotation, thinks that in favour of an apportionment of judges according to population the fact can be adduced that the interests of a world power with 65,000,000 citizens in a new truly permanent international court of justice is much greater than in the case of a small state with 4,000,000 inhabitants. Granting that the point is well taken, it must be categorically denied that the great power which has a larger interest in the existence of the international court would also have a larger interest in being specially represented upon the tribunal. Quite the contrary! The greater the advantages which a great power may expect from a truly ideal court, the more will it urge an impartial composition of that court. But this will be entirely impossible if the great power is to have a larger share in

[1] Of the same opinion is Bustamante, p. 185.

[2] Op. cit., p. 107.

[3] Op. cit., chap. iv, sec. 5.

the appointment of members, and a consequently stronger representation upon the court.

Oppenheim [1] has recently advocated the unequal apportionment of judges, and asserts that the principle of the equality of states has after all nothing to do with the composition of an international court, so long as no state is forced against its will to resort to such a court. On the contrary it must be insisted that according to the principle of the equality of states an equal right must be secured to every state upon the tribunal created by all the powers together, and the circumstance of the voluntary or enforced resort to this tribunal has nothing to do with the matter. An enforced resort to a tribunal is entirely out of the question in international law. All states must, as Schücking [2] also has pointed out, be allowed to share equally in the work of perfecting the organization of states, which urgently calls for a permanent court of justice.

In addition to Oppenheim, the majority of recent writers advocate an unequal apportionment of judges; among them are Pohl,[3] Ozanam,[4] Lawrence,[5] Despagnet,[6] Higgins,[7] de Louter,[8] Schücking,[9] and Fried.[10]

Apart from Huber, Bustamante,[11] Nippold,[12] and Fischer [13] have vigorously rejected the idea of an unequal apportionment of judges.

It is quite significant that before the meeting of the Second Hague Peace Conference the principle of the

[1] *Die Zukunft des Völkerrechts*, p. 43. [2] Op. cit.
[3] *Deutsche Prisengerichtsbarkeit*, p. 144.
[4] *La juridiction internationale des prises maritimes*, p. 171.
[5] *The Principles of International Law*, p. 583; *International Problems and Hague Conferences*, p. 75. [6] pp. 1359 et seq.
[7] p. 517. [8] Vol. ii, p. 158. [9] Op. cit.
[10] *Zweite Haager Konferenz*, p. 117. [11] pp. 184 et seq.
[12] *Die zweite Haager Konferenz*, vol. ii, p. 255. [13] p. 11.

equality of states only rarely came into question in connexion with a proposed international law court. Rather, with the exception of Schlief,[1] all writers of consequence, in particular, Kamarovski,[2] considered that this principle must necessarily be followed. It was only to the Turks[3] that certain authors, such as Kamarovski, wished to deny a vote upon the court. The possibility of an unequal representation of states upon the court was likewise considered by the commission of the Interparliamentary Union at its meeting at The Hague; but it likewise came to the conclusion, as Stanhope[4] had already done, and later Descamps also,[5] that the principle of the equal representation of all states should be upheld. Schlief is led to propose a system of unequal representation because he sets up the false principle that 'the members of the court of justice are not identical with the ordinary incumbents of a judicial office, but are merely representatives of their states. . . . The members must naturally proceed and vote strictly according to the instructions of their governments'.[6] Schlief then compares the representation upon the court of justice with the representation of the German federal states in the Bundesrat, and thus confounds judicial and legislative bodies.

The determination of a system of unequal representation

[1] *Der Friede in Europa*, pp. 292 et seq. [2] p. 499.

[3] It is interesting to note how this idea of the exclusion of the Turks from the European family of nations is constantly recurring. Compare especially Schücking, *Die Organisation der Welt*, and the plans there discussed of Dubois, Podebrad, Sully, and Alberoni. The well-known Dutch international law jurist, van Daehne van Varick, has recently again raised this question in his article, *La révolution et la question d'Orient* (Hague, 1911), and advocates a crusade against the Turks and the German protectorate over Syria.

[4] *Conférence interparlementaire*, p. 228. [5] *Denkschrift*, p. 72.

[6] pp. 298, 300.

of states upon the court would have, moreover, the further great disadvantage that the whole apportionment would be broken up if only one of the forty-five states failed to ratify the convention. Barbosa [1] pointed out this objection in 1907.

At the meeting of the American Society of International Law [2] all of the speakers except Aymar declared the equality of states to be a fiction, and laid down as postulate the principle of the inequality of states. Even if this assertion were true as a general proposition, the question must be raised whether this principle is also applicable to the composition of an international court, or whether in this particular case an exception should not be made from the principle of inequality (which is now even more than formerly an incorrect principle of international law).

After all, I can find no reason why states should be represented upon the court of justice. We need international judges who, in fulfilling the functions of their office, will consider themselves as little as possible the nationals of a particular country. If it is in fact not possible to de-nationalize the judges, as Scott [3] has rightly pointed out, yet we must do all in our power to prevent the judges from regarding themselves as representatives of their countries. But this will always be more or less the case if the individual states appoint the judges. Accordingly some other method of selecting the judges must be advocated.

In making this demand we must first of all point out that the Supreme Court of the United States of North America likewise consists of judges who are not appointed by the individual states, but by the President with the

[1] p. 618. [2] 1909, pp. 238 et seq.
[3] p. 319; *Revue générale*, 1911, p. 225.

approval of the Senate.[1] It must also be remembered that between the individual states of the United States, and indeed between the individual states of a federal state, there do not exist such important differences as between many states of the world. The Supreme Court never has to decide questions which might almost lead to war, as must the highest court of the world, even if its chief duty will be to pass upon questions more of a legal nature. Accordingly, it is all the more imperative that the composition of the world court be an impartial one. If in the case of the highest court of the United States, the appointment of judges is not in the hands of the individual states, how much more must the same principle be followed in the case of the new world court!

Many judges will be influenced by sympathies and by prejudices, not only in the case of disputes which concern their own state, but also in disputes between foreign powers. There are but very few men who are so free from passion as to be worthy of a position upon the world court of justice. For that very reason every possible guarantee must be given that only the best and the noblest men will be international judges.

All the more caution in the selection of judges will be needed because the positions upon the international court of justice are not very numerous, and each state will have at most only one judge to appoint. Can the individual states be trusted to be so unselfish as to waive in the selection of judges every test of national loyalty, and on the contrary select persons who will look only to the welfare of the community, and not merely to that of their own state?

A further reason why the judges should not be appointed

[1] See in this connexion, e.g., Montague, *Proceedings*, pp. 214 et seq.

by the states directly is this. I have in mind the case that a judge like Lammasch, who enjoys throughout the world a singular esteem as an international judge, might not be appointed as a member of the court by the ministry then in power for reasons of one kind or another, whether personal or political; or again the case that a state might not possess at the time a citizen who is specially qualified for the office of an international judge, whereas another state might have two unusually able international jurists. Look, for example, at the case of Holland, which has at its disposal a particularly large number of international jurists—men such as Asser, Loeff, and de Savornin Lohman! Now the community of states would be able to make no use of the distinguished services of these men, if each state could only appoint one judge, or if indeed the absurd system of rotation brought it about that the nationals of states such as Holland and Belgium had only temporary positions upon the court. It is no answer to these facts to point out that such distinguished men might perhaps be appointed by other states in case their own state was not in a position to appoint them. For if each state may appoint only one judge upon the tribunal, it will without any doubt designate one of its own citizens. I have before me the list of the present judges of the Hague Permanent Court, and I find that not a single state has appointed a national of a foreign power, and this in spite of the fact that each government may appoint not one but four judges.

Hence some method of selection must be introduced by which only the most distinguished men throughout the world will be appointed to positions upon the world court. without respect to the state of which they are citizens. For that reason I regard as unsuitable all the proposals of the Second Hague Peace Conference, according to which the states themselves were to appoint the judges; rather, the appointment must be made by an approved international board. This board is to be composed not of diplomats, but only of international jurists of high standing. If the list of the present Hague Permanent Court were to be so formed that no diplomats, but only international jurists, should be represented upon it (and not many changes would be necessary for this, since there are only very few diplomats represented in the list, Germany and Austria-Hungary having appointed only jurists), we would have a body of persons by whom the new court could be constituted, and there would be created at the same time a connexion between the Permanent Court of 1899 and the new court of justice. It was in this way that Bourgeois also wished to see the question settled at the Second Peace Conference.

Oppenheim [1] also makes the following very apt remark: 'It might be possible to constitute an international court which sets aside entirely the principle that definite states must be represented upon it, and perhaps such a court may be possible in the future, when more confidence is placed in the judicial settlement of international disputes.' Mougins de Roquefort [2] has already pointed out that there would be more guarantees for the impartiality of a tribunal if the states were not themselves to appoint the judges.

It appears that in America also it has been gradually perceived of late that the judges are only representatives of the international community, and accordingly should be appointed only by an organ of the body of states, and not by the individual governments. Th. R. White, in a brilliant address before the Lake Mohonk Conference of 1911,[3]

[1] *Die Zukunft des Völkerrechts*, p. 44.
[2] p. 216. [3] *Report*, pp. 104 et seq.

expressed this idea, with great earnestness. White does not want the judges of the new court to be appointed by the states themselves, and proposes the creation of a board, not too large, which will act as the agent of all the states for the purpose of selecting the body of judges.

In this connexion I think that Huber [1] is not justified when he speaks as follows of this and similar plans:

> These proposals would, to be sure, have satisfied the equality of states, but, on the other hand, would have done an injury to the unity of the court and to the uniformity of its composition. But, above all, such a method of composing the court offered to the more influential states no security for a position upon the court corresponding to their political significance.

First of all as regards the last statement of Huber, I cannot understand, as I have insisted above, why the larger states should have a larger representation upon the tribunal. The mistaken idea that the individual judges are representatives of their states and not of the international community is here in evidence.

If the judges are representatives of the community of states, it will be a matter of entire indifference to the individual state whether it has one national more or less upon the court. For this national must, in conformity with the oath which he takes as judge, regard himself only as a representative of the community of states.

I also consider Huber's criticism with respect to the uniform composition of the court unfounded. The representation of all the particular legal systems upon the tribunal is not at all necessary in the case of questions of strict international law, but only in the case of questions belonging to international private law, which must come

[1] *Jahrbuch des öffentlichen Rechts*, vol. ii, p. 511.

before another court, that is, before another jurisdiction. In the case of questions of international law it is only important that there be upon the court persons who know the legal systems which are in force in the countries of the disputants. For this reason and also because of difficulties arising from the languages used, it will often be necessary to enlarge the court *ad hoc*. Citizens of the various states will naturally be appointed as far as possible, and as it is hoped that in time a number of judicial bodies will be established, it will be quite possible that at the same time citizens of almost all the states, in so far as they can offer good international judges, will sit upon the court. A number of nationals of the same state will not be appointed without necessity, but only when such a state has at its disposal two jurists of such distinction that they cannot be denied a place upon the supreme court of the world.

6. Other Features of the Organization and Procedure of the Court of Justice

If it were decided to confer upon the judges inscribed on the list of the present Hague Court the right to appoint the members of the new court, this would secure at the same time that the judges would not be subject to removal, as has been rightly advocated by Bustamante,[1] Scott,[2] Mérey,[3] and Bourgeois.[4] For as the states would not themselves appoint the judges they would also have no right to remove them, as they have in the case of the present Hague Permanent Court. It is true that even in the case of the Permanent Court, as Schücking[5] remarks, the removal of judges cannot be justified legally. But it is

[1] *La Seconde Conférence*, p. 205. [2] p. 317. [3] pp. 615, 667.
[4] p. 615. [5] Op. cit., p. 52.

clear from the report of the First Hague Conference and from the proceedings of the Second Conference, that it was not desired to deprive states of this right in fact.

As regards the right to remove judges, the proposal of Kriege [1] that it should be turned over to a higher court could not be carried out. But there would be hardly any hesitation in entrusting this right to the Administrative Council of the present Hague Permanent Court. If a right of removal were admitted the judges would be at least removable in case of unworthiness, &c., as Lammasch and Baron Marschall [2] in particular advocated in 1907. To make the moral responsibility of the parties the sole guarantee that the judges will not be removed, as Renault [1] proposed, does not seem to me wise.

No doubt this right of removal will have the result that states need not leave their cases to be decided by unworthy persons. But further security yet must be given that the judges will enjoy the confidence of the parties in the highest degree. Fixed conditions must be laid down under which a judge may not exercise the functions of his office. I have in mind in particular a case which the International Law Association also contemplated in its project of 1895 framed to protect states from unsuitable judges, namely, the case in which the judges might have a personal interest in the result of the dispute. But when this project also held as prejudiced judges who had already in their writings and addresses, or as members of courts, expressed their opinion upon the same question of law, it went too far. Under such a rule it would be quite impossible for a judge to render his decision a number of times upon the same legal question.

But the exclusion of prejudiced judges would not give states a complete guarantee; in addition, each state which

[1] p. 667. [2] p. 615.

resorts to the court must have a right of challenge which will enable it to reject at the start two judges out of five, three judges out of seven, &c. The places of the judges thus struck out must be filled at first from the general body of the reserve judges of the international court. The challenging of a given judge need not at all imply a lack of confidence in him, and no reason must be given for so doing. I am convinced that the states will make use of this right of challenge only in special cases. In 1907 Lammasch [1] in particular proposed that a qualified right of challenge be given to the disputants. He observed very justly that 'we live to-day under a system of political alliances and it is possible that more than one of the judges of the committee will be appointed by the ally of one of the parties. . . . The right of challenge would have the further effect of removing all fear of the concentration of the powers of the committee in the hands of a very limited number of members.' Crowe was wrong in fearing that under this method the best judges would be set aside.

The right of removal and of challenge would completely prevent the judges from obtaining too great power over the states, as Barbosa [2] in particular, as well as Mérignhac,[3] feared when he declared: 'Above all think of the position of the judges of this court, which is possessed of a power unequalled in the world!'

Finally, as regards the regulation of the procedure of the new court, it follows from what has been said that the judges of the new court must render a detailed opinion in connexion with their decision. Further, a court of appeal must be set up, which is not a difficult matter in the case of a permanent court, as Descamps [4] has already shown.

[1] p. 652.
[2] p. 344.
[3] *L'Arbitrage*, p. 428.
[4] *Denkschrift*, p. 55.

The states must be secured against errors of law, as Kamarovski and Mérignhac have already clearly pointed out. Moreover, an action in annulment must be allowed in certain cases. Thus, Baron Marschall referred in 1907 to the case in which the judges might have used more or less offensive expressions against one of the parties in their decision. As a precaution against this he advocated the presence of national judges. But in my opinion the court will overstep its jurisdiction in this way so rarely that in practice an action in annulment will answer the purpose.

Special attention would further be given to the question of languages, upon which I have expressed myself on page 111 of my commentary. Here also some form of a court of changing composition appears indispensable. The entire body of the international court of justice would first have to determine which group of judges was to have jurisdiction in case a special knowledge of languages should be required in a given instance. The friends of Esperanto have repeatedly endeavoured to show that all the difficulties of language could be avoided through the use of Esperanto. Even Schlief [1] has opposed this. Until Esperanto has come into much wider use this idea cannot be given further consideration. But I will gladly repeat here the view of the treatise *Germana Esperantisto*,[2] quoted in connexion with statements made on pp. 31, 32 of my article, ‘An International Court of Justice for Private Claims’:

> These difficulties could easily be removed by the use of Esperanto. First of all there is already a considerable number of persons who know Esperanto, and who could act as arbitrators upon such a court. But as the language is exceedingly easy to learn, as soon as such a court could be erected there would soon be found the necessary number of judges, attorneys, and secretaries familiar

[1] *Der Friede in Europa*, p. 313.

[2] Berlin, 1911, p. 234.

> with Esperanto. Hence there would only remain the task of translating the necessary documents into Esperanto. All those who have looked into this question carefully know that no other language is so well adapted to reproduce the finest shades of thought of the original as is Esperanto.

That may be true. But can a translation ever supply for the original, when it is a question often of the meaning of each individual word ?

Finally, in connexion with the new court there would be established a permanent body of attorneys who would naturally be much more capable than the present agents and counsel, who vary with each case. Not only would there be the advantage that judges and attorneys would come to know one another better, but the attorneys would thenceforth hold their position for life, and accordingly would become more familiar with the methods of procedure and with the principles of international law than has hitherto been the case. Their great and valuable services, which, as Wambaugh[1] says, have not yet been properly estimated, would in a greater measure than hitherto influence the development of procedure in a manner which would contribute to the development of international law. The judges and attorneys of the supreme court would then be real practitioners. The entirely unsatisfactory condition of things, under which the most important and difficult questions of law in the world are decided by jurists who are at the same time judges and attorneys upon the Hague Court of Arbitration, must cease. The office of an international judge and of an international attorney is indeed sufficiently honourable that international jurists should devote their whole life to it.

[1] *Proceedings*, p. 142.

It is also evident that states need not turn over to the supreme court of the world all or even the most important disputes; I am convinced that states certainly need not submit questions of honour and vital interests to the supreme court in accordance with an obligatory world arbitration treaty. States must be unquestionably allowed to interpose the objection that a dispute affects their honour or their vital interests, as a final bar to an action against them.

If this guarantee be given I see no reason why we should not set aside the *compromis,* which involves a waste of time and is often difficult to draw up. On the contrary, it must be recognized that each state possesses the right upon its own motion to institute an action for the settlement of strictly legal disputes with another state in a strictly legal way. It must be understood, of course, that here also diplomatic negotiations have failed to bring about a settlement. The fact that as prudent and learned a man as Privy Councillor Zorn, in spite of his rejection of obligatory arbitral agreements, is not opposed to a unilateral right of action under these circumstances is evident from the fact that in his inaugural address as rector he supported the memorial of the leaders of the Berlin Chamber of Commerce in favour of an international court for private claims. For according to this project private persons will be given a direct right of action against debtor states. In the project of an International Prize Court, and in the Central American Court of Justice which has been long in existence, we have significant models for the creation of such a direct right of action, as has also been advocated by Dumas. Moreover, Article 53, paragraph 2, No. 2 of the Convention for the pacific settlement of international disputes has already provided that upon motion of only one of the parties the Permanent Court is competent to settle the *compromis* under certain conditions.[1] Just as this form of compulsory *compromis* is only possible by the agreement of both parties, so a unilateral right of action even in the individual concrete case would have to be in accordance with the wishes of both parties.

In case such a unilateral right of action should be granted, guarantees would have to be given that judgments by default would be excluded, since they clearly have no place in disputes between states. It must be accepted as a self-evident proposition in this connexion that no state will refuse to answer to a process without cause. It has never thus far happened that a state has evaded a decision of an international arbitration court. Accordingly we may assume that there would be an equal willingness to fulfil treaty obligations with respect to an action before the court of justice. But if a state should refuse to answer to an action, international law does not provide a formal remedy, and we must be content with the fact that such a state draws down upon itself by its manner of acting other evils of various kinds, the chief of which is that it destroys the confidence of other governments in its good faith. Why, therefore, spoil in the minds of states the whole idea of an international judiciary by creating an executory jurisdiction or by admitting the possibility of a judgment by default!

But, it may be said, are not states legally obligated to answer to an action, and in consequence must not those states recognize judgments by default which are willing to allow the arbitrators under certain circumstances the right to fix a period for the fulfilment of the arbitral award? This question must be answered in the negative, because in the case of a judgment by default there exists the

[1] See pp. 99 et seq. of my commentary.

express or implied declaration of one party that it will not answer to the action, whereas such is not the case when a period is fixed for the execution of the arbitral award. Those forms only of international procedure are inadmissible which are contrary to the express or implied consent of the parties.

But if an agreement should not be reached with respect to a unilateral right of action, in that case the new court of justice must be declared competent to settle the *compromis* at the request of only one party. But the *compromis* should not be thus formulated without previous oral negotiations between the parties, in order that through the appearance of both parties evidence may be given of their express agreement that the terms of the *compromis* shall be settled unilaterally. Should one of the parties not be willing to enter into these negotiations a *compromis* should not be drawn up. It is not well, as is provided in Article 19, paragraph 2, no. 1 of the Draft Convention relative to the creation of a Judicial Arbitration Court, to have the compulsory *compromis* drawn up unless both parties co-operate in it.[1]

However, I must not enter into further details with respect to the establishment of the permanent court of justice. For it is not for me to add a new project to the large number of those already on hand. I shall merely endeavour to clear up the more important aspects of the question. I might lay stress, however, upon the fact that I am in full accord with Schücking when he advocates the greatest possible organic connexion between the old Permanent Court of 1899 and the new court of justice.

It is not at all necessary to delay the establishment of a permanent international court of justice until an obliga-

[1] See p. 103 of my commentary.

tory arbitration treaty between all nations has been entered into. For even without such a treaty the court of justice will have enough to do if the states are willing to refer many cases to it for decision. But it is evident that an obligatory arbitration treaty between all nations would be of advantage to the permanent court of justice, and Bourgeois, d'Oliveira, de Soveral,[1] and Milovanovitch[2] are to that extent right in thinking that the question of the adjective and that of the substantive law are closely connected. Descamps in his memorandum[3] and Schücking[4] have also laid stress upon this. An arbitration treaty between all nations would bring before the new court particularly those disputes which are not overmuch political in character. For as Scott[5] says in his masterly report, disputes which are rather legal in character must be referred to the new court of justice, and, on the other hand, disputes of a political character must be referred to the old Court of Arbitration of 1899. To be sure, an objective distinction between legal and political questions is hardly possible; in the last instance the parties themselves must always decide whether they regard a dispute as legal rather than political in character. But it would be entirely satisfactory to withdraw from the competence of the court of justice those political questions which at the same time affect the vital interests of a nation. If, therefore, this clause is introduced into the arbitration treaty between all nations, such as is advocated above, all disputes can be turned over to the new court of justice. The old Court of Arbitration would then only be resorted to in accordance with special arbitration treaties.

That numerous disputes would be turned over to an

[1] p. 346. [2] p. 335. [3] p. 50.
[4] Op. cit., p. 152. [5] Vol. i, p. 348.

international court of justice has been asserted in particular by Baron Marschall at the Second Hague Conference in the following words: [1]

> A truly permanent arbitration court will automatically attract disputes to it. There are enough on hand in the archives of governments.[2]

The contrary view of Mérignhac [3] and of von Bar [4] is untenable at the present day. Fried [5] is not wrong when he says 'it is possible that while we must now look for judges in order to settle a dispute, we shall henceforth look for disputes in order to keep a body of judges busy'.

Moreover, even after an arbitration treaty between all nations has been entered into, the court of justice will still not have jurisdiction if both parties should be agreed to turn over the question to a special arbitration court.

A number of persons, e. g., Barbosa [6] and Zorn,[7] have asserted that there is no need at all for two international permanent arbitration courts. But they do not reflect that entirely different tasks will be assigned to the old Arbitration Court and to the new court of justice. Numerous disputes would without doubt be turned over to the new court of justice on account of the lesser expense, thus justifying the existence of the International Bureau, so that continued existence of the old Permanent Court, even if it should be no longer put to use by the nations, would not involve the slightest cost or trouble. If the old Court of Arbitration should no longer be put to use by the powers after the erection of the new court, there would certainly

[1] p. 595.
[2] See also Marburg, *Proceedings*, p. x; Scott, p. 317; and Choate, p. 594.
[3] *L'Arbitrage*, p. 428.
[4] *Berliner Tageblatt*, August 22, 1907.
[5] *Pester Lloyd*, 1907, vol. 8, p. viii.
[6] p. 620.
[7] *Zeitschrift für Politik*, vol. ii, p. 350.

be no need to continue it in existence. The question is, then, which court is necessary. In my opinion, the new court is in any case indispensable, and accordingly those who want only one court must forego the old Court of Arbitration.[1]

In the case of the smaller international differences it would be possible henceforth to have them settled in a strictly legal way. If a state desired to do that at the present time, and had special judges brought to The Hague for each individual case, international jurists would soon hesitate to abandon temporarily their profession on account of a trifling object, and there would soon be a lack of suitable judges. For this reason, as well as on account of the relatively great expense involved, only the most important questions have thus far been submitted to the Hague Court of Arbitration, as has already been pointed out above. Those very disputes which are in themselves insignificant often contain questions the strictly legal settlement of which would be of benefit to international law.

A permanent court of justice would by its decisions not only contribute directly to the development of international law, but by its very existence would prepare the way for a gradual codification of international law, as Marburg[2] aptly observes.

Even though the new court of justice should after all not decide any disputes which endanger the vital interests of one of the parties, that would not prevent it from settling, at the mutual request of the parties, serious dissensions endangering the peace; and it would thereby directly further the cause of world peace, instead of merely doing

[1] See the excellent remarks of Schücking, op. cit., pp. 91, 92, against Zorn's contrary opinion.

[2] *Proceedings*, p. xv.

so by hastening the organization of the community of states through the development of international law.

Finally, the great moral influence which the very existence of a permanent international court of justice exercises over nations is not to be under-estimated. Descamps [1] has already pointed this out in forcible terms. The international court of justice embodies the idea that all of the disputes which arise between nations and which are susceptible of a strictly legal settlement can best be decided by a supreme court, in case there are difficulties in the way of diplomatic settlement. The fact that in international life law is being more and more substituted for force is a much more potent sign than the present arbitration court. This is also the view of Hull [2] and of Schücking.[3]

The above statements are made upon the hypothesis that the new court will be in reality a 'judicial tribunal', and not a 'political or diplomatic Areopagus'.[4] 'Its prime function is to interpret the law and to apply it to a concrete case. . . . What we desire are fixed juridical rules and an equally certain interpretation of those rules.' Only in that case will the international court of justice serve both to secure international peace and to develop international law, and thus justify its existence side by side with the Hague Permanent Court of 1899.

[1] *Denkschrift*, p. 51.

[2] *Proceedings of the Second National Peace Congress*, 1909, p. 219.

[3] Op. cit., p. 90.

[4] Scott, p. 316.

CHAPTER V

THE TECHNICAL ADVANTAGES OF A PERMANENT COURT OF JUSTICE. GREATER PROMPTNESS AND LESS EXPENSE IN THE SETTLEMENT OF CASES

If, then, we have seen that a permanent international court can alone guarantee the development of international law, a permanent court is also for technical reasons to be preferred to a tribunal organized for each case, in so far as the settlement of strictly legal disputes is concerned.

First of all it must be observed that a court of justice, whose members know one another well and are bound together by reason of their long work in common, is of more value than a court constituted for each case. Only when the individual members are intimate with one another is real co-operation possible. It may, for example, happen that a judge, who is in other respects distinguished and possessed of a shrewd insight for important questions, is a poor speaker and cannot give proper expression to his thoughts. If a man of this kind joins in with a body of judges who do not understand his individuality, they will as a rule imagine that he is not a particularly competent jurist, and will regard his opinions sceptically. This will make co-operation more difficult. The deliberations of the body of judges will be of quite a different character if each one knows that the opinions of this judge are always very clear-cut and deserving of consideration. If the personality of an individual judge is well known his opinions will be more highly valued by his fellow judges. If the judges are well known to one another they will more readily exchange in a confidential way their private views upon the particular object of dispute. They will discuss questions with one another not only as jurists, but at the same time as men who have in view the supreme interests of mankind. All this will greatly help to bring about a good decision. I recall merely the well-known remarkable co-operation of the members of the Committee of Examination of the First Hague Peace Conference as an ever laudable example of noble work in concert. When such relations exist between them, the high ideals which certainly animate many judges will be communicated to their colleagues. This is a matter not to be under-estimated. Scott puts the case well in his report: [1]

> The intercourse of the judges in view of the analysis and of the development of international law, their co-operation in judicial decisions, will necessarily develop an *esprit de corps*, which will have an inevitable influence upon each of them in the accomplishment of his duties. In the contact of judicial responsibilities, individual opinions, and even individual prejudices, will lose something of their rigidity, and the decisions of the court will furnish the greatest guarantee of international impartiality.

Moreover, a court whose members are on intimate terms with one another will certainly be able to settle a case more quickly; and assuredly the strictly permanent nature of the court of justice will greatly contribute to this end. Esteva,[2] Bourgeois,[3] Marschall,[4] and Asser [5] in particular laid stress upon this point in 1907. The appointment of judges to the special arbitration courts has indeed repeatedly taken up too much time. For that reason Descamps [6] has already advocated a permanent tribunal.

[1] Vol. i, p. 362. [2] p. 335. [3] p. 347.
[4] p. 594. [5] p. 595. [6] *Denkschrift*, p. 46.

The judges appointed in the Muscat case were after a month unable to agree upon the selection of the umpire, and recourse had to be had to the provision of the *compromis*, according to which the King of Italy was given the right to appoint the third judge. The decision of the King of Italy likewise consumed much time. At the meeting of the Venezuelan arbitration court on September 1, 1903, only a single judge was present, because the two others could not yet be appointed. Even with the help of modern means of communication it is impossible to constitute the tribunal more quickly in such cases, as Zorn [1] mistakenly thinks. Lammasch, the most distinguished international arbitrator, clearly says: 'Without doubt the meeting of the arbitration court would have been helped, hastened and facilitated if, having been constituted in advance, it had not been necessary to constitute it at the time.' [2]

It is true that the time between the constitution of the tribunal and its meeting would only be slightly shortened by means of a permanent court of justice, because each case would require longer preparation.

The experience in strictly technical procedure which the judges upon a permanent court would have acquired by their profession, would contribute greatly to the more prompt settlement of a case.

Similarly, I insist, in opposition to Brun,[3] that the question of expense argues forcibly in favour of a permanent international court. In my commentary [4] I showed, in agreement with Mérignhac, that there is no justification for the assertion that a truly permanent court of arbitration is less expensive. But in making this statement I had in

[1] *Das Deutsche Reich und die internationale Schiedsgerichtsbarkeit*, p. 44.
[2] *Recht*, 1911, p. 146; Schücking is of the same opinion, op. cit., p. 91.
[3] p. 148.
[4] p. 157.

mind merely international arbitration, which has for its object simply to dispose of disputes, and not international judicial settlement, which at the same time contributes to the development of international law. So long as the attitude is taken that international procedure exists merely for the adjustment of disputes, only the more important disputes, which either directly endanger the peace or can in time do so, will be considered in that connexion. Smaller disputes, particularly such as are of a strictly legal nature without a political background, can practically not be considered in connexion with procedural settlement. Accordingly each state will, in the case of arbitration, be ready to pay a much bigger sum, if by the establishment of an arbitration court it can avoid a war. For such cases a special arbitration court is not too expensive. Lammasch justly observes that a few shots from a modern battery cost quite as much. As a matter of fact the states are at liberty to be represented before the tribunal by state officials; the expense of the Bureau is borne by the nations as a body, and there remains only the expense of the judges. All this is worth while in the case of more or less significant disputes, but not in the case of smaller differences which arise quite frequently between states. In this case an arbitration court constituted for each case is too costly, and states must forego on that account the development of international law which would result from a legal settlement of these disputes. Each arbitration court costs a few thousand marks, and this amount of money is often out of all proportion to the importance of the object of the dispute, however great benefit the community of states might obtain from the legal settlement of these differences. Asser's just remarks, in 1907, with regard to the Permanent Court of Arbitration, are well known: 'It is difficult, time-consuming, and costly to set it in motion.'[1] We need, therefore, a permanent court before which so many disputes will be brought for decision that the cost of each individual case will be trifling.

Moreover, such a division of the expense among the states as a body is more in accord with justice. Each state profits from the award of an international court, even though it is not itself a party to the case. It knows for the future how it must regulate its conduct legally. The international court of justice will furnish the governments of the individual states with valuable principles of conduct and contribute to the avoidance of war, both matters in which those states also which are not parties to the dispute are greatly concerned. Accordingly, the community of states must also pay the cost.

It must not be thought that individual states would suffer through this division of the total expenses, in that certain governments might perhaps make more frequent use of the court of justice at the expense of the other states, so that the total expense would grow and the share falling to each state would be greater. Such a frequent resort to the court of justice would only have the effect of inducing the other states to refer still more disputes to the court on their own part. Moreover, the fact that after all the states as a body benefit by a decision makes it a matter of indifference whether some states resort to the court more than others. Besides, it is evident that the smaller states will have a much smaller share of the cost to pay, as is the case already in the system of apportionment adopted by the Universal Postal Union.

If then the cost of the settlement of one of the numerous

[1] Of the same view were Choate (p. 311), Martens (p. 322), Marschall (p. 594), as well as Schücking (op. cit.).

legal disputes between states under the new procedure will be considerably less than formerly—indeed the trifling sum makes possible for the first time a legal decision—the total contribution of the individual states will indeed not be high in comparison with the valuable services of the new court of justice for the development of international law.

The expenses of the Permanent Bureau of the Hague Arbitration Court of 1899 amounted, for the whole body of states,

In the year	1900,	to 42,499	gulden
,,	1901,	,, 30,438	,,
,,	1902,	,, 28,768	,,
,,	1903,	,, 25,917	,,
,,	1904,	,, 24,514	,,
,,	1905,	,, 28,201	,,
,,	1906,	,, 26,011	,,
,,	1907,	,, 27,801	,,
,,	1908,	,, 28,651	,,
,,	1909,	,, 28,832	,,

The individual states had approximately the following sums to pay for the International Bureau in 1910:

Germany, Austria-Hungary, China, United States, France, England, Italy, Japan, Russia, Turkey (each 25/375) . .	1,514	gulden
Spain (20/375)	1,211	,,
Belgium, Roumania, Holland, Sweden, Brazil, Switzerland (each 15/375) .	908	,,
Denmark, Norway, Portugal (each 10/375)	605	,,
Bulgaria, Greece, Mexico, Serbia, Argentine, Chile, Columbia, Peru (each 5/375)	302	,,
Luxemburg, Siam, Bolivia, Cuba, San Domingo, Ecuador, Guatemala, Haiti, Nicaragua, Panama, Paraguay, Persia, Uruguay, Venezuela, San Salvador (each 3/375)	181	,,
Montenegro (1/375)	60	,,

Let us estimate that the new court will have fifteen judges with a yearly salary of 6,000 gulden. If we assume that the judges will be kept busy for ten months continuously, so that each judge, in addition to the yearly salary of 6,000 gulden, will receive a compensation[1] of 100 gulden for each 300 days, the cost of the permanent court will amount to a total of 15 $(6{,}000 + 100 \times 300)$ gulden, or 540,000 gulden. Accordingly the share falling to each individual state would be after this method about twenty times as great as the present contribution to the expenses of the International Bureau. For example, Germany would pay 50,000 marks instead of 2,500, and Montenegro 2,000 marks instead of 100. Are these not trifling sums in comparison with those which are expended at the present day upon armaments ? The present contribution of many states to the International Bureau is in fact smaller than the allowance granted by them to the Interparliamentary Union or to the Berne Peace Bureau. Have the great powers not 50,000 marks, and the medium-sized and smaller states, say, 30,000 marks to spare for an international court of justice ? There can hardly be any doubt that they have. With such a court the states would above all have an excellent opportunity of having even the smallest differences settled promptly and of thereby developing international law at the same time. If the international decisions of the highest court have cleared up important questions, will not the expense of international conferences be saved, since they will need to meet for a much shorter time ?

Richet[2] estimates the cost of a permanent court of

[1] This daily compensation is provided for in the project of the Judicial Arbitration Court. In truth, it would be better to do away with it altogether and settle upon a fixed sum of, say, 20,000 gulden.

[2] p. 250.

justice at about six to seven million francs a year. This sum is too high, since Richet assumes that as many as one hundred judges, with a salary of 50,000 francs each, will officiate. There is no reason, of course, why the pay of the judges should not later on reach such an amount. But that time is still far off.

Moreover, the cost of counsel would fall considerably in the case of a truly permanent court. For the attorneys in permanent residence at The Hague would certainly not demand such huge honoraria for each case as most agents and counsel do at the present time.

At the first American National Peace Congress of 1907 [1] Professor Moore, the celebrated author of the six-volume work upon international arbitration, stated that for a country which refers many cases to arbitration a permanent tribunal was cheaper :

> Since we began our national existence we alone have had with other powers more than sixty arbitrations ; and I found that the total number of years during which these tribunals had sat was a hundred and twenty-five—more than the entire duration of our national existence since the formation of the Constitution. The excess of aggregate time is explained by the fact that now and then there were two or three tribunals in session at once. It is also to be observed that the total expense of all our tribunals . . . doubtless was greater, far greater, than would have been the cost of an actual court always in session.

If, therefore, in the case of international arbitration it is true that the cost of a permanent court is smaller in case the number of disputes comes up to a certain number, how much more must this be the case with an international judicial tribunal !

[1] *Proceedings of the National Arbitration and Peace Congress*, p. 417.

CHAPTER VI

THE PERMANENT INTERNATIONAL COURT OF JUSTICE AND INTERNATIONAL LAW

DOES an international court of justice contradict, as Barbosa [1] asserted in 1907, the 'essential principles of international law'? The assertion is often made that international arbitration is in harmony with the nature of international law, whereas judicial settlement runs counter to its fundamental principles. There are two particular errors in this connexion which make an understanding of the questions here under discussion unnecessarily difficult, namely, the common assertion that international arbitration is voluntary in character, whereas international judicial settlement presupposes force, and further, that, in contrast with a court of arbitration, a court of justice is a 'supranational court'. A third objection against judicial settlement is the unfounded assertion that an international court of justice constitutes merely the first step towards a federation of states, or that it is inconceivable without some such close organization between them.

1. INTERNATIONAL JUDICIAL SETTLEMENT AND THE ENFORCEMENT OF DECISIONS

The idea that an international court of justice naturally presupposes force, an executive power which when necessary ensures respect for the decision of the court, is quite intelligible. One has always in mind the organization of

[1] p. 703.

municipal courts which, to be sure, could not act without an attendant power of enforcing their decisions. But states as subjects of international law are very different in character from persons as subjects of national law. It cannot be too clearly pointed out that in international law we are dealing with some forty-five legal persons, whereas in municipal law, for example in the German Empire, we are dealing with more than sixty million legal persons. We cannot at the present time or in the near future succeed in persuading sixty million men, whose characters are so widely varied, that they must act only in a legal way. We cannot do so because the sixty million persons are not so dependent upon one another that an injury which *A* does to *B* will to-morrow react upon *A*. But the case is quite different with states, which recognize more and more that their community life must proceed according to legal principles, and that each state at the same time suffers vital injury when it wrongs any other state. To hold that the future of states will not be generally regulated by law, even without the existence of a compulsory power, is to despair of the great work of mankind and to deny the public facts which are evidence of a continually stronger development in this direction. Thus far all of the numerous arbitral awards have been voluntarily carried out.[1]

Accordingly, there is no need for an executive power in international law, so that the ideal of an international court of justice does not require the accompaniment of force. To those who cannot free themselves from national views, and who understand by a court of justice in international law only a body of officers possessing the power of compulsion, there is under the present conditions of

[1] See, in addition, my commentary upon the Convention for the pacific settlement of international disputes, pp. 52 et seq.

international life really no such thing as international judicial settlement. But the representatives of this theory have not yet proved that judicial settlement between states is identical with judicial settlement within the state, so that a power of enforcement must be demanded as an abstract condition of judicial settlement between states. In the discussion of the concepts ' arbitration ' and ' judicial settlement ' in international life it is entirely wrong to take any other point of view than that of existing international law and, without taking into account its peculiar character, to introduce into international law the idea of judicial settlement within the state. That was done some years ago, particularly by Bluntschli, Lorimer, Fiore, Dudley Field and Mérignhac, and recently in particular by Pohl. At the Second Hague Peace Conference it was likewise asserted by a number of speakers that an international court of justice without the power of enforcing its decrees was inconceivable. Barbosa in particular confused the question when he said that the necessary consequence of the erection of the ' court of arbitral justice ' would be the creation of an executive power.[1] The same opinion was expressed on another occasion by Ordoñez.[2]

It is true that Barbosa and Ordoñez had as an excuse for their mistaken views the fact that the American Ambassador Choate, in laying down the basis of his project, let fall certain words which could only too easily be misunderstood. He cited [3] the words of Roosevelt in a letter to Carnegie in which the wish was expressed that ' in each case that may come before them [the judges], they will decide between the nations, great or small, exactly as a judge within our own limits decides between the individuals, great or small, who come before him '. The mistaken conclusion could

[1] p. 659. [2] p. 156. [3] p. 309.

have been drawn from these words that later on in the case of judicial settlement between states—as a close comparison was made in that letter between it and judicial settlement within a state—an executive power was to be introduced, especially since Carnegie, to whom the letter was addressed, was known to be an ardent advocate of forcible execution by the international court. But that was not the meaning of Choate's words. We see this when we consider the basic principles of the American Society for the Judicial Settlement of International Disputes, founded by James Brown Scott, who is of the same mind as Choate. In the programme of this society printed at the end of Baldwin's pamphlet *The New Era of International Courts*, it is said *inter alia*: 'The purpose underlying the formation of the American Society for Judicial Settlement of International Disputes is promotion of the project to establish a judicial tribunal which will do for the civilized world what the ordinary courts of justice do for the individual.' It is perfectly clear from the numerous addresses made at the first yearly meeting of this society (e.g., that of Macfarland) [1] and from the plans which I have before me of the founder and president of the society, Scott, that this new world court will rely merely upon public opinion for the enforcement of its decrees, and that it will justly observe in the exercise of its activity the marked distinction between the legal subjects of municipal and those of international law. These facts clearly show that Choate had no intention of pushing the analogy between national and international courts as far as it is frequently and mistakenly done. Had he foreseen that Barbosa and others would only be further led astray by his comparison, he would certainly have omitted it. For when Barbosa [2] said that 'the

[1] *Proceedings*, p. 166.

[2] p. 341.

judicial settlement of disputes between states cannot be carried out in the same way in which disputes between individuals are settled ', he had another idea in mind than the comparison made by Choate.

It is true that of late the movement to create an international authority to enforce decisions has, in consequence of the efforts of van Vollenhoven, unfortunately obtained influence.[1] Foulke[2] also declared in 1911, at the Lake Mohonk Conference, that a power of enforcement was desirable, although Tripp[3] at the meeting of the same body held four years earlier vigorously attacked this proposal.

2. International Judicial Settlement and the Sovereignty of States

A second assertion which makes an understanding upon the question of international judicial settlement and arbitration very difficult is based upon the belief that arbitration rests upon the consent of the parties, whereas judicial settlement presupposes a higher authority. As a matter of fact international judicial settlement is as much in keeping with present international law as is international arbitration. Both institutions rest upon the consent of the parties, and arbitral courts as well as courts of justice are both supra-national organs.[4] The sole difference is that in the one case the will of the parties is more strongly obligated than in the other. I do not understand how

[1] See in particular the collected opinions in the volume *In't Zicht der derde Vredesconferentie*, 1911 ; also den Beer Poortugael, *Le Droit des gens en marche vers la paix*, 1912.

[2] Report, p. 99.

[3] Report, 1907, pp. 49 et seq.

[4] Kohler, in *Zeitschrift für Völkerrecht und Bundesstaatsrecht*, 1912, p. 106, is of the same opinion ; Pohl, op. cit., pp. 193 et seq., is of the opposite opinion.

those who think that international arbitration in contrast with the international court of justice is not a supra-national organ mean to classify the various intermediate stages between international judicial settlement and arbitration, whether in these cases they would say that there was a supra-national authority or not. We may distinguish the following steps in the development of arbitration into judicial settlement:

1. The free choice of the arbitrators and the reference of the dispute to a court in accordance with a special *compromis*.
2. The free choice of the arbitrators and the reference of the dispute to a court in virtue of a general arbitration treaty.
3. The reference of the dispute to a permanent court in accordance with a special *compromis*.
4. The reference of the dispute to a permanent court in accordance with a general treaty of arbitration.
5. The unilateral institution of an action by the parties before a permanent court of justice.

If we consider now the above-mentioned five cases there can be no doubt that the only difference between them consists in the degree of legal obligation. The foundations of international law will not be shaken any the more whether two states refer a definite case to an arbitration court yet to be constituted, or whether the community of states establishes once for all a court of justice for the settlement of certain disputes, before which each state can by its own motion institute an action.[1] In any case, however, a state can in fact only be condemned with its express consent, whether this be granted in general or in each special case. It must be further observed that a judgment by default is not possible, and the objection can

[1] See in this connexion Hold von Ferneck in *Zeitschrift für Völkerrecht und Bundesstaatsrecht*, 1912, pp. 16 et seq.

always be raised that honour and vital interests are involved. Accordingly, even in the case of a general obligation, it is only possible for a state to be condemned when it appears in person before the court, and thereby makes it clear a second time that it gives its assent to a settlement by legal proceedings. The foundations of international law would not be altered unless an international court of justice were set up by an organ exercising authority over the states. A sharp distinction must be made between a supranational court set up with the consent of states and a tribunal imposed upon states by a higher authority.

Is there any great difference whether in the one case the judges are appointed only after a dispute has arisen between the parties or are appointed in advance by the community of states ? Does not the individual state also co-operate directly or indirectly in the latter case in the appointment of the body of judges, whether by taking part in the appointment of the judges, or, in case the selection is entrusted to other organs, by ratifying the convention and thus delegating others to appoint the judges ?

How far the states are willing to obligate themselves legally depends above all upon the community of interests between them. The greater the number of disputes which arise between states the greater the necessity there will be no longer to conclude at every turn special arbitration agreements, but to bring into existence an institution which works as it were automatically.

Undoubtedly, a state will be more and more obligated as arbitration develops into a regular judicial system, and the authority of the body placed over the states will become constantly greater, especially as the states will confer more and more jurisdiction upon it. In the first world arbitration treaty (later on the better name will be,—a treaty concerning the jurisdiction of the international court of justice) very few disputes will be entrusted to it; but the number will gradually become greater and greater. But there can be no doubt that the most significant restriction of the sovereignty of states did not come about on the day when, in the form of an International Prize Court, there was created for the first time a sort of international court of justice, but at the time when for the first time disputes were no longer without exception settled by virtue of the state's own sovereign judgment with its final appeal to the sword, but were referred to a third party as arbitrator for a settlement in accordance with international law. This transition was just as gradual as the development of international arbitration into judicial settlement, which is taking place to-day before our very eyes. It cannot of course be said that the transition began thousands of years ago, when the first isolated arbitration awards were rendered. For at that time there was no community of interests, and in consequence no international law. When three hundred years before Christ five Rhodian judges rendered a decision in a boundary dispute between Samos and Priene, there existed no guarantee that the decision would be carried out. Both parties waited to learn what the arbitration award was to be, and if it did not suit them they did not carry it out. If, however, the losing party fulfilled the obligation assumed by the arbitration treaty, it did so not out of respect for international law, but out of a fear of the punishment of the gods, who would avenge the violation of an oath, or out of fear of its more powerful opponent. In the Middle Ages as well the parties, in entering into an arbitration treaty, were scarcely acknowledging the authority of international law, as they undoubtedly do at the present day.

At best the princes had recourse to the pope, as the pope stood above the nations and no one dared defy his award. But in so doing princes were in every case yielding to the authority of the pope, not to that of international law. Accordingly, as the papal power weakened the arbitral activity of the popes likewise ceased, and arbitration only came again into favour when the foundations of international law were gradually laid. To be sure, arbitral awards were rendered also by legal bodies and by other isolated individuals. But the question was always one of disposing of the disputes by mediation, as is seen in the fact that the arbitrators of the Middle Ages often spoke of themselves as 'friendly compositors'. States did not wish to go to war for such trifles, and endeavoured to dispose of them in any manner at all. But in so doing they did not recognize the authority of international law before which it was their duty to bow. Indeed, when at last with the beginning of the nineteenth century numerous arbitration treaties were entered into merely with the object of giving the parties their rights, the arbitrators could not immediately get accustomed to this new rôle in which they were to be actual judges. The submission of the parties to a higher power, international law, appeared to them so serious a limitation of sovereignty, that they at most felt called upon under commission of the parties only to dispose of the dispute in the fairest manner, not to expound international law.

As soon, however, as this has once been recognized, it is thereupon evident that with respect to the question of the limitation of sovereignty it is incorrect to make a distinction whether the arbitrators, that is the judges, are appointed once for all or are chosen anew in each individual case, whether a unilateral right of action exists or a *compromis* is necessary. The fundamental point is merely the fact that the state bows its will to the authority of international law. What persons or what bodies are the exponents of this international law is indeed highly significant as a practical matter, but is of no importance for the legal aspect of the question.

These facts were for a long time unknown, and thus it happened that the establishment of the International Prize Court, which lies entirely in the line of development, was attributed too prominent significance. Pohl [1] and Schücking [2] are right in not assigning to the International Prize Court an essentially different legal character from that of a special arbitration court.

Von Liszt [3] puts the question differently in his treatise *Das Wesen des völkerrechtlichen Staatenverbandes und der internationale Prisenhof.*[4] He thinks that 'as a matter of fact it cannot be denied that the establishment of a permanent court of justice having jurisdiction independently of the consent of the parties involves an encroachment upon the sovereignty of states, and consequently an alteration of the foundations of international law', whereas I am of the opinion that the encroachment upon the sovereignty of states began much earlier with the authority of international law. It is merely that this fact has been brought to the knowledge of the public in consequence of the erection of the Prize Court, which is a particularly clear manifestation of that change. And if von Liszt sees in the composition of the Prize Court an alteration of the foundations of international law, I for my part regard the grouping of states in that Convention as so unusual that no further

[1] Op. cit. [2] Op. cit., pp. 119 et seq.

[3] See also Flourens in *Deutsche Revue*, April, 1912, pp. 85 et seq.

[4] pp. 11 et seq.

consequences can be attached to it. The vigorous opposition of the states to the composition of the Judicial Arbitration Court according to the system of rotation is the best sign that the establishment of an international court of justice will never take place with such disregard for the smaller states. That the position held by the judges of the Prize Court and their international oath are not in contradiction to international law also seems to me self-evident. For arbitrators have been before the present day international judges, not mere representatives of their states, and have also in previous times taken an international oath before entering upon their office.[1]

On the contrary, I agree with Nippold [2] when he thinks that the chief and most significant progress resulting from the creation of the International Prize Court consists in the organization of an international judiciary. This view is justified by the simple fact that a healthy development of international law can for the most part take place only gradually, and accordingly an international permanent court of justice, if it signified a complete setting aside of international law, could hardly be recommended.

The chief significance of the International Prize Court, in addition to the creation of a permanent judicial organization, consists in the clear recognition of the principle that 'international law is superior to national law', and in the exceedingly wide substantive jurisdiction of the court.

But it will now be asked why I speak of the Judicial

[1] See in addition the excellent remarks of Schücking, op. cit., pp. 117 et seq., with which I am for the most part in accord. A right of denunciation on the part of states seems to me of no consequence at all as regards the legal constitution of the Judicial Arbitration Court. Would not something have to be put in its place if this right were renounced?

[2] *Die zweite Haager Konferenz*, p. 185.

Arbitration Court and the Prize Court as judicial institutions, when both depend upon the consent of the parties in the same way as the separately constituted arbitration courts. On this point I have the following remarks to make: the most important distinction between judicial settlement and arbitration in *national* law is to be found in the fact that judicial settlement is carried out by a legal authority placed over the individual persons, namely, the state, whereas, on the contrary, arbitration rests upon the consent of the parties. Now in *international* law a judicial institution set up by one having legal authority over the states cannot be considered. Every legal decision must in the abstract rest upon the free consent of the states. If a court of justice were set up over states by a superior body we should then have a world federal state. This court of justice would be, therefore, not an international institution but a national one. The concept 'judicial settlement' must be taken in such a sense that it can be realized within the present international community. But there can be no question of that if the national idea of judicial settlement is without any change transferred into international law.

It follows therefore that the idea of judicial settlement must be given a different meaning in international law from that which it has in national law. Here I am in disagreement with Pohl [1] and Schücking.[2] By courts of justice in international law I understand those agencies which are set up for the strictly legal settlement of international disputes, and are permanently adapted to the purpose. In contrast to these courts are arbitral courts which merely have in view a disposal of the case on the basis of equity. Whether there is or is not behind these

[1] Op. cit., p. 195. [2] Op. cit., p. 124.

courts a power to enforce their decrees seems to me to be of no consequence. For I hold, as was shown above, that such a power of enforcement in international law is in no case necessary.

If Schücking's definition were accepted and if the Prize Court and the Judicial Arbitration Court were characterized as arbitration courts, the great contrast between arbitration courts, which are set up as the occasion calls for them, and completely organized courts, which are appointed to expound international law in a systematic way, would be entirely obliterated. On the contrary, it must be recognized that there exists a sharp distinction in the judicial organization and in the purposes of these two forms of procedure for the settlement of differences, and that this distinction should also appear in the names given them. Schücking's definition is merely negative: 'The Prize Court and the Judicial Arbitration Court are not courts of justice in the municipal sense but rest entirely upon arbitral foundations, if the municipal idea of arbitration be taken as a basis.' That is without doubt correct. But it is not enough to say what those institutions are according to a standard borrowed from municipal law and not at all suited to international law; we must approach them from a new, strictly international point of view. Why should we not introduce into international law the terms 'judicial settlement' and 'arbitration' to distinguish the two chief methods of procedure for the settlement of disputes, provided that in so doing we keep before our eyes that by international judicial settlement we understand something quite different from judicial settlement in municipal law? Certainly two other terms would perhaps better prevent a confusion of municipal and international concepts. But the distinction which I advocate is made use of in so many quarters, and the terms arbitration and judicial settlement are so current throughout the world, that more appropriate names cannot be found. Distinguished authors, such as Hold von Ferneck,[1] Huber,[2] Nippold,[3] Kohler,[4] Despagnet,[5] Davis,[6] Oppenheim,[7] von Liszt,[8] Fried,[9] and in the beginning Zorn[10] also, have spoken of the Prize Court, which is identical in principle with the Judicial Arbitration Court, as a true judicial court. Moreover, in the proceedings of the Second Hague Conference a group of delegates referred to the Judicial Arbitration Court and the Prize Court as judicial courts. Further, it seems to me worthy of remark that the numerous Americans of recognized standing who support the project of the Judicial Arbitration Court, and Scott in particular, not only in their addresses and writings speak of a 'court of justice', but have given to the society which they have formed the name 'American Society for the *Judicial* Settlement of International Disputes'. On the other hand, those who see in the Court of Arbitral Justice an arbitration court stand apart by themselves. I mention in addition Schücking, Pohl, Bustamante, Lammasch, de Louter, and von Ullmann.

The terminology proposed by me is, moreover, highly commendable for the further reason that in municipal law judicial settlement furnishes us with a strictly legal decision, whereas arbitration furnishes us with a decision in accordance with equity. Accordingly, the distinction between judicial settlement and arbitration in international law is

[1] Op. cit., pp. 13 et seq.
[2] Op. cit., p. 472.
[3] *Haager Friedenskonferenz*, vol. i, p. 185.
[4] *Zeitschrift für Völkerrecht und Bundesstaatsrecht*, vol. vi, p. 108.
[5] *Droit international public*, p. 1359.
[6] *Elements of International Law*, p. 262.
[7] Op. cit., pp. 43 et seq.
[8] *Völkerrecht*, p. 277.
[9] *Die zweite Haager Konferenz*, p. 185.
[10] *Festgabe für Laband*, pp. 179, 180.

analogous to the corresponding distinction in municipal law, even if on important points the above-mentioned differences exist.

As regards the names of the Judicial Arbitration Court and the Prize Court, I have, however, the following remarks to make:

With respect to those who have only a superior contempt for the remarkable progress of international law, and also of the peace movement which in recent years has been the result of it, and who fail to appreciate the high idealism based upon actual facts which exists in the world of pacifism,—with respect to these sceptics I assert that the splendid progress which is involved in the Judicial Arbitration Court, adopted in principle by the Second Peace Conference, and in the Prize Court, cannot be too strongly emphasized. Those who are always insisting that the occasional and the permanent agencies for the settlement of international disputes are identical, and are 'only' arbitration courts, seek to create the impression in the minds of the present generation, or create it unconsciously, that these permanent institutions signify at bottom no progress, and they depreciate the great value of the new project in the eyes of a public which has for a long time failed to recognize the true significance of the great Hague Peace Conferences. For when it is said that 'the International Prize Court and the Judicial Arbitration Court are not judicial courts', it is thereby directly asserted that there is a still higher stage in the procedural settlement of international disputes, and that a tribunal set up by a higher authority is after all conceivable in international law, and that the highest form of an international court has not yet been reached by far. But in truth a supranational court, in the sense that it derives its power from an authority higher than that of the states, is simply inconceivable under the system of international law; and the Judicial Arbitration Court and the Prize Court in their main elements (prescinding from details)[1] are indeed the highest form of possible institutions for the peaceful settlement of international disputes. Accordingly, in using the name 'court of justice' we at the same time publish to the world the splendid progress of the new organization of nations. To this end all those must co-operate who are of the belief that the true science of international law has not merely as its aim to define legal concepts in a clear and precise manner, but that it must co-operate with the strength and enthusiasm of the true disciples of international law in the reform of international anarchy, and that it must devote itself to the great questions of actual life (not those of the study room) with that same zeal of which the pacifists are continually giving us an example. That is indeed the critical question which we German international law jurists must face: Shall we, with Kohler, von Liszt, von Ullmann, Niemeyer, Lammasch, Schücking and others, agree to recognize that the fundamental idea of the peace movement is entirely sound, or shall we continue to stand in the way of a development which is fraught with such great blessings for mankind? Shall we not rather, before we reject the proposals of the pacifists with a superior disdain, give ourselves up to long and earnest study of history and of doctrine, in order to comprehend that great world of ideas?

Accordingly, the contest over the terms 'arbitration' or 'judicial settlement' is in my mind not merely a verbal dispute but a fundamental question. Precisely because of the significance of pacifism, to make known the positive

[1] See below.

foundations of which is my highest endeavour, we should not fear to call things by their right names.

And now in conclusion let us read the statements made by Barbosa[1] at the sixth meeting of the committee appointed by the Second Hague Peace Conference to consider the Judicial Arbitration Court:

> Judicial settlement and arbitration are ultimately reducible to one and the same idea, that of the recognition of right between two opposing claims. The arbitrators decide; they do justice; their decisions bear the name of awards. These are the elementary ideas with which every one is familiar. Nevertheless, in spite of this fact there is from a juristic point of view so real a distinction between the office of judge and that of arbitrator that the two will never be confounded without introducing uncertainty and confusion into the principles which are most necessary to the organization of judicial institutions and methods of procedure.
>
> Observe the laws of all countries. They prescribe judicial institutions. They authorize arbitration. The two institutions live side by side, mutually aid each other, replace each other, intertwine at times, but never destroy or become confounded with each other. This fact proves the fundamental difference between them and at the same time the necessity of their parallel activities; for if they were substantially identical the contact between them would have resulted in their being confounded with the other and they would not have continued to be employed for thousands of years as distinct agencies.
>
> Hence courts of justice and arbitral tribunals are both indispensable. The existence of each is justified and they have separate functions and a distinct character. In what do they differ? First of all, as to the source from which they are derived; next, as to the social principle which animates them; lastly, as to the juridical form in which they are clothed.
>
> In the case of the courts of justice the juridical form

[1] p. 659.

of the court is permanent and unalterable. It is fixed by law. In the case of arbitral tribunals the juridical procedure is variable and determined by the circumstances. It depends upon an agreement of the parties. Courts of justice proceed from sovereignty and are imposed upon an obedient people. Its agencies are created by the power of states. The parties are constrained to submit to them. On the contrary, arbitral courts are based upon free choice and are the result of a convention. They have no greater authority than that given them by the contracting parties, and their judges are such as are voluntarily appointed by the parties. It is for this reason that while courts of justice are preferable for the settlement of the relations between individuals, courts of arbitration are the only ones possible between nations. The latter submit only to the authorities which they have themselves created. To substitute courts of justice for those of arbitration would be in the case of nations to replace voluntary consent by constraint.

How appropriate were these words of Barbosa in 1907, when it was sought to introduce an international judicial system not based in fact upon the free consent of the states, but imposed by a higher power! But how mistaken Barbosa's remarks appear to be when one keeps in mind that the Judicial Arbitration Court of the Second Peace Conference, measured by the standard of national law, was only a truly permanent arbitration court! Barbosa overlooked the distinction between national and international judicial institutions. He took offence at the term 'court of justice', and thought that an institution bearing that name must necessarily rest upon the same foundations in international law as in national law. It seems to me, indeed, most remarkable that none of the delegates fully cleared up Barbosa's error.

3. International Judicial Settlement and World Federation

Finally, there remains to be disproved the assertion that the erection of an international court of justice would be the preliminary step to a federation of states.

Barbosa's statements at the Second Hague Conference[1] in this connexion were incorrect: 'A further step forward and there will be established an international executive as a preliminary to a universal legislature. It will be the Constitution of the United States of the World.'

The union of the community of states upon an international basis came about so unexpectedly that for a long time men could not formulate it in legal terms. Just as the jurists of private law brought false views into the field of public law, so the public law jurists confused the fundamental principles of international law with concepts derived from public law but not suited to international law. Many of them saw with dread in the ever-growing community of states the approach of a world federation. And yet nothing could be more absurd than this view. When the German Empire was a federal state it was necessary to unite the individual small states, so that a single, noble, self-conscious legal authority took the place of many small states. The individual states were at that time all German. They all wanted to stand as a powerful unit against foreign enemies. But the condition of similarity of race does not hold good in the case of the states of the world. Moreover, the states of the world are not under the necessity of arming themselves against a foreign foe. Accordingly, everything argues against a development in the direction of a world federation. It is unintelligible, therefore, why persons should regard

[1] p. 659.

the ever-growing union of states as the beginning of the development of a world federation.

In particular is it absurd to maintain that a world court of justice is only possible in the world federation of states desired by Utopian dreamers, because otherwise the decisions of the court of justice would not be executed. It has been pointed out above that the judicial settlement of international disputes is possible even without the power of enforcement possessed by a federal state.

Having thus shown that an international permanent court of justice is entirely consistent with the nature of international law, we dispose of the opinion of Zorn[1] that the old Arbitration Court of 1899 is more in keeping with international law than the project of the Second Hague Conference concerning a Judicial Arbitration Court. For the idea of the international court of justice, the establishment of which is an absolute necessity, does indeed imply a closer union of the states than is the case with arbitration, but this union is after all only one which is intimately bound up with the interests of all the members of the community of states.

If a clear grasp be had of the significance of judicial settlement under the system of international law, it will be possible to understand the meaning of the conflict at the Second Hague Peace Conference concerning the name of the truly permanent tribunal. This conflict, of which I shall speak in the following section, would not have been possible if the distinction between judicial settlement in national and judicial settlement in international law had been clearly understood.

[1] *Zeitschrift für Politik*, vol. ii, p. 351.

4. The Name of the New Court of Justice

The first name proposed in 1907 for the permanent court of justice was the 'International High Court of Justice'. As was pointed out by Scott in his report,[1] the authors of this proposal wished by that name to bring out the idea 'that the court would be an international court and that it would have for its object the settlement of all disputes whatever which might be submitted to it in a spirit dominated by the sense of judicial responsibility which is peculiar to courts of justice'.

This was opposed first of all by Lammasch,[2] who observed that if the court bore that name persons might obtain the idea that the new tribunal was merely a court of appeal. He further expressed the opinion that if the court were so named the impression would be given that it was the intention to substitute judicial settlement in place of arbitration. Barbosa[3] naturally agreed with him fully upon both points. Asser and Renault[3] also feared that the proposed name might give rise to the idea that the court was a court of appeal. In answer to this Fry[4] explained (and in my opinion he was right) that in England the term 'High Court' signified not necessarily a court of appeal, but only a court of justice of first instance for certain disputes of a particularly important significance. 'The term High Court', he insisted, 'signifies not only a court of appeal, but also a court for important cases.' Baron Marschall[3] also failed to see that the expression 'International Court of Justice' would give occasion to misunderstandings.

After these proceedings in committee *B* of the first sub-

[1] Vol. i, p. 355.
[2] pp. 596, 597.
[3] p. 597.
[4] Op. cit.

commission of the first commission, a drafting committee was appointed consisting of Lammasch, Asser, Renault, Crowe, Kriege, and Scott. This committee decided to retain the term 'International Court of Justice' for the very good reason that the question at issue was that of international judicial settlement. At the next meeting Lammasch [1] opposed the resolution of the majority of this committee. He expressed the fear that if the court bore that name the appearance might be given that the court was composed of judges having authority over the states. He advocated the name 'International Court of Arbitral Justice'. Scott, Renault, and Fry [1] were opposed to this. Mérey [2] brought forward two arguments in support of the proposal of his countryman, Lammasch—the first being that, in his view, the old and the new court had the same 'arbitral mission' (in which idea he was mistaken), and the second that he had been convinced from the preceding discussions that if the word 'arbitral' were stricken out it would be equivalent to a denial of the arbitral character of the new tribunal. Here again Mérey was wrong in thinking that the whole nature of an institution could be changed by giving it another name. He proposed instead of 'Court of Arbitral Justice' the name 'Court of Arbitral Jurisdiction'. To this proposal Lammasch [3] also gave his adherence at the next meeting.

Those who wished to see the judicial functions of the new tribunal expressed in its name yielded finally to those who held out for the alleged arbitral activity of the new court. Scott [3] also was weary of the dispute over a word. The important thing was the principle at issue, and this was not hurt by giving the new tribunal the inappropriate name of an arbitral court. The name 'Court of Arbitral Justice'

[1] p. 658. [2] p. 660. [3] p. 681.

was thereupon adopted, after Bourgeois [1] had remarked very appropriately that 'the word "justice" indicates the end which we seek more and more to attain; the primary question will not be to weigh interests, but to do justice. On the other hand, the word "arbitral" stands for the free and independent consent of the parties.'

It is first of all remarkable that the name 'Court of Arbitral Justice' (the word 'international' in the title was struck out as being too awkward) does not express the fact that the Court of Justice is a permanent one. Asser [2] and Fusinato [1] had proposed to take away the adjective 'permanent' from the Permanent Court of 1899, and to affix it to the name of the new court of justice. But on this question there was so little agreement that it was doubtful whether the new tribunal would actually come into existence. Accordingly it seemed unwise to alter the name by which the old Arbitration Court was known throughout the whole world.

There is no need to emphasize further the fact that the name 'Court of Arbitral Justice' is not a particularly appropriate one. The term 'International Court of Justice' which was first proposed would have been much more in keeping with the nature of the institution. That was pointed out in particular by Jarousse de Sillac,[3] who very justly remarked 'the word "arbitral" implies the idea that the arbitrators are chosen for each dispute which arises, and that their powers expire with the settlement of it'.[4]

[1] p. 682. [2] pp. 597, 682. [3] p. 456.

[4] TRANSLATOR'S NOTE. The French title *Cour de Justice Arbitral* has been rendered 'Judicial Arbitration Court' in the official translations of the United States and Great Britain, and that designation has therefore been employed in other parts of this work as a substitute for the unofficial title 'Court of Arbitral Justice' still in use.

CHAPTER VII

THE EFFORTS THUS FAR MADE FOR THE ESTABLISHMENT OF A PERMANENT COURT OF JUSTICE

AT the present day, when the development of international arbitration into judicial settlement appears assured, it is of particular interest to glance back at the numerous efforts made during the last decades for the creation of an international court. In so doing I shall omit the earliest projects, which have essentially only a historical value, as well as those which at the same time advocate a world federation, a European federation, &c.

1. THE PROJECTS OF INDIVIDUAL PERSONS

Until recent years all the authorities have regarded arbitration more or less exclusively as a means for the settlement of disputes. Bentham crudely expressed the theory of the settlement of disputes as follows: 'Just or unjust, the award of an arbitral tribunal will save the credit and honour of the losing state.' Accordingly, the court advocated by these authors is merely an instrument of peace and not of law. This idea is particularly prominent in the 'draft project for the creation of an international council and high court of arbitration' drawn up by Levi in 1887, and warmly recommended by the Universal Peace Congress at Paris in 1889. This project advocated the creation of an 'international council of arbitration', which was to settle disputes between the parties by means of mediation and other such measures. In case a dispute was to be settled by arbitration, the council was to appoint from among its members a 'high court of international arbitration'. Provisions of that kind by which a dispute was to be in any case settled by mediation, and only in an extreme case was to be referred to an arbitration court, are just as unsuited to the legal settlement of the greatest possible number of cases as is the commission of inquiry in the new arbitration treaty between Great Britain and the United States, of which we shall speak later.

The projects of Dudley Field and Saint-Georges d'Armstrong, which are quite similar, can just as little be regarded as models. Dudley Field desired a permanent court of arbitration, whose members were to be appointed by the parties in each case. This 'Joint High Commission' was to act as far as possible as mediator in the dispute. A court of appeal was then to be constituted of judges who were to be proposed by the neutral states. The parties were to have a limited right of challenge with regard to the judges of the court of appeal. According to the project of Saint-Georges d'Armstrong, the parties were in like manner to have complete freedom in the choice of the judges of the court of first instance, but were to have only a limited right of challenge with respect to the judges of the arbitral court of appeal, who were to be appointed from some impartial quarter. Both projects, which look to the erection of a permanent tribunal only as a court of appeal, rest upon a false basis. The distinguished character of most judges makes a revision of the decision so infrequent that international law could not be developed as it should by such a court of appeal. Accordingly, both projects were drawn up merely with the object of disposing of disputes.

Ladd, the founder of the American Peace Society, comes much nearer to our idea of legal decisions; he started out with the principle that the chief duty of an international permanent court was to expound international law, and accordingly substantive international law must first of all be defined and codified through the agency of a congress. At the same time a permanent tribunal, upon which each state was to name two judges, was to be established. In the period from 1835 to 1840 Ladd repeatedly presented petitions to the American Congress with this object in view, but met with no success. His ideas, which were taken up by Elihu Burritt, Miles, Sartorius, Larroque, Marcoartu, Bara, Pays and de Laveleye, and were represented in Germany in 1886 by the Darmstadt[1] Peace Union, do not take into consideration the fact that a general codification of international law is at present impossible and, accordingly, that the only method must be that of the gradual codification of international law by means of international conventions and the decisions of an international court. In addition, Dudley Field and Saint-Georges d'Armstrong had also advocated the codification of international law, but they did not make this a direct condition for the creation of an international court, as did the last-mentioned writers.

The projects presented by Count Kamarovski and his high-minded pupils, Mougins de Roquefort and Revon, stand upon a much higher plane. They desire an international court which would dispose of disputes according to law and equity. ' As a matter of fact,' says Mougins de Roquefort,[2] ' in international affairs arbitrators play very often the part of friendly compositors.' Accordingly, he wants a true court of justice. Nevertheless these writers practically fail to emphasize the development of inter-

[1] See Dietz, *Franz Wirth und der Frankfurter Friedensverein*, p. 22.
[2] p. 208.

national law as one of the duties of this court. It appeared perhaps self-evident to them. For a court which decides according to strict law must naturally render much more detailed decisions than an arbitration court which decides according to equity, and it would thereby of itself contribute to the development of international law. The project of a permanent tribunal as an organ of the law existing between states, which Schlief brings forward in his book *Der Friede in Europa*, is also well worthy of consideration.

Numerous projects proposed during those years, such as those of Adler, Lorimer, Gondon d'Assone, Saint-Simon, Pecqueur, Marchand, Villiaume, Trendelenburg, Sigaud, Ferrer, Dupasquier and Loewenthal, recommended an international congress, that is, an arbitration court with executive power. They belong more or less to the class of Utopian schemes. Side by side with these comprehensive projects there naturally arose others which show greater restraint and which endeavour above all to make it easier for states to take the first step towards a permanent court. Among these writers the first place is to be assigned without question to the Frenchman, Mérignhac. Mérignhac did not advocate a truly permanent court, but laid down similar principles to those which were later recognized by the First Hague Peace Conference.

Individual writers likewise presented restricted plans in advocating that the activity of the new court be limited at first to certain states. Thus de Laveleye proposed a court of justice between Great Britain and the United States, which was later on to extend its functions also to disputes arising between other nations.[1] Bentham, Schlief, and Stanhope (the latter at the meeting of the Interparlia-

[1] See also Report, 1898, p. 40.

mentary Union at The Hague in 1894) proposed a strictly European court. This idea of a court limited to certain groups of states was never abandoned, and at the Lake Mohonk Conference in 1901 certain speakers, in particular Baldwin, expressed the opinion that it would do no harm if the new court of justice did not come into use for all the states at the same time; it would then be at least, as in the case of the Central American Court of Justice, a step upon the way towards the great goal.

2. The Projects of International Unofficial Associations—Universal Peace Congresses, Interparliamentary Union, Institute of International Law, International Law Association

In the presence of so great a number of projects the international associations, which in the ten years preceding the First Hague Peace Conference occupied themselves with the plan of a permanent court of arbitration, found themselves in quite a difficult position. Stanhope [1] justly observed at the meeting of the Interparliamentary Union at The Hague that 'for one who desires to go into details of the constitution of a Central European court, it is difficult to choose among the thousand and one projects which for some years have appeared in the form of pamphlets or in the European press'.

The Universal Peace Congresses which began in 1889, occupied themselves almost every year down to 1895 with a careful consideration of the problem of an arbitration court. The Congress which met at Paris in 1889 advocated the above-mentioned project of Levi, as Darby also says in his *International Tribunals*.[2] This project met, to be

[1] *Conférence interparlementaire*, 1894, p. 225.

[2] pp. 216 et seq.

sure, with little success. The Congress further expressed its conviction that the best preparation for the creation of a permanent court of arbitration would be for each state, when concluding a permanent treaty of arbitration, to designate the jurists who, in case a dispute should arise, were to be its representatives upon the tribunal. The project of Levi was approved in the report presented to the Universal Peace Congress of 1892 at Berne by Hodgson Pratt. Mazzoleni and Kolben made a motion at Berne that the Congress should declare it desirable that the Peace Congresses should yearly appoint from among their members and the members of the parliaments an arbitration court to which the states might refer their disputes. But the proposal received no support.[1]

At the Universal Peace Congress which met at Chicago in the following year, Butler, Eaton, and Brainerd presented an elaborate project advocating a permanent court of arbitration. This is also printed in Darby.[2] A list of judges already appears in this moderate proposal as the key-note of the new idea. At the Congress at Chicago a long and important discussion concerning the court of arbitration took place, which for scientific depth and political insight has been surpassed by the deliberations of no other body. A commission appointed by the Congress at Chicago made a report at the next Universal Peace Congress at Antwerp in 1894. This report is likewise based upon actual political facts, and is in no way Utopian in

[1] The French Peace Society, *La Paix par le Droit*, made a similar motion at the general meeting of the delegates of the Berne Peace Bureau in 1898 (Bulletin, pp. 13–19, 23), but after a detailed report of Dr. Richter (Pforzheim) it was set aside on the ground that no government would have its disputes decided by persons not possessing an official character.

[2] pp. 500 et seq.

character, as can best be seen from the following statements on page 5 of the report drawn up by La Fontaine :

> In fact there seems to exist a certain confusion in the ideas of the various jurists who have occupied themselves with this question. Some speak of a 'court of arbitration', others of an 'international tribunal'. They seem to think that they are both one and the same institution. That is, however, not the case. It was generally agreed that the 'court of arbitration' was to be made up of arbitrators who were to be appointed in advance by the several states, and that from among these the contending parties were to appoint the arbitrators who were to decide in a specific case. Such an arrangement would merely have the effect of limiting the choice of the several states; they could no longer appoint as arbitrator the man who approved himself to their judgment. The procedure was to be very simple, and similar to that worked out in the code proposed by us. On the other hand, the 'international tribunal' would exercise truly permanent judicial functions, similar to those of the ordinary courts of justice of the various countries, with the sole difference that instead of private persons states would bring suit before it. Such an organization is still in the distant future.

The Universal Peace Congress at Antwerp did not pass a definite resolution, but referred the question for further study to the Commission of the Berne Bureau for International Law.

Is not the fundamental principle upon which the Hague Court rests already expressed, though with a noticeable difference, in that report of the Congress at Antwerp ? Unfortunately the material upon this subject is not easily accessible, so that it is quite intelligible how the historians of the Hague Conferences, in describing the history of the period preceding the adoption of the Hague Court of Arbitration, have failed thus far to point out the valuable preliminary work of the Universal Peace Congresses. Of all the projects of the Congress at Antwerp, Hornby's report is the most deserving of consideration. For here we find for the first time a clear exposition of the ideas which Scott advocated at The Hague in 1907.

The Universal Peace Congress at Antwerp took place some days before the meeting of the Interparliamentary Union at The Hague. At the meeting of the Interparliamentary Union Stanhope (not only an interparliamentarian in the strict sense but a pacifist) arose and proposed [1] that the meeting take into consideration the problem of the permanent court of arbitration. But Tydeman, Marcoartu, and von Bar thereupon arose and asserted that the creation of an international court of arbitration would be worthless, unless the fundamental principles of international law were previously determined. On the other hand, Gobat vigorously supported Stanhope's motion, pointing out with justice that the codification of international law and the creation of an international court of justice were entirely independent propositions. He was supported by Houzeau de Lehaie and Passy, but only after another German, Hirsch, in addition to von Bar, had first asserted that it was impossible to establish the court until international law had been codified. Hirsch went so far as to say that persons should not make themselves ridiculous by taking up such projects. From this we can easily see that at the meeting of the Interparliamentary Union it was those who were interparliamentarians in the strict sense and not pacifists, who almost defeated the motion made by Stanhope; and it was only saved by the support which it

[1] As Bertha von Suttner (*Memoirs*, pp. 289, 297, 313) points out, Stanhope was also acting under the influence of Gladstone in making his proposal.

received from those who were in the strict sense of the term pacifists. Von Bar, Hirsch, and Tydeman were merely interparliamentarians; but Passy, Houzeau de Lehaie, Stanhope, and Gobat were pacifists in the strict sense. In the commission which was appointed by the Interparliamentary Union at this meeting to examine the project of the arbitration court, four pacifists were the only ones who later presented plans—Stanhope, Rahusen, Gobat, and Houzeau de Lehaie. The last-mentioned also made the report to the Interparliamentary Union at its session at Brussels. The name of Houzeau de Lehaie, signed to the report of the commission of the Interparliamentary Union, is also signed to the report of the Universal Peace Congress at Antwerp in 1894, of which Houzeau de Lehaie was president. No further facts are needed to show the close connexion between the draft of the Interparliamentary Union at Brussels and those of the preceding Universal Peace Congresses.[1] It is a point of special importance in this connexion that the basic principle of a list of judges had already been advocated unanimously by the Antwerp Congress.

Under these circumstances it is all the more to the honour of the pacifists when we consider that the next Universal Peace Congress, which met at Budapest in 1896, set aside its own valuable preliminary labours upon an independent project, and chose rather to give its warm support to the project of the Interparliamentary Union and to recommend to the powers the acceptance of it. That act was of marked service to the cause; for by it the agitation was kept from being divided in favour of several

[1] The connexion between the Peace Congress at Antwerp and the meeting of the Interparliamentary Union at The Hague I find already pointed out in 1905 by Moch in his book, *Désarmons les Alpes*.

distinct projects, and the whole influence was brought to bear upon the accomplishment of a single project. In spite of all its own good work in connexion with the subject, the Universal Peace Congress at Budapest did not attach the slightest value to coming before the public with a project all its own, but contented itself in all modesty with recommending the project of the Interparliamentary Union which it had influenced. This unselfishness cannot be too highly estimated. Precisely on that ground the historian has all the more reason to lay special stress upon the great merit of the preceding Universal Peace Congresses. The Universal Peace Congress at Hamburg, in 1897, also adopted a resolution in favour of a permanent international court of arbitration.

This does not detract from the work of the Interparliamentary Union, which has won a place in history. This association, indeed, interceded with the whole force of its prestige in favour of the project for a court of arbitration, and succeeded in directing the attention of governments to the problem of an arbitration court to a far greater extent than ever before. Not only did the masterly memorial of Descamps contribute to that end, but likewise the two resolutions taken at Budapest in 1896 and at Brussels in 1897, in which the Union again recommended that project and exhorted its members to intercede with their governments for its realization.

In addition it must be pointed out that the Interparliamentary Union twice before 1894 took up the question of a permanent court of arbitration. The Congress at Rome in 1891 for the first time called upon the interparliamentary committees of the individual countries to place the question of the permanent court upon the programme of the next conference. In 1892 Hilty, a professor at Berne, submitted to the Interparliamentary Conference at Berne a project in which he advocated that each state should hold ready a list of five persons competent to perform the duties of arbitrator. The contending parties were then to appoint their arbitrators from the lists of five different states which were not connected with the dispute. The Interparliamentary Union referred this project to a commission of six members under the presidency of Stanhope, who made the above-mentioned motion at The Hague in 1894. There is no special connexion, however, between the motion made by Stanhope in 1894 and the work of the commission appointed in 1892. As was pointed out above, Hirsch, the German member of the commission of the Interparliamentary Union at its session at Berne, spoke against any further activity of the Interparliamentary Union in connexion with the project of an arbitration court. Hilty's project was a private endeavour which, in consequence of the inactivity of the commission appointed at Berne, produced no results. No specific plan for a permanent court of arbitration was recommended by the Interparliamentary Union before its session of 1895, as had been done by a number of previous Universal Peace Congresses.

The project of the Interparliamentary Union at its session at Brussels was in its political foresight a masterly document. It was so drawn up that without renouncing their prerogatives too far the states could create a court of justice after that fashion; and although the ideal of an international court of justice, as we have shown above, is of a different character, particularly in that it calls for a truly permanent court instead of a list of judges, yet the compromises offered by this project were always acceptable. In particular the principle of the equality of states was recognized. The questions of the burden of expense (the parties to the case and not the community of states were to pay the honoraria of the judges) as well as that of the constitution of the court were, to be sure, less commendably handled. In particular the president of the court was given too great power.

The Institute of International Law and the International Law Association have given considerably less support to the work of the Interparliamentary Union than has been given by the Universal Peace Congresses.

The Institute of International Law occupied itself with the question of an arbitration court especially in the seventies of the last century. The only practical outcome, however, was a very significant project of Goldschmidt dealing with arbitral procedure for special arbitration courts. Goldschmidt [1] rejected the idea of the creation of a permanent court of arbitration in the following words:

> This institution is unknown to existing international law, and the obstacles to its creation between absolutely sovereign states appear to be difficult to surmount.

The bold Dutch pacifist, den Beer Poortugael, was the only one at that time to propose a permanent court of justice.[2]

The International Law Association frequently occupied itself before the First Hague Conference with the question of a permanent court of arbitration. As early as 1875 Richard proposed at the meeting at The Hague the creation of a court of arbitration, but he failed to win a sufficient following. At the conference held at London in 1893, a

[1] *Revue de droit international*, 1874, p. 422.

[2] *Bulletin de l'Institut de droit international*, 1875, p. 133. As late as 1900, Anitchkow, in his work *War and Labour*, written before the Hague Peace Conference, praises the far-sighted attitude of the Institute upon this question.

special committee was entrusted with the elaboration of a project for an international arbitration court. However, as this project merely sought to regulate arbitral procedure and did not enter into the idea of a permanent court of arbitration, the meeting at Brussels in 1895 referred the question of the permanent court of arbitration to a new committee. But this committee did not present its report until 1899, at Buffalo, at the time when the Hague Peace Conference had already reached a provisional agreement upon the subject.

Finally, special mention is due to the notable and continued efforts of the Lake Mohonk Conferences, which at all of the meetings from 1895 to 1899 advocated a permanent court of arbitration. The first conference of 1895 [1] promptly adopted a resolution in which the President of the United States was requested to take the initiative in the creation of a permanent court of arbitration, and a commission consisting of Abbott, Earl, and Hale, was appointed to study the question. This commission presented a detailed report in 1896. At this second meeting Hodgson Pratt recommended the project of the Interparliamentary Union as 'the most authoritative'.[2] Furthermore, the meetings of 1898 and 1899 gave first consideration to the plan of an arbitration court.

3. The Permanent Court of Arbitration of the First Hague Conference

According to the first circular letter of the Czar the question of an arbitration court was not to figure at all in the programme of the First Hague Peace Conference. Martens, who was entrusted by the Czar with the drafting

[1] Report, p. 52. [2] Ibid., p. 70.

of the second circular letter, was the first to place the erection of a permanent court of arbitration upon the programme of the Conference, although only a general reference was made in the second circular letter to 'freely chosen arbitrators'. But under the reference to these freely chosen arbitrators Russia secretly included the question of a court of arbitration as well. Projects for a permanent court of arbitration were presented to the Conference by Great Britain and the United States as well as by Russia. While the American and British projects merely proposed the drawing up of a list of arbitral judges from among whom the contending parties in a given dispute were to appoint the requisite judges, the Russian project desired at least to determine in advance those states each of which was to appoint a member of the court when a dispute should arise. The provisions elaborated by the Hague Conference are based upon the British project, since this provided the greatest freedom for the parties.

With regard to the necessity of a strictly legal decision of disputes and the development of international law by that means, there was as little agreement at the First Hague Peace Conference as there was at the numerous international private congresses which had occupied themselves with that problem.

At the meeting of the First Commission on July 17, 1899, Martens made the following characteristic remarks: 'The aim and object of arbitration is to dispose completely of a dispute. The great utility of arbitration lies in the fact that the moment the arbitral award has been duly published the affair is at an end.' That in addition to arbitration there also existed a method of judicial settlement with the object of developing international law was entirely overlooked by Martens. With the exception of Descamps and Rahusen no other delegates at the First Hague Conference called attention to this important work of the Hague Conferences. The delegates, who it is true came in part unprepared for their difficult tasks, looked upon arbitration merely as a means of settling the particular dispute before the court. They did not recognize the necessity of the judicial settlement of international disputes, whereby future disputes as well would be prevented by reason of established legal principles. But nevertheless some of the delegates at least had doubts whether the settlement of disputes was not being accomplished at the cost of justice. Holls and Seth Low in particular called attention to the necessity that the decision should do justice to each party, and that in consequence an appeal should be allowed as a guarantee against wrong decisions. In like manner opposition was encountered upon the question of judicial opinions. Here again Martens pointed out that in view of the political nature of arbitration judicial opinions should not accompany the decisions; that to be sure in this way the science of law would be given firm foundations, but that in an international dispute the judges were at the same time representatives of their governments, and it should not be expected of them that they should set forth in detail what might possibly be a sentence against their government, since they would be forced thereby to demean its political prestige.

This view of Martens met with strong opposition from all the others. The German delegate, Zorn, was the first to reply, and he declared that a legal decision without an accompanying judicial opinion was not to be thought of. Descamps, Holls, and Asser in particular agreed with Zorn. But these speakers, instead of pointing out the necessity of developing international law by means of legal decisions, kept expressing the one idea that provision must be made for the necessary impartiality in the individual decision. Thus Descamps spoke as follows at the meeting of the Commission on July 17:

> The committee was not in agreement upon the question. As for himself he thought that the obligation of rendering a judicial opinion constituted a fundamental guarantee. There were questions which could not be answered by yes or no, and in such a case it was the judicial opinion which justified the decision. There was no case in which a decision had been rendered without being accompanied by the grounds upon which it was based.[1] He did not think that the objections pointed out by Martens were insurmountable and irreconcilable with the obligation of formulating a judicial opinion. The form and extent of these opinions were left to the judgment of the court. Moreover, if states wished to provide against a case when there would be serious objections against formulating a judicial opinion, they were free to excuse their arbitrators from this formality.

It must be conceded that in this statement the question of the development of international law received no mention at all. But nevertheless two of the speakers, Descamps (on a former occasion) and Rahusen, touched briefly upon this point in referring to the fact that unless international decisions were accompanied by judicial opinions no consistent series of legal decisions could be developed. None of the other speakers took up the ideas thus casually thrown out. It must be said that the idea of a judicial system did not receive due consideration.

If one considers without prejudice the results of the First Hague Peace Conference in the domain of arbitration, one

[1] That this has, however, happened is pointed out by Renault, p. xii of the *Recueil des arbitrages*.

must admit that in that celebrated convention strictly legal and arbitral elements were confounded. Arbitration seeks the settlement of disputes upon the basis of equity; judicial settlement, on the other hand, seeks to give to the parties their due. The equitable settlement of disputes will best be secured by a court whose members are appointed by the parties. The strictly legal settlement of a dispute, however, calls for a permanent court of justice. Accordingly, the First Hague Conference should have made its choice between arbitration and judicial settlement in international law.

It did not, however, do so, but in a general way it saw in what was at that time called international arbitration a uniform institution; it did not recognize the two constituent parts of that institution, or suspect that within a short time there would be brought about a division of the so-called international arbitration into 'international judicial settlement' and 'international arbitration in the strict sense'. That is all the more extraordinary since to the casual observer this lack of unity is at once noticeable. Meurer in his *Haager Friedenskonferenz* [1] speaks as follows, though without making clear the whole import of his statements:

> My objections can be summed up in the fact that judicial institutions can only be set up with difficulty, and hence would rarely be set up as an actual fact, and political influences are still too prominent in them.

The mistake of the First Hague Conference can, to be sure, be well understood when one remembers that up to that time arbitration alone existed in international life, and judicial settlement was not known. That was quite natural. For judicial settlement presupposes permanent

[1] Vol. i, p. 369.

judges, and up to that time no such judges had been appointed by the international community.

What the First Hague Peace Conference desired was, and here I am opposed to Scott[1] and Beernaert,[2] the further development of international arbitration, and accordingly the settlement of disputes *ex aequo et bono*. But at the same time, by making a compromise with international arbitration, it laid the foundations of international judicial settlement. It did so at the very time when it provided that disputes should be settled by the Hague Court of Arbitration 'on the basis of respect for law', instead of according to equity, and also by the provision that strictly legal disputes could best be settled by arbitration, whereas in fact they will be much better settled with the help of a judicial institution. Arbitration is especially suited to political questions, and on the other hand, judicial settlement, which by a strictly legal decision is at the same time to contribute to the development of international law, is suited above all to disputes of a more or less legal character.

Thus the Hague Permanent Court of Arbitration, while unable by its organization to serve the cause of international judicial settlement, at the same time is not adapted to the system of international arbitration, which in principle calls for the complete freedom of the parties in the choice of judges, and is opposed to the imposition of any restriction which would limit the selection of arbitrators as far as possible to the members of a small list of judges.[3] In a third of the disputes referred to the Hague Court of Arbitration the states have appointed their arbitrators from persons not inscribed upon the list of the International

[1] Vol. i, p. 348 ; vol. ii, p. 315. [2] p. 332.

[3] See Baty, *International Law*, pp. 5 et seq.

Bureau. Besides this they have had recourse more frequently to special arbitration courts than to the Hague Court. The actual practice therefore shows that the system of arbitration seeks the most complete freedom in the choice of judges, and that a limitation of this freedom is not permanently practicable. This development was dimly suspected in 1899, and in consequence no compulsion was provided for in the provision that the parties must make a selection of arbitrators from the list of the Hague Court, but the selection from the list was merely laid down as a general rule. But even in this case there was an abandonment of the fundamental principles of arbitration.[1]

As a transitional step from international arbitration to judicial settlement the Hague Permanent Court was a necessity, since it above all led the states to take the first and hardest step upon the road to judicial settlement. The fact that nearly all states are actively considering the problem here under discussion is to be ascribed not least of all to the world-wide significance of the erection of the Hague Court of Arbitration. The Hague Court of Arbitration will in the course of history come to be more and more recognized as an institution of historical significance, and those who in loyalty to the great results of the First Hague Conference opposed every development of them can take comfort above all in this thought. The creation of a truly permanent court of justice can never attain the historical significance of what was accomplished in

[1] These facts also account for the proposal which I made above that henceforth only jurists should be designated for the list of the Hague Court, in order that these jurists might choose the judges of the international court of justice. For although arbitration demands complete freedom and in my proposal diplomats are excluded, yet it must be remembered on the other hand that the appointment of arbitrators is already restricted by the mere existence of the list of judges.

1899—not, at least, what was accomplished by Zorn.[1] For what was accomplished in 1899, in carrying on the development of civilization to a higher perfection, was accomplished in the face of indescribably strong opposition, whereas the new court of justice may be regarded at the present day as already won.

It is Revon's great service to have recognized clearly in his distinguished treatise *L'arbitrage international*,[2] crowned by the Institute of France, that the world would not obtain an international court of justice all at once, but would only arrive at that goal gradually by way of arbitration. He wrote in 1892:

> My observation of facts allows me to state that arbitration may be either optional or permanent, and that it tends more and more to abandon the first form in favour

[1] The study of Zorn's remarkable achievement of world-wide importance, the particulars of which I unfortunately cannot set forth in detail because they were communicated to me confidentially, exercised in the year 1908, and later, when I first gave my attention to it, a strong influence upon my mind and made of me a pacifist. The most impressive period in the development of my mental life was the time when I studied in Meurer's treatise time and again the proceedings of the memorable arbitration committee, which Zorn, at the session of the committee on June 9, 1899, declared he had followed with the greatest attention and deep sympathy. I was fortunate enough at that time to learn of the intimate details of that event from members, then living, of the committee and of the commission of the First Hague Conference; and when I reflected upon the reasons for that resolute action of the German delegates, I came to the conclusion that Zorn's achievement was not to be explained by his patriotism alone, but that at that turning-point in the world's history Zorn obtained, under the influence of Holls, d'Estournelles, Count Nigra, Bourgeois, Asser, Lammasch, and others, a suspicion of the grandeur of the pacifist ideal. It was then made clear to me that an ideal which was capable of accomplishing so much must have a future. Whether in Zorn's case also this suspicion ever became a belief is still doubtful; in any case every new article that Zorn writes shows how gradually he is being drawn into the peace movement.

[2] p. 488.

of the second, which will lead it by gradual and imperceptible steps to become more definite and more concise, so that it will crystallize, so to speak, into a true court.

At the Second Hague Conference it was again pointed out that the Permanent Court of Arbitration of 1899 was regarded by the First Hague Conference merely as an experiment.[1]

That the formation of a permanent list of judges is merely the preliminary step to a truly permanent court of justice, and that it is precisely this which constitutes the significance of the Hague Court of Arbitration, can be seen in the development of national law in numerous countries. The eminent American, Scott, in various articles[2] has called attention to the fact that in Roman law there are three stages of development to be distinguished:

1. The parties choose for themselves definite persons as arbitrators.
2. A permanent list of arbitrators is drawn up from which the judges are selected.
3. A permanent court of justice is established.

Dr. Tettenborn likewise says, very appropriately, in her treatise upon the Hague Court of Arbitration:[3] 'The list of arbitrators is a contrivance similar to the *album iudicum selectorum* in which the names of Roman citizens were entered, from among whom the prætor selected the *iudex* in each individual case.'

In connexion with the very exhaustive article of Scott in the *American Journal*,[4] I may be allowed to add a few

[1] See Scott's report, vol. i, p. 349.

[2] 'An International Court of Arbitral Justice' in *The Outlook*, New York, June 18, 1910, p. 369; 'One Aspect of the Peace Movement' in *The Alumni Quarterly*, Illinois, 1911, p. 204; 'Some Phases of the Peace Movement,' Address at Mohonk, New York, May 24, 1911, p. 5; and 'The Evolution of a Permanent International Judiciary' in *American Journal*, 1912, pp. 316–58.

[3] p. 59.

[4] 1912, pp. 315 et seq.

further words upon the development of Roman arbitration, since it has a very instructive bearing upon our subject. But, unlike Scott, I see not merely analogies but also sharp differences in the development of procedure in Roman law and in international law.

Self-help was in the beginning the prevailing method in Rome. Then the two parties chose as arbitrator some third person who in consequence of his religious or secular office was particularly influential. It frequently happened then that the parties turned to the prætor as being specially versed in the law. As the prætor received more petitions than he could answer, he found it necessary to appoint another competent person to settle the dispute as judge. But even the judge thus appointed performed his functions merely with the consent of the parties. For if the defendant were unwilling to come before the prætor, the latter could not appoint a third person to act as judge in the dispute. The judge was only a private person and in no way an official behind whom stood the power of the Roman state. The similarity between this judge and the arbitrator who in earlier times had been appointed directly by the parties led a Roman jurist to remark: *Compromissum ad similitudinem iudiciorum redigitur.* The decision of the judge was moreover not a judgment, but merely an award (*sententia, pronuntiatio*), as the victor in a legal dispute was obliged to carry out the decision himself by means of *manus iniectio,* and the state did not give him its support in the execution of the decision. In every case a mutual agreement of the parties was required for proceedings *in iure,* and for this reason von Jhering [1] has also described the *litis contestatio* as a contract by which the parties appointed the arbitrator in the presence of the prætor.

[1] *Geist des Römischen Rechts,* vol. i, 3rd ed., p. 171.

Just as happened in the case of international arbitration at the First Hague Peace Conference in 1899, the choice of the judges was limited by the provision of the *lex Aurelia* (70 B.C.), under which the Senate drew up a list of arbitrators whose names were published by the prætor in the *album iudicum*. Side by side with the ordinary procedure, according to which the prætor appointed the judge, there later arose in special cases the extraordinary procedure under which the twofold division was discontinued and the prætor decided the case as an official judge. From the time of Diocletian and his immediate successors, there existed only the extraordinary truly judicial procedure. The judges were now permanent officials and decided in the name of the Roman state. From this we see clearly that arbitration stands half-way between self-help and judicial decision. It must, in so far as it represents a formal method of settling disputes, be regarded not as the end of a development, but merely as a transitional step.

Accordingly, if we compare briefly the development of procedure in Roman and in international law, we must take as the *tertium comparationis* merely the transition from arbitrators appointed by the parties to a list of arbitrators, and then further to a permanent tribunal, and not likewise the development of the arbitrator as a person acting with the consent of the parties into an official placed over the parties and deciding by the authority of the state. Only when we keep this before our eyes shall we rightly understand the analogy between the development of procedure in Roman and in international law.

In his article 'The Necessity of a Permanent Tribunal' [1] Nys has shown that this development took place in other

[1] Baltimore, 1910 ; also *Revue de droit international*, 1906, pp. 6, 7.

countries as well.[1] The recent development of international arbitration, and in particular the American proposal at the Second Hague Conference, has shown that this comparison is to the fullest extent justified. In addition numerous authors, e.g., Pradier-Fodéré,[2] Kamarovski,[3] and Mougins de Roquefort,[4] have as a body expressed the view that in international law, as in national law, the establishment of a judicial system has in every case had its origin in arbitration.

A particularly interesting example of this development is furnished us in the history of the formation of the Supreme Court of the United States,[5] whose special function it is to decide disputes between the individual states of the union. The Articles of Confederation of 1778 had provided that such disputes should be settled by a court of arbitration chosen in each case by the parties or by Congress from a list of arbitrators. This arrangement, however, did not stand the test. Scott [6] says that 'the net result of procedure under Article 9 was the trial and final determination of one case (Pennsylvania *v.* Connecticut); the appointment by mutual agreement of commissioners in two controversies, settled, however, out of court (Massachusetts *v.* New York; South Carolina *v.* Georgia); with petitions for the appointment of a court in some three other cases. The temporary tribunal was unsatisfactory.' Accordingly the new Constitution of 1789 established a permanent court of justice of nine members, who are appointed by the president with the approval of the Senate.[7]

[1] See, in the case of France, especially Vavasseur, *L'Organisation d'une juridiction arbitrale internationale*, 1907, pp. 11 et seq.

[2] *Cours de droit diplomatique*, vol. ii, p. 472.

[3] p. 108.

[4] p. 236.

[5] It is set forth in detail by Scott in *American Journal*, 1912, pp. 336 et seq.

[6] *The Hague Peace Conferences*, vol. i, p. 464.

[7] See also Report, 1911, p. 106.

The literature of international law in Germany has not recognized the inconsistency in modern arbitration due to the presence of strictly legal elements. A. H. Fried alone has forcibly called attention to these facts with his usual clear insight. In his valuable *Handbuch der Friedensbewegung* [1] he writes as follows:

> It becomes clearer every day that arbitration is only a transitional institution by which states are first of all to be accustomed to the legal settlement of their disputes. As a matter of fact there is also in the very nature of arbitration the inconsistency of all transitional institutions. It makes use partly of diplomatic means and partly (and at the present day to an ever greater extent) of juristic means. The arbitrator is at times not a true judge, but merely a diplomat. As he is bound by no law he will be in a position to settle a dispute by diplomatic agreement. Whereas on the other hand the ordinary municipal judge will always decide according to law only, whether according to the statute law of the state or (when there is none such at hand) according to the general legal principles determined by custom and good judgment.

Dumas spoke in similar terms at the Sixteenth Universal Peace Congress at Munich: [2]

> Arbitration is only a stage on the way towards the organization of an international judicial system, the attainment of which involves the established right of every state to institute an action directly before the competent jurisdiction, without a preliminary *compromis*.

From the provision of the First Hague Conference that international arbitration has for its object the settlement of disputes 'upon the basis of respect for law', and from the other strictly legal elements present in the Hague Convention, it cannot fail to be seen that the development of international arbitration into judicial settlement is

[1] 2nd ed., 1911, vol. i, p. 195.

[2] Bulletin, p. 54.

making ever greater progress. Certain incidents which have recently taken place point out this clearly. It is a remarkable fact that since 1899 the states have continued to reappoint as arbitrators those who have already had experience in the field of international arbitration. Of the numerous arbitration courts which have met at The Hague since 1899 in conformity with the Convention for the pacific settlement of international disputes, not one can be named in which the judges of the court were all *homines novi* in the field of international arbitration.

In every case a number of the arbitrators had obtained great experience and were the embodiment of a tradition. Men such as Gram, Matzen, and Martens were repeatedly chosen in the old period preceding 1899. The new judges who stood the test were constantly reappointed: thus Renault was reappointed five times, Lammasch and de Savornin Lohman each four times, Fusinato and Hammarskjöld each three times, Fry, Beernaert, Asser, Kriege, and Baron Taube each twice. In selecting such men who had already distinguished themselves it was doubtless the intention to secure a guarantee that the decision would be as far as possible a legal and impartial one. For the same reason the appointments were almost exclusively of jurists and not of diplomats. Of the forty-nine arbitrators appointed up to this time only one can be designated as a diplomat pure and simple, namely, the Japanese, Motono. All the others were either wholly jurists, or, if diplomats by profession, persons who had won a reputation for their knowledge of international law. One need only go back to the earlier times when recourse was much more often had to mere diplomats and in particular to irresponsible sovereigns, in order to mark the progress of the development in the direction of the judicial settlement of international disputes. Such methods are milestones upon the road towards judicial settlement, which presupposes a tradition and the exclusion of mere diplomats. In recent times the desire of the states for a permanent court is particularly manifest in the fact that the arbitrators appointed to decide the case of the *Manouba* and the *Carthage* were, with the exception of Baron Taube, identical with the arbitrators of the Casablanca arbitration court.

It would not be necessary to discuss in such detail the twofold nature of the Hague Court of Arbitration if certain scholars were not continually holding up that court as the ideal institution for the settlement of international disputes. As a matter of fact, the Hague Court of Arbitration and the above-mentioned projects of unofficial congresses, in particular the project of the Interparliamentary Union at its meeting at Brussels, are only valuable in so far as they have shown mankind the way to a higher development. On the other hand, certain other projects, in particular that of Count Kamarovski, regarded in their theoretical and logical character, stand upon a noticeably higher plane.

4. The Project of the Second Hague Peace Conference

Shortly after the First Hague Peace Conference the International Law Association at its meeting at Buffalo on August 31, 1899, brought forward a project of an international court of arbitration, in which it was advocated, for the first time after the Hague Conference, that the judges of the court of arbitration should be in permanent residence at The Hague. The arbitrators appointed by the states were to work out a 'code of international law', and the states were to have complete freedom as to which judges they desired to appoint from among the members of the court in the individual case. We have already met with similar projects above. The Interparliamentary Union at its meeting at Paris in 1900 emphasized still more sharply the necessity of establishing a truly permanent court of justice and a special resolution was passed to that effect. At the second Pan American Conference[1] of 1901–2, Guatemala made a motion for the erection of a permanent court of arbitration for claims for indemnity and other money claims, but it was not adopted.

No other projects worthy of special note were proposed from that time until the Second Hague Peace Conference.

The *Institut international de la Paix* put forth in 1907 two drafts, drawn up by Vavasseur and Lepert, for a permanent court of justice, but neither was of any special value. Vavasseur proposed a court of justice consisting of sovereigns and ambassadors, which was to have direct jurisdiction to decide serious disputes.[2] Lepert advocated, as did Duplessix later, a sort of world federation of states. Much more important are the proposals made at the same time by the Dutch lawyer, Stipriaan Luiscius. In his book *L'Avenir de l'arbitrage international* he proposed that there should be in addition to the Hague Court of Arbitration a court of appeal, which was also to have jurisdiction as a court of first instance in certain specified cases. His proposal aims rather at a development of arbitration than of judicial settlement, and on many points goes too far. It is gratifying that Stipriaan Luiscius does not give any consideration to the enforcement of judicial decisions, which is at present merely a Utopian idea.

The Judicial Arbitration Court[3] proposed by Scott and

[1] Proceedings of the Conference, pp. 290 et seq.

[2] See above, p. 53.

[3] See translator's note, p. 127.

Choate at the Second Hague Peace Conference and based upon the instructions of Root[1] was in its fundamental principles for the most part an excellent proposal. Had the principle of the equality of states been recognized in it, it is very possible that the states might have agreed upon it. But it is to be regretted that the justifiable protest of the small states received so little consideration. In Scott's excellent report the distinction between arbitration and judicial settlement in international law is brought out more clearly than ever before. Scott showed that on the one hand there are disputes which the parties regard as rather political in character, and on the other hand there are disputes which the parties regard as of a strictly legal nature, and that cases involving the first class of disputes should be settled by arbitration upon the principles of equity, and cases of the second class by a real court of justice upon principles of strict law.

In this connexion it is of interest to inquire how far the Judicial Arbitration Court of the Second Hague Peace Conference, and also the International Prize Court adopted at the same time, would have fulfilled the conditions of an ideal international court of justice in case the two Conventions had been ratified.

In chapter iv we learned to recognize the following marks as characteristic of an international court of justice:

1. No diplomats, but judges by profession.
2. The exclusion of national judges.
3. A truly permanent court of justice.
4. The exclusion of members appointed by the contending parties and above all by the states.
5. A direct right of action.
6. The creation of a right of appeal.

[1] Scott, *The Hague Peace Conferences*, vol. ii, p. 191.

In contrast with these conditions both nationals and diplomats may sit upon international courts of arbitration; the court may be constituted for each particular case; a *compromis* must be entered into; a right of appeal may not be given. For the object of arbitration, namely, the settlement of disputes according to equity, will in this way be fully attained. At the same time it must, indeed, be pointed out that what is at present called arbitration is not strictly arbitration, but contains many elements of a legal nature. In order to bring about a sound development of the present form of arbitration in the direction of judicial settlement, numerous authors have advocated the exclusion of national judges and the creation of a court to which an appeal may be taken from the present arbitration courts. These improvements in arbitration would not need to be realized if we had an international judicial system. For in that case the ideas of arbitration and judicial settlement could be kept completely separate.

Great confusion existed at the Second Hague Peace Conference over the nature of the Judicial Arbitration Court. Scott,[1] Barbosa,[2] and Carlin[3] called it a 'judicial institution'; Lammasch, Mérey, Beldiman, and Martens called it an 'arbitral institution'.[4] Renault[2] and Bourgeois[5] declared that the new court at all events was an 'approach' to a judicial institution. Fry[2] even thought that arbitration and judicial settlement were the same thing in international life. Under these circumstances Beldiman[6] was entirely right when he pointed out that 'there are differences not only as to external form but also as to principle between the opinions of Messrs. Scott and

[1] pp. 316, 658. [2] p. 658. [3] p. 146.
[4] pp. 16, 658; see Moch, *Autour de la Conférence*, p. 23.
[5] p. 615. [6] p. 660.

Lammasch, and even between the opinions of the authors of the project'. Accordingly, the diversity of opinion in 1907 became of necessity all the greater when, for example, Scott justly represented an international judicial court as an ideal worthy to be striven for by the community of states, whereas Barbosa declared that 'arbitration is the sole judicial system which can be established between nations'.

The Judicial Arbitration Court is undoubtedly in its essential nature to be designated as an international court of justice, as Davis[1] also points out. Its chief function was to render judicial decisions and to develop international law. It is true that of all the individual technical requisites of an international court of justice, the only condition which it fulfilled, to be sure a very important one, was that of a permanent institution. Moreover, the judges were regarded as essentially judges by profession, but diplomats were not in consequence excluded if they possessed a knowledge of international law,[2] nor were national judges excluded.[3] An agreement could not be reached concerning the appointment of the judges; though the majority of the projects provided that the judges were to be appointed by the individual states. The parties were not given a direct right of action nor was an appeal allowed. Accordingly it must be said that the Judicial Arbitration Court was indeed in its intrinsic nature a judicial institution, but that there were present in its exterior constitution too many of the characteristics of a court of arbitration.

To come back, therefore, to the controversy at the Second Hague Peace Conference concerning the nature of the Judicial Arbitration Court, we must concede that the view of Bourgeois and Renault that the new court was an

[1] *The Elements of International Law*, p. 262.

[2] Article 2 of the Draft.

[3] Article 20 of the Draft.

approach to a 'judicial institution' can be very well maintained, although the view held by Scott does greater justice to the intrinsic nature of the court. On one occasion Bourgeois[1] remarked very appropriately that 'the prime consideration will be not to weigh interests but to give justice'.

The project of the Second Hague Peace Conference for the creation of an International Prize Court comes much nearer to the ideal of an international judicial court. This court will not merely settle disputes, but in addition develop international law by its judicial decisions. Like the Judicial Arbitration Court, it was regarded as permanent. Moreover, the parties had a direct right of action. But although the judges were to be principally judges by profession, yet diplomats were not excluded nor strictly national judges. In like manner the appointment of the judges was made by the states. The fact that no right of appeal against the decision of the International Prize Court is granted cannot, indeed, detract from the truly judicial nature of the Prize Court, because the Prize Court was to act as a court of appeal from the decisions of national courts, and was thus itself regarded as a court of appeal.

Thus in the case of the International Prize Court there are still present isolated external characteristics of international arbitration.

5. Recent American Proposals—Central American Court, Proposal of Secretary Knox, British-American Arbitration Treaty

The Central American Court established in 1907 at the first Central American Conference between the five Central American states, Costa Rica, Honduras, Nicaragua, San

[1] p. 682.

Salvador, and Guatemala, is a truly permanent court, is composed of judges by profession, and can be resorted to by a state directly upon its own motion without the necessity of a previous *compromis.* Yet even here the ideal of an international court is not quite attained, since the judges are always in part nationals of the contending parties and are appointed by them. As is well known, the court is composed of one national from each of the five republics. Moreover, no provision is made for an appeal. The fundamental idea of the Central American Court is above all to 'dispose of disputes', as is evident from the history of its establishment under the influence, more or less, of the United States. As a matter of fact the development of international law by this court would be of little value, since this can only be done by a court embracing all the states. The Central American Court of Justice, which may serve as a model in many respects for the new world court, is a peculiar and unique phenomenon, which has only arisen in consequence of the state of unrest in Central American relations. We experience here the singular fact that a permanent judicial court exercises chiefly arbitral functions.

On October 18, 1909, Mr. Knox, the Secretary of State of the United States, surprised the world with an important circular note in which he proposed to the states participating in the Second Hague Peace Conference that the International Prize Court should be ratified under an arrangement that it should at the same time be authorized to perform the functions of a court of arbitral justice. This circular note read as follows: [1]

> It has been a subject of profound regret to the Government and people of the United States that a Court of

[1] *American Journal,* Supplement, 1910, p. 102.

Arbitral Justice,[1] composed of permanent judges and acting under a sense of judicial responsibility, representing the various judicial systems of the world and capable of insuring continuity in arbitral jurisprudence, was not established at the Second Hague Peace Conference, and the United States likewise regrets that the composition of the proposed Court of Arbitral Justice has not yet been effected through diplomatic channels, in accordance with the . . . recommendation of the Conference. . . .

A careful consideration of the project and of the difficulties preventing the constitution of the court, owing to the shortness of time at the disposal of the Conference, has led the Government of the United States to the conclusion that it is necessary in the interest of arbitration and the peaceful settlement of international disputes to take up the question of the establishment of the court as recommended by the recent Conference at The Hague and secure through diplomatic channels its institution.

The necessary and close connexion between the International Prize Court and the proposed Court of Arbitral Justice was indicated in Article 16 of the Draft Convention of the Court of Arbitral Justice as follows:

> The judges and deputy judges, members of the Judicial Arbitration Court, can also exercise the functions of judge and deputy judge in the International Prize Court.

The reason which existed in 1907 and led to the formulation of the articles still continues. It has therefore occurred to the United States that the difficulty in the way of reaching an agreement upon the composition of the court would be obviated by giving practical effect to Article 16 by an international agreement by virtue of which the judges of the International Prize Court should be competent to sit as judges of the Court of Arbitral Justice for such nations as may freely consent thereto, and that when so sitting the judges of the International Prize Court shall entertain jurisdiction of any case of arbitration submitted by a signatory for

[1] See translator's note, p. 127.

their determination and decide the same in accordance with the procedure prescribed in the Draft Convention. In proposing to invest the International Prize Court with the jurisdiction and functions of the proposed Court of Arbitral Justice the United States is actuated by the desire to establish a court of arbitration permanently in session at The Hague for the peaceful solution of controversies arising in time of peace between the nations accepting and applying in their foreign relations the principles of an enlightened and progressive international law.

It is a truism that it is easier to enlarge the jurisdiction of an existing institution than to call a new one into being, and as the judges and deputy judges of the International Prize Court must be thoroughly versed in international law and of the highest moral reputation, there can be no logical or inherent objection to enlarging their sphere of beneficent influence in vesting them with the quality of judges of the proposed Court of Arbitral Justice.

The proposal of the United States does not involve the modification either of the letter or spirit of the Draft Convention, nor would it require a change in wording of any of its articles. It would, however, secure the establishment of the Court of Arbitral Justice as a chamber of the world's first international judiciary and thus complete through diplomatic channels the work of the Second Hague Conference by giving full effect to its first recommendation.

In proposing this solution of the difficulty the United States is influenced by daily practice and procedure in its national courts of justice, where one and the same judge administers law and equity, admiralty and prize, which, under its system of procedure, are different systems of law. The United States therefore proposes that in the instrument of ratification of the International Prize Court Convention, signed at The Hague October 18, 1907, any of its signatories consenting to invest the International Prize Court with the powers of a Court of Arbitral Justice shall signify its assent thereto in the following form :

> *Whereas,* It is highly desirable that the Court of Arbitral Justice, approved and recommended by the Second Hague Peace Conference, be established through diplomatic channels; and
>
> *Whereas,* Investing the International Prize Court with the duties and functions of the proposed Court of Arbitral Justice would constitute for the consenting powers the said Court of Arbitral Justice, as recommended by the first *vœu* of the Final Act of the said conference;
>
> Therefore, the Government of . . . . . . agrees that the International Court of Prize, established by the Convention signed at The Hague October 18, 1907, and the judges thereof, shall be competent to entertain and decide any case of arbitration presented to it by a signatory of the International Court of Prize, and that when sitting as a Court of Arbitral Justice the said International Court of Prize shall conduct its proceedings in accordance with the Draft Convention for the establishment of a Court of Arbitral Justice, approved and recommended by the Second Hague Peace Conference on October 18, 1907.

At the sixteenth Lake Mohonk Conference, held in May 1910, Knox's proposal was warmly advocated by almost all of the speakers. Trueblood and Baldwin alone expressed their doubts about it. The Conference had, to be sure, a special reason for looking upon the project with great sympathy, because at the first conference at Lake Mohonk, in the year 1895, Dr. Hale had made an almost similar proposal. Hale advocated at that time a plan that Austria-Hungary, Germany, France, Great Britain, United States, and Russia should agree upon the organization of an international court of justice and invite the other states to appoint members. Thus the great powers would, by this proposal also, have imposed their will upon the other states in the appointment of the members of the new court.

At the Lake Mohonk Conference of 1910, Macfarland [1] pointed out that it was extremely probable that the new Court of Arbitral Justice would be established separately as an independent tribunal and not as a chamber of the Prize Court. From this very fact it is evident the most zealous friends of Knox's project are not all of the same opinion, and that they regard the project merely as a practical means of advancing nearer to the goal of an international court of justice.

It is clear from what has been said, that the principles which led me to reject the whole system of rotation must also lead me to regard the realization of Knox's project, which was warmly taken up by the Universal Peace Congress at Stockholm in 1910,[2] as not marking progress in the development of international law. It does not guarantee an impartial composition of the court for the great and small states alike, as Trueblood [3] and Count Apponyi [4] have rightly pointed out. At the Lake Mohonk Conference Marburg [5] adduced a quite original reason in favour of the unequal apportionment of judges. He declared that if all nations of the world were represented in equal proportion upon the court the European powers would view the new court with suspicion; for the United States was so preponderant on the American continent that the European powers would always believe that the South and Central American judges were, even though unconsciously, under the influence of the United States. If we are to act upon such conjectures we shall never succeed in reaching an agreement upon the composition of a court of justice.

[1] Report, p. 77. [2] Bulletin, p. 55. [3] Report, 1910, p. 30.
[4] *Union interparlementaire*, 1910, p. 133.
[5] Report, 1910, p. 86.

There is a further objection to Knox's proposal, which has been pointed out by von Plener also,[1] namely, that the International Prize Court itself would doubtless be called into question if the functions of an international court of justice for disputes arising in time of peace were also to be assigned to it. The distribution of seats upon the Prize Court is of such a kind as to favour the great powers. But as the Prize Court possesses a quite unique jurisdiction with regard to the laws of maritime warfare, in which the interests of the smaller states are not specially involved, they have for the most part given their assent to the apportionment of judges, even if in some cases only with reluctance. But it cannot be thought that these states would allow themselves to be degraded to the rank of second-class powers when there was a question of an institution so important and comprehensive (because not limited to a special field) as the creation of a world court of justice. Accordingly, they will positively reject the International Prize Court if the functions of an international court of justice be conferred upon it.

On the other hand, the further reasons adduced by von Plener against the proposal of Secretary Knox are not in point. In the first place the assertion that the Prize Court is a judicial court, and that the functions of an arbitration court cannot, therefore, be conferred upon it, is incorrect.[2] On the contrary, it is to be observed that according to Knox's proposal the Prize Court is to exercise judicial, not arbitral functions. In answer to von Plener's further assumption that the Prize Court is not a truly permanent one and will meet only in time of war, and that the judges are, moreover, specialists and therefore not versed in all

[1] *Union interparlementaire*, 1910, p. 99.

[2] See also Baldwin, Report, 1910, p. 81.

questions of international law,[1] it is evident from the nature of Knox's proposal that thenceforth the court would sit permanently and that, as Fried [2] has already pointed out, in the appointment of judges consideration would be given to a knowledge of international law as a whole. Hence only trifling modifications of the Prize Convention would be necessary, and these could be agreed upon in a supplemental act.

Nevertheless Knox's project must be rejected for the above-mentioned reasons. At the sixteenth meeting of the Interparliamentary Union at Brussels in September 1910 all the speakers, with the exception of those from the United States, spoke against the proposal. In particular, von Plener, Count Apponyi, and Franck opposed it vigorously. Nevertheless, as Eickhoff reports in the *Tag*,[3] he, as well as certain other continental members of the Conference, did not share in this view and were inclined to adopt Knox's proposal. That the opponents of the project won the upper hand at the Conference is evident from the resolution taken by the Conference, which reads as follows:

> The Interparliamentary Conference, in consideration of the important results to be expected from the adoption of the International Prize Court both in a general way and with respect to commercial interests, recommends that the Convention of October 18, 1907, relating thereto be ratified as quickly as possible, independently of any question relating to the organization of a permanent judicial court, although the Conference expresses its respect for the motives which have led the Government of the United States to make a proposal of this kind.

While Fried [2] thinks that the temporary nature of Knox's proposal must be borne in mind, since he only desired to

[1] See also Baldwin, op. cit.
[2] *Friedenswarte*, 1910, p. 186.
[3] September 24, 1910.

settle the question up to the time of the next Hague Peace Conference, it is my opinion that this problem is much too important to permit of a temporary settlement, and that we had better have no world court at all than a bad one.

The proposal of Secretary Knox had no prospect of being accepted after a representative of the German Foreign Office, in answer to an inquiry of Eickhoff, declared, on March 27, 1911, that the German Government could not agree to Knox's proposal. Accordingly the prophecy made by Pohl [1] has been fulfilled. In the face of the opposition of one of the great powers, so important a problem naturally cannot be solved.

As is pointed out in the *Friedenswarte*,[2] after the rejection of his first proposal Knox presented a new one to the powers. This proposal provided for the composition of the Judicial Arbitration Court in such a way that the small and medium-sized states were represented to the same extent as the great powers. This proposal must also be rejected, because it is in like manner based upon a form of rotation. If the sound principles of international law are not to be abandoned, there can be no question of any kind of unequal apportionment of judges, upon whatever ratio it is based. Only when all states are represented alike can the proposal be advocated, although, as we have pointed out, a complete guarantee for the impartiality of the judges only exists when they are not appointed directly by the states.

It would appear that Knox's proposal in its modified form has met with greater approval. However that may be, Knox had it announced at the Lake Mohonk Conference in 1911 that the International Court of Justice would be erected before the Third Hague Peace Conference,

[1] *Deutsche Prisengerichtsbarkeit*, p. 210. [2] 1911, p. 55.

because the idea had recently made great progress with all the nations. According to a report of January, 1912, Knox was said to have received favourable answers from the governments as a body. Nevertheless I have my doubts as to the realization of the project. The address of Foulke [1] at the Lake Mohonk Conference of 1911, in which he declares that Knox's project was in principle the best settlement of the problem, and then proceeded to say that greater consideration would have to be shown to China in the apportionment of judges than was the case in the project for the Prize Court, tells the real truth. It is quite impossible to find a satisfactory solution by way of the system of rotation.

The legislative factors in international law alone force us to combat vigorously the temporary composition of a court with an unequal representation of the states. For if the powers are represented equally upon the court of justice they will later on relatively easily give their consent to have the judges appointed no longer by themselves but by an impartial organ. But the situation is entirely different if the states are represented upon the tribunal unequally. In such case the final goal of a truly ideal court of justice, whose members are no longer appointed by the states but by an impartial body of officials, can only be attained by the renunciation on the part of the great powers of the position of preponderance upon the court which they had previously had. Thus the slight need of a temporary settlement of the question will be purchased at the cost of making it very much harder to attain the goal of a completely ideal judicial system.

Moreover, the states must picture to themselves the following result of the system of rotation, which is pointed

[1] Report, p. 97.

out by Th. R. White.[1] As each state is to be represented upon the international judicial court in proportion to its external strength, all the states would thenceforth increase their armies and their navies all the more, in order to obtain greater respect and a correspondingly better representation upon the court, and thus the peace of the world would be all the more endangered.

Coming now to the last American proposal in this connexion, it must be pointed out that the project of an arbitration treaty between the United States and Great Britain is from the standpoint of the development of international law open to grave doubts. According to this project all differences which are to be settled by arbitration shall, at the request of one of the parties, first be referred to a commission of inquiry which is composed of three nationals of each of the parties, without a neutral umpire. The commission of inquiry is authorized to examine into and report upon the differences referred to it, for the purpose of facilitating the solution of disputes by elucidating the facts, and further to include in its report such recommendations and conclusions as may seem appropriate to it. The findings of the commission shall not be regarded as decisions of the questions so submitted either on the facts or on the law, and shall in no way have the character of an arbitral award. The reference to such a commission of inquiry may, if either party so desires, be postponed until the expiration of one year after the date of the formal request therefor in order to afford an opportunity for diplomatic discussion and adjustment of the questions in controversy.

The second task of the commission of inquiry consists in deciding upon a difference of opinion between the parties

[1] Ibid., 1911, p. 104.

as to whether the dispute comes within the scope of the arbitration treaty, that is to say, whether it is justiciable in its nature by reason of being susceptible of decision by the application of the principles of law or equity. If in this case all the members of the commission, or all but one of them, agree and report that the difference is of the above-mentioned kind it shall be referred to arbitration.

Against the second duty of the commission of inquiry I have not the slightest objection to raise. But it seems to me that in the first duty of the commission there easily lies the danger that many disputes which by their nature can best be settled in a strictly legal way, thereby creating precedents for future decisions, will thenceforth be adjusted by diplomacy and by a sort of compromise—a result which cannot be justified. The proposal would be well founded only if difficult political disputes were at issue, the settlement of which by war could be prevented by the efforts of the commission of inquiry. But the activities of the British-American commission of inquiry are to extend also to those disputes the possible culmination of which in war is entirely out of the question. Under these circumstances the strictly legal settlement of disputes will be hindered by the efforts of the commission of inquiry, without any other valuable result being obtained. In this connexion it must be remembered that before the dispute is referred to the commission of inquiry diplomatic negotiations must have taken place and have produced no result. Accordingly, the commission of inquiry is in all cases dealing with disputes the diplomatic settlement of which is unusually difficult. It is precisely such disputes which call for a decision which will clear the way for the future. The report of the commission of inquiry will have the appearance of a decision upon grounds of equity and will be like the numerous arbitral awards of the last century. Should the parties agree upon the report of the commission of inquiry, and should a similar dispute arise later in international life, we are no wiser than before and must again first go to the trouble of diplomatic negotiations and afterwards set up another commission of inquiry.

But in spite of these doubts I must insist that I do not desire the overthrow of the principle of the British-American arbitration treaty, which in the abstract is a valuable one. In my article 'The British-American Arbitration Treaty', which appeared in the third part of the *Zeitschrift für Völkerrecht und Bundesstaatsrecht* for 1912, I warmly advocated the new project.

CHAPTER VIII

THE PROCEEDINGS OF THE SECOND HAGUE PEACE CONFERENCE WITH REGARD TO THE JUDICIAL ARBITRATION COURT

HAVING set forth the principles involved in the recent proposals for the creation of an international judiciary, it will be useful to describe the progress of the proceedings of the Second Hague Peace Conference, chiefly in order that we may obtain an idea of the position taken by the individual states towards this important problem. The project was discussed successively by the subcommission, by the committee of examination, as well as by the Commission, and at the plenary meeting of the Conference.

1. THE DISCUSSION IN THE SUBCOMMISSION

At the opening of the proceedings there were laid before the Conference two proposals, one supported by Russia and the other by the United States. The first subject for consideration was whether a truly permanent court of justice was in principle desirable.

On August 1, 1907, Choate [1] made a long introductory speech before the first subcommission in favour of the United States proposal. He insisted that the new proposal did not involve a disparagement of the work of the First Hague Peace Conference.[2] He pointed out that even in

[1] p. 309.

[2] As a matter of fact the United States delegates intended in the beginning to substitute the new judicial court for the old Court of Arbitration. See Zorn, *Zeitschrift für Politik*, vol. ii, p. 350.

1899 the Court of Arbitration was regarded merely as an experiment. Only four cases had thus far been referred to the Court of Arbitration, and only one-third of the judges inscribed upon the list of the Hague Court had exercised judicial functions. Having pointed out the defects of the old Court, he passed on to speak of the fundamental principle of the new judicial court. He asserted that the new judicial court must have at least seventeen judges, who were to be appointed by all the states great and small; and that the various legal systems, languages and the great human interests must be represented upon it. Each year the judicial court must meet for a fixed period in order to settle the cases submitted to it. Apart from a special agreement to the contrary no judge was to be allowed to act in a dispute to which the state of which he was a national was a party, for a judicial court and not a mixed commission was desired. The jurisdiction of the court, he said, extended to all disputes which the parties might submit to it. The judges who, as far as possible, were to be appointed from among the members of the old Court of Arbitration, were to be allowed to be members of the commission of inquiry also or of other special arbitration courts. The American delegation, he said, did not insist absolutely upon the details of its proposal, but merely upon the principle of the establishment of a truly permanent court; the Conference would betray the interests of mankind entrusted to it, if it did not succeed in reaching an agreement upon this proposal.

After Choate had spoken, Scott,[1] the honoured technical delegate of the same country, took the floor. He cited first of all a portion of the notable address made by Root at the opening of the New York Peace Conference on

[1] pp. 313 et seq.

April 15, 1907. He pointed out that the American proposal was based upon the express instructions of the Secretary of State, while the exposition of the general principles of the project before the committee was the work of Choate. He then in like manner criticized the Court of Arbitration of 1899, and showed the necessity for and laid out the foundations of a permanent judicial court. In particular he asserted that while a long experience in the field of international politics could only be of use to the judges, yet this experience alone, as well as a training in diplomacy, could not supply for the judicial spirit which was to be looked for above all in the new judges; accordingly it was necessary to exclude diplomats from the new judicial court. Scott then set himself[1] to the difficult task of preparing the small states for their position of inferiority upon the new court. He declared with firmness that each state must have a judge upon the court, and that the principle of the equality of states should not be set aside. But although the states were equal, he said, yet there were great differences between them in so far as their material interests were concerned. He then showed that in the United States one state had more courts than another. But when he inferred from the fact that in like manner the states with a larger population must be more strongly represented upon the international court his reasoning was not conclusive. Having set forth the distinct principles which were to prevail in the composition of the new judicial court he concluded with the words: 'Through justice towards peace.'

After him the Russian jurist, Martens,[2] made a powerful and winning address in favour of a permanent judicial court. He observed that the Russian delegation had

[1] p. 317. [2] p. 321

already, in 1899, proposed a court of arbitration with a more permanent organization; but that in the spirit of friendly concession the English draft was then agreed upon. After he had set forth separately the principal features of the Russian proposal, he concluded with the following words which inspired the Conference:

> Permit me to say a few words more which come from the bottom of my heart. There have always been in history periods when great ideas dominated and led away men's minds; at one time it was religion, at another a new philosophy, and again a political idea. The Crusades are the most striking illustration of this. In all countries was heard the cry: 'To Jerusalem; God wills it.' The great idea which dominates our time is that of arbitration. When a dispute arises between nations . . . the unanimous cry has been heard since the year 1899: 'To The Hague.'

After he had spoken, Marschall,[1] in the name of Germany, de la Barra,[1] in the name of Mexico, Fry,[2] in the name of Great Britain, Larreta [2] and Drago,[3] in the name of the Argentine Republic, expressed their sympathy with the United States proposal. Drago, however, put the condition that the composition of the judicial court should correspond to the claims of the individual powers. He asserted that, in the opinion of the Argentine Republic, each state must be given a representation upon the court corresponding to its foreign commerce, because commerce and production were the best proof of the culture and of the progress of a nation. This last remark made by Drago shows that the South American states were not in the beginning opponents of the system of rotation, but first became so when an inferior apportionment was assigned to them in that system.

[1] p. 323. [2] p. 324. [3] p. 325.

On August 3, 1907, the discussions of the first sub-commission were renewed. Beernaert, a distinguished Belgian jurist of long experience in international law, honorary president of the Interparliamentary Union, and an old champion of the cause of arbitration, delivered a long address [1] against the American proposal of a permanent court. The numerous objections brought forward by him have been discussed earlier in this work. He asserted that only judges appointed by the parties themselves were in keeping with the system of arbitration ; that Martens had pointed out that they must have jurists instead of diplomats, but that on the contrary it should be observed that there already existed in the form of the old Court of Arbitration a ' judicial institution ', for according to the Hague Convention, disputes must be settled upon the basis of respect for law. In the four cases thus far decided, he said, distinguished judges had done excellent work ; and if so few disputes had been referred to the Court of Arbitration that was the fault of the governments alone. He referred appropriately to the earlier opinions of Mérignhac and Bourgeois.

Esteva [2] next spoke in the name of the Mexican Government. He, also, expressed his approval in principle of the American proposal, but asserted that the acceptability of the new court depended upon its composition. Milovanovitch,[2] in the name of the Serbian Government, also imposed this last condition. He asserted that either the absolute equality of states must be preserved or else the plan of a representation according to states must be abandoned altogether. Further, Serbia could only find the new court necessary after an obligatory world arbitration treaty had first been entered into.

[1] p. 331. [2] p. 335.

Porras,[1] the delegate of the little American Republic of Panama, assented to the plan of a permanent court of justice in a noteworthy address. He pointed out that Panama had an interest in the decision of international disputes by judges instead of by governments ; for it had more confidence in honourable judges than in a government, which at best furnishes a political settlement of the dispute instead of a legal decision. The view had often been expressed, he said, that arbitration was made use of by the great powers to meddle with the disputes of smaller states. In this respect a judicial court offered other guarantees.

Léger,[2] the delegate from Haiti, likewise advocated a permanent judicial court. He laid special stress upon the fact that the judges would possess the confidence of the powers, and proposed with that object in view that the judges should be forbidden to accept decorations or rewards from any other governments than their own. He was, moreover, more in favour of the Russian than of the United States proposal. He proposed that each state should appoint one judge ; that the forty-five judges should then together choose the judicial court ; and that the members thus chosen should themselves choose their successors, and should be divided into sections which should sit for longer or shorter periods. In this way the whole body of judges would never be replaced by new ones all at once, so that the constant presence upon the court of judges who had previously sat upon it would establish a tradition and would make it possible to create and to develop international law. Finally, Léger [3] proposed to entrust the judicial court with the codification of international law according to one of the numerous plans mentioned in chapter vii.

[1] p. 335. [2] p. 336. [3] p. 338.

Fortoul,[1] the representative of Venezuela, made objection to the principles in accordance with which the United States draft sought to determine the apportionment of judges. He called attention to the fact that the population, commerce, manufactures, &c., of a country were subject to change, and he advocated that the separate continents of America, Europe and Asia should each appoint an equal number of judges for the court, taking indeed into consideration the languages most widely in use.

After him Barbosa,[1] the impassioned representative of Brazil, in a long address set forth the point of view of his country. There were two distinct methods, he said, which it was absolutely necessary to separate. Obligatory arbitration could be accepted for all international disputes without the states binding themselves in any way with respect to a court before which disputes should be brought for decision; on the other hand, states might subject themselves to a judicial court and yet limit this subjection to a trifling number of cases. Thus far arbitration had always been spoken of as the sole method of the juristic settlement of international differences, and the free choice of the judges had been left to the states; it was now sought to change that. It was sought to create an international judiciary of the same kind which exists for the legal relations between private persons; yes, a further step still was advocated. Private individuals could by mutual agreement refer an affair to a court of arbitration instead of a judicial court; but in the case of states it was sought to make a judicial court obligatory. In contradiction to Porras he insisted that it was precisely sovereigns who had rendered many excellent decisions, whereas the decisions of

[1] p. 339.

arbitration courts composed of reputable jurists had often been criticized by the whole world. The establishment of a permanent court of arbitration was, he said, inconsistent with the sovereignty of states.

Karandjouloff,[1] in the name of Bulgaria, expressed himself heartily in favour of a permanent judicial court. He advocated that each state should appoint one judge, and that the judges so appointed should choose the members of the judicial court. Moreover, he did not wish to see nationals excluded from the judicial court.

Thereupon Fry [2] said a few words in reply to Beernaert. Beernaert, so Fry asserted, would be right in his opposition to the new arbitration court if it were desired to set up the new court in place of the existing one; but that was not the case; the two courts would be equally open to the use of the nations.

The Portuguese, Marquis de Soveral,[2] said that it had been asserted by Choate that the creation of a permanent judicial court was the chief work before the Conference; but that was not true; the question of an obligatory world arbitration treaty was more important.

Momtaz-es-Saltaneh,[2] the representative of Persia, voiced the adherence of his government to the new project.

Thereupon Bourgeois [3] yielded the chair of the subcommission to de Beaufort in order to speak in his character as first delegate of France. He said in particular that there was no question of suppressing the old Court of Arbitration which existed essentially for political differences, but that it was a question of creating a court of justice in the true sense of the word, which would act within a limited field, namely, that of strictly legal disputes. There were even disputes, he said, for which the Court of

[1] pp. 344 et seq. [2] p. 346. [3] p. 347.

Arbitration of 1899 was preferable, and others which would be better settled by the new judicial court; in the one case arbitrators were wanted, in the other judges. In conclusion, Bourgeois professed his adherence to modern pacifism in the following noble words:

> For centuries men believed only in the rule *si vis pacem, para bellum,* that is to say, they limited themselves to the military organization of peace. We have gone beyond that, but it is not enough for us to establish a more humane organization, I was about to say, the peaceful organization of war.[1]

Beldiman,[1] the Roumanian delegate, desired that before the proposal be referred to the committee of examination the United States should give more information as to the manner in which the seventeen judges of the judicial court were to be appointed, and he proposed that it should be expressly provided in the Convention that the states should have complete freedom to choose between the old and the new Court of Arbitration.

Castro [2] accepted the proposal in the name of Uruguay, but declared that he wished to withhold his vote until the composition of the Judicial Court was known.

Réchid Bey,[2] the Turkish representative, said that he would withhold his vote in the absence of instructions from his government.

Beldiman [2] declared that as he had received no answer to his inquiry from the delegation of the United States, he desired to abstain from voting.

A vote was thereupon taken upon the abstract question of the creation of a permanent judicial court, and although the details of the American proposal with respect to the

[1] p. 349. [2] p. 350.

manner in which the court was to be constituted were not yet known, the proposal was adopted by twenty-eight votes with twelve abstentions. The states abstaining were Austria-Hungary, Belgium, Denmark, Spain, Greece, Norway, Roumania, Serbia, Siam, Sweden, Switzerland, and Turkey.

The United States and Russian proposals were thereupon referred to committee *B* of the first subcommission.

2. The Discussion in Committee of Examination *B*

The committee of examination had before it at its first meeting on August 13, 1907, in addition to the Russian proposal which was not further discussed, merely a common British-German-United States proposal, which was taken as the basis for further discussion.

Choate [1] first of all pointed out once more the importance of the new project and proposed that an agreement should first be reached upon the principles and that the question of the determination of the judges should be left to the end.

Marschall [1] supported the remarks of Choate and referred to a statement made by one of the speakers that it was the fault of the governments if too few cases had been referred to the old Court of Arbitration of 1899. He did not intend, he said, to examine into the justice of that reproach; but if that was the case, it must be made as difficult as possible for governments to fall back into their old faults. Finally, he insisted that optional arbitration with an excellent court was better than obligatory arbitration necessarily limited to a small field.

[1] p. 594.

Fry [1] assented to the remarks of both speakers.

Asser [1] dwelt particularly upon the time consumed by and the cost involved in the old Court of Arbitration, and called attention to the fact that Holland also had had the intention of introducing a proposal with the object of perfecting a court of arbitration, but had abandoned the idea in view of the other proposals which went further.

Bourgeois [1] directed his remarks chiefly against the statements made by Marschall. He insisted that the chief problem before the Conference was indeed that of obligatory arbitration. Marschall had, he said, compared the new Court of Arbitration to a magnificent frame; he begged them all to put a picture into that frame.

Beldiman [2] renewed the request which he had already made in the subcommission to be informed as to the method of appointing the judges; otherwise it would be impossible to form any picture of the British-German-United States proposal.

Scott [2] promised that the composition of the new court would be made known at the next meeting.

In the subsequent discussion of the first six articles of the project there were differences of opinion as to the name of the new judicial court.[3]

Choate [4] finally declared that the name was a matter of secondary importance; when the child was once baptized, he said, it would not be its name but its deeds which would determine its career. But Bourgeois [4] thought that it was not the name only, but rather the sex of the child, which was at issue.

At the second meeting of committee *B* on August 17,

[1] p. 595.
[2] p. 596.
[3] See above, p. 125.
[4] p. 598.

1907, Lou Tseng-tsiang [1] declared in the name of China that the standing of the powers in the Universal Postal Union should be taken as the basis of the new judicial court. If China were not given due consideration, he said, his government would withhold its vote.

After the discussion of some details, Scott [2] set forth the scheme of the British-German-United States project with regard to the apportionment of judges. He said that the question of the composition of the court of justice would be a simple one if each country were to appoint one judge and the judges so appointed were to form the court; but in that case they would have a court of forty-five members, which would be no longer a 'court', but a 'judicial assembly'. He developed the factors which were to be taken into consideration in the distribution of judges, and insisted that the principle of the equality of states would not be abandoned by making distinctions of any kind in the length of tenure of the judges. In addition, he said, each nation would be at liberty to combine with other nations in order to appoint one judge in common, who would be allowed to sit for a longer period. The plan proposed was the result of long work; it was not perfect, but could well furnish a basis for discussion.

It has been shown above in chapter iv, section 5, how many years out of a period of twelve years the great powers and the other states were to be represented upon the tribunal. As is well known, the great powers are to be represented for twelve years continuously by a judge and a deputy judge, whereas the other states are to be represented for only ten years, four years, two years, and one year. The rotation according to which the medium-sized and small states were to be represented during the

[1] p. 602. [2] p. 606.

twelve years by a judge and a deputy judge was as follows:

First year	*Second year*	*Third year*	*Fourth year*
1. Argentine.	1. Argentine.	1. Brazil.	1. Brazil.
2. Belgium.	2. Belgium.	2. Chile.	2. Chile.
3. Bolivia.	3. China.	3. Costa Rica.	3. Cuba.
4. China.	4. Columbia.	4. Denmark.	4. Denmark.
5. Spain.	5. Spain.	5. Spain.	5. Greece.
6. Netherlands.	6. Netherlands.	6. Greece.	6. Netherlands.
7. Roumania.	7. Roumania.	7. Netherlands.	7. Portugal.
8. Sweden.	8. Sweden.	8. Portugal.	8. Siam.
9. Turkey.	9. Turkey.	9. Turkey.	9. Turkey.
Fifth year	*Sixth year*	*Seventh year*	*Eighth year*
1. San Domingo.	1. Bulgaria.	1. Argentine.	1. Argentine.
2. Ecuador.	2. Spain.	2. Belgium.	2. Belgium.
3. Spain.	3. Guatemala.	3. China.	3. China.
4. Mexico.	4. Haiti.	4. Spain.	4. Spain.
5. Norway.	5. Luxemburg.	5. Honduras.	5. Nicaragua.
6. Netherlands.	6. Mexico.	6. Netherlands.	6. Netherlands.
7. Servia.	7. Norway.	7. Roumania.	7. Roumania.
8. Switzerland.	8. Persia.	8. Sweden.	8. Sweden.
9. Turkey.	9. Switzerland.	9. Turkey.	9. Turkey.
Ninth year	*Tenth year*	*Eleventh year*	*Twelfth year*
1. Brazil.	1. Brazil.	1. Spain.	1. Bulgaria.
2. Chile.	2. Chile.	2. Mexico.	2. Spain.
3. Denmark.	3. Denmark.	3. Norway.	3. Mexico.
4. Spain.	4. Greece.	4. Netherlands.	4. Montenegro.
5. Greece.	5. Paraguay.	5. Peru.	5. Norway.
6. Panama.	6. Netherlands.	6. Salvador.	6. Persia.
7. Netherlands.	7. Portugal.	7. Servia.	7. Switzerland.
8. Portugal.	8. Siam.	8. Switzerland.	8. Uruguay.
9. Turkey.	9. Turkey.	9. Turkey.	9. Venezuela.

Kriege[1] thereupon gave some explanations. Each state could, he said, have its own deputy judge, but several states could combine together in order to appoint one judge and one deputy judge; likewise Belgium's judge, for example, could at the same time be Switzerland's deputy

[1] p. 613.

judge, and Switzerland's judge could be Belgium's deputy judge; the table might even designate the states whose judges, in case of need, were to act as the deputy judges of other states.

Barbosa [1] observed that it had been agreed that the discussion of the scheme of rotation should not take place until forty-eight hours after the table had been distributed, and yet the discussion had already begun. But Bourgeois [1] insisted that the agreement had not been violated.

After Lou Tseng-tsiang [1] had next pointed out in the name of China that in the apportionment of judges due consideration had not been given to the population of that country, the discussion was confined to the details of the project.

At the third meeting of the committee on August 20, 1907, Barbosa [2] spoke with great vehemence against the system of rotation and introduced as an offset to it the proposal of Brazil above described.[3] He demanded that the British-German-United States proposal should be brought before the subcommission for discussion, because it set aside the principle of the equality of states expressly adopted by the First Hague Peace Conference, and the committee was not authorized to make such a project the basis of its deliberations.

Beldiman [4] and Fry [4] likewise said that the committee was not competent to pass upon a new proposal, while Nelidow, Marschall, Martens, and Choate [4] were of the opposite opinion and desired to see the question settled by reference to the precedents in the committee of 1899. Choate [4] and Scott [5] contradicted the assertion made by

[1] p. 613. [2] p. 618. [3] See pp. 72 et seq
[4] p. 623. [5] p. 624.

Barbosa that the United States proposal was inconsistent with the equality of states.

Barbosa [1] did not desire to insist absolutely upon his point that the affair should be referred to the subcommission. He said that he knew well what were the views of the majority of the committee. Nevertheless, he once more supported his point of view in an elaborate address. He remarked with great irony that persons thought they were acting in accordance with the principle of equality of states when one state was represented for twelve years, another for ten, four, and one year upon the court; why were the poor South American states not given a judge for only twenty-four hours? Was not the difference between twelve years and twenty-four hours a small one? It was a question, he said, of the practical application of the principle of the equality of states, and the present system ran diametrically contrary to it. A distinction should be made between the two different rights, that of appointing and that of sitting. In the right of appointment all the states were fully equal. But in the right of sitting upon the court there was an absolute inequality; and it was this very inequality to which he was opposed.

After Bourgeois and Scott [2] had once more spoken in favour of the competency of the committee, details of the project were taken up and passed upon.

At the next meeting of the committee on August 24, 1907, the Persian delegate, Momtaz-es-Saltaneh,[3] protested that his country, which could look back upon a glorious past of a thousand years, had been degraded to the rank of a power of the fourth class. The Conference had met, he said, with the object of securing the triumph of right, not of might, and Persia would vote against the project because it

[1] p. 624. [2] p. 627. [3] p. 632.

violated the principle of the equality of states. The rest of the meeting was taken up with discussions over particular points of the draft.

At the opening of the fifth meeting of the committee on August 27, 1907, Barbosa[1] vigorously renewed his attack upon the system of rotation. He declared that it did away with the principle of the equality of states, which alone up to that time had been the 'last check upon the ambition and the pride of nations' and which had been expressly confirmed by the First Hague Conference. The British-German-United States proposal embodied nothing but injustice; it degraded the South and Central American states to powers of the third, fourth, and fifth class. If Europe and the United States, he said, had a better knowledge of his continent they would not have made such a proposal; the South American states had in no way arrived at the end of their development, but were in the prime of their strength and had a great future before them. Brazil was, he said, a sovereign state, and would have nothing to do with the grades proposed by the system of rotation. Root had once said at the Third Pan American Congress that the United States regarded the rights of the smallest and weakest states as equal with those of the strongest nations. Now was the opportunity, he said, to put that principle into practice. If it were violated, the cause of peace would not be advanced. It was true that some states were weaker than others, but was not that all the more reason for protecting the weak against the strong? Disputes between the great powers were rare; it generally happened that differences arose between small and large states. The whole project was arbitrary, he said; a new court was unnecessary, and the committee should confine its

[1] pp. 643 et seq.

efforts to improving the old court. The number of judges was first fixed arbitrarily at seventeen, and now it was sought to adjust the sovereignty of states to that number, instead of starting out with the equality of states and determining the number of judges upon that basis. Moreover, judicial settlement and arbitration were being confused with each other; in international life arbitration was the only possible method; for judicial settlement presupposed a superior authority. The Brazilian draft, he said, unlike that of the United States, was in keeping with the principles of international law.

Esteva [1] protested in the name of Mexico against the draft, which he said violated the principle of the equality of states. The discussion was then given over to details.

At the sixth meeting of the committee on September 2, 1907, the project was merely discussed in its details. The fundamental significance of the proposal was touched upon only once by Barbosa,[2] when Lammasch brought up the question of the name of the new court.

At the seventh meeting of the committee, on September 5, 1907, the greater part of the discussion was likewise taken up with details of the draft; it was only at the end that Choate [3] and Barbosa [4] delivered lengthy addresses. Choate appealed to the good-will of the committee, gave a review of all the proposals thus far made with regard to the composition of the judicial court, and expressed the hope that an agreement would be reached upon some basis or other. Barbosa replied to the attacks which the Brazilian proposal had met with from the press.

The committee then appointed a subcommittee consist-

[1] p. 650.
[2] p. 658. The greater part of his address has been given above, p. 121.
[3] p. 683.
[4] p. 687.

ing of Bourgeois, as president, Nelidow, Tornielli, Choate, Marschall, Barbosa, Fry, and Mérey,[1] in order to reach an agreement upon the question of the composition of the judicial court.

At the eighth (final) meeting of committee *B* the delegates of Roumania, Belgium, Mexico, Greece, and Brazil [2] declared that Choate had asserted at the last meeting that the committee had unanimously declared itself in favour of the new court of arbitration; that this was not at all the case, since they had as a body withheld their assent.

Nelidow [2] thereupon presented a report upon the work of the subcommittee. He said that the British-German-United States project had received no support, and the system of rotation had accordingly been dropped; that a new system had thereupon been proposed according to which the members of the Court of Arbitration of 1899 were to choose from among themselves fifteen to seventeen judges; that the objection had been made that the members of the old Court of Arbitration were not exclusively jurists, and accordingly did not offer the necessary guarantees for a proper selection. It had also been proposed, he said, that each government should appoint four candidates, and that out of the list thus composed the members of the old Court of Arbitration of 1899 should make their choice; but that it had been objected that this selection would be too complicated, and would leave it to the states to appoint the members of the new judicial court. It was, therefore, impossible to submit a proposal to the committee.

Barbosa [3] then declared that the Brazilian Government did not see the necessity of a new court of arbitration; but that if it was desired to create one the principle of the equality of states must be secured; that this had only

[1] pp. 688, 694. [2] p. 694. [3] p. 695.

been done in the Brazilian proposal; and that those proposals which were consistent with the principle of the equality of states did not take into consideration that it was essential to arbitration that the states themselves, and not a third body, should appoint the judges. So long as his statements were not refuted, he said, he must in the name of Brazil withhold his approval.

Fry [1] thought that an agreement upon the composition of the court was impossible, and proposed a resolution according to which the Conference was to declare it desirable that the powers should accept the project submitted to them without the insertion of any provision as to the appointment of the judges.

Choate,[1] however, declared that there was no reason for despair; perhaps an agreement might yet be reached; he regretted that Barbosa rejected every other project but the Brazilian one. It must be borne in mind, he said, that it was necessary to create something new; he for his part considered the system of rotation the best one; the fears of the small states were, he thought, not justifiable; he wanted, however, to meet all objections, and now offered a proposal which would certainly meet with general approval. Each state was to appoint one judge and one deputy judge of any nationality; the list thus made up was to be communicated by the International Bureau to the states; thereupon each state was to communicate to the Bureau the names of the fifteen judges and fifteen deputy judges whom it desired to have appointed; those judges who received the most votes were to be regarded as elected; if a vacancy should occur, the state which had chosen the judge should appoint the new one.

Barbosa [2] replied in a long address that he did not insist

[1] p. 697. [2] p. 699.

absolutely upon the adoption of the Brazilian proposal, but that he could not accept any project which did not embody the principle which he had so often explained. As for the resolution proposed by Fry, he could not, he said, assent to it; if the Conference could not agree upon the basis of a new court of arbitration, the proposal had failed; it was not correct to assert that the Conference would disappoint the world if it did not create the court of arbitration; for the world was not expecting such a court; in the programme of the Conference no mention was made of the establishment of a new court of arbitration, but merely of improvements to be made in the old Arbitration Court of 1899.

Nelidow [1] thought that he could not wholly agree with Choate when the latter said that something new must be created; the truth was, he said, that something good should be created or nothing at all. At this Conference they had not succeeded in reaching an agreement upon the new court of arbitration, and there was nothing else left to do but to agree upon Fry's proposal.

Asser [1] declared that the Dutch delegation had had from the start a proposal of its own, but that it had not brought it forward because a great number of other projects had been introduced; as these projects appeared to have failed, the Dutch delegation reserved the right to introduce its proposal subsequently.

Bourgeois,[1] the president of the committee, declared that they had two proposals before them, that of Choate and that of Fry. He asked Barbosa whether he insisted upon his proposal.

Barbosa [1] answered that he had no interest in having his proposal adopted; he had merely wished to indicate in his draft how a permanent court of arbitration could be erected without violating the principle of the equality of

[1] p. 702.

states. He placed no value, he said, upon the creation of a new court of arbitration; he concluded, moreover, that the principle of the equality of states which he had fought for had been recognized by the Conference, since the system of rotation had been rejected on all sides.

Bourgeois [1] then proposed that Choate's proposal should first be voted upon, and he reminded the committee of the great responsibility which it was under towards the Commission.

After some remarks upon the form of the draft the votes were cast.[2] Five powers—United States, France, Greece, Netherlands, and Peru—voted for the proposal.

Nine powers—Germany, Austria-Hungary, Belgium, Brazil, Great Britain, Italy, Portugal, Roumania, and Russia—voted against it.

Choate [2] expressed his regret over the rejection of his proposal.

After further debate, Fry's proposal was voted upon.[3]

Eight states—Netherlands, Germany, Great Britain, United States, Italy, Portugal, Russia, and France—voted in favour of it.

Five powers—Greece, Peru, Brazil, Roumania, and Belgium—voted against it.

Austria-Hungary and Luxemburg refrained from voting.

Fry's proposal was accordingly adopted.

3. The Discussion in the Commission and in the Plenary Session

Scott was appointed to report upon the work of the subcommission concerning the court of arbitral justice. On October 9, 1907, the First Commission discussed the report presented by Scott.

[1] p. 703. [2] p. 704. [3] p. 707.

Esteva [1] first of all declared in the name of Mexico that he adopted Fry's proposal now that Articles 6, 7, and 8 of the British-German-United States proposal, which were inconsistent with the principle of the equality of states, had been eliminated.

Carlin,[2] the Swiss delegate, set forth the reasons why he could not accept Fry's proposal. He said that although the principle of the equality of states was, indeed, not violated by it, yet the fundamental principle of arbitration, namely, the free choice of the judges by the states in each individual case, was violated by it. Even if the old Court of Arbitration remained in existence, still the new court would obtain the upper hand, and moral pressure would be exercised upon the states to refer their disputes to it, and the most important feature of arbitration, the free choice of judges, would be lost to the states.

Henriquez i Carvajal [3] expressed in the name of San Domingo his approval of Fry's proposal.

Brun [4] declared in the name of Denmark that he must vote against the proposal, since it violated the most essential characteristic of arbitration, the free choice of judges.

Barbosa [5] then delivered the longest of the addresses which he made before the Conference upon the question of the Judicial Arbitration Court. He said that his government would in the spirit of friendly understanding vote in favour of the proposal. He then once more gave in great detail the reasons for his previous opposition. He recalled the rôle which Brazil had played up to that time in the matter of arbitration. Nationals of Brazil, he said, had often been appointed as arbitrators in important cases; for example, in the *Alabama* dispute.

[1] p. 144. [2] p. 145. [3] p. 146. [4] p. 147. [5] p. 148.

In 1870, 1871, 1880, and 1884 Germany and Italy had each appointed a Brazilian national as arbitrator once, and France, Great Britain, and the United States each twice; the South American states could not be pushed aside as it had been attempted to do at the Conference. Moreover, it was not only these states which rejected the British-German-United States proposal, but the great powers as well. At the last meeting of the subcommittee even Germany and Great Britain had abandoned the system of rotation. It was entirely unfair, he said, to say that the smaller states had caused the failure of the most important projects; on the contrary, it was the opposition of the great powers which had prevented an agreement upon the questions of capture at sea, contraband of war, blockade, and obligatory arbitration. The South American states were entirely willing to co-operate with the projects looking to peace. Japan had entered into the European concert some years before through the gate of war, whereas South America now appeared to be entering into it through the gate of peace.

Ordoñez [1] declared in the name of Uruguay that he desired to withhold his vote. He said that the free choice of judges was the only method consistent with the nature of arbitration.

Machado,[2] in the name of Guatemala, expressed his adherence to the statements made by Barbosa; he too pointed out the development of the South American states and protested against the statements made by the press, which, he said, greatly under-estimated the South American states in asserting that the great powers could not let their disputes be decided by judges from Brazil, Guatemala, &c.

Hagerup [3] voted for Fry's proposal and declared that

[1] p. 156. [2] p. 158. [3] p. 159.

the principle of the equality of states was a fundamental principle of international law.

Guillaume and Beldiman,[1] in the name of Belgium and of Serbia, rejected the proposal as inconsistent with the fundamental principles of arbitration.

Martens [1] observed that the Russian proposal had not yet been withdrawn.

Lou Tseng-tsiang and Momtaz-es-Saltaneh [1] declared in the name of China and of Persia that they desired to vote for the British proposal, but that they entered a solemn protest against any attempt to set aside the principle of the equality of states.

At the next meeting of the Commission on October 10, 1907, Beldiman [2] explained the reasons for the attitude of Serbia towards the British proposal. He further attacked a remark made by Scott that the Peace Palace had been built, so that it was only necessary for the judges to take their seats within it. The palace had not at all been built, he said, since no agreement had been reached upon the fundamental question, the constitution of the court.

Corragioni d'Orelli [3] declared himself in favour of the British proposal, as did also Matte,[3] the representative of Chile. The latter directed a forcible attack against the overthrow of the principle of the equality of states.

Thereupon the President of the Second Conference, the Russian Ambassador Nelidow,[4] exhorted the delegates to unite in a spirit of friendly understanding upon the British proposal. The Conference greeted his words with applause.

Hudicourt,[5] in the name of Haiti, voted for the proposal after he had obtained an assurance from Scott that the clause 'based upon the equality of states', which by

[1] p. 160. [2] p. 178. [3] p. 180.
[4] p. 181. [5] p. 182.

a mistake on the part of the printer had been introduced in the original draft of the proposal, had not been left out because the principle had been abandoned.

Fortoul and Carvajal,[1] in the name of Venezuela and of San Domingo, announced that they withheld their votes.

Rangabé [1] declared in the name of Greece that he withheld his vote, since an agreement had not been reached upon the most important points.

A vote was then taken upon the draft. Thirty-seven states voted for it; Belgium, Roumania, and Switzerland voted against it. There were four abstentions: Denmark, Greece, Uruguay, and Venezuela. The proposal was accordingly adopted.[2]

A discussion then followed as to whether Fry's proposal should be given the name of a *vœu* or a 'resolution'. Count Tornielli [3] declared that in accordance with the earlier attitude of the governments unanimity was not necessary in the case of a *vœu*, but was necessary in the case of a 'resolution'. Accordingly, in the present case the first term must be preferred.

At the plenary meeting of the Conference, on October 16, 1907, Scott,[4] in presenting his report, recommended the adoption of the *vœu*. Thereupon certain states had it recorded in the minutes of the meeting that they would vote for the *vœu* only upon the express condition that the principle of the equality of states would not be violated in the subsequent erection of the court.

A vote was finally taken upon Fry's proposal, which read as follows:

> The Conference calls the attention of the signatory powers to the advisability of adopting the annexed draft

[1] p. 182. [2] p. 189. [3] p. 190.
[4] Vol. i, p. 332.

Convention for the creation of a Judicial Arbitration Court, and of bringing it into force as soon as an agreement has been reached respecting the selection of the judges and the constitution of the court.

Thirty-eight states voted for the *vœu,* while six refrained from voting, namely, Belgium, Denmark, Greece, Roumania, Switzerland, Uruguay. Accordingly, the *vœu* was adopted by the Conference.

CHAPTER IX

COMMENTARY UPON THE DRAFT CONVENTION OF THE SECOND HAGUE PEACE CONFERENCE RELATIVE TO THE JUDICIAL ARBITRATION COURT

In the preceding chapters we have merely been able to discuss the leading principles which, in our opinion, are involved in the new Judicial Court. It may also be of interest, however, to learn the details of the new project as they were laid down by the Second Hague Peace Conference. The best method of doing this seems to me to present a brief commentary upon the draft convention of the Second Hague Peace Conference. In doing so I shall keep closely to Scott's report, but avoid as far as possible the repetition of points already discussed above.

Annexe au Ier vœu émis par la Deuxième Conférence de la Paix

Projet d'une Convention relative à l'établissement d'une Cour de Justice arbitrale

Titre Ier. Organisation de la Cour de Justice arbitrale

Article 1er

Dans le but de faire progresser la cause de l'arbitrage, les puissances contractantes conviennent d'organiser, sans porter atteinte à la cour permanente d'arbitrage, une Cour de justice arbitrale,

Annex to the first vœu expressed by the Second Peace Conference

Draft Convention relative to the creation of a Judicial Arbitration Court

Part I. Constitution of the Judicial Arbitration Court

Article 1

With a view to promoting the cause of arbitration, the contracting powers agree to constitute, without altering the status of the Permanent Court of Arbitration, a Judicial Arbitration Court, d'un accès libre et facile, réunissant des juges représentant les divers systèmes juridiques du monde, et capable d'assurer la continuité de la jurisprudence arbitrale.

of free and easy access, composed of judges representing the various juridical systems of the world, and capable of ensuring continuity in arbitral jurisprudence.

The Convention, in truth, has in view not the advancement of arbitration, but merely the judicial settlement of international disputes, as was pointed out above. But in order to meet the unfounded fears of certain states the Convention retains the expression, ' arbitration court '.

The Convention has two objects in view: the advancement of arbitration and the securing of the continuity of judicial decisions. For the fulfilment of this second duty two conditions, in Scott's opinion, appear to be necessary: the permanence of the tribunal and the representation upon it of the various legal systems. We have seen above that this second condition hardly possesses the great importance attached to it by Scott, because we would like to see cases of international private law referred to a special court, that is, to a special chamber. A series of cases would thus be withdrawn from the jurisdiction of the international court.[1] But Scott is correct when he points out that an international interpretation of treaties must be substituted for the mainly national interpretation thus far given.[2] Yet this will happen of itself if nationals of the various states are represented upon the court. There is no need for that purpose to make a special selection according to nationalities.

The new court is to sit side by side with the old Court of Arbitration of 1899. The text first read: 'To erect a Judicial Arbitration Court side by side with the old

[1] See in this connexion Meili, *Das internationale Zivilprozessrecht auf Grund der Theorie, Gesetzgebung und Praxis*, Zürich, 1904, pt. 2, p. 290.

[2] See also Oppenheim, *Die Zukunft des Völkerrechts*, p. 38.

Court of Arbitration.' Bourgeois [1] objected to these words because, he said, the impression would be given that the old and the new courts were entirely independent of each other, whereas on the contrary it should be pointed out that there existed a common bond between the two. Upon motion of Mérey [2] the following words were substituted: 'Without altering the status of the Permanent Court of Arbitration.'

The Judicial Court is to be 'of free and easy access'. The text at first read: 'Easy of access and free of charge.' By the expression 'free of charge' it was meant that only the resort to the court should be free, and that the honoraria of the counsel, &c., should be borne by the parties. [3] Martens [4] thought that this expression was not clear enough. In consequence the present wording was adopted. According to the article the costs of the decision are to be borne by the contracting states, not by the parties to the particular case, since the court exists in the interest of the states as a body.

Article 2

La Cour de justice arbitrale se compose de juges et de juges suppléants choisis parmi les personnes jouissant de la plus haute considération morale et qui tous devront remplir les conditions requises, dans leurs pays respectifs, pour l'admission dans la haute magistrature, ou être des jurisconsultes d'une compétence notoire en matière de droit international.

Article 2

The Judicial Arbitration Court is composed of judges and deputy judges chosen from persons of the highest moral reputation, and all fulfilling conditions qualifying them, in their respective countries, to occupy high legal posts, or be jurists of recognized competence in matters of international law.

[1] p. 598. [2] p. 661. [3] p. 596. [4] p. 657.

Les juges et les juges suppléants de la Cour sont choisis, autant que possible, parmi les membres de la Cour permanente d'arbitrage. Le choix sera fait dans les six mois qui suivront la ratification de la présente Convention.

The judges and deputy judges of the Court are appointed, as far as possible, from the members of the Permanent Court of Arbitration. The appointment shall be made within the six months following the ratification of the present Convention.

The first paragraph of this article secures the exclusion of diplomats and the presence upon the court of true jurists only. Kriege [1] desired that only those judges should be appointed who had fulfilled the conditions requisite for admission to the highest judicial offices of their own states. He said that in many countries the conditions for admission to the lowest judicial offices were trifling, and that not even university work was always required. But the delegates hesitated to adopt this proposal, because to do so would be to exclude many distinguished jurists from a seat upon the court. Martens [2] proposed that the governments should make a definite declaration to the International Bureau concerning the qualifications of the judges appointed by them. Nelidow agreed to this proposal. But in order to avoid any hurt feelings, the wording of the text was agreed upon. According to this, the capacity to hold the highest judicial offices within the state is demanded as a general rule, but at the same time an acknowledged experience in the field of international law is sufficient.[3] The important point is, as Scott well observed, not the title, but the character and personality of the judges.[4]

With regard to the words 'as far as possible' in the

[1] p. 661. [2] p. 662.

[3] See p. 71 of my commentary upon the Convention for the pacific settlement of international disputes. [4] See Schücking, op. cit., p. 91.

second paragraph, a discussion took place at the first meeting of the committee.[1] Barbosa said that it was inconsistent with a convention to insert in it any other obligation than a legal one, and pointed out that nothing more than a wish was expressed in the words 'as far as possible' in the second paragraph. Hence it would be better to say 'may be chosen'. This suggestion, however, was not taken up, and the discussion merely turned upon the point whether the states should have complete freedom with respect to the choice of judges, as Fry proposed, or not. Bourgeois urged in favour of the latter rule above all the fact that in this way a connexion would be established between the two courts. He desired, he said, that the judges of the new court should not only be appointed from among the members, but also by the members, of the old Court of Arbitration of 1899. Asser thought that if the states were to be obliged to appoint the judges from the list of the Permanent Court the powers should be permitted to put five persons instead of four upon the list. But Choate and Bourgeois regarded this number as too great. The latter thought, moreover, that a clause could be inserted to the effect that the judges might be freely chosen in case suitable judges could not be obtained from the list of the Hague Court of Arbitration. But Howard[2] thought that this would be a reflection upon the members of the old Court of Arbitration. Baron Marschall declared that he was in favour of the appointment of the judges from among the members of the old Court of Arbitration. The question was then referred to a drafting committee, which adopted the orginal phrasing, so that the connexion is clearly expressed between the new Judicial Court and the old Court of Arbitration.

[1] p. 599. [2] p. 600.

Attention was called to the fact that in the second paragraph of this article and in subsequent articles the new court is repeatedly termed a ‘judicial court’. Likewise ‘judges’ and ‘deputy judges’ are spoken of instead of ‘arbitrators’.

Article 3

Les juges et les juges suppléants sont nommés pour une période de douze ans à compter de la date où la nomination aura été notifiée au Conseil administratif institué par la Convention pour le règlement pacifique des conflits internationaux. Leur mandat peut être renouvelé.

En cas de décès ou de démission d'un juge ou d'un juge suppléant, il est pourvu à son remplacement selon le mode fixé pour sa nomination. Dans ce cas, la nomination est faite pour une nouvelle période de douze ans.

Article 3

The judges and deputy judges are appointed for a period of twelve years, counting from the date on which the appointment is notified to the Administrative Council created by the Convention for the pacific settlement of international disputes. Their appointments can be renewed.

Should a judge or deputy judge die or retire, the vacancy is filled in the manner in which his appointment was made. In this case, the appointment is made for a fresh period of twelve years.

Underlying the first paragraph is the idea that the regularity and continuity of judicial decisions must be secured. But this can only be when the period during which the individual judges exercise their functions is not too brief. Only then can the experience obtained by the judges be of advantage to the community of states. With this idea particularly in view, it was provided that a judge might be reappointed, thus making it possible for him to sit for a period exceeding twelve years.

Whereas the names of the members of the Court of

Arbitration of 1899 were notified to the International Bureau, the names of the judges of the new Judicial Court are notified to the Administrative Council, which is composed of the diplomatic representatives of the contracting powers accredited to The Hague and of the Netherland Minister for Foreign Affairs, who acts as president.[1] This distinction was made on the ground that the appointment of a judge to the new Judicial Court was a more important affair than the designation of a person upon the list of the old Permanent Court, since in the latter case nothing more was created than the possibility of the exercise of judicial functions; and also upon the ground that the International Bureau is strictly a bureau, whereas the Administrative Council is diplomatic in character.

Paragraph 2 is analogous to Article 44, paragraphs 7 and 59, of the Convention for the pacific settlement of international disputes. The attempt was made at first to secure that the judges be irremovable by providing that the reasons for their recall should be given, but it did not succeed. Mérey [2] in particular had demanded every guarantee in that respect; but Kriege [2] answered that this principle had already been recognized in the provision that the length of judicial tenure should be a period of twelve years, but that a judge could nevertheless be recalled by his country for special reasons, namely, those reasons which were provided for in the domestic legislation of his country. To formulate general rules in this respect would be difficult, he said. Renault also [2] pointed out that each new appointment of a government and the recall of a judge was valid from an international point of view, and that the powers

[1] See Article 49 of the Convention for the pacific settlement of international disputes.

[2] p. 667.

were morally obligated not to remove judges without justifying reasons. Nevertheless Mérey[1] insisted that it should be laid down as a principle that the judges were not subject to removal, and Martens agreed with him. But Mérey's proposal was rejected, as was that of Marschall,[1] according to which the states were to communicate to the other states the grounds for the removal of a given judge. Likewise Choate's proposal[2] that the judges should only be removed in case of permanent physical incapacity found no support.

The recall of the judges is accordingly left to the good faith of the states. If a judge is recalled, the decision in which he has taken part is in every case valid.

Article 4

Les juges de la Cour de justice arbitrale sont égaux entre eux et prennent rang d'après la date de la notification de leur nomination. La préséance appartient au plus âgé, au cas où la date est la même.

Les juges suppléants sont, dans l'exercice de leurs fonctions, assimilés aux juges titulaires. Toutefois, ils prennent rang après ceux-ci.

Article 4

The judges of the Judicial Arbitration Court are equal, and rank according to the date on which their appointment was notified. The judge who is senior in point of age takes precedence when the date of notification is the same.

The deputy judges are assimilated, in the exercise of their functions, with the judges. They rank, however, below the latter.

These provisions were made very precise in order to avoid any differences of interpretation.

The last sentence of the first paragraph is of importance in connexion with Article 26, paragraph 1.

[1] p. 668.

[2] p. 669.

Article 5

Les juges jouissent des privilèges et immunités diplomatiques dans l'exercice de leurs fonctions et en dehors de leurs pays.

Avant de prendre possession de leur siège, les juges et les juges suppléants doivent, devant le Conseil administratif, prêter serment ou faire une affirmation solennelle d'exercer leurs fonctions avec impartialité et en toute conscience.

Article 5

The judges enjoy diplomatic privileges and immunities in the exercise of their functions, outside their own country.

Before taking their seat, the judges and deputy judges must, before the Administrative Council, swear or make a solemn affirmation to exercise their functions impartially and conscientiously.

The first paragraph of Article 5 corresponds to Article 46, paragraph 4, of the Convention for the pacific settlement of international disputes.[1] The words ' outside their own country ' provide for the case in which a state should appoint a national of another state as judge. Such a judge would naturally not enjoy diplomatic privileges in his own country.

Consideration is made for the fact that many judges might refuse to take an oath because they considered that act as inconsistent with their religious convictions. For this reason a solemn affirmation is made equivalent to an oath. This arrangement was not made, however, without a struggle. Many desired that an oath should be made under all circumstances. Kriege [2] declared that an oath bound in a special way the conscience of the one taking it, and accordingly that particular form must be chosen. The provision of the text was advocated in particular by Martens.

[1] See p. 83 of my commentary upon this Convention.

[2] p. 614.

De Beaufort [1] proposed that it should be provided that the judges should take an oath that they had received no remuneration from private persons, because Article 10 makes mention only of remuneration from governments. But it was thought that such a provision would be offensive.

The Administrative Council must put it on record that the oath has been taken or the solemn affirmation made.

In response to an inquiry made by Eyschen [2] it was unanimously agreed that the members of the new Judicial Court were under obligation to exercise the duties of their office. 'They must sit except in cases where they are legally excused,' said Bourgeois.[2] The contrary rule holds good for the judges of the Permanent Court of Arbitration of 1899.[3]

Article 6

La Cour désigne annuellement trois juges qui forment une délégation spéciale et trois autres destinés à les remplacer en cas d'empêchement. Ils peuvent être réélus. L'élection se fait au scrutin de liste. Sont considérés comme élus ceux qui réunissent le plus grand nombre de voix. La délégation élit elle-même son président, qui, à défaut d'une majorité, est désigné par le sort.

Un membre de la délégation ne peut exercer ses fonctions quand la puissance qui l'a nommé, ou dont il est le national, est une des parties.

Article 6

The Court annually nominates three judges to form a special delegation, and three more to replace them should the necessity arise. They may be re-elected. They are balloted for. The persons who secure the largest number of votes are considered elected. The delegation itself elects its president, who, in default of a majority, is appointed by lot.

A member of the delegation cannot exercise his duties when the power which appointed him, or of which he is a national, is one of the parties.

[1] p. 615.

[2] p. 637.

[3] See p. 71 of my commentary upon the Convention for the pacific settlement of international disputes.

Les membres de la délégation terminent les affaires qui leur ont été soumises, même au cas où la période pour laquelle ils ont été nommés juges serait expirée.	The members of the delegation are to conclude all matters submitted to them, even if the period for which they have been appointed judges has expired.

In place of the word 'delegation' the word 'committee' was first used in the text. Martens [1] objected to this term on the ground that it was too weak an expression in consideration of the work to be done by the delegation. He said that in no other country except Great Britain would the name 'committee' be conferred upon a body performing legal functions. He proposed the expression 'permanent arbitral tribunal', and observed that the committee had not only the duty of deciding disputes, but also that of calling together the judicial court. Later on Kriege [2] thought that the expression 'special commission' might be selected, as this name is also adopted in the Prize Court Convention. An agreement seemed to be reached upon this term at first. But Martens [3] disliked this expression as well and proposed 'special tribunal'. Thereupon Crowe [3] called attention to the fact that this expression was already employed in the Peace Convention of 1899. As a result the term 'delegation' was selected. Against the use of the word 'tribunal' was the fact that it would give the appearance that there were in reality two judicial courts, whereas the delegation is only a body of judges taken from the court as a whole.

Paragraph 2 of the article expressly excludes national judges from the delegation.[4] The effort was made, so Scott says in his report,[5] to prevent the delegation from becoming arbitral rather than judicial in character.

[1] p. 616. [2] p. 634. [3] p. 664.
[4] See my remarks on p. 75 of my commentary. [5] Vol. i, p. 365.

Paragraph 3 makes it possible for two delegations to be in session at the same time. It would evidently delay the settlement of a case if the members of the delegation after the expiration of a year could not bring to an end suits begun during their term, but would be obliged to turn them over to a new delegation. If the judges exercise their functions professionally cases will for the most part not last very long, and the delegation of the preceding year will accordingly bring to an end as quickly as possible cases which have begun under them.

Asser[1] thought it better that the election should not take place every year, but only every second year. International cases, he said, often last a long time, and in addition distinguished jurists would not readily give up the work of their permanent profession, if they were members of the Judicial Court for only one year. But Kriege[1] thought differently, holding that it was precisely the short term of their commission which would induce many respected jurists to accept it. Under the present method, he said, jurists could easily be reappointed, and it was, moreover, possible to remove incapable persons at the end of a year.

Concerning the jurisdiction of the delegation see Articles 18 and 19.

Article 7

L'exercice des fonctions judiciaires est interdit au juge dans les affaires au sujet desquelles il aura, à un titre quelconque, concouru à la décision d'un tribunal national, d'un tribunal d'arbitrage ou d'une commission

Article 7

A judge may not exercise his judicial functions in any case in which he has, in any way whatever, taken part in the decision of a national tribunal, of a tribunal of arbitration, or of a commission of inquiry, or has

[1] p. 665.

d'enquête, ou figuré dans l'instance comme conseil ou avocat d'une partie.

Aucun juge ne peut intervenir comme agent ou comme avocat devant la Cour de justice arbitrale ou la Cour permanente d'arbitrage, devant un tribunal spécial d'arbitrage ou une commission d'enquête, ni y agir pour une partie en quelque qualité que ce soit, pendant toute la durée de son mandat.

figured in the suit as counsel or advocate for one of the parties.

A judge cannot act as agent or advocate before the Judicial Arbitration Court or the Permanent Court of Arbitration, before a special tribunal of arbitration or a commission of inquiry, nor act for one of the parties in any capacity whatsoever so long as his appointment lasts.

This article aims at securing the impartiality of the judges. In the cases mentioned in Article 7 the mere possibility that such a judge is not impartial is sufficient to call for a positive prohibition. Even if a judge has the best will to decide fairly in such matters, that is not enough to secure a satisfactory result. The very effort of the judge to be entirely fair in such cases will often be prejudicial to the parties.

On account of the close connexion between the old Court of Arbitration of 1899 and the new Judicial Court the judges are forbidden to act as advocates not only before the new Court but also before the Court of Arbitration.

Nay, the article goes even further and has extended this prohibition even to acting as advocate before special arbitration courts and commissions of inquiry, because the duties of an advocate, as Scott says in his report,[1] are not consistent with the equanimity and dignity of a judge. This provision is preferable to the one in Article 62, paragraph 3 of the Convention for the pacific settlement of international disputes.

[1] Vol. i, p. 366.

In order to meet all possible cases it is even provided that a judge is disqualified if he has previously acted for one of the parties under the specified conditions. In this way, contrary to the provision of the Convention for the pacific settlement of international disputes,[1] a judge is forbidden to furnish legal opinions or give advice to either of the parties.

Great Britain and the United States at first advocated that nationals of the parties should not be allowed to sit upon the Court; just as they are excluded from the delegation so they should be excluded from the deliberations of the Court itself. But Marschall[2] was opposed to this, because he expected greater impartiality from the presence of national judges. He said that a distinction must be made between national courts of justice and international courts of arbitration: in national courts in no instance can one be a judge in his own case, whereas in international arbitration this may well happen. Choate and Fry[3] said that they could not agree with this view, but that they would yield in a spirit of friendly concession.

In his excellent report,[4] Scott offered the following reasons in explanation of the non-exclusion of national judges: in a small court, he said, the presence of national judges might arouse a suspicion of partiality; in such cases it is to be feared that each national judge will win over the umpire to his side. But in a large court it is impossible that the few national judges should win over the whole body, and there are not the same objections to admitting to the court judges who are nationals of the parties. Moreover, disputes will be referred to the new Court, he said, in accordance with general or special arbitration treaties;

[1] See my commentary, p. 119.
[2] p. 604.
[3] p. 605.
[4] Vol. i, p. 367.

now the free choice of judges was a characteristic of arbitration; accordingly the parties should not be limited in the exercise of this freedom of choice. This last reason is not at all conclusive. In any case a dispute would be submitted to the new court in accordance with a general or special arbitration treaty, since a direct right of action is not granted; but we should not let it escape our attention that we are dealing here with an institution which is a cross between a judicial and an arbitral court. Now, since the object of the whole Convention is the furtherance of judicial settlement and the repudiation of arbitration, it is hard to understand how the presence of national judges upon the court can be justified by the principles of arbitration when the Convention seeks to get rid of that very method of procedure.

Scott[1] advances the following additional reasons in favour of the presence of national judges: the parties have a greater assurance that the decision will not contain any offensive expressions; finally, the decision has not only for its object, he said, the settlement of the dispute, but also the development of international law; and in this last work national judges could not be forbidden to co-operate. This last observation does not take into consideration that the settlement of a dispute and the development of international law are closely connected, so that a judge who has any interest in the result of the case would seem in this respect unsuited to develop international law.

Unfortunately the Conference neglected to decide the question as to what would happen if by the system of rotation only one party were represented upon the Court. Was the other party in this case, as it was originally

[1] Vol. i, p. 368.

intended, to be allowed to appoint a representative of its own? Asser [1] called attention to the case in which more than two parties should come before the Court. Should each be then allowed to appoint a national judge? In answer Marschall [2] said that if *A* were the plaintiff and *B*, *C*, and *D* the defendants, *B*, *C*, and *D* would have to agree together upon the appointment of one judge.[3]

Asser [4] also desired to know what would happen if *B*, *C*, and *D* already had each a representative upon the Court. Kriege answered evasively that this question could not be specially regulated in the Convention, because otherwise the composition of the Court would be too complicated.

Should the national judge in such cases be appointed *ad hoc* if the parties were at the time not represented upon the Court by a judge? This question, which was raised by Asser, was answered by Bourgeois and Marschall [2] to the effect that the judge appointed by the parties in question would sit upon the Court even if by the system of rotation he was not at the time a member of it.

All these questions were left undecided in the article. Accordingly it is to be assumed that the system of rotation is to be introduced, no matter what hardships result from it.

Article 8

La Cour élit son président et son vice-président à la majorité absolue des suffrages exprimés. Après deux tours de scrutin, l'élection se fait à la majorité relative et, en cas de partage des voix, le sort décide.

Article 8

The Court elects its president and vice-president by an absolute majority of the votes cast. After two ballots, the election is made by a bare majority, and, in case the votes are even, by lot.

[1] p. 605.

[2] p. 606.

[3] See p. 75 of my commentary upon the Convention for the pacific settlement of international disputes.

[4] p. 663.

As the appointment of the chief dignitaries is of great importance, it was regulated in detail. Bourgeois, Martens, and Crowe[1] had advocated that in case the votes were equal the election should be not by lot but by age, and this was at first agreed to. But later on doubts arose, and the wording of the text was adopted.

It was at first provided that the appointment of the president should take place every three years. According to the present wording the period for which the president is elected is fixed by the rules of the Court.

Article 9

Les juges de la Cour de justice arbitrale reçoivent une indemnité annuelle de six mille florins néerlandais. Cette indemnité est payée à l'expiration de chaque semestre à dater du jour de la première réunion de la Cour.

Pendant l'exercice de leurs fonctions au cours des sessions ou dans les cas spéciaux prévus par la présente Convention, ils touchent une somme de cent florins par jour. Il leur est alloué, en outre, une indemnité de voyage fixée d'après les règlements de leur pays. Les dispositions du présent alinéa s'appliquent aussi aux juges suppléants remplaçant les juges.

Article 9

The judges of the Judicial Arbitration Court receive an annual salary of 6,000 Netherland florins. This salary is paid at the end of each half-year, reckoned from the date on which the Court meets for the first time.

In the exercise of their duties during the sessions or in the special cases covered by the present Convention, they receive the sum of 100 florins *per diem*. They are further entitled to receive a travelling allowance fixed in accordance with regulations existing in their own country. The provisions of the present paragraph are applicable also to a deputy judge when acting for a judge.

[1] p. 613.

Ces allocations, comprises dans les frais généraux de la Cour, prévus par l'article 31, sont versées par l'entremise du Bureau international institué par la Convention pour le règlement pacifique des conflits internationaux.

These emoluments are included in the general expenses of the Court dealt with in Article 31, and are paid through the International Bureau created by the Convention for the pacific settlement of international disputes.

In the discussions at The Hague a dispute arose as to the honoraria to be given to the judges. Choate [1] held that the sum of 6,000 florins a year was too small in consideration of the fact that many of the judges would be obliged to come from China, Japan and South America. But Crowe [1] observed that travelling expenses were especially provided for. Later on, Choate [2] returned once more to the question and wanted to have the honorarium fixed at 10,000 florins. He pointed out that people would lose respect for the Court if the judges were too poorly paid. Thereupon Crowe observed that most of the judges would receive in addition the salaries which they drew as judges in their own country. Lammasch [3] pointed out that if the salary was too high unworthy persons might be attracted, and that, above all, the high honour of the position of an international judge should be taken into consideration. The other delegates voted with Lammasch.

The honorarium of 6,000 florins goes only to the judges, not to the deputy judges. It must be regarded as a requital for the necessity under which the judges are to hold themselves ready at all times to go to The Hague, should a case come up for decision.

In addition to the 6,000 florins the judges receive during the exercise of their functions the sum of 100 florins *per diem*. Moreover, their travelling expenses are paid. The

[1] p. 666. [2] p. 680. [3] p. 681.

amount of the latter is determined in accordance with the regulations of their own country. The same provisions hold good for the deputy judges when acting as judges.

In the first draft the above-mentioned sums were to be paid by the contracting states according to the proportion in which they shared the expenses of the Universal Postal Union; as the text reads the payments are to be made by the International Bureau in accordance with a special agreement to be later entered into between the contracting parties.

Article 10

Les juges ne peuvent recevoir de leur propre Gouvernement ou de celui d'une autre puissance aucune rémunération pour des services rentrant dans leurs devoirs comme membres de la Cour.

Article 10

The judges may not accept from their own Government or from that of any other power any remuneration for services connected with their duties in their capacity of members of the Court.

This article constitutes a further guarantee for the impartiality of the judges. In so far as the judges are national officials or university professors they draw naturally their salary as such. Only that remuneration from their government is forbidden which would be given in return for work connected with their duties upon the international court.

The provisions of the article are applicable not only to the duties of the judges upon the Court, but also to the services which they render as members of the delegation or of a commission of inquiry.

Article 11

La Cour de justice arbitrale a son siège à La Haye et ne peut, sauf le cas de force majeure, le transporter ailleurs.

Article 11

The seat of the Judicial Arbitration Court is at The Hague, and cannot be transferred, unless absolutely obliged by circumstances, elsewhere.

La délégation peut, avec l'assentiment des parties, choisir un autre lieu pour ses réunions si des circonstances particulières l'exigent.

The delegation may choose, with the assent of the parties concerned, another site for its meetings, if special circumstances render such a step necessary.

A permanent judicial court is inconceivable without a fixed place where all its sessions are held. Only in cases of *force majeure*, e. g., in the case of an epidemic, a war, &c., may the seat of the Court be transferred.[1] It is otherwise with respect to the delegation, the purpose of which is the quickest possible settlement especially of the smaller disputes. It can transfer its seat if (1) the parties agree and (2) special circumstances require it. In this case there need be no question of *force majeure*; mere convenience is enough.

The delegation has complete freedom to hold its sessions where it will, particularly when it acts as a commission of inquiry in accordance with Article 20. It is precisely the determination of facts which frequently requires local investigation, &c. The agreement of the parties is, however, necessary.

Article 12

Le Conseil administratif remplit à l'égard de la Cour de justice arbitrale les fonctions qu'il remplit à l'égard de la Cour permanente d'arbitrage.

Article 12

The Administrative Council fulfils with regard to the Judicial Arbitration Court the same functions as to the Permanent Court of Arbitration.

This article also indicates the connexion between the old Permanent Court and the new Judicial Court.

Concerning the duties of the Administrative Council,

[1] See Article 60 of the Convention for the pacific settlement of international disputes.

see pages 89 et seq. of my commentary upon the Convention for the pacific settlement of international disputes.

Article 13

Le Bureau international sert de greffe à la Cour de justice arbitrale et doit mettre ses locaux et son organisation à la disposition de la Cour. Il a la garde des archives et la gestion des affaires administratives.

Le secrétaire-général du Bureau remplit les fonctions de greffier.

Les secrétaires adjoints au greffier, les traducteurs et les sténographes nécessaires sont désignés et assermentés par la Cour.

Article 13

The International Bureau acts as registry to the Judicial Arbitration Court, and must place its offices and staff at the disposal of the Court. It has charge of the archives and carries out the administrative work.

The secretary-general of the Bureau discharges the functions of registrar.

The necessary secretaries to assist the registrar, translators and shorthand writers are appointed and sworn in by the Court.

Like the Administrative Council, the International Bureau is common to the old Permanent Court and the new Judicial Court. Concerning the International Bureau see p. 66 of my commentary upon the Convention for the pacific settlement of international disputes. As in numerous cases thus far translators and stenographers have been shown to be necessary, the present Convention provides expressly for the presence of these assistants. The fact that the Convention considers translators necessary is a proof that the knowledge of foreign languages possessed by the judges has not always been adequate to meet the difficulties arising from them.

Article 14

La Cour se réunit en session une fois par an. La session commence le troisième

Article 14

The Court meets in session once a year. The session opens the third Wednesday mercredi de juin et dure tant que l'ordre du jour n'aura pas été épuisé.

La Cour ne se réunit pas en session, si la délégation estime que cette réunion n'est pas nécessaire. Toutefois, si une puissance est partie à un litige actuellement pendant devant la Cour et dont l'instruction est terminée ou va être terminée, elle a le droit d'exiger que la session ait lieu.

En cas de nécessité, la délégation peut convoquer la Cour en session extraordinaire.

in June, and lasts until all the business on the agenda has been transacted.

The Court does not meet in session if the delegation considers that such meeting is unnecessary. However, when a power is party in a case actually pending before the Court, the pleadings in which are closed, or about to be closed, it may insist that the session should be held.

When necessary, the delegation may summon the Court in extraordinary session.

As Scott points out in his report[1] the provisions of this article have an important bearing upon the permanence and impartiality of the Court, which are its chief attributes.

The United States draft had provided that the judges should not only be permanently ready to settle international disputes, but should also be in residence at The Hague. It was thought that in this way the judges could best be gradually freed from national prejudices. But many feared that a regulation of that kind would be very inconvenient for the judges, and that it would contribute very little to increase the respect for the Court if in the first years only a few disputes should be referred to it. But Marschall declared that, on the contrary, it was to be hoped that too many disputes would be referred to the Court, because there were numerous cases reposing in the chancelleries of the governments. Choate thought that the

[1] Vol. i, p. 373.

fact should, after all, be reckoned with that many disputes would not be referred to the Court until it had won general confidence; but this was no objection to having the judges already in permanent residence at The Hague. Nevertheless, the suggestion that the judges should not permanently reside at The Hague was set aside, and it was decided to have plenary sessions only.[1]

It was at first sought to provide for two plenary sessions of the Court each year. But Asser and d'Oliveira considered that two sessions were too many.[2] Asser thought that most disputes would be referred to the delegation, and that frequent sessions of the court would be accordingly unnecessary. A yearly session was determined upon, but it was insisted that difficult disputes should in no case be entrusted to the delegation, but to the Court itself.

D'Oliveira thereupon proposed that special sessions of the Court should be held every five or six years, at which all the powers should be represented. At these sessions the members of the Court were to exchange their views upon questions of international law. But Marschall, Fry and Lammasch[3] considered it dangerous to have the judges express abstract opinions upon questions upon which they would later on be obliged to give an opinion in concrete cases.

The opening session of the Court was fixed for the day which would coincide as far as possible with the date of the opening of the Second Hague Peace Conference.

The draft provided that the Court should not meet in session if the delegation considered that the cases on hand did not justify having the Court meet. But Tornielli expressed the fear that the delegation might misuse its

[1] See Schücking, *op. cit.*, p. 156. [2] p. 616. [3] p. 617.

power and prevent the meeting of the Court. Accordingly, every state which has a case pending before the Court, the pleadings in which are closed or are about to be closed, is given the right to insist that the session be held, if the delegation has not thought it necessary. Provision is made therefore in this article that the Court shall meet yearly if necessary. It depends only upon the number of disputes whether the Court shall be in continuous session at The Hague.

Article 15

Un compte-rendu des travaux de la Cour sera dressé chaque année par la délégation. Ce compte-rendu sera transmis aux puissances contractantes par l'intermédiaire du Bureau international. Il sera communiqué aussi à tous les juges et juges suppléants de la Cour.

Article 15

A report of the doings of the Court shall be drawn up every year by the delegation. This report shall be forwarded to the contracting powers through the International Bureau. It shall also be communicated to the judges and the deputy judges of the Court.

Whereas the report of the work of the old Permanent Court is prepared by the Administrative Council, the work of drawing up the report of the new Court is in the hands of the delegation.

The report will include in particular the decisions of the Court; and for this very reason it will be of special importance.

The draft at first provided that the report should be turned over by the delegation to the Administrative Council; but it was feared that if this were done the Judicial Court would appear to be subordinate to some extent to the Administrative Council.

The report may not criticize the work of the Court.[1]

[1] p. 670.

Article 16

Les juges et les juges suppléants, membres de la Cour de justice arbitrale, peuvent aussi être nommés aux fonctions de juge et de juge suppléant dans la Cour internationale des prises.

Article 16

The judges and deputy judges, members of the Judicial Arbitration Court, can also exercise the functions of judge and deputy judge in the International Prize Court.

As there is a connexion between the new Judicial Court and the old Permanent Court, so also is there to be a connexion between the new Judicial Court and the International Prize Court. Article 16 has its origin in this idea.[1]

Titre II. Compétence et procédure

Article 17

La Cour de justice arbitrale est compétente pour tous les cas qui sont portés devant elle, en vertu d'une stipulation générale d'arbitrage ou d'un accord spécial.

Part II. Competence and Procedure

Article 17

The Judicial Arbitration Court is competent to deal with all cases submitted to it, in virtue either of a general undertaking to have recourse to arbitration or of a special agreement.

The British-German-United States proposal desired to make the Court obligatory for all disputes which, in accordance with a general treaty of arbitration entered into before the ratification of this Convention, were to be submitted to the Permanent Court of Arbitration. But Fusinato[2] urged the objection that this would constitute a preference for the new Judicial Court over the old Court of Arbitration; and it also involved an alteration of the provisions of the general arbitration treaties without the consent of the parties. Martens, Renault, and Fry agreed

[1] See Schücking, *op. cit.*, pp. 104 et seq.

[2] p. 628.

with him, while Asser supported the above-mentioned proposal. It was finally decided to suppress that provision, the committee being of the opinion of Renault that the new Judicial Court, when it had won confidence, would come to be preferred on its own account.

A German proposal had sought to have the Court declared competent 'to revise the decisions of arbitration courts and the reports of commissions of inquiry, as well as to determine the rights and duties resulting therefrom, in all cases in which the parties, in accordance with a general treaty or a special agreement, resort to the Court for that purpose'. In this proposal also Fry,[1] Bourgeois,[2] and Choate [2] saw a preference for the Judicial Court over the Permanent Court of Arbitration, and their views were warmly supported by Martens,[3] the vigorous opponent of the idea of the revision of decisions. The latter feared that such a provision would only encourage the parties to question the revised decisions. Renault [4] objected to the proposal that the report of a commission of inquiry should be reviewed by a court. But Kriege thought that it need only be said that the Judicial Court should in that instance act as a commission of inquiry on appeal, not as a court. Apropos of the remark made by Fusinato [5] that the revision of an arbitral award should be made by the same court which rendered the award, Kriege declared [2] that it might happen that the judges of the first court would be prevented from acting upon the second court; moreover, the parties could enter into an express agreement in the *compromis* looking to a revision of the decision by the new Judicial Court. The German proposal was finally struck out.

[1] pp. 628, 630. [2] p. 630. [3] pp. 629 et seq.
[4] p. 629. [5] p. 628.

It was expressly laid down by Bourgeois [1] that 'a general undertaking to have recourse to arbitration' was to be understood to include arbitral clauses in treaties.

Article 18

La délégation est compétente :

1° pour juger les cas d'arbitrage visés à l'article précédent, si les parties sont d'accord pour réclamer l'application de la procédure sommaire, réglée au Titre IV, Chapitre 4 de la Convention pour le règlement pacifique des conflits internationaux.

2° pour procéder à une enquête en vertu et en conformité du Titre III de la dite Convention en tant que la délégation en est chargée par les parties agissant d'un commun accord. Avec l'assentiment des parties et par dérogation à l'article 7, alinéa 1, les membres de la délégation ayant pris part à l'enquête peuvent siéger comme juges, si le litige est soumis à l'arbitrage de la Cour ou de la délégation elle-même.

Article 18

The delegation is competent:

1. To decide the arbitrations referred to in the preceding article, if the parties concerned are agreed that the summary procedure, laid down in Part IV, Chapter IV, of the Convention for the pacific settlement of international disputes is to be applied.

2. To hold an inquiry under and in accordance with Part III of the said Convention, in so far as the delegation is entrusted with such inquiry by the parties acting in common agreement. With the assent of the parties concerned, and as an exception to Article 7, paragraph 1, the members of the delegation who have taken part in the inquiry may sit as judges, if the case in dispute is submitted to the arbitration of the Court or of the delegation itself.

Whereas the preceding article deals with the jurisdiction of the Judicial Court, the present article determines the jurisdiction of the delegation. The summary procedure can be undertaken in connexion with the old Court of

[1] p. 631.

Arbitration as well as with the new Judicial Court. The parties have complete freedom in this respect, and no preference for the delegation is to be created. With regard to the summary procedure the rules of the Convention for the pacific settlement of international disputes (Articles 86–90) are to be applied, unless the parties make some other provision.

A proposal made by Asser and Martens,[1] according to which all other disputes were to be referred to the delegation under agreement of the parties, was rejected, since in the interest of the Judicial Court it was not desired to extend too far the jurisdiction of the delegation; the Judicial Court, so it was pointed out—in particular by Crowe and Bourgeois, was to develop international law, and it would not contribute to that end if the settlement of the most important questions were to be turned over to the delegation, which consisted only of three persons.

With respect to No. 2, Lammasch urged the objection that the functions of a commission of inquiry were being assigned to a judicial court. He pointed out that a court was in the habit of deciding legal disputes, and would with difficulty restrict itself to the determination of facts. But Fry, Kriege, Renault, and Martens would not admit that, and Lammasch's proposal was not adopted.

Asser desired that the members of the delegation should receive a special honorarium for their work upon the commission of inquiry.[2] But this was objected to in particular by Choate. He said that a special compensation was justly granted only to those who were not appointed from among the members of the Judicial Court.[3] These last must be paid by the parties themselves. On the

[1] p. 673. [2] p. 635. [3] See Article 20, paragraph 2.

contrary, the other members of a commission of inquiry, who belong to the delegation of the Judicial Court, could demand no further honoraria than the salary fixed in Article 8.

Renault and Martens [1] raised the question whether the reference of an affair to the new Judicial Court after the report of the commission of inquiry would not operate to disqualify the members of the commission from becoming members of the Court. Renault advocated the presence of the judges who had taken part in the commission of inquiry, since they could render valuable service upon the Court. Asser also agreed to this, referring to the recent legislation in certain countries which provided that the judge who had undertaken an inquiry should sit upon the Court. The committee was in general of the opinion that the duties of a judge were entirely distinct from those of a commissioner. Accordingly the last paragraph of the article which, when the parties are agreed upon it, makes an exception of the provision of Article 7, paragraph 1, was allowed to stand.

Article 19	Article 19
La délégation est, en outre, compétente pour l'établissement du compromis visé par l'article 52 de la Convention pour le règlement pacifique des conflits internationaux, si les parties sont d'accord pour s'en remettre à la Cour.	The delegation is also competent to settle the *compromis* referred to in Article 52 of the Convention for the pacific settlement of international disputes if the parties are agreed to leave it to the Court.
Elle est également compétente, même si la demande est faite seulement par l'une des parties, après qu'un accord par la voie diplomatique	It is equally competent to do so, even when the request is only made by one of the parties concerned, if all attempts have failed to reach

[1] p. 635.

a été vainement essayé, quand il s'agit :

1° d'un différend rentrant dans un traité d'arbitrage général conclu ou renouvelé après la mise en vigueur de cette Convention et qui prévoit pour chaque différend un compromis, et n'exclut pour l'établissement de ce dernier ni explicitement ni implicitement la compétence de la délégation. Toutefois, le recours à la Cour n'a pas lieu si l'autre partie déclare qu'à son avis le différend n'appartient pas à la catégorie des questions à soumettre à un arbitrage obligatoire, à moins que le traité d'arbitrage ne confère au tribunal arbitral le pouvoir de décider cette question préalable ;

2° d'un différend provenant de dettes contractuelles réclamées à une puissance par une autre puissance comme dues à ses nationaux, et pour la solution duquel l'offre d'arbitrage a été acceptée. Cette disposition n'est pas applicable si l'acceptation a été subordonnée à la condition que le compromis soit établi selon un autre mode.

an understanding through the diplomatic channel, in the case of :

1. A dispute covered by a general treaty of arbitration concluded or renewed after the present Convention has come into force, providing for a *compromis* in all disputes, and not either explicitly or implicitly excluding the settlement of the *compromis* from the competence of the delegation. Recourse cannot, however, be had to the Court if the other party declares that in its opinion the dispute does not belong to the category of questions to be submitted to compulsory arbitration, unless the treaty of arbitration confers upon the arbitration tribunal the power of deciding this preliminary question.

2. A dispute arising from contract debts claimed from one power by another power as due to its nationals, and for the settlement of which the offer of arbitration has been accepted. This arrangement is not applicable if acceptance is subject to the condition that the *compromis* should be settled in some other way.

See in connexion with this article, pages 96 et seq. of my commentary upon the Convention for the pacific settlement of international disputes ; also page 92 above.

In paragraph one, Asser [1] wished to have inserted the additional provision that the delegation should also be competent to regulate special points of the *compromis.* But it was thought that it could do so even without such a provision.

Paragraph two, no. 1, had its origin in a German proposal. Marschall [2] called attention to the case in which a dispute might in the opinion of both parties come under a permanent treaty of arbitration, but in which an agreement upon the *compromis* could not be reached ; under such conditions, he said, an obligatory *compromis* was the necessary complement of the obligatory arbitration treaty; but that such a compulsory *compromis* could not naturally be enforced, if one of the parties asserted that the case affected its honour or its vital interests. In consequence of this limitation Fry [3] and Martens [4] considered that a legal obligation was entirely out of place. Choate [4] declared that he could not accept the German proposal because the Senate of the United States would reject such an obligation, which would take from it its right to ratify each individual *compromis.* The German proposal was, however, agreed to, subject to the exception that the delegation should not be competent to settle a compulsory *compromis* in those cases in which a general arbitration treaty should have been entered into before the present Convention came into force.[5] Thereupon Choate,[6] in the name of the United States, declared himself in favour of the new provision.

According to the Convention respecting the limitation of the employment of force for the recovery of contract debts the use of force is only permitted when, in consequence

[1] p. 638. [2] p. 639. [3] p. 640.
[4] p. 641. [5] pp. 651, 675. [6] p. 676.

of the conduct of the debtor state, a *compromis* cannot be agreed upon. In order to facilitate the settlement of the *compromis* in such cases, the delegation was declared competent to settle the *compromis* in case diplomatic negotiations should have produced no result and one of the parties should make a request to that effect.

Article 20

Chacune des parties a le droit de désigner un juge de la Cour pour prendre part, avec voix délibérative, à l'examen de l'affaire soumise à la délégation.

Si la délégation fonctionne en qualité de commission d'enquête, ce mandat peut être confié à des personnes prises en dehors des juges de la Cour. Les frais de déplacement et la rétribution à allouer aux dites personnes sont fixés et supportés par les puissances qui les ont nommés.

Article 20

Each of the parties concerned may nominate a judge of the Court to take part, with power to vote, in the examination of the case submitted to the delegation.

If the delegation acts as a commission of inquiry, this task may be entrusted to persons other than the judges of the Court. The travelling expenses and remuneration to be given to the said persons are fixed and borne by the powers appointing them.

The first paragraph contemplates the possibility of establishing without great difficulty a larger court of five judges, by having each of the parties appoint a judge from among the members of the Judicial Court. It is not stated clearly whether these judges may be nationals of the parties. This question can scarcely be answered in the negative.[1]

As persons who are not jurists can also be of service upon a commission of inquiry, paragraph 2 provides that persons who do not belong to the Judicial Court may be chosen for the commission of inquiry.

[1] See Article 30, paragraph 2.

With regard to question of cost in the second sentence of the last paragraph, see the concluding remarks upon Article 18.

Article 21

L'accès de la Cour de justice arbitrale, instituée par la présente Convention, n'est ouvert qu'aux puissances contractantes.

Article 21

The contracting powers only may have access to the Judicial Arbitration Court set up by the present Convention.

Asser [1] had inquired whether the International Judicial Court should be open only to contracting states or to other powers as well, and he recommended that they should restrict it to the contracting states. This proposal was agreed to, but it was specially laid down that only those states should be considered as contracting states which should later on ratify the Convention, and not those which had merely signed it. This was due especially to the fact that the access to the Court was gratuitous, and accordingly it did not seem fair that the contracting states should also bear the burden of the expenses incurred by the Court in the settlement of disputes between states not parties to the Convention.[2]

Article 22

La Cour de justice arbitrale suit les règles de procédure édictées par la Convention pour le règlement pacifique des conflits internationaux, sauf ce qui est prescrit par la présente Convention.

Article 22

The Judicial Arbitration Court follows the rules of procedure laid down in the Convention for the pacific settlement of international disputes, except in so far as the procedure is laid down in the present Convention.

The preceding article also creates a connexion between the old Court of Arbitration and the new Judicial Court,

[1] p. 633. [2] p. 677.

by making the same rules of procedure applicable to both.

See in addition Articles 51 et seq. of the Convention for the pacific settlement of international disputes.

Article 23

La Cour décide du choix de la langue dont elle fera usage, et des langues dont l'emploi sera autorisé devant elle.

Article 23

The Court determines what language it will itself use and what languages may be used before it.

According to Article 52 of the Convention for the pacific settlement of international disputes, the parties themselves determine what languages are to be used. Such a procedure, however, would in the case of a judicial court of seventeen members lead to difficulties. For this reason the Court is given exclusive right to decide the question of languages.

Upon the question of languages, compare in addition the statements made on pages 88 et seq. above, and also pages 110 et seq. of my commentary upon the Convention for the pacific settlement of international disputes.

Article 24

Le Bureau international sert d'intermédiaire pour toutes les communications à faire aux juges au cours de l'instruction prévue à l'article 63, alinéa 2, de la Convention pour le règlement pacifique des conflits internationaux.

Article 24

The International Bureau serves as channel for all communications to be made to the judges during the interchange of pleadings provided for in Article 63, paragraph 2, of the Convention for the pacific settlement of international disputes.

According to Article 52 of the Convention for the pacific settlement of international disputes, the form, the order, and the time for the communication of the cases, counter-cases, &c., of the parties are fixed in the *compromis*. This provision did not appear applicable in connexion with the judicial court, because it could not be decided by a court of seventeen members until the individual cases were carefully studied. Accordingly the above general provision was adopted.

Article 25

Pour toutes les notifications à faire, notamment aux parties, aux témoins et aux experts, la Cour peut s'adresser directement au Gouvernement de la puissance sur le territoire de laquelle la notification doit être effectuée. Il en est de même s'il s'agit de faire procéder à l'établissement de tout moyen de preuve.

Les requêtes adressées à cet effet ne peuvent être refusées que si la puissance requise les juge de nature à porter atteinte à sa souveraineté ou à sa sécurité. S'il est donné suite à la requête, les frais ne comprennent que les dépenses d'exécution réellement effectuées.

La Cour a également la faculté de recourir à l'intermédiaire de la puissance sur le territoire de laquelle elle a son siège.

Les notifications à faire aux parties dans le lieu où siège la Cour peuvent être exécutées par le Bureau international.

Article 25

For all notices to be served, in particular on the parties, witnesses, or experts, the Court may apply direct to the Government of the state on whose territory the service is to be carried out. The same rule applies in the case of steps being taken to procure evidence.

The requests addressed for this purpose can only be rejected when the power applied to considers them likely to impair its sovereign rights or its safety. If the request is complied with, the fees charged must only comprise the expenses actually incurred.

The Court is equally entitled to act through the power on whose territory it sits.

Notices to be given to parties in the place where the Court sits may be served through the International Bureau.

This article is an amplification of Article 24 of the Convention for the pacific settlement of international disputes. The contracting states are to assist the Judicial Court in the accomplishment of its purpose. Accordingly the fees must represent only expenses actually incurred, for recourse to the Court must not be made a source of revenue.

Article 26

Les débats sont dirigés par le président ou le vice-président et, en cas d'absence ou d'empêchement de l'un et de l'autre, par le plus ancien des juges présents.

Le juge nommé par une des parties ne peut siéger comme président.

Article 26

The discussions are under the control of the president or vice-president, or, in case they are absent or cannot act, of the senior judge present.

The judge appointed by one of the parties cannot preside.

There are no remarks to be made upon paragraph one.

Although it was not desired to prohibit the presence of nationals upon the large Court of seventeen members, yet they were rightly denied the competency to direct the discussions, since the regularity of the procedure would otherwise be disturbed. Such is the origin of the second paragraph.

Article 27

Les délibérations de la Cour ont lieu à huis clos et restent secrètes.

Toute décision est prise à la majorité des juges présents. Si la Cour siège en nombre pair et qu'il y ait partage des voix, la voix du dernier des juges, dans l'ordre de préséance établi d'après l'article 4, alinéa 1, ne sera pas comptée.

Article 27

The Court considers its decisions in private, and the proceedings are secret.

All decisions are arrived at by a majority of the judges present. If the number of judges is even and equally divided, the vote of the junior judge, in the order of precedence laid down in Article 4, paragraph 1, is not counted.

Paragraph 1 follows word for word Article 78, paragraph 1, of the Convention for the pacific settlement of international disputes. In paragraph 2, it is expressly said, 'of the judges present'. This is equivalent to saying that the decision can also be rendered even if all the judges are not present, contrary to the provision of Article 78, paragraph 2, of the Convention for the pacific settlement of international disputes. The fact that the number of judges was very great was sufficient reason for not placing special importance upon the presence of the whole body of judges. But if all the judges do not take part in a case, it may happen that the number of judges will be even and a tie vote result. It was desired that in such a case the vote of the president should not be the deciding one, because he would otherwise have too great power ; it was provided, therefore, that the vote of the youngest judge should not be counted.[1]

Article 28

Les arrêts de la Cour doivent être motivés. Ils mentionnent les noms des juges qui y ont participé ; ils sont signés par le président et par le greffier.

Article 28

The judgment of the Court must give the reasons on which it is based. It contains the names of the judges taking part in it ; it is signed by the president and registrar.

This article corresponds to Article 79 of the Convention for the pacific settlement of international disputes ; see page 137 of my commentary. Those judges who do not agree with the majority are not called upon to record their dissenting opinion. But in order not to force these judges to sign the decision, provision is made that it is to be signed only by the president and the registrar. The signature of the president does not express his opinion, but merely

[1] See also Article 30, paragraph 2.

proves the document to be authentic. Copies of the decision will merely be signed by the registrar.

Article 29

Chaque partie supporte ses propres frais et une part égale des frais spéciaux de l'instance.

Article 29

Each party pays its own costs and an equal share of the costs of the trial.

This article corresponds to Article 85 of the Convention for the pacific settlement of international disputes and calls for no special remarks.

Article 30

Les dispositions des articles 21 à 29 seront appliquées par analogie dans la procédure devant la délégation.

Lorsque le droit d'adjoindre un membre à la délégation n'a été exercé que par une seule partie, la voix du membre adjoint n'est pas comptée, s'il y a partage de voix.

Article 30

The provisions of Articles 21 to 29 are applicable by analogy to the procedure before the delegation.

When the right of attaching a member to the delegation has been exercised by one of the parties only, the vote of the member attached is not recorded if the votes are evenly divided.

As the delegation is only a branch of the Judicial Court, the same rules of procedure must be applied by it as by the Court.

In case only one of the parties makes use of the right granted by Article 20, paragraph 1, it may easily happen that the votes will be equally divided. In that case the provision of Article 27, paragraph 2, that the vote of the youngest judge shall be set aside, does not hold good, but rather the vote of the member appointed by the one party is excluded. This provision is a just one, as it is to be assumed that the member appointed by state *A* will be very reluctant to vote against state *A*, and accordingly will be more or less biased.

Article 31

Les frais généraux de la Cour sont supportés par les puissances contractantes.

Le Conseil administratif s'adresse aux puissances pour obtenir les fonds nécessaires au fonctionnement de la Cour.

Article 31

The general expenses of the Court are borne by the contracting powers.

The Administrative Council applies to the powers to obtain the funds requisite for the working of the Court.

As the Judicial Court is established in the interest of the contracting parties as a body, it is just that the expenses of the Court shall be borne by them in common. The Second Hague Peace Conference was of the opinion that those states which had more judges upon the Court should bear the larger share of the expenses; but as an agreement could not be reached upon the composition of the Court, no method of apportionment was determined upon.

Compare Article 50 of the Convention for the pacific settlement of international disputes.

Article 32

La Cour fait elle-même son règlement d'ordre intérieur qui doit être communiqué aux puissances contractantes.

Après la ratification de la présente Convention, la Cour se réunira aussitôt que possible pour élaborer ce règlement, pour élire le président et le vice-président, ainsi que pour désigner les membres de la délégation.

Article 32

The Court itself draws up its own rules of procedure, which must be communicated to the contracting powers.

After the ratification of the present Convention the Court shall meet as early as possible in order to elaborate these rules, elect the president and vice-president, and appoint the members of the delegation.

As the rules of procedure of this Convention, as well as those referred to in Article 52 of the Convention for the pacific settlement of international disputes, only comprise the main outlines of a procedure, it was necessary to provide that the Judicial Court should promptly draw up in detail its own rules of procedure. Accordingly, as soon as it has obtained some experience, it must draw up its own rules of procedure. These rules are to be communicated to the contracting powers, in order that they may be able to regulate their procedure accordingly.

Paragraph 2 provides that the Judicial Court shall meet as soon as possible.

Article 33

La Cour peut proposer des modifications à apporter aux dispositions de la présente Convention qui concernent la procédure. Ces propositions sont communiquées par l'intermédiaire du Gouvernement des Pays-Bas aux puissances contractantes qui se concerteront sur la suite à y donner.

Article 33

The Court may propose modifications in the provisions of the present Convention concerning procedure. These proposals are communicated through the Netherland Government to the contracting powers, which will consider together as to the measures to be taken.

Article 32 makes provision for supplementing the general rules of procedure, and for modifying them as well. These modifications cannot be made by the Judicial Court of its own accord, since it possesses no legislative power. It may, however, make proposals, the adoption of which is left to the decision of the powers.

Titre III. Dispositions finales

Article 34

La présente Convention sera ratifiée dans le plus bref délai possible.

Part III. Final Provisions

Article 34

The present Convention shall be ratified as soon as possible.

Les ratifications seront déposées à La Haye.

Il sera dressé du dépôt de chaque ratification un procès-verbal, dont une copie, certifiée conforme, sera remise par la voie diplomatique à toutes les puissances signataires.

The ratification shall be deposited at The Hague.

A *procès-verbal* of the deposit of each ratification shall be drawn up, of which a duly certified copy shall be sent through the diplomatic channel to all the signatory powers.

Article 35

La Convention entrera en vigueur six mois après sa ratification.

Elle aura une durée de douze ans, et sera renouvelée tacitement de douze ans en douze ans, sauf dénonciation.

La dénonciation devra être notifiée, au moins deux ans avant l'expiration de chaque période, au Gouvernement des Pays-Bas, qui en donnera connaissance aux autres puissances.

La dénonciation ne produira effet qu'à l'égard de la puissance qui l'aura notifiée. La Convention restera exécutoire dans les rapports entre les autres puissances.

Article 35

The Convention shall come into force six months after its ratification.

It shall remain in force for twelve years, and shall be tacitly renewed for periods of twelve years, unless denounced.

The denunciation must be notified, at least two years before the expiration of each period, to the Netherland Government, which will inform the other powers.

The denunciation shall only have effect in regard to the notifying power. The Convention shall continue in force as far as the other powers are concerned.

CHAPTER X

THE PROSPECT FOR THE CREATION OF AN INTERNATIONAL COURT OF JUSTICE

THE almost unanimous acceptance of the fundamental principles of the Draft Convention for the creation of a Judicial Arbitration Court adopted at the Second Hague Peace Conference, a project which was also advocated by the Universal Peace Congress at Munich in 1907, indicates that the possibility of the creation of an international court of justice at the next Hague Peace Conference is exceedingly great.

The Government and the people of the United States have since the time of the Second Hague Peace Conference fought their hardest for the establishment of an international court of justice. In this connexion the project of Secretary Knox is once more particularly deserving of attention. All American writers of repute advocate the new Judicial Court, and in the year 1910 the partisans of this great idea formed the American Society for the Judicial Settlement of International Disputes. This society has among its members many distinguished citizens of the United States. Its honorary president is President Taft, its president, Scott, its vice-president, Hammond, and its secretary, Marburg. The object of the society is the furtherance of the project of a judicial tribunal, which shall do for the civilized world what the ordinary courts of justice do for the citizens of a state. Moreover, the society will, after the creation of the judicial tribunal, labour with all its strength to have the states resort to the Judicial

Court as frequently as possible. The society has, besides, nothing to do with peace propaganda. It does not strive after disarmament, but is convinced that, as the number of international disputes settled by judicial methods becomes more and more numerous, the advent of the era of disarmament will be hastened. The first meeting of the society at Washington from December 15–17, 1910, was a great demonstration. In the table of contents of the *Proceedings* of this meeting, which cover 400 pages, it can be seen that the President of the United States, the French Ambassador, two former Secretaries of State, the presidents of three of the leading universities of the United States, many former diplomats, &c., were among the speakers at the various sessions.

During the period preceding the founding of the American Society for Judicial Settlement of International Disputes, the American Society of International Law, as well as the Lake Mohonk Conferences, discussed the project of a permanent international court of justice.

As has been justly pointed out, in particular by Nys [1] and Balch,[2] no state of the world has shown greater zeal in the cause of arbitration than has the United States. The great perseverance which this nation has shown in the pursuit of its high and admirable ideal justifies the hope that in the future as well it will continue to pursue the same noble path for the good of mankind.

The words of President Butler, spoken at the opening of the Fifteenth Lake Mohonk Conference on May 19, 1909, have particular application to an international judiciary:

> We Americans have a peculiar responsibility toward the political organization of the world. Whether we

[1] *The Necessity of a Permanent Tribunal*, p. 27.

[2] *Revue de droit international*, 1908, p. 398.

recognize it or not we are universally looked to, if not to lead in this undertaking, at least to contribute powerfully toward it.

In doing so the United States will have all the fewer difficulties to overcome, in that writers outside of the United States are in favour of the permanent court. In particular, the French are entirely in favour of the project.

Apart from Germany, I know of but two writers who are opposed to the new Judicial Court—namely, the Englishman, Baty,[1] and the Cuban, Bustamante.[2] But these writers stand alone in their view upon this question. In Great Britain the idea of an international judicial court is advocated in particular by Oppenheim,[3] Lawrence,[4] Higgins,[5] Whittuck,[6] Darby, Barclay, Stead and others. Of the Dutchmen who have expressed themselves in this connexion, in addition to Asser I may mention den Beer Poortugael, who in the *Onze Eeuw*[7] advocates with much enthusiasm the Judicial Arbitration Court, van der Mandere,[8] de Louter, who, in his article 'The Future of International Law',[9] advocated the Permanent Court of Justice and abandoned his earlier contrary opinion,[10] and finally van Karnebeek, the honoured vice-president of the First

[1] *International Law*, pp. 9 et seq.; see *La Paix par le Droit*, 1911, p. 193, where Baty also objects to the Knox proposal upon general grounds. He further says in the same article that the National Peace Council in Great Britain and Tryon in the United States have called attention to the dangers of the new Judicial Court.

[2] *La Seconde Conférence*, pp. 204 et seq.

[3] *Zukunft des Völkerrechtes*, p. 47.

[4] *International Problems and Hague Conferences*, p. 75.

[5] *The Hague Peace Conferences*, p. 517.

[6] *International Documents*, p. xvii.

[7] April, 1911, p. 46,

[8] *Vragen des Tijds*, 1910, pp. 376 et seq.

[9] 1912.

[10] *Het stellig volkenrecht*, vol. ii, p. 158.

Hague Conference, who, in an address on the occasion of the laying of the corner-stone of the Peace Palace, expressed the wish that a permanent court in the true sense of the word would soon be established. In Germany a violent conflict of opinion exists upon the question of a permanent court of justice; but the number of those in favour of it outweighs the others by far. I shall place the names of the German-Austrian writers who are in favour of the Judicial Court side by side with the names of those who are opposed to it.

In Favour of the Judicial Court

1. Eickhoff: *Die internationale Schiedsgerichtsbewegung,* 1910, p. 30.
2. Fried: In all his books and articles.
3. Harmening: *Weltparlament,* pp. 20 et seq.
4. Kohler: In an address delivered on October 20, 1911, before the Cologne Society for the Development of Legal and Political Sciences.
5. Meurer: See Year Book of the Carnegie Endowment for International Peace, 1911, p. 133.
6. Lammasch: In *Das Recht,* 1911, p. 151.
7. Von Liszt: *Völkerrecht,* 1911, p. 151; *Das Wesen des völkerrechtlichen Staatenverbandes,* 1911, pp. 10 et seq.
8. Von Martitz: *Internationale Monatsschrift,* 1911, p. 162.

Opposed to the Judicial Court

1. Von Bar: *Berliner Tageblatt,* August 22, 1907.
2. Nippold: *Fortbildung des Verfahrens,* pp. 150 et seq.; *Zweite Haager Konferenz,* pp. 221 et seq.
3. Von Plener: *Interparliamentary Union,* 1908, p. 46.
4. Pohl: *Deutsche Prisengerichtsbarkeit,* p. 210.
5. Tettenborn: *Das Haager Schiedsgericht,* p. 55.
6. Zorn: *Zeitschrift für Politik,* vol. ii, p. 351; *Das Deutsche Reich und die internationale Schiedsgerichtsbarkeit,* p. 44.

9. Schücking: op. cit., pp. 90 et seq.
10. Schlief: *Der Friede in Europa*, p. 275.
11. Sturm: *Psychologische Grundlage des Rechts*, p. 370.
12. Von Ullmann: *Völkerrecht*, p. 448; but see also p. 39, note 1.
13. Wehberg: *Kommentar zu dem Haager Abkommen*, p. 48.

It is noteworthy that, of all the members of the committee of examination of the arbitration commission of the First Hague Conference, Zorn was the only one to vote against the truly permanent Judicial Court.[1]

A commission of the Institute of International Law, consisting of Renault, Hagerup, Rolin, Scott, Westlake, Fauchille, Fromageot, Holland and von Bar, which met in 1911, adopted a resolution to the effect that the next Peace Conference should carry out the *vœu* of the Second Hague Conference relative to the creation of a judicial arbitration court.[2] The result of the meeting of the Interparliamentary Union at Ghent in September, 1912, will be awaited with great interest. A proposal concerning obligatory arbitration,[3] made by Zorn and already approved by the Interparliamentary Council, reads as follows:

The Hague Permanent Court of Arbitration has stood

[1] Shortly before the opening of the Second Peace Conference, White, the American delegate to the First Hague Peace Conference, declared at the Lake Mohonk Conference of 1907 that a truly permanent arbitration court had not been created in 1899 because it was believed that there was not enough work for it to do, and that he was of the conviction that the same reasons would prevent the adoption of the project of a permanent court of justice at the Second Conference (*Report*, pp. 30, 31).

[2] See *Revue de droit international*, 1911, p. 596.

[3] Article 3.

> the test in the form given it by the First Hague Peace Conference. However, the erection of a permanent judicial court for all questions of international private law is to be recommended. . . .[1]

In this resolution the need of an international court of justice is therefore impliedly denied. The fact that three such important opponents of the international court of justice as Beernaert, von Plener and Zorn will play an important part in the deliberations of the Interparliamentary Union makes the adoption of this resolution not impossible. It is all the more necessary, therefore, to point out distinctly that neither Beernaert, von Plener nor Zorn have been able to adduce even the slightest valid reasons against the project here discussed, at least no reasons which have not been proved to be completely without foundation in the remarks we have made above. If Zorn's resolution were to be adopted the Interparliamentary Union would prove to the world that a majority of its members have failed to comprehend the significance of the latest development of international law.

Just as Martens once pointed out that there were three stages in the development of international arbitration, so it seems to me three stages can be distinguished in the progress towards an international judiciary. In the first stage disputes were merely disposed of in accordance with equity. In the second stage in which we now stand the demand that all disputes, as far as possible, shall be decided in a strictly legal way is calling for recognition, but there is still lacking a suitable organ to assist in the triumph of this idea. In the last period the institution that is needed for this purpose will be created in the form of the International Permanent Court of Justice.

[1] *Annuaire*, 1912, p. 77

To be sure, the International Court of Justice will not be a symbol of perpetual peace. We must reckon with the fact that the foundations of the old political international law will only break down slowly, and that the moral elements in the community of states will only gradually prevail with the progress of the economic and cultural unity of nations. Above all we must have patience and not be discouraged by the fact that great ideals can only be reached at a later day.

But the Permanent International Court of Justice will represent a very significant progress towards the goal of international organization. Its creation will necessarily be regarded as a striking victory of the modern peace movement, which has advocated it incessantly for decades. As the Hague Permanent Court of 1899 traces back to the report of La Fontaine on the occasion of the Universal Peace Congress at Antwerp, so the American proposal of an international court of justice found its first advocate in Hornby at that same Universal Peace Congress. The establishment of the International Court of Justice will strengthen the belief of thousands that the nations must hasten to an ever higher stage of their development and that their policies must assume gradually a nobler character. In this way it will once more be shown that all great ideas prevail in the course of the centuries, and mankind must only have perseverance in order to reach its highest goal.

SUBJECT INDEX

(See also Contents)

INDEX OF PERSONS